"Non-asymptotic, high-dimensional theory is critical for modern statistics and machine learning. This book is unique in providing a crystal clear, complete, and unified treatment of the area. With topics ranging from concentration of measure to graphical models, the author weaves together probability theory and its applications to statistics. Ideal for graduate students and researchers. This will surely be the standard reference on the topic for many years."

— Larry Wasserman, *Carnegie Mellon University*

"Martin Wainwright brings his large box of analytical power tools to bear on the problems of the day—the analysis of models for wide data. A broad knowledge of this new area combines with his powerful analytical skills to deliver this impressive and intimidating work—bound to be an essential reference for all the brave souls that try their hand."

— Trevor Hastie, *Stanford University*

"This book provides an excellent treatment of perhaps the fastest growing area within high-dimensional theoretical statistics—non-asymptotic theory that seeks to provide probabilistic bounds on estimators as a function of sample size and dimension. It offers the most thorough, clear, and engaging coverage of this area to date, and is thus poised to become the definitive reference and textbook on this topic."

— Genevera Allen, *Rice University*

"Statistical theory and practice have undergone a renaissance in the past two decades, with intensive study of high-dimensional data analysis. No researcher has deepened our understanding of high-dimensional statistics more than Martin Wainwright. This book brings the signature clarity and incisiveness of his published research into book form. It will be a fantastic resource for both beginning students and seasoned researchers, as the field continues to make exciting breakthroughs."

— John Lafferty, *Yale University*

"This is an outstanding book on high-dimensional statistics, written by a creative and celebrated researcher in the field. It gives comprehensive treatments of many important topics in statistical machine learning and, furthermore, is self-contained, from introductory material to the most up-to-date results on various research frontiers. This book is a must-read for those who wish to learn and to develop modern statistical machine theory, methods and algorithms."

— Jianqing Fan, *Princeton University*

"This book provides an in-depth mathematical treatment and methodological intuition for high-dimensional statistics. The main technical tools from probability theory are carefully developed and the construction and analysis of statistical methods and algorithms for high-dimensional problems are presented in an outstandingly clear way. Martin Wainwright has written a truly exceptional, inspiring, and beautiful masterpiece!"

— Peter Bühlmann, *ETH Zurich*

"This new book by Martin Wainwright covers modern topics in high-dimensional statistical inference, and focuses primarily on explicit non-asymptotic results related to sparsity and non-parametric estimation. This is a must-read for all graduate students in mathematical statistics and theoretical machine learning, both for the breadth of recent advances it covers and the depth of results which are presented. The exposition is outstandingly clear, starting from the first introductory chapters on the necessary probabilistic tools. Then, the book covers state-of-the-art advances in high-dimensional statistics, with always a clever choice of results which have the perfect mix of significance and mathematical depth."

— Francis Bach, *INRIA Paris*

"Wainwright's book on those parts of probability theory and mathematical statistics critical to understanding of the new phenomena encountered in high dimensions is marked by the clarity of its presentation and the depth to which it travels. In every chapter he starts with intuitive examples and simulations which are systematically developed either into powerful mathematical tools or complete answers to fundamental questions of inference. It is not easy, but elegant and rewarding whether read systematically or dipped into as a reference."

— Peter Bickel, *UC Berkeley*

High-Dimensional Statistics

Recent years have witnessed an explosion in the volume and variety of data collected in all scientific disciplines and industrial settings. Such massive data sets present a number of challenges to researchers in statistics and machine learning. This book provides a self-contained introduction to the area of high-dimensional statistics, aimed at the first-year graduate level. It includes chapters that are focused on core methodology and theory—including tail bounds, concentration inequalities, uniform laws and empirical process, and random matrices—as well as chapters devoted to in-depth exploration of particular model classes—including sparse linear models, matrix models with rank constraints, graphical models, and various types of non-parametric models.

With hundreds of worked examples and exercises, this text is intended both for courses and for self-study by graduate students and researchers in statistics, machine learning, and related fields who must understand, apply, and adapt modern statistical methods suited to large-scale data.

MARTIN J. WAINWRIGHT is a Chancellor's Professor at the University of California, Berkeley, with a joint appointment between the Department of Statistics and the Department of Electrical Engineering and Computer Sciences. His research lies at the nexus of statistics, machine learning, optimization, and information theory, and he has published widely in all of these disciplines. He has written two other books, one on graphical models together with Michael I. Jordan, and one on sparse learning together with Trevor Hastie and Robert Tibshirani. Among other awards, he has received the COPSS Presidents' Award, has been a Medallion Lecturer and Blackwell Lecturer for the Institute of Mathematical Statistics, and has received Best Paper Awards from the NIPS, ICML, and UAI conferences, as well as from the IEEE Information Theory Society.

CAMBRIDGE SERIES IN STATISTICAL AND PROBABILISTIC MATHEMATICS

This series of high-quality upper-division textbooks and expository monographs covers all aspects of stochastic applicable mathematics. The topics range from pure and applied statistics to probability theory, operations research, optimization, and mathematical programming. The books contain clear presentations of new developments in the field and also of the state of the art in classical methods. While emphasizing rigorous treatment of theoretical methods, the books also contain applications and discussions of new techniques made possible by advances in computational practice.

A complete list of books in the series can be found at www.cambridge.org/statistics.
Recent titles include the following:

23. *Applied Asymptotics,* by A. R. Brazzale, A. C. Davison and N. Reid
24. *Random Networks for Communication,* by Massimo Franceschetti and Ronald Meester
25. *Design of Comparative Experiments,* by R. A. Bailey
26. *Symmetry Studies,* by Marlos A. G. Viana
27. *Model Selection and Model Averaging,* by Gerda Claeskens and Nils Lid Hjort
28. *Bayesian Nonparametrics,* edited by Nils Lid Hjort *et al.*
29. *From Finite Sample to Asymptotic Methods in Statistics,* by Pranab K. Sen, Julio M. Singer and Antonio C. Pedrosa de Lima
30. *Brownian Motion,* by Peter Mörters and Yuval Peres
31. *Probability: Theory and Examples (Fourth Edition),* by Rick Durrett
33. *Stochastic Processes,* by Richard F. Bass
34. *Regression for Categorical Data,* by Gerhard Tutz
35. *Exercises in Probability (Second Edition),* by Loïc Chaumont and Marc Yor
36. *Statistical Principles for the Design of Experiments,* by R. Mead, S. G. Gilmour and A. Mead
37. *Quantum Stochastics,* by Mou-Hsiung Chang
38. *Nonparametric Estimation under Shape Constraints,* by Piet Groeneboom and Geurt Jongbloed
39. *Large Sample Covariance Matrices and High-Dimensional Data Analysis,* by Jianfeng Yao, Shurong Zheng and Zhidong Bai
40. *Mathematical Foundations of Infinite-Dimensional Statistical Models,* by Evarist Giné and Richard Nickl
41. *Confidence, Likelihood, Probability,* by Tore Schweder and Nils Lid Hjort
42. *Probability on Trees and Networks,* by Russell Lyons and Yuval Peres
43. *Random Graphs and Complex Networks (Volume 1),* by Remco van der Hofstad
44. *Fundamentals of Nonparametric Bayesian Inference,* by Subhashis Ghosal and Aad van der Vaart
45. *Long-Range Dependence and Self-Similarity,* by Vladas Pipiras and Murad S. Taqqu
46. *Predictive Statistics,* by Bertrand S. Clarke and Jennifer L. Clarke
47. *High-Dimensional Probability,* by Roman Vershynin
48. *High-Dimensional Statistics,* by Martin J. Wainwright
49. *Probability: Theory and Examples (Fifth Edition),* by Rick Durrett

High-Dimensional Statistics

A Non-Asymptotic Viewpoint

Martin J. Wainwright
University of California, Berkeley

CAMBRIDGE
UNIVERSITY PRESS

CAMBRIDGE
UNIVERSITY PRESS

University Printing House, Cambridge CB2 8BS, United Kingdom

One Liberty Plaza, 20th Floor, New York, NY 10006, USA

477 Williamstown Road, Port Melbourne, VIC 3207, Australia

314-321, 3rd Floor, Plot 3, Splendor Forum, Jasola District Centre, New Delhi - 110025, India

79 Anson Road, #06-04/06, Singapore 079906

Cambridge University Press is part of the University of Cambridge.

It furthers the University's mission by disseminating knowledge in the pursuit of education, learning and research at the highest international levels of excellence.

www.cambridge.org
Information on this title: www.cambridge.org/9781108498029
DOI: 10.1017/9781108627771

First published 2019

A catalogue record for this publication is available from the British Library

Library of Congress Cataloging in Publication data
Names: Wainwright, Martin (Martin J.), author.
Title: High-dimensional statistics : a non-asymptotic viewpoint / Martin J. Wainwright (University of California, Berkeley).
Description: Cambridge ; New York, NY : Cambridge University Press, 2019. |
Series: Cambridge series in statistical and probabilistic mathematics ; 48 |
Includes bibliographical references and indexes.
Identifiers: LCCN 2018043475 | ISBN 9781108498029 (hardback)
Subjects: LCSH: Mathematical statistics–Textbooks. | Big data.
Classification: LCC QA276.18 .W35 2019 | DDC 519.5–dc23
LC record available at https://lccn.loc.gov/2018043475

ISBN 978-1-108-49802-9 Hardback

List of chapters

Contents

Illustrations

Acknowledgements

This book would not exist without the help of many people in my life. I thank my parents, John and Patricia, for their nurture and support over the years, and my family, Haruko, Hana, Mina, and Kento for bringing daily joy. In developing my own understanding of and perspective on the broad area of high-dimensional statistics, I have been fortunate to interact with and learn from many wonderful colleagues, both here at Berkeley and elsewhere. For valuable discussions, insights and feedback, I would like to thank, among others, Bryon Aragam, Alex d'Asprémont, Francis Bach, Peter Bickel, Peter Bühlmann, Tony Cai, Emmanuel Candès, Constantine Caramanis, David Donoho, Noureddine El Karoui, Jianqing Fan, Aditya Guntuboyina, Trevor Hastie, Iain Johnstone, Michael Jordan, John Lafferty, Eliza Levina, Zongming Ma, Nicolai Meinshausen, Andrea Montanari, Axel Munk, Richard Nickl, Eric Price, Philippe Rigollet, Alessandro Rinaldo, Richard Samworth, Robert Tibshirani, Ryan Tibshirani, Alexander Tsybakov, Sara Van de Geer, Larry Wasserman, Frank Werner, Bin Yu, Ming Yuan, and Harry Zhou. The Statistics and EECS departments, staff, faculty and students at UC Berkeley have provided a wonderful intellectual environment for this work; I also thank the Statistics Group at ETH Zurich, as well as the Laboratory for Information and Decision Sciences (LIDS) at MIT for their support during my visiting professor stays.

Over the years, I have also had the pleasure of working with many outstanding students and postdoctoral associates. Our discussions and research together as well as their feedback have been instrumental in shaping and improving this book. I would like to thank my current and former students, postdocs and visting fellows—many of whom are now current colleagues—including Alekh Agarwal, Arash Amini, Sivaraman Balakrishan, Merle Behr, Joseph Bradley, Yudong Chen, Alex Dimakis, John Duchi, Reinhard Heckel, Nhat Ho, Johannes Lederer, Po-Ling Loh, Sahand Negahban, Xuanlong Nguyen, Nima Noorshams, Jonas Peters, Mert Pilanci, Aaditya Ramdas, Garvesh Raskutti, Pradeep Ravikumar, Prasad Santhanam, Nihar Shah, Yuting Wei, Fanny Yang, Yun Yang, and Yuchen Zhang. Yuting Wei put her artistic skills to excellent use in designing the cover for this book. Last but not least, I would like to thank Cambridge University Press, and in particular the support and encouragement throughout the process of Senior Editor Diana Gillooly. My apologies to anyone whose help I may have failed inadvertently to acknowledge.

1

Introduction

The focus of this book is non-asymptotic theory in high-dimensional statistics. As an area of intellectual inquiry, high-dimensional statistics is not new: it has roots going back to the seminal work of Rao, Wigner, Kolmogorov, Huber and others, from the 1950s onwards. What is new—and very exciting—is the dramatic surge of interest and activity in high-dimensional analysis over the past two decades. The impetus for this research is the nature of data sets arising in modern science and engineering: many of them are extremely large, often with the dimension of the same order as, or possibly even larger than, the sample size. In such regimes, classical asymptotic theory often fails to provide useful predictions, and standard methods may break down in dramatic ways. These phenomena call for the development of new theory as well as new methods. Developments in high-dimensional statistics have connections with many areas of applied mathematics—among them machine learning, optimization, numerical analysis, functional and geometric analysis, information theory, approximation theory and probability theory. The goal of this book is to provide a coherent introduction to this body of work.

1.1 Classical versus high-dimensional theory

What is meant by the term "high-dimensional", and why is it important and interesting to study high-dimensional problems? In order to answer these questions, we first need to understand the distinction between classical as opposed to high-dimensional theory.

Classical theory in probability and statistics provides statements that apply to a fixed class of models, parameterized by an index n that is allowed to increase. In statistical settings, this integer-valued index has an interpretation as a sample size. The canonical instance of such a theoretical statement is the *law of large numbers*. In its simplest instantiation, it concerns the limiting behavior of the sample mean of n independent and identically distributed d-dimensional random vectors $\{X_i\}_{i=1}^n$, say, with mean $\mu = \mathbb{E}[X_1]$ and a finite variance. The law of large numbers guarantees that the sample mean $\hat{\mu}_n := \frac{1}{n}\sum_{i=1}^n X_i$ converges in probability to μ. Consequently, the sample mean $\hat{\mu}_n$ is a consistent estimator of the unknown population mean. A more refined statement is provided by the *central limit theorem*, which guarantees that the rescaled deviation $\sqrt{n}(\hat{\mu}_n - \mu)$ converges in distribution to a centered Gaussian with covariance matrix $\mathbf{\Sigma} = \text{cov}(X_1)$. These two theoretical statements underlie the analysis of a wide range of classical statistical estimators—in particular, ensuring their consistency and asymptotic normality, respectively.

In a classical theoretical framework, the ambient dimension d of the data space is typically viewed as fixed. In order to appreciate the motivation for high-dimensional statistics, it is worthwhile considering the following:

Question Suppose that we are given $n = 1000$ samples from a statistical model in $d = 500$ dimensions. Will theory that requires $n \to +\infty$ with the dimension d remaining fixed provide useful predictions?

Of course, this question cannot be answered definitively without further details on the model under consideration. Some essential facts that motivate our discussion in this book are the following:

1. The data sets arising in many parts of modern science and engineering have a "high-dimensional flavor", with d on the same order as, or possibly larger than, the sample size n.
2. For many of these applications, classical "large n, fixed d" theory fails to provide useful predictions.
3. Classical methods can break down dramatically in high-dimensional regimes.

These facts motivate the study of high-dimensional statistical models, as well as the associated methodology and theory for estimation, testing and inference in such models.

1.2 What can go wrong in high dimensions?

In order to appreciate the challenges associated with high-dimensional problems, it is worthwhile considering some simple problems in which classical results break down. Accordingly, this section is devoted to three brief forays into some examples of high-dimensional phenomena.

1.2.1 Linear discriminant analysis

In the problem of binary hypothesis testing, the goal is to determine whether an observed vector $x \in \mathbb{R}^d$ has been drawn from one of two possible distributions, say $\mathbb{P}_1$ versus $\mathbb{P}_2$. When these two distributions are known, then a natural decision rule is based on thresholding the log-likelihood ratio $\log \frac{\mathbb{P}_2[x]}{\mathbb{P}_1[x]}$; varying the setting of the threshold allows for a principled trade-off between the two types of errors—namely, deciding $\mathbb{P}_1$ when the true distribution is $\mathbb{P}_2$, and vice versa. The celebrated Neyman–Pearson lemma guarantees that this family of decision rules, possibly with randomization, are optimal in the sense that they trace out the curve giving the best possible trade-off between the two error types.

As a special case, suppose that the two classes are distributed as multivariate Gaussians, say $N(\mu_1, \Sigma)$ and $N(\mu_2, \Sigma)$, respectively, differing only in their mean vectors. In this case, the log-likelihood ratio reduces to the linear statistic

$$\Psi(x) := \left\langle \mu_1 - \mu_2, \, \Sigma^{-1}\left(x - \frac{\mu_1 + \mu_2}{2}\right)\right\rangle, \tag{1.1}$$

where $\langle \cdot, \cdot \rangle$ denotes the Euclidean inner product in $\mathbb{R}^d$. The optimal decision rule is based on thresholding this statistic. We can evaluate the quality of this decision rule by computing the probability of incorrect classification. Concretely, if the two classes are equally likely, this probability is given by

$$\mathrm{Err}(\Psi) := \tfrac{1}{2}\mathbb{P}_1[\Psi(X') \le 0] + \tfrac{1}{2}\mathbb{P}_2[\Psi(X'') > 0],$$

where X' and X'' are random vectors drawn from the distributions $\mathbb{P}_1$ and $\mathbb{P}_2$, respectively. Given our Gaussian assumptions, some algebra shows that the error probability can be written in terms of the Gaussian cumulative distribution function Φ as

$$\mathrm{Err}(\Psi) = \underbrace{\frac{1}{\sqrt{2\pi}}\int_{-\infty}^{-\gamma/2} e^{-t^2/2}\,dt}_{\Phi(-\gamma/2)}, \qquad \text{where } \gamma = \sqrt{(\mu_1 - \mu_2)^{\mathrm{T}}\boldsymbol{\Sigma}^{-1}(\mu_1 - \mu_2)}. \tag{1.2}$$

In practice, the class conditional distributions are not known, but instead one observes a collection of labeled samples, say $\{x_1, \ldots, x_{n_1}\}$ drawn independently from $\mathbb{P}_1$, and $\{x_{n_1+1}, \ldots, x_{n_1+n_2}\}$ drawn independently from $\mathbb{P}_2$. A natural approach is to use these samples in order to estimate the class conditional distributions, and then "plug" these estimates into the log-likelihood ratio. In the Gaussian case, estimating the distributions is equivalent to estimating the mean vectors μ_1 and μ_2, as well as the covariance matrix $\boldsymbol{\Sigma}$, and standard estimates are the samples means

$$\hat{\mu}_1 := \frac{1}{n_1}\sum_{i=1}^{n_1} x_i \quad \text{and} \quad \hat{\mu}_2 := \frac{1}{n_2}\sum_{i=n_1+1}^{n_1+n_2} x_i, \tag{1.3a}$$

as well as the pooled sample covariance matrix

$$\widehat{\boldsymbol{\Sigma}} := \frac{1}{n_1 - 1}\sum_{i=1}^{n_1}(x_i - \hat{\mu}_1)(x_i - \hat{\mu}_1)^{\mathrm{T}} + \frac{1}{n_2 - 1}\sum_{i=n_1+1}^{n_1+n_2}(x_i - \hat{\mu}_2)(x_i - \hat{\mu}_2)^{\mathrm{T}}. \tag{1.3b}$$

Substituting these estimates into the log-likelihood ratio (1.1) yields the *Fisher linear discriminant function*

$$\widehat{\Psi}(x) = \left\langle \hat{\mu}_1 - \hat{\mu}_2,\ \widehat{\boldsymbol{\Sigma}}^{-1}\left(x - \frac{\hat{\mu}_1 + \hat{\mu}_2}{2}\right)\right\rangle. \tag{1.4}$$

Here we have assumed that the sample covariance is invertible, and hence are assuming implicitly that $n_i > d$.

Let us assume that the two classes are equally likely *a priori*. In this case, the error probability obtained by using a zero threshold is given by

$$\mathrm{Err}(\widehat{\Psi}) := \tfrac{1}{2}\mathbb{P}_1[\widehat{\Psi}(X') \le 0] + \tfrac{1}{2}\mathbb{P}_2[\widehat{\Psi}(X'') > 0],$$

where X' and X'' are samples drawn independently from the distributions $\mathbb{P}_1$ and $\mathbb{P}_2$, respectively. Note that the error probability is itself a random variable, since the discriminant function $\widehat{\Psi}$ is a function of the samples $\{X_i\}_{i=1}^{n_1+n_2}$.

In the 1960s, Kolmogorov analyzed a simple version of the Fisher linear discriminant, in which the covariance matrix $\boldsymbol{\Sigma}$ is known *a priori* to be the identity, so that the linear statistic (1.4) simplifies to

$$\widehat{\Psi}_{\mathrm{id}}(x) = \left\langle \hat{\mu}_1 - \hat{\mu}_2,\ x - \frac{\hat{\mu}_1 + \hat{\mu}_2}{2}\right\rangle. \tag{1.5}$$

Working under an assumption of Gaussian data, he analyzed the behavior of this method under a form of high-dimensional asymptotics, in which the triple (n_1, n_2, d) all tend to infinity, with the ratios d/n_i, for $i = 1, 2$, converging to some non-negative fraction $\alpha > 0$, and the Euclidean[1] distance $\|\mu_1 - \mu_2\|_2$ converging to a constant $\gamma > 0$. Under this type of high-dimensional scaling, he showed that the error $\mathrm{Err}(\widehat{\Psi}_{\mathrm{id}})$ converges in probability to a fixed number—in particular,

$$\mathrm{Err}(\widehat{\Psi}_{\mathrm{id}}) \xrightarrow{\text{prob.}} \Phi\Big(-\frac{\gamma^2}{2\sqrt{\gamma^2 + 2\alpha}}\Big), \tag{1.6}$$

where $\Phi(t) := \mathbb{P}[Z \le t]$ is the cumulative distribution function of a standard normal variable. Thus, if $d/n_i \to 0$, then the asymptotic error probability is simply $\Phi(-\gamma/2)$, as is predicted by classical scaling (1.2). However, when the ratios d/n_i converge to a strictly positive number $\alpha > 0$, then the asymptotic error probability is strictly larger than the classical prediction, since the quantity $\frac{\gamma^2}{2\sqrt{\gamma^2+2\alpha}}$ is shifted towards zero.

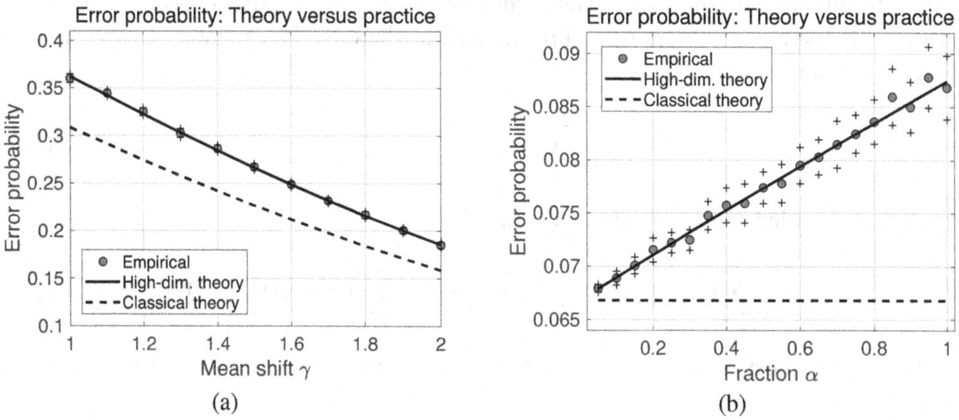

Figure 1.1 (a) Plots of the error probability $\mathrm{Err}(\widehat{\Psi}_{\mathrm{id}})$ versus the mean shift parameter $\gamma \in [1, 2]$ for $d = 400$ and fraction $\alpha = 0.5$, so that $n_1 = n_2 = 800$. Gray circles correspond to the empirical error probabilities, averaged over 50 trials and confidence bands shown with plus signs, as defined by three times the standard error. The solid curve gives the high-dimensional prediction (1.6), whereas the dashed curve gives the classical prediction (1.2). (b) Plots of the error probability $\mathrm{Err}(\widehat{\Psi}_{\mathrm{id}})$ versus the fraction $\alpha \in [0, 1]$ for $d = 400$ and $\gamma = 2$. In this case, the classical prediction $\Phi(-\gamma/2)$ plotted as a dashed line remains flat, since it is independent of α.

Recalling our original motivating question from Section 1.1, it is natural to ask whether the error probability of the test $\widehat{\Psi}_{\mathrm{id}}$, for some finite triple (d, n_1, n_2), is better described by the classical prediction (1.2), or the high-dimensional analog (1.6). In Figure 1.1, we plot comparisons between the empirical behavior and theoretical predictions for different choices of the mean shift parameter γ and limiting fraction α. Figure 1.1(a) shows plots of the error probability $\mathrm{Err}(\widehat{\Psi}_{\mathrm{id}})$ versus the mean shift parameter γ for dimension $d = 400$ and fraction $\alpha = 0.5$, meaning that $n_1 = n_2 = 800$. Gray circles correspond to the empirical

[1] We note that the Mahalanobis distance from equation (1.2) reduces to the Euclidean distance when $\boldsymbol{\Sigma} = \mathbf{I}_d$.

performance averaged over 50 trials, whereas the solid and dashed lines correspond to the high-dimensional and classical predictions, respectively. Note that the high-dimensional prediction (1.6) with $\alpha = 0.5$ shows excellent agreement with the behavior in practice, whereas the classical prediction $\Phi(-\gamma)$ drastically underestimates the error rate. Figure 1.1(b) shows a similar plot, again with dimension $d = 400$ but with $\gamma = 2$ and the fraction α ranging in the interval $[0.05, 1]$. In this case, the classical prediction is flat, since it has no dependence on α. Once again, the empirical behavior shows good agreement with the high-dimensional prediction.

A failure to take into account high-dimensional effects can also lead to sub-optimality. A simple instance of this phenomenon arises when the two fractions d/n_i, $i = 1, 2$, converge to possibly different quantities $\alpha_i \geq 0$ for $i = 1, 2$. For reasons to become clear shortly, it is natural to consider the behavior of the discriminant function $\widehat{\Psi}_{\text{id}}$ for a general choice of threshold $t \in \mathbb{R}$, in which case the associated error probability takes the form

$$\text{Err}_t(\widehat{\Psi}_{\text{id}}) = \tfrac{1}{2}\mathbb{P}_1[\widehat{\Psi}_{\text{id}}(X') \leq t] + \tfrac{1}{2}\mathbb{P}_2[\widehat{\Psi}_{\text{id}}(X'') > t], \tag{1.7}$$

where X' and X'' are again independent samples from $\mathbb{P}_1$ and $\mathbb{P}_2$, respectively. For this setup, it can be shown that

$$\text{Err}_t(\widehat{\Psi}_{\text{id}}) \xrightarrow{\text{prob.}} \frac{1}{2}\Phi\left(-\frac{\gamma^2 + 2t + (\alpha_1 - \alpha_2)}{2\sqrt{\gamma^2 + \alpha_1 + \alpha_2}}\right) + \frac{1}{2}\Phi\left(-\frac{\gamma^2 - 2t - (\alpha_1 - \alpha_2)}{2\sqrt{\gamma^2 + \alpha_1 + \alpha_2}}\right),$$

a formula which reduces to the earlier expression (1.6) in the special case when $\alpha_1 = \alpha_2 = \alpha$ and $t = 0$. Due to the additional term $\alpha_1 - \alpha_2$, whose sign differs between the two terms, the choice $t = 0$ is no longer asymptotically optimal, even though we have assumed that the two classes are equally likely *a priori*. Instead, the optimal choice of the threshold is $t = \frac{\alpha_2 - \alpha_1}{2}$, a choice that takes into account the different sample sizes between the two classes.

1.2.2 Covariance estimation

We now turn to an exploration of high-dimensional effects for the problem of covariance estimation. In concrete terms, suppose that we are given a collection of random vectors $\{x_1, \ldots, x_n\}$, where each x_i is drawn in an independent and identically distributed (i.i.d.) manner from some zero-mean distribution in $\mathbb{R}^d$, and our goal is to estimate the unknown covariance matrix $\mathbf{\Sigma} = \text{cov}(X)$. A natural estimator is the *sample covariance matrix*

$$\widehat{\mathbf{\Sigma}} := \frac{1}{n}\sum_{i=1}^{n} x_i x_i^{\mathrm{T}}, \tag{1.8}$$

a $d \times d$ random matrix corresponding to the sample average of the outer products $x_i x_i^T \in \mathbb{R}^{d \times d}$. By construction, the sample covariance $\widehat{\mathbf{\Sigma}}$ is an unbiased estimate, meaning that $\mathbb{E}[\widehat{\mathbf{\Sigma}}] = \mathbf{\Sigma}$.

A classical analysis considers the behavior of the sample covariance matrix $\widehat{\mathbf{\Sigma}}$ as the sample size n increases while the ambient dimension d stays fixed. There are different ways in which to measure the distance between the random matrix $\widehat{\mathbf{\Sigma}}$ and the population covariance matrix $\mathbf{\Sigma}$, but, regardless of which norm is used, the sample covariance is a consistent estimate. One useful matrix norm is the ℓ_2-operator norm, given by

$$|||\widehat{\Sigma} - \Sigma|||_2 := \sup_{u \neq 0} \frac{\|(\widehat{\Sigma} - \Sigma)u\|_2}{\|u\|_2}. \tag{1.9}$$

Under mild moment conditions, an argument based on the classical law of large numbers can be used to show that the difference $|||\widehat{\Sigma} - \Sigma|||_2$ converges to zero almost surely as $n \to \infty$. Consequently, the sample covariance is a strongly consistent estimate of the population covariance in the classical setting.

Is this type of consistency preserved if we also allow the dimension d to tend to infinity? In order to pose the question more crisply, let us consider sequences of problems $(\widehat{\Sigma}, \Sigma)$ indexed by the pair (n, d), and suppose that we allow both n and d to increase with their ratio remaining fixed—in particular, say $d/n = \alpha \in (0, 1)$. In Figure 1.2, we plot the results of simulations for a random ensemble $\Sigma = \mathbf{I}_d$, with each $X_i \sim N(0, \mathbf{I}_d)$ for $i = 1, \ldots, n$. Using these n samples, we generated the sample covariance matrix (1.8), and then computed its vector of eigenvalues $\gamma(\widehat{\Sigma}) \in \mathbb{R}^d$, say arranged in non-increasing order as

$$\gamma_{\max}(\widehat{\Sigma}) = \gamma_1(\widehat{\Sigma}) \geq \gamma_2(\widehat{\Sigma}) \geq \cdots \geq \gamma_d(\widehat{\Sigma}) = \gamma_{\min}(\widehat{\Sigma}) \geq 0.$$

Each plot shows a histogram of the vector $\gamma(\widehat{\Sigma}) \in \mathbb{R}^d$ of eigenvalues: Figure 1.2(a) corresponds to the case $(n, d) = (4000, 800)$ or $\alpha = 0.2$, whereas Figure 1.2(b) shows the pair $(n, d) = (4000, 2000)$ or $\alpha = 0.5$. If the sample covariance matrix were converging to the identity matrix, then the vector of eigenvalues $\gamma(\widehat{\Sigma})$ should converge to the all-ones vector, and the corresponding histograms should concentrate around 1. Instead, the histograms in both plots are highly dispersed around 1, with differing shapes depending on the aspect ratios.

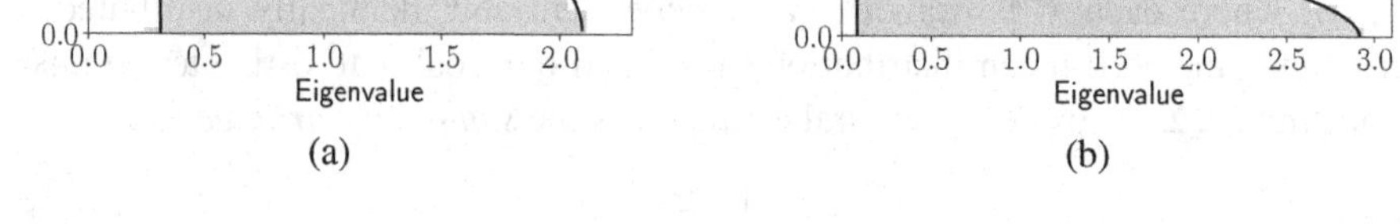

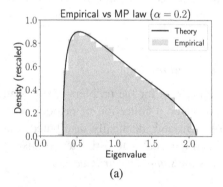

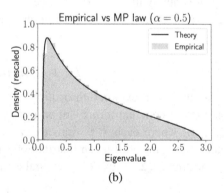

Figure 1.2 Empirical distribution of the eigenvalues of a sample covariance matrix $\widehat{\Sigma}$ versus the asymptotic prediction of the Marčenko–Pastur law. It is specified by a density of the form $f_{\text{MP}}(\gamma) \propto \sqrt{\frac{(t_{\max}(\alpha)-\gamma)(\gamma-t_{\min}(\alpha))}{\gamma}}$, supported on the interval $[t_{\min}(\alpha), t_{\max}(\alpha)] = [(1 - \sqrt{\alpha})^2, (1 + \sqrt{\alpha})^2]$. (a) Aspect ratio $\alpha = 0.2$ and $(n, d) = (4000, 800)$. (b) Aspect ratio $\alpha = 0.5$ and $(n, d) = (4000, 2000)$. In both cases, the maximum eigenvalue $\gamma_{\max}(\widehat{\Sigma})$ is very close to $(1 + \sqrt{\alpha})^2$, consistent with theory.

These shapes—if we let both the sample size and dimension increase in such a way that

$d/n \to \alpha \in (0, 1)$—are characterized by an asymptotic distribution known as the Marčenko–Pastur law. Under some mild moment conditions, this theory predicts convergence to a strictly positive density supported on the interval $[t_{\min}(\alpha), t_{\max}(\alpha)]$, where

$$t_{\min}(\alpha) := (1 - \sqrt{\alpha})^2 \quad \text{and} \quad t_{\max}(\alpha) := (1 + \sqrt{\alpha})^2. \tag{1.10}$$

See the caption of Figure 1.2 for more details.

The Marčenko–Pastur law is an asymptotic statement, albeit of a non-classical flavor since it allows both the sample size and dimension to diverge. By contrast, the primary focus of this book are results that are non-asymptotic in nature—that is, in the current context, we seek results that hold for *all* choices of the pair (n, d), and that provide explicit bounds on the events of interest. For example, as we discuss at more length in Chapter 6, in the setting of Figure 1.2, it can be shown that the maximum eigenvalue $\gamma_{\max}(\widehat{\boldsymbol{\Sigma}})$ satisfies the upper deviation inequality

$$\mathbb{P}[\gamma_{\max}(\widehat{\boldsymbol{\Sigma}}) \geq (1 + \sqrt{d/n} + \delta)^2] \leq e^{-n\delta^2/2} \qquad \text{for all } \delta \geq 0, \tag{1.11}$$

with an analogous lower deviation inequality for the minimum eigenvalue $\gamma_{\min}(\widehat{\boldsymbol{\Sigma}})$ in the regime $n \geq d$. This result gives us more refined information about the maximum eigenvalue, showing that the probability that it deviates above $(1 + \sqrt{d/n})^2$ is exponentially small in the sample size n. In addition, this inequality (and related results) can be used to show that the sample covariance matrix $\widehat{\boldsymbol{\Sigma}}$ is an operator-norm-consistent estimate of the population covariance matrix $\boldsymbol{\Sigma}$ as long as $d/n \to 0$.

1.2.3 Nonparametric regression

The effects of high dimensions on regression problems can be even more dramatic. In one instance of the problem known as *nonparametric regression*, we are interested in estimating a function from the unit hypercube $[0, 1]^d$ to the real line $\mathbb{R}$; this function can be viewed as mapping a vector $x \in [0, 1]^d$ of predictors or covariates to a scalar response variable $y \in \mathbb{R}$. If we view the pair (X, Y) as random variables, then we can ask for the function f that minimizes the least-squares prediction error $\mathbb{E}[(Y - f(X))^2]$. An easy calculation shows that the optimal such function is defined by the conditional expectation $f(x) = \mathbb{E}[Y \mid x]$, and it is known as the regression function.

In practice, the joint distribution $\mathbb{P}_{X,Y}$ of (X, Y) is unknown, so that computing f directly is not possible. Instead, we are given samples (X_i, Y_i) for $i = 1, \ldots, n$, drawn in an i.i.d. manner from $\mathbb{P}_{X,Y}$, and our goal is to find a function $\widehat{f}$ for which the mean-squared error (MSE)

$$\|\widehat{f} - f\|_{L^2}^2 := \mathbb{E}_X[(\widehat{f}(X) - f(X))^2] \tag{1.12}$$

is as small as possible.

It turns out that this problem becomes extremely difficult in high dimensions, a manifestation of what is known as the *curse of dimensionality*. This notion will be made precise in our discussion of nonparametric regression in Chapter 13. Here, let us do some simple simulations to address the following question: How many samples n should be required as a function of the problem dimension d? For concreteness, let us suppose that the covariate vector X is uniformly distributed over $[0, 1]^d$, so that $\mathbb{P}_X$ is the uniform distribution, denoted by $\text{Uni}([0, 1]^d)$. If we are able to generate a good estimate of $\widehat{f}$ based on the samples

$X_1, \ldots, X_n$, then it should be the case that a typical vector $X' \in [0,1]^d$ is relatively close to at least one of our samples. To formalize this notation, we might study the quantity

$$\rho_\infty(n,d) := \mathbb{E}_{X',X}\Big[\min_{i=1,\ldots,n} \|X' - X_i\|_\infty\Big], \tag{1.13}$$

which measures the average distance between an independently drawn sample X', again from the uniform distribution $\text{Uni}([0,1]^d)$, and our original data set $\{X_1, \ldots, X_n\}$.

How many samples n do we need to collect as a function of the dimension d so as to ensure that $\rho_\infty(n,d)$ falls below some threshold δ? For illustrative purposes, we use $\delta = 1/3$ in the simulations to follow. As in the previous sections, let us first consider a scaling in which the ratio d/n converges to some constant $\alpha > 0$, say $\alpha = 0.5$ for concreteness, so that $n = 2d$. Figure 1.3(a) shows the results of estimating the quantity $\rho_\infty(2d,d)$ on the basis of 20 trials. As shown by the gray circles, in practice, the closest point (on average) to a data set based on $n = 2d$ samples tends to increase with dimension, and certainly stays bounded above $1/3$. What happens if we try a more aggressive scaling of the sample size? Figure 1.3(b) shows the results of the same experiments with $n = d^2$ samples; again, the minimum distance tends to increase as the dimension increases, and stays bounded well above $1/3$.

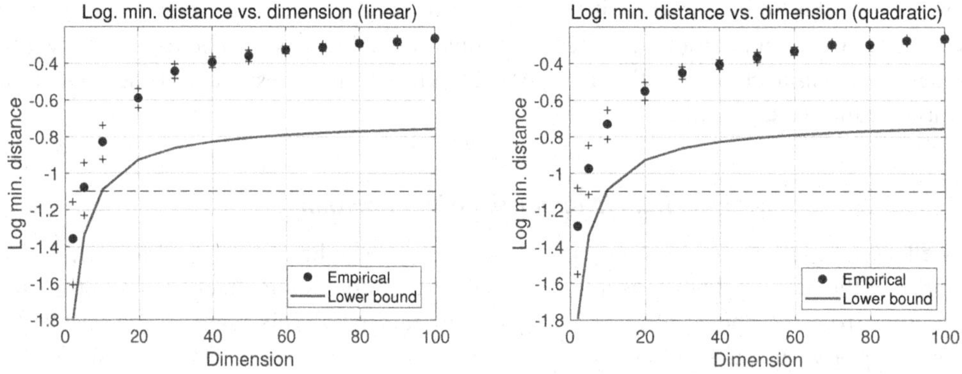

(a) (b)

Figure 1.3 Behavior of the quantity $\rho_\infty(n,d)$ versus the dimension d, for different scalings of the pair (n,d). Full circles correspond to the average over 20 trials, with confidence bands shown with plus signs, whereas the solid curve provides the theoretical lower bound (1.14). (a) Behavior of the variable $\rho_\infty(2d,d)$. (b) Behavior of the variable $\rho_\infty(d^2,d)$. In both cases, the expected minimum distance remains bounded above $1/3$, corresponding to $\log(1/3) \approx -1.1$ (horizontal dashed line) on this logarithmic scale.

In fact, we would need to take an *exponentially large* sample size in order to ensure that $\rho_\infty(n,d)$ remained below δ as the dimension increased. This fact can be confirmed by proving the lower bound

$$\log \rho_\infty(n,d) \geq \log \frac{d}{2(d+1)} - \frac{\log n}{d}, \tag{1.14}$$

which implies that a sample size $n > (1/\delta)^d$ is required to ensure that the upper bound $\rho_\infty(n,d) \leq \delta$ holds. We leave the proof of the bound (1.14) as an exercise for the reader.

We have chosen to illustrate this exponential explosion in a randomized setting, where the covariates X are drawn uniformly from the hypercube $[0,1]^d$. But the curse of dimensionality manifests itself with equal ferocity in the deterministic setting, where we are given the freedom of choosing some collection $\{x_i\}_{i=1}^n$ of vectors in the hypercube $[0,1]^d$. Let us investigate the minimal number n required to ensure that any vector $x' \in [0,1]^d$ is at most distance δ in the ℓ_∞-norm to some vector in our collection—that is, such that

$$\sup_{x' \in [0,1]^d} \min_{i=1,\ldots,n} \|x' - x_i\|_\infty \leq \delta. \tag{1.15}$$

The most straightforward way of ensuring this approximation quality is by a uniform gridding of the unit hypercube: in particular, suppose that we divide each of the d sides of the cube into $\lceil 1/(2\delta) \rceil$ sub-intervals,[2] each of length 2δ. Taking the Cartesian products of these sub-intervals yields a total of $\lceil 1/(2\delta) \rceil^d$ boxes. Placing one of our points x_i at the center of each of these boxes yields the desired approximation (1.15).

This construction provides an instance of what is known as a δ-covering of the unit hypercube in the ℓ_∞-norm, and we see that its size must grow exponentially in the dimension. By studying a related quantity known as a δ-packing, this exponential scaling can be shown to be inescapable—that is, there is not a covering set with substantially fewer elements. See Chapter 5 for a much more detailed treatment of the notions of packing and covering.

1.3 What can help us in high dimensions?

An important fact is that the high-dimensional phenomena described in the previous sections are *all unavoidable*. Concretely, for the classification problem described in Section 1.2.1, if the ratio d/n stays bounded strictly above zero, then it is not possible to achieve the optimal classification rate (1.2). For the covariance estimation problem described in Section 1.2.2, there is no consistent estimator of the covariance matrix in ℓ_2-operator norm when d/n remains bounded away from zero. Finally, for the nonparametric regression problem in Section 1.2.3, given the goal of estimating a differentiable regression function f, no consistent procedure is possible unless the sample size n grows exponentially in the dimension d. All of these statements can be made rigorous via the notions of metric entropy and minimax lower bounds, to be developed in Chapters 5 and 15, respectively.

Given these "no free lunch" guarantees, what can help us in the high-dimensional setting? Essentially, our only hope is that the data is endowed with some form of *low-dimensional structure,* one which makes it simpler than the high-dimensional view might suggest. Much of high-dimensional statistics involves constructing models of high-dimensional phenomena that involve some implicit form of low-dimensional structure, and then studying the statistical and computational gains afforded by exploiting this structure. In order to illustrate, let us revisit our earlier three vignettes, and show how the behavior can change dramatically when low-dimensional structure is present.

[2] Here $\lceil a \rceil$ denotes the ceiling of a, or the smallest integer greater than or equal to a.

1.3.1 Sparsity in vectors

Recall the simple classification problem described in Section 1.2.1, in which, for $j = 1, 2$, we observe n_j samples of a multivariate Gaussian with mean $\mu_j \in \mathbb{R}^d$ and identity covariance matrix $\mathbf{I}_d$. Setting $n = n_1 = n_2$, let us recall the scaling in which the ratios d/n_j are fixed to some number $\alpha \in (0, \infty)$. What is the underlying cause of the inaccuracy of the classical prediction shown in Figure 1.1? Recalling that $\hat{\mu}_j$ denotes the sample mean of the n_j samples, the squared Euclidean error $\|\hat{\mu}_j - \mu_j\|_2^2$ turns out to concentrate sharply around $\frac{d}{n_j} = \alpha$. This fact is a straightforward consequence of the chi-squared (χ^2) tail bounds to be developed in Chapter 2—in particular, see Example 2.11. When $\alpha > 0$, there is a constant level of error, for which reason the classical prediction (1.2) of the error rate is overly optimistic.

But the sample mean is not the only possible estimate of the true mean: when the true mean vector is equipped with some type of low-dimensional structure, there can be much better estimators. Perhaps the simplest form of structure is sparsity: suppose that we knew that each mean vector μ_j were relatively sparse, with only s of its d entries being non-zero, for some sparsity parameter $s \ll d$. In this case, we can obtain a substantially better estimator by applying some form of thresholding to the sample means. As an example, for a given threshold level $\lambda > 0$, the hard-thresholding estimator is given by

$$H_\lambda(x) = x\mathbb{I}[|x| > \lambda] = \begin{cases} x & \text{if } |x| > \lambda, \\ 0 & \text{otherwise,} \end{cases} \tag{1.16}$$

where $\mathbb{I}[|x| > \lambda]$ is a 0–1 indicator for the event $\{|x| > \lambda\}$. As shown by the solid curve in Figure 1.4(a), it is a "keep-or-kill" function that zeroes out x whenever its absolute value falls below the threshold λ, and does nothing otherwise. A closely related function is the soft-thresholding operator

$$T_\lambda(x) = \mathbb{I}[|x| > \lambda](x - \lambda\,\text{sign}(x)) = \begin{cases} x - \lambda\,\text{sign}(x) & \text{if } |x| > \lambda, \\ 0 & \text{otherwise.} \end{cases} \tag{1.17}$$

As shown by the dashed line in Figure 1.4(a), it has been shifted so as to be continuous, in contrast to the hard-thresholding function.

In the context of our classification problem, instead of using the sample means $\hat{\mu}_j$ in the plug-in classification rule (1.5), suppose that we used hard-thresholded versions of the sample means—namely

$$\widetilde{\mu}_j = H_{\lambda_n}(\hat{\mu}_j) \quad \text{for } j = 1, 2 \qquad \text{where } \lambda_n := \sqrt{\frac{2\log d}{n}}. \tag{1.18}$$

Standard tail bounds to be developed in Chapter 2—see Exercise 2.12 in particular—will illuminate why this particular choice of threshold λ_n is a good one. Using these thresholded estimates, we can then implement a classifier based on the linear discriminant

$$\widetilde{\Psi}(x) := \left\langle \widetilde{\mu}_1 - \widetilde{\mu}_2,\ x - \frac{\widetilde{\mu}_1 + \widetilde{\mu}_2}{2} \right\rangle. \tag{1.19}$$

In order to explore the performance of this classifier, we performed simulations using the same parameters as those in Figure 1.1(a); Figure 1.4(b) gives a plot of the error $\text{Err}(\widetilde{\Psi})$

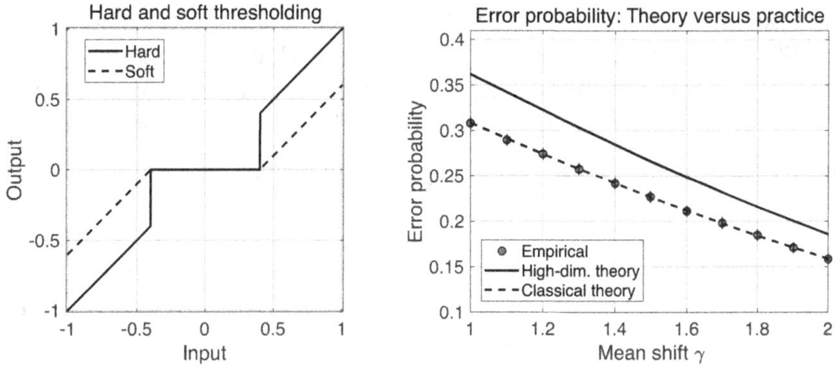

(a) (b)

Figure 1.4 (a) Plots of the hard-thresholding and soft-thresholding functions at some level $\lambda > 0$. (b) Plots of the error probability $\mathrm{Err}(\widehat{\Psi}_{\mathrm{id}})$ versus the mean shift parameter $\gamma \in [1, 2]$ with the same set-up as the simulations in Figure 1.1: dimension $d = 400$, and sample sizes $n = n_1 = n_2 = 800$. In this case, the mean vectors μ_1 and μ_2 each had $s = 5$ non-zero entries, and the classification was based on hard-thresholded versions of the sample means at the level $\lambda_n = \sqrt{\frac{2 \log d}{n}}$. Gray circles correspond to the empirical error probabilities, averaged over 50 trials and confidence intervals defined by three times the standard error. The solid curve gives the high-dimensional prediction (1.6), whereas the dashed curve gives the classical prediction (1.2). In contrast to Figure 1.1(a), the classical prediction is now accurate.

versus the mean shift γ. Overlaid for comparison are both the classical (1.2) and high-dimensional (1.6) predictions. In contrast to Figure 1.1(a), the classical prediction now gives an excellent fit to the observed behavior. In fact, the classical limit prediction is exact whenever the ratio $\log \binom{d}{s}/n$ approaches zero. Our theory on sparse vector estimation in Chapter 7 can be used to provide a rigorous justification of this claim.

1.3.2 Structure in covariance matrices

In Section 1.2.2, we analyzed the behavior of the eigenvalues of a sample covariance matrix $\widehat{\mathbf{\Sigma}}$ based on n samples of a d-dimensional random vector with the identity matrix as its covariance. As shown in Figure 1.2, when the ratio d/n remains bounded away from zero, the sample eigenspectrum $\gamma(\widehat{\mathbf{\Sigma}})$ remains highly dispersed around 1, showing that $\widehat{\mathbf{\Sigma}}$ is not a good estimate of the population covariance matrix $\mathbf{\Sigma} = \mathbf{I}_d$. Again, we can ask the questions: What types of low-dimensional structure might be appropriate for modeling covariance matrices? And how can they can be exploited to construct better estimators?

As a very simple example, suppose that our goal is to estimate a covariance matrix known to be diagonal. It is then intuitively clear that the sample covariance matrix can be improved by zeroing out its non-diagonal entries, leading to the diagonal covariance estimate $\widehat{\mathbf{D}}$. A little more realistically, if the covariance matrix $\mathbf{\Sigma}$ were assumed to be sparse but the positions were unknown, then a reasonable estimator would be the hard-thresholded version $\widetilde{\mathbf{\Sigma}} := T_{\lambda_n}(\widehat{\mathbf{\Sigma}})$ of the sample covariance, say with $\lambda_n = \sqrt{\frac{2 \log d}{n}}$ as before. Figure 1.5(a) shows the resulting eigenspectrum $\gamma(\widetilde{\boldsymbol{\Sigma}})$ of this estimator with aspect ratio $\alpha = 0.2$ and $(n, d) = (4000, 800)$—that is, the same settings as Figure 1.2(a). In contrast to the Marčenko–Pastur behavior shown in the former figure, we now see that the eigenspectrum $\gamma(\widetilde{\boldsymbol{\Sigma}})$ is sharply concentrated around the point mass at 1. Tail bounds and theory from Chapters 2 and 6 can be used to show that $|||\widetilde{\boldsymbol{\Sigma}} - \boldsymbol{\Sigma}|||_2 \precsim \sqrt{\frac{\log d}{n}}$ with high probability.

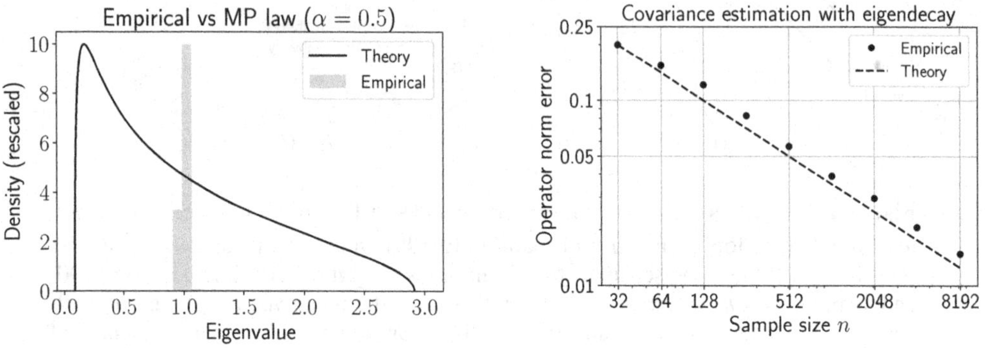

(a) (b)

Figure 1.5 (a) Behavior of the eigenspectrum $\gamma(\widetilde{\boldsymbol{\Sigma}})$ for a hard-thresholded version of the sample covariance matrix. Unlike the sample covariance matrix itself, it can be a consistent estimator of a sparse covariance matrix even for scalings such that $d/n = \alpha > 0$. (b) Behavior of the sample covariance matrix for estimating sequences of covariance matrices of increasing dimension but all satisfying the constraint $\text{trace}(\boldsymbol{\Sigma}) \leq 20$. Consistent with theoretical predictions, the operator norm error $|||\widehat{\boldsymbol{\Sigma}} - \boldsymbol{\Sigma}|||_2$ for this sequence decays at the rate $1/\sqrt{n}$, as shown by the solid line on the log–log plot.

An alternative form of low-dimensional structure for symmetric matrices is that of fast decay in their eigenspectra. If we again consider sequences of problems indexed by (n, d), suppose that our sequence of covariance matrices have a bounded trace—that is, $\text{trace}(\boldsymbol{\Sigma}) \leq R$, independent of the dimension d. This requirement means that the ordered eigenvalues $\gamma_j(\boldsymbol{\Sigma})$ must decay a little more quickly than j^{-1}. As we discuss in Chapter 10, these types of eigendecay conditions hold in a variety of applications. Figure 1.5(b) shows a log–log plot of the operator norm error $|||\widehat{\boldsymbol{\Sigma}} - \boldsymbol{\Sigma}|||_2$ over a range of pairs (n, d), all with the fixed ratio $d/n = 0.2$, for a sequence of covariance matrices that all satisfy the constraint $\text{trace}(\boldsymbol{\Sigma}) \leq 20$. Theoretical results to be developed in Chapter 6 predict that, for such a sequence of covariance matrices, the error $|||\widehat{\boldsymbol{\Sigma}} - \boldsymbol{\Sigma}|||_2$ should decay as $n^{-1/2}$, even if the dimension d grows in proportion to the sample size n. See also Chapters 8 and 10 for discussion of other forms of matrix estimation in which these types of rank or eigendecay constraints play a role.

1.3.3 Structured forms of regression

As discussed in Section 1.2.3, a generic regression problem in high dimensions suffers from a severe curse of dimensionality. What type of structure can alleviate this curse? There are various forms of low-dimensional structure that have been studied in past and on-going work on high-dimensional regression.

One form of structure is that of an *additive decomposition* in the regression function—say of the form

$$f(x_1, \ldots, x_d) = \sum_{j=1}^{d} g_j(x_j), \tag{1.20}$$

where each univariate function $g_j \colon \mathbb{R} \to \mathbb{R}$ is chosen from some base class. For such functions, the problem of regression is reduced to estimating a collection of d separate univariate functions. The general theory developed in Chapters 13 and 14 can be used to show how the additive assumption (1.20) largely circumvents[3] the curse of dimensionality. A very special case of the additive decomposition (1.20) is the classical linear model, in which, for each $j = 1, \ldots, d$, the univariate function takes the form $g_j(x_j) = \theta_j x_j$ for some coefficients $\theta_j \in \mathbb{R}$. More generally, we might assume that each g_j belongs to a reproducing kernel Hilbert space, a class of function spaces studied at length in Chapter 12.

Assumptions of sparsity also play an important role in the regression setting. The *sparse additive model* (SPAM) is based on positing the existence of some subset $S \subset \{1, 2, \ldots, d\}$ of cardinality $s = |S|$ such that the regression function can be decomposed as

$$f(x_1, \ldots, x_d) = \sum_{j \in S} g_j(x_j). \tag{1.21}$$

In this model, there are two different classes of objects to be estimated: (i) the unknown subset S that ranges over all $\binom{d}{s}$ possible subsets of size s; and (ii) the univariate functions $\{g_j, j \in S\}$ associated with this subset. A special case of the SPAM decomposition (1.21) is the *sparse linear model*, in which $f(x) = \sum_{j=1}^{d} \theta_j x_j$ for some vector $\theta \in \mathbb{R}^d$ that is s-sparse. See Chapter 7 for a detailed discussion of this class of models, and the conditions under which accurate estimation is possible even when $d \gg n$.

There are a variety of other types of structured regression models to which the methods and theory developed in this book can be applied. Examples include the *multiple-index model*, in which the regression function takes the form

$$f(x_1, \ldots, x_d) = h(\mathbf{A}x), \tag{1.22}$$

for some matrix $\mathbf{A} \in \mathbb{R}^{s \times d}$, and function $h \colon \mathbb{R}^s \to \mathbb{R}$. The single-index model is the special case of this model with $s = 1$, so that $f(x) = h(\langle a, x \rangle)$ for some vector $a \in \mathbb{R}^d$. Another special case of this more general family is the SPAM class (1.21): it can be obtained by letting the rows of $\mathbf{A}$ be the standard basis vectors $\{e_j, j \in S\}$, and letting the function h belong to the additive class (1.20).

Taking sums of single-index models leads to a method known as *projection pursuit regression*, involving functions of the form

$$f(x_1, \ldots, x_d) = \sum_{j=1}^{M} g_j(\langle a_j, x \rangle), \tag{1.23}$$

for some collection of univariate functions $\{g_j\}_{j=1}^{M}$, and a collection of d vectors $\{a_j\}_{j=1}^{M}$. Such

[3] In particular, see Exercise 13.9, as well as Examples 14.11 and 14.14.

models can also help alleviate the curse of dimensionality, as long as the number of terms M can be kept relatively small while retaining a good fit to the regression function.

1.4 What is the non-asymptotic viewpoint?

As indicated by its title, this book emphasizes non-asymptotic results in high-dimensional statistics. In order to put this emphasis in context, we can distinguish between at least three types of statistical analysis, depending on how the sample size behaves relative to the dimension and other problem parameters:

- *Classical asymptotics.* The sample size n is taken to infinity, with the dimension d and all other problem parameters remaining fixed. The standard laws of large numbers and central limit theorem are examples of this type of theory.
- *High-dimensional asymptotics.* The pair (n, d) is taken to infinity simultaneously, while enforcing that, for some scaling function Ψ, the sequence $\Psi(n, d)$ remains fixed, or converges to some value $\alpha \in [0, \infty]$. For example, in our discussions of linear discriminant analysis (Section 1.2.1) and covariance estimation (Section 1.2.2), we considered such scalings with the function $\Psi(n, d) = d/n$. More generally, the scaling function might depend on other problem parameters in addition to (n, d). For example, in studying vector estimation problems involving a sparsity parameter s, the scaling function $\Psi(n, d, s) = \log \binom{d}{s}/n$ might be used. Here the numerator reflects that there are $\binom{d}{s}$ possible subsets of cardinality s contained in the set of all possible indices $\{1, 2, \ldots, d\}$.
- *Non-asymptotic bounds.* The pair (n, d), as well as other problem parameters, are viewed as fixed, and high-probability statements are made as a function of them. The previously stated bound (1.11) on the maximum eigenvalue of a sample covariance matrix is a standard example of such a result. Results of this type—that is, tail bounds and concentration inequalities on the performance of statistical estimators—are the primary focus of this book.

To be clear, these modes of analysis are closely related. Tail bounds and concentration inequalities typically underlie the proofs of classical asymptotic theorems, such as almost sure convergence of a sequence of random variables. Non-asymptotic theory can be used to predict some aspects of high-dimensional asymptotic phenomena—for instance, it can be used to derive the limiting forms of the error probabilities (1.6) for linear discriminant analysis. In random matrix theory, it can be used to establish that the sample eigenspectrum of a sample covariance matrix with $d/n = \alpha$ lies within[4] the interval $[(1 - \sqrt{\alpha})^2, (1 + \sqrt{\alpha})^2]$ with probability one as (n, d) grow—cf. Figure 1.2. Finally, the functions that arise in a non-asymptotic analysis can suggest appropriate forms of scaling functions Ψ suitable for performing a high-dimensional asymptotic analysis so as to unveil limiting distributional behavior.

One topic *not* covered in this book—due to space constraints—is an evolving line of work that seeks to characterize the asymptotic behavior of low-dimensional functions of a given high-dimensional estimator; see the bibliography in Section 1.6 for some references.

[4] To be clear, it does not predict the precise shape of the distribution on this interval, as given by the Marčenko–Pastur law.

For instance, in sparse vector estimation, one natural goal is to seek a confidence interval for a given coordinate of the d-dimensional vector. At the heart of such analyses are non-asymptotic tail bounds, which allow for control of residuals within the asymptotics. Consequently, the reader who has mastered the techniques laid out in this book will be well equipped to follow these types of derivations.

1.5 Overview of the book

With this motivation in hand, let us now turn to a broad overview of the structure of this book, as well as some suggestions regarding its potential use in a teaching context.

1.5.1 Chapter structure and synopses

The chapters follow a rough division into two types: material on *Tools and techniques* (TT), and material on *Models and estimators* (ME). Chapters of the TT type are foundational in nature, meant to develop techniques and derive theory that is broadly applicable in high-dimensional statistics. The ME chapters are meant to be complementary in nature: each such chapter focuses on a particular class of statistical estimation problems, and brings to bear the methods developed in the foundational chapters.

Tools and techniques

- Chapter 2: This chapter provides an introduction to standard techniques in deriving tail bounds and concentration inequalities. It is required reading for all other chapters in the book.
- Chapter 3: Following directly from Chapter 2, this chapter is devoted to more advanced material on concentration of measure, including the entropic method, log-Sobolev inequalities, and transportation cost inequalities. It is meant for the reader interested in a deeper understanding of the concentration phenomenon, but is not required reading for the remaining chapters. The concentration inequalities in Section 3.4 for empirical processes are used in later analysis of nonparametric models.
- Chapter 4: This chapter is again required reading for most other chapters, as it introduces the foundational ideas of uniform laws of large numbers, along with techniques such as symmetrization, which leads naturally to the Rademacher complexity of a set. It also covers the notion of Vapnik–Chervonenkis (VC) dimension as a particular way of bounding the Rademacher complexity.
- Chapter 5: This chapter introduces the geometric notions of covering and packing in metric spaces, along with the associated discretization and chaining arguments that underlie proofs of uniform laws via entropic arguments. These arguments, including Dudley's entropy integral, are required for later study of nonparametric models in Chapters 13 and 14. Also covered in this chapter are various connections to Gaussian processes, including the Sudakov–Fernique and Gordon–Slepian bounds, as well as Sudakov's lower bound.
- Chapter 12: This chapter provides a self-contained introduction to reproducing kernel Hilbert spaces, including material on kernel functions, Mercer's theorem and eigenvalues, the representer theorem, and applications to function interpolation and estimation via kernel ridge regression. This material is not a prerequisite for reading Chapters 13 and 14,

but is required for understanding the kernel-based examples covered in these chapters on nonparametric problems.

- Chapter 14: This chapter follows the material from Chapters 4 and 13, and is devoted to more advanced material on uniform laws, including an in-depth analysis of two-sided and one-sided uniform laws for the population and empirical L^2-norms. It also includes some extensions to certain Lipschitz cost functions, along with applications to nonparametric density estimation.
- Chapter 15: This chapter provides a self-contained introduction to techniques for proving minimax lower bounds, including in-depth discussions of Le Cam's method in both its naive and general forms, the local and Yang–Barron versions of the Fano method, along with various examples. It can be read independently of any other chapter, but does make reference (for comparison) to upper bounds proved in other chapters.

Models and estimators

- Chapter 6: This chapter is devoted to the problem of covariance estimation. It develops various non-asymptotic bounds for the singular values and operator norms of random matrices, using methods based on comparison inequalities for Gaussian matrices, discretization methods for sub-Gaussian and sub-exponential variables, as well as tail bounds of the Ahlswede–Winter type. It also covers the estimation of sparse and structured covariance matrices via thresholding and related techniques. Material from Chapters 2, 4 and 5 is needed for a full understanding of the proofs in this chapter.
- Chapter 7: The sparse linear model is possibly the most widely studied instance of a high-dimensional statistical model, and arises in various applications. This chapter is devoted to theoretical results on the behavior of ℓ_1-relaxations for estimating sparse vectors, including results on exact recovery for noiseless models, estimation in ℓ_2-norm and prediction semi-norms for noisy models, as well as results on variable selection. It makes substantial use of various tail bounds from Chapter 2.
- Chapter 8: Principal component analysis is a standard method in multivariate data analysis, and exhibits a number of interesting phenomena in the high-dimensional setting. This chapter is devoted to a non-asymptotic study of its properties, in both its unstructured and sparse versions. The underlying analysis makes use of techniques from Chapters 2 and 6.
- Chapter 9: This chapter develops general techniques for analyzing estimators that are based on decomposable regularizers, including the ℓ_1-norm and nuclear norm as special cases. It builds on the material on sparse linear regression from Chapter 7, and makes uses of techniques from Chapters 2 and 4.
- Chapter 10: There are various applications that involve the estimation of low-rank matrices in high dimensions, and this chapter is devoted to estimators based on replacing the rank constraint with a nuclear norm penalty. It makes direct use of the framework from Chapter 9, as well as tail bounds and random matrix theory from Chapters 2 and 6.
- Chapter 11: Graphical models combine ideas from probability theory and graph theory, and are widely used in modeling high-dimensional data. This chapter addresses various types of estimation and model selection problems that arise in graphical models. It requires background from Chapters 2 and 7.
- Chapter 13: This chapter is devoted to an in-depth analysis of least-squares estimation

in the general nonparametric setting, with a broad range of examples. It exploits techniques from Chapters 2, 4 and 5, along with some concentration inequalities for empirical processes from Chapter 3.

1.5.2 Recommended background

This book is targeted at graduate students with an interest in applied mathematics broadly defined, including mathematically oriented branches of statistics, computer science, electrical engineering and econometrics. As such, it assumes a strong undergraduate background in basic aspects of mathematics, including the following:

- A course in linear algebra, including material on matrices, eigenvalues and eigendecompositions, singular values, and so on.
- A course in basic real analysis, at the level of Rudin's elementary book (Rudin, 1964), covering convergence of sequences and series, metric spaces and abstract integration.
- A course in probability theory, including both discrete and continuous variables, laws of large numbers, as well as central limit theory. A measure-theoretic version is not required, but the ability to deal with the abstraction of this type is useful. Some useful books include Breiman (1992), Chung (1991), Durrett (2010) and Williams (1991).
- A course in classical mathematical statistics, including some background on decision theory, basics of estimation and testing, maximum likelihood estimation and some asymptotic theory. Some standard books at the appropriate level include Keener (2010), Bickel and Doksum (2015) and Shao (2007).

Probably the most subtle requirement is a certain degree of mathematical maturity on the part of the reader. This book is meant for the person who is interested in gaining a deep understanding of the core issues in high-dimensional statistics. As with anything worthwhile in life, doing so requires effort. This basic fact should be kept in mind while working through the proofs, examples and exercises in this book.

At the same time, this book has been written with self-study and/or teaching in mind. To wit, we have often sacrificed generality or sharpness in theorem statements for the sake of proof clarity. In lieu of an exhaustive treatment, our primary emphasis is on developing techniques that can be used to analyze many different problems. To this end, each chapter is seeded with a large number of examples, in which we derive specific consequences of more abstract statements. Working through these examples in detail, as well as through some of the many exercises at the end of each chapter, is the best way to gain a robust grasp of the material. As a warning to the reader: the exercises range in difficulty from relatively straightforward to extremely challenging. *Don't be discouraged* if you find an exercise to be challenging; some of them are meant to be!

1.5.3 Teaching possibilities and a flow diagram

This book has been used for teaching one-semester graduate courses on high-dimensional statistics at various universities, including the University of California Berkeley, Carnegie

Mellon University, Massachusetts Institute of Technology and Yale University. The book has far too much material for a one-semester class, but there are various ways of working through different subsets of chapters over time periods ranging from five to 15 weeks. See Figure 1.6 for a flow diagram that illustrates some of these different pathways through the book.

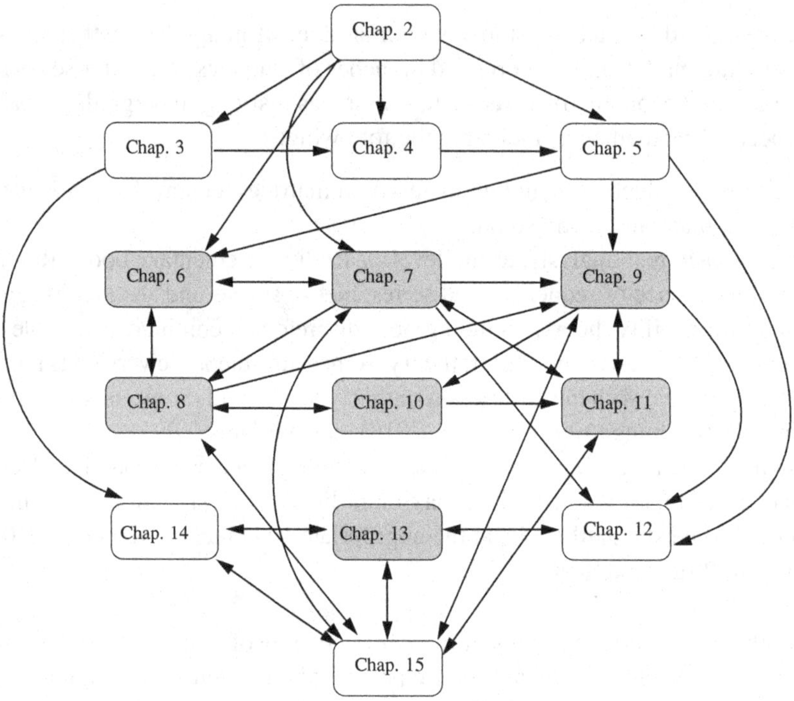

Figure 1.6 A flow diagram of Chapters 2–15 and some of their dependence structure. Various tours of subsets of chapters are possible; see the text for more details.

A short introduction. Given a shorter period of a few weeks, it would be reasonable to cover Chapter 2 followed by Chapter 7 on sparse linear regression, followed by parts of Chapter 6 on covariance estimation. Other brief tours beginning with Chapter 2 are also possible.

A longer look. Given a few more weeks, a longer look could be obtained by supplementing the short introduction with some material from Chapter 5 on metric entropy and Dudley's entropy integral, followed by Chapter 13 on nonparametric least squares. This supplement would give a taste of the nonparametric material in the book. Alternative additions are possible, depending on interests.

A full semester course. A semester-length tour through the book could include Chapter 2 on tail bounds, Chapter 4 on uniform laws, the material in Sections 5.1 through 5.3.3 on metric entropy through to Dudley's entropy integral, followed by parts of Chapter 6 on covariance estimation, Chapter 7 on sparse linear regression, and Chapter 8 on principal component analysis. A second component of the course could consist of Chapter 12 on reproducing kernel Hilbert spaces, followed by Chapter 13 on nonparametric least squares. Depending on the semester length, it could also be possible to cover some material on minimax lower bounds from Chapter 15.

1.6 Bibliographic details and background

Rao (1949) was one of the first authors to consider high-dimensional effects in two-sample testing problems. The high-dimensional linear discriminant problem discussed in Section 1.2.1 was first proposed and analyzed by Kolmogorov in the 1960s. Deev, working in the group of Kolmogorov, analyzed the high-dimensional asymptotics of the general Fisher linear discriminant for fractions $\alpha_i \in [0, 1)$. See the book by Serdobolskii (2000) and the survey paper by Raudys and Young (2004) for further detail on this early line of Russian research in high-dimensional classification.

The study of high-dimensional random matrices, as treated briefly in Section 1.2.2, also has deep roots, dating back to the seminal work from the 1950s onwards (e.g., Wigner, 1955, 1958; Marčenko and Pastur, 1967; Pastur, 1972; Wachter, 1978; Geman, 1980). The high-dimensional asymptotic law for the eigenvalues of a sample covariance matrix illustrated in Figure 1.2 is due to Marčenko and Pastur (1967); this asymptotic prediction has been shown to be a remarkably robust phenomenon, requiring only mild moment conditions (e.g., Silverstein, 1995; Bai and Silverstein, 2010). See also the paper by Götze and Tikhomirov (2004) for quantitative bounds on the distance to this limiting distribution.

In his Wald Memorial Lecture, Huber (1973) studied the asymptotics of robust regression under a high-dimensional scaling with d/n constant. Portnoy (1984; 1985) studied M-estimators for high-dimensional linear regression models, proving consistency when the ratio $\frac{d \log d}{n}$ goes to zero, and asymptotic normality under somewhat more stringent conditions. See also Portnoy (1988) for extensions to more general exponential family models. The high-dimensional asymptotics of various forms of robust regression estimators have been studied in recent work by El Karoui and co-authors (e.g., Bean et al., 2013; El Karoui, 2013; El Karoui et al., 2013), as well as by Donoho and Montanari (2013).

Thresholding estimators are widely used in statistical problems in which the estimand is expected to be sparse. See the book by Johnstone (2015) for an extensive discussion of thresholding estimators in the context of the normal sequence model, with various applications in nonparametric estimation and density estimation. See also Chapters 6 and 7 for some discussion and analysis of thresholding estimators. Soft thresholding is very closely related to ℓ_1-regularization, a method with a lengthy history (e.g., Levy and Fullagar, 1981; Santosa and Symes, 1986; Tibshirani, 1996; Chen et al., 1998; Juditsky and Nemirovski, 2000; Donoho and Huo, 2001; Elad and Bruckstein, 2002; Candès and Tao, 2005; Donoho, 2006b; Bickel et al., 2009); see Chapter 7 for an in-depth discussion.

Stone (1985) introduced the class of additive models (1.20) for nonparametric regression; see the book by Hastie and Tibshirani (1990) for more details. The SPAM class (1.21) has been studied by many researchers (e.g., Meier et al., 2009; Ravikumar et al., 2009; Koltchinskii and Yuan, 2010; Raskutti et al., 2012). The single-index model (1.22), as a particular instance of a semiparametric model, has also been widely studied; for instance, see the various papers (Härdle and Stoker, 1989; Härdle et al., 1993; Ichimura, 1993; Hristache et al., 2001) and references therein for further details. Friedman and Stuetzle (1981) introduced the idea of projection pursuit regression (1.23). In broad terms, projection pursuit methods are based on seeking "interesting" projections of high-dimensional data (Kruskal, 1969; Huber, 1985; Friedman and Tukey, 1994), and projection pursuit regression is based on this idea in the context of regression.

2

Basic tail and concentration bounds

In a variety of settings, it is of interest to obtain bounds on the tails of a random variable, or two-sided inequalities that guarantee that a random variable is close to its mean or median. In this chapter, we explore a number of elementary techniques for obtaining both deviation and concentration inequalities. This chapter serves as an entry point to more advanced literature on large-deviation bounds and concentration of measure.

2.1 Classical bounds

One way in which to control a tail probability $\mathbb{P}[X \geq t]$ is by controlling the moments of the random variable X. Gaining control of higher-order moments leads to correspondingly sharper bounds on tail probabilities, ranging from Markov's inequality (which requires only existence of the first moment) to the Chernoff bound (which requires existence of the moment generating function).

2.1.1 From Markov to Chernoff

The most elementary tail bound is *Markov's inequality*: given a non-negative random variable X with finite mean, we have

$$\mathbb{P}[X \geq t] \leq \frac{\mathbb{E}[X]}{t} \qquad \text{for all } t > 0. \tag{2.1}$$

This is a simple instance of an upper tail bound. For a random variable X that also has a finite variance, we have *Chebyshev's inequality*:

$$\mathbb{P}[|X - \mu| \geq t] \leq \frac{\text{var}(X)}{t^2} \qquad \text{for all } t > 0. \tag{2.2}$$

This is a simple form of concentration inequality, guaranteeing that X is close to its mean $\mu = \mathbb{E}[X]$ whenever its variance is small. Observe that Chebyshev's inequality follows by applying Markov's inequality to the non-negative random variable $Y = (X - \mu)^2$. Both Markov's and Chebyshev's inequalities are sharp, meaning that they cannot be improved in general (see Exercise 2.1).

There are various extensions of Markov's inequality applicable to random variables with higher-order moments. For instance, whenever X has a central moment of order k, an application of Markov's inequality to the random variable $|X - \mu|^k$ yields that

$$\mathbb{P}[|X - \mu| \geq t] \leq \frac{\mathbb{E}[|X - \mu|^k]}{t^k} \qquad \text{for all } t > 0. \tag{2.3}$$

Of course, the same procedure can be applied to functions other than polynomials $|X - \mu|^k$. For instance, suppose that the random variable X has a moment generating function in a neighborhood of zero, meaning that there is some constant $b > 0$ such that the function $\varphi(\lambda) = \mathbb{E}[e^{\lambda(X-\mu)}]$ exists for all $\lambda \leq |b|$. In this case, for any $\lambda \in [0, b]$, we may apply Markov's inequality to the random variable $Y = e^{\lambda(X-\mu)}$, thereby obtaining the upper bound

$$\mathbb{P}[(X-\mu) \geq t] = \mathbb{P}[e^{\lambda(X-\mu)} \geq e^{\lambda t}] \leq \frac{\mathbb{E}[e^{\lambda(X-\mu)}]}{e^{\lambda t}}. \tag{2.4}$$

Optimizing our choice of λ so as to obtain the tightest result yields the *Chernoff bound*—namely, the inequality

$$\log \mathbb{P}[(X-\mu) \geq t] \leq \inf_{\lambda \in [0,b]} \left\{ \log \mathbb{E}[e^{\lambda(X-\mu)}] - \lambda t \right\}. \tag{2.5}$$

As we explore in Exercise 2.3, the moment bound (2.3) with an optimal choice of k is never worse than the bound (2.5) based on the moment generating function. Nonetheless, the Chernoff bound is most widely used in practice, possibly due to the ease of manipulating moment generating functions. Indeed, a variety of important tail bounds can be obtained as particular cases of inequality (2.5), as we discuss in examples to follow.

2.1.2 Sub-Gaussian variables and Hoeffding bounds

The form of tail bound obtained via the Chernoff approach depends on the growth rate of the moment generating function. Accordingly, in the study of tail bounds, it is natural to classify random variables in terms of their moment generating functions. For reasons to become clear in the sequel, the simplest type of behavior is known as sub-Gaussian. In order to motivate this notion, let us illustrate the use of the Chernoff bound (2.5) in deriving tail bounds for a Gaussian variable.

Example 2.1 (Gaussian tail bounds) Let $X \sim \mathcal{N}(\mu, \sigma^2)$ be a Gaussian random variable with mean μ and variance σ^2. By a straightforward calculation, we find that X has the moment generating function

$$\mathbb{E}[e^{\lambda X}] = e^{\mu\lambda + \frac{\sigma^2\lambda^2}{2}}, \qquad \text{valid for all } \lambda \in \mathbb{R}. \tag{2.6}$$

Substituting this expression into the optimization problem defining the optimized Chernoff bound (2.5), we obtain

$$\inf_{\lambda \geq 0} \left\{ \log \mathbb{E}[e^{\lambda(X-\mu)}] - \lambda t \right\} = \inf_{\lambda \geq 0} \left\{ \frac{\lambda^2 \sigma^2}{2} - \lambda t \right\} = -\frac{t^2}{2\sigma^2},$$

where we have taken derivatives in order to find the optimum of this quadratic function. Returning to the Chernoff bound (2.5), we conclude that any $\mathcal{N}(\mu, \sigma^2)$ random variable satisfies the *upper deviation inequality*

$$\mathbb{P}[X \geq \mu + t] \leq e^{-\frac{t^2}{2\sigma^2}} \qquad \text{for all } t \geq 0. \tag{2.7}$$

In fact, this bound is sharp up to polynomial-factor corrections, as shown by our exploration of the Mills ratio in Exercise 2.2. ♣

Motivated by the structure of this example, we are led to introduce the following definition.

Definition 2.2 A random variable X with mean $\mu = \mathbb{E}[X]$ is *sub-Gaussian* if there is a positive number σ such that

$$\mathbb{E}[e^{\lambda(X-\mu)}] \leq e^{\sigma^2\lambda^2/2} \quad \text{for all } \lambda \in \mathbb{R}. \tag{2.8}$$

The constant σ is referred to as the *sub-Gaussian parameter*; for instance, we say that X is sub-Gaussian with parameter σ when the condition (2.8) holds. Naturally, any Gaussian variable with variance σ^2 is sub-Gaussian with parameter σ, as should be clear from the calculation described in Example 2.1. In addition, as we will see in the examples and exercises to follow, a large number of non-Gaussian random variables also satisfy the condition (2.8).

The condition (2.8), when combined with the Chernoff bound as in Example 2.1, shows that, if X is sub-Gaussian with parameter σ, then it satisfies the *upper deviation inequality* (2.7). Moreover, by the symmetry of the definition, the variable $-X$ is sub-Gaussian if and only if X is sub-Gaussian, so that we also have the *lower deviation inequality* $\mathbb{P}[X \leq \mu - t] \leq e^{-\frac{t^2}{2\sigma^2}}$, valid for all $t \geq 0$. Combining the pieces, we conclude that any sub-Gaussian variable satisfies the *concentration inequality*

$$\mathbb{P}[|X - \mu| \geq t] \leq 2\,e^{-\frac{t^2}{2\sigma^2}} \quad \text{for all } t \in \mathbb{R}. \tag{2.9}$$

Let us consider some examples of sub-Gaussian variables that are non-Gaussian.

Example 2.3 (Rademacher variables) A Rademacher random variable ε takes the values $\{-1, +1\}$ equiprobably. We claim that it is sub-Gaussian with parameter $\sigma = 1$. By taking expectations and using the power-series expansion for the exponential, we obtain

$$\begin{aligned}
\mathbb{E}[e^{\lambda\varepsilon}] = \frac{1}{2}\{e^{-\lambda} + e^{\lambda}\} &= \frac{1}{2}\left\{\sum_{k=0}^{\infty} \frac{(-\lambda)^k}{k!} + \sum_{k=0}^{\infty} \frac{(\lambda)^k}{k!}\right\} \\
&= \sum_{k=0}^{\infty} \frac{\lambda^{2k}}{(2k)!} \\
&\leq 1 + \sum_{k=1}^{\infty} \frac{\lambda^{2k}}{2^k\, k!} \\
&= e^{\lambda^2/2},
\end{aligned}$$

which shows that ε is sub-Gaussian with parameter $\sigma = 1$ as claimed. ♣

We now generalize the preceding example to show that any bounded random variable is also sub-Gaussian.

Example 2.4 (Bounded random variables) Let X be zero-mean, and supported on some interval $[a, b]$. Letting X' be an independent copy, for any $\lambda \in \mathbb{R}$, we have

$$\mathbb{E}_X[e^{\lambda X}] = \mathbb{E}_X[e^{\lambda(X - \mathbb{E}_{X'}[X'])}] \leq \mathbb{E}_{X,X'}[e^{\lambda(X-X')}],$$

where the inequality follows from the convexity of the exponential, and Jensen's inequality. Letting ε be an independent Rademacher variable, note that the distribution of $(X - X')$ is the same as that of $\varepsilon(X - X')$, so that we have

$$\mathbb{E}_{X,X'}[e^{\lambda(X-X')}] = \mathbb{E}_{X,X'}[\mathbb{E}_\varepsilon[e^{\lambda\varepsilon(X-X')}]] \overset{(i)}{\leq} \mathbb{E}_{X,X'}[e^{\frac{\lambda^2(X-X')^2}{2}}],$$

where step (i) follows from the result of Example 2.3, applied conditionally with (X, X') held fixed. Since $|X - X'| \leq b - a$, we are guaranteed that

$$\mathbb{E}_{X,X'}[e^{\frac{\lambda^2(X-X')^2}{2}}] \leq e^{\frac{\lambda^2(b-a)^2}{2}}.$$

Putting together the pieces, we have shown that X is sub-Gaussian with parameter at most $\sigma = b - a$. This result is useful but can be sharpened. In Exercise 2.4, we work through a more involved argument to show that X is sub-Gaussian with parameter at most $\sigma = \frac{b-a}{2}$.

Remark: The technique used in Example 2.4 is a simple example of a *symmetrization argument*, in which we first introduce an independent copy X', and then symmetrize the problem with a Rademacher variable. Such symmetrization arguments are useful in a variety of contexts, as will be seen in later chapters.

Just as the property of Gaussianity is preserved by linear operations, so is the property of sub-Gaussianity. For instance, if X_1 and X_2 are independent sub-Gaussian variables with parameters σ_1 and σ_2, then $X_1 + X_2$ is sub-Gaussian with parameter $\sqrt{\sigma_1^2 + \sigma_2^2}$. See Exercise 2.13 for verification of this fact, as well as some related properties. As a consequence of this fact and the basic sub-Gaussian tail bound (2.7), we obtain an important result, applicable to sums of independent sub-Gaussian random variables, and known as the *Hoeffding bound*:

Proposition 2.5 (Hoeffding bound) *Suppose that the variables X_i, $i = 1, \ldots, n$, are independent, and X_i has mean μ_i and sub-Gaussian parameter σ_i. Then for all $t \geq 0$, we have*

$$\mathbb{P}\left[\sum_{i=1}^{n}(X_i - \mu_i) \geq t\right] \leq \exp\left\{-\frac{t^2}{2\sum_{i=1}^{n}\sigma_i^2}\right\}. \tag{2.10}$$

The Hoeffding bound is often stated only for the special case of bounded random variables. In particular, if $X_i \in [a, b]$ for all $i = 1, 2, \ldots, n$, then from the result of Exercise 2.4, it is sub-Gaussian with parameter $\sigma = \frac{b-a}{2}$, so that we obtain the bound

$$\mathbb{P}\Big[\sum_{i=1}^{n}(X_i - \mu_i) \geq t\Big] \leq e^{-\frac{2t^2}{n(b-a)^2}}. \tag{2.11}$$

Although the Hoeffding bound is often stated in this form, the basic idea applies somewhat more generally to sub-Gaussian variables, as we have given here.

We conclude our discussion of sub-Gaussianity with a result that provides three different characterizations of sub-Gaussian variables. First, the most direct way in which to establish sub-Gaussianity is by computing or bounding the moment generating function, as we have done in Example 2.1. A second intuition is that any sub-Gaussian variable is dominated in a certain sense by a Gaussian variable. Third, sub-Gaussianity also follows by having suitably tight control on the moments of the random variable. The following result shows that all three notions are equivalent in a precise sense.

Theorem 2.6 (Equivalent characterizations of sub-Gaussian variables) *Given any zero-mean random variable X, the following properties are equivalent:*

(I) *There is a constant $\sigma \geq 0$ such that*

$$\mathbb{E}[e^{\lambda X}] \leq e^{\frac{\lambda^2 \sigma^2}{2}} \quad \text{for all } \lambda \in \mathbb{R}. \tag{2.12a}$$

(II) *There is a constant $c \geq 0$ and Gaussian random variable $Z \sim \mathcal{N}(0, \tau^2)$ such that*

$$\mathbb{P}[|X| \geq s] \leq c\,\mathbb{P}[|Z| \geq s] \quad \text{for all } s \geq 0. \tag{2.12b}$$

(III) *There is a constant $\theta \geq 0$ such that*

$$\mathbb{E}[X^{2k}] \leq \frac{(2k)!}{2^k k!}\theta^{2k} \quad \text{for all } k = 1, 2, \ldots. \tag{2.12c}$$

(IV) *There is a constant $\sigma \geq 0$ such that*

$$\mathbb{E}[e^{\frac{\lambda X^2}{2\sigma^2}}] \leq \frac{1}{\sqrt{1-\lambda}} \quad \text{for all } \lambda \in [0, 1). \tag{2.12d}$$

See Appendix A (Section 2.4) for the proof of these equivalences.

2.1.3 Sub-exponential variables and Bernstein bounds

The notion of sub-Gaussianity is fairly restrictive, so that it is natural to consider various relaxations of it. Accordingly, we now turn to the class of sub-exponential variables, which are defined by a slightly milder condition on the moment generating function:

Definition 2.7 A random variable X with mean $\mu = \mathbb{E}[X]$ is *sub-exponential* if there are non-negative parameters (ν, α) such that

$$\mathbb{E}[e^{\lambda(X-\mu)}] \leq e^{\frac{\nu^2\lambda^2}{2}} \qquad \text{for all } |\lambda| < \frac{1}{\alpha}. \tag{2.13}$$

It follows immediately from this definition that any sub-Gaussian variable is also sub-exponential—in particular, with $\nu = \sigma$ and $\alpha = 0$, where we interpret $1/0$ as being the same as $+\infty$. However, the converse statement is not true, as shown by the following calculation:

Example 2.8 (Sub-exponential but not sub-Gaussian) Let $Z \sim \mathcal{N}(0, 1)$, and consider the random variable $X = Z^2$. For $\lambda < \frac{1}{2}$, we have

$$\begin{aligned}\mathbb{E}[e^{\lambda(X-1)}] &= \frac{1}{\sqrt{2\pi}} \int_{-\infty}^{+\infty} e^{\lambda(z^2-1)} e^{-z^2/2}\, dz \\ &= \frac{e^{-\lambda}}{\sqrt{1-2\lambda}}.\end{aligned}$$

For $\lambda > \frac{1}{2}$, the moment generating function is infinite, which reveals that X is *not* sub-Gaussian.

As will be seen momentarily, the existence of the moment generating function in a neighborhood of zero is actually an equivalent definition of a sub-exponential variable. Let us verify directly that condition (2.13) is satisfied. Following some calculus, we find that

$$\frac{e^{-\lambda}}{\sqrt{1-2\lambda}} \leq e^{2\lambda^2} = e^{4\lambda^2/2}, \qquad \text{for all } |\lambda| < \frac{1}{4}, \tag{2.14}$$

which shows that X is sub-exponential with parameters $(\nu, \alpha) = (2, 4)$. ♣

As with sub-Gaussianity, the control (2.13) on the moment generating function, when combined with the Chernoff technique, yields deviation and concentration inequalities for sub-exponential variables. When t is small enough, these bounds are sub-Gaussian in nature (i.e., with the exponent quadratic in t), whereas for larger t, the exponential component of the bound scales linearly in t. We summarize in the following:

Proposition 2.9 (Sub-exponential tail bound) *Suppose that X is sub-exponential with parameters (ν, α). Then*

$$\mathbb{P}[X - \mu \geq t] \leq \begin{cases} e^{-\frac{t^2}{2\nu^2}} & \text{if } 0 \leq t \leq \frac{\nu^2}{\alpha}, \\ e^{-\frac{t}{2\alpha}} & \text{for } t > \frac{\nu^2}{\alpha}. \end{cases}$$

As with the Hoeffding inequality, similar bounds can be derived for the left-sided event $\{X - \mu \leq -t\}$, as well as the two-sided event $\{|X - \mu| \geq t\}$, with an additional factor of 2 in the latter case.

Proof By recentering as needed, we may assume without loss of generality that $\mu = 0$. We follow the usual Chernoff-type approach: combining it with the definition (2.13) of a sub-exponential variable yields the upper bound

$$\mathbb{P}[X \geq t] \leq e^{-\lambda t}\, \mathbb{E}[e^{\lambda X}] \leq \exp\underbrace{\left(-\lambda t + \frac{\lambda^2 \nu^2}{2}\right)}_{g(\lambda,t)}, \qquad \text{valid for all } \lambda \in [0, \alpha^{-1}).$$

In order to complete the proof, it remains to compute, for each fixed $t \geq 0$, the quantity $g^*(t) := \inf_{\lambda \in [0,\alpha^{-1})} g(\lambda, t)$. Note that the unconstrained minimum of the function $g(\cdot, t)$ occurs at $\lambda^* = t/\nu^2$. If $0 \leq t < \frac{\nu^2}{\alpha}$, then this unconstrained optimum corresponds to the constrained minimum as well, so that $g^*(t) = -\frac{t^2}{2\nu^2}$ over this interval.

Otherwise, we may assume that $t \geq \frac{\nu^2}{\alpha}$. In this case, since the function $g(\cdot, t)$ is monotonically decreasing in the interval $[0, \lambda^*)$, the constrained minimum is achieved at the boundary point $\lambda^\dagger = \alpha^{-1}$, and we have

$$g^*(t) = g(\lambda^\dagger, t) = -\frac{t}{\alpha} + \frac{1}{2\alpha}\frac{\nu^2}{\alpha} \overset{\text{(i)}}{\leq} -\frac{t}{2\alpha},$$

where inequality (i) uses the fact that $\frac{\nu^2}{\alpha} \leq t$. □

As shown in Example 2.8, the sub-exponential property can be verified by explicitly computing or bounding the moment generating function. This direct calculation may be impracticable in many settings, so it is natural to seek alternative approaches. One such method is based on control of the polynomial moments of X. Given a random variable X with mean $\mu = \mathbb{E}[X]$ and variance $\sigma^2 = \mathbb{E}[X^2] - \mu^2$, we say that *Bernstein's condition* with parameter b holds if

$$|\mathbb{E}[(X-\mu)^k]| \leq \tfrac{1}{2} k! \sigma^2 b^{k-2} \qquad \text{for } k = 2, 3, 4, \ldots. \tag{2.15}$$

One sufficient condition for Bernstein's condition to hold is that X be bounded; in particular, if $|X - \mu| \leq b$, then it is straightforward to verify that condition (2.15) holds. Even for bounded variables, our next result will show that the Bernstein condition can be used to obtain tail bounds that may be tighter than the Hoeffding bound. Moreover, Bernstein's condition is also satisfied by various unbounded variables, a property which lends it much broader applicability.

When X satisfies the Bernstein condition, then it is sub-exponential with parameters determined by σ^2 and b. Indeed, by the power-series expansion of the exponential, we have

$$\begin{aligned}
\mathbb{E}[e^{\lambda(X-\mu)}] &= 1 + \frac{\lambda^2\sigma^2}{2} + \sum_{k=3}^{\infty} \lambda^k \frac{\mathbb{E}[(X-\mu)^k]}{k!} \\
&\overset{\text{(i)}}{\leq} 1 + \frac{\lambda^2\sigma^2}{2} + \frac{\lambda^2\sigma^2}{2} \sum_{k=3}^{\infty} (|\lambda|\, b)^{k-2},
\end{aligned}$$

where the inequality (i) makes use of the Bernstein condition (2.15). For any $|\lambda| < 1/b$, we can sum the geometric series so as to obtain

$$\mathbb{E}[e^{\lambda(X-\mu)}] \leq 1 + \frac{\lambda^2\sigma^2/2}{1-b|\lambda|} \stackrel{\text{(ii)}}{\leq} e^{\frac{\lambda^2\sigma^2/2}{1-b|\lambda|}}, \tag{2.16}$$

where inequality (ii) follows from the bound $1 + t \leq e^t$. Consequently, we conclude that

$$\mathbb{E}[e^{\lambda(X-\mu)}] \leq e^{\frac{\lambda^2(\sqrt{2}\sigma)^2}{2}} \qquad \text{for all } |\lambda| < \tfrac{1}{2b},$$

showing that X is sub-exponential with parameters $(\sqrt{2}\sigma, 2b)$.

As a consequence, an application of Proposition 2.9 leads directly to tail bounds on a random variable satisfying the Bernstein condition (2.15). However, the resulting tail bound can be sharpened slightly, at least in terms of constant factors, by making direct use of the upper bound (2.16). We summarize in the following:

Proposition 2.10 (Bernstein-type bound) *For any random variable satisfying the Bernstein condition* (2.15), *we have*

$$\mathbb{E}[e^{\lambda(X-\mu)}] \leq e^{\frac{\lambda^2\sigma^2/2}{1-b|\lambda|}} \qquad \textit{for all } |\lambda| < \tfrac{1}{b}, \tag{2.17a}$$

and, moreover, the concentration inequality

$$\mathbb{P}[|X-\mu| \geq t] \leq 2e^{-\frac{t^2}{2(\sigma^2+bt)}} \qquad \textit{for all } t \geq 0. \tag{2.17b}$$

We proved inequality (2.17a) in the discussion preceding this proposition. Using this bound on the moment generating function, the tail bound (2.17b) follows by setting $\lambda = \frac{t}{bt+\sigma^2} \in [0, \frac{1}{b})$ in the Chernoff bound, and then simplifying the resulting expression.

Remark: Proposition 2.10 has an important consequence even for bounded random variables (i.e., those satisfying $|X - \mu| \leq b$). The most straightforward way to control such variables is by exploiting the boundedness to show that $(X-\mu)$ is sub-Gaussian with parameter b (see Exercise 2.4), and then applying a Hoeffding-type inequality (see Proposition 2.5). Alternatively, using the fact that any bounded variable satisfies the Bernstein condition (2.16), we can also apply Proposition 2.10, thereby obtaining the tail bound (2.17b), that involves *both* the variance σ^2 and the bound b. This tail bound shows that for suitably small t, the variable X has sub-Gaussian behavior with parameter σ, as opposed to the parameter b that would arise from a Hoeffding approach. Since $\sigma^2 = \mathbb{E}[(X-\mu)^2] \leq b^2$, this bound is never worse; moreover, it is substantially better when $\sigma^2 \ll b^2$, as would be the case for a random variable that occasionally takes on large values, but has relatively small variance. Such variance-based control frequently plays a key role in obtaining optimal rates in statistical problems, as will be seen in later chapters. For bounded random variables, Bennett's inequality can be used to provide sharper control on the tails (see Exercise 2.7).

Like the sub-Gaussian property, the sub-exponential property is preserved under summation for independent random variables, and the parameters (ν, α) transform in a simple way. In particular, consider an independent sequence $\{X_k\}_{k=1}^n$ of random variables, such that X_k has mean μ_k, and is sub-exponential with parameters (ν_k, α_k). We compute the moment generating function

$$\mathbb{E}[e^{\lambda \sum_{k=1}^n (X_k-\mu_k)}] \stackrel{\text{(i)}}{=} \prod_{k=1}^n \mathbb{E}[e^{\lambda(X_k-\mu_k)}] \stackrel{\text{(ii)}}{\leq} \prod_{k=1}^n e^{\lambda^2 \nu_k^2/2},$$

valid for all $|\lambda| < \left(\max_{k=1,\ldots,n} \alpha_k\right)^{-1}$, where equality (i) follows from independence, and inequality (ii) follows since X_k is sub-exponential with parameters (ν_k, α_k). Thus, we conclude that the variable $\sum_{k=1}^n (X_k - \mu_k)$ is sub-exponential with the parameters (ν_*, α_*), where

$$\alpha_* := \max_{k=1,\ldots,n} \alpha_k \quad \text{and} \quad \nu_* := \sqrt{\sum_{k=1}^n \nu_k^2}.$$

Using the same argument as in Proposition 2.9, this observation leads directly to the upper tail bound

$$\mathbb{P}\left[\frac{1}{n}\sum_{i=1}^n (X_k - \mu_k) \geq t\right] \leq \begin{cases} e^{-\frac{nt^2}{2(\nu_*^2/n)}} & \text{for } 0 \leq t \leq \frac{\nu_*^2}{n\alpha_*}, \\ e^{-\frac{nt}{2\alpha_*}} & \text{for } t > \frac{\nu_*^2}{n\alpha_*}, \end{cases} \tag{2.18}$$

along with similar two-sided tail bounds. Let us illustrate our development thus far with some examples.

Example 2.11 (χ^2-variables) A chi-squared (χ^2) random variable with n degrees of freedom, denoted by $Y \sim \chi_n^2$, can be represented as the sum $Y = \sum_{k=1}^n Z_k^2$ where $Z_k \sim \mathcal{N}(0, 1)$ are i.i.d. variates. As discussed in Example 2.8, the variable Z_k^2 is sub-exponential with parameters $(2, 4)$. Consequently, since the variables $\{Z_k\}_{k=1}^n$ are independent, the χ^2-variate Y is sub-exponential with parameters $(\nu, \alpha) = (2\sqrt{n}, 4)$, and the preceding discussion yields the two-sided tail bound

$$\mathbb{P}\left[\left|\frac{1}{n}\sum_{k=1}^n Z_k^2 - 1\right| \geq t\right] \leq 2e^{-nt^2/8}, \qquad \text{for all } t \in (0, 1). \tag{2.19}$$

♣

The concentration of χ^2-variables plays an important role in the analysis of procedures based on taking random projections. A classical instance of the random projection method is the Johnson–Lindenstrauss analysis of metric embedding.

Example 2.12 (Johnson–Lindenstrauss embedding) As one application of the concentration of χ^2-variables, consider the following problem. Suppose that we are given $N \geq 2$ distinct vectors $\{u^1, \ldots, u^N\}$, with each vector lying in $\mathbb{R}^d$. If the data dimension d is large, then it might be expensive to store and manipulate the data set. The idea of dimensionality reduction is to construct a mapping $F\colon \mathbb{R}^d \to \mathbb{R}^m$—with the projected dimension m substantially smaller than d—that preserves some "essential" features of the data set. What features should we try to preserve? There is not a unique answer to this question but, as one interesting example, we might consider preserving pairwise distances, or equivalently norms and inner products. Many algorithms are based on such pairwise quantities, including linear regression, methods for principal components, the k-means algorithm for clustering, and nearest-neighbor algorithms for density estimation. With these motivations in mind, given some tolerance $\delta \in (0, 1)$, we might be interested in a mapping F with the guarantee that

$$(1-\delta) \le \frac{\|F(u^i) - F(u^j)\|_2^2}{\|u^i - u^j\|_2^2} \le (1+\delta) \qquad \text{for all pairs } u^i \neq u^j. \tag{2.20}$$

In words, the projected data set $\{F(u^1), \ldots, F(u^N)\}$ preserves all pairwise squared distances up to a multiplicative factor of δ. Of course, this is always possible if the projected dimension m is large enough, but the goal is to do it with relatively small m.

Constructing such a mapping that satisfies the condition (2.20) with high probability turns out to be straightforward as long as the projected dimension is lower bounded as $m \gtrsim \frac{1}{\delta^2} \log N$. Observe that the projected dimension is independent of the ambient dimension d, and scales only logarithmically with the number of data points N.

The construction is probabilistic: first form a random matrix $\mathbf{X} \in \mathbb{R}^{m \times d}$ filled with independent $\mathcal{N}(0, 1)$ entries, and use it to define a linear mapping $F: \mathbb{R}^d \to \mathbb{R}^m$ via $u \mapsto \mathbf{X}u/\sqrt{m}$. We now verify that F satisfies condition (2.20) with high probability. Let $x_i \in \mathbb{R}^d$ denote the ith row of $\mathbf{X}$, and consider some fixed $u \neq 0$. Since x_i is a standard normal vector, the variable $\langle x_i, u/\|u\|_2 \rangle$ follows a $\mathcal{N}(0, 1)$ distribution, and hence the quantity

$$Y := \frac{\|\mathbf{X}u\|_2^2}{\|u\|_2^2} = \sum_{i=1}^m \langle x_i, u/\|u\|_2 \rangle^2,$$

follows a χ^2 distribution with m degrees of freedom, using the independence of the rows. Therefore, applying the tail bound (2.19), we find that

$$\mathbb{P}\left[\left|\frac{\|\mathbf{X}u\|_2^2}{m\,\|u\|_2^2} - 1\right| \ge \delta\right] \le 2e^{-m\delta^2/8} \qquad \text{for all } \delta \in (0, 1).$$

Rearranging and recalling the definition of F yields the bound

$$\mathbb{P}\left[\frac{\|F(u)\|_2^2}{\|u\|_2^2} \notin [(1-\delta), (1+\delta)]\right] \le 2e^{-m\delta^2/8}, \qquad \text{for any fixed } 0 \neq u \in \mathbb{R}^d.$$

Noting that there are $\binom{N}{2}$ distinct pairs of data points, we apply the union bound to conclude that

$$\mathbb{P}\left[\frac{\|F(u^i - u^j)\|_2^2}{\|u^i - u^j\|_2^2} \notin [(1-\delta), (1+\delta)] \text{ for some } u^i \neq u^j\right] \le 2\binom{N}{2} e^{-m\delta^2/8}.$$

For any $\epsilon \in (0, 1)$, this probability can be driven below ϵ by choosing $m > \frac{16}{\delta^2} \log(N/\epsilon)$. ♣

In parallel to Theorem 2.13, there are a number of equivalent ways to characterize a sub-exponential random variable. The following theorem provides a summary:

Theorem 2.13 (Equivalent characterizations of sub-exponential variables) *For a zero-mean random variable X, the following statements are equivalent:*

(I) *There are non-negative numbers* (ν, α) *such that*

$$\mathbb{E}[e^{\lambda X}] \leq e^{\frac{\nu^2 \lambda^2}{2}} \qquad \text{for all } |\lambda| < \tfrac{1}{\alpha}. \tag{2.21a}$$

(II) *There is a positive number* $c_0 > 0$ *such that* $\mathbb{E}[e^{\lambda X}] < \infty$ *for all* $|\lambda| \leq c_0$.

(III) *There are constants* $c_1, c_2 > 0$ *such that*

$$\mathbb{P}[|X| \geq t] \leq c_1 \, e^{-c_2 t} \qquad \text{for all } t > 0. \tag{2.21b}$$

(IV) *The quantity* $\gamma := \sup_{k \geq 2} \left[\frac{\mathbb{E}[X^k]}{k!}\right]^{1/k}$ *is finite.*

See Appendix B (Section 2.5) for the proof of this claim.

2.1.4 Some one-sided results

Up to this point, we have focused on two-sided forms of Bernstein's condition, which yields bounds on both the upper and lower tails. As we have seen, one sufficient condition for Bernstein's condition to hold is a bound on the absolute value, say $|X| \leq b$ almost surely. Of course, if such a bound only holds in a one-sided way, it is still possible to derive one-sided bounds. In this section, we state and prove one such result.

Proposition 2.14 (One-sided Bernstein's inequality) *If* $X \leq b$ *almost surely, then*

$$\mathbb{E}[e^{\lambda(X - \mathbb{E}[X])}] \leq \exp\left(\frac{\frac{\lambda^2}{2}\mathbb{E}[X^2]}{1 - \frac{b\lambda}{3}}\right) \qquad \text{for all } \lambda \in [0, 3/b). \tag{2.22a}$$

Consequently, given n independent random variables such that $X_i \leq b$ *almost surely, we have*

$$\mathbb{P}\left[\sum_{i=1}^{n}(X_i - \mathbb{E}[X_i]) \geq n\delta\right] \leq \exp\left(-\frac{n\delta^2}{2\left(\frac{1}{n}\sum_{i=1}^{n}\mathbb{E}[X_i^2] + \frac{b\delta}{3}\right)}\right). \tag{2.22b}$$

Of course, if a random variable is bounded from below, then the same result can be used to derive bounds on its lower tail; we simply apply the bound (2.22b) to the random variable $-X$. In the special case of independent non-negative random variables $Y_i \geq 0$, we find that

$$\mathbb{P}\left[\sum_{i=1}^{n}(Y_i - \mathbb{E}[Y_i]) \leq -n\delta\right] \leq \exp\left(-\frac{n\delta^2}{\frac{2}{n}\sum_{i=1}^{n}\mathbb{E}[Y_i^2]}\right). \tag{2.23}$$

Thus, we see that the lower tail of *any* non-negative random variable satisfies a bound of the sub-Gaussian type, albeit with the second moment instead of the variance.

The proof of Proposition 2.14 is quite straightforward given our development thus far.

Proof Defining the function

$$h(u) := 2\frac{e^u - u - 1}{u^2} = 2\sum_{k=2}^{\infty} \frac{u^{k-2}}{k!},$$

we have the expansion

$$\mathbb{E}[e^{\lambda X}] = 1 + \lambda \mathbb{E}[X] + \tfrac{1}{2}\lambda^2 \mathbb{E}[X^2 h(\lambda X)].$$

Observe that for all scalars $x < 0$, $x' \in [0, b]$ and $\lambda > 0$, we have

$$h(\lambda x) \leq h(0) \leq h(\lambda x') \leq h(\lambda b).$$

Consequently, since $X \leq b$ almost surely, we have $\mathbb{E}[X^2 h(\lambda X)] \leq \mathbb{E}[X^2] h(\lambda b)$, and hence

$$\begin{aligned}\mathbb{E}[e^{\lambda(X - \mathbb{E}[X])}] &\leq e^{-\lambda \mathbb{E}[X]}\{1 + \lambda \mathbb{E}[X] + \tfrac{1}{2}\lambda^2 \mathbb{E}[X^2] h(\lambda b)\} \\ &\leq \exp\left\{\frac{\lambda^2 \mathbb{E}[X^2]}{2} h(\lambda b)\right\}.\end{aligned}$$

Consequently, the bound (2.22a) will follow if we can show that $h(\lambda b) \leq (1 - \frac{\lambda b}{3})^{-1}$ for $\lambda b < 3$. By applying the inequality $k! \geq 2(3^{k-2})$, valid for all $k \geq 2$, we find that

$$h(\lambda b) = 2\sum_{k=2}^{\infty} \frac{(\lambda b)^{k-2}}{k!} \leq \sum_{k=2}^{\infty} \left(\frac{\lambda b}{3}\right)^{k-2} = \frac{1}{1 - \frac{\lambda b}{3}},$$

where the condition $\frac{\lambda b}{3} \in [0, 1)$ allows us to sum the geometric series.

In order to prove the upper tail bound (2.22b), we apply the Chernoff bound, exploiting independence to apply the moment generating function bound (2.22a) separately, and thereby find that

$$\mathbb{P}\left[\sum_{i=1}^{n}(X_i - \mathbb{E}[X_i]) \geq n\delta\right] \leq \exp\left(-\lambda n\delta + \frac{\frac{\lambda^2}{2}\sum_{i=1}^{n} \mathbb{E}[X_i^2]}{1 - \frac{b\lambda}{3}}\right), \qquad \text{valid for } b\lambda \in [0, 3).$$

Substituting

$$\lambda = \frac{n\delta}{\sum_{i=1}^{n} \mathbb{E}[X_i^2] + \frac{n\delta b}{3}} \in [0, 3/b)$$

and simplifying yields the bound. □

2.2 Martingale-based methods

Up until this point, our techniques have provided various types of bounds on sums of independent random variables. Many problems require bounds on more general functions of random variables, and one classical approach is based on martingale decompositions. In this section, we describe some of the results in this area along with some examples. Our treatment is quite brief, so we refer the reader to the bibliographic section for additional references.

2.2.1 Background

Let us begin by introducing a particular case of a martingale sequence that is especially relevant for obtaining tail bounds. Let $\{X_k\}_{k=1}^n$ be a sequence of independent random variables, and consider the random variable $f(X) = f(X_1 \ldots, X_n)$, for some function $f\colon \mathbb{R}^n \to \mathbb{R}$. Suppose that our goal is to obtain bounds on the deviations of f from its mean. In order to do so, we consider the sequence of random variables given by $Y_0 = \mathbb{E}[f(X)]$, $Y_n = f(X)$, and

$$Y_k = \mathbb{E}[f(X) \mid X_1, \ldots, X_k] \qquad \text{for } k = 1, \ldots, n-1, \tag{2.24}$$

where we assume that all conditional expectations exist. Note that Y_0 is a constant, and the random variables Y_k will tend to exhibit more fluctuations as we move along the sequence from Y_0 to Y_n. Based on this intuition, the martingale approach to tail bounds is based on the telescoping decomposition

$$f(X) - \mathbb{E}[f(X)] = Y_n - Y_0 = \sum_{k=1}^{n} \underbrace{(Y_k - Y_{k-1})}_{D_k},$$

in which the deviation $f(X) - \mathbb{E}[f(X)]$ is written as a sum of increments $\{D_k\}_{k=1}^n$. As we will see, the sequence $\{Y_k\}_{k=1}^n$ is a particular example of a martingale sequence, known as the *Doob martingale*, whereas the sequence $\{D_k\}_{k=1}^n$ is an example of a martingale difference sequence.

With this example in mind, we now turn to the general definition of a martingale sequence. Let $\{\mathcal{F}_k\}_{k=1}^\infty$ be a sequence of σ-fields that are nested, meaning that $\mathcal{F}_k \subseteq \mathcal{F}_{k+1}$ for all $k \geq 1$; such a sequence is known as a *filtration*. In the Doob martingale described above, the σ-field $\sigma(X_1, \ldots, X_k)$ generated by the first k variables plays the role of $\mathcal{F}_k$. Let $\{Y_k\}_{k=1}^\infty$ be a sequence of random variables such that Y_k is measurable with respect to the σ-field $\mathcal{F}_k$. In this case, we say that $\{Y_k\}_{k=1}^\infty$ is adapted to the filtration $\{\mathcal{F}_k\}_{k=1}^\infty$. In the Doob martingale, the random variable Y_k is a measurable function of $(X_1, \ldots, X_k)$, and hence the sequence is adapted to the filtration defined by the σ-fields. We are now ready to define a general martingale:

Definition 2.15 Given a sequence $\{Y_k\}_{k=1}^\infty$ of random variables adapted to a filtration $\{\mathcal{F}_k\}_{k=1}^\infty$, the pair $\{(Y_k, \mathcal{F}_k)\}_{k=1}^\infty$ is a *martingale* if, for all $k \geq 1$,

$$\mathbb{E}[|Y_k|] < \infty \quad \text{and} \quad \mathbb{E}[Y_{k+1} \mid \mathcal{F}_k] = Y_k. \tag{2.25}$$

It is frequently the case that the filtration is defined by a second sequence of random variables $\{X_k\}_{k=1}^\infty$ via the canonical σ-fields $\mathcal{F}_k := \sigma(X_1, \ldots, X_k)$. In this case, we say that $\{Y_k\}_{k=1}^\infty$ is a martingale sequence with respect to $\{X_k\}_{k=1}^\infty$. The Doob construction is an instance of such a martingale sequence. If a sequence is martingale with respect to itself (i.e., with $\mathcal{F}_k = \sigma(Y_1, \ldots, Y_k)$), then we say simply that $\{Y_k\}_{k=1}^\infty$ forms a martingale sequence.

Let us consider some examples to illustrate:

Example 2.16 (Partial sums as martingales) Perhaps the simplest instance of a martingale is provided by considering partial sums of an i.i.d. sequence. Let $\{X_k\}_{k=1}^\infty$ be a sequence of i.i.d. random variables with mean μ, and define the partial sums $S_k := \sum_{j=1}^k X_j$. Defining $\mathcal{F}_k = \sigma(X_1, \ldots, X_k)$, the random variable S_k is measurable with respect to $\mathcal{F}_k$, and, moreover, we have

$$\begin{aligned}\mathbb{E}[S_{k+1} \mid \mathcal{F}_k] &= \mathbb{E}[X_{k+1} + S_k \mid X_1, \ldots, X_k] \\ &= \mathbb{E}[X_{k+1}] + S_k \\ &= \mu + S_k.\end{aligned}$$

Here we have used the facts that X_{k+1} is independent of $X_1^k := (X_1, \ldots, X_k)$, and that S_k is a function of X_1^k. Thus, while the sequence $\{S_k\}_{k=1}^\infty$ itself is not a martingale unless $\mu = 0$, the recentered variables $Y_k := S_k - k\mu$ for $k \geq 1$ define a martingale sequence with respect to $\{X_k\}_{k=1}^\infty$. ♣

Let us now show that the Doob construction does lead to a martingale, as long as the underlying function f is absolutely integrable.

Example 2.17 (Doob construction) Given a sequence of independent random variables $\{X_k\}_{k=1}^n$, recall the sequence $Y_k = \mathbb{E}[f(X) \mid X_1, \ldots, X_k]$ previously defined, and suppose that $\mathbb{E}[|f(X)|] < \infty$. We claim that $\{Y_k\}_{k=0}^n$ is a martingale with respect to $\{X_k\}_{k=1}^n$. Indeed, in terms of the shorthand $X_1^k = (X_1, X_2, \ldots, X_k)$, we have

$$\mathbb{E}[|Y_k|] = \mathbb{E}[|\mathbb{E}[f(X) \mid X_1^k]|] \leq \mathbb{E}[|f(X)|] < \infty,$$

where the bound follows from Jensen's inequality. Turning to the second property, we have

$$\mathbb{E}[Y_{k+1} \mid X_1^k] = \mathbb{E}[\mathbb{E}[f(X) \mid X_1^{k+1}] \mid X_1^k] \stackrel{\text{(i)}}{=} \mathbb{E}[f(X) \mid X_1^k] = Y_k,$$

where we have used the tower property of conditional expectation in step (i). ♣

The following martingale plays an important role in analyzing stopping rules for sequential hypothesis tests:

Example 2.18 (Likelihood ratio) Let f and g be two mutually absolutely continuous densities, and let $\{X_k\}_{k=1}^\infty$ be a sequence of random variables drawn i.i.d. according to f. For each $k \geq 1$, let $Y_k := \prod_{\ell=1}^k \frac{g(X_\ell)}{f(X_\ell)}$ be the likelihood ratio based on the first k samples. Then the sequence $\{Y_k\}_{k=1}^\infty$ is a martingale with respect to $\{X_k\}_{k=1}^\infty$. Indeed, we have

$$\mathbb{E}[Y_{n+1} \mid X_1, \ldots, X_n] = \mathbb{E}\left[\frac{g(X_{n+1})}{f(X_{n+1})}\right] \prod_{k=1}^n \frac{g(X_k)}{f(X_k)} = Y_n,$$

using the fact that $\mathbb{E}\left[\frac{g(X_{n+1})}{f(X_{n+1})}\right] = 1$. ♣

A closely related notion is that of *martingale difference sequence*, meaning an adapted sequence $\{(D_k, \mathcal{F}_k)\}_{k=1}^\infty$ such that, for all $k \geq 1$,

$$\mathbb{E}[|D_k|] < \infty \quad \text{and} \quad \mathbb{E}[D_{k+1} \mid \mathcal{F}_k] = 0. \tag{2.26}$$

As suggested by their name, such difference sequences arise in a natural way from martingales. In particular, given a martingale $\{(Y_k, \mathcal{F}_k)\}_{k=0}^\infty$, let us define $D_k = Y_k - Y_{k-1}$ for $k \geq 1$.

We then have

$$\begin{aligned}\mathbb{E}[D_{k+1} \mid \mathcal{F}_k] &= \mathbb{E}[Y_{k+1} \mid \mathcal{F}_k] - \mathbb{E}[Y_k \mid \mathcal{F}_k] \\ &= \mathbb{E}[Y_{k+1} \mid \mathcal{F}_k] - Y_k = 0,\end{aligned}$$

using the martingale property (2.25) and the fact that Y_k is measurable with respect to $\mathcal{F}_k$. Thus, for any martingale sequence $\{Y_k\}_{k=0}^{\infty}$, we have the telescoping decomposition

$$Y_n - Y_0 = \sum_{k=1}^{n} D_k, \tag{2.27}$$

where $\{D_k\}_{k=1}^{\infty}$ is a martingale difference sequence. This decomposition plays an important role in our development of concentration inequalities to follow.

2.2.2 Concentration bounds for martingale difference sequences

We now turn to the derivation of concentration inequalities for martingales. These inequalities can be viewed in one of two ways: either as bounds for the difference $Y_n - Y_0$, or as bounds for the sum $\sum_{k=1}^{n} D_k$ of the associated martingale difference sequence. Throughout this section, we present results mainly in terms of martingale differences, with the understanding that such bounds have direct consequences for martingale sequences. Of particular interest to us is the Doob martingale described in Example 2.17, which can be used to control the deviations of a function from its expectation.

We begin by stating and proving a general Bernstein-type bound for a martingale difference sequence, based on imposing a sub-exponential condition on the martingale differences.

Theorem 2.19 *Let $\{(D_k, \mathcal{F}_k)\}_{k=1}^{\infty}$ be a martingale difference sequence, and suppose that $\mathbb{E}[e^{\lambda D_k} \mid \mathcal{F}_{k-1}] \leq e^{\lambda^2 \nu_k^2/2}$ almost surely for any $|\lambda| < 1/\alpha_k$. Then the following hold:*

(a) *The sum $\sum_{k=1}^{n} D_k$ is sub-exponential with parameters $\left(\sqrt{\sum_{k=1}^{n} \nu_k^2},\ \alpha_*\right)$ where $\alpha_* := \max_{k=1,\ldots,n} \alpha_k$.*

(b) *The sum satisfies the concentration inequality*

$$\mathbb{P}\left[\left|\sum_{k=1}^{n} D_k\right| \geq t\right] \leq \begin{cases} 2e^{-\frac{t^2}{2\sum_{k=1}^{n} \nu_k^2}} & \text{if } 0 \leq t \leq \frac{\sum_{k=1}^{n} \nu_k^2}{\alpha_*}, \\ 2e^{-\frac{t}{2\alpha_*}} & \text{if } t > \frac{\sum_{k=1}^{n} \nu_k^2}{\alpha_*}. \end{cases} \tag{2.28}$$

Proof We follow the standard approach of controlling the moment generating function of $\sum_{k=1}^{n} D_k$, and then applying the Chernoff bound. For any scalar λ such that $|\lambda| < \frac{1}{\alpha_*}$, conditioning on $\mathcal{F}_{n-1}$ and applying iterated expectation yields

$$\mathbb{E}[e^{\lambda\left(\sum_{k=1}^{n} D_k\right)}] = \mathbb{E}[e^{\lambda\left(\sum_{k=1}^{n-1} D_k\right)}\mathbb{E}[e^{\lambda D_n} \mid \mathcal{F}_{n-1}]]$$
$$\leq \mathbb{E}[e^{\lambda \sum_{k=1}^{n-1} D_k}]e^{\lambda^2 \nu_n^2/2}, \tag{2.29}$$

where the inequality follows from the stated assumption on D_n. Iterating this procedure yields the bound $\mathbb{E}[e^{\lambda \sum_{k=1}^{n} D_k}] \leq e^{\lambda^2 \sum_{k=1}^{n} \nu_k^2/2}$, valid for all $|\lambda| < \frac{1}{\alpha_*}$. By definition, we conclude that $\sum_{k=1}^{n} D_k$ is sub-exponential with parameters $\left(\sqrt{\sum_{k=1}^{n} \nu_k^2}, \alpha_*\right)$, as claimed. The tail bound (2.28) follows by applying Proposition 2.9. □

In order for Theorem 2.19 to be useful in practice, we need to isolate sufficient and easily checkable conditions for the differences D_k to be almost surely sub-exponential (or sub-Gaussian when $\alpha = 0$). As discussed previously, bounded random variables are sub-Gaussian, which leads to the following corollary:

Corollary 2.20 (Azuma–Hoeffding) *Let $(\{(D_k, \mathcal{F}_k)\}_{k=1}^{\infty})$ be a martingale difference sequence for which there are constants $\{(a_k, b_k)\}_{k=1}^{n}$ such that $D_k \in [a_k, b_k]$ almost surely for all $k = 1, \ldots, n$. Then, for all $t \geq 0$,*

$$\mathbb{P}\left[\left|\sum_{k=1}^{n} D_k\right| \geq t\right] \leq 2e^{-\frac{2t^2}{\sum_{k=1}^{n}(b_k-a_k)^2}}. \tag{2.30}$$

Proof Recall the decomposition (2.29) in the proof of Theorem 2.19; from the structure of this argument, it suffices to show that $\mathbb{E}[e^{\lambda D_k} \mid \mathcal{F}_{k-1}] \leq e^{\lambda^2(b_k-a_k)^2/8}$ almost surely for each $k = 1, 2, \ldots, n$. But since $D_k \in [a_k, b_k]$ almost surely, the conditioned variable $(D_k \mid \mathcal{F}_{k-1})$ also belongs to this interval almost surely, and hence from the result of Exercise 2.4, it is sub-Gaussian with parameter at most $\sigma = (b_k - a_k)/2$. □

An important application of Corollary 2.20 concerns functions that satisfy a bounded difference property. Let us first introduce some convenient notation. Given vectors $x, x' \in \mathbb{R}^n$ and an index $k \in \{1, 2, \ldots, n\}$, we define a new vector $x^{\backslash k} \in \mathbb{R}^n$ via

$$x_j^{\backslash k} := \begin{cases} x_j & \text{if } j \neq k, \\ x'_k & \text{if } j = k. \end{cases} \tag{2.31}$$

With this notation, we say that $f\colon \mathbb{R}^n \to \mathbb{R}$ satisfies the bounded difference inequality with parameters $(L_1, \ldots, L_n)$ if, for each index $k = 1, 2, \ldots, n$,

$$|f(x) - f(x^{\backslash k})| \leq L_k \qquad \text{for all } x, x' \in \mathbb{R}^n. \tag{2.32}$$

For instance, if the function f is L-Lipschitz with respect to the Hamming norm $d_H(x, y) = \sum_{i=1}^{n} \mathbb{I}[x_i \neq y_i]$, which counts the number of positions in which x and y differ, then the bounded difference inequality holds with parameter L uniformly across all coordinates.

Corollary 2.21 (Bounded differences inequality) *Suppose that f satisfies the bounded difference property (2.32) with parameters $(L_1, \ldots, L_n)$ and that the random vector $X = (X_1, X_2, \ldots, X_n)$ has independent components. Then*

$$\mathbb{P}[|f(X) - \mathbb{E}[f(X)]| \geq t] \leq 2e^{-\frac{2t^2}{\sum_{k=1}^n L_k^2}} \qquad \text{for all } t \geq 0. \tag{2.33}$$

Proof Recalling the Doob martingale introduced in Example 2.17, consider the associated martingale difference sequence

$$D_k = \mathbb{E}[f(X) \mid X_1, \ldots, X_k] - \mathbb{E}[f(X) \mid X_1, \ldots, X_{k-1}]. \tag{2.34}$$

We claim that D_k lies in an interval of length at most L_k almost surely. In order to prove this claim, define the random variables

$$A_k := \inf_x \mathbb{E}[f(X) \mid X_1, \ldots, X_{k-1}, x] - \mathbb{E}[f(X) \mid X_1, \ldots, X_{k-1}]$$

and

$$B_k := \sup_x \mathbb{E}[f(X) \mid X_1, \ldots, X_{k-1}, x] - \mathbb{E}[f(X) \mid X_1, \ldots, X_{k-1}].$$

On one hand, we have

$$D_k - A_k = \mathbb{E}[f(X) \mid X_1, \ldots, X_k] - \inf_x \mathbb{E}[f(X) \mid X_1, \ldots, X_{k-1}, x],$$

so that $D_k \geq A_k$ almost surely. A similar argument shows that $D_k \leq B_k$ almost surely.

We now need to show that $B_k - A_k \leq L_k$ almost surely. Observe that by the independence of $\{X_k\}_{k=1}^n$, we have

$$\mathbb{E}[f(X) \mid x_1, \ldots, x_k] = \mathbb{E}_{k+1}[f(x_1, \ldots, x_k, X_{k+1}^n)] \qquad \text{for any vector } (x_1, \ldots, x_k),$$

where $\mathbb{E}_{k+1}$ denotes expectation over $X_{k+1}^n := (X_{k+1}, \ldots, X_n)$. Consequently, we have

$$\begin{aligned} B_k - A_k &= \sup_x \mathbb{E}_{k+1}[f(X_1, \ldots, X_{k-1}, x, X_{k+1}^n] - \inf_x \mathbb{E}_{k+1}[f(X_1, \ldots, X_{k-1}, x, X_{k+1}^n] \\ &\leq \sup_{x,y} \left|\mathbb{E}_{k+1}[f(X_1, \ldots, X_{k-1}, x, X_{k+1}^n) - f(X_1, \ldots, X_{k-1}, y, X_{k+1}^n)]\right| \\ &\leq L_k, \end{aligned}$$

using the bounded differences assumption. Thus, the variable D_k lies within an interval of length L_k at most surely, so that the claim follows as a corollary of the Azuma–Hoeffding inequality. □

Remark: In the special case when f is L-Lipschitz with respect to the Hamming norm, Corollary 2.21 implies that

$$\mathbb{P}[|f(X) - \mathbb{E}[f(X)]| \geq t] \leq 2e^{-\frac{2t^2}{nL^2}} \qquad \text{for all } t \geq 0. \tag{2.35}$$

Let us consider some examples to illustrate.

Example 2.22 (Classical Hoeffding from bounded differences) As a warm-up, let us show how the classical Hoeffding bound (2.11) for bounded variables—say $X_i \in [a, b]$ almost surely—follows as an immediate corollary of the bound (2.35). Consider the function $f(x_1, \ldots, x_n) = \sum_{i=1}^n (x_i - \mu_i)$, where $\mu_i = \mathbb{E}[X_i]$ is the mean of the ith random variable. For any index $k \in \{1, \ldots, n\}$, we have

$$\begin{aligned} |f(x) - f(x^{\backslash k})| &= |(x_k - \mu_k) - (x'_k - \mu_k)| \\ &= |x_k - x'_k| \leq b - a, \end{aligned}$$

showing that f satisfies the bounded difference inequality in each coordinate with parameter $L = b - a$. Consequently, it follows from the bounded difference inequality (2.35) that

$$\mathbb{P}\left[\left|\sum_{i=1}^n (X_i - \mu_i)\right| \geq t\right] \leq 2e^{-\frac{2t^2}{n(b-a)^2}},$$

which is the classical Hoeffding bound for independent random variables. ♣

The class of U-statistics frequently arise in statistical problems; let us now study their concentration properties.

Example 2.23 (U-statistics) Let $g \colon \mathbb{R}^2 \to \mathbb{R}$ be a symmetric function of its arguments. Given an i.i.d. sequence X_k, $k \geq 1$, of random variables, the quantity

$$U := \frac{1}{\binom{n}{2}} \sum_{j<k} g(X_j, X_k) \tag{2.36}$$

is known as a pairwise *U-statistic*. For instance, if $g(s, t) = |s - t|$, then U is an unbiased estimator of the mean absolute pairwise deviation $\mathbb{E}[|X_1 - X_2|]$. Note that, while U is *not* a sum of independent random variables, the dependence is relatively weak, and this fact can be revealed by a martingale analysis. If g is bounded (say $\|g\|_\infty \leq b$), then Corollary 2.21 can be used to establish the concentration of U around its mean. Viewing U as a function $f(x) = f(x_1, \ldots, x_n)$, for any given coordinate k, we have

$$\begin{aligned} |f(x) - f(x^{\backslash k})| &\leq \frac{1}{\binom{n}{2}} \sum_{j \neq k} |g(x_j, x_k) - g(x_j, x'_k)| \\ &\leq \frac{(n-1)(2b)}{\binom{n}{2}} = \frac{4b}{n}, \end{aligned}$$

so that the bounded differences property holds with parameter $L_k = \frac{4b}{n}$ in each coordinate. Thus, we conclude that

$$\mathbb{P}[|U - \mathbb{E}[U]| \geq t] \leq 2e^{-\frac{nt^2}{8b^2}}.$$

This tail inequality implies that U is a consistent estimate of $\mathbb{E}[U]$, and also yields finite sample bounds on its quality as an estimator. Similar techniques can be used to obtain tail bounds on U-statistics of higher order, involving sums over k-tuples of variables. ♣

Martingales and the bounded difference property also play an important role in analyzing the properties of random graphs, and other random combinatorial structures.

Example 2.24 (Clique number in random graphs) An undirected graph is a pair $G = (V, E)$, composed of a vertex set $V = \{1, \ldots, d\}$ and an edge set E, where each edge $e = (i, j)$ is an unordered pair of distinct vertices ($i \neq j$). A graph clique C is a subset of vertices such that $(i, j) \in E$ for all $i, j \in C$. The clique number $C(G)$ of the graph is the cardinality of the largest clique—note that $C(G) \in [1, d]$. When the edges E of the graph are drawn according to some random process, then the clique number $C(G)$ is a random variable, and we can study its concentration around its mean $\mathbb{E}[C(G)]$.

The *Erdös–Rényi* ensemble of random graphs is one of the most well-studied models: it is defined by a parameter $p \in (0, 1)$ that specifies the probability with which each edge (i, j) is included in the graph, independently across all $\binom{d}{2}$ edges. More formally, for each $i < j$, let us introduce a Bernoulli *edge-indicator variable* X_{ij} with parameter p, where $X_{ij} = 1$ means that edge (i, j) is included in the graph, and $X_{ij} = 0$ means that it is not included.

Note that the $\binom{d}{2}$-dimensional random vector $Z := \{X_{ij}\}_{i<j}$ specifies the edge set; thus, we may view the clique number $C(G)$ as a function $Z \mapsto f(Z)$. Let Z' denote a vector in which a single coordinate of Z has been changed, and let G' and G be the associated graphs. It is easy to see that $C(G')$ can differ from $C(G)$ by at most 1, so that $|f(Z') - f(Z)| \leq 1$. Thus, the function $C(G) = f(Z)$ satisfies the bounded difference property in each coordinate with parameter $L = 1$, so that

$$\mathbb{P}[\tfrac{1}{n}|C(G) - \mathbb{E}[C(G)]| \geq \delta] \leq 2e^{-2n\delta^2}.$$

Consequently, we see that the clique number of an Erdös–Rényi random graph is very sharply concentrated around its expectation. ♣

Finally, let us study concentration of the Rademacher complexity, a notion that plays a central role in our subsequent development in Chapters 4 and 5.

Example 2.25 (Rademacher complexity) Let $\{\varepsilon_k\}_{k=1}^n$ be an i.i.d. sequence of Rademacher variables (i.e., taking the values $\{-1, +1\}$ equiprobably, as in Example 2.3). Given a collection of vectors $\mathcal{A} \subset \mathbb{R}^n$, define the random variable[1]

$$Z := \sup_{a \in \mathcal{A}} \left[\sum_{k=1}^{n} a_k \varepsilon_k \right] = \sup_{a \in \mathcal{A}} [\langle a, \varepsilon \rangle]. \tag{2.37}$$

The random variable Z measures the size of $\mathcal{A}$ in a certain sense, and its expectation $\mathcal{R}(\mathcal{A}) := \mathbb{E}[Z(\mathcal{A})]$ is known as the *Rademacher complexity* of the set $\mathcal{A}$.

Let us now show how Corollary 2.21 can be used to establish that $Z(\mathcal{A})$ is sub-Gaussian. Viewing $Z(\mathcal{A})$ as a function $(\varepsilon_1, \ldots, \varepsilon_n) \mapsto f(\varepsilon_1, \ldots, \varepsilon_n)$, we need to bound the maximum change when coordinate k is changed. Given two Rademacher vectors $\varepsilon, \varepsilon' \in \{-1, +1\}^n$, recall our definition (2.31) of the modified vector $\varepsilon^{\setminus k}$. Since $f(\varepsilon^{\setminus k}) \geq \langle a, \varepsilon^{\setminus k} \rangle$ for any $a \in \mathcal{A}$, we have

$$\langle a, \varepsilon \rangle - f(\varepsilon^{\setminus k}) \leq \langle a, \varepsilon - \varepsilon^{\setminus k} \rangle = a_k(\varepsilon_k - \varepsilon'_k) \leq 2|a_k|.$$

Taking the supremum over $\mathcal{A}$ on both sides, we obtain the inequality

$$f(\varepsilon) - f(\varepsilon^{\setminus k}) \leq 2 \sup_{a \in \mathcal{A}} |a_k|.$$

[1] For the reader concerned about measurability, see the bibliographic discussion in Chapter 4.

Since the same argument applies with the roles of ε and $\varepsilon^{\backslash k}$ reversed, we conclude that f satisfies the bounded difference inequality in coordinate k with parameter $2 \sup_{a \in \mathcal{A}} |a_k|$. Consequently, Corollary 2.21 implies that the random variable $Z(\mathcal{A})$ is sub-Gaussian with parameter at most $2\sqrt{\sum_{k=1}^{n} \sup_{a \in \mathcal{A}} a_k^2}$. This sub-Gaussian parameter can be reduced to the (potentially much) smaller quantity $\sqrt{\sup_{a \in \mathcal{A}} \sum_{k=1}^{n} a_k^2}$ using alternative techniques; in particular, see Example 3.5 in Chapter 3 for further details. ♣

2.3 Lipschitz functions of Gaussian variables

We conclude this chapter with a classical result on the concentration properties of Lipschitz functions of Gaussian variables. These functions exhibit a particularly attractive form of dimension-free concentration. Let us say that a function $f : \mathbb{R}^n \to \mathbb{R}$ is L-Lipschitz with respect to the Euclidean norm $\|\cdot\|_2$ if

$$|f(x) - f(y)| \leq L \|x - y\|_2 \quad \text{for all } x, y \in \mathbb{R}^n. \tag{2.38}$$

The following result guarantees that any such function is sub-Gaussian with parameter at most L:

Theorem 2.26 *Let $(X_1, \ldots, X_n)$ be a vector of i.i.d. standard Gaussian variables, and let $f : \mathbb{R}^n \to \mathbb{R}$ be L-Lipschitz with respect to the Euclidean norm. Then the variable $f(X) - \mathbb{E}[f(X)]$ is sub-Gaussian with parameter at most L, and hence*

$$\mathbb{P}[|f(X) - \mathbb{E}[f(X)]| \geq t] \leq 2e^{-\frac{t^2}{2L^2}} \quad \textit{for all } t \geq 0. \tag{2.39}$$

Note that this result is truly remarkable: it guarantees that any L-Lipschitz function of a standard Gaussian random vector, regardless of the dimension, exhibits concentration like a scalar Gaussian variable with variance L^2.

Proof With the aim of keeping the proof as simple as possible, let us prove a version of the concentration bound (2.39) with a weaker constant in the exponent. (See the bibliographic notes for references to proofs of the sharpest results.) We also prove the result for a function that is both Lipschitz *and* differentiable; since any Lipschitz function is differentiable almost everywhere,[2] it is then straightforward to extend this result to the general setting. For a differentiable function, the Lipschitz property guarantees that $\|\nabla f(x)\|_2 \leq L$ for all $x \in \mathbb{R}^n$. In order to prove this version of the theorem, we begin by stating an auxiliary technical lemma:

[2] This fact is a consequence of Rademacher's theorem.

Lemma 2.27 *Suppose that $f\colon \mathbb{R}^n \to \mathbb{R}$ is differentiable. Then for any convex function $\phi\colon \mathbb{R} \to \mathbb{R}$, we have*

$$\mathbb{E}[\phi(f(X) - \mathbb{E}[f(X)])] \le \mathbb{E}\Big[\phi\Big(\frac{\pi}{2}\langle \nabla f(X), Y\rangle\Big)\Big], \tag{2.40}$$

where $X, Y \sim \mathcal{N}(0, \mathbf{I}_n)$ are standard multivariate Gaussian, and independent.

We now prove the theorem using this lemma. For any fixed $\lambda \in \mathbb{R}$, applying inequality (2.40) to the convex function $t \mapsto e^{\lambda t}$ yields

$$\begin{aligned}\mathbb{E}_X\Big[\exp\big(\lambda\{f(X) - \mathbb{E}[f(X)]\}\big)\Big] &\le \mathbb{E}_{X,Y}\Big[\exp\Big(\frac{\lambda\pi}{2}\langle Y, \nabla f(X)\rangle\Big)\Big]\\ &= \mathbb{E}_X\Big[\exp\Big(\frac{\lambda^2\pi^2}{8}\|\nabla f(X)\|_2^2\Big)\Big],\end{aligned}$$

where we have used the independence of X and Y to first take the expectation over Y marginally, and the fact that $\langle Y, \nabla f(x)\rangle$ is a zero-mean Gaussian variable with variance $\|\nabla f(x)\|_2^2$. Due to the Lipschitz condition on f, we have $\|\nabla f(x)\|_2 \le L$ for all $x \in \mathbb{R}^n$, whence

$$\mathbb{E}\Big[\exp\big(\lambda\{f(X) - \mathbb{E}[f(X)]\}\big)\Big] \le e^{\frac{1}{8}\lambda^2\pi^2L^2},$$

which shows that $f(X) - \mathbb{E}[f(X)]$ is sub-Gaussian with parameter at most $\frac{\pi L}{2}$. The tail bound

$$\mathbb{P}[|f(X) - \mathbb{E}[f(X)]| \ge t] \le 2\exp\Big(-\frac{2t^2}{\pi^2L^2}\Big) \qquad \text{for all } t \ge 0$$

follows from Proposition 2.5.

It remains to prove Lemma 2.27, and we do so via a classical interpolation method that exploits the rotation invariance of the Gaussian distribution. For each $\theta \in [0, \pi/2]$, consider the random vector $Z(\theta) \in \mathbb{R}^n$ with components

$$Z_k(\theta) := X_k \sin\theta + Y_k \cos\theta \qquad \text{for } k = 1, 2, \ldots, n.$$

By the convexity of ϕ, we have

$$\mathbb{E}_X[\phi(f(X) - \mathbb{E}_Y[f(Y)])] \le \mathbb{E}_{X,Y}[\phi(f(X) - f(Y))]. \tag{2.41}$$

Now since $Z_k(0) = Y_k$ and $Z_k(\pi/2) = X_k$ for all $k = 1, \ldots, n$, we have

$$f(X) - f(Y) = \int_0^{\pi/2} \frac{d}{d\theta} f(Z(\theta))\, d\theta = \int_0^{\pi/2} \langle \nabla f(Z(\theta)), Z'(\theta)\rangle\, d\theta, \tag{2.42}$$

where $Z'(\theta) \in \mathbb{R}^n$ denotes the elementwise derivative, a vector with the components $Z_k'(\theta) = X_k\cos\theta - Y_k\sin\theta$. Substituting the integral representation (2.42) into our earlier bound (2.41) yields

$$\begin{aligned}
\mathbb{E}_X[\phi(f(X) - \mathbb{E}_Y[f(Y)])] &\le \mathbb{E}_{X,Y}\left[\phi\left(\int_0^{\pi/2} \langle \nabla f(Z(\theta)), Z'(\theta)\rangle \, d\theta\right)\right] \\
&= \mathbb{E}_{X,Y}\left[\phi\left(\frac{1}{\pi/2}\int_0^{\pi/2} \frac{\pi}{2}\langle \nabla f(Z(\theta)), Z'(\theta)\rangle \, d\theta\right)\right] \\
&\le \frac{1}{\pi/2}\int_0^{\pi/2} \mathbb{E}_{X,Y}\left[\phi\left(\frac{\pi}{2}\langle \nabla f(Z(\theta)), Z'(\theta)\rangle\right)\right] d\theta, \qquad (2.43)
\end{aligned}$$

where the final step again uses convexity of ϕ. By the rotation invariance of the Gaussian distribution, for each $\theta \in [0, \pi/2]$, the pair $(Z_k(\theta), Z_k'(\theta))$ is a jointly Gaussian vector, with zero mean and identity covariance $\mathbf{I}_2$. Therefore, the expectation inside the integral in equation (2.43) does not depend on θ, and hence

$$\frac{1}{\pi/2}\int_0^{\pi/2} \mathbb{E}_{X,Y}\left[\phi\left(\frac{\pi}{2}\langle \nabla f(Z(\theta)), Z'(\theta)\rangle\right)\right] d\theta = \mathbb{E}\left[\phi\left(\frac{\pi}{2}\left\langle \nabla f(\widetilde{X}), \widetilde{Y}\right\rangle\right)\right],$$

where $(\widetilde{X}, \widetilde{Y})$ are independent standard Gaussian n-vectors. This completes the proof of the bound (2.40). □

Note that the proof makes essential use of various properties specific to the standard Gaussian distribution. However, similar concentration results hold for other non-Gaussian distributions, including the uniform distribution on the sphere and any strictly log-concave distribution (see Chapter 3 for further discussion of such distributions). However, without additional structure of the function f (such as convexity), dimension-free concentration for Lipschitz functions need not hold for an arbitrary sub-Gaussian distribution; see the bibliographic section for further discussion of this fact.

Theorem 2.26 is useful for a broad range of problems; let us consider some examples to illustrate.

Example 2.28 (χ^2 concentration) For a given sequence $\{Z_k\}_{k=1}^n$ of i.i.d. standard normal variates, the random variable $Y := \sum_{k=1}^n Z_k^2$ follows a χ^2-distribution with n degrees of freedom. The most direct way to obtain tail bounds on Y is by noting that Z_k^2 is sub-exponential, and exploiting independence (see Example 2.11). In this example, we pursue an alternative approach—namely, via concentration for Lipschitz functions of Gaussian variates. Indeed, defining the variable $V = \sqrt{Y}/\sqrt{n}$, we can write $V = \|(Z_1, \ldots, Z_n)\|_2/\sqrt{n}$, and since the Euclidean norm is a 1-Lipschitz function, Theorem 2.26 implies that

$$\mathbb{P}[V \ge \mathbb{E}[V] + \delta] \le e^{-n\delta^2/2} \qquad \text{for all } \delta \ge 0.$$

Using concavity of the square-root function and Jensen's inequality, we have

$$\mathbb{E}[V] \le \sqrt{\mathbb{E}[V^2]} = \left\{\frac{1}{n}\sum_{i=1}^n \mathbb{E}[Z_k^2]\right\}^{1/2} = 1.$$

Recalling that $V = \sqrt{Y}/\sqrt{n}$ and putting together the pieces yields

$$\mathbb{P}[Y/n \ge (1+\delta)^2] \le e^{-n\delta^2/2} \qquad \text{for all } \delta \ge 0.$$

Since $(1+\delta)^2 = 1 + 2\delta + \delta^2 \leq 1 + 3\delta$ for all $\delta \in [0, 1]$, we conclude that

$$\mathbb{P}[Y \geq n(1+t)] \leq e^{-nt^2/18} \qquad \text{for all } t \in [0, 3], \tag{2.44}$$

where we have made the substitution $t = 3\delta$. It is worthwhile comparing this tail bound to those that can be obtained by using the fact that each Z_k^2 is sub-exponential, as discussed in Example 2.11. ♣

Example 2.29 (Order statistics) Given a random vector $(X_1, X_2, \ldots, X_n)$, its order statistics are obtained by reordering its entries in a non-decreasing manner—namely as

$$X_{(1)} \leq X_{(2)} \leq \cdots \leq X_{(n-1)} \leq X_{(n)}. \tag{2.45}$$

As particular cases, we have $X_{(n)} = \max_{k=1,\ldots,n} X_k$ and $X_{(1)} = \min_{k=1,\ldots,n} X_k$. Given another random vector $(Y_1, \ldots, Y_n)$, it can be shown that $|X_{(k)} - Y_{(k)}| \leq \|X - Y\|_2$ for all $k = 1, \ldots, n$, so that each order statistic is a 1-Lipschitz function. (We leave the verification of this inequality as an exercise for the reader.) Consequently, when X is a Gaussian random vector, Theorem 2.26 implies that

$$\mathbb{P}[|X_{(k)} - \mathbb{E}[X_{(k)}]| \geq \delta] \leq 2e^{-\frac{\delta^2}{2}} \qquad \text{for all } \delta \geq 0.$$

♣

Example 2.30 (Gaussian complexity) This example is closely related to our earlier discussion of Rademacher complexity in Example 2.25. Let $\{W_k\}_{k=1}^n$ be an i.i.d. sequence of $\mathcal{N}(0, 1)$ variables. Given a collection of vectors $\mathcal{A} \subset \mathbb{R}^n$, define the random variable[3]

$$Z := \sup_{a \in \mathcal{A}} \left[\sum_{k=1}^n a_k W_k \right] = \sup_{a \in \mathcal{A}} \langle a, W \rangle. \tag{2.46}$$

As with the Rademacher complexity, the variable $Z = Z(\mathcal{A})$ is one way of measuring the size of the set $\mathcal{A}$, and will play an important role in later chapters. Viewing Z as a function $(w_1, \ldots, w_n) \mapsto f(w_1, \ldots, w_n)$, let us verify that f is Lipschitz (with respect to Euclidean norm) with parameter $\sup_{a \in \mathcal{A}} \|a\|_2$. Let $w, w' \in \mathbb{R}^n$ be arbitrary, and let $a^* \in \mathcal{A}$ be any vector that achieves the maximum defining $f(w)$. Following the same argument as Example 2.25, we have the upper bound

$$f(w) - f(w') \leq \langle a^*, w - w' \rangle \leq D(\mathcal{A}) \, \|w - w'\|_2,$$

where $D(\mathcal{A}) = \sup_{a \in \mathcal{A}} \|a\|_2$ is the Euclidean width of the set. The same argument holds with the roles of w and w' reversed, and hence

$$|f(w) - f(w')| \leq D(\mathcal{A}) \, \|w - w'\|_2.$$

Consequently, Theorem 2.26 implies that

$$\mathbb{P}[|Z - \mathbb{E}[Z]| \geq \delta] \leq 2 \exp\left(-\frac{\delta^2}{2D^2(\mathcal{A})} \right). \tag{2.47}$$

♣

[3] For measurability concerns, see the bibliographic discussion in Chapter 4.

Example 2.31 (Gaussian chaos variables) As a generalization of the previous example, let $\mathbf{Q} \in \mathbb{R}^{n\times n}$ be a symmetric matrix, and let $w, \widetilde{w}$ be independent zero-mean Gaussian random vectors with covariance matrix $\mathbf{I}_n$. The random variable

$$Z := \sum_{i,j=1}^{n} Q_{ij} w_i \widetilde{w}_j = w^{\mathrm{T}} \mathbf{Q} \widetilde{w}$$

is known as a (decoupled) Gaussian chaos. By the independence of w and $\widetilde{w}$, we have $\mathbb{E}[Z] = 0$, so it is natural to seek a tail bound on Z.

Conditioned on $\widetilde{w}$, the variable Z is a zero-mean Gaussian variable with variance $\|\mathbf{Q}\widetilde{w}\|_2^2 = \widetilde{w}^{\mathrm{T}} \mathbf{Q}^2 \widetilde{w}$, whence

$$\mathbb{P}[|Z| \geq \delta \mid \widetilde{w}] \leq 2e^{-\frac{\delta^2}{2\|\mathbf{Q}\widetilde{w}\|_2^2}}. \tag{2.48}$$

Let us now control the random variable $Y := \|\mathbf{Q}\widetilde{w}\|_2$. Viewed as a function of the Gaussian vector $\widetilde{w}$, it is Lipschitz with constant

$$|||\mathbf{Q}|||_2 := \sup_{\|u\|_2=1} \|\mathbf{Q}u\|_2, \tag{2.49}$$

corresponding to the *ℓ_2-operator norm* of the matrix $\mathbf{Q}$. Moreover, by Jensen's inequality, we have $\mathbb{E}[Y] \leq \sqrt{\mathbb{E}[\widetilde{w}^{\mathrm{T}} \mathbf{Q}^2 \widetilde{w}]} = |||\mathbf{Q}|||_{\mathrm{F}}$, where

$$|||\mathbf{Q}|||_{\mathrm{F}} := \sqrt{\sum_{i=1}^{n} \sum_{j=1}^{n} Q_{ij}^2} \tag{2.50}$$

is the *Frobenius norm* of the matrix $\mathbf{Q}$. Putting together the pieces yields the tail bound

$$\mathbb{P}[\|\mathbf{Q}\widetilde{w}\|_2 \geq |||\mathbf{Q}|||_{\mathrm{F}} + t] \leq 2 \exp\Big(-\frac{t^2}{2|||\mathbf{Q}|||_2^2}\Big).$$

Note that $(|||\mathbf{Q}|||_{\mathrm{F}} + t)^2 \leq 2|||\mathbf{Q}|||_{\mathrm{F}}^2 + 2t^2$. Consequently, setting $t^2 = \delta|||\mathbf{Q}|||_2$ and simplifying yields

$$\mathbb{P}[\widetilde{w}^T \mathbf{Q}^2 \widetilde{w} \geq 2|||\mathbf{Q}|||_{\mathrm{F}}^2 + 2\delta|||\mathbf{Q}|||_2] \leq 2 \exp\Big(-\frac{\delta}{2|||\mathbf{Q}|||_2}\Big).$$

Putting together the pieces, we find that

$$\begin{aligned}\mathbb{P}[|Z| \geq \delta] &\leq 2 \exp\Big(-\frac{\delta^2}{4|||\mathbf{Q}|||_{\mathrm{F}}^2 + 4\delta|||\mathbf{Q}|||_2}\Big) + 2 \exp\Big(-\frac{\delta}{2|||\mathbf{Q}|||_2}\Big) \\ &\leq 4 \exp\Big(-\frac{\delta^2}{4|||\mathbf{Q}|||_{\mathrm{F}}^2 + 4\delta|||\mathbf{Q}|||_2}\Big).\end{aligned}$$

We have thus shown that the Gaussian chaos variable satisfies a sub-exponential tail bound. ♣

Example 2.32 (Singular values of Gaussian random matrices) For integers $n > d$, let $\mathbf{X} \in \mathbb{R}^{n\times d}$ be a random matrix with i.i.d. $\mathcal{N}(0, 1)$ entries, and let

$$\sigma_1(\mathbf{X}) \geq \sigma_2(\mathbf{X}) \geq \cdots \geq \sigma_d(\mathbf{X}) \geq 0$$

denote its ordered singular values. By Weyl's theorem (see Exercise 8.3), given another matrix $\mathbf{Y} \in \mathbb{R}^{n \times d}$, we have

$$\max_{k=1,\ldots,d} |\sigma_k(\mathbf{X}) - \sigma_k(\mathbf{Y})| \leq |||\mathbf{X} - \mathbf{Y}|||_2 \leq |||\mathbf{X} - \mathbf{Y}|||_{\mathrm{F}}, \tag{2.51}$$

where $|||\cdot|||_{\mathrm{F}}$ denotes the Frobenius norm. The inequality (2.51) shows that each singular value $\sigma_k(\mathbf{X})$ is a 1-Lipschitz function of the random matrix, so that Theorem 2.26 implies that, for each $k = 1, \ldots, d$, we have

$$\mathbb{P}[|\sigma_k(\mathbf{X}) - \mathbb{E}[\sigma_k(\mathbf{X})]| \geq \delta] \leq 2e^{-\frac{\delta^2}{2}} \qquad \text{for all } \delta \geq 0. \tag{2.52}$$

Consequently, even though our techniques are not yet powerful enough to characterize the expected value of these random singular values, we are guaranteed that the expectations are representative of the typical behavior. See Chapter 6 for a more detailed discussion of the singular values of random matrices. ♣

2.4 Appendix A: Equivalent versions of sub-Gaussian variables

In this appendix, we prove Theorem 2.6. We establish the equivalence by proving the circle of implications (I) $\Rightarrow$ (II) $\Rightarrow$ (III) $\Rightarrow$ (I), followed by the equivalence (I) $\Leftrightarrow$ (IV).

Implication (I) $\Rightarrow$ (II): If X is zero-mean and sub-Gaussian with parameter σ, then we claim that, for $Z \sim \mathcal{N}(0, 2\sigma^2)$,

$$\frac{\mathbb{P}[X \geq t]}{\mathbb{P}[Z \geq t]} \leq \sqrt{8}e \qquad \text{for all } t \geq 0,$$

showing that X is majorized by Z with constant $c = \sqrt{8}e$. On one hand, by the sub-Gaussianity of X, we have $\mathbb{P}[X \geq t] \leq \exp(-\frac{t^2}{2\sigma^2})$ for all $t \geq 0$. On the other hand, by the Mills ratio for Gaussian tails, if $Z \sim \mathcal{N}(0, 2\sigma^2)$, then we have

$$\mathbb{P}[Z \geq t] \geq \left(\frac{\sqrt{2}\sigma}{t} - \frac{(\sqrt{2}\sigma)^3}{t^3}\right) e^{-\frac{t^2}{4\sigma^2}} \qquad \text{for all } t > 0. \tag{2.53}$$

(See Exercise 2.2 for a derivation of this inequality.) We split the remainder of our analysis into two cases.

Case 1: First, suppose that $t \in [0, 2\sigma]$. Since the function $\Phi(t) = \mathbb{P}[Z \geq t]$ is decreasing, for all t in this interval,

$$\mathbb{P}[Z \geq t] \geq \mathbb{P}[Z \geq 2\sigma] \geq \left(\frac{1}{\sqrt{2}} - \frac{1}{2\sqrt{2}}\right) e^{-1} = \frac{1}{\sqrt{8}e}.$$

Since $\mathbb{P}[X \geq t] \leq 1$, we conclude that $\frac{\mathbb{P}[X \geq t]}{\mathbb{P}[Z \geq t]} \leq \sqrt{8}e$ for all $t \in [0, 2\sigma]$.

Case 2: Otherwise, we may assume that $t > 2\sigma$. In this case, by combining the Mills ratio (2.53) and the sub-Gaussian tail bound and making the substitution $s = t/\sigma$, we find

that

$$\sup_{t>2\sigma} \frac{\mathbb{P}[X \geq t]}{\mathbb{P}[Z \geq t]} \leq \sup_{s>2} \frac{e^{-\frac{s^2}{4}}}{\left(\frac{\sqrt{2}}{s} - \frac{(\sqrt{2})^3}{s^3}\right)}$$
$$\leq \sup_{s>2} s^3 e^{-\frac{s^2}{4}}$$
$$\leq \sqrt{8}e,$$

where the last step follows from a numerical calculation.

Implication (II) ⇒ (III): Suppose that X is majorized by a zero-mean Gaussian with variance τ^2. Since X^{2k} is a non-negative random variable, we have

$$\mathbb{E}[X^{2k}] = \int_0^\infty \mathbb{P}[X^{2k} > s]\, ds = \int_0^\infty \mathbb{P}[|X| > s^{1/(2k)}]\, ds.$$

Under the majorization assumption, there is some constant $c \geq 1$ such that

$$\int_0^\infty \mathbb{P}[|X| > s^{1/(2k)}]\, ds \leq c \int_0^\infty \mathbb{P}[|Z| > s^{1/(2k)}]\, ds = c\mathbb{E}[Z^{2k}],$$

where $Z \sim \mathcal{N}(0, \tau^2)$. The polynomial moments of Z are given by

$$\mathbb{E}[Z^{2k}] = \frac{(2k)!}{2^k k!}\tau^{2k}, \qquad \text{for } k = 1, 2, \ldots, \tag{2.54}$$

whence

$$\mathbb{E}[X^{2k}] \leq c\mathbb{E}[Z^{2k}] = c\frac{(2k)!}{2^k k!}\tau^{2k} \leq \frac{(2k)!}{2^k k!}(c\tau)^{2k}, \qquad \text{for all } k = 1, 2, \ldots.$$

Consequently, the moment bound (2.12c) holds with $\theta = c\tau$.

Implication (III) ⇒ (I): For each $\lambda \in \mathbb{R}$, we have

$$\mathbb{E}[e^{\lambda X}] \leq 1 + \sum_{k=2}^{\infty} \frac{|\lambda|^k \mathbb{E}[|X|^k]}{k!}, \tag{2.55}$$

where we have used the fact $\mathbb{E}[X] = 0$ to eliminate the term involving $k = 1$. If X is symmetric around zero, then all of its odd moments vanish, and by applying our assumption on $\theta(X)$, we obtain

$$\mathbb{E}[e^{\lambda X}] \leq 1 + \sum_{k=1}^{\infty} \frac{\lambda^{2k}}{(2k)!} \frac{(2k)!\theta^{2k}}{2^k k!} = e^{\frac{\lambda^2\theta^2}{2}},$$

which shows that X is sub-Gaussian with parameter θ.

When X is not symmetric, we can bound the odd moments in terms of the even ones as

$$\mathbb{E}[|\lambda X|^{2k+1}] \stackrel{\text{(i)}}{\leq} (\mathbb{E}[|\lambda X|^{2k}]\mathbb{E}[|\lambda X|^{2k+2}])^{1/2} \stackrel{\text{(ii)}}{\leq} \tfrac{1}{2}(\lambda^{2k}\mathbb{E}[X^{2k}] + \lambda^{2k+2}\mathbb{E}[X^{2k+2}]), \tag{2.56}$$

where step (i) follows from the Cauchy–Schwarz inequality; and step (ii) follows from the arithmetic–geometric mean inequality. Applying this bound to the power-series expansion (2.55), we obtain

$$\begin{aligned}\mathbb{E}[e^{\lambda X}] &\le 1 + \Big(\frac{1}{2} + \frac{1}{2\cdot 3!}\Big)\lambda^2\mathbb{E}[X^2] + \sum_{k=2}^{\infty}\Big(\frac{1}{(2k)!} + \frac{1}{2}\Big[\frac{1}{(2k-1)!} + \frac{1}{(2k+1)!}\Big]\Big)\lambda^{2k}\mathbb{E}[X^{2k}] \\ &\le \sum_{k=0}^{\infty} 2^k \frac{\lambda^{2k}\mathbb{E}[X^{2k}]}{(2k)!} \\ &\le e^{\frac{(\sqrt{2}\lambda\theta)^2}{2}},\end{aligned}$$

which establishes the claim.

Implication (I) $\Rightarrow$ (IV): This result is obvious for $s = 0$. For $s \in (0, 1)$, we begin with the sub-Gaussian inequality $\mathbb{E}[e^{\lambda X}] \le e^{\frac{\lambda^2\sigma^2}{2}}$, and multiply both sides by $e^{-\frac{\lambda^2\sigma^2}{2s}}$, thereby obtaining

$$\mathbb{E}[e^{\lambda X - \frac{\lambda^2\sigma^2}{2s}}] \le e^{\frac{\lambda^2\sigma^2(s-1)}{2s}}.$$

Since this inequality holds for all $\lambda \in \mathbb{R}$, we may integrate both sides over $\lambda \in \mathbb{R}$, using Fubini's theorem to justify exchanging the order of integration. On the right-hand side, we have

$$\int_{-\infty}^{\infty} \exp\Big(\frac{\lambda^2\sigma^2\,(s-1)}{2s}\Big)\,d\lambda = \frac{1}{\sigma}\sqrt{\frac{2\pi s}{1-s}}.$$

Turning to the left-hand side, for each fixed $x \in \mathbb{R}$, we have

$$\int_{-\infty}^{\infty} \exp\Big(\lambda x - \frac{\lambda^2\sigma^2}{2s}\Big)\,d\lambda = \frac{\sqrt{2\pi s}}{\sigma}e^{\frac{sx^2}{2\sigma^2}}.$$

Taking expectations with respect to X, we conclude that

$$\mathbb{E}[e^{\frac{sX^2}{2\sigma^2}}] \le \frac{\sigma}{\sqrt{2\pi s}}\,\frac{1}{\sigma}\sqrt{\frac{2\pi s}{1-s}} = \frac{1}{\sqrt{1-s}},$$

which establishes the claim.

Implication (IV) $\Rightarrow$ (I): Applying the bound $e^u \le u + e^{9u^2/16}$ with $u = \lambda X$ and then taking expectations, we find that

$$\mathbb{E}[e^{\lambda X}] \le \mathbb{E}[\lambda X] + \mathbb{E}[e^{\frac{9\lambda^2 X^2}{16}}] = \mathbb{E}[e^{\frac{sX^2}{2\sigma^2}}] \le \frac{1}{\sqrt{1-s}},$$

valid whenever $s = \frac{9}{8}\lambda^2\sigma^2$ is strictly less than 1. Noting that $\frac{1}{\sqrt{1-s}} \le e^s$ for all $s \in [0, \frac{1}{2}]$ and that $s < \frac{1}{2}$ whenever $|\lambda| < \frac{2}{3\sigma}$, we conclude that

$$\mathbb{E}[e^{\lambda X}] \le e^{\frac{9}{8}\lambda^2\sigma^2} \qquad \text{for all } |\lambda| < \tfrac{2}{3\sigma}. \tag{2.57a}$$

It remains to establish a similar upper bound for $|\lambda| \ge \frac{2}{3\sigma}$. Note that, for any $\alpha > 0$, the functions $f(u) = \frac{u^2}{2\alpha}$ and $f^*(v) = \frac{\alpha v^2}{2}$ are conjugate duals. Thus, the Fenchel–Young inequality implies that $uv \leq \frac{u^2}{2\alpha} + \frac{\alpha v^2}{2}$, valid for all $u, v \in \mathbb{R}$ and $\alpha > 0$. We apply this inequality with $u = \lambda$, $v = X$ and $\alpha = c/\sigma^2$ for a constant $c > 0$ to be chosen; doing so yields

$$\mathbb{E}[e^{\lambda X}] \leq \mathbb{E}[e^{\frac{\lambda^2\sigma^2}{2c} + \frac{cX^2}{2\sigma^2}}] = e^{\frac{\lambda^2\sigma^2}{2c}} \mathbb{E}[e^{\frac{cX^2}{2\sigma^2}}] \stackrel{\text{(ii)}}{\leq} e^{\frac{\lambda^2\sigma^2}{2c}} e^c,$$

where step (ii) is valid for any $c \in (0, 1/2)$, using the same argument that led to the bound (2.57a). In particular, setting $c = 1/4$ yields $\mathbb{E}[e^{\lambda X}] \leq e^{2\lambda^2\sigma^2} e^{1/4}$.

Finally, when $|\lambda| \geq \frac{2}{3\sigma}$, then we have $\frac{1}{4} \leq \frac{9}{16}\lambda^2\sigma^2$, and hence

$$\mathbb{E}[e^{\lambda X}] \leq e^{2\lambda^2\sigma^2 + \frac{9}{16}\lambda^2\sigma^2} \leq e^{3\lambda^2\sigma^2}. \tag{2.57b}$$

This inequality, combined with the bound (2.57a), completes the proof.

2.5 Appendix B: Equivalent versions of sub-exponential variables

This appendix is devoted to the proof of Theorem 2.13. In particular, we prove the chain of equivalences I ⇔ II ⇔ III, followed by the equivalence II ⇔ IV.

(II) ⇒ (I): The existence of the moment generating function for $|\lambda| < c_0$ implies that $\mathbb{E}[e^{\lambda X}] = 1 + \frac{\lambda^2 \mathbb{E}[X^2]}{2} + o(\lambda^2)$ as $\lambda \to 0$. Moreover, an ordinary Taylor-series expansion implies that $e^{\frac{\sigma^2\lambda^2}{2}} = 1 + \frac{\sigma^2\lambda^2}{2} + o(\lambda^2)$ as $\lambda \to 0$. Therefore, as long as $\sigma^2 > \mathbb{E}[X^2]$, there exists some $b \geq 0$ such that $\mathbb{E}[e^{\lambda X}] \leq e^{\frac{\sigma^2\lambda^2}{2}}$ for all $|\lambda| \leq \frac{1}{b}$.

(I) ⇒ (II): This implication is immediate.

(III) ⇒ (II): For an exponent $a > 0$ and truncation level $T > 0$ to be chosen, we have

$$\mathbb{E}[e^{a|X|}\mathbb{I}[e^{a|X|} \leq e^{aT}]] \leq \int_0^{e^{aT}} \mathbb{P}[e^{a|X|} \geq t]\, dt \leq 1 + \int_1^{e^{aT}} \mathbb{P}\Big[|X| \geq \frac{\log t}{a}\Big] dt.$$

Applying the assumed tail bound, we obtain

$$\mathbb{E}[e^{a|X|}\mathbb{I}[e^{a|X|} \leq e^{aT}]] \leq 1 + c_1 \int_1^{e^{aT}} e^{-\frac{c_2 \log t}{a}}\, dt = 1 + c_1 \int_1^{e^{aT}} t^{-c_2/a}\, dt.$$

Thus, for any $a \in [0, \frac{c_2}{2}]$, we have

$$\mathbb{E}[e^{a|X|}\mathbb{I}[e^{a|X|} \leq e^{aT}]] \leq 1 + \frac{c_1}{2}(1 - e^{-aT}) \leq 1 + \frac{c_1}{2}.$$

By taking the limit as $T \to \infty$, we conclude that $\mathbb{E}[e^{a|X|}]$ is finite for all $a \in [0, \frac{c_2}{2}]$. Since both e^{aX} and e^{-aX} are upper bounded by $e^{|a|\,|X|}$, it follows that $\mathbb{E}[e^{aX}]$ is finite for all $|a| \leq \frac{c_2}{2}$.

(II) ⇒ (III): By the Chernoff bound with $\lambda = c_0/2$, we have

$$\mathbb{P}[X \geq t] \leq \mathbb{E}[e^{\frac{c_0 X}{2}}] e^{-\frac{c_0 t}{2}}.$$

Applying a similar argument to $-X$, we conclude that $\mathbb{P}[|X| \geq t] \leq c_1 e^{-c_2 t}$ with $c_1 = \mathbb{E}[e^{c_0 X/2}] + \mathbb{E}[e^{-c_0 X/2}]$ and $c_2 = c_0/2$.

(II) $\Leftrightarrow$ (IV): Since the moment generating function exists in an open interval around zero, we can consider the power-series expansion

$$\mathbb{E}[e^{\lambda X}] = 1 + \sum_{k=2}^{\infty} \frac{\lambda^k \mathbb{E}[X^k]}{k!} \qquad \text{for all } |\lambda| < a. \tag{2.58}$$

By definition, the quantity $\gamma(X)$ is the radius of convergence of this power series, from which the equivalence between (II) and (IV) follows.

2.6 Bibliographic details and background

Further background and details on tail bounds can be found in various books (e.g., Saulis and Statulevicius, 1991; Petrov, 1995; Buldygin and Kozachenko, 2000; Boucheron et al., 2013). Classic papers on tail bounds include those of Bernstein (1937), Chernoff (1952), Bahadur and Ranga Rao (1960), Bennett (1962), Hoeffding (1963) and Azuma (1967). The idea of using the cumulant function to bound the tails of a random variable was first introduced by Bernstein (1937), and further developed by Chernoff (1952), whose name is now frequently associated with the method. The book by Saulis and Statulevicius (1991) provides a number of more refined results that can be established using cumulant-based techniques. The original work of Hoeffding (1963) gives results both for sums of independent random variables, assumed to be bounded from above, as well as certain types of dependent random variables, including U-statistics. The work of Azuma (1967) applies to general martingales that are sub-Gaussian in a conditional sense, as in Theorem 2.19.

The book by Buldygin and Kozachenko (2000) provides a range of results on sub-Gaussian and sub-exponential variates. In particular, Theorems 2.6 and 2.13 are based on results from this book. The Orlicz norms, discussed briefly in Exercises 2.18 and 2.19, provide an elegant generalization of the sub-exponential and sub-Gaussian families. See Section 5.6 and the books (Ledoux and Talagrand, 1991; Buldygin and Kozachenko, 2000) for further background on Orlicz norms.

The Johnson–Lindenstrauss lemma, discussed in Example 2.12, was originally proved by Johnson and Lindenstrauss (1984) as an intermediate step in a more general result about Lipschitz embeddings. The original proof of the lemma was based on random matrices with orthonormal rows, as opposed to the standard Gaussian random matrix used here. The use of random projection for dimension reduction and algorithmic speed-ups has a wide range of applications; see the sources (Vempala, 2004; Mahoney, 2011; Cormode, 2012; Kane and Nelson, 2014; Woodruff, 2014; Bourgain et al., 2015; Pilanci and Wainwright, 2015) for further details.

Tail bounds for U-statistics, as sketched out in Example 2.23, were derived by Hoeffding (1963). The book by de la Peña and Giné (1999) provides more advanced results, including extensions to uniform laws for U-processes and decoupling results. The bounded differences inequality (Corollary 2.21) and extensions thereof have many applications in the study of randomized algorithms as well as random graphs and other combinatorial objects. A number of such applications can be found in the survey by McDiarmid (1989), and the book by Boucheron et al. (2013).

Milman and Schechtman (1986) provide the short proof of Gaussian concentration for

Lipschitz functions, on which Theorem 2.26 is based. Ledoux (2001) provides an example of a Lipschitz function of an i.i.d. sequence of Rademacher variables (i.e., taking values $\{-1, +1\}$ equiprobably) for which sub-Gaussian concentration fails to hold (cf. p. 128 in his book). However, sub-Gaussian concentration does hold for Lipschitz functions of bounded random variables with an additional convexity condition; see Section 3.3.5 for further details.

The kernel density estimation problem from Exercise 2.15 is a particular form of non-parametric estimation; we return to such problems in Chapters 13 and 14. Although we have focused exclusively on tail bounds for real-valued random variables, there are many generalizations to random variables taking values in Hilbert and other function spaces, as considered in Exercise 2.16. The books (Ledoux and Talagrand, 1991; Yurinsky, 1995) contain further background on such results. We also return to consider some versions of these bounds in Chapter 14. The Hanson–Wright inequality discussed in Exercise 2.17 was proved in the papers (Hanson and Wright, 1971; Wright, 1973); see the papers (Hsu et al., 2012b; Rudelson and Vershynin, 2013) for more modern treatments. The moment-based tail bound from Exercise 2.20 relies on a classical inequality due to Rosenthal (1970). Exercise 2.21 outlines the proof of the rate-distortion theorem for the Bernoulli source. It is a particular instance of more general information-theoretic results that are proved using probabilistic techniques; see the book by Cover and Thomas (1991) for further reading. The Ising model (2.74) discussed in Exercise 2.22 has a lengthy history dating back to Ising (1925). The book by Talagrand (2003) contains a wealth of information on spin glass models and their mathematical properties.

2.7 Exercises

Exercise 2.1 (Tightness of inequalities) The Markov and Chebyshev inequalities cannot be improved in general.

(a) Provide a non-negative random variable X for which Markov's inequality (2.1) is met with equality.
(b) Provide a random variable Y for which Chebyshev's inequality (2.2) is met with equality.

Exercise 2.2 (Mills ratio) Let $\phi(z) = \frac{1}{\sqrt{2\pi}} e^{-z^2/2}$ be the density function of a standard normal $Z \sim \mathcal{N}(0, 1)$ variate.

(a) Show that $\phi'(z) + z\phi(z) = 0$.
(b) Use part (a) to show that

$$\phi(z)\left(\frac{1}{z} - \frac{1}{z^3}\right) \leq \mathbb{P}[Z \geq z] \leq \phi(z)\left(\frac{1}{z} - \frac{1}{z^3} + \frac{3}{z^5}\right) \qquad \text{for all } z > 0. \tag{2.59}$$

Exercise 2.3 (Polynomial Markov versus Chernoff) Suppose that $X \geq 0$, and that the moment generating function of X exists in an interval around zero. Given some $\delta > 0$ and integer $k = 1, 2, \ldots$, show that

$$\inf_{k=0,1,2,\ldots} \frac{\mathbb{E}[|X|^k]}{\delta^k} \leq \inf_{\lambda>0} \frac{\mathbb{E}[e^{\lambda X}]}{e^{\lambda\delta}}. \tag{2.60}$$

Consequently, an optimized bound based on polynomial moments is always at least as good as the Chernoff upper bound.

Exercise 2.4 (Sharp sub-Gaussian parameter for bounded random variable) Consider a random variable X with mean $\mu = \mathbb{E}[X]$, and such that, for some scalars $b > a$, $X \in [a, b]$ almost surely.

(a) Defining the function $\psi(\lambda) = \log \mathbb{E}[e^{\lambda X}]$, show that $\psi(0) = 0$ and $\psi'(0) = \mu$.
(b) Show that $\psi''(\lambda) = \mathbb{E}_\lambda[X^2] - (\mathbb{E}_\lambda[X])^2$, where we define $\mathbb{E}_\lambda[f(X)] := \frac{\mathbb{E}[f(X)e^{\lambda X}]}{\mathbb{E}[e^{\lambda X}]}$. Use this fact to obtain an upper bound on $\sup_{\lambda \in \mathbb{R}} |\psi''(\lambda)|$.
(c) Use parts (a) and (b) to establish that X is sub-Gaussian with parameter at most $\sigma = \frac{b-a}{2}$.

Exercise 2.5 (Sub-Gaussian bounds and means/variances) Consider a random variable X such that

$$\mathbb{E}[e^{\lambda X}] \leq e^{\frac{\lambda^2 \sigma^2}{2} + \lambda \mu} \qquad \text{for all } \lambda \in \mathbb{R}. \tag{2.61}$$

(a) Show that $\mathbb{E}[X] = \mu$.
(b) Show that $\text{var}(X) \leq \sigma^2$.
(c) Suppose that the smallest possible σ satisfying the inequality (2.61) is chosen. Is it then true that $\text{var}(X) = \sigma^2$? Prove or disprove.

Exercise 2.6 (Lower bounds on squared sub-Gaussians) Letting $\{X_i\}_{i=1}^n$ be an i.i.d. sequence of zero-mean sub-Gaussian variables with parameter σ, consider the normalized sum $Z_n := \frac{1}{n} \sum_{i=1}^n X_i^2$. Prove that

$$\mathbb{P}[Z_n \leq \mathbb{E}[Z_n] - \sigma^2 \delta] \leq e^{-n\delta^2/16} \qquad \text{for all } \delta \geq 0.$$

This result shows that the lower tail of a sum of squared sub-Gaussian variables behaves in a sub-Gaussian way.

Exercise 2.7 (Bennett's inequality) This exercise is devoted to a proof of a strengthening of Bernstein's inequality, known as Bennett's inequality.

(a) Consider a zero-mean random variable such that $|X_i| \leq b$ for some $b > 0$. Prove that

$$\log \mathbb{E}[e^{\lambda X_i}] \leq \sigma_i^2 \lambda^2 \left\{ \frac{e^{\lambda b} - 1 - \lambda b}{(\lambda b)^2} \right\} \qquad \text{for all } \lambda \in \mathbb{R},$$

where $\sigma_i^2 = \text{var}(X_i)$.
(b) Given independent random variables $X_1, \ldots, X_n$ satisfying the condition of part (a), let $\sigma^2 := \frac{1}{n} \sum_{i=1}^n \sigma_i^2$ be the average variance. Prove *Bennett's inequality*

$$\mathbb{P}\left[\sum_{i=1}^n X_i \geq n\delta \right] \leq \exp\left\{ -\frac{n\sigma^2}{b^2} h\left(\frac{b\delta}{\sigma^2} \right) \right\}, \tag{2.62}$$

where $h(t) := (1 + t) \log(1 + t) - t$ for $t \geq 0$.
(c) Show that Bennett's inequality is at least as good as Bernstein's inequality.

Exercise 2.8 (Bernstein and expectations) Consider a non-negative random variable that satisfies a concentration inequality of the form

$$\mathbb{P}[Z \geq t] \leq C e^{-\frac{t^2}{2(\nu^2+bt)}} \tag{2.63}$$

for positive constants (ν, b) and $C \geq 1$.

(a) Show that $\mathbb{E}[Z] \leq 2\nu(\sqrt{\pi} + \sqrt{\log C}) + 4b(1 + \log C)$.
(b) Let $\{X_i\}_{i=1}^n$ be an i.i.d. sequence of zero-mean variables satisfying the Bernstein condition (2.15). Use part (a) to show that

$$\mathbb{E}\left[\left|\frac{1}{n}\sum_{i=1}^n X_i\right|\right] \leq \frac{2\sigma}{\sqrt{n}}\left(\sqrt{\pi} + \sqrt{\log 2}\right) + \frac{4b}{n}(1 + \log 2).$$

Exercise 2.9 (Sharp upper bounds on binomial tails) Let $\{X_i\}_{i=1}^n$ be an i.i.d. sequence of Bernoulli variables with parameter $\alpha \in (0, 1/2]$, and consider the binomial random variable $Z_n = \sum_{i=1}^n X_i$. The goal of this exercise is to prove, for any $\delta \in (0, \alpha)$, a sharp upper bound on the tail probability $\mathbb{P}[Z_n \leq \delta n]$.

(a) Show that $\mathbb{P}[Z_n \leq \delta n] \leq e^{-nD(\delta \,\|\, \alpha)}$, where the quantity

$$D(\delta \,\|\, \alpha) := \delta \log \frac{\delta}{\alpha} + (1 - \delta) \log \frac{(1 - \delta)}{(1 - \alpha)} \tag{2.64}$$

is the Kullback–Leibler divergence between the Bernoulli distributions with parameters δ and α, respectively.
(b) Show that the bound from part (a) is strictly better than the Hoeffding bound for all $\delta \in (0, \alpha)$.

Exercise 2.10 (Lower bounds on binomial tail probabilities) Let $\{X_i\}_{i=1}^n$ be a sequence of i.i.d. Bernoulli variables with parameter $\alpha \in (0, 1/2]$, and consider the binomial random variable $Z_n = \sum_{i=1}^n X_i$. In this exercise, we establish a *lower bound* on the probability $\mathbb{P}[Z_n \leq \delta n]$ for each fixed $\delta \in (0, \alpha)$, thereby establishing that the upper bound from Exercise 2.9 is tight up to a polynomial pre-factor. Throughout the analysis, we define $m = \lfloor n\delta \rfloor$, the largest integer less than or equal to $n\delta$, and set $\widetilde{\delta} = \frac{m}{n}$.

(a) Prove that $\frac{1}{n} \log \mathbb{P}[Z_n \leq \delta n] \geq \frac{1}{n} \log \binom{n}{m} + \widetilde{\delta} \log \alpha + (1 - \widetilde{\delta}) \log(1 - \alpha)$.
(b) Show that

$$\frac{1}{n} \log \binom{n}{m} \geq \phi(\widetilde{\delta}) - \frac{\log(n + 1)}{n}, \tag{2.65a}$$

where $\phi(\widetilde{\delta}) = -\widetilde{\delta} \log(\widetilde{\delta}) - (1 - \widetilde{\delta}) \log(1 - \widetilde{\delta})$ is the binary entropy. (*Hint:* Let Y be a binomial random variable with parameters $(n, \widetilde{\delta})$ and show that $\mathbb{P}[Y = \ell]$ is maximized when $\ell = m = \widetilde{\delta} n$.)
(c) Show that

$$\mathbb{P}[Z_n \leq \delta n] \geq \frac{1}{n + 1} e^{-nD(\delta \,\|\, \alpha)}, \tag{2.65b}$$

where the Kullback–Leibler divergence $D(\delta \,\|\, \alpha)$ was previously defined (2.64).

Exercise 2.11 (Upper and lower bounds for Gaussian maxima) Let $\{X_i\}_{i=1}^n$ be an i.i.d. sequence of $\mathcal{N}(0, \sigma^2)$ variables, and consider the random variable $Z_n := \max_{i=1,\ldots,n} |X_i|$.

(a) Prove that

$$\mathbb{E}[Z_n] \leq \sqrt{2\sigma^2 \log n} + \frac{4\sigma}{\sqrt{2 \log n}} \qquad \text{for all } n \geq 2.$$

(*Hint:* You may use the tail bound $\mathbb{P}[U \geq \delta] \leq \sqrt{\frac{2}{\pi}} \frac{1}{\delta} e^{-\delta^2/2}$, valid for any standard normal variate.)

(b) Prove that

$$\mathbb{E}[Z_n] \geq (1 - 1/e) \sqrt{2\sigma^2 \log n} \qquad \text{for all } n \geq 5.$$

(c) Prove that $\frac{\mathbb{E}[Z_n]}{\sqrt{2\sigma^2 \log n}} \to 1$ as $n \to +\infty$.

Exercise 2.12 (Upper bounds for sub-Gaussian maxima) Let $\{X_i\}_{i=1}^n$ be a sequence of zero-mean random variables, each sub-Gaussian with parameter σ. (No independence assumptions are needed.)

(a) Prove that

$$\mathbb{E}\Big[\max_{i=1,\ldots,n} X_i\Big] \leq \sqrt{2\sigma^2 \log n} \qquad \text{for all } n \geq 1. \tag{2.66}$$

(*Hint:* The exponential is a convex function.)

(b) Prove that the random variable $Z = \max_{i=1,\ldots,n} |X_i|$ satisfies

$$\mathbb{E}[Z] \leq \sqrt{2\sigma^2 \log(2n)} \leq 2\sqrt{\sigma^2 \log n}, \tag{2.67}$$

valid for all $n \geq 2$.

Exercise 2.13 (Operations on sub-Gaussian variables) Suppose that X_1 and X_2 are zero-mean and sub-Gaussian with parameters σ_1 and σ_2, respectively.

(a) If X_1 and X_2 are independent, show that the random variable $X_1 + X_2$ is sub-Gaussian with parameter $\sqrt{\sigma_1^2 + \sigma_2^2}$.

(b) Show that, in general (without assuming independence), the random variable $X_1 + X_2$ is sub-Gaussian with parameter at most $\sqrt{2}\sqrt{\sigma_1^2 + \sigma_2^2}$.

(c) In the same setting as part (b), show that $X_1 + X_2$ is sub-Gaussian with parameter at most $\sigma_1 + \sigma_2$.

(d) If X_1 and X_2 are independent, show that $X_1 X_2$ is sub-exponential with parameters $(\nu, b) = (\sqrt{2}\sigma_1\sigma_2, \sqrt{2}\sigma_1\sigma_2)$.

Exercise 2.14 (Concentration around medians and means) Given a scalar random variable X, suppose that there are positive constants c_1, c_2 such that

$$\mathbb{P}[|X - \mathbb{E}[X]| \geq t] \leq c_1 e^{-c_2 t^2} \qquad \text{for all } t \geq 0. \tag{2.68}$$

(a) Prove that $\text{var}(X) \leq \frac{c_1}{c_2}$.

(b) A median m_X is any number such that $\mathbb{P}[X \geq m_X] \geq 1/2$ and $\mathbb{P}[X \leq m_X] \geq 1/2$. Show by example that the median need not be unique.

(c) Show that whenever the mean concentration bound (2.68) holds, then for any median m_X, we have

$$\mathbb{P}[|X - m_X| \geq t] \leq c_3 e^{-c_4 t^2} \qquad \text{for all } t \geq 0, \tag{2.69}$$

where $c_3 := 4c_1$ and $c_4 := \frac{c_2}{8}$.

(d) Conversely, show that whenever the median concentration bound (2.69) holds, then mean concentration (2.68) holds with $c_1 = 2c_3$ and $c_2 = \frac{c_4}{4}$.

Exercise 2.15 (Concentration and kernel density estimation) Let $\{X_i\}_{i=1}^n$ be an i.i.d. sequence of random variables drawn from a density f on the real line. A standard estimate of f is the *kernel density estimate*

$$\widehat{f}_n(x) := \frac{1}{nh} \sum_{i=1}^n K\Big(\frac{x - X_i}{h}\Big),$$

where $K : \mathbb{R} \to [0, \infty)$ is a kernel function satisfying $\int_{-\infty}^{\infty} K(t)\, dt = 1$, and $h > 0$ is a bandwidth parameter. Suppose that we assess the quality of $\widehat{f}_n$ using the L^1-norm $\|\widehat{f}_n - f\|_1 := \int_{-\infty}^{\infty} |\widehat{f}_n(t) - f(t)|\, dt$. Prove that

$$\mathbb{P}[\|\widehat{f}_n - f\|_1 \geq \mathbb{E}[\|\widehat{f}_n - f\|_1] + \delta] \leq e^{-\frac{n\delta^2}{8}}.$$

Exercise 2.16 (Deviation inequalities in a Hilbert space) Let $\{X_i\}_{i=1}^n$ be a sequence of independent random variables taking values in a Hilbert space $\mathbb{H}$, and suppose that $\|X_i\|_{\mathbb{H}} \leq b_i$ almost surely. Consider the real-valued random variable $S_n = \big\| \sum_{i=1}^n X_i \big\|_{\mathbb{H}}$.

(a) Show that, for all $\delta \geq 0$,

$$\mathbb{P}[|S_n - \mathbb{E}[S_n]| \geq n\delta] \leq 2e^{-\frac{n\delta^2}{8b^2}}, \qquad \text{where } b^2 = \tfrac{1}{n} \textstyle\sum_{i=1}^n b_i^2.$$

(b) Show that $\mathbb{P}[\frac{S_n}{n} \geq a + \delta] \leq e^{-\frac{n\delta^2}{8b^2}}$, where $a := \sqrt{\frac{1}{n^2} \sum_{i=1}^n \mathbb{E}[\|X_i\|_{\mathbb{H}}^2]}$.

(*Note:* See Chapter 12 for basic background on Hilbert spaces.)

Exercise 2.17 (Hanson–Wright inequality) Given random variables $\{X_i\}_{i=1}^n$ and a positive semidefinite matrix $\mathbf{Q} \in \mathcal{S}_+^{n \times n}$, consider the random quadratic form

$$Z = \sum_{i=1}^n \sum_{j=1}^n \mathbf{Q}_{ij} X_i X_j. \tag{2.70}$$

The *Hanson–Wright inequality* guarantees that whenever the random variables $\{X_i\}_{i=1}^n$ are i.i.d. with mean zero, unit variance, and σ-sub-Gaussian, then there are universal constants (c_1, c_2) such that

$$\mathbb{P}[Z \geq \operatorname{trace}(\mathbf{Q}) + \sigma t] \leq 2 \exp\Big\{ -\min\Big(\frac{c_1 t}{|||\mathbf{Q}|||_2}, \frac{c_2 t^2}{|||\mathbf{Q}|||_F^2} \Big) \Big\}, \tag{2.71}$$

where $|||\mathbf{Q}|||_2$ and $|||\mathbf{Q}|||_F$ denote the operator and Frobenius norms, respectively. Prove this inequality in the special case $X_i \sim N(0, 1)$. (*Hint:* The rotation invariance of the Gaussian distribution and sub-exponential nature of χ^2-variates could be useful.)

Exercise 2.18 (Orlicz norms) Let $\psi : \mathbb{R}_+ \to \mathbb{R}_+$ be a strictly increasing convex function that satisfies $\psi(0) = 0$. The ψ-*Orlicz* norm of a random variable X is defined as

$$\|X\|_\psi := \inf\{t > 0 \mid \mathbb{E}[\psi(t^{-1}|X|)] \leq 1\}, \tag{2.72}$$

where $\|X\|_\psi$ is infinite if there is no finite t for which the expectation $\mathbb{E}[\psi(t^{-1}|X|)]$ exists. For the functions $u \mapsto u^q$ for some $q \in [1, \infty]$, then the Orlicz norm is simply the usual ℓ_q-norm $\|X\|_q = (\mathbb{E}[|X|^q])^{1/q}$. In this exercise, we consider the Orlicz norms $\|\cdot\|_{\psi_q}$ defined by the convex functions $\psi_q(u) = \exp(u^q) - 1$, for $q \geq 1$.

(a) If $\|X\|_{\psi_q} < +\infty$, show that there exist positive constants c_1, c_2 such that

$$\mathbb{P}[|X| > t] \leq c_1 \exp(-c_2 t^q) \qquad \text{for all } t > 0. \tag{2.73}$$

(In particular, you should be able to show that this bound holds with $c_1 = 2$ and $c_2 = \|X\|_{\psi_q}^{-q}$.)

(b) Suppose that a random variable Z satisfies the tail bound (2.73). Show that $\|X\|_{\psi_q}$ is finite.

Exercise 2.19 (Maxima of Orlicz variables) Recall the definition of Orlicz norm from Exercise 2.18. Let $\{X_i\}_{i=1}^n$ be an i.i.d. sequence of zero-mean random variables with finite Orlicz norm $\sigma = \|X_i\|_\psi$. Show that

$$\mathbb{E}\Big[\max_{i=1,\ldots,n} |X_i|\Big] \leq \sigma \psi^{-1}(n).$$

Exercise 2.20 (Tail bounds under moment conditions) Suppose that $\{X_i\}_{i=1}^n$ are zero-mean and independent random variables such that, for some fixed integer $m \geq 1$, they satisfy the moment bound $\|X_i\|_{2m} := (\mathbb{E}[X_i^{2m}])^{\frac{1}{2m}} \leq C_m$. Show that

$$\mathbb{P}\Bigg[\Bigg|\frac{1}{n}\sum_{i=1}^n X_i\Bigg| \geq \delta\Bigg] \leq B_m \left(\frac{1}{\sqrt{n}\delta}\right)^{2m} \qquad \text{for all } \delta > 0,$$

where B_m is a universal constant depending only on C_m and m.

Hint: You may find the following form of Rosenthal's inequality to be useful. Under the stated conditions, there is a universal constant R_m such that

$$\mathbb{E}\Bigg[\Bigg(\sum_{i=1}^n X_i\Bigg)^{2m}\Bigg] \leq R_m \Bigg\{\sum_{i=1}^n \mathbb{E}[X_i^{2m}] + \Bigg(\sum_{i=1}^n \mathbb{E}[X_i^2]\Bigg)^m\Bigg\}.$$

Exercise 2.21 (Concentration and data compression) Let $X = (X_1, X_2, \ldots, X_n)$ be a vector of i.i.d. Bernoulli variables with parameter $1/2$. The goal of lossy data compression is to represent X using a collection of binary vectors, say $\{z^1, \ldots, z^N\}$, such that the *rescaled Hamming distortion*

$$d(X) := \min_{j=1,\ldots,N} \rho_H(X, z^j) = \min_{j=1,\ldots,N} \Bigg\{\frac{1}{n}\sum_{i=1}^n \mathbb{I}[X_i \neq z_i^j]\Bigg\}$$

is as small as possible. Each binary vector z^j is known as a codeword, and the full collection is called a codebook. Of course, one can always achieve zero distortion using a codebook with $N = 2^n$ codewords, so the goal is to use $N = 2^{Rn}$ codewords for some rate $R < 1$. In this exercise, we use tail bounds to study the trade-off between the rate R and the distortion δ.

(a) Suppose that the rate R is upper bounded as

$$R < D_2(\delta \,\|\, 1/2) = \delta \log_2 \frac{\delta}{1/2} + (1-\delta) \log_2 \frac{1-\delta}{1/2}.$$

Show that, for any codebook $\{z^1, \ldots, z^N\}$ with $N \leq 2^{nR}$ codewords, the probability of the event $\{d(X) \leq \delta\}$ goes to zero as n goes to infinity. (*Hint:* Let V^j be a $\{0,1\}$-valued indicator variable for the event $\rho_H(X, z^j) \leq \delta$, and define $V = \sum_{j=1}^N V^j$. The tail bounds from Exercise 2.9 could be useful in bounding the probability $\mathbb{P}[V \geq 1]$.)

(b) We now show that, if $\Delta R := R - D_2(\delta \,\|\, 1/2) > 0$, then there exists a codebook that achieves distortion δ. In order to do so, consider a random codebook $\{Z^1, \ldots, Z^N\}$, formed by generating each codeword Z^j independently, and with all i.i.d. Ber$(1/2)$ entries. Let V^j be an indicator for the event $\rho_H(X, Z^j) \leq \delta$, and define $V = \sum_{j=1}^N V^j$.

   (i) Show that $\mathbb{P}[V \geq 1] \geq \frac{(\mathbb{E}[V])^2}{\mathbb{E}[V^2]}$.
   (ii) Use part (i) to show that $\mathbb{P}[V \geq 1] \to +\infty$ as $n \to +\infty$. (*Hint:* The tail bounds from Exercise 2.10 could be useful.)

Exercise 2.22 (Concentration for spin glasses) For some positive integer $d \geq 2$, consider a collection $\{\theta_{jk}\}_{j \neq k}$ of weights, one for each distinct pair $j \neq k$ of indices in $\{1, 2, \ldots, d\}$. We can then define a probability distribution over the Boolean hypercube $\{-1, +1\}^d$ via the mass function

$$\mathbb{P}_\theta(x_1, \ldots, x_d) = \exp\left\{ \frac{1}{\sqrt{d}} \sum_{i \neq j} \theta_{jk} x_j x_k - F_d(\theta) \right\}, \tag{2.74}$$

where the function $F_d : \mathbb{R}^{\binom{d}{2}} \to \mathbb{R}$, known as the *free energy*, is given by

$$F_d(\theta) := \log\left(\sum_{x \in \{-1,+1\}^d} \exp\left\{ \frac{1}{\sqrt{d}} \sum_{j \neq k} \theta_{jk} x_j x_k \right\} \right) \tag{2.75}$$

serves to normalize the distribution. The probability distribution (2.74) was originally used to describe the behavior of magnets in statistical physics, in which context it is known as the *Ising model.* Suppose that the weights are chosen as i.i.d. random variables, so that equation (2.74) now describes a random family of probability distributions. This family is known as the Sherrington–Kirkpatrick model in statistical physics.

(a) Show that F_d is a convex function.
(b) For any two vectors $\theta, \theta' \in \mathbb{R}^{\binom{d}{2}}$, show that $\|F_d(\theta) - F_d(\theta')\|_2 \leq \sqrt{d}\, \|\theta - \theta'\|_2$.
(c) Suppose that the weights are chosen in an i.i.d. manner as $\theta_{jk} \sim \mathcal{N}(0, \beta^2)$ for each $j \neq k$. Use the previous parts and Jensen's inequality to show that

$$\mathbb{P}\left[\frac{F_d(\theta)}{d} \geq \log 2 + \frac{\beta^2}{4} + t \right] \leq 2e^{-\beta d t^2/2} \qquad \text{for all } t > 0. \tag{2.76}$$

Remark: Interestingly, it is known that, for any $\beta \in [0, 1)$, this upper tail bound captures the asymptotic behavior of $F_d(\theta)/d$ accurately, in that $\frac{F_d(\theta)}{d} \xrightarrow{a.s.} \log 2 + \beta^2/4$ as $d \to \infty$. By contrast, for $\beta \geq 1$, the behavior of this spin glass model is much more subtle; we refer the reader to the bibliographic section for additional reading.

3

Concentration of measure

Building upon the foundation of Chapter 2, this chapter is devoted to an exploration of more advanced material on the concentration of measure. In particular, our goal is to provide an overview of the different types of methods available to derive tail bounds and concentration inequalities. We begin in Section 3.1 with a discussion of the entropy method for concentration, and illustrate its use in deriving tail bounds for Lipschitz functions of independent random variables. In Section 3.2, we turn to some geometric aspects of concentration inequalities, a viewpoint that is historically among the oldest. Section 3.3 is devoted to the use of transportation cost inequalities for deriving concentration inequalities, a method that is in some sense dual to the entropy method, and well suited to certain types of dependent random variables. We conclude in Section 3.4 by deriving some tail bounds for empirical processes, including versions of the functional Hoeffding and Bernstein inequalities. These inequalities play an especially important role in our later treatment of nonparametric problems.

3.1 Concentration by entropic techniques

We begin our exploration with the entropy method and related techniques for deriving concentration inequalities.

3.1.1 Entropy and its properties

Given a convex function $\phi \colon \mathbb{R} \to \mathbb{R}$, it can be used to define a functional on the space of probability distributions via

$$\mathbb{H}_\phi(X) := \mathbb{E}[\phi(X)] - \phi(\mathbb{E}[X]),$$

where $X \sim \mathbb{P}$. This quantity, which is well defined for any random variable such that both X and $\phi(X)$ have finite expectations, is known as the *ϕ-entropy*[1] of the random variable X. By Jensen's inequality and the convexity of ϕ, the ϕ-entropy is always non-negative. As the name suggests, it serves as a measure of variability. For instance, in the most extreme case, we have $\mathbb{H}_\phi(X) = 0$ for any random variable such that X is equal to its expectation $\mathbb{P}$-almost-everywhere.

[1] The notation $\mathbb{H}_\phi(X)$ has the potential to mislead, since it suggests that the entropy is a function of X, and hence a random variable. To be clear, the entropy $\mathbb{H}_\phi$ is a functional that acts on the probability measure $\mathbb{P}$, as opposed to the random variable X.

There are various types of entropies, depending on the choice of the underlying convex function ϕ. Some of these entropies are already familiar to us. For example, the convex function $\phi(u) = u^2$ yields

$$\mathbb{H}_\phi(X) = \mathbb{E}[X^2] - (\mathbb{E}[X])^2 = \text{var}(X),$$

corresponding to the usual variance of the random variable X. Another interesting choice is the convex function $\phi(u) = -\log u$ defined on the positive real line. When applied to the positive random variable $Z := e^{\lambda X}$, this choice of ϕ yields

$$\mathbb{H}_\phi(e^{\lambda X}) = -\lambda \mathbb{E}[X] + \log \mathbb{E}[e^{\lambda X}] = \log \mathbb{E}[e^{\lambda(X - \mathbb{E}[X])}],$$

a type of entropy corresponding to the centered cumulant generating function. In Chapter 2, we have seen how both the variance and the cumulant generating function are useful objects for obtaining concentration inequalities—in particular, in the form of Chebyshev's inequality and the Chernoff bound, respectively.

Throughout the remainder of this chapter, we focus on a slightly different choice of entropy functional, namely the convex function $\phi\colon [0, \infty) \to \mathbb{R}$ defined as

$$\phi(u) := u \log u \quad \text{for } u > 0, \qquad \text{and} \quad \phi(0) := 0. \tag{3.1}$$

For any non-negative random variable $Z \geq 0$, it defines the ϕ-entropy given by

$$\mathbb{H}(Z) = \mathbb{E}[Z \log Z] - \mathbb{E}[Z] \log \mathbb{E}[Z], \tag{3.2}$$

assuming that all relevant expectations exist. In the remainder of this chapter, we omit the subscript ϕ, since the choice (3.1) is to be implicitly understood.

The reader familiar with information theory may observe that the entropy (3.2) is closely related to the Shannon entropy, as well as the Kullback–Leibler divergence; see Exercise 3.1 for an exploration of this connection. As will be clarified in the sequel, the most attractive property of the ϕ-entropy (3.2) is its so-called tensorization when applied to functions of independent random variables.

For the random variable $Z := e^{\lambda X}$, the entropy has an explicit expression as a function of the moment generating function $\varphi_x(\lambda) = \mathbb{E}[e^{\lambda X}]$ and its first derivative. In particular, a short calculation yields

$$\mathbb{H}(e^{\lambda X}) = \lambda \varphi'_x(\lambda) - \varphi_x(\lambda) \log \varphi_x(\lambda). \tag{3.3}$$

Consequently, if we know the moment generating function of X, then it is straightforward to compute the entropy $\mathbb{H}(e^{\lambda X})$. Let us consider a simple example to illustrate:

Example 3.1 (Entropy of a Gaussian random variable) For the scalar Gaussian variable $X \sim \mathcal{N}(0, \sigma^2)$, we have $\varphi_x(\lambda) = e^{\lambda^2\sigma^2/2}$. By taking derivatives, we find that $\varphi'_x(\lambda) = \lambda\sigma^2\varphi_x(\lambda)$, and hence

$$\mathbb{H}(e^{\lambda X}) = \lambda^2\sigma^2\varphi_x(\lambda) - \tfrac{1}{2}\lambda^2\sigma^2 \varphi_x(\lambda) = \tfrac{1}{2}\lambda^2\sigma^2 \varphi_x(\lambda). \tag{3.4}$$

♣

Given that the moment generating function can be used to obtain concentration inequalities via the Chernoff method, this connection suggests that there should also be a connection between the entropy (3.3) and tail bounds. It is the goal of the following sections to make this connection precise for various classes of random variables. We then show how the entropy based on $\phi(u) = u \log u$ has a certain tensorization property that makes it particularly well suited to dealing with general Lipschitz functions of collections of random variables.

3.1.2 Herbst argument and its extensions

Intuitively, the entropy is a measure of the fluctuations in a random variable, so that control on the entropy should translate into bounds on its tails. The Herbst argument makes this intuition precise for a certain class of random variables. In particular, suppose that there is a constant $\sigma > 0$ such that the entropy of $e^{\lambda X}$ satisfies an upper bound of the form

$$\mathbb{H}(e^{\lambda X}) \leq \tfrac{1}{2}\sigma^2\lambda^2\,\varphi_x(\lambda). \tag{3.5}$$

Note that by our earlier calculation in Example 3.1, any Gaussian variable $X \sim \mathcal{N}(0, \sigma^2)$ satisfies this condition *with equality* for all $\lambda \in \mathbb{R}$. Moreover, as shown in Exercise 3.7, any bounded random variable satisfies an inequality of the form (3.5).

Of interest here is the other implication: What does the entropy bound (3.5) imply about the tail behavior of the random variable? The classical Herbst argument answers this question, in particular showing that any such variable must have sub-Gaussian tail behavior.

Proposition 3.2 (Herbst argument) *Suppose that the entropy $\mathbb{H}(e^{\lambda X})$ satisfies inequality (3.5) for all $\lambda \in I$, where I can be either of the intervals $[0, \infty)$ or $\mathbb{R}$. Then X satisfies the bound*

$$\log \mathbb{E}[e^{\lambda(X - \mathbb{E}[X])}] \leq \tfrac{1}{2}\lambda^2\sigma^2 \qquad \textit{for all } \lambda \in I. \tag{3.6}$$

Remarks: When $I = \mathbb{R}$, then the inequality (3.6) is equivalent to asserting that the centered variable $X - \mathbb{E}[X]$ is sub-Gaussian with parameter σ. Via an application of the usual Chernoff argument, the bound (3.6) with $I = [0, \infty)$ implies the one-sided tail bound

$$\mathbb{P}[X \geq \mathbb{E}[X] + t] \leq e^{-\frac{t^2}{2\sigma^2}}, \tag{3.7}$$

and with $I = \mathbb{R}$, it implies the two-sided bound $\mathbb{P}[|X - \mathbb{E}[X]| \geq t] \leq 2e^{-\frac{t^2}{2\sigma^2}}$. Of course, these are the familiar tail bounds for sub-Gaussian variables discussed previously in Chapter 2.

Proof Recall the representation (3.3) of entropy in terms of the moment generating function. Combined with the assumed upper bound (3.5), we conclude that the moment generating function $\varphi \equiv \varphi_x$ satisfies the differential inequality

$$\lambda\varphi'(\lambda) - \varphi(\lambda)\log\varphi(\lambda) \leq \tfrac{1}{2}\sigma^2\lambda^2\,\varphi(\lambda), \qquad \text{valid for all } \lambda \geq 0. \tag{3.8}$$

Define the function $G(\lambda) = \frac{1}{\lambda}\log\varphi(\lambda)$ for $\lambda \neq 0$, and extend the definition by continuity to

$$G(0) := \lim_{\lambda \to 0} G(\lambda) = \mathbb{E}[X]. \tag{3.9}$$

Note that we have $G'(\lambda) = \frac{1}{\lambda}\frac{\varphi'(\lambda)}{\varphi(\lambda)} - \frac{1}{\lambda^2}\log\varphi(\lambda)$, so that the inequality (3.8) can be rewritten in the simple form $G'(\lambda) \leq \frac{1}{2}\sigma^2$ for all $\lambda \in I$. For any $\lambda_0 > 0$, we can integrate both sides of the inequality to obtain

$$G(\lambda) - G(\lambda_0) \leq \tfrac{1}{2}\sigma^2(\lambda - \lambda_0).$$

Letting $\lambda_0 \to 0^+$ and using the relation (3.9), we conclude that

$$G(\lambda) - \mathbb{E}[X] \leq \tfrac{1}{2}\sigma^2\lambda,$$

which is equivalent to the claim (3.6). We leave the extension of this proof to the case $I = \mathbb{R}$ as an exercise for the reader. □

Thus far, we have seen how a particular upper bound (3.5) on the entropy $\mathbb{H}(e^{\lambda X})$ translates into a bound on the cumulant generating function (3.6), and hence into sub-Gaussian tail bounds via the usual Chernoff argument. It is natural to explore to what extent this approach may be generalized. As seen previously in Chapter 2, a broader class of random variables are those with sub-exponential tails, and the following result is the analog of Proposition 3.2 in this case.

Proposition 3.3 (Bernstein entropy bound) *Suppose that there are positive constants b and σ such that the entropy $\mathbb{H}(e^{\lambda X})$ satisfies the bound*

$$\mathbb{H}(e^{\lambda X}) \leq \lambda^2\{b\varphi_x'(\lambda) + \varphi_x(\lambda)(\sigma^2 - b\mathbb{E}[X])\} \quad \textit{for all } \lambda \in [0, 1/b). \tag{3.10}$$

Then X satisfies the bound

$$\log \mathbb{E}[e^{\lambda(X-\mathbb{E}[X])}] \leq \sigma^2\lambda^2(1 - b\lambda)^{-1} \quad \textit{for all } \lambda \in [0, 1/b). \tag{3.11}$$

Remarks: As a consequence of the usual Chernoff argument, Proposition 3.3 implies that X satisfies the upper tail bound

$$\mathbb{P}[X \geq \mathbb{E}[X] + \delta] \leq \exp\left(-\frac{\delta^2}{4\sigma^2 + 2b\delta}\right) \quad \text{for all } \delta \geq 0, \tag{3.12}$$

which (modulo non-optimal constants) is the usual Bernstein-type bound to be expected for a variable with sub-exponential tails. See Proposition 2.10 from Chapter 2 for further details on such Bernstein bounds.

We now turn to the proof of Proposition 3.3.

Proof As before, we omit the dependence of φ_x on X throughout this proof so as to simplify notation. By rescaling and recentering arguments sketched out in Exercise 3.6, we may assume without loss of generality that $\mathbb{E}[X] = 0$ and $b = 1$, in which case the inequality (3.10) simplifies to

$$\mathbb{H}(e^{\lambda X}) \leq \lambda^2\{\varphi'(\lambda) + \varphi(\lambda)\sigma^2\} \quad \text{for all } \lambda \in [0, 1). \tag{3.13}$$

Recalling the function $G(\lambda) = \frac{1}{\lambda}\log\varphi(\lambda)$ from the proof of Proposition 3.2, a little bit of algebra shows that condition (3.13) is equivalent to the differential inequality $G' \leq \sigma^2 + \frac{\varphi'}{\varphi}$. Letting $\lambda_0 > 0$ be arbitrary and integrating both sides of this inequality over the interval (λ_0, λ), we obtain

$$G(\lambda) - G(\lambda_0) \leq \sigma^2(\lambda - \lambda_0) + \log \varphi(\lambda) - \log \varphi(\lambda_0).$$

Since this inequality holds for all $\lambda_0 > 0$, we may take the limit as $\lambda_0 \to 0^+$. Doing so and using the facts that $\lim_{\lambda_0 \to 0^+} G(\lambda_0) = G(0) = \mathbb{E}[X]$ and $\log \varphi(0) = 0$, we obtain the bound

$$G(\lambda) - \mathbb{E}[X] \leq \sigma^2 \lambda + \log \varphi(\lambda). \tag{3.14}$$

Substituting the definition of G and rearranging yields the claim (3.11).

□

3.1.3 Separately convex functions and the entropic method

Thus far, we have seen how the entropic method can be used to derive sub-Gaussian and sub-exponential tail bounds for scalar random variables. If this were the only use of the entropic method, then we would have gained little beyond what can be done via the usual Chernoff bound. The real power of the entropic method—as we now will see—manifests itself in dealing with concentration for functions of many random variables.

As an illustration, we begin by stating a deep result that can be proven in a relatively direct manner using the entropy method. We say that a function $f\colon \mathbb{R}^n \to \mathbb{R}$ is *separately convex* if, for each index $k \in \{1, 2, \ldots, n\}$, the univariate function

$$y_k \mapsto f(x_1, x_2, \ldots, x_{k-1}, y_k, x_{k+1}, \ldots, x_n)$$

is convex for each fixed vector $(x_1, x_2, \ldots, x_{k-1}, x_{k+1}, \ldots, x_n) \in \mathbb{R}^{n-1}$. A function f is L-Lipschitz with respect to the Euclidean norm if

$$|f(x) - f(x')| \leq L\|x - x'\|_2 \qquad \text{for all } x, x' \in \mathbb{R}^n. \tag{3.15}$$

The following result applies to separately convex and L-Lipschitz functions.

Theorem 3.4 *Let $\{X_i\}_{i=1}^n$ be independent random variables, each supported on the interval $[a, b]$, and let $f\colon \mathbb{R}^n \to \mathbb{R}$ be separately convex, and L-Lipschitz with respect to the Euclidean norm. Then, for all $\delta > 0$, we have*

$$\mathbb{P}[f(X) \geq \mathbb{E}[f(X)] + \delta] \leq \exp\left(-\frac{\delta^2}{4L^2(b-a)^2}\right). \tag{3.16}$$

Remarks: This result is the analog of the upper tail bound for Lipschitz functions of Gaussian variables (cf. Theorem 2.26 in Chapter 2), but applicable to independent and bounded variables instead. In contrast to the Gaussian case, the additional assumption of separate convexity cannot be eliminated in general; see the bibliographic section for further discussion. When f is jointly convex, other techniques can be used to obtain the lower tail bound as well; see Theorem 3.24 in the sequel for one such example.

Theorem 3.4 can be used to obtain order-optimal bounds for a number of interesting problems. As one illustration, we return to the Rademacher complexity, first introduced in Example 2.25 of Chapter 2.

Example 3.5 (Sharp bounds on Rademacher complexity) Given a bounded subset $\mathcal{A} \subset \mathbb{R}^n$, consider the random variable $Z = \sup_{a\in\mathcal{A}} \sum_{k=1}^n a_k \varepsilon_k$, where $\varepsilon_k \in \{-1, +1\}$ are i.i.d. Rademacher variables. Let us view Z as a function of the random signs, and use Theorem 3.4 to bound the probability of the tail event $\{Z \geq \mathbb{E}[Z] + t\}$.

It suffices to verify the convexity and Lipschitz conditions of the theorem. First, since $Z = Z(\varepsilon_1, \ldots, \varepsilon_n)$ is the maximum of a collection of linear functions, it is jointly (and hence separately) convex. Let $Z' = Z(\varepsilon'_1, \ldots, \varepsilon'_n)$ where $\varepsilon' \in \{-1, +1\}^n$ is a second vector of sign variables. For any $a \in \mathcal{A}$, we have

$$\underbrace{\langle a,\, \varepsilon\rangle}_{\sum_{k=1}^n a_k\varepsilon_k} - Z' = \langle a,\, \varepsilon\rangle - \sup_{a'\in\mathcal{A}} \langle a',\, \varepsilon'\rangle \leq \langle a,\, \varepsilon - \varepsilon'\rangle \leq \|a\|_2\, \|\varepsilon - \varepsilon'\|_2.$$

Taking suprema over $a \in \mathcal{A}$ yields that $Z - Z' \leq (\sup_{a\in\mathcal{A}} \|a\|_2)\, \|\varepsilon - \varepsilon'\|_2$. Since the same argument may be applied with the roles of ε and ε' reversed, we conclude that Z is Lipschitz with parameter $\mathcal{W}(\mathcal{A}) := \sup_{a\in\mathcal{A}} \|a\|_2$, corresponding to the Euclidean width of the set. Putting together the pieces, Theorem 3.4 implies that

$$\mathbb{P}[Z \geq \mathbb{E}[Z] + t] \leq \exp\left(-\frac{t^2}{16\mathcal{W}^2(\mathcal{A})}\right). \tag{3.17}$$

Note that parameter $\mathcal{W}^2(\mathcal{A})$ may be substantially smaller than the quantity $\sum_{k=1}^n \sup_{a\in\mathcal{A}} a_k^2$ —indeed, possibly as much as a factor of n smaller! In such cases, Theorem 3.4 yields a much sharper tail bound than our earlier tail bound from Example 2.25, which was obtained by applying the bounded differences inequality. ♣

Another use of Theorem 3.4 is in random matrix theory.

Example 3.6 (Operator norm of a random matrix) Let $\mathbf{X} \in \mathbb{R}^{n\times d}$ be a random matrix, say with X_{ij} drawn i.i.d. from some zero-mean distribution supported on the unit interval $[-1, +1]$. The spectral or ℓ_2-operator norm of X, denoted by $|||\mathbf{X}|||_2$, is its maximum singular value, given by

$$|||\mathbf{X}|||_2 = \max_{\substack{v\in\mathbb{R}^d \\ \|v\|_2=1}} \|\mathbf{X}v\|_2 = \max_{\substack{v\in\mathbb{R}^d \\ \|v\|_2=1}} \max_{\substack{u\in\mathbb{R}^n \\ \|u\|_2=1}} u^{\mathrm{T}}\mathbf{X}v. \tag{3.18}$$

Let us view the mapping $\mathbf{X} \mapsto |||\mathbf{X}|||_2$ as a function f from $\mathbb{R}^{nd}$ to $\mathbb{R}$. In order to apply Theorem 3.4, we need to show that f is both Lipschitz and convex. From its definition (3.18), the operator norm is the supremum of a collection of functions that are linear in the entries $\mathbf{X}$; any such supremum is a convex function. Moreover, we have

$$\Big| |||\mathbf{X}|||_2 - |||\mathbf{X}'|||_2 \Big| \overset{\text{(i)}}{\leq} |||\mathbf{X} - \mathbf{X}'|||_2 \overset{\text{(ii)}}{\leq} |||\mathbf{X} - \mathbf{X}'|||_{\mathrm{F}}, \tag{3.19}$$

where step (i) follows from the triangle inequality, and step (ii) follows since the Frobenius norm of a matrix always upper bounds the operator norm. (The Frobenius norm $|||\mathbf{M}|||_F$ of a matrix $\mathbf{M} \in \mathbb{R}^{n\times d}$ is simply the Euclidean norm of all its entries; see equation (2.50).) Consequently, the operator norm is Lipschitz with parameter $L = 1$, and thus Theorem 3.4 implies that

$$\mathbb{P}[|||\mathbf{X}|||_2 \geq \mathbb{E}[|||\mathbf{X}|||_2] + \delta] \leq e^{-\frac{\delta^2}{16}}.$$

It is worth observing that this bound is the analog of our earlier bound (2.52) on the operator norm of a Gaussian random matrix, albeit with a worse constant. See Example 2.32 in Chapter 2 for further details on this Gaussian case. ♣

3.1.4 Tensorization and separately convex functions

We now return to prove Theorem 3.4. The proof is based on two lemmas, both of which are of independent interest. Here we state these results and discuss some of their consequences, deferring their proofs to the end of this section. Our first lemma establishes an entropy bound for univariate functions:

Lemma 3.7 (Entropy bound for univariate functions) *Let $X, Y \sim \mathbb{P}$ be a pair of i.i.d. variates. Then for any function $g\colon \mathbb{R} \to \mathbb{R}$, we have*

$$\mathbb{H}(e^{\lambda g(X)}) \leq \lambda^2 \mathbb{E}[(g(X) - g(Y))^2 e^{\lambda g(X)}\, \mathbb{I}[g(X) \geq g(Y)]] \qquad \textit{for all } \lambda > 0. \tag{3.20a}$$

If in addition X is supported on $[a, b]$, and g is convex and Lipschitz, then

$$\mathbb{H}(e^{\lambda g(X)}) \leq \lambda^2 (b-a)^2\, \mathbb{E}[(g'(X))^2 e^{\lambda g(X)}] \qquad \textit{for all } \lambda > 0, \tag{3.20b}$$

where g' is the derivative.

In stating this lemma, we have used the fact that any convex and Lipschitz function has a derivative defined almost everywhere, a result known as Rademacher's theorem. Moreover, note that if g is Lipschitz with parameter L, then we are guaranteed that $\|g'\|_\infty \leq L$, so that inequality (3.20b) implies an entropy bound of the form

$$\mathbb{H}(e^{\lambda g(X)}) \leq \lambda^2 L^2 (b-a)^2\, \mathbb{E}[e^{\lambda g(X)}] \qquad \text{for all } \lambda > 0.$$

In turn, by an application of Proposition 3.2, such an entropy inequality implies the upper tail bound

$$\mathbb{P}[g(X) \geq \mathbb{E}[g(X)] + \delta] \leq e^{-\frac{\delta^2}{4L^2(b-a)^2}}.$$

Thus, Lemma 3.7 implies the univariate version of Theorem 3.4. However, the inequality (3.20b) is sharper, in that it involves $g'(X)$ as opposed to the worst-case bound L, and this distinction will be important in deriving the sharp result of Theorem 3.4. The more general inequality (3.20b) will be useful in deriving functional versions of the Hoeffding and Bernstein inequalities (see Section 3.4).

Returning to the main thread, it remains to extend this univariate result to the multivariate setting, and the so-called *tensorization property* of entropy plays a key role here. Given a function $f\colon \mathbb{R}^n \to \mathbb{R}$, an index $k \in \{1, 2, \ldots, n\}$ and a vector $x_{\backslash k} = (x_i, i \neq k) \in \mathbb{R}^{n-1}$, we define the conditional entropy in coordinate k via

$$\mathbb{H}(e^{\lambda f_k(X_k)} \mid x_{\backslash k}) := \mathbb{H}(e^{\lambda f(x_1,\ldots,x_{k-1},X_k,x_{k+1},\ldots,x_n)}),$$

where $f_k\colon \mathbb{R} \to \mathbb{R}$ is the coordinate function $x_k \mapsto f(x_1, \ldots, x_k, \ldots, x_n)$. To be clear, for a random vector $X^{\backslash k} \in \mathbb{R}^{n-1}$, the entropy $\mathbb{H}(e^{\lambda f_k(X_k)} \mid X^{\backslash k})$ is a random variable, and its expectation is often referred to as the conditional entropy.) The following result shows that the joint entropy can be upper bounded by a sum of univariate entropies, suitably defined.

Lemma 3.8 (Tensorization of entropy) *Let $f\colon \mathbb{R}^n \to \mathbb{R}$, and let $\{X_k\}_{k=1}^n$ be independent random variables. Then*

$$\mathbb{H}(e^{\lambda f(X_1,\ldots,X_n)}) \le \mathbb{E}\left[\sum_{k=1}^n \mathbb{H}(e^{\lambda f_k(X_k)} \mid X^{\backslash k})\right] \quad \textit{for all } \lambda > 0. \tag{3.21}$$

Equipped with these two results, we are now ready to prove Theorem 3.4.

Proof of Theorem 3.4 For any $k \in \{1, 2, \ldots, n\}$ and fixed vector $x_{\backslash k} \in \mathbb{R}^{n-1}$, our assumptions imply that the coordinate function f_k is convex, and hence Lemma 3.7 implies that, for all $\lambda > 0$, we have

$$\begin{aligned}
\mathbb{H}(e^{\lambda f_k(X_k)} \mid x_{\backslash k}) &\le \lambda^2 (b-a)^2\, \mathbb{E}_{X_k}[(f_k'(X_k))^2 e^{\lambda f_k(X_k)} \mid x_{\backslash k}] \\
&= \lambda^2 (b-a)^2\, \mathbb{E}_{X_k}\left[\left(\frac{\partial f(x_1, \ldots, X_k, \ldots, x_n)}{\partial x_k}\right)^2 e^{\lambda f(x_1,\ldots,X_k,\ldots,x_n)}\right],
\end{aligned}$$

where the second line involves unpacking the definition of the conditional entropy.

Combined with Lemma 3.8, we find that the unconditional entropy is upper bounded as

$$\mathbb{H}(e^{\lambda f(X)}) \le \lambda^2 (b-a)^2\, \mathbb{E}\left[\sum_{k=1}^n \left(\frac{\partial f(X)}{\partial x_k}\right)^2 e^{\lambda f(X)}\right] \overset{\text{(i)}}{\le} \lambda^2 (b-a)^2 L^2\, \mathbb{E}[e^{\lambda f(X)}].$$

Here step (i) follows from the Lipschitz condition, which guarantees that

$$\|\nabla f(x)\|_2^2 = \sum_{k=1}^n \left(\frac{\partial f(x)}{\partial x_k}\right)^2 \le L^2$$

almost surely. Thus, the tail bound (3.16) follows from an application of Proposition 3.2. □

It remains to prove the two auxiliary lemmas used in the preceding proof—namely, Lemma 3.7 on entropy bounds for univariate Lipschitz functions, and Lemma 3.8 on the tensorization of entropy. We begin with the former property.

Proof of Lemma 3.7

By the definition of entropy, we can write

$$\begin{aligned}\mathbb{H}(e^{\lambda g(X)}) &= \mathbb{E}_X[\lambda g(X)e^{\lambda g(X)}] - \mathbb{E}_X[e^{\lambda g(X)}]\log(\mathbb{E}_Y[e^{\lambda g(Y)}])\\ &\overset{\text{(i)}}{\leq} \mathbb{E}_X[\lambda g(X)e^{\lambda g(X)}] - \mathbb{E}_{X,Y}[e^{\lambda g(X)}\lambda g(Y)]\\ &= \tfrac{1}{2}\mathbb{E}_{X,Y}\Big[\lambda\{g(X)-g(Y)\}\{e^{\lambda g(X)} - e^{\lambda g(Y)}\}\Big]\\ &\overset{\text{(ii)}}{=} \lambda\mathbb{E}\Big[\{g(X)-g(Y)\}\{e^{\lambda g(X)} - e^{\lambda g(Y)}\}\,\mathbb{I}[g(X)\geq g(Y)]\Big],\end{aligned} \tag{3.22}$$

where step (i) follows from Jensen's inequality, and step (ii) follows from symmetry of X and Y.

By convexity of the exponential, we have $e^s - e^t \leq e^s(s-t)$ for all $s, t \in \mathbb{R}$. For $s \geq t$, we can multiply both sides by $(s-t) \geq 0$, thereby obtaining

$$(s-t)(e^s - e^t)\,\mathbb{I}[s \geq t] \leq (s-t)^2 e^s\,\mathbb{I}[s\geq t].$$

Applying this bound with $s = \lambda g(X)$ and $t = \lambda g(Y)$ to the inequality (3.22) yields

$$\mathbb{H}(e^{\lambda g(X)}) \leq \lambda^2\,\mathbb{E}[(g(X)-g(Y))^2 e^{\lambda g(X)}\,\mathbb{I}[g(X)\geq g(Y)]], \tag{3.23}$$

where we have recalled the assumption that $\lambda > 0$.

If in addition g is convex, then we have the upper bound $g(x) - g(y) \leq g'(x)(x-y)$, and hence, for $g(x) \geq g(y)$,

$$(g(x)-g(y))^2 \leq (g'(x))^2(x-y)^2 \leq (g'(x))^2(b-a)^2,$$

where the final step uses the assumption that $x, y \in [a, b]$. Combining the pieces yields the claim.

We now turn to the tensorization property of entropy.

Proof of Lemma 3.8

The proof makes use of the following variational representation for entropy:

$$\mathbb{H}(e^{\lambda f(X)}) = \sup_g\{\mathbb{E}[g(X)e^{\lambda f(X)}] \mid \mathbb{E}[e^{g(X)}] \leq 1\}. \tag{3.24}$$

This equivalence follows by a duality argument that we explore in Exercise 3.9.

For each $j \in \{1, 2, \ldots, n\}$, define $X_j^n = (X_j, \ldots, X_n)$. Let g be any function that satisfies $\mathbb{E}[e^{g(X)}] \leq 1$. We can then define an auxiliary sequence of functions $\{g^1, \ldots, g^n\}$ via

$$g^1(X_1, \ldots, X_n) := g(X) - \log\mathbb{E}[e^{g(X)} \mid X_2^n]$$

and

$$g^k(X_k, \ldots, X_n) := \log\frac{\mathbb{E}[e^{g(X)} \mid X_k^n]}{\mathbb{E}[e^{g(X)} \mid X_{k+1}^n]} \qquad \text{for } k = 2, \ldots, n.$$

By construction, we have

$$\sum_{k=1}^n g^k(X_k, \ldots, X_n) = g(X) - \log\mathbb{E}[e^{g(X)}] \geq g(X) \tag{3.25}$$

and moreover $\mathbb{E}[\exp(g^k(X_k, X_{k+1}, \ldots, X_n)) \mid X_{k+1}^n] = 1$.

We now use this decomposition within the variational representation (3.24), thereby obtaining the chain of upper bounds

$$\begin{aligned}
\mathbb{E}[g(X)e^{\lambda f(X)}] &\overset{\text{(i)}}{\leq} \sum_{k=1}^{n} \mathbb{E}[g^k(X_k, \ldots, X_n)e^{\lambda f(X)}] \\
&= \sum_{k=1}^{n} \mathbb{E}_{X_{\setminus k}}[\mathbb{E}_{X_k}[g^k(X_k, \ldots, X_n)e^{\lambda f(X)} \mid X_{\setminus k}]] \\
&\overset{\text{(ii)}}{\leq} \sum_{k=1}^{n} \mathbb{E}_{X_{\setminus k}}[\mathbb{H}(e^{\lambda f_k(X_k)} \mid X_{\setminus k})],
\end{aligned}$$

where inequality (i) uses the bound (3.25), and inequality (ii) applies the variational representation (3.24) to the univariate functions, and also makes use of the fact that $\mathbb{E}[g^k(X_k, \ldots, X_n) \mid X_{\setminus k}] = 1$. Since this argument applies to any function g such that $\mathbb{E}[e^{g(X)}] \leq 1$, we may take the supremum over the left-hand side, and combined with the variational representation (3.24), we conclude that

$$\mathbb{H}(e^{\lambda f(X)}) \leq \sum_{k=1}^{n} \mathbb{E}_{X_{\setminus k}}[\mathbb{H}(e^{\lambda f_k(X_k)} \mid X_{\setminus k})],$$

as claimed.

3.2 A geometric perspective on concentration

We now turn to some geometric aspects of the concentration of measure. Historically, this geometric viewpoint is among the oldest, dating back to the classical result of Lévy on concentration of measure for Lipschitz functions of Gaussians. It also establishes deep links between probabilistic concepts and high-dimensional geometry.

The results of this section are most conveniently stated in terms of a *metric measure space*—namely, a metric space $(\mathcal{X}, \rho)$ endowed with a probability measure $\mathbb{P}$ on its Borel sets. Some canonical examples of metric spaces for the reader to keep in mind are the set $\mathcal{X} = \mathbb{R}^n$ equipped with the usual Euclidean metric $\rho(x, y) := \|x - y\|_2$, and the discrete cube $\mathcal{X} = \{0, 1\}^n$ equipped with the Hamming metric $\rho(x, y) = \sum_{j=1}^{n} \mathbb{I}[x_j \neq y_j]$.

Associated with any metric measure space is an object known as its *concentration function*, which is defined in a geometric manner via the ϵ-enlargements of sets. The concentration function specifies how rapidly, as a function of ϵ, the probability of any ϵ-enlargement increases towards one. As we will see, this function is intimately related to the concentration properties of Lipschitz functions on the metric space.

3.2.1 Concentration functions

Given a set $A \subseteq \mathcal{X}$ and a point $x \in \mathcal{X}$, define the quantity

$$\rho(x, A) := \inf_{y \in A} \rho(x, y), \tag{3.26}$$

which measures the distance between the point x and the closest point in the set A. Given a parameter $\epsilon > 0$, the *ϵ-enlargement* of A is given by

$$A^\epsilon := \{x \in \mathcal{X} \mid \rho(x, A) < \epsilon\}. \tag{3.27}$$

In words, the set A^ϵ corresponds to the open neighborhood of points lying at distance less than ϵ from A. With this notation, the concentration function of the metric measure space $(\mathcal{X}, \rho, \mathbb{P})$ is defined as follows:

Definition 3.9 The *concentration function* $\alpha\colon [0, \infty) \to \mathbb{R}_+$ associated with metric measure space $(\mathbb{P}, \mathcal{X}, \rho)$ is given by

$$\alpha_{\mathbb{P},(\mathcal{X},\rho)}(\epsilon) := \sup_{A\subseteq\mathcal{X}}\{1 - \mathbb{P}[A^\epsilon] \mid \mathbb{P}[A] \geq \tfrac{1}{2}\}, \tag{3.28}$$

where the supremum is taken over all measurable subsets A.

When the underlying metric space $(\mathcal{X}, \rho)$ is clear from the context, we frequently use the abbreviated notation $\alpha_{\mathbb{P}}$. It follows immediately from the definition (3.28) that $\alpha_{\mathbb{P}}(\epsilon) \in [0, \frac{1}{2}]$ for all $\epsilon \geq 0$. Of primary interest is the behavior of the concentration function as ϵ increases, and, more precisely, how rapidly it approaches zero. Let us consider some examples to illustrate.

Example 3.10 (Concentration function for sphere) Consider the metric measure space defined by the uniform distribution over the n-dimensional Euclidean sphere

$$\mathbb{S}^{n-1} := \{x \in \mathbb{R}^n \mid \|x\|_2 = 1\}, \tag{3.29}$$

equipped with the geodesic distance $\rho(x, y) := \arccos\langle x,\ y\rangle$. Let us upper bound the concentration function $\alpha_{\mathbb{S}^{n-1}}$ defined by the triplet $(\mathbb{P}, \mathbb{S}^{n-1}, \rho)$, where $\mathbb{P}$ is the uniform distribution over the sphere. For each $y \in \mathbb{S}^{n-1}$, we can define the hemisphere

$$H_y := \{x \in \mathbb{S}^{n-1} \mid \rho(x, y) \geq \pi/2\} = \{x \in \mathbb{S}^{n-1} \mid \langle x,\ y\rangle \leq 0\}, \tag{3.30}$$

as illustrated in Figure 3.1(a). With some simple geometry, it can be shown that its ϵ-enlargement corresponds to the set

$$H_y^\epsilon = \{z \in \mathbb{S}^{n-1} \mid \langle z,\ y\rangle < \sin(\epsilon)\}, \tag{3.31}$$

as illustrated in Figure 3.1(b). Note that $\mathbb{P}[H_y] = 1/2$, so that the hemisphere (3.30) is a candidate set for the supremum defining the concentration function (3.28). The classical isoperimetric theorem of Lévy asserts that these hemispheres are *extremal*, meaning that they achieve the supremum, viz.

$$\alpha_{\mathbb{S}^{n-1}}(\epsilon) = 1 - \mathbb{P}[H_y^\epsilon]. \tag{3.32}$$

Let us take this fact as given, and use it to compute an upper bound on the concentration

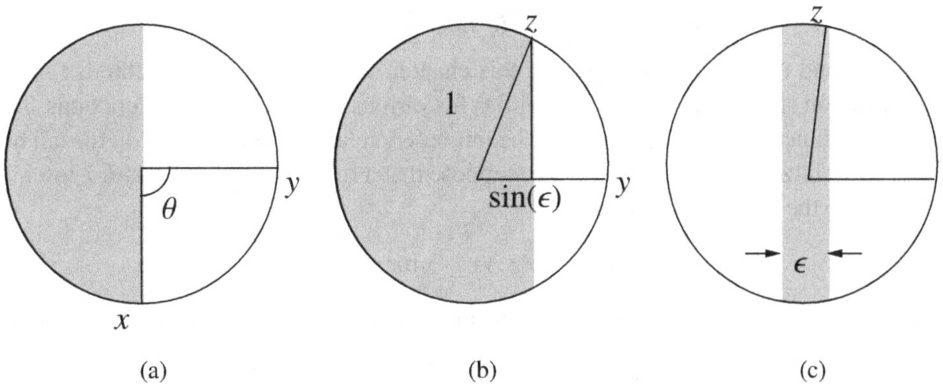

Figure 3.1 (a) Idealized illustration of the sphere $\mathbb{S}^{n-1}$. Any vector $y \in \mathbb{S}^{n-1}$ defines a hemisphere $H_y = \{x \in \mathbb{S}^{n-1} \mid \langle x, y \rangle \leq 0\}$, corresponding to those vectors whose angle $\theta = \arccos \langle x, y \rangle$ with y is at least $\pi/2$ radians. (b) The ϵ-enlargement of the hemisphere H_y. (c) A central slice $T_y(\epsilon)$ of the sphere of width ϵ.

function. In order to do so, we need to lower bound the probability $\mathbb{P}[H_y^\epsilon]$. Since $\sin(\epsilon) \geq \epsilon/2$ for all $\epsilon \in (0, \pi/2]$, the enlargement contains the set

$$\widetilde{H}_y^\epsilon := \{z \in \mathbb{S}^{n-1} \mid \langle z, y \rangle \leq \tfrac{1}{2}\epsilon\},$$

and hence $\mathbb{P}[H_y^\epsilon] \geq \mathbb{P}[\widetilde{H}_y^\epsilon]$. Finally, a geometric calculation, left as an exercise for the reader, yields that, for all $\epsilon \in (0, \sqrt{2})$, we have

$$\mathbb{P}[\widetilde{H}_y^\epsilon] \geq 1 - \left(1 - \left(\frac{\epsilon}{2}\right)^2\right)^{n/2} \geq 1 - e^{-n\epsilon^2/8}, \tag{3.33}$$

where we have used the inequality $(1 - t) \leq e^{-t}$ with $t = \epsilon^2/4$. We thus obtain that the concentration function is upper bounded as $\alpha_{\mathbb{S}^{n-1}}(\epsilon) \leq e^{-n\epsilon^2/8}$. A similar but more careful approach to bounding $\mathbb{P}[H_y]$ can be used to establish the sharper upper bound

$$\alpha_{\mathbb{S}^{n-1}}(\epsilon) \leq \sqrt{\frac{\pi}{2}}\, e^{-\frac{n\epsilon^2}{2}}. \tag{3.34}$$

The bound (3.34) is an extraordinary conclusion, originally due to Lévy, and it is worth pausing to think about it in more depth. Among other consequences, it implies that, if we consider a central slice of the sphere of width ϵ, say a set of the form

$$T_y(\epsilon) := \{z \in \mathbb{S}^{n-1} \mid |\langle z, y \rangle| \leq \epsilon/2\}, \tag{3.35}$$

as illustrated in Figure 3.1(c), then it occupies a huge fraction of the total volume: in particular, we have $\mathbb{P}[T_y(\epsilon)] \geq 1 - \sqrt{2\pi} \exp(-\frac{n\epsilon^2}{2})$. Moreover, this conclusion holds for *any* such slice. To be clear, the two-dimensional instance shown in Figure 3.1(c)—like any low-dimensional example—fails to capture the behavior of high-dimensional spheres. In general, our low-dimensional intuition can be *very* misleading when applied to high-dimensional settings. ♣

3.2.2 Connection to Lipschitz functions

In Chapter 2 and the preceding section of this chapter, we explored some methods for obtaining deviation and concentration inequalities for various types of Lipschitz functions. The concentration function $\alpha_{\mathbb{P},(\mathcal{X},\rho)}$ turns out to be intimately related to such results on the tail behavior of Lipschitz functions. In particular, suppose that a function $f\colon \mathcal{X} \to \mathbb{R}$ is L-Lipschitz with respect to the metric ρ—that is,

$$|f(x) - f(y)| \leq L\rho(x, y) \qquad \text{for all } x, y \in \mathcal{X}. \tag{3.36}$$

Given a random variable $X \sim \mathbb{P}$, let m_f be any median of $f(X)$, meaning a number such that

$$\mathbb{P}[f(X) \geq m_f] \geq 1/2 \quad \text{and} \quad \mathbb{P}[f(X) \leq m_f] \geq 1/2. \tag{3.37}$$

Define the set $A = \{x \in \mathcal{X} \mid f(x) \leq m_f\}$, and consider its $\frac{\epsilon}{L}$-enlargement $A^{\epsilon/L}$. For any $x \in A^{\epsilon/L}$, there exists some $y \in A$ such that $\rho(x, y) < \epsilon/L$. Combined with the Lipschitz property, we conclude that $|f(y) - f(x)| \leq L\rho(x, y) < \epsilon$, and hence that

$$A^{\epsilon/L} \subseteq \{x \in \mathcal{X} \mid f(x) < m_f + \epsilon\}. \tag{3.38}$$

Consequently, we have

$$\mathbb{P}[f(X) \geq m_f + \epsilon] \overset{\text{(i)}}{\leq} 1 - \mathbb{P}[A^{\epsilon/L}] \overset{\text{(ii)}}{\leq} \alpha_{\mathbb{P}}(\epsilon/L),$$

where inequality (i) follows from the inclusion (3.38), and inequality (ii) uses the fact $\mathbb{P}[A] \geq 1/2$, and the definition (3.28). Applying a similar argument to $-f$ yields an analogous left-sided deviation inequality $\mathbb{P}[f(X) \leq m_f - \epsilon] \leq \alpha_{\mathbb{P}}(\epsilon/L)$, and putting together the pieces yields the concentration inequality

$$\mathbb{P}[|f(X) - m_f| \geq \epsilon] \leq 2\alpha_{\mathbb{P}}(\epsilon/L).$$

As shown in Exercise 2.14 from Chapter 2, such sharp concentration around the median is equivalent (up to constant factors) to concentration around the mean. Consequently, we have shown that bounds on the concentration function (3.28) imply concentration inequalities for any Lipschitz function. This argument can also be reversed, yielding the following equivalence between control on the concentration function, and the behavior of Lipschitz functions.

Proposition 3.11 *Given a random variable $X \sim \mathbb{P}$ and concentration function $\alpha_{\mathbb{P}}$, any 1-Lipschitz function on $(\mathcal{X}, \rho)$ satisfies*

$$\mathbb{P}[|f(X) - m_f| \geq \epsilon] \leq 2\alpha_{\mathbb{P}}(\epsilon), \tag{3.39a}$$

where m_f is any median of f. Conversely, suppose that there is a function $\beta\colon \mathbb{R}_+ \to \mathbb{R}_+$ such that, for any 1-Lipschitz function on $(\mathcal{X}, \rho)$,

$$\mathbb{P}[f(X) \geq \mathbb{E}[f(X)] + \epsilon] \leq \beta(\epsilon) \qquad \text{for all } \epsilon \geq 0. \tag{3.39b}$$

Then the concentration function satisfies the bound $\alpha_{\mathbb{P}}(\epsilon) \leq \beta(\epsilon/2)$.

Proof It remains to prove the converse claim. Fix some $\epsilon \geq 0$, and let A be an arbitrary measurable set with $\mathbb{P}[A] \geq 1/2$. Recalling the definition of $\rho(x, A)$ from equation (3.26), let us consider the function $f(x) := \min\{\rho(x, A), \epsilon\}$. It can be seen that f is 1-Lipschitz, and moreover that $1 - \mathbb{P}[A^\epsilon] = \mathbb{P}[f(X) \geq \epsilon]$. On the other hand, our construction guarantees that

$$\mathbb{E}[f(X)] \leq (1 - \mathbb{P}[A])\epsilon \leq \epsilon/2,$$

whence we have

$$\mathbb{P}[f(X) \geq \epsilon] \leq \mathbb{P}[f(X) \geq \mathbb{E}[f(X)] + \epsilon/2] \leq \beta(\epsilon/2),$$

where the final inequality uses the assumed condition (3.39b). □

Proposition 3.11 has a number of concrete interpretations in specific settings.

Example 3.12 (Lévy concentration on $\mathbb{S}^{n-1}$) From our earlier discussion in Example 3.10, the concentration function for the uniform distribution over the sphere $\mathbb{S}^{n-1}$ can be upper bounded as

$$\alpha_{\mathbb{S}^{n-1}}(\epsilon) \leq \sqrt{\frac{\pi}{2}}\, e^{-\frac{n\epsilon^2}{2}}.$$

Consequently, for any 1-Lipschitz function f defined on the sphere $\mathbb{S}^{n-1}$, we have the two-sided bound

$$\mathbb{P}[|f(X) - m_f| \geq \epsilon] \leq \sqrt{2\pi}\, e^{-\frac{n\epsilon^2}{2}}, \tag{3.40}$$

where m_f is any median of f. Moreover, by the result of Exercise 2.14(d), we also have

$$\mathbb{P}[|f(X) - \mathbb{E}[f(X)]| \geq \epsilon] \leq 2\sqrt{2\pi}\, e^{-\frac{n\epsilon^2}{8}}. \tag{3.41}$$

♣

Example 3.13 (Concentration for Boolean hypercube) Consider the Boolean hypercube $\mathcal{X} = \{0, 1\}^n$ equipped with the usual Hamming metric

$$\rho_H(x, y) := \sum_{j=1}^{n} \mathbb{I}[x_j \neq y_j].$$

Given this metric, we can define the Hamming ball

$$\mathbb{B}_H(r; x) = \{y \in \{0, 1\}^n \mid \rho_H(y, x) \leq r\}$$

of radius r centered at some $x \in \{0, 1\}^n$. Of interest here are the Hamming balls centered at the all-zeros vector 0 and all-ones vector 1, respectively. In particular, in this example, we show how a classical combinatorial result due to Harper can be used to bound the concentration function of the metric measure space consisting of the Hamming metric along with the uniform distribution $\mathbb{P}$.

Given two non-empty subsets A and B of the binary hypercube, one consequence of Harper's theorem is that we can always find two positive integers r_A and r_B, and associated subsets A' and B', with the following properties:

- the sets A' and B' are sandwiched as

$$\mathbb{B}_H(r_A - 1; 0) \subseteq A' \subseteq \mathbb{B}_H(r_A; 0) \quad \text{and} \quad \mathbb{B}_H(r_B - 1; 1) \subseteq B' \subseteq \mathbb{B}_H(r_B; 1);$$

- the cardinalities are matched as $\text{card}(A) = \text{card}(A')$ and $\text{card}(B) = \text{card}(B')$;
- we have the lower bound $\rho_H(A', B') \geq \rho_H(A, B)$.

Let us now show that this combinatorial theorem implies that

$$\alpha_{\mathbb{P}}(\epsilon) \leq e^{-\frac{2\epsilon^2}{n}} \qquad \text{for all } n \geq 3. \tag{3.42}$$

Consider any subset such that $\mathbb{P}[A] = \frac{\text{card}(A)}{2^n} \geq \frac{1}{2}$. For any $\epsilon > 0$, define the set $B = \{0, 1\}^n \setminus A^\epsilon$. In order to prove the bound (3.42), it suffices to show that $\mathbb{P}[B] \leq e^{-\frac{2\epsilon^2}{n}}$. Since we always have $\mathbb{P}[B] \leq \frac{1}{2} \leq e^{-\frac{2}{n}}$ for $n \geq 3$, it suffices to restrict our attention to $\epsilon > 1$. By construction, we have

$$\rho_H(A, B) = \min_{a \in A, b \in B} \rho_H(a, b) \geq \epsilon.$$

Let A' and B' denote the subsets guaranteed by Harper's theorem. Since A has cardinality at least 2^{n-1}, the set A', which has the same cardinality as A, must contain all vectors with at most $n/2$ ones. Moreover, by the cardinality matching condition and our choice of the uniform distribution, we have $\mathbb{P}[B] = \mathbb{P}[B']$. On the other hand, the set B' is contained within a Hamming ball centered at the all-ones vector, and we have $\rho_H(A', B') \geq \epsilon > 1$. Consequently, any vector $b \in B'$ must contain at least $\frac{n}{2} + \epsilon$ ones. Thus, if we let $\{X_i\}_{i=1}^n$ be a sequence of i.i.d. Bernoulli variables, we have $\mathbb{P}[B'] \leq \mathbb{P}[\sum_{i=1}^n X_i \geq \frac{n}{2} + \epsilon] \leq e^{-\frac{2\epsilon^2}{n}}$, where the final inequality follows from the Hoeffding bound.

Since A was an arbitrary set with $\mathbb{P}[A] \geq \frac{1}{2}$, we have shown that the concentration function satisfies the bound (3.42). Applying Proposition 3.11, we conclude that any 1-Lipschitz function on the Boolean hypercube satisfies the concentration bound

$$\mathbb{P}[|f(X) - m_f| \geq \epsilon] \leq 2e^{-\frac{2\epsilon^2}{n}}.$$

Thus, modulo the negligible difference between the mean and median (see Exercise 2.14), we have recovered the bounded differences inequality (2.35) for Lipschitz functions on the Boolean hypercube. ♣

3.2.3 From geometry to concentration

The geometric perspective suggests the possibility of a variety of connections between convex geometry and the concentration of measure. Consider, for instance, the Brunn–Minkowski inequality: in one of its formulations, it asserts that, for any two convex bodies[2] C and D in $\mathbb{R}^n$, we have

$$[\text{vol}(\lambda C + (1 - \lambda)D)]^{1/n} \geq \lambda[\text{vol}(C)]^{1/n} + (1 - \lambda)[\text{vol}(D)]^{1/n} \qquad \text{for all } \lambda \in [0, 1]. \tag{3.43}$$

Here we use

$$\lambda C + (1 - \lambda)D := \{\lambda c + (1 - \lambda)d \mid c \in C, d \in D\}$$

to denote the Minkowski sum of the two sets. The Brunn–Minkowski inequality and its variants are intimately connected to concentration of measure. To appreciate the connection,

[2] A convex body in $\mathbb{R}^n$ is a compact and closed set.

observe that the concentration function (3.28) defines a notion of extremal sets—namely, those that minimize the measure $\mathbb{P}[A^\epsilon]$ subject to a constraint on the size of $\mathbb{P}[A]$. Viewing the volume as a type of unnormalized probability measure, the Brunn–Minkowski inequality (3.43) can be used to prove a classical result of this type:

Example 3.14 (Classical isoperimetric inequality in $\mathbb{R}^n$) Consider the Euclidean sphere $\mathbb{B}_2^n := \{x \in \mathbb{R}^n \mid \|x\|_2 \leq 1\}$ in $\mathbb{R}^n$. The classical isoperimetric inequality asserts that, for any set $A \subset \mathbb{R}^n$ such that $\mathrm{vol}(A) = \mathrm{vol}(\mathbb{B}_2^n)$, the volume of its ϵ-enlargement A^ϵ is lower bounded as

$$\mathrm{vol}(A^\epsilon) \geq \mathrm{vol}([\mathbb{B}_2^n]^\epsilon), \tag{3.44}$$

showing that the ball $\mathbb{B}_2^n$ is extremal. In order to verify this bound, we note that

$$[\mathrm{vol}(A^\epsilon)]^{1/n} = [\mathrm{vol}(A + \epsilon\mathbb{B}_2^n)]^{1/n} \geq [\mathrm{vol}(A)]^{1/n} + [\mathrm{vol}(\epsilon\mathbb{B}_2^n)]^{1/n},$$

where the lower bound follows by applying the Brunn–Minkowski inequality (3.43) with appropriate choices of (λ, C, D); see Exercise 3.10 for the details. Since $\mathrm{vol}(A) = \mathrm{vol}(\mathbb{B}_2^n)$ and $[\mathrm{vol}(\epsilon\mathbb{B}_2^n)]^{1/n} = \epsilon\,\mathrm{vol}(\mathbb{B}_2^n)$, we see that

$$\mathrm{vol}(A^\epsilon)^{1/n} \geq (1+\epsilon)\,\mathrm{vol}(\mathbb{B}_2^n)^{1/n} = [\mathrm{vol}((\mathbb{B}_2^n)^\epsilon)]^{1/n},$$

which establishes the claim. ♣

The Brunn–Minkowski inequality has various equivalent formulations. For instance, it can also be stated as

$$\mathrm{vol}(\lambda C + (1-\lambda)D) \geq [\mathrm{vol}(C)]^\lambda [\mathrm{vol}(D)]^{1-\lambda} \qquad \text{for all } \lambda \in [0,1]. \tag{3.45}$$

This form of the Brunn–Minkowski inequality can be used to establish Lévy-type concentration for the uniform measure on the sphere, albeit with slightly weaker constants than the derivation in Example 3.10. In Exercise 3.10, we explore the equivalence between inequality (3.45) and our original statement (3.43) of the Brunn–Minkowski inequality.

The modified form (3.45) of the Brunn–Minkowski inequality also leads naturally to a functional-analytic generalization, due to Prékopa and Leindler. In turn, this generalized inequality can be used to derive concentration inequalities for strongly log-concave measures.

Theorem 3.15 (Prékopa–Leindler inequality) *Let u, v, w be non-negative integrable functions such that, for some $\lambda \in [0,1]$, we have*

$$w(\lambda x + (1-\lambda)y) \geq [u(x)]^\lambda [v(y)]^{1-\lambda} \qquad \textit{for all } x, y \in \mathbb{R}^n. \tag{3.46}$$

Then

$$\int w(x)\,dx \geq \left(\int u(x)\,dx\right)^\lambda \left(\int v(x)\,dx\right)^{1-\lambda}. \tag{3.47}$$

In order to see how this claim implies the classical Brunn–Minkowski inequality (3.45), consider the choices

$$u(x) = \mathbb{I}_C(x), \quad v(x) = \mathbb{I}_D(x) \quad \text{and} \quad w(x) = \mathbb{I}_{\lambda C+(1-\lambda)D}(x),$$

respectively. Here $\mathbb{I}_C$ denotes the binary-valued indicator function for the event $\{x \in C\}$, with the other indicators defined in an analogous way. In order to show that the classical inequality (3.45) follows as a consequence of Theorem 3.15, we need to verify that

$$\mathbb{I}_{\lambda C+(1-\lambda)D}(\lambda x + (1-\lambda)y) \geq [\mathbb{I}_C(x)]^{\lambda}[\mathbb{I}_D(y)]^{1-\lambda} \qquad \text{for all } x, y \in \mathbb{R}^n.$$

For $\lambda = 0$ or $\lambda = 1$, the claim is immediate. For any $\lambda \in (0, 1)$, if either $x \notin C$ or $y \notin D$, the right-hand side is zero, so the statement is trivial. Otherwise, if $x \in C$ and $y \in D$, then both sides are equal to one.

The Prékopa–Leindler inequality can be used to establish some interesting concentration inequalities of Lipschitz functions for a particular subclass of distributions, one which allows for some dependence. In particular, we say that a distribution $\mathbb{P}$ with a density p (with respect to the Lebesgue measure) is a *strongly log-concave distribution* if the function $\log p$ is strongly concave. Equivalently stated, this condition means that the density can be written in the form $p(x) = \exp(-\psi(x))$, where the function $\psi\colon \mathbb{R}^n \to \mathbb{R}$ is strongly convex, meaning that there is some $\gamma > 0$ such that

$$\lambda\psi(x) + (1-\lambda)\psi(y) - \psi(\lambda x + (1-\lambda)y) \geq \frac{\gamma}{2}\lambda(1-\lambda)\,\|x-y\|_2^2 \tag{3.48}$$

for all $\lambda \in [0, 1]$, and $x, y \in \mathbb{R}^n$. For instance, it is easy to verify that the distribution of a standard Gaussian vector in n dimensions is strongly log-concave with parameter $\gamma = 1$. More generally, any Gaussian distribution with covariance matrix $\mathbf{\Sigma} > 0$ is strongly log-concave with parameter $\gamma = \gamma_{\min}(\mathbf{\Sigma}^{-1}) = (\gamma_{\max}(\mathbf{\Sigma}))^{-1}$. In addition, there are a variety of non-Gaussian distributions that are also strongly log-concave. For any such distribution, Lipschitz functions are guaranteed to concentrate, as summarized in the following:

Theorem 3.16 *Let $\mathbb{P}$ be any strongly log-concave distribution with parameter $\gamma > 0$. Then for any function $f\colon \mathbb{R}^n \to \mathbb{R}$ that is L-Lipschitz with respect to Euclidean norm, we have*

$$\mathbb{P}[|f(X) - \mathbb{E}[f(X)]| \geq t] \leq 2e^{-\frac{\gamma t^2}{4L^2}}. \tag{3.49}$$

Remark: Since the standard Gaussian distribution is log-concave with parameter $\gamma = 1$, this theorem implies our earlier result (Theorem 2.26), albeit with a sub-optimal constant in the exponent.

Proof Let h be an arbitrary zero-mean function with Lipschitz constant L with respect to the Euclidean norm. It suffices to show that $\mathbb{E}[e^{h(X)}] \leq e^{\frac{L^2}{\gamma}}$. Indeed, if this inequality holds, then, given an arbitrary function f with Lipschitz constant K and $\lambda \in \mathbb{R}$, we can apply this inequality to the zero-mean function $h := \lambda(f - \mathbb{E}[f(X)])$, which has Lipschitz constant $L = \lambda K$. Doing so yields the bound

$$\mathbb{E}[e^{\lambda(f(X)-\mathbb{E}[f(X)])}] \leq e^{\frac{\lambda^2 K^2}{\gamma}} \qquad \text{for all } \lambda \in \mathbb{R},$$

which shows that $f(X) - \mathbb{E}[f(X)]$ is a sub-Gaussian random variable. As shown in Chapter 2, this type of uniform control on the moment generating function implies the claimed tail bound.

Accordingly, for a given zero-mean function h that is L-Lipschitz and for given $\lambda \in (0, 1)$ and $x, y \in \mathbb{R}^n$, define the function

$$g(y) := \inf_{x \in \mathbb{R}^n} \left\{ h(x) + \frac{\gamma}{4} \|x - y\|_2^2 \right\},$$

known as the inf-convolution of h with the rescaled Euclidean norm. With this definition, the proof is based on applying the Prékopa–Leindler inequality with $\lambda = 1/2$ to the triplet of functions $w(z) \equiv p(z) = \exp(-\psi(z))$, the density of $\mathbb{P}$, and the pair of functions

$$u(x) := \exp(-h(x) - \psi(x)) \quad \text{and} \quad v(y) := \exp(g(y) - \psi(y)).$$

We first need to verify that the inequality (3.46) holds with $\lambda = 1/2$. By the definitions of u and v, the logarithm of the right-hand side of inequality (3.46)—call it R for short—is given by

$$R = \tfrac{1}{2}\{g(y) - h(x)\} - \tfrac{1}{2}\psi(x) - \tfrac{1}{2}\psi(y) = \tfrac{1}{2}\{g(y) - h(x) - 2E(x, y)\} - \psi(x/2 + y/2),$$

where $E(x, y) := \frac{1}{2}\psi(x) + \frac{1}{2}\psi(y) - \psi(x/2 + y/2)$. Since $\mathbb{P}$ is a γ-log-concave distribution, the function ψ is γ strongly convex, and hence $2E(x, y) \geq \frac{\gamma}{4} \|x - y\|_2^2$. Substituting into the earlier representation of R, we find that

$$R \leq \frac{1}{2}\left\{ g(y) - h(x) - \frac{\gamma}{4} \|x - y\|_2^2 \right\} - \psi(x/2 + y/2) \leq -\psi(x/2 + y/2),$$

where the final inequality follows from the definition of the inf-convolution g. We have thus verified condition (3.46) with $\lambda = 1/2$.

Now since $\int w(x)\, dx = \int p(x)\, dx = 1$ by construction, the Prékopa–Leindler inequality implies that

$$0 \geq \frac{1}{2} \log \int e^{-h(x)-\psi(x)}\, dx + \frac{1}{2} \log \int e^{g(y)-\psi(y)}\, dy.$$

Rewriting the integrals as expectations and rearranging yields

$$\mathbb{E}[e^{g(Y)}] \leq \frac{1}{\mathbb{E}[e^{-h(X)}]} \overset{\text{(i)}}{\leq} \frac{1}{e^{\mathbb{E}[-h(X)]}} \overset{\text{(ii)}}{=} 1, \tag{3.50}$$

where step (i) follows from Jensen's inequality, and convexity of the function $t \mapsto \exp(-t)$, and step (ii) uses the fact that $\mathbb{E}[-h(X)] = 0$ by assumption. Finally, since h is an L-Lipschitz function, we have $|h(x) - h(y)| \leq L\,\|x - y\|_2$, and hence

$$g(y) = \inf_{x \in \mathbb{R}^n} \Big\{ h(x) + \frac{\gamma}{4}\,\|x - y\|_2^2 \Big\} \geq h(y) + \inf_{x \in \mathbb{R}^n} \Big\{ -L\,\|x - y\|_2 + \frac{\gamma}{4}\,\|x - y\|_2^2 \Big\}$$
$$= h(y) - \frac{L^2}{\gamma}.$$

Combined with the bound (3.50), we conclude that $\mathbb{E}[e^{h(Y)}] \leq \exp(\frac{L^2}{\gamma})$, as claimed. □

3.3 Wasserstein distances and information inequalities

We now turn to the topic of Wasserstein distances and information inequalities, also known as *transportation cost inequalities*. On one hand, the transportation cost approach can be used to obtain some sharp results for Lipschitz functions of independent random variables. Perhaps more importantly, it is especially well suited to certain types of dependent random variables, such as those arising in Markov chains and other types of mixing processes.

3.3.1 Wasserstein distances

We begin by defining the notion of a Wasserstein distance. Given a metric space $(\mathcal{X}, \rho)$, a function $f\colon \mathcal{X} \to \mathbb{R}$ is L-Lipschitz with respect to the metric ρ if

$$|f(x) - f(x')| \leq L\rho(x, x') \qquad \text{for all } x, x' \in \mathcal{X}, \tag{3.51}$$

and we use $\|f\|_{\mathrm{Lip}}$ to denote the smallest L for which this inequality holds. Given two probability distributions $\mathbb{Q}$ and $\mathbb{P}$ on $\mathcal{X}$, we can then measure the distance between them via

$$W_\rho(\mathbb{Q}, \mathbb{P}) = \sup_{\|f\|_{\mathrm{Lip}} \leq 1} \Big[\int f\, d\mathbb{Q} - \int f\, d\mathbb{P} \Big], \tag{3.52}$$

where the supremum ranges over all 1-Lipschitz functions. This distance measure is referred to as the *Wasserstein metric induced by* ρ. It can be verified that, for each choice of the metric ρ, this definition defines a distance on the space of probability measures.

Example 3.17 (Hamming metric and total variation distance) Consider the Hamming metric $\rho(x, x') = \mathbb{I}[x \neq x']$. We claim that, in this case, the associated Wasserstein distance is equivalent to the *total variation distance*

$$\|\mathbb{Q} - \mathbb{P}\|_{\mathrm{TV}} := \sup_{A \subseteq \mathcal{X}} |\mathbb{Q}(A) - \mathbb{P}(A)|, \tag{3.53}$$

where the supremum ranges over all measurable subsets A. To see this equivalence, note that any function that is 1-Lipschitz with respect to the Hamming distance satisfies the bound $|f(x) - f(x')| \leq 1$. Since the supremum (3.52) is invariant to constant offsets of the function, we may restrict the supremum to functions such that $f(x) \in [0, 1]$ for all $x \in \mathcal{X}$, thereby obtaining

$$W_{\mathrm{Ham}}(\mathbb{Q}, \mathbb{P}) = \sup_{f\colon \mathcal{X} \to [0,1]} \int f\,(d\mathbb{Q} - d\mathbb{P}) \overset{\text{(i)}}{=} \|\mathbb{Q} - \mathbb{P}\|_{\mathrm{TV}},$$

where equality (i) follows from Exercise 3.13.

In terms of the underlying densities[3] p and q taken with respect to a base measure ν, we can write

$$W_{\mathrm{Ham}}(\mathbb{Q}, \mathbb{P}) = \|\mathbb{Q} - \mathbb{P}\|_{\mathrm{TV}} = \frac{1}{2} \int |p(x) - q(x)| \nu(dx),$$

corresponding to (one half) the $L^1(\nu)$-norm between the densities. Again, see Exercise 3.13 for further details on this equivalence. ♣

By a classical and deep result in duality theory (see the bibliographic section for details), any Wasserstein distance has an equivalent definition as a type of coupling-based distance. A distribution $\mathbb{M}$ on the product space $\mathcal{X} \otimes \mathcal{X}$ is a *coupling* of the pair $(\mathbb{Q}, \mathbb{P})$ if its marginal distributions in the first and second coordinates coincide with $\mathbb{Q}$ and $\mathbb{P}$, respectively. In order to see the relation to the Wasserstein distance, let $f\colon \mathcal{X} \to \mathbb{R}$ be any 1-Lipschitz function, and let $\mathbb{M}$ be any coupling. We then have

$$\int \rho(x, x')\, d\mathbb{M}(x, x') \stackrel{\text{(i)}}{\geq} \int (f(x) - f(x'))\, d\mathbb{M}(x, x') \stackrel{\text{(ii)}}{=} \int f\,(d\mathbb{P} - d\mathbb{Q}), \tag{3.54}$$

where the inequality (i) follows from the 1-Lipschitz nature of f, and the equality (ii) follows since $\mathbb{M}$ is a coupling. The *Kantorovich–Rubinstein duality* guarantees the following important fact: if we minimize over all possible couplings, then this argument can be reversed, and in fact we have the equivalence

$$\underbrace{\sup_{\|f\|_{\mathrm{Lip}} \leq 1} \int f\,(d\mathbb{Q} - d\mathbb{P})}_{W_\rho(\mathbb{P}, \mathbb{Q})} = \inf_{\mathbb{M}} \int_{\mathcal{X} \times \mathcal{X}} \rho(x, x')\, d\mathbb{M}(x, x') = \inf_{\mathbb{M}} \mathbb{E}_{\mathbb{M}}[\rho(X, X')], \tag{3.55}$$

where the infimum ranges over all couplings $\mathbb{M}$ of the pair $(\mathbb{P}, \mathbb{Q})$. This coupling-based representation of the Wasserstein distance plays an important role in many of the proofs to follow.

The term "transportation cost" arises from the following interpretation of coupling-based representation (3.55). For concreteness, let us consider the case where $\mathbb{P}$ and $\mathbb{Q}$ have densities p and q with respect to Lebesgue measure on $\mathcal{X}$, and the coupling $\mathbb{M}$ has density m with respect to Lebesgue measure on the product space. The density p can be viewed as describing some initial distribution of mass over the space $\mathcal{X}$, whereas the density q can be interpreted as some desired distribution of the mass. Our goal is to shift mass so as to transform the initial distribution p to the desired distribution q. The quantity $\rho(x, x')\, dx\, dx'$ can be interpreted as the cost of transporting a small increment of mass dx to the new increment dx'. The joint distribution $m(x, x')$ is known as a *transportation plan*, meaning a scheme for shifting mass so that p is transformed to q. Combining these ingredients, we conclude that the transportation cost associated with the plan m is given by

$$\int_{\mathcal{X} \times \mathcal{X}} \rho(x, x') m(x, x')\, dx\, dx',$$

and minimizing over all admissible plans—that is, those that marginalize down to p and q,

[3] This assumption entails no loss of generality, since $\mathbb{P}$ and $\mathbb{Q}$ both have densities with respect to $\nu = \frac{1}{2}(\mathbb{P} + \mathbb{Q})$.

respectively—yields the Wasserstein distance.

3.3.2 Transportation cost and concentration inequalities

Let us now turn to the notion of a transportation cost inequality, and its implications for the concentration of measure. Transportation cost inequalities are based on upper bounding the Wasserstein distance $W_\rho(\mathbb{Q}, \mathbb{P})$ in terms of the *Kullback–Leibler (KL) divergence*. Given two distributions $\mathbb{Q}$ and $\mathbb{P}$, the KL divergence between them is given by

$$D(\mathbb{Q}\,\|\,\mathbb{P}) := \begin{cases} \mathbb{E}_{\mathbb{Q}}\big[\log \frac{d\mathbb{Q}}{d\mathbb{P}}\big] & \text{when } \mathbb{Q} \text{ is absolutely continuous with respect to } \mathbb{P}, \\ +\infty & \text{otherwise.} \end{cases} \tag{3.56}$$

If the measures have densities[4] with respect to some underlying measure ν—say q and p—then the Kullback–Leibler divergence can be written in the form

$$D(\mathbb{Q}\,\|\,\mathbb{P}) = \int_{\mathcal{X}} q(x) \log \frac{q(x)}{p(x)} \nu(dx). \tag{3.57}$$

Although the KL divergence provides a measure of distance between distributions, it is not actually a metric (since, for instance, it is not symmetric in general).

We say that a transportation cost inequality is satisfied when the Wasserstein distance is upper bounded by a multiple of the square-root KL divergence.

Definition 3.18 For a given metric ρ, the probability measure $\mathbb{P}$ is said to satisfy a *ρ-transportation cost inequality* with parameter $\gamma > 0$ if

$$W_\rho(\mathbb{Q}, \mathbb{P}) \leq \sqrt{2\gamma D(\mathbb{Q}\,\|\,\mathbb{P})} \tag{3.58}$$

for all probability measures $\mathbb{Q}$.

Such results are also known as *information inequalities*, due to the role of the Kullback–Leibler divergence in information theory. A classical example of an information inequality is the *Pinsker–Csiszár–Kullback inequality*, which relates the total variation distance with the KL divergence. More precisely, for all probability distributions $\mathbb{P}$ and $\mathbb{Q}$, we have

$$\|\mathbb{P} - \mathbb{Q}\|_{\mathrm{TV}} \leq \sqrt{\tfrac{1}{2} D(\mathbb{Q}\,\|\,\mathbb{P})}. \tag{3.59}$$

From our development in Example 3.17, this inequality corresponds to a transportation cost inequality, in which $\gamma = 1/4$ and the Wasserstein distance is based on the Hamming norm $\rho(x, x') = \mathbb{I}[x \neq x']$. As will be seen shortly, this inequality can be used to recover the bounded differences inequality, corresponding to a concentration statement for functions that are Lipschitz with respect to the Hamming norm. See Exercise 15.6 in Chapter 15 for

[4] In the special case of a discrete space $\mathcal{X}$, and probability mass functions q and p, we have $D(\mathbb{Q}\,\|\,\mathbb{P}) = \sum_{x \in \mathcal{X}} q(x) \log \frac{q(x)}{p(x)}$.

the proof of this bound.

By the definition (3.52) of the Wasserstein distance, the transportation cost inequality (3.58) can be used to upper bound the deviation $\int f\, d\mathbb{Q} - \int f\, d\mathbb{P}$ in terms of the Kullback–Leibler divergence $D(\mathbb{Q}\,\|\,\mathbb{P})$. As shown by the following result, a particular choice of distribution $\mathbb{Q}$ can be used to derive a concentration bound for f under $\mathbb{P}$. In this way, a transportation cost inequality leads to concentration bounds for Lipschitz functions:

Theorem 3.19 (From transportation cost to concentration) *Consider a metric measure space $(\mathbb{P}, \mathcal{X}, \rho)$, and suppose that $\mathbb{P}$ satisfies the ρ-transportation cost inequality* (3.58). *Then its concentration function satisfies the bound*

$$\alpha_{\mathbb{P},(\mathcal{X},\rho)}(t) \le 2\exp\left(-\frac{t^2}{2\gamma}\right). \tag{3.60}$$

Moreover, for any $X \sim \mathbb{P}$ and any L-Lipschitz function $f\colon \mathcal{X} \to \mathbb{R}$, we have the concentration inequality

$$\mathbb{P}[|f(X) - \mathbb{E}[f(X)]| \ge t] \le 2\exp\left(-\frac{t^2}{2\gamma L^2}\right). \tag{3.61}$$

Remarks: By Proposition 3.11, the bound (3.60) implies that

$$\mathbb{P}[|f(X) - m_f| \ge t] \le 2\exp\left(-\frac{t^2}{2\gamma L^2}\right), \tag{3.62}$$

where m_f is any median of f. In turn, this bound can be used to establish concentration around the mean, albeit with worse constants than the bound (3.61). (See Exercise 2.14 for details on this equivalence.) In our proof, we make use of separate arguments for the median and mean, so as to obtain sharp constants.

Proof We begin by proving the bound (3.60). For any set A with $\mathbb{P}[A] \ge 1/2$ and a given $\epsilon > 0$, consider the set

$$B := (A^\epsilon)^c = \{y \in \mathcal{X} \mid \rho(x, y) \ge \epsilon \quad \forall\, x \in A\}.$$

If $\mathbb{P}(A^\epsilon) = 1$, then the proof is complete, so that we may assume that $\mathbb{P}(B) > 0$.

By construction, we have $\rho(A, B) := \inf_{x\in A} \inf_{y\in B} \rho(x, y) \ge \epsilon$. On the other hand, let $\mathbb{P}_A$ and $\mathbb{P}_B$ denote the distributions of $\mathbb{P}$ conditioned on A and B, and let $\mathbb{M}$ denote any coupling of this pair. Since the marginals of $\mathbb{M}$ are supported on A and B, respectively, we have $\rho(A, B) \le \int \rho(x, x')\, d\mathbb{M}(x, x')$. Taking the infimum over all couplings, we conclude that $\epsilon \le \rho(A, B) \le W_\rho(\mathbb{P}_A, \mathbb{P}_B)$.

Now applying the triangle inequality, we have

$$\begin{aligned} \epsilon \le W_\rho(\mathbb{P}_A, \mathbb{P}_B) \le W_\rho(\mathbb{P}, \mathbb{P}_A) + W_\rho(\mathbb{P}, \mathbb{P}_B) &\overset{\text{(ii)}}{\le} \sqrt{\gamma D(\mathbb{P}_A\,\|\,\mathbb{P})} + \sqrt{\gamma D(\mathbb{P}_B\,\|\,\mathbb{P})} \\ &\overset{\text{(iii)}}{\le} \sqrt{2\gamma}\,\{D(\mathbb{P}_A\,\|\,\mathbb{P}) + D(\mathbb{P}_B\,\|\,\mathbb{P})\}^{1/2}, \end{aligned}$$

where step (ii) follows from the transportation cost inequality, and step (iii) follows from the inequality $(a + b)^2 \leq 2a^2 + 2b^2$.

It remains to compute the Kullback–Leibler divergences. For any measurable set C, we have $\mathbb{P}_A(C) = \mathbb{P}(C \cap A)/\mathbb{P}(A)$, so that $D(\mathbb{P}_A \,||\, \mathbb{P}) = \log \frac{1}{\mathbb{P}(A)}$. Similarly, we have $D(\mathbb{P}_B \,||\, \mathbb{P}) = \log \frac{1}{\mathbb{P}(B)}$. Combining the pieces, we conclude that

$$\epsilon^2 \leq 2\gamma\{\log(1/\mathbb{P}(A)) + \log(1/\mathbb{P}(B))\} = 2\gamma \log\left(\frac{1}{\mathbb{P}(A)\mathbb{P}(B)}\right),$$

or equivalently $\mathbb{P}(A)\mathbb{P}(B) \leq \exp\left(-\frac{\epsilon^2}{2\gamma}\right)$. Since $\mathbb{P}(A) \geq 1/2$ and $B = (A^\epsilon)^c$, we conclude that $\mathbb{P}(A^\epsilon) \geq 1 - 2\exp\left(-\frac{\epsilon^2}{2\gamma}\right)$. Since A was an arbitrary set with $\mathbb{P}(A) \geq 1/2$, the bound (3.60) follows.

We now turn to the proof of the concentration statement (3.61) for the mean. If one is not concerned about constants, such a bound follows immediately by combining claim (3.60) with the result of Exercise 2.14. Here we present an alternative proof with the dual goals of obtaining the sharp result and illustrating a different proof technique. Throughout this proof, we use $\mathbb{E}_{\mathbb{Q}}[f]$ and $\mathbb{E}_{\mathbb{P}}[f]$ to denote the mean of the random variable $f(X)$ when $X \sim \mathbb{Q}$ and $X \sim \mathbb{P}$, respectively. We begin by observing that

$$\int f\,(d\mathbb{Q} - d\mathbb{P}) \overset{\text{(i)}}{\leq} L W_\rho(\mathbb{Q}, \mathbb{P}) \overset{\text{(ii)}}{\leq} \sqrt{2L^2\gamma D(\mathbb{Q} \,||\, \mathbb{P})},$$

where step (i) follows from the L-Lipschitz condition on f and the definition (3.52); and step (ii) follows from the information inequality (3.58). For any positive numbers (u, v, λ), we have $\sqrt{2uv} \leq \frac{u}{2}\lambda + \frac{v}{\lambda}$. Applying this inequality with $u = L^2\gamma$ and $v = D(\mathbb{Q} \,||\, \mathbb{P})$ yields

$$\int f\,(d\mathbb{Q} - d\mathbb{P}) \leq \frac{\lambda\gamma L^2}{2} + \frac{1}{\lambda} D(\mathbb{Q} \,||\, \mathbb{P}), \tag{3.63}$$

valid for all $\lambda > 0$.

Now define a distribution $\mathbb{Q}$ with Radon–Nikodym derivative $\frac{d\mathbb{Q}}{d\mathbb{P}}(x) = e^{g(x)}/\mathbb{E}_{\mathbb{P}}[e^{g(X)}]$, where $g(x) := \lambda(f(x) - \mathbb{E}_{\mathbb{P}}(f)) - \frac{L^2\gamma\lambda^2}{2}$. (Note that our proof of the bound (3.61) ensures that $\mathbb{E}_{\mathbb{P}}[e^{g(X)}]$ exists.) With this choice, we have

$$D(\mathbb{Q} \,||\, \mathbb{P}) = \mathbb{E}_{\mathbb{Q}} \log\left(\frac{e^{g(X)}}{\mathbb{E}_{\mathbb{P}}[e^{g(X)}]}\right) = \lambda\{\mathbb{E}_{\mathbb{Q}}(f(X)) - \mathbb{E}_{\mathbb{P}}(f(X))\} - \frac{\gamma L^2\lambda^2}{2} - \log \mathbb{E}_{\mathbb{P}}[e^{g(X)}].$$

Combining with inequality (3.63) and performing some algebra (during which the reader should recall that $\lambda > 0$), we find that $\log \mathbb{E}_{\mathbb{P}}[e^{g(X)}] \leq 0$, or equivalently

$$\mathbb{E}_{\mathbb{P}}[e^{\lambda(f(X) - \mathbb{E}_{\mathbb{P}}[f(X')])}] \leq e^{\frac{\lambda^2\gamma L^2}{2}}.$$

The upper tail bound thus follows by the Chernoff bound. The same argument can be applied to $-f$, which yields the lower tail bound. □

3.3.3 Tensorization for transportation cost

Based on Theorem 3.19, we see that transportation cost inequalities can be translated into concentration inequalities. Like entropy, transportation cost inequalities behave nicely for product measures, and can be combined in an additive manner. Doing so yields concentration inequalities for Lipschitz functions in the higher-dimensional space. We summarize in the following:

Proposition 3.20 *Suppose that, for each $k = 1, 2, \ldots, n$, the univariate distribution $\mathbb{P}_k$ satisfies a ρ_k-transportation cost inequality with parameter γ_k. Then the product distribution $\mathbb{P} = \bigotimes_{k=1}^n \mathbb{P}_k$ satisfies the transportation cost inequality*

$$W_\rho(\mathbb{Q}, \mathbb{P}) \le \sqrt{2\Big(\sum_{k=1}^n \gamma_k\Big) D(\mathbb{Q} \,\|\, \mathbb{P})} \quad \text{for all distributions } \mathbb{Q}, \tag{3.64}$$

where the Wasserstein metric is defined using the distance $\rho(x, y) := \sum_{k=1}^n \rho_k(x_k, y_k)$.

Before turning to the proof of Proposition 3.20, it is instructive to see how, in conjunction with Theorem 3.19, it can be used to recover the bounded differences inequality.

Example 3.21 (Bounded differences inequality) Suppose that f satisfies the bounded differences inequality with parameter L_k in coordinate k. Then using the triangle inequality and the bounded differences property, it can be verified that f is a 1-Lipschitz function with respect to the rescaled Hamming metric

$$\rho(x, y) := \sum_{k=1}^n \rho_k(x_k, y_k), \qquad \text{where } \rho_k(x_k, y_k) := L_k \, \mathbb{I}[x_k \neq y_k].$$

By the Pinsker–Csiszár–Kullback inequality (3.59), each univariate distribution $\mathbb{P}_k$ satisfies a ρ_k-transportation cost inequality with parameter $\gamma_k = \frac{L_k^2}{4}$, so that Proposition 3.20 implies that $\mathbb{P} = \bigotimes_{k=1}^n \mathbb{P}_k$ satisfies a ρ-transportation cost inequality with parameter $\gamma := \frac{1}{4} \sum_{k=1}^n L_k^2$. Since f is 1-Lipschitz with respect to the metric ρ, Theorem 3.19 implies that

$$\mathbb{P}[|f(X) - \mathbb{E}[f(X)]| \ge t] \le 2 \exp\left(-\frac{2t^2}{\sum_{k=1}^n L_k^2}\right). \tag{3.65}$$

In this way, we recover the bounded differences inequality from Chapter 2 from a transportation cost argument. ♣

Our proof of Proposition 3.20 is based on the coupling-based characterization (3.55) of Wasserstein distances.

Proof Letting $\mathbb{Q}$ be an arbitrary distribution over the product space $\mathcal{X}^n$, we construct a coupling $\mathbb{M}$ of the pair $(\mathbb{P}, \mathbb{Q})$. For each $j = 2, \ldots, n$, let $\mathbb{M}_1^j$ denote the joint distribution over the pair $(X_1^j, Y_1^j) = (X_1, \ldots, X_j, Y_1, \ldots, Y_j)$, and let $\mathbb{M}_{j|j-1}$ denote the conditional distribution of (X_j, Y_j) given (X_1^{j-1}, Y_1^{j-1}). By the dual representation (3.55), we have

$$W_\rho(\mathbb{Q}, \mathbb{P}) \le \mathbb{E}_{\mathbb{M}_1}[\rho_1(X_1, Y_1)] + \sum_{j=2}^n \mathbb{E}_{\mathbb{M}_1^{j-1}}[\mathbb{E}_{\mathbb{M}_{j|j-1}}[\rho_j(X_j, Y_j)]],$$

where $\mathbb{M}_j$ denotes the marginal distribution over the pair (X_j, Y_j). We now define our coupling $\mathbb{M}$ in an inductive manner as follows. First, choose $\mathbb{M}_1$ to be an optimal coupling of the pair $(\mathbb{P}_1, \mathbb{Q}_1)$, thereby ensuring that

$$\mathbb{E}_{\mathbb{M}_1}[\rho_1(X_1, Y_1)] \stackrel{\text{(i)}}{=} W_\rho(\mathbb{Q}_1, \mathbb{P}_1) \stackrel{\text{(ii)}}{\leq} \sqrt{2\gamma_1 D(\mathbb{Q}_1 \,\|\, \mathbb{P}_1)},$$

where equality (i) follows by the optimality of the coupling, and inequality (ii) follows from the assumed transportation cost inequality for $\mathbb{P}_1$. Now assume that the joint distribution over (X_1^{j-1}, Y_1^{j-1}) has been defined. We choose conditional distribution $\mathbb{M}_{j|j-1}(\cdot \mid x_1^{j-1}, y_1^{j-1})$ to be an optimal coupling for the pair $(\mathbb{P}_j, \mathbb{Q}_{j|j-1}(\cdot \mid y_1^{j-1}))$, thereby ensuring that

$$\mathbb{E}_{\mathbb{M}_{j|j-1}}[\rho_j(X_j, Y_j)] \leq \sqrt{2\gamma_j D(\mathbb{Q}_{j|j-1}(\cdot \mid y_1^{j-1}) \,\|\, \mathbb{P}_j)},$$

valid for each y_1^{j-1}. Taking averages over Y_1^{j-1} with respect to the marginal distribution $\mathbb{M}_1^{j-1}$—or, equivalently, the marginal $\mathbb{Q}_1^{j-1}$—the concavity of the square-root function and Jensen's inequality implies that

$$\mathbb{E}_{\mathbb{M}_1^{j-1}}[\mathbb{E}_{\mathbb{M}_{j|j-1}}[\rho_j(X_j, Y_j)]] \leq \sqrt{2\gamma_j \mathbb{E}_{\mathbb{Q}_1^{j-1}} D(\mathbb{Q}_{j|j-1}(\cdot \mid Y_1^{j-1}) \,\|\, \mathbb{P}_j)}.$$

Combining the ingredients, we obtain

$$\begin{aligned} W_\rho(\mathbb{Q}, \mathbb{P}) &\leq \sqrt{2\gamma_1 D(\mathbb{Q}_1 \,\|\, \mathbb{P}_1)} + \sum_{j=2}^{n} \sqrt{2\gamma_j \mathbb{E}_{\mathbb{Q}_1^{j-1}}[D(\mathbb{Q}_{j|j-1}(\cdot \mid Y_1^{j-1}) \,\|\, \mathbb{P}_j)]} \\ &\stackrel{\text{(i)}}{\leq} \sqrt{2\Big(\sum_{j=1}^{n} \gamma_j\Big)} \sqrt{D(\mathbb{Q}_1 \,\|\, \mathbb{P}_1) + \sum_{j=2}^{n} \mathbb{E}_{\mathbb{Q}_1^{j-1}}[D(\mathbb{Q}_{j|j-1}(\cdot \mid Y_1^{j-1}) \,\|\, \mathbb{P}_j)]} \\ &\stackrel{\text{(ii)}}{=} \sqrt{2\Big(\sum_{j=1}^{n} \gamma_j\Big) D(\mathbb{Q} \,\|\, \mathbb{P})}, \end{aligned}$$

where step (i) by follows the Cauchy–Schwarz inequality, and equality (ii) uses the chain rule for Kullback–Leibler divergence from Exercise 3.2. □

In Exercise 3.14, we sketch out an alternative proof of Proposition 3.20, one which makes direct use of the Lipschitz characterization of the Wasserstein distance.

3.3.4 Transportation cost inequalities for Markov chains

As mentioned previously, the transportation cost approach has some desirable features in application to Lipschitz functions involving certain types of dependent random variables. Here we illustrate this type of argument for the case of a Markov chain. (See the bibliographic section for references to more general results on concentration for dependent random variables.)

More concretely, let $(X_1, \ldots, X_n)$ be a random vector generated by a Markov chain, where each X_i takes values in a countable space $\mathcal{X}$. Its distribution $\mathbb{P}$ over $\mathcal{X}^n$ is defined by an initial distribution $X_1 \sim \mathbb{P}_1$, and the transition kernels

$$\mathbb{K}_{i+1}(x_{i+1} \mid x_i) = \mathbb{P}_{i+1}(X_{i+1} = x_{i+1} \mid X_i = x_i). \tag{3.66}$$

Here we focus on discrete state Markov chains that are β-contractive, meaning that there exists some $\beta \in [0, 1)$ such that

$$\max_{i=1,\ldots,n-1} \sup_{x_i, x_i'} \|\mathbb{K}_{i+1}(\cdot \mid x_i) - \mathbb{K}_{i+1}(\cdot \mid x_i')\|_{\mathrm{TV}} \leq \beta, \tag{3.67}$$

where the total variation norm (3.53) was previously defined.

Theorem 3.22 *Let $\mathbb{P}$ be the distribution of a β-contractive Markov chain* (3.67) *over the discrete space $\mathcal{X}^n$. Then for any other distribution $\mathbb{Q}$ over $\mathcal{X}^n$, we have*

$$W_\rho(\mathbb{Q}, \mathbb{P}) \leq \frac{1}{1-\beta} \sqrt{\frac{n}{2} D(\mathbb{Q} \,\|\, \mathbb{P})}, \tag{3.68}$$

where the Wasserstein distance is defined with respect to the Hamming norm $\rho(x, y) = \sum_{i=1}^{n} \mathbb{I}[x_i \neq y_i]$.

Remark: See the bibliography section for references to proofs of this result. Using Theorem 3.19, an immediate corollary of the bound (3.68) is that for any function $f \colon \mathcal{X}^n \to \mathbb{R}$ that is L-Lipschitz with respect to the Hamming norm, we have

$$\mathbb{P}[|f(X) - \mathbb{E}[f(X)]| \geq t] \leq 2 \exp\left(-\frac{2(1-\beta)^2 t^2}{nL^2}\right). \tag{3.69}$$

Note that this result is a strict generalization of the bounded difference inequality for independent random variables, to which it reduces when $\beta = 0$.

Example 3.23 (Parameter estimation for a binary Markov chain) Consider a Markov chain over binary variables $X_i \in \{0, 1\}^2$ specified by an initial distribution $\mathbb{P}_1$ that is uniform, and the transition kernel

$$\mathbb{K}_{i+1}(x_{i+1} \mid x_i) = \begin{cases} \frac{1}{2}(1+\delta) & \text{if } x_{i+1} = x_i, \\ \frac{1}{2}(1-\delta) & \text{if } x_{i+1} \neq x_i, \end{cases}$$

where $\delta \in [0, 1]$ is a "stickiness" parameter. Suppose that our goal is to estimate the parameter δ based on an n-length vector $(X_1, \ldots, X_n)$ drawn according to this chain. An unbiased estimate of $\frac{1}{2}(1+\delta)$ is given by the function

$$f(X_1, \ldots, X_n) := \frac{1}{n-1} \sum_{i=1}^{n-1} \mathbb{I}[X_i = X_{i+1}],$$

corresponding to the fraction of times that successive samples take the same value. We claim that f satisfies the concentration inequality

$$\mathbb{P}[|f(X) - \tfrac{1}{2}(1+\delta)| \geq t] \leq 2e^{-\frac{(n-1)^2(1-\delta)^2 t^2}{2n}} \leq 2e^{-\frac{(n-1)(1-\delta)^2 t^2}{4}}. \tag{3.70}$$

Following some calculation, we find that the chain is β-contractive with $\beta = \delta$. Moreover, the function f is $\frac{2}{n-1}$-Lipschitz with respect to the Hamming norm. Consequently, the bound (3.70) follows as a consequence of our earlier general result (3.69). ♣

3.3.5 Asymmetric coupling cost

Thus far, we have considered various types of Wasserstein distances, which can be used to obtain concentration for Lipschitz functions. However, this approach—as with most methods that involve Lipschitz conditions with respect to ℓ_1-type norms—typically does not yield dimension-independent bounds. By contrast, as we have seen previously, Lipschitz conditions based on the ℓ_2-norm often do lead to dimension-independent results.

With this motivation in mind, this section is devoted to consideration of another type of coupling-based distance between probability distributions, but one that is asymmetric in its two arguments, and of a quadratic nature. In particular, we define

$$C(\mathbb{Q}, \mathbb{P}) := \inf_{\mathbb{M}} \sqrt{\int \sum_{i=1}^{n} (\mathbb{M}[Y_i \neq x_i \mid X_i = x_i])^2 \, d\mathbb{P}(x)}, \tag{3.71}$$

where once again the infimum ranges over all couplings $\mathbb{M}$ of the pair $(\mathbb{P}, \mathbb{Q})$. This distance is relatively closely related to the total variation distance; in particular, it can be shown that an equivalent representation for this asymmetric distance is

$$C(\mathbb{Q}, \mathbb{P}) = \sqrt{\int \left|1 - \frac{d\mathbb{Q}}{d\mathbb{P}}(x)\right|_+^2 d\mathbb{P}(x)}, \tag{3.72}$$

where $t_+ := \max\{0, t\}$. We leave this equivalence as an exercise for the reader. This representation reveals the close link to the total variation distance, for which

$$\|\mathbb{P} - \mathbb{Q}\|_{\mathrm{TV}} = \int \left|1 - \frac{d\mathbb{Q}}{d\mathbb{P}}\right| d\mathbb{P}(x) = 2 \int \left|1 - \frac{d\mathbb{Q}}{d\mathbb{P}}\right|_+ d\mathbb{P}(x).$$

An especially interesting aspect of the asymmetric coupling distance is that it satisfies a Pinsker-type inequality for product distributions. In particular, given any product distribution $\mathbb{P}$ in n variables, we have

$$\max\{C(\mathbb{Q}, \mathbb{P}), C(\mathbb{P}, \mathbb{Q})\} \leq \sqrt{2D(\mathbb{Q} \,\|\, \mathbb{P})} \tag{3.73}$$

for all distributions $\mathbb{Q}$ in n dimensions. This deep result is due to Samson; see the bibliographic section for further discussion. While simple to state, it is non-trivial to prove, and has some very powerful consequences for the concentration of convex and Lipschitz functions, as summarized in the following:

Theorem 3.24 *Consider a vector of independent random variables $(X_1, \ldots, X_n)$, each taking values in $[0, 1]$, and let $f\colon \mathbb{R}^n \to \mathbb{R}$ be convex, and L-Lipschitz with respect to the Euclidean norm. Then for all $t \geq 0$, we have*

$$\mathbb{P}[|f(X) - \mathbb{E}[f(X)]| \geq t] \leq 2e^{-\frac{t^2}{2L^2}}. \tag{3.74}$$

Remarks: Note that this is the analog of Theorem 2.26—namely, a dimension-independent form of concentration for Lipschitz functions of independent Gaussian variables, but formulated for *Lipschitz and convex* functions of bounded random variables.

Of course, the same bound also applies to a concave and Lipschitz function. Earlier, we saw that upper tail bounds can obtained under a slightly milder condition, namely that of separate convexity (see Theorem 3.4). However, two-sided tail bounds (or concentration inequalities) require these stronger convexity or concavity conditions, as imposed here.

Example 3.25 (Rademacher revisited) As previously introduced in Example 3.5, the Rademacher complexity of a set $\mathcal{A} \subseteq \mathbb{R}^n$ is defined in terms of the random variable

$$Z \equiv Z(\varepsilon_1, \ldots, \varepsilon_n) := \sup_{a \in \mathcal{A}} \sum_{k=1}^{n} a_k \varepsilon_k,$$

where $\{\varepsilon_k\}_{k=1}^n$ is an i.i.d. sequence of Rademacher variables. As shown in Example 3.5, the function $(\varepsilon_1, \ldots, \varepsilon_n) \mapsto Z(\varepsilon_1, \ldots, \varepsilon_n)$ is jointly convex, and Lipschitz with respect to the Euclidean norm with parameter $\mathcal{W}(\mathcal{A}) := \sup_{a \in \mathcal{A}} \|a\|_2$. Consequently, Theorem 3.24 implies that

$$\mathbb{P}[|Z - \mathbb{E}[Z]| \geq t] \leq 2 \exp\left(-\frac{t^2}{2\mathcal{W}^2(\mathcal{A})}\right). \tag{3.75}$$

Note that this bound sharpens our earlier inequality (3.17), both in terms of the exponent and in providing a two-sided result. ♣

Let us now prove Theorem 3.24.

Proof As defined, any Wasserstein distance immediately yields an upper bound on a quantity of the form $\int f(d\mathbb{Q} - d\mathbb{P})$, where f is a Lipschitz function. Although the asymmetric coupling-based distance is not a Wasserstein distance, the key fact is that it can be used to upper bound such differences when $f\colon [0,1]^n \to \mathbb{R}$ is Lipschitz and convex. Indeed, for a convex f, we have the lower bound $f(x) \geq f(y) + \langle \nabla f(y), x - y \rangle$, which implies that

$$f(y) - f(x) \leq \sum_{j=1}^{n} \left| \frac{\partial f}{\partial y_j}(y) \right| \mathbb{I}[x_j \neq y_j].$$

Here we have also used the fact that $|x_j - y_j| \leq \mathbb{I}[x_j \neq y_j]$ for variables taking values in the unit interval $[0, 1]$. Consequently, for any coupling $\mathbb{M}$ of the pair $(\mathbb{P}, \mathbb{Q})$, we have

$$\begin{aligned}
\int f(y)\, d\mathbb{Q}(y) - \int f(x)\, d\mathbb{P}(x) &\leq \sum_{j=1}^{n} \left| \frac{\partial f}{\partial y_j}(y) \right| \mathbb{I}[x_j \neq y_j]\, d\mathbb{M}(x, y) \\
&= \int \sum_{j=1}^{n} \left| \frac{\partial f}{\partial y_j}(y) \right| \mathbb{M}[X_j \neq y_j \mid Y_j = y_j]\, d\mathbb{Q}(y) \\
&\leq \int \|\nabla f(y)\|_2 \sqrt{\sum_{j=1}^{n} \mathbb{M}^2[X_j \neq y_j \mid Y_j = y_j]}\, d\mathbb{Q}(y),
\end{aligned}$$

where we have applied the Cauchy–Schwarz inequality. By the Lipschitz condition and convexity, we have $\|\nabla f(y)\|_2 \leq L$ almost everywhere, and hence

$$\begin{aligned}\int f(y)\,d\mathbb{Q}(y) - \int f(x)\,d\mathbb{P}(x) &\leq L \int \left\{ \sum_{j=1}^{n} \mathbb{M}^2[X_j \neq y_j \mid Y_j = y_j] \right\}^{1/2} d\mathbb{Q}(y) \\ &\leq L \left[\int \sum_{j=1}^{n} \mathbb{M}^2[X_j \neq y_j \mid Y_j = y_j]\, d\mathbb{Q}(y) \right]^{1/2} \\ &= L C(\mathbb{P}, \mathbb{Q}).\end{aligned}$$

Consequently, the upper tail bound follows by a combination of the information inequality (3.73) and Theorem 3.19.

To obtain the lower bound for a convex Lipschitz function, it suffices to establish an upper bound for a concave Lipschitz function, say $g\colon [0,1]^n \to \mathbb{R}$. In this case, we have the upper bound

$$g(y) \leq g(x) + \langle \nabla g(x),\, y - x \rangle \leq g(x) + \sum_{j=1}^{n} \left| \frac{\partial g(x)}{\partial x_j} \right| \mathbb{I}[x_j \neq y_j],$$

and consequently

$$\int g\, d\mathbb{Q}(y) - \int g\, d\mathbb{P}(x) \leq \sum_{j=1}^{n} \left| \frac{\partial g(x)}{\partial x_j} \right| \mathbb{I}[x_j \neq y_j]\, d\mathbb{M}(x, y).$$

The same line of reasoning then shows that $\int g\, d\mathbb{Q}(y) - \int g\, d\mathbb{P}(x) \leq L C(\mathbb{Q}, \mathbb{P})$, from which the claim then follows as before. □

We have stated Theorem 3.24 for the familiar case of independent random variables. However, a version of the underlying information inequality (3.73) holds for many collections of random variables. In particular, consider an n-dimensional distribution $\mathbb{P}$ for which there exists some $\gamma > 0$ such that the following inequality holds:

$$\max\{C(\mathbb{Q}, \mathbb{P}),\, C(\mathbb{P}, \mathbb{Q})\} \leq \sqrt{2\gamma D(\mathbb{Q} \,\|\, \mathbb{P})} \qquad \text{for all distributions } \mathbb{Q}. \tag{3.76}$$

The same proof then shows that any L-Lipschitz function satisfies the concentration inequality

$$\mathbb{P}[|f(X) - \mathbb{E}[f(X)]| \geq t] \leq 2 \exp\left(-\frac{t^2}{2\gamma L^2} \right). \tag{3.77}$$

For example, for a Markov chain that satisfies the β-contraction condition (3.67), it can be shown that the information inequality (3.76) holds with $\gamma = (\frac{1}{1-\sqrt{\beta}})^2$. Consequently, any L-Lipschitz function (with respect to the Euclidean norm) of a β-contractive Markov chain satisfies the concentration inequality

$$\mathbb{P}[|f(X) - \mathbb{E}[f(X)]| \geq t] \leq 2 \exp\left(-\frac{(1 - \sqrt{\beta})^2 t^2}{2L^2} \right). \tag{3.78}$$

This bound is a dimension-independent analog of our earlier bound (3.69) for a contractive Markov chain. We refer the reader to the bibliographic section for further discussion of results of this type.

3.4 Tail bounds for empirical processes

In this section, we illustrate the use of concentration inequalities in application to empirical processes. We encourage the interested reader to look ahead to Chapter 4 so as to acquire the statistical motivation for the classes of problems studied in this section. Here we use the entropy method to derive various tail bounds on the suprema of empirical processes—in particular, for random variables that are generated by taking suprema of sample averages over function classes. More precisely let $\mathscr{F}$ be a class of functions (each of the form $f: \mathcal{X} \to \mathbb{R}$), and let $(X_1, \ldots, X_n)$ be drawn from a product distribution $\mathbb{P} = \bigotimes_{i=1}^n \mathbb{P}_i$, where each $\mathbb{P}_i$ is supported on some set $\mathcal{X}_i \subseteq \mathcal{X}$. We then consider the random variable[5]

$$Z = \sup_{f \in \mathscr{F}} \left\{ \frac{1}{n} \sum_{i=1}^n f(X_i) \right\}. \tag{3.79}$$

The primary goal of this section is to derive a number of upper bounds on the tail event $\{Z \geq \mathbb{E}[Z] + \delta\}$.

As a passing remark, we note that, if the goal is to obtain bounds on the random variable $\sup_{f \in \mathscr{F}} \left|\frac{1}{n} \sum_{i=1}^n f(X_i)\right|$, then it can be reduced to an instance of the variable (3.79) by considering the augmented function class $\widetilde{\mathscr{F}} = \mathscr{F} \cup \{-\mathscr{F}\}$.

3.4.1 A functional Hoeffding inequality

We begin with the simplest type of tail bound for the random variable Z, namely one of the Hoeffding type. The following result is a generalization of the classical Hoeffding theorem for sums of bounded random variables.

Theorem 3.26 (Functional Hoeffding theorem) *For each $f \in \mathscr{F}$ and $i = 1, \ldots, n$, assume that there are real numbers $a_{i,f} \leq b_{i,f}$ such that $f(x) \in [a_{i,f}, b_{i,f}]$ for all $x \in \mathcal{X}_i$. Then for all $\delta \geq 0$, we have*

$$\mathbb{P}[Z \geq \mathbb{E}[Z] + \delta] \leq \exp\left(-\frac{n\delta^2}{4L^2}\right), \tag{3.80}$$

where $L^2 := \sup_{f \in \mathscr{F}} \{\frac{1}{n} \sum_{i=1}^n (b_{i,f} - a_{i,f})^2\}$.

Remark: In a very special case, Theorem 3.26 can be used to recover the classical Hoeffding inequality in the case of bounded random variables, albeit with a slightly worse constant. Indeed, if we let $\mathscr{F}$ be a singleton consisting of the identity function $f(x) = x$, then we have $Z = \frac{1}{n} \sum_{i=1}^n X_i$. Consequently, as long as $x_i \in [a_i, b_i]$, Theorem 3.26 implies that

$$\mathbb{P}\left[\frac{1}{n} \sum_{i=1}^n (X_i - \mathbb{E}[X_i]) \geq \delta\right] \leq e^{-\frac{n\delta^2}{4L^2}},$$

[5] Note that there can be measurability problems associated with this definition if $\mathscr{F}$ is not countable. See the bibliographic discussion in Chapter 4 for more details on how to resolve them.

where $L^2 = \frac{1}{n}\sum_{i=1}^n (b_i - a_i)^2$. We thus recover the classical Hoeffding theorem, although the constant $1/4$ in the exponent is not optimal.

More substantive implications of Theorem 3.26 arise when it is applied to a larger function class $\mathscr{F}$. In order to appreciate its power, let us compare the upper tail bound (3.80) to the corresponding bound that can be derived from the bounded differences inequality, as applied to the function $(x_1, \ldots, x_n) \mapsto Z(x_1, \ldots, x_n)$. With some calculation, it can be seen that this function satisfies the bounded difference inequality with constant $L_i := \sup_{f \in \mathscr{F}} |b_{i,f} - a_{i,f}|$ in coordinate i. Consequently, the bounded differences method (Corollary 2.21) yields a sub-Gaussian tail bound, analogous to the bound (3.80), but with the parameter

$$\widetilde{L}^2 = \frac{1}{n}\sum_{i=1}^n \sup_{f \in \mathscr{F}} (b_{i,f} - a_{i,f})^2.$$

Note that the quantity $\widetilde{L}$—since it is defined by applying the supremum separately to each coordinate—can be substantially larger than the constant L defined in the theorem statement.

Proof It suffices to prove the result for a finite class of functions $\mathscr{F}$; the general result can be recovered by taking limits over an increasing sequence of such finite classes. Let us view Z as a function of the random variables $(X_1, \ldots, X_n)$. For each index $j = 1, \ldots, n$, define the random function

$$x_j \mapsto Z_j(x_j) = Z(X_1, \ldots, X_{j-1}, x_j, X_{j+1}, \ldots, X_n).$$

In order to avoid notational clutter, we work throughout this proof with the *unrescaled* version of Z, namely $Z = \sup_{f \in \mathscr{F}} \sum_{i=1}^n f(X_i)$. Combining the tensorization Lemma 3.8 with the bound (3.20a) from Lemma 3.7, we obtain

$$\mathbb{H}(e^{\lambda Z(X)}) \le \lambda^2 \mathbb{E}\Bigg[\sum_{j=1}^n \mathbb{E}[(Z_j(X_j) - Z_j(Y_j))^2\, \mathbb{I}[Z_j(X_j) \ge Z_j(Y_j)]\, e^{\lambda Z(X)} \mid X^{\setminus j}]\Bigg]. \tag{3.81}$$

For each $f \in \mathscr{F}$, define the set $\mathcal{A}(f) := \{(x_1, \ldots, x_n) \in \mathbb{R}^n \mid Z = \sum_{i=1}^n f(x_i)\}$, corresponding to the set of realizations for which the maximum defining Z is achieved by f. (If there are ties, then we resolve them arbitrarily so as to make the sets $\mathcal{A}(f)$ disjoint.) For any $x \in \mathcal{A}(f)$, we have

$$Z_j(x_j) - Z_j(y_j) = f(x_j) + \sum_{i \ne j}^n f(x_i) - \max_{\widetilde{f} \in \mathscr{F}}\Big\{\widetilde{f}(y_j) + \sum_{i \ne j}^n \widetilde{f}(x_i)\Big\} \le f(x_j) - f(y_j).$$

As long as $Z_j(x_j) \ge Z_j(y_j)$, this inequality still holds after squaring both sides. Considering all possible sets $\mathcal{A}(f)$, we arrive at the upper bound

$$(Z_j(x_j) - Z_j(y_j))^2\, \mathbb{I}[Z_j(x_j) \ge Z_j(y_j)] \le \sum_{f \in \mathscr{F}} \mathbb{I}[x \in \mathcal{A}(f)](f(x_j) - f(y_j))^2. \tag{3.82}$$

Since $(f(x_j) - f(y_j))^2 \leq (b_{j,f} - a_{j,f})^2$ by assumption, summing over the indices j yields

$$\begin{aligned}\sum_{j=1}^{n}(Z_j(x_j) - Z_j(y_j))^2\, \mathbb{I}[Z_k(x_k) \geq Z_k(y_k)]\, e^{\lambda Z(x)} &\leq \sum_{h\in\mathscr{F}} \mathbb{I}[x \in \mathcal{A}(h)] \sum_{k=1}^{n}(b_{k,h} - a_{k,h})^2 e^{\lambda Z(x)} \\ &\leq \sup_{f\in\mathscr{F}} \sum_{j=1}^{n}(b_{j,f} - a_{j,f})^2 e^{\lambda Z(x)} \\ &= nL^2 e^{\lambda Z(x)}.\end{aligned}$$

Substituting back into our earlier inequality (3.81), we find that

$$\mathbb{H}(e^{\lambda Z(X)}) \leq nL^2\lambda^2\, \mathbb{E}[e^{\lambda Z(X)}].$$

This is a sub-Gaussian entropy bound (3.5) with $\sigma = \sqrt{2n}\,L$, so that Proposition 3.2 implies that the unrescaled version of Z satisfies the tail bound

$$\mathbb{P}[Z \geq \mathbb{E}[Z] + t] \leq e^{-\frac{t^2}{4nL^2}}.$$

Setting $t = n\delta$ yields the claim (3.80) for the rescaled version of Z. □

3.4.2 A functional Bernstein inequality

In this section, we turn to the Bernstein refinement of the functional Hoeffding inequality from Theorem 3.26. As opposed to control only in terms of bounds on the function values, it also brings a notion of variance into play. As will be discussed at length in later chapters, this type of variance control plays a key role in obtaining sharp bounds for various types of statistical estimators.

Theorem 3.27 (Talagrand concentration for empirical processes) *Consider a countable class of functions $\mathscr{F}$ uniformly bounded by b. Then for all $\delta > 0$, the random variable* (3.79) *satisfies the upper tail bound*

$$\mathbb{P}[Z \geq \mathbb{E}[Z] + \delta] \leq 2\exp\left(\frac{-n\delta^2}{8e\,\mathbb{E}[\Sigma^2] + 4b\delta}\right), \tag{3.83}$$

where $\Sigma^2 = \sup_{f\in\mathscr{F}} \frac{1}{n}\sum_{i=1}^{n} f^2(X_i)$.

In order to obtain a simpler bound, the expectation $\mathbb{E}[\Sigma^2]$ can be upper bounded. Using symmetrization techniques to be developed in Chapter 4, it can be shown that

$$\mathbb{E}[\Sigma^2] \leq \sigma^2 + 2b\,\mathbb{E}[Z], \tag{3.84}$$

where $\sigma^2 = \sup_{f\in\mathscr{F}} \mathbb{E}[f^2(X)]$. Using this upper bound on $\mathbb{E}[\Sigma^2]$ and performing some algebra, we obtain that there are universal positive constants (c_0, c_1) such that

$$\mathbb{P}[Z \geq \mathbb{E}[Z] + c_0\gamma\sqrt{t} + c_1 bt] \leq e^{-nt} \qquad \text{for all } t > 0, \tag{3.85}$$

where $\gamma^2 = \sigma^2 + 2b\,\mathbb{E}[Z]$. See Exercise 3.16 for the derivation of this inequality from Theorem 3.27 and the upper bound (3.84). Although the proof outlined here leads to poor constants, the best known are $c_0 = \sqrt{2}$ and $c_1 = 1/3$; see the bibliographic section for further details.

In certain settings, it can be useful to exploit the bound (3.85) in an alternative form: in particular, for any $\epsilon > 0$, it implies the upper bound

$$\mathbb{P}[Z \geq (1+\epsilon)\mathbb{E}[Z] + c_0\sigma\sqrt{t} + (c_1 + c_0^2/\epsilon)bt] \leq e^{-nt}. \tag{3.86}$$

Conversely, we can recover the tail bound (3.85) by optimizing over $\epsilon > 0$ in the family of bounds (3.86); see Exercise 3.16 for the details of this equivalence.

Proof We assume without loss of generality that $b = 1$, since the general case can be reduced to this one. Moreover, as in the proof of Theorem 3.26, we work with the unrescaled version—namely, the variable $Z = \sup_{f\in\mathscr{F}} \sum_{i=1}^n f(X_i)$—and then translate our results back. Recall the definition of the sets $\mathcal{A}(f)$, and the upper bound (3.82) from the previous proof; substituting it into the entropy bound (3.81) yields the upper bound

$$\mathbb{H}(e^{\lambda Z}) \leq \lambda^2\, \mathbb{E}\left[\sum_{j=1}^n \mathbb{E}\left[\sum_{f\in\mathscr{F}} \mathbb{I}[x \in \mathcal{A}(f)](f(X_j) - f(Y_j))^2 e^{\lambda Z} \mid X^{\backslash j}\right]\right].$$

Now we have

$$\sum_{i=1}^n \sum_{f\in\mathscr{F}} \mathbb{I}[X \in \mathcal{A}(f)](f(X_j) - f(Y_j))^2 \leq 2\sup_{f\in\mathscr{F}} \sum_{i=1}^n f^2(X_i) + 2\sup_{f\in\mathscr{F}} \sum_{i=1}^n f^2(Y_i)$$
$$= 2\{\Gamma(X) + \Gamma(Y)\},$$

where $\Gamma(X) := \sup_{f\in\mathscr{F}} \sum_{i=1}^n f^2(X_i)$ is the unrescaled version of Σ^2. Combined with our earlier inequality, we see that the entropy satisfies the upper bound

$$\mathbb{H}(e^{\lambda Z}) \leq 2\lambda^2\{\mathbb{E}[\Gamma e^{\lambda Z}] + \mathbb{E}[\Gamma]\,\mathbb{E}[e^{\lambda Z}]\}. \tag{3.87}$$

From the result of Exercise 3.4, we have $\mathbb{H}(e^{\lambda(Z+c)}) = e^{\lambda c}\mathbb{H}(e^{\lambda Z})$ for any constant $c \in \mathbb{R}$. Since the right-hand side also contains a term $e^{\lambda Z}$ in each component, we see that the same upper bound holds for $\mathbb{H}(e^{\lambda\widetilde{Z}})$, where $\widetilde{Z} = Z - \mathbb{E}[Z]$ is the centered version. We now introduce a lemma to control the term $\mathbb{E}[\Gamma e^{\lambda\widetilde{Z}}]$.

Lemma 3.28 (Controlling the random variance) *For all $\lambda > 0$, we have*

$$\mathbb{E}[\Gamma e^{\lambda\widetilde{Z}}] \leq (e-1)\mathbb{E}[\Gamma]\mathbb{E}[e^{\lambda\widetilde{Z}}] + \mathbb{E}[\widetilde{Z}e^{\lambda\widetilde{Z}}]. \tag{3.88}$$

Combining the upper bound (3.88) with the entropy upper bound (3.87) for $\widetilde{Z}$, we obtain

$$\mathbb{H}(e^{\lambda\widetilde{Z}}) \leq \lambda^2\{2e\,\mathbb{E}[\Gamma]\varphi(\lambda) + 2\varphi'(\lambda)\} \qquad \text{for all } \lambda > 0,$$

where $\varphi(\lambda) := \mathbb{E}[e^{\lambda\widetilde{Z}}]$ is the moment generating function of $\widetilde{Z}$. Since $\mathbb{E}[\widetilde{Z}] = 0$, we recognize this as an entropy bound of the Bernstein form (3.10) with $b = 2$ and $\sigma^2 = 2e\,\mathbb{E}[\Gamma]$.

Consequently, by the consequence (3.12) stated following Proposition 3.3, we conclude that

$$\mathbb{P}[\widetilde{Z} \geq \mathbb{E}[\widetilde{Z}] + \delta] \leq \exp\left(-\frac{\delta^2}{8e\,\mathbb{E}[\Gamma] + 4\delta}\right) \qquad \text{for all } \delta \geq 0.$$

Recalling the definition of Γ and rescaling by $1/n$, we obtain the stated claim of the theorem with $b = 1$.

It remains to prove Lemma 3.28. Consider the function $g(t) = e^t$ with conjugate dual $g^*(s) = s \log s - s$ for $s > 0$. By the definition of conjugate duality (also known as Young's inequality), we have $st \leq s \log s - s + e^t$ for all $s > 0$ and $t \in \mathbb{R}$. Applying this inequality with $s = e^{\lambda \widetilde{Z}}$ and $t = \Gamma - (e-1)\mathbb{E}[\Gamma]$ and then taking expectations, we find that

$$\mathbb{E}[\Gamma e^{\lambda \widetilde{Z}}] - (e-1)\mathbb{E}[e^{\lambda \widetilde{Z}}]\,\mathbb{E}[\Gamma] \leq \lambda \mathbb{E}[\widetilde{Z} e^{\lambda \widetilde{Z}}] - \mathbb{E}[e^{\lambda \widetilde{Z}}] + \mathbb{E}[e^{\Gamma - (e-1)\mathbb{E}[\Gamma]}].$$

Note that Γ is defined as a supremum of a class of functions taking values in $[0, 1]$. Therefore, by the result of Exercise 3.15, we have $\mathbb{E}[e^{\Gamma - (e-1)\mathbb{E}[\Gamma]}] \leq 1$. Moreover, by Jensen's inequality, we have $\mathbb{E}[e^{\lambda \widetilde{Z}}] \geq e^{\lambda \mathbb{E}[\widetilde{Z}]} = 1$. Putting together the pieces yields the claim (3.88). □

3.5 Bibliographic details and background

Concentration of measure is an extremely rich and deep area with an extensive literature; we refer the reader to the books by Ledoux (2001) and Boucheron et al. (2013) for more comprehensive treatments. Logarithmic Sobolev inequalities were introduced by Gross (1975) in a functional-analytic context. Their dimension-free nature makes them especially well suited for controlling infinite-dimensional stochastic processes (e.g., Holley and Stroock, 1987). The argument underlying the proof of Proposition 3.2 is based on the unpublished notes of Herbst. Ledoux (1996; 2001) pioneered the entropy method in application to a wider range of problems. The proof of Theorem 3.4 is based on Ledoux (1996), whereas the proofs of Lemmas 3.7 and 3.8 follow the book (Ledoux, 2001). A result of the form in Theorem 3.4 was initially proved by Talagrand (1991; 1995; 1996b) using his convex distance inequalities.

The Brunn–Minkowski theorem is a classical result from geometry and real analysis; see Gardner (2002) for a survey of its history and connections. Theorem 3.15 was proved independently by Prékopa (1971; 1973) and Leindler (1972). Brascamp and Lieb (1976) developed various connections between log-concavity and log-Sobolev inequalities; see the paper by Bobkov (1999) for further discussion. The inf-convolution argument underlying the proof of Theorem 3.16 was initiated by Maurey (1991), and further developed by Bobkov and Ledoux (2000). The lecture notes by Ball (1997) contain a wealth of information on geometric aspects of concentration, including spherical sections of convex bodies. Harper's theorem quoted in Example 3.13 is proven in the paper (Harper, 1966); it is a special case of a more general class of results known as discrete isoperimetric inequalities.

The Kantorovich–Rubinstein duality (3.55) was established by Kantorovich and Rubinstein (1958); it is a special case of more general results in optimal transport theory (e.g., Villani, 2008; Rachev and Ruschendorf, 1998). Marton (1996a) pioneered the use of the transportation cost method for deriving concentration inequalities, with subsequent contributions from various researchers (e.g., Dembo and Zeitouni, 1996; Dembo, 1997; Bobkov and Götze, 1999; Ledoux, 2001). See Marton's paper (1996b) for a proof of Theorem 3.22.

The information inequality (3.73) was proved by Samson (2000). As noted following the statement of Theorem 3.24, he actually proves a much more general result, applicable to various types of dependent random variables. Other results on concentration for dependent random variables include the papers (Marton, 2004; Kontorovich and Ramanan, 2008).

Upper tail bounds on the suprema of empirical processes can be proved using chaining methods; see Chapter 5 for more details. Talagrand (1996a) initiated the use of concentration techniques to control deviations above the mean, as in Theorems 3.26 and 3.27. The theorems and entropy-based arguments given here are based on Chapter 7 of Ledoux (2001); the sketch in Exercise 3.15 is adapted from arguments in the same chapter. Sharper forms of Theorem 3.27 have been established by various authors (e.g., Massart, 2000; Bousquet, 2002, 2003; Klein and Rio, 2005). In particular, Bousquet (2003) proved that the bound (3.85) holds with constants $c_0 = \sqrt{2}$ and $c_1 = 1/3$. There are also various results on concentration of empirical processes for unbounded and/or dependent random variables (e.g., Adamczak, 2008; Mendelson, 2010); see also Chapter 14 for some one-sided results in this direction.

3.6 Exercises

Exercise 3.1 (Shannon entropy and Kullback–Leibler divergence) Given a discrete random variable $X \in \mathcal{X}$ with probability mass function p, its Shannon entropy is given by $H(X) := -\sum_{x\in\mathcal{X}} p(x)\log p(x)$. In this exercise, we explore the connection between the entropy functional $\mathbb{H}$ based on $\phi(u) = u\log u$ (see equation (3.2)) and the Shannon entropy.

(a) Consider the random variable $Z = p(U)$, where U is uniformly distributed over $\mathcal{X}$. Show that

$$\mathbb{H}(Z) = \frac{1}{|\mathcal{X}|}\{\log|\mathcal{X}| - H(X)\}.$$

(b) Use part (a) to show that Shannon entropy for a discrete random variable is maximized by a uniform distribution.

(c) Given two probability mass functions p and q, specify a choice of random variable Y such that $\mathbb{H}(Y) = D(p \,\|\, q)$, corresponding to the Kullback–Leibler divergence between p and q.

Exercise 3.2 (Chain rule and Kullback–Leibler divergence) Given two n-variate distributions $\mathbb{Q}$ and $\mathbb{P}$, show that the Kullback–Leibler divergence can be decomposed as

$$D(\mathbb{Q} \,\|\, \mathbb{P}) = D(\mathbb{Q}_1 \,\|\, \mathbb{P}_1) + \sum_{j=2}^{n} \mathbb{E}_{\mathbb{Q}_1^{j-1}}[D(\mathbb{Q}_j(\cdot \mid X_1^{j-1}) \,\|\, \mathbb{P}_j(\cdot \mid X_1^{j-1}))],$$

where $\mathbb{Q}_j(\cdot \mid X_1^{j-1})$ denotes the conditional distribution of X_j given $(X_1, \ldots, X_{j-1})$ under $\mathbb{Q}$, with a similar definition for $\mathbb{P}_j(\cdot \mid X_1^{j-1})$.

Exercise 3.3 (Variational representation for entropy) Show that the entropy has the variational representation

$$\mathbb{H}(e^{\lambda X}) = \inf_{t\in\mathbb{R}} \mathbb{E}[\psi(\lambda(X-t))e^{\lambda X}], \tag{3.89}$$

where $\psi(u) := e^{-u} - 1 + u$.

Exercise 3.4 (Entropy and constant shifts) In this exercise, we explore some properties of the entropy.

(a) Show that for any random variable X and constant $c \in \mathbb{R}$,

$$\mathbb{H}(e^{\lambda(X+c)}) = e^{\lambda c}\,\mathbb{H}(e^{\lambda X}).$$

(b) Use part (a) to show that, if X satisfies the entropy bound (3.5), then so does $X + c$ for any constant c.

Exercise 3.5 (Equivalent forms of entropy) Let $\mathbb{H}_\varphi$ denote the entropy defined by the convex function $\varphi(u) = u \log u - u$. Show that $\mathbb{H}_\varphi(e^{\lambda X}) = \mathbb{H}(e^{\lambda X})$, where $\mathbb{H}$ denotes the usual entropy (defined by $\phi(u) = u \log u$).

Exercise 3.6 (Entropy rescaling) In this problem, we develop recentering and rescaling arguments used in the proof of Proposition 3.3.

(a) Show that a random variable X satisfies the Bernstein entropy bound (3.10) if and only if $\widetilde{X} = X - \mathbb{E}[X]$ satisfies the inequality

$$\mathbb{H}(e^{\lambda X}) \leq \lambda^2\{b\varphi_x'(\lambda) + \varphi_x(\lambda)\sigma^2\} \qquad \text{for all } \lambda \in [0, 1/b). \tag{3.90}$$

(b) Show that a zero-mean random variable X satisfies inequality (3.90) if and only if $\widetilde{X} = X/b$ satisfies the bound

$$\mathbb{H}(e^{\lambda \widetilde{X}}) \leq \lambda^2\{\varphi_{\widetilde{X}}'(\lambda) + \tilde{\sigma}^2 \varphi_{\widetilde{X}}(\lambda)\} \qquad \text{for all } \lambda \in [0, 1),$$

where $\tilde{\sigma}^2 = \sigma^2/b^2$.

Exercise 3.7 (Entropy for bounded variables) Consider a zero-mean random variable X taking values in a finite interval $[a, b]$ almost surely. Show that its entropy satisfies the bound $\mathbb{H}(e^{\lambda X}) \leq \frac{\lambda^2\sigma^2}{2}\varphi_x(\lambda)$ with $\sigma := (b - a)/2$. (*Hint:* You may find the result of Exercise 3.3 useful.)

Exercise 3.8 (Exponential families and entropy) Consider a random variable $Y \in \mathcal{Y}$ with an exponential family distribution of the form

$$p_\theta(y) = h(y)e^{\langle \theta, T(y)\rangle - \Phi(\theta)},$$

where $T\colon \mathcal{Y} \to \mathbb{R}^d$ defines the vector of sufficient statistics, the function h is fixed, and the density p_θ is taken with respect to base measure μ. Assume that the log normalization term $\Phi(\theta) = \log \int_{\mathcal{Y}} \exp(\langle \theta, T(y)\rangle) h(y)\mu(dy)$ is finite for all $\theta \in \mathbb{R}^d$, and suppose moreover that ∇A is Lipschitz with parameter L, meaning that

$$\|\nabla\Phi(\theta) - \nabla\Phi(\theta')\|_2 \leq L\|\theta - \theta'\|_2 \qquad \text{for all } \theta, \theta' \in \mathbb{R}^d. \tag{3.91}$$

(a) For fixed unit-norm vector $v \in \mathbb{R}^d$, consider the random variable $X = \langle v, T(Y) \rangle$. Show that

$$\mathbb{H}(e^{\lambda X}) \leq L\lambda^2 \varphi_x(\lambda) \qquad \text{for all } \lambda \in \mathbb{R}.$$

Conclude that X is sub-Gaussian with parameter $\sqrt{2L}$.

(b) Apply part (a) to establish the sub-Gaussian property for:

(i) the univariate Gaussian distribution $Y \sim \mathcal{N}(\mu, \sigma^2)$ (*Hint:* Viewing σ^2 as fixed, write it as a one-dimensional exponential family.)

(ii) the Bernoulli variable $Y \in \{0, 1\}$ with $\theta = \frac{1}{2} \log \frac{\mathbb{P}[Y=1]}{\mathbb{P}[Y=0]}$.

Exercise 3.9 (Another variational representation) Prove the following variational representation:

$$\mathbb{H}(e^{\lambda f(X)}) = \sup_g \{\mathbb{E}[g(X)e^{\lambda f(X)}] \mid \mathbb{E}[e^{g(X)}] \leq 1\},$$

where the supremum ranges over all measurable functions. Exhibit a function g at which the supremum is obtained. (*Hint:* The result of Exercise 3.5 and the notion of conjugate duality could be useful.)

Exercise 3.10 (Brunn–Minkowski and classical isoperimetric inequality) In this exercise, we explore the connection between the Brunn–Minkowski (BM) inequality and the classical isoperimetric inequality.

(a) Show that the BM inequality (3.43) holds if and only if

$$\text{vol}(A + B)^{1/n} \geq \text{vol}(A)^{1/n} + \text{vol}(B)^{1/n} \tag{3.92}$$

for all convex bodies A and B.

(b) Show that the BM inequality (3.43) implies the "weaker" inequality (3.45).

(c) Conversely, show that inequality (3.45) also implies the original BM inequality (3.43). (*Hint:* From part (a), it suffices to prove the inequality (3.92) for bodies A and B with strictly positive volumes. Consider applying inequality (3.45) to the rescaled bodies $C := \frac{A}{\text{vol}(A)}$ and $D := \frac{B}{\text{vol}(B)}$, and a suitable choice of λ.)

Exercise 3.11 (Concentration on the Euclidean ball) Consider the uniform measure $\mathbb{P}$ over the Euclidean unit ball $\mathbb{B}_2^n = \{x \in \mathbb{R}^n \mid \|x\|_2 \leq 1\}$. In this example, we bound its concentration function using the Brunn–Minkowski inequality (3.45).

(a) Given any subset $A \subseteq \mathbb{B}_2^n$, show that

$$\frac{1}{2}\|a + b\|_2 \leq 1 - \frac{\epsilon^2}{8} \qquad \text{for all } a \in A \text{ and } b \in (A^\epsilon)^c.$$

To be clear, here we define $(A^\epsilon)^c := \mathbb{B}_2^n \backslash A^\epsilon$.

(b) Use the BM inequality (3.45) to show that $\mathbb{P}[A](1 - \mathbb{P}[A^\epsilon]) \leq (1 - \frac{\epsilon^2}{8})^{2n}$.

(c) Conclude that

$$\alpha_{\mathbb{P},(\mathcal{X},\rho)}(\epsilon) \leq 2e^{-n\epsilon^2/4} \qquad \text{for } \mathcal{X} = \mathbb{B}_2^n \text{ with } \rho(\cdot) = \|\cdot\|_2.$$

Exercise 3.12 (Rademacher chaos variables) A symmetric positive semidefinite matrix $\mathbf{Q} \in \mathcal{S}_+^{d\times d}$ can be used to define a Rademacher chaos variable $X = \sum_{i,j=1}^d Q_{ij}\varepsilon_i\varepsilon_j$, where $\{\varepsilon_i\}_{i=1}^d$ are i.i.d. Rademacher variables.

(a) Prove that

$$\mathbb{P}[X \geq (\sqrt{\operatorname{trace}\mathbf{Q}} + t)^2] \leq 2\exp\left(-\frac{t^2}{16\,|||\mathbf{Q}|||_2}\right). \tag{3.93}$$

(b) Given an arbitrary symmetric matrix $\mathbf{M} \in \mathcal{S}^{d\times d}$, consider the decoupled Rademacher chaos variable $Y = \sum_{i=1,j=1}^d M_{ij}\varepsilon_i\varepsilon'_j$, where $\{\varepsilon'_j\}_{j=1}^d$ is a second i.i.d. Rademacher sequence, independent of the first. Show that

$$\mathbb{P}[Y \geq \delta] \leq 2\exp\left(-\frac{\delta^2}{4\,|||\mathbf{M}|||_F^2 + 16\delta\,|||\mathbf{M}|||_2}\right).$$

(*Hint:* Part (a) could be useful in an intermediate step.)

Exercise 3.13 (Total variation and Wasserstein) Consider the Wasserstein distance based on the Hamming metric, namely $W_\rho(\mathbb{P}, \mathbb{Q}) = \inf_{\mathbb{M}} \mathbb{M}[X \neq Y]$, where the infimum is taken over all couplings $\mathbb{M}$—that is, distributions on the product space $\mathcal{X}\times\mathcal{X}$ with marginals $\mathbb{P}$ and $\mathbb{Q}$, respectively. Show that

$$\inf_{\mathbb{M}} \mathbb{M}[X \neq Y] = \|\mathbb{P} - \mathbb{Q}\|_{\text{TV}} = \sup_A |\mathbb{P}(A) - \mathbb{Q}(A)|,$$

where the supremum ranges over all measurable subsets A of $\mathcal{X}$.

Exercise 3.14 (Alternative proof) In this exercise, we work through an alternative proof of Proposition 3.20. As noted, it suffices to consider the case $n = 2$. Let $\mathbb{P} = \mathbb{P}_1 \otimes \mathbb{P}_2$ be a product distribution, and let $\mathbb{Q}$ be an arbitrary distribution on $\mathcal{X} \times \mathcal{X}$.

(a) Show that the Wasserstein distance $W_\rho(\mathbb{Q}, \mathbb{P})$ is upper bounded by

$$\sup_{\|f\|_{\text{Lip}}\leq 1} \left\{\int\left[\int f(x_1, x_2)\,(d\mathbb{Q}_{2|1} - d\mathbb{P}_2)\right] d\mathbb{Q}_1 + \int\left[\int f(x_1, x_2)\,d\mathbb{P}_2\right](d\mathbb{Q}_1 - d\mathbb{P}_1)\right\},$$

where the supremum ranges over all functions that are 1-Lipschitz with respect to the metric $\rho(x, x') = \sum_{i=1}^2 \rho_i(x_i, x'_i)$.

(b) Use part (a) to show that

$$W_\rho(\mathbb{Q}, \mathbb{P}) \leq \left[\int \sqrt{2\gamma_2 D(\mathbb{Q}_{2|1} \,\|\, \mathbb{P}_2)}\, d\mathbb{Q}_1\right] + \sqrt{2\gamma_1 D(\mathbb{Q}_1 \,\|\, \mathbb{P}_1)}.$$

(c) Complete the proof using part (b). (*Hint:* Cauchy–Schwarz and Exercise 3.2 could be useful.

Exercise 3.15 (Bounds for suprema of non-negative functions) Consider a random variable of the form $Z = \sup_{f\in\mathscr{F}} \sum_{i=1}^n f(V_i)$ where $\{V_i\}_{i=1}^n$ is an i.i.d. sequence of random variables, and $\mathscr{F}$ is a class of functions taking values in the interval $[0, 1]$. In this exercise, we prove that

$$\log \mathbb{E}[e^{\lambda Z}] \leq (e^\lambda - 1)\mathbb{E}[Z] \qquad \text{for any } \lambda \geq 0. \tag{3.94}$$

As in our main development, we can reduce the problem to a finite class of functions $\mathscr{F}$, say with M functions $\{f^1, \ldots, f^M\}$. Defining the random vectors $X_i = (f^1(V_i), \ldots, f^M(V_i)) \in \mathbb{R}^M$ for $i = 1, \ldots, n$, we can then consider the function $Z(X) = \max_{j=1,\ldots,M} \sum_{i=1}^n X_i^j$. We let Z_k denote the function $X_k \mapsto Z(X)$ with all other X_i for $i \neq k$ fixed.

(a) Define $Y_k(X) := (X_1, \ldots, X_{k-1}, 0, X_{k+1}, X_n)$. Explain why $Z(X) - Z(Y_k(X)) \geq 0$.

(b) Use the tensorization approach and the variational representation from Exercise 3.3 to show that

$$\mathbb{H}(e^{\lambda Z(X)}) \leq \mathbb{E}\left[\sum_{k=1}^n \mathbb{E}[\psi(\lambda(Z(X) - Z(Y_k(X))))e^{\lambda Z(X)} \mid X^{\backslash k}]\right] \qquad \text{for all } \lambda > 0.$$

(c) For each $\ell = 1, \ldots, M$, let

$$\mathbb{A}_\ell = \left\{ x = (x_1, \ldots, x_n) \in \mathbb{R}^{M \times n} \;\middle|\; \sum_{i=1}^n x_i^\ell = \max_{j=1,\ldots,M} \sum_{i=1}^n x_i^j \right\}.$$

Prove that

$$0 \leq \lambda\{Z(X) - Z(Y_k(X))\} \leq \lambda \sum_{\ell=1}^M \mathbb{I}[X \in \mathbb{A}_\ell] X_k^\ell \qquad \text{valid for all } \lambda \geq 0.$$

(d) Noting that $\psi(t) = e^{-t} + 1 - t$ is non-negative with $\psi(0) = 0$, argue by the convexity of ψ that

$$\psi(\lambda(Z(X) - Z(Y_k(X)))) \leq \psi(\lambda)\left[\sum_{\ell=1}^M \mathbb{I}[X \in \mathbb{A}_\ell] X_k^\ell\right] \qquad \text{for all } \lambda \geq 0.$$

(e) Combining with previous parts, prove that

$$\mathbb{H}(e^{\lambda Z}) \leq \psi(\lambda) \sum_{k=1}^n \mathbb{E}\left[\sum_{\ell=1}^M \mathbb{I}[X \in \mathbb{A}_\ell] X_k^\ell e^{\lambda Z(X)}\right] = \psi(\lambda)\mathbb{E}[Z(X)e^{\lambda Z(X)}].$$

(*Hint:* Observe that $\sum_{k=1}^n \sum_{\ell=1}^M \mathbb{I}[X \in \mathbb{A}_\ell] X_k^\ell = Z(X)$ by definition of the sets $\mathbb{A}_\ell$.)

(f) Use part (e) to show that $\varphi_Z(\lambda) = \mathbb{E}[e^{\lambda Z}]$ satisfies the differential inequality

$$[\log \varphi_Z(\lambda)]' \leq \frac{e^\lambda}{e^\lambda - 1} \log \varphi_Z(\lambda) \qquad \text{for all } \lambda > 0,$$

and use this to complete the proof.

Exercise 3.16 (Different forms of functional Bernstein) Consider a random variable Z that satisfies a Bernstein tail bound of the form

$$\mathbb{P}[Z \geq \mathbb{E}[Z] + \delta] \leq \exp\left(-\frac{n\delta^2}{c_1\gamma^2 + c_2 b\delta}\right) \qquad \text{for all } \delta \geq 0,$$

where c_1 and c_2 are universal constants.

(a) Show that

$$\mathbb{P}\left[Z \geq \mathbb{E}[Z] + \gamma\sqrt{\frac{c_1 t}{n}} + \frac{c_2 b t}{n}\right] \leq e^{-t} \qquad \text{for all } t \geq 0. \tag{3.95a}$$

(b) If, in addition, $\gamma^2 \leq \sigma^2 + c_3 b \mathbb{E}[Z]$, we have

$$\mathbb{P}\left[Z \geq (1+\epsilon)\mathbb{E}[Z] + \sigma\sqrt{\frac{c_1 t}{n}} + \left(c_2 + \frac{c_1 c_3}{2\epsilon}\right)\frac{bt}{n}\right] \leq e^{-t} \quad \text{for all } t \geq 0 \text{ and } \epsilon > 0. \tag{3.95b}$$

4

Uniform laws of large numbers

The focus of this chapter is a class of results known as uniform laws of large numbers. As suggested by their name, these results represent a strengthening of the usual law of large numbers, which applies to a fixed sequence of random variables, to related laws that hold uniformly over collections of random variables. On one hand, such uniform laws are of theoretical interest in their own right, and represent an entry point to a rich area of probability and statistics known as empirical process theory. On the other hand, uniform laws also play a key role in more applied settings, including understanding the behavior of different types of statistical estimators. The classical versions of uniform laws are of an asymptotic nature, whereas more recent work in the area has emphasized non-asymptotic results. Consistent with the overall goals of this book, this chapter will follow the non-asymptotic route, presenting results that apply to all sample sizes. In order to do so, we make use of the tail bounds and the notion of Rademacher complexity previously introduced in Chapter 2.

4.1 Motivation

We begin with some statistical motivations for deriving laws of large numbers, first for the case of cumulative distribution functions and then for more general function classes.

4.1.1 Uniform convergence of cumulative distribution functions

The law of any scalar random variable X can be fully specified by its cumulative distribution function (CDF), whose value at any point $t \in \mathbb{R}$ is given by $F(t) := \mathbb{P}[X \leq t]$. Now suppose that we are given a collection $\{X_i\}_{i=1}^n$ of n i.i.d. samples, each drawn according to the law specified by F. A natural estimate of F is the *empirical CDF* given by

$$\widehat{F}_n(t) := \frac{1}{n} \sum_{i=1}^{n} \mathbb{I}_{(-\infty,t]}[X_i], \tag{4.1}$$

where $\mathbb{I}_{(-\infty,t]}[x]$ is a $\{0,1\}$-valued indicator function for the event $\{x \leq t\}$. Since the population CDF can be written as $F(t) = \mathbb{E}[\mathbb{I}_{(-\infty,t]}[X]]$, the empirical CDF is an unbiased estimate.

Figure 4.1 provides some illustrations of empirical CDFs for the uniform distribution on the interval $[0,1]$ for two different sample sizes. Note that $\widehat{F}_n$ is a random function, with the value $\widehat{F}_n(t)$ corresponding to the fraction of samples that lie in the interval $(-\infty, t]$. As the sample size n grows, we see that $\widehat{F}_n$ approaches F—compare the plot for $n = 10$ in Figure 4.1(a) to that for $n = 100$ in Figure 4.1(b). It is easy to see that $\widehat{F}_n$ converges to F in

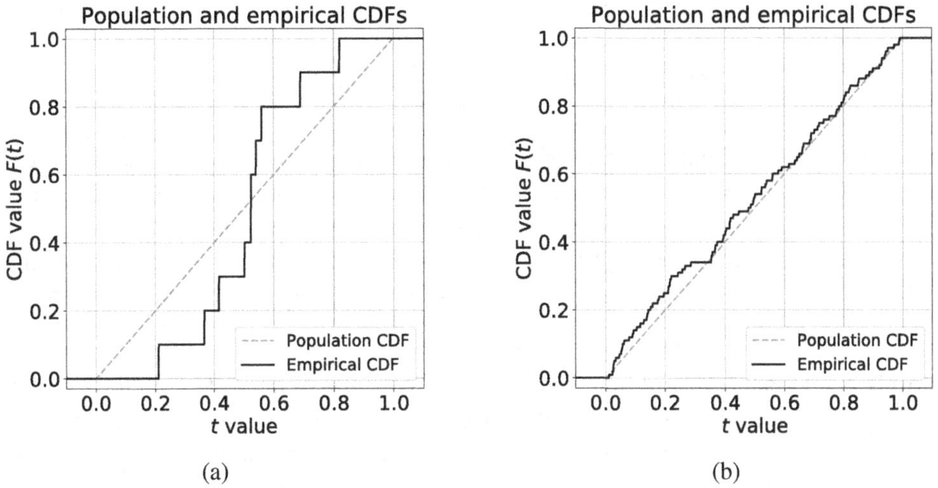

Figure 4.1 Plots of population and empirical CDF functions for the uniform distribution on $[0, 1]$. (a) Empirical CDF based on $n = 10$ samples. (b) Empirical CDF based on $n = 100$ samples.

a pointwise sense. Indeed, for any fixed $t \in \mathbb{R}$, the random variable $\widehat{F}_n(t)$ has mean $F(t)$, and moments of all orders, so that the strong law of large numbers implies that $\widehat{F}_n(t) \xrightarrow{a.s.} F(t)$. A natural goal is to strengthen this pointwise convergence to a form of uniform convergence.

Why are uniform convergence results interesting and important? In statistical settings, a typical use of the empirical CDF is to construct estimators of various quantities associated with the population CDF. Many such estimation problems can be formulated in a terms of functional γ that maps any CDF F to a real number $\gamma(F)$—that is, $F \mapsto \gamma(F)$. Given a set of samples distributed according to F, the *plug-in principle* suggests replacing the unknown F with the empirical CDF $\widehat{F}_n$, thereby obtaining $\gamma(\widehat{F}_n)$ as an estimate of $\gamma(F)$. Let us illustrate this procedure via some examples.

Example 4.1 (Expectation functionals) Given some integrable function g, we may define the *expectation functional* γ_g via

$$\gamma_g(F) := \int g(x)\, dF(x). \tag{4.2}$$

For instance, for the function $g(x) = x$, the functional γ_g maps F to $\mathbb{E}[X]$, where X is a random variable with CDF F. For any g, the plug-in estimate is given by $\gamma_g(\widehat{F}_n) = \frac{1}{n}\sum_{i=1}^n g(X_i)$, corresponding to the sample mean of $g(X)$. In the special case $g(x) = x$, we recover the usual sample mean $\frac{1}{n}\sum_{i=1}^n X_i$ as an estimate for the mean $\mu = \mathbb{E}[X]$. A similar interpretation applies to other choices of the underlying function g. ♣

Example 4.2 (Quantile functionals) For any $\alpha \in [0, 1]$, the *quantile functional* Q_α is given by

$$Q_\alpha(F) := \inf\{t \in \mathbb{R} \mid F(t) \geq \alpha\}. \tag{4.3}$$

The median corresponds to the special case $\alpha = 0.5$. The plug-in estimate is given by

$$Q_\alpha(\widehat{F}_n) := \inf\left\{ t \in \mathbb{R} \;\middle|\; \frac{1}{n}\sum_{i=1}^{n} \mathbb{I}_{(-\infty,t]}[X_i] \geq \alpha \right\}, \tag{4.4}$$

and corresponds to estimating the αth quantile of the distribution by the αth sample quantile. In the special case $\alpha = 0.5$, this estimate corresponds to the sample median. Again, it is of interest to determine in what sense (if any) the random variable $Q_\alpha(\widehat{F}_n)$ approaches $Q_\alpha(F)$ as n becomes large. In this case, $Q_\alpha(\widehat{F}_n)$ is a fairly complicated, nonlinear function of all the variables, so that this convergence does not follow immediately by a classical result such as the law of large numbers. ♣

Example 4.3 (Goodness-of-fit functionals) It is frequently of interest to test the hypothesis of whether or not a given set of data has been drawn from a known distribution F_0. For instance, we might be interested in assessing departures from uniformity, in which case F_0 would be a uniform distribution on some interval, or departures from Gaussianity, in which case F_0 would specify a Gaussian with a fixed mean and variance. Such tests can be performed using functionals that measure the distance between F and the target CDF F_0, including the sup-norm distance $\|F - F_0\|_\infty$, or other distances such as the Cramér–von Mises criterion based on the functional $\gamma(F) := \int_{-\infty}^{\infty} [F(x) - F_0(x)]^2 \, dF_0(x)$. ♣

For any plug-in estimator $\gamma(\widehat{F}_n)$, an important question is to understand when it is consistent—that is, when does $\gamma(\widehat{F}_n)$ converge to $\gamma(F)$ in probability (or almost surely)? This question can be addressed in a unified manner for many functionals by defining a notion of continuity. Given a pair of CDFs F and G, let us measure the distance between them using the sup-norm

$$\|G - F\|_\infty := \sup_{t \in \mathbb{R}} |G(t) - F(t)|. \tag{4.5}$$

We can then define the continuity of a functional γ with respect to this norm: more precisely, we say that the functional γ is *continuous at F in the sup-norm* if, for all $\epsilon > 0$, there exists a $\delta > 0$ such that $\|G - F\|_\infty \leq \delta$ implies that $|\gamma(G) - \gamma(F)| \leq \epsilon$.

As we explore in Exercise 4.1, this notion is useful, because for any continuous functional, it reduces the consistency question for the plug-in estimator $\gamma(\widehat{F}_n)$ to the issue of whether or not the random variable $\|\widehat{F}_n - F\|_\infty$ converges to zero. A classical result, known as the Glivenko–Cantelli theorem, addresses the latter question:

Theorem 4.4 (Glivenko–Cantelli) *For any distribution, the empirical CDF $\widehat{F}_n$ is a strongly consistent estimator of the population CDF in the uniform norm, meaning that*

$$\|\widehat{F}_n - F\|_\infty \xrightarrow{a.s.} 0. \tag{4.6}$$

We provide a proof of this claim as a corollary of a more general result to follow (see Theorem 4.10). For statistical applications, an important consequence of Theorem 4.4 is that the plug-in estimate $\gamma(\widehat{F}_n)$ is almost surely consistent as an estimator of $\gamma(F)$ for any functional γ that is continuous with respect to the sup-norm. See Exercise 4.1 for further exploration of this connection.

4.1.2 Uniform laws for more general function classes

We now turn to more general consideration of uniform laws of large numbers. Let $\mathscr{F}$ be a class of integrable real-valued functions with domain $\mathcal{X}$, and let $\{X_i\}_{i=1}^n$ be a collection of i.i.d. samples from some distribution $\mathbb{P}$ over $\mathcal{X}$. Consider the random variable

$$\|\mathbb{P}_n - \mathbb{P}\|_{\mathscr{F}} := \sup_{f \in \mathscr{F}} \left| \frac{1}{n} \sum_{i=1}^n f(X_i) - \mathbb{E}[f(X)] \right|, \tag{4.7}$$

which measures the absolute deviation between the sample average $\frac{1}{n}\sum_{i=1}^n f(X_i)$ and the population average $\mathbb{E}[f(X)]$, uniformly over the class $\mathscr{F}$. Note that there can be measurability concerns associated with the definition (4.7); see the bibliographic section for discussion of different ways in which to resolve them.

Definition 4.5 We say that $\mathscr{F}$ is a *Glivenko–Cantelli* class for $\mathbb{P}$ if $\|\mathbb{P}_n - \mathbb{P}\|_{\mathscr{F}}$ converges to zero in probability as $n \to \infty$.

This notion can also be defined in a stronger sense, requiring almost sure convergence of $\|\mathbb{P}_n - \mathbb{P}\|_{\mathscr{F}}$, in which case we say that $\mathscr{F}$ satisfies a *strong Glivenko–Cantelli law.* The classical result on the empirical CDF (Theorem 4.4) can be reformulated as a particular case of this notion:

Example 4.6 (Empirical CDFs and indicator functions) Consider the function class

$$\mathscr{F} = \{\mathbb{I}_{(-\infty,t]}(\cdot) \mid t \in \mathbb{R}\}, \tag{4.8}$$

where $\mathbb{I}_{(-\infty,t]}$ is the $\{0,1\}$-valued indicator function of the interval $(-\infty, t]$. For each fixed $t \in \mathbb{R}$, we have the equality $\mathbb{E}[\mathbb{I}_{(-\infty,t]}(X)] = \mathbb{P}[X \le t] = F(t)$, so that the classical Glivenko–Cantelli theorem is equivalent to a strong uniform law for the class (4.8). ♣

Not all classes of functions are Glivenko–Cantelli, as illustrated by the following example.

Example 4.7 (Failure of uniform law) Let $\mathcal{S}$ be the class of all subsets S of $[0,1]$ such that the subset S has a finite number of elements, and consider the function class $\mathscr{F}_{\mathcal{S}} = \{\mathbb{I}_S(\cdot) \mid S \in \mathcal{S}\}$ of ($\{0\text{-}1\}$-valued) indicator functions of such sets. Suppose that samples X_i are drawn from some distribution over $[0,1]$ that has no atoms (i.e., $\mathbb{P}(\{x\}) = 0$ for all $x \in [0,1]$); this class includes any distribution that has a density with respect to Lebesgue measure. For any such distribution, we are guaranteed that $\mathbb{P}[S] = 0$ for all $S \in \mathcal{S}$. On the other hand, for any positive integer $n \in \mathbb{N}$, the discrete set $\{X_1, \ldots, X_n\}$ belongs to $\mathcal{S}$, and moreover, by definition of the empirical distribution, we have $\mathbb{P}_n[X_1^n] = 1$. Putting together the pieces, we conclude that

$$\sup_{S \in \mathcal{S}} |\mathbb{P}_n[S] - \mathbb{P}[S]| = 1 - 0 = 1, \tag{4.9}$$

so that the function class $\mathscr{F}_{\mathcal{S}}$ is *not* a Glivenko–Cantelli class for $\mathbb{P}$. ♣

We have seen that the classical Glivenko–Cantelli law—which guarantees convergence of a special case of the variable $\|\mathbb{P}_n - \mathbb{P}\|_{\mathscr{F}}$—is of interest in analyzing estimators based on "plugging in" the empirical CDF. It is natural to ask in what other statistical contexts do these quantities arise? In fact, variables of the form $\|\mathbb{P}_n - \mathbb{P}\|_{\mathscr{F}}$ are ubiquitous throughout statistics—in particular, they lie at the heart of methods based on empirical risk minimization. In order to describe this notion more concretely, let us consider an indexed family of probability distributions $\{\mathbb{P}_\theta \mid \theta \in \Omega\}$, and suppose that we are given n samples $\{X_i\}_{i=1}^n$, each sample lying in some space $\mathcal{X}$. Suppose that the samples are drawn i.i.d. according to a distribution $\mathbb{P}_{\theta^*}$, for some fixed but unknown $\theta^* \in \Omega$. Here the index θ^* could lie within a finite-dimensional space, such as $\Omega = \mathbb{R}^d$ in a vector estimation problem, or could lie within some function class $\Omega = \mathscr{G}$, in which case the problem is of the nonparametric variety.

In either case, a standard decision-theoretic approach to estimating θ^* is based on minimizing a cost function of the form $\theta \mapsto \mathcal{L}_\theta(X)$, which measures the "fit" between a parameter $\theta \in \Omega$ and the sample $X \in \mathcal{X}$. Given the collection of n samples $\{X_i\}_{i=1}^n$, the principle of empirical risk minimization is based on the objective function

$$\widehat{R}_n(\theta, \theta^*) := \frac{1}{n} \sum_{i=1}^n \mathcal{L}_\theta(X_i).$$

This quantity is known as the *empirical risk*, since it is defined by the samples X_1^n, and our notation reflects the fact that these samples depend—in turn—on the unknown distribution $\mathbb{P}_{\theta^*}$. This empirical risk should be contrasted with the *population risk*,

$$R(\theta, \theta^*) := \mathbb{E}_{\theta^*}[\mathcal{L}_\theta(X)],$$

where the expectation $\mathbb{E}_{\theta^*}$ is taken over a sample $X \sim \mathbb{P}_{\theta^*}$.

In practice, one minimizes the empirical risk over some subset Ω_0 of the full space Ω, thereby obtaining some estimate $\widehat{\theta}$. The statistical question is how to bound the *excess risk*, measured in terms of the population quantities—namely the difference

$$E(\widehat{\theta}, \theta^*) := R(\widehat{\theta}, \theta^*) - \inf_{\theta \in \Omega_0} R(\theta, \theta^*).$$

Let us consider some examples to illustrate.

Example 4.8 (Maximum likelihood) Consider a parameterized family of distributions—say $\{\mathbb{P}_\theta, \theta \in \Omega\}$—each with a strictly positive density p_θ defined with respect to a common underlying measure. Now suppose that we are given n i.i.d. samples from an unknown distribution $\mathbb{P}_{\theta^*}$, and we would like to estimate the unknown parameter θ^*. In order to do so, we consider the cost function

$$\mathcal{L}_\theta(x) := \log\left[\frac{p_{\theta^*}(x)}{p_\theta(x)}\right].$$

The term $p_{\theta^*}(x)$, which we have included for later theoretical convenience, has no effect on the minimization over θ. Indeed, the maximum likelihood estimate is obtained by minimizing the empirical risk defined by this cost function—that is

$$\widehat{\theta} \in \arg\min_{\theta \in \Omega_0} \underbrace{\left\{ \frac{1}{n} \sum_{i=1}^{n} \log \frac{p_{\theta^*}(X_i)}{p_\theta(X_i)} \right\}}_{\widehat{R}_n(\theta,\theta^*)} = \arg\min_{\theta \in \Omega_0} \left\{ \frac{1}{n} \sum_{i=1}^{n} \log \frac{1}{p_\theta(X_i)} \right\}.$$

The population risk is given by $R(\theta, \theta^*) = \mathbb{E}_{\theta^*}\big[\log \frac{p_{\theta^*}(X)}{p_\theta(X)}\big]$, a quantity known as the *Kullback–Leibler divergence* between p_{θ^*} and p_θ. In the special case that $\theta^* \in \Omega_0$, the excess risk is simply the Kullback–Leibler divergence between the true density p_{θ^*} and the fitted model $p_{\widehat{\theta}}$. See Exercise 4.3 for some concrete examples. ♣

Example 4.9 (Binary classification) Suppose that we observe n pairs of samples, each of the form $(X_i, Y_i) \in \mathbb{R}^d \times \{-1, +1\}$, where the vector X_i corresponds to a set of d predictors or features, and the binary variable Y_i corresponds to a label. We can view such data as being generated by some distribution $\mathbb{P}_X$ over the features, and a conditional distribution $\mathbb{P}_{Y|X}$. Since Y takes binary values, the conditional distribution is fully specified by the likelihood ratio $\psi(x) = \frac{\mathbb{P}[Y=+1 \mid X=x]}{\mathbb{P}[Y=-1 \mid X=x]}$.

The goal of binary classification is to estimate a function $f\colon \mathbb{R}^d \to \{-1, +1\}$ that minimizes the probability of misclassification $\mathbb{P}[f(X) \neq Y]$, for an independently drawn pair (X, Y). Note that this probability of error corresponds to the population risk for the cost function

$$\mathcal{L}_f(X, Y) := \begin{cases} 1 & \text{if } f(X) \neq Y, \\ 0 & \text{otherwise.} \end{cases} \tag{4.10}$$

A function that minimizes this probability of error is known as a *Bayes classifier* f^*; in the special case of equally probable classes—that is, when $\mathbb{P}[Y = +1] = \mathbb{P}[Y = -1] = \frac{1}{2}$—a Bayes classifier is given by

$$f^*(x) = \begin{cases} +1 & \text{if } \psi(x) \geq 1, \\ -1 & \text{otherwise.} \end{cases}$$

Since the likelihood ratio ψ (and hence f^*) is unknown, a natural approach to approximating the Bayes rule is based on choosing $\widehat{f}$ to minimize the empirical risk

$$\widehat{R}_n(f, f^*) := \frac{1}{n} \sum_{i=1}^{n} \underbrace{\mathbb{I}[f(X_i) \neq Y_i]}_{\mathcal{L}_f(X_i, Y_i)},$$

corresponding to the fraction of training samples that are misclassified. Typically, the minimization over f is restricted to some subset of all possible decision rules. See Chapter 14 for some further discussion of how to analyze such methods for binary classification. ♣

Returning to the main thread, our goal is to develop methods for controlling the excess risk. For simplicity, let us assume[1] that there exists some $\theta_0 \in \Omega_0$ such that $R(\theta_0, \theta^*) =$

[1] If the infimum is not achieved, then we choose an element θ_0 for which this equality holds up to some arbitrarily small tolerance $\epsilon > 0$, and the analysis to follow holds up to this tolerance.

$\inf_{\theta \in \Omega_0} R(\theta, \theta^*)$. With this notation, the excess risk can be decomposed as

$$E(\widehat{\theta}, \theta^*) = \underbrace{\{R(\widehat{\theta}, \theta^*) - \widehat{R}_n(\widehat{\theta}, \theta^*)\}}_{T_1} + \underbrace{\{\widehat{R}_n(\widehat{\theta}, \theta^*) - \widehat{R}_n(\theta_0, \theta^*)\}}_{T_2 \leq 0} + \underbrace{\{\widehat{R}_n(\theta_0, \theta^*) - R(\theta_0, \theta^*)\}}_{T_3}.$$

Note that T_2 is non-positive, since $\widehat{\theta}$ minimizes the empirical risk over Ω_0.

The third term T_3 can be dealt with in a relatively straightforward manner, because θ_0 is an unknown but non-random quantity. Indeed, recalling the definition of the empirical risk, we have

$$T_3 = \left[\frac{1}{n} \sum_{i=1}^{n} \mathcal{L}_{\theta_0}(X_i)\right] - \mathbb{E}_X[\mathcal{L}_{\theta_0}(X)],$$

corresponding to the deviation of a sample mean from its expectation for the random variable $\mathcal{L}_{\theta_0}(X)$. This quantity can be controlled using the techniques introduced in Chapter 2—for instance, via the Hoeffding bound when the samples are independent and the cost function is bounded.

Finally, returning to the first term, it can be written in a similar way, namely as the difference

$$T_1 = \mathbb{E}_X[\mathcal{L}_{\widehat{\theta}}(X)] - \left[\frac{1}{n} \sum_{i=1}^{n} \mathcal{L}_{\widehat{\theta}}(X_i)\right].$$

This quantity is more challenging to control, because the parameter $\widehat{\theta}$—in contrast to the deterministic quantity θ_0—is now random, and moreover depends on the samples $\{X_i\}_{i=1}^n$, since it was obtained by minimizing the empirical risk. For this reason, controlling the first term requires a stronger result, such as a uniform law of large numbers over the cost function class $\mathfrak{L}(\Omega_0) := \{x \mapsto \mathcal{L}_\theta(x), \theta \in \Omega_0\}$. With this notation, we have

$$T_1 \leq \sup_{\theta \in \Omega_0} \left|\frac{1}{n} \sum_{i=1}^{n} \mathcal{L}_\theta(X_i) - \mathbb{E}_X[\mathcal{L}_\theta(X)]\right| = \|\mathbb{P}_n - \mathbb{P}\|_{\mathfrak{L}(\Omega_0)}.$$

Since T_3 is also dominated by this same quantity, we conclude that the excess risk is at most $2\|\mathbb{P}_n - \mathbb{P}\|_{\mathfrak{L}(\Omega_0)}$. This derivation demonstrates that the central challenge in analyzing estimators based on empirical risk minimization is to establish a uniform law of large numbers for the loss class $\mathfrak{L}(\Omega_0)$. We explore various concrete examples of this procedure in the exercises.

4.2 A uniform law via Rademacher complexity

Having developed various motivations for studying uniform laws, let us now turn to the technical details of deriving such results. An important quantity that underlies the study of uniform laws is the *Rademacher complexity* of the function class $\mathscr{F}$. For any fixed collection $x_1^n := (x_1, \ldots, x_n)$ of points, consider the subset of $\mathbb{R}^n$ given by

$$\mathscr{F}(x_1^n) := \{(f(x_1), \ldots, f(x_n)) \mid f \in \mathscr{F}\}. \tag{4.11}$$

The set $\mathscr{F}(x_1^n)$ corresponds to all those vectors in $\mathbb{R}^n$ that can be realized by applying a function $f \in \mathscr{F}$ to the collection $(x_1, \ldots, x_n)$, and the *empirical Rademacher complexity* is given by

$$\mathcal{R}(\mathscr{F}(x_1^n)/n) := \mathbb{E}_\varepsilon\left[\sup_{f\in\mathscr{F}}\left|\frac{1}{n}\sum_{i=1}^n \varepsilon_i f(x_i)\right|\right]. \tag{4.12}$$

Note that this definition coincides with our earlier definition of the Rademacher complexity of a set (see Example 2.25).

Given a collection $X_1^n := \{X_i\}_{i=1}^n$ of random samples, then the empirical Rademacher complexity $\mathcal{R}(\mathscr{F}(X_1^n)/n)$ is a random variable. Taking its expectation yields the *Rademacher complexity of the function class* $\mathscr{F}$—namely, the deterministic quantity

$$\mathcal{R}_n(\mathscr{F}) := \mathbb{E}_X[\mathcal{R}(\mathscr{F}(X_1^n)/n)] = \mathbb{E}_{X,\varepsilon}\left[\sup_{f\in\mathscr{F}}\left|\frac{1}{n}\sum_{i=1}^n \varepsilon_i f(X_i)\right|\right]. \tag{4.13}$$

Note that the Rademacher complexity is the average of the maximum correlation between the vector $(f(X_1),\ldots,f(X_n))$ and the "noise vector" $(\varepsilon_1,\ldots,\varepsilon_n)$, where the maximum is taken over all functions $f \in \mathscr{F}$. The intuition is a natural one: a function class is extremely large—and, in fact, "too large" for statistical purposes—if we can always find a function that has a high correlation with a randomly drawn noise vector. Conversely, when the Rademacher complexity decays as a function of sample size, then it is impossible to find a function that correlates very highly in expectation with a randomly drawn noise vector.

We now make precise the connection between Rademacher complexity and the Glivenko–Cantelli property, in particular by showing that, for any bounded function class $\mathscr{F}$, the condition $\mathcal{R}_n(\mathscr{F}) = o(1)$ implies the Glivenko–Cantelli property. More precisely, we prove a non-asymptotic statement, in terms of a tail bound for the probability that the random variable $\|\mathbb{P}_n - \mathbb{P}\|_{\mathscr{F}}$ deviates substantially above a multiple of the Rademacher complexity. It applies to a function class $\mathscr{F}$ that is b-uniformly bounded, meaning that $\|f\|_\infty \leq b$ for all $f \in \mathscr{F}$.

Theorem 4.10 *For any b-uniformly bounded class of functions $\mathscr{F}$, any positive integer $n \geq 1$ and any scalar $\delta \geq 0$, we have*

$$\|\mathbb{P}_n - \mathbb{P}\|_{\mathscr{F}} \leq 2\mathcal{R}_n(\mathscr{F}) + \delta \tag{4.14}$$

with $\mathbb{P}$-probability at least $1 - \exp\left(-\frac{n\delta^2}{2b^2}\right)$. Consequently, as long as $\mathcal{R}_n(\mathscr{F}) = o(1)$, we have $\|\mathbb{P}_n - \mathbb{P}\|_{\mathscr{F}} \xrightarrow{a.s.} 0$.

In order for Theorem 4.10 to be useful, we need to obtain upper bounds on the Rademacher complexity. There are a variety of methods for doing so, ranging from direct calculations to alternative complexity measures. In Section 4.3, we develop some techniques for upper bounding the Rademacher complexity for indicator functions of half-intervals, as required for the classical Glivenko–Cantelli theorem (see Example 4.6); we also discuss the notion of Vapnik–Chervonenkis dimension, which can be used to upper bound the Rademacher complexity for other function classes. In Chapter 5, we introduce more advanced techniques based on metric entropy and chaining for controlling Rademacher complexity and related sub-Gaussian processes. In the meantime, let us turn to the proof of Theorem 4.10.

Proof We first note that if $\mathcal{R}_n(\mathscr{F}) = o(1)$, then the almost-sure convergence follows from the tail bound (4.14) and the Borel–Cantelli lemma. Accordingly, the remainder of the argument is devoted to proving the tail bound (4.14).

Concentration around mean: We first claim that, when $\mathscr{F}$ is uniformly bounded, then the random variable $\|\mathbb{P}_n - \mathbb{P}\|_{\mathscr{F}}$ is sharply concentrated around its mean. In order to simplify notation, it is convenient to define the recentered functions $\bar{f}(x) := f(x) - \mathbb{E}[f(X)]$, and to write $\|\mathbb{P}_n - \mathbb{P}\|_{\mathscr{F}} = \sup_{f\in\mathscr{F}} \left|\frac{1}{n}\sum_{i=1}^n \bar{f}(X_i)\right|$. Thinking of the samples as fixed for the moment, consider the function

$$G(x_1, \ldots, x_n) := \sup_{f\in\mathscr{F}} \left|\frac{1}{n}\sum_{i=1}^n \bar{f}(x_i)\right|.$$

We claim that G satisfies the Lipschitz property required to apply the bounded differences method (recall Corollary 2.21). Since the function G is invariant to permutation of its coordinates, it suffices to bound the difference when the first coordinate x_1 is perturbed. Accordingly, we define the vector $y \in \mathbb{R}^n$ with $y_i = x_i$ for all $i \neq 1$, and seek to bound the difference $|G(x) - G(y)|$. For any function $\bar{f} = f - \mathbb{E}[f]$, we have

$$\begin{aligned}
\left|\frac{1}{n}\sum_{i=1}^n \bar{f}(x_i)\right| - \sup_{h\in\mathscr{F}} \left|\frac{1}{n}\sum_{i=1}^n \bar{h}(y_i)\right| &\leq \left|\frac{1}{n}\sum_{i=1}^n \bar{f}(x_i)\right| - \left|\frac{1}{n}\sum_{i=1}^n \bar{f}(y_i)\right| \\
&\leq \frac{1}{n}\left|\bar{f}(x_1) - \bar{f}(y_1)\right| \\
&\leq \frac{2b}{n},
\end{aligned} \tag{4.15}$$

where the final inequality uses the fact that

$$|\bar{f}(x_1) - \bar{f}(y_1)| = |f(x_1) - f(y_1)| \leq 2b,$$

which follows from the uniform boundedness condition $\|f\|_\infty \leq b$. Since the inequality (4.15) holds for any function f, we may take the supremum over $f \in \mathscr{F}$ on both sides; doing so yields the inequality $G(x) - G(y) \leq \frac{2b}{n}$. Since the same argument may be applied with the roles of x and y reversed, we conclude that $|G(x) - G(y)| \leq \frac{2b}{n}$. Therefore, by the bounded differences method (see Corollary 2.21), we have

$$\|\mathbb{P}_n - \mathbb{P}\|_{\mathscr{F}} - \mathbb{E}[\|\mathbb{P}_n - \mathbb{P}\|_{\mathscr{F}}] \leq t \qquad \text{with } \mathbb{P}\text{-prob. at least } 1 - \exp\left(-\tfrac{nt^2}{2b^2}\right), \tag{4.16}$$

valid for all $t \geq 0$.

Upper bound on mean: It remains to show that $\mathbb{E}[\|\mathbb{P}_n - \mathbb{P}\|_{\mathscr{F}}]$ is upper bounded by $2\mathcal{R}_n(\mathscr{F})$, and we do so using a classical symmetrization argument. Letting $(Y_1, \ldots, Y_n)$ be a second i.i.d. sequence, independent of $(X_1, \ldots, X_n)$, we have

$$
\begin{aligned}
\mathbb{E}[\|\mathbb{P}_n - \mathbb{P}\|_{\mathscr{F}}] &= \mathbb{E}_X\left[\sup_{f\in\mathscr{F}}\left|\frac{1}{n}\sum_{i=1}^{n}\{f(X_i) - \mathbb{E}_{Y_i}[f(Y_i)]\}\right|\right] \\
&= \mathbb{E}_X\left[\sup_{f\in\mathscr{F}}\left|\mathbb{E}_Y[\frac{1}{n}\sum_{i=1}^{n}\{f(X_i) - f(Y_i)\}]\right|\right] \\
&\stackrel{(i)}{\leq} \mathbb{E}_{X,Y}\left[\sup_{f\in\mathscr{F}}\left|\frac{1}{n}\sum_{i=1}^{n}\{f(X_i) - f(Y_i)\}\right|\right],
\end{aligned} \tag{4.17}
$$

where the upper bound (i) follows from the calculation of Exercise 4.4.

Now let $(\varepsilon_1, \ldots, \varepsilon_n)$ be an i.i.d. sequence of Rademacher variables, independent of X and Y. Given our independence assumptions, for any function $f \in \mathscr{F}$, the random vector with components $\varepsilon_i(f(X_i) - f(Y_i))$ has the same joint distribution as the random vector with components $f(X_i) - f(Y_i)$, whence

$$
\begin{aligned}
\mathbb{E}_{X,Y}\left[\sup_{f\in\mathscr{F}}\left|\frac{1}{n}\sum_{i=1}^{n}\{f(X_i) - f(Y_i)\}\right|\right] &= \mathbb{E}_{X,Y,\varepsilon}\left[\sup_{f\in\mathscr{F}}\left|\frac{1}{n}\sum_{i=1}^{n}\varepsilon_i(f(X_i) - f(Y_i))\right|\right] \\
&\leq 2\,\mathbb{E}_{X,\varepsilon}\left[\sup_{f\in\mathscr{F}}\left|\frac{1}{n}\sum_{i=1}^{n}\varepsilon_i f(X_i)\right|\right] = 2\mathcal{R}_n(\mathscr{F}).
\end{aligned} \tag{4.18}
$$

Combining the upper bound (4.18) with the tail bound (4.16) yields the claim. □

4.2.1 Necessary conditions with Rademacher complexity

The proof of Theorem 4.10 illustrates an important technique known as symmetrization, which relates the random variable $\|\mathbb{P}_n - \mathbb{P}\|_{\mathscr{F}}$ to its symmetrized version

$$
\|\mathbb{S}_n\|_{\mathscr{F}} := \sup_{f\in\mathscr{F}}\left|\frac{1}{n}\sum_{i=1}^{n}\varepsilon_i f(X_i)\right|. \tag{4.19}
$$

Note that the expectation of $\|\mathbb{S}_n\|_{\mathscr{F}}$ corresponds to the Rademacher complexity, which plays a central role in Theorem 4.10. It is natural to wonder whether much was lost in moving from the variable $\|\mathbb{P}_n - \mathbb{P}\|_{\mathscr{F}}$ to its symmetrized version. The following "sandwich" result relates these quantities.

Proposition 4.11 *For any convex non-decreasing function $\Phi\colon \mathbb{R} \to \mathbb{R}$, we have*

$$
\mathbb{E}_{X,\varepsilon}[\Phi(\tfrac{1}{2}\|\mathbb{S}_n\|_{\bar{\mathscr{F}}})] \stackrel{(a)}{\leq} \mathbb{E}_X[\Phi(\|\mathbb{P}_n - \mathbb{P}\|_{\mathscr{F}})] \stackrel{(b)}{\leq} \mathbb{E}_{X,\varepsilon}[\Phi(2\|\mathbb{S}_n\|_{\mathscr{F}})], \tag{4.20}
$$

where $\bar{\mathscr{F}} = \{f - \mathbb{E}[f], f \in \mathscr{F}\}$ is the recentered function class.

When applied with the convex non-decreasing function $\Phi(t) = t$, Proposition 4.11 yields the inequalities

$$
\tfrac{1}{2}\mathbb{E}_{X,\varepsilon}\|\mathbb{S}_n\|_{\bar{\mathscr{F}}} \leq \mathbb{E}_X[\|\mathbb{P}_n - \mathbb{P}\|_{\mathscr{F}}] \leq 2\mathbb{E}_{X,\varepsilon}\|\mathbb{S}_n\|_{\mathscr{F}}, \tag{4.21}
$$

with the only differences being the constant pre-factors, and the use of $\mathscr{F}$ in the upper bound, and the recentered class $\overline{\mathscr{F}}$ in the lower bound.

Other choices of interest include $\Phi(t) = e^{\lambda t}$ for some $\lambda > 0$, which can be used to control the moment generating function.

Proof Beginning with bound (b), we have

$$\begin{aligned}
\mathbb{E}_X[\Phi(\|\mathbb{P}_n - \mathbb{P}\|_{\mathscr{F}})] &= \mathbb{E}_X\left[\Phi\left(\sup_{f\in\mathscr{F}}\left|\frac{1}{n}\sum_{i=1}^n f(X_i) - \mathbb{E}_Y[f(Y_i)]\right|\right)\right] \\
&\overset{\text{(i)}}{\leq} \mathbb{E}_{X,Y}\left[\Phi\left(\sup_{f\in\mathscr{F}}\left|\frac{1}{n}\sum_{i=1}^n f(X_i) - f(Y_i)\right|\right)\right] \\
&\overset{\text{(ii)}}{=} \underbrace{\mathbb{E}_{X,Y,\varepsilon}\left[\Phi\left(\sup_{f\in\mathscr{F}}\left|\frac{1}{n}\sum_{i=1}^n \varepsilon_i\{f(X_i) - f(Y_i)\}\right|\right)\right]}_{:=T_1},
\end{aligned}$$

where inequality (i) follows from Exercise 4.4, using the convexity and non-decreasing properties of Φ, and equality (ii) follows since the random vector with components $\varepsilon_i(f(X_i) - f(Y_i))$ has the same joint distribution as the random vector with components $f(X_i) - f(Y_i)$. By the triangle inequality, we have

$$\begin{aligned}
T_1 &\leq \mathbb{E}_{X,Y,\varepsilon}\left[\Phi\left(\sup_{f\in\mathscr{F}}\left|\frac{1}{n}\sum_{i=1}^n \varepsilon_i f(X_i)\right| + \left|\frac{1}{n}\sum_{i=1}^n \varepsilon_i f(Y_i)\right|\right)\right] \\
&\overset{\text{(iii)}}{\leq} \frac{1}{2}\mathbb{E}_{X,\varepsilon}\left[\Phi\left(2\sup_{f\in\mathscr{F}}\left|\frac{1}{n}\sum_{i=1}^n \varepsilon_i f(X_i)\right|\right)\right] + \frac{1}{2}\mathbb{E}_{Y,\varepsilon}\left[\Phi\left(2\sup_{f\in\mathscr{F}}\left|\frac{1}{n}\sum_{i=1}^n \varepsilon_i f(Y_i)\right|\right)\right] \\
&\overset{\text{(iv)}}{=} \mathbb{E}_{X,\varepsilon}\left[\Phi\left(2\sup_{f\in\mathscr{F}}\left|\frac{1}{n}\sum_{i=1}^n \varepsilon_i f(X_i)\right|\right)\right],
\end{aligned}$$

where step (iii) follows from Jensen's inequality and the convexity of Φ, and step (iv) follows since X and Y are i.i.d. samples.

Turning to the bound (a), we have

$$\begin{aligned}
\mathbb{E}_{X,\varepsilon}[\Phi(\tfrac{1}{2}\|\mathbb{S}_n\|_{\overline{\mathscr{F}}})] &= \mathbb{E}_{X,\varepsilon}\left[\Phi\left(\frac{1}{2}\sup_{f\in\mathscr{F}}\left|\frac{1}{n}\sum_{i=1}^n \varepsilon_i\{f(X_i) - \mathbb{E}_{Y_i}[f(Y_i)]\}\right|\right)\right] \\
&\overset{\text{(i)}}{\leq} \mathbb{E}_{X,Y,\varepsilon}\left[\Phi\left(\frac{1}{2}\sup_{f\in\mathscr{F}}\left|\frac{1}{n}\sum_{i=1}^n \varepsilon_i\{f(X_i) - f(Y_i)\}\right|\right)\right] \\
&\overset{\text{(ii)}}{=} \mathbb{E}_{X,Y}\left[\Phi\left(\frac{1}{2}\sup_{f\in\mathscr{F}}\left|\frac{1}{n}\sum_{i=1}^n \{f(X_i) - f(Y_i)\}\right|\right)\right],
\end{aligned}$$

where inequality (i) follows from Jensen's inequality and the convexity of Φ; and equality (ii) follows since for each $i = 1, 2, \ldots, n$ and $f \in \mathscr{F}$, the variables $\varepsilon_i\{f(X_i) - f(Y_i)\}$ and $f(X_i) - f(Y_i)$ have the same distribution.

Now focusing on the quantity $T_2 := \frac{1}{2}\sup_{f\in\mathscr{F}}\left|\frac{1}{n}\sum_{i=1}^n \{f(X_i) - f(Y_i)\}\right|$, we add and subtract a term of the form $\mathbb{E}[f]$, and then apply the triangle inequality, thereby obtaining the upper bound

$$T_2 \leq \frac{1}{2} \sup_{f \in \mathscr{F}} \left| \frac{1}{n} \sum_{i=1}^{n} \{f(X_i) - \mathbb{E}[f]\} \right| + \frac{1}{2} \sup_{f \in \mathscr{F}} \left| \frac{1}{n} \sum_{i=1}^{n} \{f(Y_i) - \mathbb{E}[f]\} \right|.$$

Since Φ is convex and non-decreasing, we are guaranteed that

$$\Phi(T_2) \leq \frac{1}{2}\Phi\left(\sup_{f \in \mathscr{F}} \left| \frac{1}{n} \sum_{i=1}^{n} \{f(X_i) - \mathbb{E}[f]\} \right| \right) + \frac{1}{2}\Phi\left(\sup_{f \in \mathscr{F}} \left| \frac{1}{n} \sum_{i=1}^{n} \{f(Y_i) - \mathbb{E}[f]\} \right| \right).$$

The claim follows by taking expectations and using the fact that X and Y are identically distributed. □

A consequence of Proposition 4.11 is that the random variable $\|\mathbb{P}_n - \mathbb{P}\|_{\mathscr{F}}$ can be lower bounded by a multiple of Rademacher complexity, and some fluctuation terms. This fact can be used to prove the following:

Proposition 4.12 *For any b-uniformly bounded function class $\mathscr{F}$, any integer $n \geq 1$ and any scalar $\delta \geq 0$, we have*

$$\|\mathbb{P}_n - \mathbb{P}\|_{\mathscr{F}} \geq \frac{1}{2}\mathcal{R}_n(\mathscr{F}) - \frac{\sup_{f \in \mathscr{F}} |\mathbb{E}[f]|}{2\sqrt{n}} - \delta \tag{4.22}$$

with $\mathbb{P}$-probability at least $1 - e^{-\frac{n\delta^2}{2b^2}}$.

We leave the proof of this result for the reader (see Exercise 4.5). As a consequence, if the Rademacher complexity $\mathcal{R}_n(\mathscr{F})$ remains bounded away from zero, then $\|\mathbb{P}_n - \mathbb{P}\|_{\mathscr{F}}$ cannot converge to zero in probability. We have thus shown that, for a uniformly bounded function class $\mathscr{F}$, the Rademacher complexity provides a necessary and sufficient condition for it to be Glivenko–Cantelli.

4.3 Upper bounds on the Rademacher complexity

Obtaining concrete results using Theorem 4.10 requires methods for upper bounding the Rademacher complexity. There are a variety of such methods, ranging from simple union bound methods (suitable for finite function classes) to more advanced techniques involving the notion of metric entropy and chaining arguments. We explore the latter techniques in Chapter 5 to follow. This section is devoted to more elementary techniques, including those required to prove the classical Glivenko–Cantelli result, and, more generally, those that apply to function classes with polynomial discrimination, as well as associated Vapnik–Chervonenkis classes.

4.3.1 Classes with polynomial discrimination

For a given collection of points $x_1^n = (x_1, \ldots, x_n)$, the "size" of the set $\mathscr{F}(x_1^n)$ provides a sample-dependent measure of the complexity of $\mathscr{F}$. In the simplest case, the set $\mathscr{F}(x_1^n)$ contains only a finite number of vectors for all sample sizes, so that its "size" can be measured via its cardinality. For instance, if $\mathscr{F}$ consists of a family of decision rules taking binary values (as in Example 4.9), then $\mathscr{F}(x_1^n)$ can contain at most 2^n elements. Of interest to us are function classes for which this cardinality grows only as a polynomial function of n, as formalized in the following:

Definition 4.13 (Polynomial discrimination) A class $\mathscr{F}$ of functions with domain $\mathcal{X}$ has polynomial discrimination of order $\nu \geq 1$ if, for each positive integer n and collection $x_1^n = \{x_1, \ldots, x_n\}$ of n points in $\mathcal{X}$, the set $\mathscr{F}(x_1^n)$ has cardinality upper bounded as

$$\text{card}(\mathscr{F}(x_1^n)) \leq (n+1)^\nu. \tag{4.23}$$

The significance of this property is that it provides a straightforward approach to controlling the Rademacher complexity. For any set $\mathbb{S} \subset \mathbb{R}^n$, we use $D(\mathbb{S}) := \sup_{x \in \mathbb{S}} \|x\|_2$ to denote its maximal width in the ℓ_2-norm.

Lemma 4.14 *Suppose that $\mathscr{F}$ has polynomial discrimination of order ν. Then for all positive integers n and any collection of points $x_1^n = (x_1, \ldots, x_n)$,*

$$\underbrace{\mathbb{E}_\varepsilon\left[\sup_{f \in \mathscr{F}} \left|\frac{1}{n}\sum_{i=1}^n \varepsilon_i f(x_i)\right|\right]}_{\mathcal{R}(\mathscr{F}(x_1^n)/n))} \leq 4D(x_1^n)\sqrt{\frac{\nu \log(n+1)}{n}},$$

where $D(x_1^n) := \sup_{f \in \mathscr{F}} \sqrt{\frac{\sum_{i=1}^n f^2(x_i)}{n}}$ is the ℓ_2-radius of the set $\mathscr{F}(x_1^n)/\sqrt{n}$.

We leave the proof of this claim for the reader (see Exercise 4.9).

Although Lemma 4.14 is stated as an upper bound on the empirical Rademacher complexity, it yields as a corollary an upper bound on the Rademacher complexity $\mathcal{R}_n(\mathscr{F}) = \mathbb{E}_X[\mathcal{R}(\mathscr{F}(X_1^n)/n)]$, one which involves the expected ℓ_2-width $\mathbb{E}_{X_1^n}[D(X)]$. An especially simple case is when the function class is b uniformly bounded, so that $D(x_1^n) \leq b$ for all samples. In this case, Lemma 4.14 implies that

$$\mathcal{R}_n(\mathscr{F}) \leq 2b\sqrt{\frac{\nu \log(n+1)}{n}} \quad \text{for all } n \geq 1. \tag{4.24}$$

Combined with Theorem 4.10, we conclude that any bounded function class with polynomial discrimination is Glivenko–Cantelli.

What types of function classes have polynomial discrimination? As discussed previously in Example 4.6, the classical Glivenko–Cantelli law is based on indicator functions of the left-sided intervals $(-\infty, t]$. These functions are uniformly bounded with $b = 1$, and moreover, as shown in the following proof, this function class has polynomial discrimination of order $\nu = 1$. Consequently, Theorem 4.10 combined with Lemma 4.14 yields a quantitative version of Theorem 4.4 as a corollary.

Corollary 4.15 (Classical Glivenko–Cantelli) *Let $F(t) = \mathbb{P}[X \leq t]$ be the CDF of a random variable X, and let $\widehat{F}_n$ be the empirical CDF based on n i.i.d. samples $X_i \sim \mathbb{P}$. Then*

$$\mathbb{P}\left[\|\widehat{F}_n - F\|_\infty \geq 8\sqrt{\frac{\log(n+1)}{n}} + \delta\right] \leq e^{-\frac{n\delta^2}{2}} \quad \textit{for all } \delta \geq 0, \tag{4.25}$$

and hence $\|\widehat{F}_n - F\|_\infty \xrightarrow{a.s.} 0$.

Proof For a given sample $x_1^n = (x_1, \ldots, x_n) \in \mathbb{R}^n$, consider the set $\mathscr{F}(x_1^n)$, where $\mathscr{F}$ is the set of all $\{0\text{-}1\}$-valued indicator functions of the half-intervals $(-\infty, t]$ for $t \in \mathbb{R}$. If we order the samples as $x_{(1)} \leq x_{(2)} \leq \cdots \leq x_{(n)}$, then they split the real line into at most $n + 1$ intervals (including the two end-intervals $(-\infty, x_{(1)})$ and $[x_{(n)}, \infty)$). For a given t, the indicator function $\mathbb{I}_{(-\infty,t]}$ takes the value one for all $x_{(i)} \leq t$, and the value zero for all other samples. Thus, we have shown that, for any given sample x_1^n, we have $\text{card}(\mathscr{F}(x_1^n)) \leq n + 1$. Applying Lemma 4.14, we obtain

$$\mathbb{E}_\varepsilon\left[\sup_{f\in\mathscr{F}}\left|\frac{1}{n}\sum_{i=1}^n \varepsilon_i f(X_i)\right|\right] \leq 4\sqrt{\frac{\log(n+1)}{n}},$$

and taking averages over the data X_i yields the upper bound $\mathcal{R}_n(\mathscr{F}) \leq 4\sqrt{\frac{\log(n+1)}{n}}$. The claim (4.25) then follows from Theorem 4.10. □

Although the exponential tail bound (4.25) is adequate for many purposes, it is far from the tightest possible. Using alternative methods, we provide a sharper result that removes the $\sqrt{\log(n+1)}$ factor in Chapter 5. See the bibliographic section for references to the sharpest possible results, including control of the constants in the exponent and the pre-factor.

4.3.2 Vapnik–Chervonenkis dimension

Thus far, we have seen that it is relatively straightforward to establish uniform laws for function classes with polynomial discrimination. In certain cases, such as in our proof of the classical Glivenko–Cantelli law, we can verify by direct calculation that a given function class has polynomial discrimination. More broadly, it is of interest to develop techniques for certifying this property in a less laborious manner. The theory of Vapnik–Chervonenkis (VC) dimension provides one such class of techniques. Accordingly, we now turn to defining the notions of shattering and VC dimension.

Let us consider a function class $\mathscr{F}$ in which each function f is binary-valued, taking the values $\{0, 1\}$ for concreteness. In this case, the set $\mathscr{F}(x_1^n)$ from equation (4.11) can have at most 2^n elements.

Definition 4.16 (Shattering and VC dimension) Given a class $\mathscr{F}$ of binary-valued functions, we say that the set $x_1^n = (x_1, \ldots, x_n)$ is *shattered* by $\mathscr{F}$ if $\text{card}(\mathscr{F}(x_1^n)) = 2^n$. The *VC dimension* $\nu(\mathscr{F})$ is the largest integer n for which there is *some* collection $x_1^n = (x_1, \ldots, x_n)$ of n points that is shattered by $\mathscr{F}$.

When the quantity $\nu(\mathscr{F})$ is finite, then the function class $\mathscr{F}$ is said to be a *VC class*. We will frequently consider function classes $\mathscr{F}$ that consist of indicator functions $\mathbb{I}_S[\cdot]$, for sets S ranging over some class of sets $\mathcal{S}$. In this case, we use $\mathcal{S}(x_1^n)$ and $\nu(\mathcal{S})$ as shorthands for the sets $\mathscr{F}(x_1^n)$ and the VC dimension of $\mathscr{F}$, respectively. For a given set class $\mathcal{S}$, the shatter coefficient of order n is given by $\max_{x_1^n} \text{card}(\mathcal{S}(x_1^n))$.

Let us illustrate the notions of shattering and VC dimension with some examples:

Example 4.17 (Intervals in $\mathbb{R}$) Consider the class of all indicator functions for left-sided half-intervals on the real line—namely, the class $\mathcal{S}_{\text{left}} := \{(-\infty, a] \mid a \in \mathbb{R}\}$. Implicit in the proof of Corollary 4.15 is a calculation of the VC dimension for this class. We first note that, for any single point x_1, both subsets ($\{x_1\}$ and the empty set $\emptyset$) can be picked out by the class of left-sided intervals $\{(-\infty, a] \mid a \in \mathbb{R}\}$. But given two distinct points $x_1 < x_2$, it is impossible to find a left-sided interval that contains x_2 but not x_1. Therefore, we conclude that $\nu(\mathcal{S}_{\text{left}}) = 1$. In the proof of Corollary 4.15, we showed more specifically that, for any collection $x_1^n = \{x_1, \ldots, x_n\}$, we have $\text{card}(\mathcal{S}_{\text{left}}(x_1^n)) \leq n + 1$.

Now consider the class of all two-sided intervals over the real line—namely, the class $\mathcal{S}_{\text{two}} := \{(b, a] \mid a, b \in \mathbb{R} \text{ such that } b < a\}$. The class $\mathcal{S}_{\text{two}}$ can shatter any two-point set. However, given three distinct points $x_1 < x_2 < x_3$, it cannot pick out the subset $\{x_1, x_3\}$, showing that $\nu(\mathcal{S}_{\text{two}}) = 2$. For future reference, let us also upper bound the shatter coefficients of $\mathcal{S}_{\text{two}}$. Note that any collection of n distinct points $x_1 < x_2 < \cdots < x_{n-1} < x_n$ divides up the real line into $(n + 1)$ intervals. Thus, any set of the form $(-b, a]$ can be specified by choosing one of $(n + 1)$ intervals for b, and a second interval for a. Thus, a crude upper bound on the shatter coefficient of order n is

$$\text{card}(\mathcal{S}_{\text{two}}(x_1^n)) \leq (n + 1)^2,$$

showing that this class has polynomial discrimination with degree $\nu = 2$. ♣

Thus far, we have seen two examples of function classes with finite VC dimension, both of which turned out also to have polynomial discrimination. Is there a general connection between the VC dimension and polynomial discriminability? Indeed, it turns out that any finite VC class has polynomial discrimination with degree at most the VC dimension; this fact is a deep result that was proved independently (in slightly different forms) in papers by Vapnik and Chervonenkis, Sauer and Shelah.

In order to understand why this fact is surprising, note that, for a given set class $\mathcal{S}$, the definition of VC dimension implies that, for all $n > \nu(\mathcal{S})$, then it must be the case that

$\mathrm{card}(\mathcal{S}(x_1^n)) < 2^n$ *for all* collections x_1^n of n samples. However, at least in principle, there could exist some subset with

$$\mathrm{card}(\mathcal{S}(x_1^n)) = 2^n - 1,$$

which is not significantly different from 2^n. The following result shows that this is *not* the case; indeed, for any VC class, the cardinality of $\mathcal{S}(x_1^n)$ can grow at most polynomially in n.

Proposition 4.18 (Vapnik–Chervonenkis, Sauer and Shelah) *Consider a set class $\mathcal{S}$ with $\nu(\mathcal{S}) < \infty$. Then for any collection of points $P = (x_1, \ldots, x_n)$ with $n \geq \nu(\mathcal{S})$, we have*

$$\mathrm{card}(\mathcal{S}(P)) \overset{\text{(i)}}{\leq} \sum_{i=0}^{\nu(\mathcal{S})} \binom{n}{i} \overset{\text{(ii)}}{\leq} (n+1)^{\nu(\mathcal{S})}. \tag{4.26}$$

Given inequality (i), inequality (ii) can be established by elementary combinatorial arguments, so we leave it to the reader (in particular, see part (a) of Exercise 4.11). Part (b) of the same exercise establishes a sharper upper bound.

Proof Given a subset of points Q and a set class $\mathcal{T}$, we let $\nu(\mathcal{T}; Q)$ denote the VC dimension of $\mathcal{T}$ when considering only whether or not subsets of Q can be shattered. Note that $\nu(\mathcal{T}) \leq k$ implies that $\nu(\mathcal{T}; Q) \leq k$ for all point sets Q. For positive integers (n, k), define the functions

$$\Phi_k(n) := \sup_{\substack{\text{point sets } Q \\ \mathrm{card}(Q) \leq n}} \; \sup_{\substack{\text{set classes } \mathcal{T} \\ \nu(\mathcal{T}; Q) \leq k}} \mathrm{card}(\mathcal{T}(Q)) \quad \text{and} \quad \Psi_k(n) := \sum_{i=0}^{k} \binom{n}{i}.$$

Here we agree that $\binom{n}{i} = 0$ whenever $i > n$. In terms of this notation, we claim that it suffices to prove that

$$\Phi_k(n) \leq \Psi_k(n). \tag{4.27}$$

Indeed, suppose there were some set class $\mathcal{S}$ with $\nu(\mathcal{S}) = k$ and collection $P = \{x_1, \ldots, x_n\}$ of n distinct points for which $\mathrm{card}(\mathcal{S}(P)) > \Psi_k(n)$. By the definition $\Phi_k(n)$, we would then have

$$\Phi_k(n) \overset{\text{(i)}}{\geq} \sup_{\substack{\text{set classes } \mathcal{T} \\ \nu(\mathcal{T}; P) \leq k}} \mathrm{card}(\mathcal{T}(P)) \overset{\text{(ii)}}{\geq} \mathrm{card}(\mathcal{S}(P)) > \Psi_k(n), \tag{4.28}$$

which contradicts the claim (4.27). Here inequality (i) follows because P is feasible for the supremum over Q that defines $\Phi_k(n)$; and inequality (ii) follows because $\nu(\mathcal{S}) = k$ implies that $\nu(\mathcal{S}; P) \leq k$.

We now prove the claim (4.27) by induction on the sum $n + k$ of the pairs (n, k).

Base case: To start, we claim that inequality (4.27) holds for all pairs with $n + k = 2$.

The claim is trivial if either $n = 0$ or $k = 0$. Otherwise, for $(n,k) = (1,1)$, both sides of inequality (4.27) are equal to 2.

Induction step: Now assume that, for some integer $\ell > 2$, the inequality (4.27) holds for all pairs with $n + k < \ell$. We claim that it then holds for all pairs with $n + k = \ell$. Fix an arbitrary pair (n,k) such that $n + k = \ell$, a point set $P = \{x_1, \ldots, x_n\}$ and a set class $\mathcal{S}$ such that $\nu(\mathcal{S}; P) = k$. Define the point set $P' = P \setminus \{x_1\}$, and let $\mathcal{S}_0 \subseteq \mathcal{S}$ be the smallest collection of subsets that labels the point set P' in the maximal number of different ways. Let $\mathcal{S}_1$ be the smallest collection of subsets inside $\mathcal{S} \setminus \mathcal{S}_0$ that produce binary labelings of the point set P that are not in $\mathcal{S}_0(P)$. (The choices of $\mathcal{S}_0$ and $\mathcal{S}_1$ need not be unique.)

As a concrete example, given a set class $\mathcal{S} = \{s_1, s_2, s_3, s_4\}$ and a point set $P = \{x_1, x_2, x_3\}$, suppose that the sets generated the binary labelings

$$s_1 \leftrightarrow (0,1,1), \quad s_2 \leftrightarrow (1,1,1), \quad s_3 \leftrightarrow (0,1,0), \quad s_4 \leftrightarrow (0,1,1).$$

In this particular case, we have $\mathcal{S}(P) = \{(0,1,1), (1,1,1), (0,1,0)\}$, and one valid choice of the pair $(\mathcal{S}_0, \mathcal{S}_1)$ would be $\mathcal{S}_0 = \{s_1, s_3\}$ and $\mathcal{S}_1 = \{s_2\}$, generating the labelings $\mathcal{S}_0(P) = \{(0,1,1), (0,1,0)\}$ and $\mathcal{S}_1(P) = \{(1,1,1)\}$.

Using this decomposition, we claim that

$$\mathrm{card}(\mathcal{S}(P)) = \mathrm{card}(\mathcal{S}_0(P')) + \mathrm{card}(\mathcal{S}_1(P')).$$

Indeed, any binary labeling in $\mathcal{S}(P)$ is either mapped to a member of $\mathcal{S}_0(P')$, or in the case that its labeling on P' corresponds to a duplicate, it can be uniquely identified with a member of $\mathcal{S}_1(P')$. This can be verified in the special case described above.

Now since P' is a subset of P and $\mathcal{S}_0$ is a subset of $\mathcal{S}$, we have

$$\nu(\mathcal{S}_0; P') \leq \nu(\mathcal{S}_0; P) \leq k.$$

Since the cardinality of P' is equal to $n - 1$, the induction hypothesis thus implies that $\mathrm{card}(\mathcal{S}_0(P')) \leq \Psi_k(n-1)$.

On the other hand, we claim that the set class $\mathcal{S}_1$ satisfies the upper bound $\nu(\mathcal{S}_1; P') \leq k - 1$. Suppose that $\mathcal{S}_1$ shatters some subset $Q' \subseteq P'$ of cardinality m; it suffices to show that $m \leq k-1$. If $\mathcal{S}_1$ shatters such a set Q', then $\mathcal{S}$ would shatter the set $Q = Q' \cup \{x_1\} \subseteq P$. (This fact follows by construction of $\mathcal{S}_1$: for every binary vector in the set $\mathcal{S}_1(P)$, the set $\mathcal{S}(P)$ must contain a binary vector with the label for x_1 flipped; see the concrete example given above for an illustration.) Since $\nu(\mathcal{S}; P) \leq k$, it must be the case that $\mathrm{card}(Q) = m + 1 \leq k$, which implies that $\nu(\mathcal{S}_1; P') \leq k - 1$. Consequently, the induction hypothesis implies that $\mathrm{card}(\mathcal{S}_1(P')) \leq \Psi_{k-1}(n-1)$.

Putting together the pieces, we have shown that

$$\mathrm{card}(\mathcal{S}(P)) \leq \Psi_k(n-1) + \Psi_{k-1}(n-1) \overset{\text{(i)}}{=} \Psi_k(n), \tag{4.29}$$

where the equality (i) follows from an elementary combinatorial argument (see Exercise 4.10). This completes the proof. □

4.3.3 Controlling the VC dimension

Since classes with finite VC dimension have polynomial discrimination, it is of interest to develop techniques for controlling the VC dimension.

Basic operations

The property of having finite VC dimension is preserved under a number of basic operations, as summarized in the following.

Proposition 4.19 *Let $\mathcal{S}$ and $\mathcal{T}$ be set classes, each with finite VC dimensions $\nu(\mathcal{S})$ and $\nu(\mathcal{T})$, respectively. Then each of the following set classes also have finite VC dimension:*

(a) *The set class $\mathcal{S}^c := \{S^c \mid S \in \mathcal{S}\}$, where S^c denotes the complement of S.*
(b) *The set class $\mathcal{S} \sqcup \mathcal{T} := \{S \cup T \mid S \in \mathcal{S},\ T \in \mathcal{T}\}$.*
(c) *The set class $\mathcal{S} \sqcap \mathcal{T} := \{S \cap T \mid S \in \mathcal{S},\ T \in \mathcal{T}\}$.*

We leave the proof of this result as an exercise for the reader (Exercise 4.8).

Vector space structure

Any class $\mathscr{G}$ of real-valued functions defines a class of sets by the operation of taking subgraphs. In particular, given a real-valued function $g\colon \mathcal{X} \to \mathbb{R}$, its subgraph at level zero is the subset $S_g := \{x \in \mathcal{X} \mid g(x) \leq 0\}$. In this way, we can associate to $\mathscr{G}$ the collection of subsets $\mathcal{S}(\mathscr{G}) := \{S_g, g \in \mathscr{G}\}$, which we refer to as the subgraph class of $\mathscr{G}$. Many interesting classes of sets are naturally defined in this way, among them half-spaces, ellipsoids and so on. In many cases, the underlying function class $\mathscr{G}$ is a vector space, and the following result allows us to upper bound the VC dimension of the associated set class $\mathcal{S}(\mathscr{G})$.

Proposition 4.20 (Finite-dimensional vector spaces) *Let $\mathscr{G}$ be a vector space of functions $g\colon \mathbb{R}^d \to \mathbb{R}$ with dimension* $\dim(\mathscr{G}) < \infty$. *Then the subgraph class $\mathcal{S}(\mathscr{G})$ has VC dimension at most* $\dim(\mathscr{G})$.

Proof By the definition of VC dimension, we need to show that no collection of $n = \dim(\mathscr{G}) + 1$ points in $\mathbb{R}^d$ can be shattered by $\mathcal{S}(\mathscr{G})$. Fix an arbitrary collection $x_1^n = \{x_1, \ldots, x_n\}$ of n points in $\mathbb{R}^d$, and consider the linear map $L\colon \mathscr{G} \to \mathbb{R}^n$ given by $L(g) = (g(x_1), \ldots, g(x_n))$. By construction, the range of the mapping L is a linear subspace of $\mathbb{R}^n$ with dimension at most $\dim(\mathscr{G}) = n - 1 < n$. Therefore, there must exist a non-zero vector $\gamma \in \mathbb{R}^n$ such that $\langle \gamma, L(g) \rangle = 0$ for all $g \in \mathscr{G}$. We may assume without loss of generality that at least one coordinate is positive, and then write

$$\sum_{\{i \mid \gamma_i \leq 0\}} (-\gamma_i) g(x_i) = \sum_{\{i \mid \gamma_i > 0\}} \gamma_i g(x_i) \qquad \text{for all } g \in \mathscr{G}. \tag{4.30}$$

Proceeding via proof by contradiction, suppose that there were to exist some $g \in \mathscr{G}$ such that the associated subgraph set $S_g = \{x \in \mathbb{R}^d \mid g(x) \leq 0\}$ included only the subset $\{x_i \mid \gamma_i \leq 0\}$. For such a function g, the right-hand side of equation (4.30) would be strictly positive while the left-hand side would be non-positive, which is a contradiction. We conclude that $\mathcal{S}(\mathscr{G})$ fails to shatter the set $\{x_1, \ldots, x_n\}$, as claimed. □

Let us illustrate the use of Proposition 4.20 with some examples:

Example 4.21 (Linear functions in $\mathbb{R}^d$) For a pair $(a, b) \in \mathbb{R}^d \times \mathbb{R}$, define the function $f_{a,b}(x) := \langle a, x \rangle + b$, and consider the family $\mathscr{L}^d := \{f_{a,b} \mid (a, b) \in \mathbb{R}^d \times \mathbb{R}\}$ of all such linear functions. The associated subgraph class $\mathcal{S}(\mathscr{L}^d)$ corresponds to the collection of all half-spaces of the form $H_{a,b} := \{x \in \mathbb{R}^d \mid \langle a, x \rangle + b \leq 0\}$. Since the family $\mathscr{L}^d$ forms a vector space of dimension $d + 1$, we obtain as an immediate consequence of Proposition 4.20 that $\mathcal{S}(\mathscr{L}^d)$ has VC dimension at most $d + 1$.

For the special case $d = 1$, let us verify this statement by a more direct calculation. In this case, the class $\mathcal{S}(\mathscr{L}^1)$ corresponds to the collection of all left-sided or right-sided intervals—that is,

$$\mathcal{S}(\mathscr{L}^1) = \{(-\infty, t] \mid t \in \mathbb{R}\} \cup \{[t, \infty) \mid t \in \mathbb{R}\}.$$

Given any two distinct points $x_1 < x_2$, the collection of all such intervals can pick out all possible subsets. However, given any three points $x_1 < x_2 < x_3$, there is no interval contained in $\mathcal{S}(\mathscr{L}^1)$ that contains x_2 while excluding both x_1 and x_3. This calculation shows that $\nu(\mathcal{S}(\mathscr{L}^1)) = 2$, which matches the upper bound obtained from Proposition 4.20. More generally, it can be shown that the VC dimension of $\mathcal{S}(\mathscr{L}^d)$ is $d + 1$, so that Proposition 4.20 yields a sharp result in all dimensions. ♣

Example 4.22 (Spheres in $\mathbb{R}^d$) Consider the sphere $S_{a,b} := \{x \in \mathbb{R}^d \mid \|x - a\|_2 \leq b\}$, where $(a, b) \in \mathbb{R}^d \times \mathbb{R}_+$ specify its center and radius, respectively, and let $\mathcal{S}^d_{\text{sphere}}$ denote the collection of all such spheres. If we define the function

$$f_{a,b}(x) := \|x\|_2^2 - 2 \sum_{j=1}^{d} a_j x_j + \|a\|_2^2 - b^2,$$

then we have $S_{a,b} = \{x \in \mathbb{R}^d \mid f_{a,b}(x) \leq 0\}$, so that the sphere $S_{a,b}$ is a subgraph of the function $f_{a,b}$.

In order to leverage Proposition 4.20, we first define a feature map $\phi \colon \mathbb{R}^d \to \mathbb{R}^{d+2}$ via $\phi(x) := (1, x_1, \ldots, x_d, \|x\|_2^2)$, and then consider functions of the form

$$g_c(x) := \langle c, \phi(x) \rangle \qquad \text{where } c \in \mathbb{R}^{d+2}.$$

The family of functions $\{g_c, c \in \mathbb{R}^{d+1}\}$ is a vector space of dimension $d + 2$, and it contains the function class $\{f_{a,b}, (a, b) \in \mathbb{R}^d \times \mathbb{R}_+\}$. Consequently, by applying Proposition 4.20 to this larger vector space, we conclude that $\nu(\mathcal{S}^d_{\text{sphere}}) \leq d + 2$. This bound is adequate for many purposes, but is not sharp: a more careful analysis shows that the VC dimension of spheres in $\mathbb{R}^d$ is actually $d + 1$. See Exercise 4.13 for an in-depth exploration of the case $d = 2$. ♣

4.4 Bibliographic details and background

First, a technical remark regarding measurability: in general, the normed difference $\|\mathbb{P}_n - \mathbb{P}\|_{\mathscr{F}}$ need not be measurable, since the function class $\mathscr{F}$ may contain an uncountable number of elements. If the function class is separable, then we may simply take the supremum over the countable dense basis. Otherwise, for a general function class, there are various ways of dealing with the issue of measurability, including the use of outer probability (cf. van der Vaart and Wellner (1996)). Here we instead adopt the following convention, suitable for defining expectations of any function ϕ of $\|\mathbb{P}_n - \mathbb{P}\|_{\mathscr{F}}$. For any finite class of functions $\mathscr{G}$ contained within $\mathscr{F}$, the random variable $\|\mathbb{P}_n - \mathbb{P}\|_{\mathscr{G}}$ is well defined, so that it is sensible to define

$$\mathbb{E}[\phi(\|\mathbb{P}_n - \mathbb{P}\|_{\mathscr{F}})] := \sup\{\mathbb{E}[\phi(\|\mathbb{P}_n - \mathbb{P}\|_{\mathscr{G}})] \mid \mathscr{G} \subset \mathscr{F},\ \mathscr{G} \text{ has finite cardinality}\}.$$

By using this definition, we can always think instead of expectations defined via suprema over finite sets.

Theorem 4.4 was originally proved by Glivenko (1933) for the continuous case, and by Cantelli (1933) in the general setting. The non-asymptotic form of the Glivenko–Cantelli theorem given in Corollary 4.15 can be sharpened substantially. For instance, Dvoretsky, Kiefer and Wolfowitz (1956) prove that there is a constant C independent of F and n such that

$$\mathbb{P}[\|\widehat{F}_n - F\|_\infty \geq \delta] \leq Ce^{-2n\delta^2} \qquad \text{for all } \delta \geq 0. \tag{4.31}$$

Massart (1990) establishes the sharpest possible result, with the leading constant $C = 2$.

The Rademacher complexity, and its relative the Gaussian complexity, have a lengthy history in the study of Banach spaces using probabilistic methods; for instance, see the books (Milman and Schechtman, 1986; Pisier, 1989; Ledoux and Talagrand, 1991). Rademacher and Gaussian complexities have also been studied extensively in the specific context of uniform laws of large numbers and empirical risk minimization (e.g. van der Vaart and Wellner, 1996; Koltchinskii and Panchenko, 2000; Koltchinskii, 2001, 2006; Bartlett and Mendelson, 2002; Bartlett et al., 2005). In Chapter 5, we develop further connections between these two forms of complexity, and the related notion of metric entropy.

Exercise 5.4 is adapted from Problem 2.6.3 from van der Vaart and Wellner (1996). The proof of Proposition 4.20 is adapted from Pollard (1984), who credits it to Steele (1978) and Dudley (1978).

4.5 Exercises

Exercise 4.1 (Continuity of functionals) Recall that the functional γ is *continuous in the sup-norm at* F if for all $\epsilon > 0$, there exists a $\delta > 0$ such that $\|G - F\|_\infty \leq \delta$ implies that $|\gamma(G) - \gamma(F)| \leq \epsilon$.

(a) Given n i.i.d. samples with law specified by F, let $\widehat{F}_n$ be the empirical CDF. Show that if γ is continuous in the sup-norm at F, then $\gamma(\widehat{F}_n) \xrightarrow{\text{prob.}} \gamma(F)$.

(b) Which of the following functionals are continuous with respect to the sup-norm? Prove or disprove.

(i) The mean functional $F \mapsto \int x\, dF(x)$.
(ii) The Cramér–von Mises functional $F \mapsto \int [F(x) - F_0(x)]^2\, dF_0(x)$.
(iii) The quantile functional $Q_\alpha(F) = \inf\{t \in \mathbb{R} \mid F(t) \geq \alpha\}$.

Exercise 4.2 (Failure of Glivenko–Cantelli) Recall from Example 4.7 the class $\mathcal{S}$ of all subsets S of $[0, 1]$ for which S has a finite number of elements. Prove that the Rademacher complexity satisfies the lower bound

$$\mathcal{R}_n(\mathcal{S}) = \mathbb{E}_{X,\varepsilon}\left[\sup_{S \in \mathcal{S}} \left|\frac{1}{n}\sum_{i=1}^{n} \varepsilon_i \mathbb{I}_S[X_i]\right|\right] \geq \frac{1}{2}. \tag{4.32}$$

Discuss the connection to Theorem 4.10.

Exercise 4.3 (Maximum likelihood and uniform laws) Recall from Example 4.8 our discussion of empirical and population risks for maximum likelihood over a family of densities $\{p_\theta, \theta \in \Omega\}$.

(a) Compute the population risk $R(\theta, \theta^*) = \mathbb{E}_{\theta^*}\left[\log \frac{p_{\theta^*}(X)}{p_\theta(X)}\right]$ in the following cases:

  (i) Bernoulli: $p_\theta(x) = \frac{e^{\theta x}}{1+e^{\theta x}}$ for $x \in \{0, 1\}$;
  (ii) Poisson: $p_\theta(x) = \frac{e^{\theta x} e^{-\exp(\theta)}}{x!}$ for $x \in \{0, 1, 2, \ldots\}$;
  (iii) multivariate Gaussian: p_θ is the density of an $\mathcal{N}(\theta, \mathbf{\Sigma})$ vector, where the covariance matrix $\mathbf{\Sigma}$ is known and fixed.

(b) For each of the above cases:

  (i) Letting $\widehat{\theta}$ denote the maximum likelihood estimate, give an explicit expression for the excess risk $E(\widehat{\theta}, \theta^*) = R(\widehat{\theta}, \theta^*) - \inf_{\theta \in \Omega} R(\theta, \theta^*)$.
  (ii) Give an upper bound on the excess risk in terms of an appropriate Rademacher complexity.

Exercise 4.4 (Details of symmetrization argument)

(a) Prove that

$$\sup_{g \in \mathscr{G}} \mathbb{E}[g(X)] \leq \mathbb{E}\left[\sup_{g \in \mathscr{G}} |g(X)|\right].$$

Use this to complete the proof of inequality (4.17).

(b) Prove that for any convex and non-decreasing function Φ,

$$\sup_{g \in \mathscr{G}} \Phi(\mathbb{E}[|g(X)|]) \leq \mathbb{E}\left[\Phi\left(\sup_{g \in \mathscr{G}} |g(X)|\right)\right].$$

Use this bound to complete the proof of Proposition 4.11.

Exercise 4.5 (Necessity of vanishing Rademacher complexity) In this exercise, we work through the proof of Proposition 4.12.

(a) Recall the recentered function class $\overline{\mathscr{F}} = \{f - \mathbb{E}[f] \mid f \in \mathscr{F}\}$. Show that

$$\mathbb{E}_{X,\varepsilon}[\|\mathbb{S}_n\|_{\overline{\mathscr{F}}}] \geq \mathbb{E}_{X,\varepsilon}[\|\mathbb{S}_n\|_{\mathscr{F}}] - \frac{\sup_{f \in \mathscr{F}} |\mathbb{E}[f]|}{\sqrt{n}}.$$

(b) Use concentration results to complete the proof of Proposition 4.12.

Exercise 4.6 (Too many linear classifiers) Consider the function class

$$\mathscr{F} = \{x \mapsto \operatorname{sign}(\langle \theta, x \rangle) \mid \theta \in \mathbb{R}^d, \ \|\theta\|_2 = 1\},$$

corresponding to the $\{-1, +1\}$-valued classification rules defined by linear functions in $\mathbb{R}^d$. Supposing that $d \geq n$, let $x_1^n = \{x_1, \ldots, x_n\}$ be a collection of vectors in $\mathbb{R}^d$ that are linearly independent. Show that the empirical Rademacher complexity satisfies

$$\mathcal{R}(\mathscr{F}(x_1^n)/n) = \mathbb{E}_\varepsilon\left[\sup_{f \in \mathscr{F}} \left|\frac{1}{n}\sum_{i=1}^n \varepsilon_i f(x_i)\right|\right] = 1.$$

Discuss the consequences for empirical risk minimization over the class $\mathscr{F}$.

Exercise 4.7 (Basic properties of Rademacher complexity) Prove the following properties of the Rademacher complexity.

(a) $\mathcal{R}_n(\mathscr{F}) = \mathcal{R}_n(\operatorname{conv}(\mathscr{F}))$.
(b) Show that $\mathcal{R}_n(\mathscr{F} + \mathscr{G}) \leq \mathcal{R}_n(\mathscr{F}) + \mathcal{R}_n(\mathscr{G})$. Give an example to demonstrate that this bound cannot be improved in general.
(c) Given a fixed and uniformly bounded function g, show that

$$\mathcal{R}_n(\mathscr{F} + g) \leq \mathcal{R}_n(\mathscr{F}) + \frac{\|g\|_\infty}{\sqrt{n}}. \tag{4.33}$$

Exercise 4.8 (Operations on VC classes) Let $\mathcal{S}$ and $\mathcal{T}$ be two classes of sets with finite VC dimensions. Show that each of the following operations lead to a new set class also with finite VC dimension.

(a) The set class $\mathcal{S}^c := \{S^c \mid S \in \mathcal{S}\}$, where S^c denotes the complement of the set S.
(b) The set class $\mathcal{S} \sqcap \mathcal{T} := \{S \cap T \mid S \in \mathcal{S}, \ T \in \mathcal{T}\}$.
(c) The set class $\mathcal{S} \sqcup \mathcal{T} := \{S \cup T \mid S \in \mathcal{S}, \ T \in \mathcal{T}\}$.

Exercise 4.9 Prove Lemma 4.14.

Exercise 4.10 Prove equality (i) in equation (4.29), namely that

$$\binom{n-1}{k} + \binom{n-1}{k-1} = \binom{n}{k}.$$

Exercise 4.11 In this exercise, we complete the proof of Proposition 4.18.

(a) Prove inequality (ii) in (4.26).
(b) For $n \geq \nu$, prove the sharper upper bound $\operatorname{card}(\mathcal{S}(x_1^n)) \leq (\frac{en}{\nu})^\nu$. (*Hint:* You might find the result of Exercise 2.9 useful.)

Exercise 4.12 (VC dimension of left-sided intervals) Consider the class of left-sided half-intervals in $\mathbb{R}^d$:

$$\mathcal{S}^d_{\text{left}} := \{(-\infty, t_1] \times (-\infty, t_2] \times \cdots \times (-\infty, t_d] \mid (t_1, \ldots, t_d) \in \mathbb{R}^d\}.$$

Show that for any collection of n points, we have $\text{card}(\mathcal{S}^d_{\text{left}}(x_1^n)) \leq (n+1)^d$ and $\nu(\mathcal{S}^d_{\text{left}}) = d$.

Exercise 4.13 (VC dimension of spheres) Consider the class of all spheres in $\mathbb{R}^2$—that is

$$\mathcal{S}^2_{\text{sphere}} := \{S_{a,b},\ (a,b) \in \mathbb{R}^2 \times \mathbb{R}_+\}, \tag{4.34}$$

where $S_{a,b} := \{x \in \mathbb{R}^2 \mid \|x - a\|_2 \leq b\}$ is the sphere of radius $b \geq 0$ centered at $a = (a_1, a_2)$.

(a) Show that $\mathcal{S}^2_{\text{sphere}}$ can shatter any subset of three points that are not collinear.
(b) Show that no subset of four points can be shattered, and conclude that the VC dimension is $\nu(\mathcal{S}^2_{\text{sphere}}) = 3$.

Exercise 4.14 (VC dimension of monotone Boolean conjunctions) For a positive integer $d \geq 2$, consider the function $h_S : \{0,1\}^d \to \{0,1\}$ of the form

$$h_s(x_1, \ldots, x_d) = \begin{cases} 1 & \text{if } x_j = 1 \text{ for all } j \in S, \\ 0 & \text{otherwise.} \end{cases}$$

The set of all Boolean monomials $\mathfrak{B}_d$ consists of all such functions as S ranges over all subsets of $\{1, 2, \ldots, d\}$, along with the constant functions $h \equiv 0$ and $h \equiv 1$. Show that the VC dimension of $\mathfrak{B}_d$ is equal to d.

Exercise 4.15 (VC dimension of closed and convex sets) Show that the class C^d_{cc} of all closed and convex sets in $\mathbb{R}^d$ does *not* have finite VC dimension. (*Hint:* Consider a set of n points on the boundary of the unit ball.)

Exercise 4.16 (VC dimension of polygons) Compute the VC dimension of the set of all polygons in $\mathbb{R}^2$ with at most four vertices.

Exercise 4.17 (Infinite VC dimension) For a scalar $t \in \mathbb{R}$, consider the function $f_t(x) = \text{sign}(\sin(tx))$. Prove that the function class $\{f_t : [-1,1] \to \mathbb{R} \mid t \in \mathbb{R}\}$ has infinite VC dimension. (*Note:* This shows that VC dimension is *not* equivalent to the number of parameters in a function class.)

5

Metric entropy and its uses

Many statistical problems require manipulating and controlling collections of random variables indexed by sets with an infinite number of elements. There are many examples of such stochastic processes. For instance, a continuous-time random walk can be viewed as a collection of random variables indexed by the unit interval $[0, 1]$. Other stochastic processes, such as those involved in random matrix theory, are indexed by vectors that lie on the Euclidean sphere. Empirical process theory, a broad area that includes the Glivenko–Cantelli laws discussed in Chapter 4, is concerned with stochastic processes that are indexed by sets of functions.

Whereas any finite set can be measured in terms of its cardinality, measuring the "size" of a set with infinitely many elements requires more delicacy. The concept of metric entropy, which dates back to the seminal work of Kolmogorov, Tikhomirov and others in the Russian school, provides one way in which to address this difficulty. Though defined in a purely deterministic manner, in terms of packing and covering in a metric space, it plays a central role in understanding the behavior of stochastic processes. Accordingly, this chapter is devoted to an exploration of metric entropy, and its various uses in the context of stochastic processes.

5.1 Covering and packing

We begin by defining the notions of packing and covering a set in a metric space. Recall that a metric space $(\mathbb{T}, \rho)$ consists of a non-empty set $\mathbb{T}$, equipped with a mapping $\rho\colon \mathbb{T} \times \mathbb{T} \to \mathbb{R}$ that satisfies the following properties:

(a) It is non-negative: $\rho(\theta, \widetilde{\theta}) \geq 0$ for all pairs $(\theta, \widetilde{\theta})$, with equality if and only if $\theta = \widetilde{\theta}$.
(b) It is symmetric: $\rho(\theta, \widetilde{\theta}) = \rho(\widetilde{\theta}, \theta)$ for all pairs $(\widetilde{\theta}, \theta)$.
(c) The triangle inequality holds: $\rho(\theta, \widetilde{\theta}) \leq \rho(\theta, \tilde{\theta}) + \rho(\tilde{\theta}, \widetilde{\theta})$ for all triples $(\theta, \widetilde{\theta}, \tilde{\theta})$.

Familiar examples of metric spaces include the real space $\mathbb{R}^d$ with the *Euclidean metric*

$$\rho(\theta, \widetilde{\theta}) = \|\theta - \widetilde{\theta}\|_2 := \sqrt{\sum_{j=1}^{d} (\theta_j - \theta'_j)^2}, \tag{5.1a}$$

and the discrete cube $\{0, 1\}^d$ with the *rescaled Hamming metric*

$$\rho_H(\theta, \widetilde{\theta}) := \frac{1}{d} \sum_{j=1}^{d} \mathbb{I}[\theta_j \neq \theta'_j]. \tag{5.1b}$$

Also of interest are various metric spaces of functions, among them the usual spaces $L^2(\mu, [0,1])$ with its metric

$$\|f - g\|_2 := \left[\int_0^1 (f(x) - g(x))^2 \, d\mu(x) \right]^{1/2}, \tag{5.1c}$$

as well as the space $C[0,1]$ of all continuous functions on $[0,1]$ equipped with the sup-norm metric

$$\|f - g\|_\infty = \sup_{x \in [0,1]} |f(x) - g(x)|. \tag{5.1d}$$

Given a metric space $(\mathbb{T}, \rho)$, a natural way in which to measure its size is in terms of number of balls of a fixed radius δ required to cover it, a quantity known as the covering number.

Definition 5.1 (Covering number) A δ-*cover* of a set $\mathbb{T}$ with respect to a metric ρ is a set $\{\theta^1, \ldots, \theta^N\} \subset \mathbb{T}$ such that for each $\theta \in \mathbb{T}$, there exists some $i \in \{1, \ldots, N\}$ such that $\rho(\theta, \theta^i) \le \delta$. The δ-*covering number* $N(\delta\,;\,\mathbb{T}, \rho)$ is the cardinality of the smallest δ-cover.

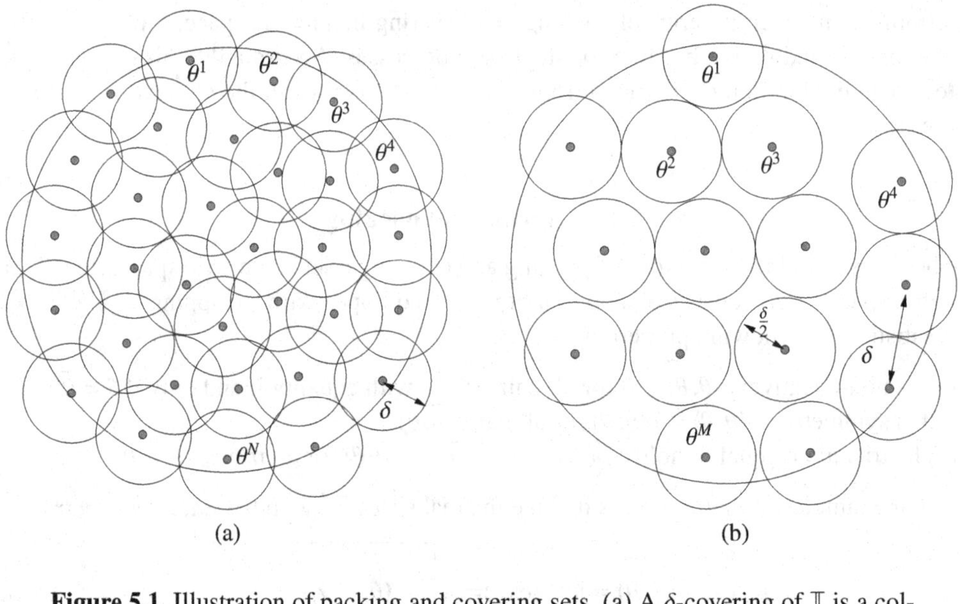

(a) (b)

Figure 5.1 Illustration of packing and covering sets. (a) A δ-covering of $\mathbb{T}$ is a collection of elements $\{\theta^1, \ldots, \theta^N\} \subset \mathbb{T}$ such that for each $\theta \in \mathbb{T}$, there is some element $j \in \{1, \ldots, N\}$ such that $\rho(\theta, \theta^j) \le \delta$. Geometrically, the union of the balls with centers θ^j and radius δ cover the set $\mathbb{T}$. (b) A δ-packing of a set $\mathbb{T}$ is a collection of elements $\{\theta^1, \ldots, \theta^M\} \subset \mathbb{T}$ such that $\rho(\theta^j, \theta^k) > \delta$ for all $j \neq k$. Geometrically, it is a collection of balls of radius $\delta/2$ with centers contained in $\mathbb{T}$ such that no pair of balls have a non-empty intersection.

As illustrated in Figure 5.1(a), a δ-covering can be visualized as a collection of balls of radius δ that cover the set $\mathbb{T}$. When discussing metric entropy, we restrict our attention to metric spaces $(\mathbb{T}, \rho)$ that are *totally bounded*, meaning that the covering number $N(\delta) = N(\delta\,;\,\mathbb{T},\,\rho)$ is finite for all $\delta > 0$. See Exercise 5.1 for an example of a metric space that is *not* totally bounded.

It is easy to see that the covering number is non-increasing in δ, meaning that $N(\delta) \geq N(\delta')$ for all $\delta \leq \delta'$. Typically, the covering number diverges as $\delta \to 0^+$, and of interest to us is this growth rate on a logarithmic scale. More specifically, the quantity $\log N(\delta\,;\,\mathbb{T},\,\rho)$ is known as the *metric entropy* of the set $\mathbb{T}$ with respect to ρ.

Example 5.2 (Covering numbers of unit cubes) Let us begin with a simple example of how covering numbers can be bounded. Consider the interval $[-1, 1]$ in $\mathbb{R}$, equipped with the metric $\rho(\theta, \theta') = |\theta - \theta'|$. Suppose that we divide the interval $[-1, 1]$ into $L := \lfloor \frac{1}{\delta} \rfloor + 1$ sub-intervals,[1] centered at the points $\theta^i = -1 + 2(i-1)\delta$ for $i \in [L] := \{1, 2, \ldots, L\}$, and each of length at most 2δ. By construction, for any point $\widetilde{\theta} \in [0, 1]$, there is some $j \in [L]$ such that $|\theta^j - \widetilde{\theta}| \leq \delta$, which shows that

$$N(\delta\,;\,[-1, 1],\,|\cdot|) \leq \frac{1}{\delta} + 1. \tag{5.2}$$

As an exercise, the reader should generalize this analysis, showing that, for the d-dimensional cube $[-1, 1]^d$, we have $N(\delta\,;\,[-1, 1]^d,\,\|\cdot\|_\infty) \leq (1 + \frac{1}{\delta})^d$. ♣

Example 5.3 (Covering of the binary hypercube) Consider the binary hypercube $\mathbb{H}^d := \{0, 1\}^d$ equipped with the rescaled Hamming metric (5.1b). First, let us upper bound its δ-covering number. Let $S = \{1, 2, \ldots, \lceil (1-\delta)d \rceil\}$, where $\lceil (1-\delta)d \rceil$ denotes the smallest integer larger than or equal to $(1 - \delta)d$. Consider the set of binary vectors

$$\mathbb{T}(\delta) := \{\theta \in \mathbb{H}^d \mid \theta_j = 0 \qquad \text{for all } j \notin S\}.$$

By construction, for any binary vector $\widetilde{\theta} \in \mathbb{H}^d$, we can find a vector $\theta \in \mathbb{T}(\delta)$ such that $\rho_H(\theta, \widetilde{\theta}) \leq \delta$. (Indeed, we can match $\widetilde{\theta}$ exactly on all entries $j \in S$, and, in the worst case, disagree on all the remaining $\lfloor \delta d \rfloor$ positions.) Since $\mathbb{T}(\delta)$ contains $2^{\lceil (1-\delta)d \rceil}$ vectors, we conclude that

$$\frac{\log N_H(\delta; \mathbb{H}^d)}{\log 2} \leq \lceil d(1 - \delta) \rceil.$$

This bound is useful but can be sharpened considerably by using a more refined argument, as discussed in Exercise 5.3.

Let us lower bound its δ-covering number, where $\delta \in (0, \frac{1}{2})$. If $\{\theta^1, \ldots, \theta^N\}$ is a δ-covering, then the (unrescaled) Hamming balls of radius $s = \delta d$ around each θ^ℓ must contain all 2^d vectors in the binary hypercube. Let $s = \lfloor \delta d \rfloor$ denote the largest integer less than or equal to δd. For each θ^ℓ, there are exactly $\sum_{j=0}^{s} \binom{d}{j}$ binary vectors lying within distance δd from it, and hence we must have $N \left\{ \sum_{j=0}^{s} \binom{d}{j} \right\} \geq 2^d$. Now let $X_i \in \{0, 1\}$ be i.i.d. Bernoulli variables

[1] For a scalar $a \in \mathbb{R}$, the notation $\lfloor a \rfloor$ denotes the greatest integer less than or equal to a.

with parameter 1/2. Rearranging the previous inequality, we have

$$\frac{1}{N} \leq \sum_{j=0}^{s} \binom{d}{j} 2^{-d} = \mathbb{P}\left[\sum_{i=1}^{d} X_i \leq \delta d\right] \overset{\text{(i)}}{\leq} e^{-2d(\frac{1}{2}-\delta)^2},$$

where inequality (i) follows by applying Hoeffding's bound to the sum of d i.i.d. Bernoulli variables. Following some algebra, we obtain the lower bound

$$\log N_H(\delta; \mathbb{H}^d) \geq 2d\left(\frac{1}{2} - \delta\right)^2, \qquad \text{valid for } \delta \in (0, \tfrac{1}{2}).$$

This lower bound is qualitatively correct, but can be tightened by using a better upper bound on the binomial tail probability. For instance, from the result of Exercise 2.9, we have $\frac{1}{d} \log \mathbb{P}[\sum_{i=1}^d X_i \leq s] \leq -D(\delta \,\|\, \frac{1}{2})$, where $D(\delta \,\|\, \frac{1}{2})$ is the Kullback–Leibler divergence between the Bernoulli distributions with parameters δ and $\frac{1}{2}$, respectively. Using this tail bound within the same argument leads to the improved lower bound

$$\log N_H(\delta; \mathbb{H}^d) \geq dD(\delta \,\|\, \tfrac{1}{2}), \qquad \text{valid for } \delta \in (0, \tfrac{1}{2}). \tag{5.3}$$

♣

In the preceding examples, we used different techniques to upper and lower bound the covering number. A complementary way in which to measure the massiveness of sets, also useful for deriving bounds on the metric entropy, is known as the packing number.

Definition 5.4 (Packing number) A *δ-packing* of a set $\mathbb{T}$ with respect to a metric ρ is a set $\{\theta^1, \ldots, \theta^M\} \subset \mathbb{T}$ such that $\rho(\theta^i, \theta^j) > \delta$ for all distinct $i, j \in \{1, 2, \ldots, M\}$. The *$\delta$-packing number* $M(\delta\,; \mathbb{T}, \rho)$ is the cardinality of the largest δ-packing.

As illustrated in Figure 5.1(b), a δ-packing can be viewed as a collection of balls of radius $\delta/2$, each centered at an element contained in $\mathbb{T}$, such that no two balls intersect. What is the relation between the covering number and packing numbers? Although not identical, they provide essentially the same measure of the massiveness of a set, as summarized in the following:

Lemma 5.5 *For all $\delta > 0$, the packing and covering numbers are related as follows:*

$$M(2\delta\,; \mathbb{T}, \rho) \overset{\text{(a)}}{\leq} N(\delta\,; \mathbb{T}, \rho) \overset{\text{(b)}}{\leq} M(\delta\,; \mathbb{T}, \rho). \tag{5.4}$$

We leave the proof of Lemma 5.5 for the reader (see Exercise 5.2). It shows that, at least up to constant factors, the packing and covering numbers exhibit the same scaling behavior as $\delta \to 0$.

Example 5.6 (Packing of unit cubes) Returning to Example 5.2, we observe that the points

$\{\theta^j,\ j = 1, \ldots, L-1\}$ are separated as $|\theta^j - \theta^k| \geq 2\delta > \delta$ for all $j \neq k$, which implies that $M(2\delta\,;[-1,1], |\cdot|) \geq \lfloor \frac{1}{\delta} \rfloor$. Combined with Lemma 5.5 and our previous upper bound (5.2), we conclude that $\log N(\delta; [-1,1], |\cdot|) \asymp \log(1/\delta)$ for $\delta > 0$ sufficiently small. This argument can be extended to the d-dimensional cube with the sup-norm $\|\cdot\|_\infty$, showing that

$$\log N(\delta\,; [0,1]^d, \|\cdot\|_\infty) \asymp d \log(1/\delta) \qquad \text{for } \delta > 0 \text{ sufficiently small.} \tag{5.5}$$

Thus, we see how an explicit construction of a packing set can be used to lower bound the metric entropy. ♣

In Exercise 5.3, we show how a packing argument can be used to obtain a refined upper bound on the covering number of the Boolean hypercube from Example 5.3.

We now seek some more general understanding of what geometric properties govern metric entropy. Since covering is defined in terms of the number of balls—each with a fixed radius and hence volume—one would expect to see connections between covering numbers and volumes of these balls. The following lemma provides a precise statement of this connection in the case of norms on $\mathbb{R}^d$ with open unit balls, for which the volume can be taken with respect to Lebesgue measure. Important examples are the usual ℓ_q-balls, defined for $q \in [1, \infty]$ via

$$\mathbb{B}_q^d(1) := \{x \in \mathbb{R}^d \mid \|x\|_q \leq 1\}, \tag{5.6}$$

where for $q \in [1, \infty)$, the ℓ_q-norm is given by

$$\|x\|_q := \begin{cases} \left(\sum_{i=1}^d |x_i|^q\right)^{1/q} & \text{for } q \in [1, \infty), \\ \max_{i=1,\ldots,d} |x_i| & \text{for } q = \infty. \end{cases} \tag{5.7}$$

The following lemma relates the metric entropy to the so-called volume ratio. It involves the Minkowski sum $A + B := \{a + b \mid a \in A,\ b \in B\}$ of two sets.

Lemma 5.7 (Volume ratios and metric entropy) *Consider a pair of norms $\|\cdot\|$ and $\|\cdot\|'$ on $\mathbb{R}^d$, and let $\mathbb{B}$ and $\mathbb{B}'$ be their corresponding unit balls (i.e., $\mathbb{B} = \{\theta \in \mathbb{R}^d \mid \|\theta\| \leq 1\}$, with $\mathbb{B}'$ similarly defined). Then the δ-covering number of $\mathbb{B}$ in the $\|\cdot\|'$-norm obeys the bounds*

$$\left(\frac{1}{\delta}\right)^d \frac{\operatorname{vol}(\mathbb{B})}{\operatorname{vol}(\mathbb{B}')} \overset{\text{(a)}}{\leq} N(\delta\,; \mathbb{B}, \|\cdot\|') \overset{\text{(b)}}{\leq} \frac{\operatorname{vol}(\frac{2}{\delta}\mathbb{B} + \mathbb{B}')}{\operatorname{vol}(\mathbb{B}')}. \tag{5.8}$$

Whenever $\mathbb{B}' \subseteq \mathbb{B}$, the upper bound (b) may be simplified by observing that

$$\operatorname{vol}\left(\frac{2}{\delta}\mathbb{B} + \mathbb{B}'\right) \leq \operatorname{vol}\left(\left(\frac{2}{\delta} + 1\right)\mathbb{B}\right) = \left(\frac{2}{\delta} + 1\right)^d \operatorname{vol}(\mathbb{B}),$$

which implies that $N(\delta\,; \mathbb{B}, \|\cdot\|') \leq (1 + \frac{2}{\delta})^d \frac{\operatorname{vol}(\mathbb{B})}{\operatorname{vol}(\mathbb{B}')}$.

Proof On one hand, if $\{\theta^1, \ldots, \theta^N\}$ is a δ-covering of $\mathbb{B}$, then we have

$$\mathbb{B} \subseteq \bigcup_{j=1}^{N} \{\theta^j + \delta\mathbb{B}'\},$$

which implies that $\text{vol}(\mathbb{B}) \leq N \,\text{vol}(\delta\mathbb{B}') = N\delta^d \,\text{vol}(\mathbb{B}')$, thus establishing inequality (a) in the claim (5.8).

In order to establish inequality (b) in (5.8), let $\{\theta^1, \ldots, \theta^M\}$ be a maximal $(\delta/2)$-packing of $\mathbb{B}$ in the $\|\cdot\|'$-norm; by maximality, this set must also be a δ-covering of $\mathbb{B}$ under the $\|\cdot\|'$-norm. The balls $\{\theta^j + \frac{\delta}{2}\mathbb{B}',\ j = 1, \ldots, M\}$ are all disjoint and contained within $\mathbb{B} + \frac{\delta}{2}\mathbb{B}'$. Taking volumes, we conclude that $\sum_{j=1}^{M} \text{vol}(\theta^j + \frac{\delta}{2}\mathbb{B}') \leq \text{vol}(\mathbb{B} + \frac{\delta}{2}\mathbb{B}')$, and hence

$$M \,\text{vol}\left(\frac{\delta}{2}\mathbb{B}'\right) \leq \text{vol}\left(\mathbb{B} + \frac{\delta}{2}\mathbb{B}'\right).$$

Finally, we have $\text{vol}(\frac{\delta}{2}\mathbb{B}') = (\frac{\delta}{2})^d \,\text{vol}(\mathbb{B}')$ and $\text{vol}(\mathbb{B} + \frac{\delta}{2}\mathbb{B}') = (\frac{\delta}{2})^d \,\text{vol}(\frac{2}{\delta}\mathbb{B} + \mathbb{B}')$, from which the claim (b) in equation (5.8) follows. □

Let us illustrate Lemma 5.7 with an example.

Example 5.8 (Covering unit balls in their own metrics) As an important special case, if we take $\mathbb{B} = \mathbb{B}'$ in Lemma 5.7, then we obtain upper and lower bounds on the metric entropy of a given unit ball in terms of its own norm—namely, we have

$$d \log(1/\delta) \leq \log N(\delta\,;\, \mathbb{B}, \|\cdot\|) \leq d \log\left(1 + \frac{2}{\delta}\right). \tag{5.9}$$

When applied to the ℓ_∞-norm, this result shows that the $\|\cdot\|_\infty$-metric entropy of $\mathbb{B}_\infty^d = [-1, 1]^d$ scales as $d \log(1/\delta)$, so that we immediately recover the end result of our more direct analysis in Examples 5.2 and 5.6. As another special case, we also find that the Euclidean unit ball $\mathbb{B}_2^d$ can be covered by at most $(1+2/\delta)^d$ balls with radius δ in the norm $\|\cdot\|_2$. In Example 5.12 to follow in the sequel, we use Lemma 5.7 to bound the metric entropy of certain ellipsoids in $\ell^2(\mathbb{N})$. ♣

Thus far, we have studied the metric entropy of various subsets of $\mathbb{R}^d$. We now turn to the metric entropy of some function classes, beginning with a simple parametric class of functions.

Example 5.9 (A parametric class of functions) For any fixed θ, define the real-valued function $f_\theta(x) := 1 - e^{-\theta x}$, and consider the function class

$$\mathscr{P} := \{f_\theta \colon [0, 1] \to \mathbb{R} \mid \theta \in [0, 1]\}.$$

The set $\mathscr{P}$ is a metric space under the uniform norm (also known as the sup-norm) given by $\|f - g\|_\infty := \sup_{x\in[0,1]} |f(x) - g(x)|$. We claim that its covering number in terms of the sup-norm is bounded above and below as

$$1 + \left\lfloor \frac{1 - 1/e}{2\delta} \right\rfloor \overset{\text{(i)}}{\leq} N_\infty(\delta; \mathscr{P}) \overset{\text{(ii)}}{\leq} \frac{1}{2\delta} + 2. \tag{5.10}$$

We first establish the upper bound given in inequality (ii) of (5.10). For a given $\delta \in (0, 1)$, let us set $T = \lfloor \frac{1}{2\delta} \rfloor$, and define $\theta^i := 2\delta i$ for $i = 0, 1, \ldots, T$. By also adding the point $\theta^{T+1} = 1$, we obtain a collection of points $\{\theta^0, \ldots, \theta^T, \theta^{T+1}\}$ contained within $[0, 1]$. We claim that the associated functions $\{f_{\theta^0}, \ldots, f_{\theta^{T+1}}\}$ form a δ-cover for $\mathscr{P}$. Indeed, for any $f_\theta \in \mathscr{P}$, we can find some θ^i in our cover such that $|\theta^i - \theta| \leq \delta$. We then have

$$\|f_{\theta^i} - f_\theta\|_\infty = \max_{x \in [0,1]} |e^{-\theta^i |x|} - e^{-\theta |x|}| \leq |\theta^i - \theta| \leq \delta,$$

which implies that $N_\infty(\delta; \mathscr{P}) \leq T + 2 \leq \frac{1}{2\delta} + 2$.

In order to prove the lower bound on the covering number, as stated in inequality (i) in (5.10), we proceed by first lower bounding the packing number, and then applying Lemma 5.5. An explicit packing can be constructed as follows: first set $\theta^0 = 0$, and then define $\theta^i = -\log(1 - \delta i)$ for all i such that $\theta^i \leq 1$. We can define θ^i in this way until $1/e = 1 - T\delta$, or $T \geq \lfloor \frac{1-1/e}{\delta} \rfloor$. Moreover, note that for any $i \neq j$ in the resulting set of functions, we have $\|f_{\theta^i} - f_{\theta^j}\|_\infty \geq |f_{\theta^i}(1) - f_{\theta^j}(1)| \geq \delta$, by definition of θ^i. Therefore, we conclude that $M_\infty(\delta; \mathscr{P}) \geq \lfloor \frac{1-1/e}{\delta} \rfloor + 1$, and hence that

$$N_\infty(\delta; \mathscr{P}) \geq M_\infty(2\delta; \mathscr{P}) \geq \left\lfloor \frac{1 - 1/e}{2\delta} \right\rfloor + 1,$$

as claimed. We have thus established the scaling $\log N(\delta\,; \mathscr{P}, \|\cdot\|_\infty) \asymp \log(1/\delta)$ as $\delta \to 0^+$. This rate is the typical one to be expected for a scalar parametric class. ♣

A function class with a metric entropy that scales as $\log(1/\delta)$ as $\delta \to 0^+$ is relatively small. Indeed, as shown in Example 5.2, the interval $[-1, 1]$ has metric entropy of this order, and the function class $\mathscr{P}$ from Example 5.9 is not essentially different. Other function classes are much richer, and so their metric entropy exhibits a correspondingly faster growth, as shown by the following example.

Example 5.10 (Lipschitz functions on the unit interval) Now consider the class of Lipschitz functions

$$\mathscr{F}_L := \{g\colon [0, 1] \to \mathbb{R} \mid g(0) = 0, \text{ and } |g(x) - g(x')| \leq L|x - x'| \quad \forall\, x, x' \in [0, 1]\}. \tag{5.11}$$

Here $L > 0$ is a fixed constant, and all of the functions in the class obey the Lipschitz bound, uniformly over all of $[0, 1]$. Note that the function class $\mathscr{P}$ from Example 5.9 is contained within the class $\mathscr{F}_L$ with $L = 1$. It is known that the metric entropy of the class $\mathscr{F}_L$ with respect to the sup-norm scales as

$$\log N_\infty(\delta\,; \mathscr{F}_L) \asymp (L/\delta) \qquad \text{for suitably small } \delta > 0. \tag{5.12}$$

Consequently, the set of Lipschitz functions is a *much* larger class than the parametric function class from Example 5.9, since its metric entropy grows as $1/\delta$ as $\delta \to 0$, as compared to $\log(1/\delta)$.

Let us prove the lower bound in equation (5.12); via Lemma 5.5, it suffices to construct a sufficiently large packing of the set $\mathscr{F}_L$. For a given $\epsilon > 0$, define $M = \lfloor 1/\epsilon \rfloor$, and consider the points in $[0, 1]$ given by

$$x_i = (i - 1)\epsilon, \qquad \text{for } i = 1, \ldots, M, \quad \text{and} \quad x_{M+1} = M\epsilon \leq 1.$$

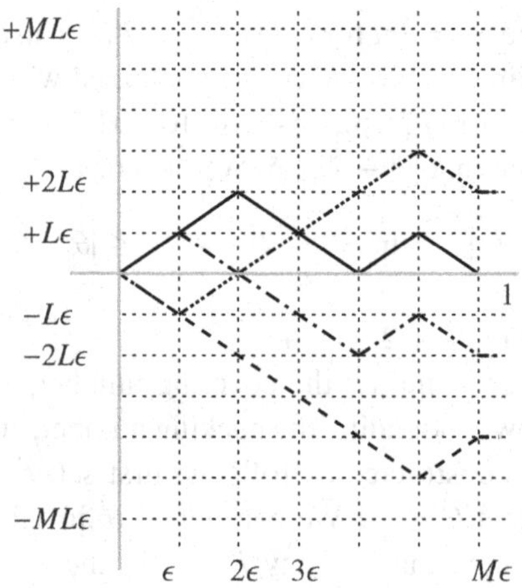

Figure 5.2 The function class $\{f_\beta, \beta \in \{-1,+1\}^M\}$ used to construct a packing of the Lipschitz class $\mathscr{F}_L$. Each function is piecewise linear over the intervals $[0,\epsilon]$, $[\epsilon, 2\epsilon]$, $\ldots$, $[(M-1)\epsilon, M\epsilon]$ with slope either $+L$ or $-L$. There are 2^M functions in total, where $M = \lfloor 1/\epsilon \rfloor$.

Moreover, define the function $\phi\colon \mathbb{R} \to \mathbb{R}_+$ via

$$\phi(u) := \begin{cases} 0 & \text{for } u < 0, \\ u & \text{for } u \in [0,1], \\ 1 & \text{otherwise.} \end{cases} \tag{5.13}$$

For each binary sequence $\beta \in \{-1,+1\}^M$, we may then define a function f_β mapping the unit interval $[0,1]$ to $[-L,+L]$ via

$$f_\beta(y) = \sum_{i=1}^{M} \beta_i L\epsilon\, \phi\left(\frac{y - x_i}{\epsilon}\right). \tag{5.14}$$

By construction, each function f_β is piecewise linear and continuous, with slope either $+L$ or $-L$ over each of the intervals $[\epsilon(i-1), \epsilon i]$ for $i = 1, \ldots, M$, and constant on the remaining interval $[M\epsilon, 1]$; see Figure 5.2 for an illustration. Moreover, it is straightforward to verify that $f_\beta(0) = 0$ and that f_β is Lipschitz with constant L, which ensures that $f_\beta \in \mathscr{F}_L$.

Given a pair of distinct binary strings $\beta \neq \beta'$ and the two functions f_β and $f_{\beta'}$, there is at least one interval where the functions start at the same point, and have the opposite slope over an interval of length ϵ. Since the functions have slopes $+L$ and $-L$, respectively, we are guaranteed that $\|f_\beta - f_{\beta'}\|_\infty \geq 2L\epsilon$, showing that the set $\{f_\beta, \beta \in \{-1,+1\}^M\}$ forms a $2L\epsilon$ packing in the sup-norm. Since this set has cardinality $2^M = 2^{\lfloor 1/\epsilon \rfloor}$, after making the substitution $\epsilon = \delta/L$ and using Lemma 5.5, we conclude that

$$\log N(\delta\,;\, \mathscr{F}_L, \|\cdot\|_\infty) \gtrsim L/\delta.$$

With a little more effort, it can also be shown that the collection of functions $\{f_\beta, \beta \in \{-1,+1\}^M\}$ defines a suitable covering of the set $\mathscr{F}_L$, which establishes the overall claim (5.12). ♣

The preceding example can be extended to Lipschitz functions on the unit cube in higher dimensions, meaning real-valued functions on $[0,1]^d$ such that

$$|f(x) - f(y)| \leq L \|x - y\|_\infty \qquad \text{for all } x, y \in [0,1]^d, \tag{5.15}$$

a class that we denote by $\mathscr{F}_L([0,1]^d)$. An extension of our argument can then be used to show that

$$\log N_\infty(\delta\,;\, \mathscr{F}_L([0,1]^d) \asymp (L/\delta)^d.$$

It is worth contrasting the *exponential dependence* of this metric entropy on the dimension d, as opposed to the linear dependence that we saw earlier for simpler sets (e.g., such as d-dimensional unit balls). This is a dramatic manifestation of the curse of dimensionality.

Another direction in which Example 5.10 can be extended is to classes of functions that have higher-order derivatives.

Example 5.11 (Higher-order smoothness classes) We now consider an example of a function class based on controlling higher-order derivatives. For a suitably differentiable function f, let us adopt the notation $f^{(k)}$ to mean the kth derivative. (Of course, $f^{(0)} = f$ in this notation.) For some integer α and parameter $\gamma \in (0,1]$, consider the class of functions $f\colon [0,1] \to \mathbb{R}$ such that

$$|f^{(j)}(x)| \leq C_j \qquad \text{for all } x \in [0,1],\ j = 0, 1, \ldots, \alpha, \tag{5.16a}$$

$$|f^{(\alpha)}(x) - f^{(\alpha)}(x')| \leq L\,|x - x'|^\gamma, \qquad \text{for all } x, x' \in [0,1]. \tag{5.16b}$$

We claim that the metric entropy of this function class $\mathscr{F}_{\alpha,\gamma}$ scales as

$$\log N(\delta\,;\, \mathscr{F}_{\alpha,\gamma}, \|\cdot\|_\infty) \asymp \left(\frac{1}{\delta}\right)^{\frac{1}{\alpha+\gamma}}. \tag{5.17}$$

(Here we have absorbed the dependence on the constants C_j and L into the order notation.) Note that this claim is consistent with our calculation in Example 5.10, which is essentially the same as the class $\mathscr{F}_{0,1}$.

Let us prove the lower bound in the claim (5.17). As in the previous example, we do so by constructing a packing $\{f_\beta,\ \beta \in \{-1,+1\}^M\}$ for a suitably chosen integer M. Define the function

$$\phi(y) := \begin{cases} c\, 2^{2(\alpha+\gamma)} y^{\alpha+\gamma} (1-y)^{\alpha+\gamma} & \text{for } y \in [0,1], \\ 0 & \text{otherwise.} \end{cases} \tag{5.18}$$

If the pre-factor c is chosen small enough (as a function of the constants C_j and L), it can be seen that the function ϕ satisfies the conditions (5.16). Now for some $\epsilon > 0$, let us set $\delta = (\epsilon/c)^{1/(\alpha+\gamma)}$. By adjusting c as needed, this can be done such that $M := \lfloor 1/\delta \rfloor < 1/\delta$, so that we consider the points in $[0,1]$ given by

$$x_i = (i-1)\delta, \quad \text{for } i = 1, \ldots, M, \quad \text{and} \quad x_{M+1} = M\delta < 1.$$

For each $\beta \in \{-1, +1\}^M$, let us define the function

$$f_\beta(x) := \sum_{i=1}^{M} \beta_i \delta^{1/(\alpha+\gamma)} \phi\left(\frac{x - x_i}{\delta}\right), \tag{5.19}$$

and note that it also satisfies the conditions (5.16). Finally, for two binary strings $\beta \neq \beta'$, there must exist some $i \in \{1, \ldots, M\}$ and an associated interval $I_{i-1} = [x_{i-1}, x_i]$ such that

$$|f_\beta(x) - f_{\beta'}(x)| = 2^{1+2(\alpha+\gamma)} c \delta^{1/(\alpha+\gamma)} \phi\left(\frac{x - x_i}{\delta}\right) \qquad \text{for all } x \in I_{i-1}.$$

By setting $x = x_i + \delta/2$, we see that

$$\|f_\beta - f_{\beta'}\|_\infty \geq 2c\, \delta^{\alpha+\gamma} = 2\epsilon,$$

so that the set $\{f_\beta, \beta \in \{-1, +1\}^M\}$ is a 2ϵ-packing. Thus, we conclude that

$$\log N(\delta\, ; \mathscr{F}_{\alpha,\gamma}, \|\cdot\|_\infty) \gtrsim (1/\delta) \asymp (1/\epsilon)^{1/(\alpha+\gamma)},$$

as claimed. ♣

Various types of function classes can be defined in terms of orthogonal expansions. Concretely, suppose that we are given a sequence of functions $(\phi_j)_{j=1}^\infty$ belonging to $L^2[0, 1]$ and such that

$$\langle \phi_i, \phi_j \rangle_{L^2[0,1]} := \int_0^1 \phi_i(x)\phi_j(x)\, dx = \begin{cases} 1 & \text{if } i = j, \\ 0 & \text{otherwise.} \end{cases}$$

For instance, the cosine basis is one such orthonormal basis, and there are many other interesting ones. Given such a basis, any function $f \in L^2[0, 1]$ can be expanded in the form $f = \sum_{j=1}^\infty \theta_j \phi_j$, where the expansion coefficients are given by the inner products $\theta_j = \langle f, \phi_j \rangle$. By Parseval's theorem, we have $\|f\|_2^2 = \sum_{j=1}^\infty \theta_j^2$ so that $\|f\|_2 < \infty$ if and only if $(\theta_j)_{j=1}^\infty \in \ell^2(\mathbb{N})$, the space of all square summable sequences. Various interesting classes of functions can be obtained by imposing additional constraints on the class of sequences, and one example is that of an ellipsoid constraint.

Example 5.12 (Function classes based on ellipsoids in $\ell^2(\mathbb{N})$) Given a sequence of non-negative real numbers $(\mu_j)_{j=1}^\infty$ such that $\sum_{j=1}^\infty \mu_j < \infty$, consider the ellipsoid

$$\mathcal{E} = \left\{ (\theta_j)_{j=1}^\infty \;\middle|\; \sum_{j=1}^\infty \frac{\theta_j^2}{\mu_j} \leq 1 \right\} \subset \ell^2(\mathbb{N}). \tag{5.20}$$

Such ellipsoids play an important role in our discussion of reproducing kernel Hilbert spaces (see Chapter 12). In this example, we study the ellipsoid specified by the sequence $\mu_j = j^{-2\alpha}$ for some parameter $\alpha > 1/2$. Ellipsoids of this type arise from certain classes of α-times-differentiable functions; see Chapter 12 for details.

We claim that the metric entropy of the associated ellipsoid with respect to the norm $\|\cdot\|_2 = \|\cdot\|_{\ell^2(\mathbb{N})}$ scales as

$$\log N(\delta; \mathcal{E}, \|\cdot\|_2) \asymp \left(\frac{1}{\delta}\right)^{1/\alpha} \qquad \text{for all suitably small } \delta > 0. \tag{5.21}$$

We begin by proving the upper bound—in particular, for a given $\delta > 0$, let us upper bound[2] the covering number $N(\sqrt{2}\delta)$. Let d be the smallest integer such that $\mu_d \leq \delta^2$, and consider the truncated ellipsoid

$$\widetilde{\mathcal{E}} := \{\theta \in \mathcal{E} \mid \theta_j = 0 \qquad \text{for all } j \geq d+1\}.$$

We claim that any δ-cover of this truncated ellipsoid, say $\{\theta^1, \ldots, \theta^N\}$, forms a $\sqrt{2}\delta$-cover of the full ellipsoid. Indeed, for any $\theta \in \mathcal{E}$, we have

$$\sum_{j=d+1}^{\infty} \theta_j^2 \leq \mu_d \sum_{j=d+1}^{\infty} \frac{\theta_j^2}{\mu_j} \leq \delta^2,$$

and hence

$$\min_{k \in [N]} \|\theta - \theta^k\|_2^2 = \min_{k \in [N]} \sum_{j=1}^{d} (\theta_j - \theta_j^k)^2 + \sum_{j=d+1}^{\infty} \theta_j^2 \leq 2\delta^2.$$

Consequently, it suffices to upper bound the cardinality N of this covering of $\widetilde{\mathcal{E}}$. Since $\delta^2 \leq \mu_j$ for all $j \in \{1, \ldots, d\}$, if we view $\widetilde{\mathcal{E}}$ as a subset of $\mathbb{R}^d$, then it contains the ball $\mathbb{B}_2^d(\delta)$, and hence $\text{vol}(\widetilde{\mathcal{E}} + \mathbb{B}_2^d(\delta/2)) \leq \text{vol}(2\widetilde{\mathcal{E}})$. Consequently, by Lemma 5.7, we have

$$N \leq \left(\frac{2}{\delta}\right)^d \frac{\text{vol}(\widetilde{\mathcal{E}} + \mathbb{B}_2^d(\delta/2))}{\text{vol}(\mathbb{B}_2^d(1))} \leq \left(\frac{4}{\delta}\right)^d \frac{\text{vol}(\widetilde{\mathcal{E}})}{\text{vol}(\mathbb{B}_2^d(1))}.$$

By standard formulae for the volume of ellipsoids, we have $\frac{\text{vol}(\widetilde{\mathcal{E}})}{\text{vol}(\mathbb{B}_2^d(1))} = \prod_{j=1}^d \sqrt{\mu_j}$. Putting together the pieces, we find that

$$\log N \leq d \log(4/\delta) + \frac{1}{2} \sum_{j=1}^{d} \log \mu_j \stackrel{\text{(i)}}{=} d \log(4/\delta) - \alpha \sum_{j=1}^{d} \log j,$$

where step (i) follows from the substitution $\mu_j = j^{-2\alpha}$. Using the elementary inequality $\sum_{j=1}^d \log j \geq d \log d - d$, we have

$$\log N \leq d(\log 4 + \alpha) + d\{\log(1/\delta) - \alpha \log d\} \leq d(\log 4 + \alpha),$$

where the final inequality follows since $\mu_d = d^{-2\alpha} \leq \delta^2$, which is equivalent to $\log(\frac{1}{\delta}) \leq \alpha \log d$. Since $(d-1)^{-2\alpha} \geq \delta^2$, we have $d \leq (1/\delta)^{1/\alpha} + 1$, and hence

$$\log N \leq \left\{\left(\frac{1}{\delta}\right)^{\frac{1}{\alpha}} + 1\right\} (\log 4 + \alpha),$$

which completes the proof of the upper bound.

For the lower bound, we note that the ellipsoid $\mathcal{E}$ contains the truncated ellipsoid $\widetilde{\mathcal{E}}$, which (when viewed as a subset of $\mathbb{R}^d$) contains the ball $\mathbb{B}_2^d(\delta)$. Thus, we have

$$\log N\left(\frac{\delta}{2}; \mathcal{E}, \|\cdot\|_2\right) \geq \log N\left(\frac{\delta}{2}; \mathbb{B}_2^d(\delta), \|\cdot\|_2\right) \geq d \log 2,$$

where the final inequality uses the lower bound (5.9) from Example 5.8. Given the inequality $d \geq (1/\delta)^{1/\alpha}$, we have established the lower bound in our original claim (5.21). ♣

[2] The additional factor of $\sqrt{2}$ is irrelevant for the purposes of establishing the claimed scaling (5.21).

5.2 Gaussian and Rademacher complexity

Although metric entropy is a purely deterministic concept, it plays a fundamental role in understanding the behavior of stochastic processes. Given a collection of random variables $\{X_\theta, \theta \in \mathbb{T}\}$ indexed by $\mathbb{T}$, it is frequently of interest to analyze how the behavior of this stochastic process depends on the structure of the set $\mathbb{T}$. In the other direction, given knowledge of a stochastic process indexed by $\mathbb{T}$, it is often possible to infer certain properties of the set $\mathbb{T}$. In our treatment to follow, we will see instances of both directions of this interplay.

An important example of this interplay is provided by the stochastic processes that define the Gaussian and Rademacher complexities. Given a set $\mathbb{T} \subseteq \mathbb{R}^d$, the family of random variables $\{G_\theta, \theta \in \mathbb{T}\}$, where

$$G_\theta := \langle w, \theta \rangle = \sum_{i=1}^{d} w_i \theta_i, \qquad \text{with } w_i \sim \mathcal{N}(0,1), \text{ i.i.d.}, \tag{5.22}$$

defines a stochastic process is known as the *canonical Gaussian process* associated with $\mathbb{T}$. As discussed earlier in Chapter 2, its expected supremum

$$\mathcal{G}(\mathbb{T}) := \mathbb{E}\Big[\sup_{\theta \in \mathbb{T}} \langle \theta, w \rangle\Big] \tag{5.23}$$

is known as the *Gaussian complexity* of $\mathbb{T}$. Like the metric entropy, the functional $\mathcal{G}(\mathbb{T})$ measures the size of the set $\mathbb{T}$ in a certain sense. Replacing the standard Gaussian variables with random signs yields the *Rademacher process* $\{R_\theta, \theta \in \mathbb{T}\}$, where

$$R_\theta := \langle \varepsilon, \theta \rangle = \sum_{i=1}^{d} \varepsilon_i \theta_i, \qquad \text{with } \varepsilon_i \text{ uniform over } \{-1,+1\}, \text{ i.i.d.} \tag{5.24}$$

Its expectation $\mathcal{R}(\mathbb{T}) := \mathbb{E}[\sup_{\theta \in \mathbb{T}} \langle \theta, \varepsilon \rangle]$ is known as the *Rademacher complexity* of $\mathbb{T}$. As shown in Exercise 5.5, we have $\mathcal{R}(\mathbb{T}) \leq \sqrt{\frac{\pi}{2}}\,\mathcal{G}(\mathbb{T})$ for any set $\mathbb{T}$, but there are sets for which the Gaussian complexity is substantially larger than the Rademacher complexity.

Example 5.13 (Rademacher/Gaussian complexity of Euclidean ball $\mathbb{B}_2^d$) Let us compute the Rademacher and Gaussian complexities of the Euclidean ball of unit norm—that is, $\mathbb{B}_2^d = \{\theta \in \mathbb{R}^d \mid \|\theta\|_2 \leq 1\}$. Computing the Rademacher complexity is straightforward: indeed, the Cauchy–Schwarz inequality implies that

$$\mathcal{R}(\mathbb{B}_2^d) = \mathbb{E}\Big[\sup_{\|\theta\|_2 \leq 1} \langle \theta, \varepsilon \rangle\Big] = \mathbb{E}\Big[\Big(\sum_{i=1}^{d} \varepsilon_i^2\Big)^{1/2}\Big] = \sqrt{d}.$$

The same argument shows that $\mathcal{G}(\mathbb{B}_2^d) = \mathbb{E}[\|w\|_2]$ and by concavity of the square-root function and Jensen's inequality, we have

$$\mathbb{E}\|w\|_2 \leq \sqrt{\mathbb{E}[\|w\|_2^2]} = \sqrt{d},$$

so that we have the upper bound $\mathcal{G}(\mathbb{B}_2^d) \leq \sqrt{d}$. On the other hand, it can be shown that $\mathbb{E}\|w\|_2 \geq \sqrt{d}\,(1 - o(1))$. This is a good exercise to work through, using concentration bounds for χ^2 variates from Chapter 2. Combining these upper and lower bounds, we conclude that

$$\mathcal{G}(\mathbb{B}_2^d)/\sqrt{d} = 1 - o(1), \tag{5.25}$$

so that the Rademacher and Gaussian complexities of $\mathbb{B}_2^d$ are essentially equivalent. ♣

Example 5.14 (Rademacher/Gaussian complexity of $\mathbb{B}_1^d$) As a second example, let us consider the ℓ_1-ball in d dimensions, denoted by $\mathbb{B}_1^d$. By the duality between the ℓ_1- and ℓ_∞-norms (or equivalently, using Hölder's inequality), we have

$$\mathcal{R}(\mathbb{B}_1^d) = \mathbb{E}\left[\sup_{\|\theta\|_1 \leq 1} \langle \theta, \varepsilon \rangle\right] = \mathbb{E}[\|\varepsilon\|_\infty] = 1.$$

Similarly, we have $\mathcal{G}(\mathbb{B}_1^d) = \mathbb{E}[\|w\|_\infty]$, and using the result of Exercise 2.11 on Gaussian maxima, we conclude that

$$\mathcal{G}(\mathbb{B}_1^d)/\sqrt{2\log d} = 1 \pm o(1). \tag{5.26}$$

Thus, we see that the Rademacher and Gaussian complexities can differ by a factor of the order $\sqrt{\log d}$; as shown in Exercise 5.5, this difference turns out to be the worst possible. But in either case, comparing with the Rademacher/Gaussian complexity of the Euclidean ball (5.25) shows that the ℓ_1-ball is a much smaller set. ♣

Example 5.15 (Gaussian complexity of ℓ_0-balls) We now turn to the Gaussian complexity of a set defined in a combinatorial manner. As we explore at more length in later chapters, sparsity plays an important role in many classes of high-dimensional statistical models. The ℓ_1-norm, as discussed in Example 5.14, is a convex constraint used to enforce sparsity. A more direct and combinatorial way[3] is by limiting the number $\|\theta\|_0 := \sum_{j=1}^d \mathbb{I}[\theta_j \neq 0]$ of non-zero entries in θ. For some integer $s \in \{1, 2, \ldots, d\}$, the ℓ_0-"ball" of radius s is given by

$$\mathbb{B}_0^d(s) := \{\theta \in \mathbb{R}^d \mid \|\theta\|_0 \leq s\}. \tag{5.27}$$

This set is non-convex, corresponding to the union of $\binom{d}{s}$ subspaces, one for each of the possible s-sized subsets of d coordinates. Since it contains these subspaces, it is also an unbounded set, so that, in computing any type of complexity measure, it is natural to impose an additional constraint. For instance, let us consider the Gaussian complexity of the set

$$\mathbb{S}^d(s) := \mathbb{B}_0^d(s) \cap \mathbb{B}_2^d(1) = \{\theta \in \mathbb{R}^d \mid \|\theta\|_0 \leq s, \text{ and } \|\theta\|_2 \leq 1\}. \tag{5.28}$$

Exercise 5.7 leads the reader through the steps required to establish the upper bound

$$\mathcal{G}(\mathbb{S}^d(s)) \precsim \sqrt{s \log \frac{ed}{s}}, \tag{5.29}$$

where $e \approx 2.7183$ is defined as usual. Moreover, we show in Exercise 5.8 that this bound is tight up to constant factors. ♣

The preceding examples focused on subsets of vectors in $\mathbb{R}^d$. Gaussian complexity also plays an important role in measuring the size of different classes of functions. For a given

[3] Despite our notation, the ℓ_0-"norm" is not actually a norm in the usual sense of the word.

class $\mathscr{F}$ of real-valued functions with domain $\mathcal{X}$, let $x_1^n := \{x_1, \ldots, x_n\}$ be a collection of n points within the domain, known as the *design points*. We can then define the set

$$\mathscr{F}(x_1^n) := \{(f(x_1), f(x_2), \ldots, f(x_n)) \mid f \in \mathscr{F}\} \subseteq \mathbb{R}^n. \tag{5.30}$$

Bounding the Gaussian complexity of this subset of $\mathbb{R}^n$ yields a measure of the complexity of $\mathscr{F}$ at scale n; this measure plays an important role in our analysis of nonparametric least squares in Chapter 13.

It is most natural to analyze a version of the set $\mathscr{F}(x_1^n)$ that is rescaled, either by $n^{-1/2}$ or by n^{-1}. It is useful to observe that the Euclidean metric on the rescaled set $\frac{\mathscr{F}(x_1^n)}{\sqrt{n}}$ corresponds to the *empirical $L^2(\mathbb{P}_n)$-metric* on the function space $\mathscr{F}$—viz.

$$\|f - g\|_n := \sqrt{\frac{1}{n}\sum_{i=1}^{n}(f(x_i) - g(x_i))^2}. \tag{5.31}$$

Note that, if the function class $\mathscr{F}$ is uniformly bounded (i.e., $\|f\|_\infty \leq b$ for all $f \in \mathscr{F}$), then we also have $\|f\|_n \leq b$ for all $f \in \mathscr{F}$. In this case, we always have the following (trivial) upper bound

$$\mathcal{G}\left(\frac{\mathscr{F}(x_1^n)}{n}\right) = \mathbb{E}\left[\sup_{f\in\mathscr{F}}\sum_{i=1}^{n}\frac{w_i}{\sqrt{n}}\frac{f(x_i)}{\sqrt{n}}\right] \leq b\,\frac{\mathbb{E}[\|w\|_2]}{\sqrt{n}} \leq b,$$

where we have recalled our analysis of $\mathbb{E}[\|w\|_2]$ from Example 5.13. Thus, a bounded function class (evaluated at n points) has Gaussian complexity that is never larger than a (scaled) Euclidean ball in $\mathbb{R}^n$. A more refined analysis will show that the Gaussian complexity of $\frac{\mathscr{F}(x_1^n)}{n}$ is often substantially smaller, depending on the structure of $\mathscr{F}$. We will study many instances of such refined bounds in the sequel.

5.3 Metric entropy and sub-Gaussian processes

Both the canonical Gaussian process (5.22) and the Rademacher process (5.24) are particular examples of sub-Gaussian processes, which we now define in more generality.

Definition 5.16 A collection of zero-mean random variables $\{X_\theta, \theta \in \mathbb{T}\}$ is a *sub-Gaussian process* with respect to a metric ρ_X on $\mathbb{T}$ if

$$\mathbb{E}[e^{\lambda(X_\theta - X_{\widetilde{\theta}})}] \leq e^{\frac{\lambda^2\rho_X^2(\theta,\widetilde{\theta})}{2}} \quad \text{for all } \theta, \widetilde{\theta} \in \mathbb{T}, \text{ and } \lambda \in \mathbb{R}. \tag{5.32}$$

By the results of Chapter 2, the bound (5.32) implies the tail bound

$$\mathbb{P}[|X_\theta - X_{\widetilde{\theta}}| \geq t] \leq 2e^{-\frac{t^2}{2\rho_X^2(\theta,\widetilde{\theta})}},$$

and imposing such a tail bound is an equivalent way in which to define a sub-Gaussian process. It is easy to see that the canonical Gaussian and Rademacher processes are both sub-Gaussian with respect to the Euclidean metric $\|\theta - \widetilde{\theta}\|_2$.

Given a sub-Gaussian process, we use the notation $N_X(\delta\,;\,\mathbb{T})$ to denote the δ-covering number of $\mathbb{T}$ with respect to ρ_X, and $N_2(\delta\,;\,\mathbb{T})$ to denote the covering number with respect to the Euclidean metric $\|\cdot\|_2$, corresponding to the case of a canonical Gaussian process. As we now discuss, these metric entropies can be used to construct upper bounds on various expected suprema involving the process.

5.3.1 Upper bound by one-step discretization

We begin with a simple upper bound obtained via a discretization argument. The basic idea is natural: by approximating the set $\mathbb{T}$ up to some accuracy δ, we may replace the supremum over $\mathbb{T}$ by a finite maximum over the δ-covering set, plus an approximation error that scales proportionally with δ. We let $D := \sup_{\theta,\widetilde{\theta}\in\mathbb{T}} \rho_X(\theta,\widetilde{\theta})$ denote the diameter of $\mathbb{T}$, and let $N_X(\delta\,;\,\mathbb{T})$ denote the δ-covering number of $\mathbb{T}$ in the ρ_X-metric.

Proposition 5.17 (One-step discretization bound) *Let $\{X_\theta,\,\theta\in\mathbb{T}\}$ be a zero-mean sub-Gaussian process with respect to the metric ρ_X. Then for any $\delta\in[0,D]$ such that $N_X(\delta\,;\,\mathbb{T})\geq 10$, we have*

$$\mathbb{E}\Big[\sup_{\theta,\widetilde{\theta}\in\mathbb{T}}(X_\theta - X_{\widetilde{\theta}})\Big] \leq 2\,\mathbb{E}\Bigg[\sup_{\substack{\gamma,\gamma'\in\mathbb{T}\\ \rho_X(\gamma,\gamma')\leq\delta}}(X_\gamma - X_{\gamma'})\Bigg] + 4\sqrt{D^2\log N_X(\delta\,;\,\mathbb{T})}. \tag{5.33}$$

Remarks: It is convenient to state the upper bound in terms of the increments $X_\theta - X_{\widetilde{\theta}}$ so as to avoid issues of considering where the set $\mathbb{T}$ is centered. However, the claim (5.33) always implies an upper bound on $\mathbb{E}[\sup_{\theta\in\mathbb{T}} X_\theta]$, since the zero-mean condition means that

$$\mathbb{E}\Big[\sup_{\theta\in\mathbb{T}} X_\theta\Big] = \mathbb{E}\Big[\sup_{\theta\in\mathbb{T}}(X_\theta - X_{\theta_0})\Big] \leq \mathbb{E}\Big[\sup_{\theta,\widetilde{\theta}\in\mathbb{T}}(X_\theta - X_{\widetilde{\theta}})\Big].$$

For each $\delta\in[0,D]$, the upper bound (5.33) consists of two quantities, corresponding to approximation error and estimation error, respectively. As $\delta\to 0^+$, the approximation error (involving the constraint $\rho_X(\gamma,\gamma')\leq\delta$) shrinks to zero, whereas the estimation error (involving the metric entropy) grows. In practice, one chooses δ so as to achieve the optimal trade-off between these two terms.

Proof For a given $\delta\geq 0$ and associated covering number $N = N_X(\delta\,;\,\mathbb{T})$, let $\{\theta^1,\ldots,\theta^N\}$ be a δ-cover of $\mathbb{T}$. For any $\theta\in\mathbb{T}$, we can find some θ^i such that $\rho_X(\theta,\theta^i)\leq\delta$, and hence

$$\begin{aligned} X_\theta - X_{\theta^1} &= (X_\theta - X_{\theta^i}) + (X_{\theta^i} - X_{\theta^1}) \\ &\leq \sup_{\substack{\gamma,\gamma'\in\mathbb{T}\\ \rho_X(\gamma,\gamma')\leq\delta}}(X_\gamma - X_{\gamma'}) + \max_{i=1,2,\ldots,N}|X_{\theta^i} - X_{\theta^1}|. \end{aligned}$$

Given some other arbitrary $\widetilde{\theta}\in\mathbb{T}$, the same upper bound holds for $X_{\theta^1} - X_{\widetilde{\theta}}$, so that adding together the bounds, we obtain

$$\sup_{\theta,\widetilde{\theta}\in\mathbb{T}}(X_\theta - X_{\widetilde{\theta}}) \leq 2 \sup_{\substack{\gamma,\gamma'\in\mathbb{T}\\ \rho_X(\gamma,\gamma')\leq\delta}} (X_\gamma - X_{\gamma'}) + 2 \max_{i=1,2,\ldots,N} |X_{\theta^i} - X_{\theta^1}|. \tag{5.34}$$

Now by assumption, for each $i = 1, 2, \ldots, N$, the random variable $X_{\theta^i} - X_{\theta^1}$ is zero-mean and sub-Gaussian with parameter at most $\rho_X(\theta^i, \theta^1) \leq D$. Consequently, by the behavior of sub-Gaussian maxima (see Exercise 2.12(c)), we are guaranteed that

$$\mathbb{E}\Big[\max_{i=1,\ldots,N} |X_{\theta^i} - X_{\theta^1}|\Big] \leq 2\sqrt{D^2 \log N},$$

which yields the claim. □

In order to gain intuition, it is worth considering the special case of the canonical Gaussian (or Rademacher) process, in which case the relevant metric is the Euclidean norm $\|\theta - \widetilde{\theta}\|_2$. In order to reduce to the essential aspects of the problem, consider a set $\mathbb{T}$ that contains the origin. The arguments[4] leading to the bound (5.33) imply that the Gaussian complexity $\mathcal{G}(\mathbb{T})$ is upper bounded as

$$\mathcal{G}(\mathbb{T}) \leq \min_{\delta\in[0,D]} \Big\{\mathcal{G}(\widetilde{\mathbb{T}}(\delta)) + 2\sqrt{D^2 \log N_2(\delta\,;\,\mathbb{T})}\Big\}, \tag{5.35}$$

where $N_2(\delta\,;\,\mathbb{T})$ is the δ-covering number in the ℓ_2-norm, and

$$\widetilde{\mathbb{T}}(\delta) := \{\gamma - \gamma' \mid \gamma, \gamma' \in \mathbb{T},\ \|\gamma - \gamma'\|_2 \leq \delta\}.$$

The quantity $\mathcal{G}(\widetilde{\mathbb{T}}(\delta))$ is referred to as a *localized Gaussian complexity*, since it measures the complexity of the set $\mathbb{T}$ within an ℓ_2-ball of radius δ. This idea of localization plays an important role in obtaining optimal rates for statistical problems; see Chapters 13 and 14 for further discussion. We note also that analogous upper bounds hold for the Rademacher complexity $\mathcal{R}(\mathbb{T})$ in terms of a localized Rademacher complexity.

In order to obtain concrete results from the discretization bound (5.35), it remains to upper bound the localized Gaussian complexity, and then optimize the choice of δ. When $\mathbb{T}$ is a subset of $\mathbb{R}^d$, the Cauchy–Schwarz inequality yields

$$\mathcal{G}(\widetilde{\mathbb{T}}(\delta)) = \mathbb{E}\Big[\sup_{\theta\in\widetilde{\mathbb{T}}(\delta)} \langle\theta,\, w\rangle\Big] \leq \delta\,\mathbb{E}[\|w\|_2] \leq \delta\sqrt{d},$$

which leads to the *naive discretization bound*

$$\mathcal{G}(\mathbb{T}) \leq \min_{\delta\in[0,D]} \Big\{\delta\sqrt{d} + 2\sqrt{D^2 \log N_2(\delta\,;\,\mathbb{T})}\Big\}. \tag{5.36}$$

For some sets, this simple bound can yield useful results, whereas for other sets, the local Gaussian (or Rademacher) complexity needs to be controlled with more care.

[4] In this case, the argument can be refined so as to remove a factor of 2.

5.3.2 Some examples of discretization bounds

Let us illustrate the use of the bounds (5.33), (5.35) and (5.36) with some examples.

Example 5.18 (Gaussian complexity of unit ball) Recall our discussion of the Gaussian complexity of the Euclidean ball $\mathbb{B}_2^d$ from Example 5.13: using direct methods, we proved the scaling $\mathcal{G}(\mathbb{B}_2^d) = \sqrt{d}\,(1 - o(1))$. The purpose of this example is to show that Proposition 5.17 yields an upper bound with this type of scaling (albeit with poor control of the pre-factor). In particular, recall from Example 5.8 that the metric entropy number of the Euclidean ball is upper bounded as $\log N_2(\delta; \mathbb{B}_2^d) \leq d \log(1 + \frac{2}{\delta})$. Thus, setting $\delta = 1/2$ in the naive discretization bound (5.36), we obtain

$$\mathcal{G}(\mathbb{B}_2^d) \leq \sqrt{d}\{\tfrac{1}{2} + 2\sqrt{2 \log 5}\,\}.$$

Relative to the exact result, the constant in this result is sub-optimal, but it does have the correct scaling as a function of d. ♣

Example 5.19 (Maximum singular value of sub-Gaussian random matrix) As a more substantive demonstration of Proposition 5.17, let us show how it can be used to control the expected ℓ_2-operator norm of a sub-Gaussian random matrix. Let $\mathbf{W} \in \mathbb{R}^{n\times d}$ be a random matrix with zero-mean i.i.d. entries W_{ij}, each sub-Gaussian with parameter $\sigma = 1$. Examples include the standard Gaussian ensemble $W_{ij} \sim \mathcal{N}(0, 1)$, and the Rademacher ensemble $W_{ij} \in \{-1, +1\}$ equiprobably. The ℓ_2-operator norm (or spectral norm) of the matrix $\mathbf{W}$ is given by its maximum singular value; equivalently, it is defined as $|||\mathbf{W}|||_2 := \sup_{v\in\mathbb{S}^{d-1}} \|\mathbf{W}v\|_2$, where $\mathbb{S}^{d-1} = \{v \in \mathbb{R}^d \mid \|v\|_2 = 1\}$ is the Euclidean unit sphere in $\mathbb{R}^d$. Here we sketch out an approach for proving the bound

$$\mathbb{E}[|||\mathbf{W}|||_2/\sqrt{n}] \precsim 1 + \sqrt{\frac{d}{n}},$$

leaving certain details for the reader in Exercise 5.11.

Let us define the class of matrices

$$\mathbb{M}^{n,d}(1) := \{\boldsymbol{\Theta} \in \mathbb{R}^{n\times d} \mid \operatorname{rank}(\boldsymbol{\Theta}) = 1,\ |||\boldsymbol{\Theta}|||_F = 1\}, \tag{5.37}$$

corresponding to the set of $n \times d$ matrices of rank one with unit Frobenius norm $|||\boldsymbol{\Theta}|||_F^2 = \sum_{i=1}^n \sum_{j=1}^d \Theta_{ij}^2$. As verified in Exercise 5.11(a), we then have the variational representation

$$|||\mathbf{W}|||_2 = \sup_{\boldsymbol{\Theta}\in\mathbb{M}^{n,d}(1)} X_{\boldsymbol{\Theta}}, \quad \text{where} \quad X_{\boldsymbol{\Theta}} := \langle\langle \mathbf{W},\, \boldsymbol{\Theta}\rangle\rangle = \sum_{i=1}^n \sum_{j=1}^d W_{ij}\Theta_{ij}. \tag{5.38}$$

In the Gaussian case, this representation shows that $\mathbb{E}[|||\mathbf{W}|||_2]$ is equal to the Gaussian complexity $\mathcal{G}(\mathbb{M}^{n,d}(1))$. For any sub-Gaussian random matrix, we show in part (b) of Exercise 5.11 that the stochastic process $\{X_{\boldsymbol{\Theta}}, \boldsymbol{\Theta} \in \mathbb{M}^{n,d}(1)\}$ is zero-mean, and sub-Gaussian with respect to the Frobenius norm $|||\boldsymbol{\Theta} - \boldsymbol{\Theta}'|||_F$. Consequently, Proposition 5.17 implies that, for all $\delta \in [0, 1]$, we have the upper bound

$$\mathbb{E}[|||\mathbf{W}|||_2] \leq 2\,\mathbb{E}\left[\sup_{\substack{\operatorname{rank}(\boldsymbol{\Gamma})=\operatorname{rank}(\boldsymbol{\Gamma}')=1\\ |||\boldsymbol{\Gamma}-\boldsymbol{\Gamma}'|||_F\leq\delta}} \langle\langle \boldsymbol{\Gamma} - \boldsymbol{\Gamma}',\, \mathbf{W}\rangle\rangle\right] + 6\sqrt{\log N_F(\delta; \mathbb{M}^{n,d}(1))}, \tag{5.39}$$

where $N_F(\delta; \mathbb{M}^{n,d}(1))$ denotes the δ-covering number in Frobenius norm. In part (c) of Exercise 5.11, we prove the upper bound

$$\mathbb{E}\left[\sup_{\substack{\operatorname{rank}(\boldsymbol{\Gamma})=\operatorname{rank}(\boldsymbol{\Gamma}')=1 \\ |||\boldsymbol{\Gamma}-\boldsymbol{\Gamma}'|||_F \leq \delta}} \langle\langle \boldsymbol{\Gamma} - \boldsymbol{\Gamma}', \, \mathbf{W} \rangle\rangle\right] \leq \sqrt{2}\,\delta\,\mathbb{E}[|||\mathbf{W}|||_2], \tag{5.40}$$

and in part (d), we upper bound the metric entropy as

$$\log N_F(\delta; \mathbb{M}^{n,d}(1)) \leq (n+d)\log\left(1 + \frac{2}{\delta}\right). \tag{5.41}$$

Substituting these upper bounds into inequality (5.39), we obtain

$$\mathbb{E}[|||\mathbf{W}|||_2] \leq \min_{\delta \in [0,1]} \left\{2\sqrt{2}\,\delta\;\mathbb{E}[|||\mathbf{W}|||_2] + 6\sqrt{(n+d)\log\left(1 + \frac{2}{\delta}\right)}\right\}.$$

Fixing $\delta = \frac{1}{4\sqrt{2}}$ (as one particular choice) and rearranging terms yields the upper bound

$$\frac{1}{\sqrt{n}}\,\mathbb{E}[|||\mathbf{W}|||_2] \leq c_1\left\{1 + \sqrt{\frac{d}{n}}\right\}$$

for some universal constant $c_1 > 1$. Again, this yields the correct scaling of $\mathbb{E}[|||\mathbf{W}|||_2]$ as a function of (n, d). As we explore in Exercise 5.14, for Gaussian random matrices, a more refined argument using the Sudakov–Fernique comparison can be used to prove the upper bound with $c_1 = 1$, which is the best possible. In Example 5.33 to follow, we establish a matching lower bound of the same order. ♣

Let us now turn to some examples of Gaussian complexity involving function spaces. Recall the definition (5.30) of the set $\mathscr{F}(x_1^n)$ as well as the empirical L^2-norm (5.31). As a consequence of the inequalities

$$\|f - g\|_n \leq \max_{i=1,\ldots,n} |f(x_i) - g(x_i)| \leq \|f - g\|_\infty,$$

we have the following relations among metric entropies:

$$\log N_2(\delta; \mathscr{F}(x_1^n)/\sqrt{n}) \leq \log N_\infty(\delta\,;\, \mathscr{F}(x_1^n)) \leq \log N(\delta\,;\, \mathscr{F},\, \|\cdot\|_\infty), \tag{5.42}$$

which will be useful in our development.

Example 5.20 (Empirical Gaussian complexity for a parametric function class) Let us bound the Gaussian complexity of the set $\mathscr{P}(x_1^n)/n$ generated by the simple parametric function class $\mathscr{P}$ from Example 5.9. Using the bound (5.42), it suffices to control the ℓ_∞-covering number of $\mathscr{P}$. From our previous calculations, it can be seen that, as long as $\delta \leq 1/4$, we have $\log N_\infty(\delta\,;\,\mathscr{P}) \leq \log(1/\delta)$. Moreover, since the function class is uniformly bounded (i.e., $\|f\|_\infty \leq 1$ for all $f \in \mathscr{P}$), the diameter in empirical L^2-norm is also well-controlled—in particular, we have $D^2 = \sup_{f \in \mathscr{P}} \frac{1}{n}\sum_{i=1}^n f^2(x_i) \leq 1$. Consequently, the discretization bound (5.33) implies that

$$\mathcal{G}(\mathscr{P}(x_1^n)/n) \leq \frac{1}{\sqrt{n}} \inf_{\delta \in (0,1/4]} \left\{\delta\sqrt{n} + 3\sqrt{\log(1/\delta)}\right\}.$$

In order to optimize the scaling of the bound, we set $\delta = 1/(4\sqrt{n})$, and thereby obtain the upper bound

$$\mathcal{G}(\mathscr{P}(x_1^n)/n) \precsim \sqrt{\frac{\log n}{n}}. \tag{5.43}$$

As we will see later, the Gaussian complexity for this function class is actually upper bounded by $1/\sqrt{n}$, so that the crude bound from Proposition 5.17 captures the correct behavior only up to a logarithmic factor. We will later develop more refined techniques that remove this logarithmic factor. ♣

Example 5.21 (Gaussian complexity for smoothness classes) Now recall the class $\mathscr{F}_L$ of Lipschitz functions from Example 5.10. From the bounds on metric entropy given there, as long as $\delta \in (0, \delta_0)$ for a sufficiently small $\delta_0 > 0$, we have $\log N_\infty(\delta\,; \mathscr{F}_L) \leq \frac{cL}{\delta}$ for some constant c. Since the functions in $\mathscr{F}_L$ are uniformly bounded by one, the discretization bound implies that

$$\mathcal{G}(\mathscr{F}_L(x_1^n)/n) \leq \frac{1}{\sqrt{n}} \inf_{\delta \in (0,\delta_0)} \left\{ \delta\sqrt{n} + 3\sqrt{\frac{cL}{\delta}} \right\}.$$

To obtain the tightest possible upper bound (up to constant factors), we set $\delta = n^{-1/3}$, and hence find that

$$\mathcal{G}(\mathscr{F}_L(x_1^n)/n) \precsim n^{-1/3}. \tag{5.44}$$

By comparison to the parametric scaling (5.43), this upper bound decays much more slowly. ♣

5.3.3 Chaining and Dudley's entropy integral

In this section, we introduce an important method known as chaining, and show how it can be used to obtain tighter bounds on the expected suprema of sub-Gaussian processes. Recall the discretization bound from Proposition 5.17: it was based on a simple one-step discretization in which we replaced the supremum over a large set with a finite maximum over a δ-cover plus an approximation error. We then bounded the finite maximum by combining the union bound with a sub-Gaussian tail bound. In this section, we describe a substantial refinement of this procedure, in which we decompose the supremum into a sum of finite maxima over sets that are successively refined. The resulting procedure is known as the *chaining method.*

In this section, we show how chaining can be used to derive a classical upper bound, originally due to Dudley (1967), on the expected supremum of a sub-Gaussian process. In Section 5.6, we show how related arguments can be used to control the probability of a deviation above this expectation. Let $\{X_\theta,\ \theta \in \mathbb{T}\}$ be a zero-mean sub-Gaussian process with respect to the (pseudo)metric ρ_X (see Definition 5.16). Define $D = \sup_{\theta,\widetilde{\theta}\in\mathbb{T}} \rho_X(\theta, \widetilde{\theta})$, and the δ-truncated *Dudley's entropy integral*

$$\mathcal{J}(\delta; D) := \int_\delta^D \sqrt{\log N_X(u\,; \mathbb{T})}\, du, \tag{5.45}$$

where we recall that $N_X(u\,;\,\mathbb{T})$ denotes the δ-covering number of $\mathbb{T}$ with respect to ρ_X.

Theorem 5.22 (Dudley's entropy integral bound) *Let $\{X_\theta,\ \theta \in \mathbb{T}\}$ be a zero-mean sub-Gaussian process with respect to the induced pseudometric ρ_X from Definition 5.16. Then for any $\delta \in [0, D]$, we have*

$$\mathbb{E}\Big[\sup_{\theta,\widetilde{\theta}\in\mathbb{T}}(X_\theta - X_{\widetilde{\theta}})\Big] \leq 2\,\mathbb{E}\Big[\sup_{\substack{\gamma,\gamma'\in\mathbb{T}\\ \rho_X(\gamma,\gamma')\leq\delta}}(X_\gamma - X_{\gamma'})\Big] + 32\,\mathcal{J}(\delta/4; D). \tag{5.46}$$

Remarks: There is no particular significance to the constant 32, which could be improved with a more careful analysis. We have stated the bound in terms of the increment $X_\theta - X_{\widetilde{\theta}}$, but it can easily be translated into an upper bound on $\mathbb{E}[\sup_{\theta\in\mathbb{T}} X_\theta]$. (See the discussion following Proposition 5.17.) The usual form of Dudley's bound corresponds to the case $\delta = 0$, and so is in terms of the entropy integral $\mathcal{J}(0; D)$. The additional flexibility to choose $\delta \in [0, D]$ can be useful in certain problems.

Proof We begin with the inequality (5.34) previously established in the proof of Proposition 5.17—namely,

$$\sup_{\theta,\widetilde{\theta}\in\mathbb{T}}(X_\theta - X_{\widetilde{\theta}}) \leq 2 \sup_{\substack{\gamma,\gamma'\in\mathbb{T}\\ \rho_X(\gamma,\gamma')\leq\delta}}(X_\gamma - X_{\gamma'}) + 2 \max_{i=1,2,\ldots,N} |X_{\theta^i} - X_{\theta^1}|.$$

In the proof of Proposition 5.17, we simply upper bounded the maximum over $i = 1, \ldots, N$ using the union bound. In this proof, we pursue a more refined chaining argument. Define $\mathbb{U} = \{\theta^1, \ldots, \theta^N\}$, and for each integer $m = 1, 2, \ldots, L$, let $\mathbb{U}_m$ be a minimal $\epsilon_m = D2^{-m}$ covering set of $\mathbb{U}$ in the metric ρ_X, where we allow for any element of $\mathbb{T}$ to be used in forming the cover. Since $\mathbb{U}$ is a subset of $\mathbb{T}$, each set has cardinality $N_m := |\mathbb{U}_m|$ upper bounded as $N_m \leq N_X(\epsilon_m\,;\,\mathbb{T})$. Since $\mathbb{U}$ is finite, there is some finite integer L for which $\mathbb{U}_L = \mathbb{U}$. (In particular, for the smallest integer such that $N_L = |\mathbb{U}|$, we can simply choose $\mathbb{U}_L = \mathbb{U}$.) For each $m = 1, \ldots, L$, define the mapping $\pi_m\colon \mathbb{U} \to \mathbb{U}_m$ via

$$\pi_m(\theta) = \arg\min_{\beta\in\mathbb{U}_m} \rho_X(\theta, \beta),$$

so that $\pi_m(\theta)$ is the best approximation of $\theta \in \mathbb{U}$ from the set $\mathbb{U}_m$. Using this notation, we can decompose the random variable X_θ into a sum of increments in terms of an associated sequence $(\gamma^1, \ldots, \gamma^L)$, where we define $\gamma^L = \theta$ and $\gamma^{m-1} := \pi_{m-1}(\gamma^m)$ recursively for $m = L, L-1, \ldots, 2$. By construction, we then have the *chaining relation*

$$X_\theta - X_{\gamma^1} = \sum_{m=2}^{L}(X_{\gamma^m} - X_{\gamma^{m-1}}), \tag{5.47}$$

and hence $|X_\theta - X_{\gamma^1}| \leq \sum_{m=2}^{L} \max_{\beta\in\mathbb{U}_m} |X_\beta - X_{\pi_{m-1}(\beta)}|$. See Figure 5.3 for an illustration of this set-up.

Thus, we have decomposed the difference between X_θ and the final element X_{γ^1} in its

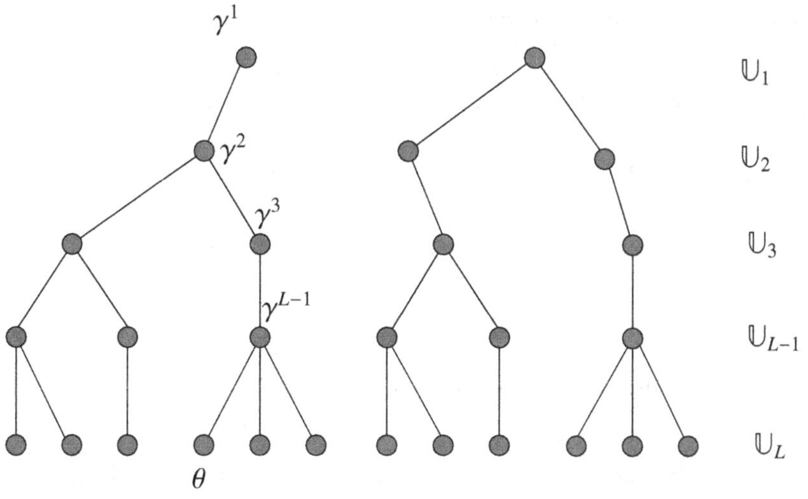

Figure 5.3 Illustration of the chaining relation in the case $L = 5$. The set $\mathbb{U}$, shown at the bottom of the figure, has a finite number of elements. For each $m = 1, \ldots, 5$, we let $\mathbb{U}_m$ be a $D\epsilon^{-m}$-cover of the set $\mathbb{U}$; the elements of the cover are shaded in gray at each level. For each element $\theta \in \mathbb{U}$, we construct the chain by setting $\gamma^5 = \theta$, and then recursively $\gamma^{m-1} = \pi_{m-1}(\gamma^m)$ for $m = 5, \ldots, 2$. We can then decompose the difference $X_\theta - X_{\gamma^1}$ as a sum (5.47) of terms along the edges of a tree.

associated chain as a sum of increments. Given any other $\widetilde{\theta} \in \mathbb{U}$, we can define the chain $\{\widetilde{\gamma}^1, \ldots, \widetilde{\gamma}^L\}$, and then derive an analogous bound for the increment $|X_{\widetilde{\theta}} - X_{\widetilde{\gamma}^1}|$. By appropriately adding and subtracting terms and then applying the triangle inequality, we obtain

$$
\begin{aligned}
|X_\theta - X_{\widetilde{\theta}}| &= |X_{\gamma^1} - X_{\widetilde{\gamma}^1} + (X_\theta - X_{\gamma^1}) + (X_{\widetilde{\gamma}^1} - X_{\widetilde{\theta}})| \\
&\leq |X_{\gamma^1} - X_{\widetilde{\gamma}^1}| + |X_\theta - X_{\gamma^1}| + |X_{\widetilde{\theta}} - X_{\widetilde{\gamma}^1}|.
\end{aligned}
$$

Taking maxima over $\theta, \widetilde{\theta} \in \mathbb{U}$ on the left-hand side and using our upper bounds on the right-hand side, we obtain

$$
\max_{\theta,\widetilde{\theta}\in\mathbb{U}} |X_\theta - X_{\widetilde{\theta}}| \leq \max_{\gamma,\widetilde{\gamma}\in\mathbb{U}_1} |X_\gamma - X_{\widetilde{\gamma}}| + 2 \sum_{m=2}^{L} \max_{\beta\in\mathbb{U}_m} |X_\beta - X_{\pi_{m-1}(\beta)}|.
$$

We first upper bound the finite maximum over $\mathbb{U}_1$, which has $N(\frac{D}{2}) := N_X(\frac{D}{2}; \mathbb{T})$ elements. By the sub-Gaussian nature of the process, the increment $X_\gamma - X_{\widetilde{\gamma}}$ is sub-Gaussian with parameter at most $\rho_X(\gamma, \widetilde{\gamma}) \leq D$. Consequently, by our earlier results on finite Gaussian maxima (see Exercise 2.12), we have

$$
\mathbb{E}\left[\max_{\gamma,\widetilde{\gamma}\in\mathbb{U}_1} |X_\gamma - X_{\widetilde{\gamma}}|\right] \leq 2D\sqrt{\log N(D/2)}.
$$

Similarly, for each $m = 2, 3, \ldots, L$, the set $\mathbb{U}_m$ has $N(D2^{-m})$ elements, and, moreover, $\max_{\beta\in\mathbb{U}_m} \rho_X(\beta, \pi_{m-1}(\beta)) \leq D2^{-(m-1)}$, whence

$$
\mathbb{E}\left[\max_{\beta\in\mathbb{U}_m} |X_\beta - X_{\pi_{m-1}(\beta)}|\right] \leq 2\,D2^{-(m-1)}\sqrt{\log N(D2^{-m})}.
$$

Combining the pieces, we conclude that

$$\mathbb{E}\Big[\max_{\theta,\widetilde{\theta}\in\mathbb{U}}|X_\theta - X_{\widetilde{\theta}}|\Big] \le 4\sum_{m=1}^{L} D2^{-(m-1)}\sqrt{\log N(D2^{-m})}.$$

Since the metric entropy $\log N(t)$ is non-increasing in t, we have

$$D2^{-(m-1)}\sqrt{\log N(D2^{-m})} \le 4\int_{D2^{-(m+1)}}^{D2^{-m}} \sqrt{\log N(u)}\,du,$$

and hence $2\,\mathbb{E}[\max_{\theta,\widetilde{\theta}\in\mathbb{U}}|X_\theta - X_{\widetilde{\theta}}|] \le 32\int_{\delta/4}^{D}\sqrt{\log N(u)}\,du$. □

Let us illustrate the Dudley entropy bound with some examples.

Example 5.23 In Example 5.20, we showed that the Gaussian complexity of the parametric function class $\mathscr{P}$ was upper bounded by $O\big(\sqrt{\frac{\log n}{n}}\big)$, a result obtained by the naive discretization bound. Here we show that the Dudley entropy integral yields the sharper upper bound $O(1/\sqrt{n})$. In particular, since the L_∞-norm metric entropy is upper bounded as $\log N(\delta\,;\,\mathscr{P},\|\cdot\|_\infty) = O(\log(1+1/\delta))$, the Dudley bound implies that

$$\mathcal{G}\Big(\frac{\mathscr{P}(x_1^n)}{n}\Big) \le \frac{c}{\sqrt{n}}\int_0^2 \sqrt{\log(1+1/u)}\,du = \frac{c'}{\sqrt{n}}.$$

Thus, we have removed the logarithmic factor from the naive discretization bound. ♣

Recall from Chapter 4 our discussion of the Vapnik–Chervonenkis dimension. As we now show, the Dudley integral can be used to obtain a sharp result for any finite VC class.

Example 5.24 (Bounds for Vapnik–Chervonenkis classes) Let $\mathscr{F}$ be a b-uniformly bounded class of functions with finite VC dimension ν, and suppose that we are interested in establishing a uniform law of large numbers for $\mathscr{F}$—that is, in controlling the random variable $\sup_{f\in\mathscr{F}}\big|\frac{1}{n}\sum_{i=1}^n f(X_i) - \mathbb{E}[f]\big|$, where $X_i \sim \mathbb{P}$ are i.i.d. samples. As discussed in Chapter 4, by exploiting concentration and symmetrization results, the study of this random variable can be reduced to controlling the expectation $\mathbb{E}_\varepsilon\big[\sup_{f\in\mathscr{F}}\big|\frac{1}{n}\sum_{i=1}^n \varepsilon_i f(x_i)\big|\big]$, where ε_i are i.i.d. Rademacher variables (random signs), and the observations x_i are fixed for the moment.

In order to see how Dudley's entropy integral may be applied, define the zero-mean random variable $Z_f := \frac{1}{\sqrt{n}}\sum_{i=1}^n \varepsilon_i f(x_i)$, and consider the stochastic process $\{Z_f \mid f\in\mathscr{F}\}$. It is straightforward to verify that the increment $Z_f - Z_g$ is sub-Gaussian with parameter

$$\|f-g\|_{\mathbb{P}_n}^2 := \frac{1}{n}\sum_{i=1}^n (f(x_i)-g(x_i))^2.$$

Consequently, by Dudley's entropy integral, we have

$$\mathbb{E}_\varepsilon\Big[\sup_{f\in\mathscr{F}}\Big|\frac{1}{n}\sum_{i=1}^n \varepsilon_i f(x_i)\Big|\Big] \le \frac{24}{\sqrt{n}}\int_0^{2b}\sqrt{\log N(t;\mathscr{F},\|\cdot\|_{\mathbb{P}_n})}\,dt, \tag{5.48}$$

where we have used the fact that $\sup_{f,g\in\mathscr{F}}\|f-g\|_{\mathbb{P}_n} \le 2b$. Now by known results on VC classes and metric entropy, there is a universal constant C such that

$$N(\epsilon; \mathscr{F}, \|\cdot\|_{\mathbb{P}_n}) \leq C\nu(16e)^{\nu}\left(\frac{b}{\epsilon}\right)^{2\nu}. \tag{5.49}$$

See Exercise 5.4 for the proof of a weaker claim of this form, and the bibliographic section for further discussion of such bounds.

Substituting the metric entropy bound (5.49) into the entropy integral (5.48), we find that there are universal constants c_0 and c_1, depending on b but not on (ν, n), such that

$$\begin{aligned}\mathbb{E}_{\varepsilon}\left[\sup_{f\in\mathscr{F}}\left|\frac{1}{n}\sum_{i=1}^{n}\varepsilon_i f(x_i)\right|\right] &\leq c_0\sqrt{\frac{\nu}{n}}\left[1+\int_0^{2b}\sqrt{\log(b/t)}\,dt\right] \\ &= c_0'\sqrt{\frac{\nu}{n}},\end{aligned} \tag{5.50}$$

since the integral is finite. ♣

Note that the bound (5.50) is sharper than the earlier $\sqrt{\frac{\nu\log(n+1)}{n}}$ bound that we proved in Lemma 4.14. It leads to various improvements of previous results that we have stated. For example, consider the classical Glivenko–Cantelli setting, which amounts to bounding $\|\widehat{F}_n - F\|_\infty = \sup_{u\in\mathbb{R}}|\widehat{F}_n(u) - F(u)|$. Since the set of indicator functions has VC dimension $\nu = 1$, the bound (5.50), combined with Theorem 4.10, yields that

$$\mathbb{P}\left[\|\widehat{F}_n - F\|_\infty \geq \frac{c}{\sqrt{n}} + \delta\right] \leq 2e^{-\frac{n\delta^2}{8}} \quad \text{for all } \delta \geq 0, \tag{5.51}$$

where c is a universal constant. Apart from better constants, this bound is unimprovable.

5.4 Some Gaussian comparison inequalities

Suppose that we are given a pair of Gaussian processes, say $\{Y_\theta,\ \theta\in\mathbb{T}\}$ and $\{Z_\theta,\ \theta\in\mathbb{T}\}$, both indexed by the same set $\mathbb{T}$. It is often useful to compare the two processes in some sense, possibly in terms of the expected value of some real-valued function F defined on the processes. One important example is the supremum $F(X) := \sup_{\theta\in\mathbb{T}} X_\theta$. Under what conditions can we say that $F(X)$ is larger (or smaller) than $F(Y)$? Results that allow us to deduce such properties are known as *Gaussian comparison inequalities*, and there are a large number of them. In this section, we derive a few of the standard ones, and illustrate them via a number of examples.

Recall that we have defined the suprema of Gaussian processes by taking limits of maxima over finite subsets. For this reason, it is sufficient to consider the case where $\mathbb{T}$ is finite, say $\mathbb{T} = \{1,\ldots,N\}$ for some integer N. We focus on this case throughout this section, adopting the notation $[N] = \{1,\ldots,N\}$ as a convenient shorthand.

5.4.1 A general comparison result

We begin by stating and proving a fairly general Gaussian comparison principle:

Theorem 5.25 *Let $(X_1, \ldots, X_N)$ and $(Y_1, \ldots, Y_N)$ be a pair of centered Gaussian random vectors, and suppose that there exist disjoint subsets A and B of $[N] \times [N]$ such that*

$$\mathbb{E}[X_i X_j] \leq \mathbb{E}[Y_i Y_j] \qquad \text{for all } (i, j) \in A, \tag{5.52a}$$

$$\mathbb{E}[X_i X_j] \geq \mathbb{E}[Y_i Y_j] \qquad \text{for all } (i, j) \in B, \tag{5.52b}$$

$$\mathbb{E}[X_i X_j] = \mathbb{E}[Y_i Y_j] \qquad \text{for all } (i, j) \notin A \cup B. \tag{5.52c}$$

Let $F \colon \mathbb{R}^N \to \mathbb{R}$ be a twice-differentiable function, and suppose that

$$\frac{\partial^2 F}{\partial u_i \partial u_j}(u) \geq 0 \qquad \text{for all } (i, j) \in A, \tag{5.53a}$$

$$\frac{\partial^2 F}{\partial u_i \partial u_j}(u) \leq 0 \qquad \text{for all } (i, j) \in B. \tag{5.53b}$$

Then we are guaranteed that

$$\mathbb{E}[F(X)] \leq \mathbb{E}[F(Y)]. \tag{5.54}$$

Proof We may assume without loss of generality that X and Y are independent. We proceed via a classical interpolation argument: define the Gaussian random vector

$$Z(t) = \sqrt{1-t}\, X + \sqrt{t}\, Y, \qquad \text{for each } t \in [0, 1], \tag{5.55}$$

and consider the function $\phi \colon [0, 1] \to \mathbb{R}$ given by $\phi(t) = \mathbb{E}[F(Z(t))]$. If we can show that $\phi'(t) \geq 0$ for all $t \in (0, 1)$, then we may conclude that

$$\mathbb{E}[F(Y)] = \phi(1) \geq \phi(0) = \mathbb{E}[F(X)].$$

With this goal in mind, for a given $t \in (0, 1)$, we begin by using the chain rule to compute the first derivative

$$\phi'(t) = \sum_{j=1}^{N} \mathbb{E}\left[\frac{\partial F}{\partial z_j}(Z(t))\, Z_j'(t)\right],$$

where $Z_j'(t) := \frac{d}{dt} Z_j(t) = -\frac{1}{2\sqrt{1-t}} X_j + \frac{1}{2\sqrt{t}} Y_j$. Computing the expectation, we find that

$$\begin{aligned}\mathbb{E}[Z_i(t)\, Z_j'(t)] &= \mathbb{E}\left[\left(\sqrt{1-t}\, X_i + \sqrt{t}\, Y_i\right)\left(-\frac{1}{2\sqrt{1-t}} X_j + \frac{1}{2\sqrt{t}} Y_j\right)\right] \\ &= \frac{1}{2}\{\mathbb{E}[Y_i Y_j] - \mathbb{E}[X_i X_j]\}.\end{aligned}$$

Consequently, for each $i = 1, \ldots, N$, we can write[5] $Z_i(t) = \alpha_{ij} Z_j'(t) + W_{ij}$, where $\alpha_{ij} \geq 0$ for $(i, j) \in A$, $\alpha_{ij} \leq 0$ for $(i, j) \in B$, and $\alpha_{ij} = 0$ if $(i, j) \notin A \cup B$. Moreover, due to the Gaussian assumption, we are guaranteed that the random vector $W(j) := (W_{1j}, \ldots, W_{Nj})$ is independent of $Z_j'(t)$.

[5] The variable W_{ij} does depend on t, but we omit this dependence so as to simplify notation.

Since F is twice differentiable, we may apply a first-order Taylor series to the function $\partial F/\partial z_j$ between the points $W(j)$ and $Z(t)$, thereby obtaining

$$\frac{\partial F}{\partial z_j}(Z(t)) = \frac{\partial F}{\partial z_j}(W(j)) + \sum_{i=1}^{N} \frac{\partial^2 F}{\partial z_j \partial z_i}(U)\, \alpha_{ij} Z_j'(t),$$

where $U \in \mathbb{R}^N$ is some intermediate point between $W(j)$ and $Z(t)$. Taking expectations then yields

$$\begin{aligned}\mathbb{E}\left[\frac{\partial F}{\partial z_j}(Z(t)) Z_j'(t)\right] &= \mathbb{E}\left[\frac{\partial F}{\partial z_j}(W(j)) Z_j'(t)\right] + \sum_{i=1}^{N} \mathbb{E}\left[\frac{\partial^2 F}{\partial z_j \partial z_i}(U)\, \alpha_{ij} (Z_j'(t))^2\right] \\ &= \sum_{i=1}^{n} \mathbb{E}\left[\frac{\partial^2 F}{\partial z_j \partial z_i}(U)\, \alpha_{ij} (Z_j'(t))^2\right],\end{aligned}$$

where the first term vanishes since $W(j)$ and $Z_j'(t)$ are independent, and $Z_j'(t)$ is zero-mean. By our assumptions on the second derivatives of f and the previously stated conditions on α_{ij}, we have $\frac{\partial^2 F}{\partial z_j \partial z_i}(U)\, \alpha_{ij} \geq 0$, so that we may conclude that $\phi'(t) \geq 0$ for all $t \in (0, 1)$, which completes the proof. □

5.4.2 Slepian and Sudakov–Fernique inequalities

An important corollary of Theorem 5.25 is *Slepian's inequality*.

Corollary 5.26 (Slepian's inequality) *Let $X \in \mathbb{R}^N$ and $Y \in \mathbb{R}^N$ be zero-mean Gaussian random vectors such that*

$$\mathbb{E}[X_i X_j] \geq \mathbb{E}[Y_i Y_j] \quad \text{for all } i \neq j, \tag{5.56a}$$

$$\mathbb{E}[X_i^2] = \mathbb{E}[Y_i^2] \quad \text{for all } i = 1, 2, \ldots, N. \tag{5.56b}$$

Then we are guaranteed

$$\mathbb{E}[\max_{i=1,\ldots,N} X_i] \leq \mathbb{E}[\max_{i=1,\ldots,N} Y_i]. \tag{5.57}$$

Proof In order to study the maximum, let us introduce, for each $\beta > 0$, a real-valued function on $\mathbb{R}^N$ via $F_\beta(x) := \beta^{-1} \log\{\sum_{j=1}^N \exp(\beta x_j)\}$. By a straightforward calculation, we find the useful relation

$$\max_{j=1,\ldots,N} x_j \leq F_\beta(x) \leq \max_{j=1,\ldots,N} x_j + \frac{\log N}{\beta}, \qquad \text{valid for all } \beta > 0, \tag{5.58}$$

so that bounds on F_β can be used to control the maximum by taking $\beta \to +\infty$. Note that F_β is twice differentiable for each $\beta > 0$; moreover, some calculus shows that $\frac{\partial^2 F_\beta}{\partial x_i \partial x_j} \leq 0$ for all $i \neq j$. Consequently, applying Theorem 5.25 with $A = \emptyset$ and $B = \{(i, j), i \neq j\}$ yields that $\mathbb{E}[F_\beta(X)] \leq \mathbb{E}[F_\beta(Y)]$. Combining this inequality with the sandwich relation (5.58), we conclude that

$$\mathbb{E}\Big[\max_{j=1,\ldots,N} X_j\Big] \leq \mathbb{E}\Big[\max_{j=1,\ldots,N} Y_j\Big] + \frac{\log N}{\beta},$$

and taking the limit $\beta \to +\infty$ yields the claim. □

Note that Theorem 5.25 and Corollary 5.26 are stated in terms of the variances and correlations of the random vector. In many cases, it is more convenient to compare two Gaussian processes in terms of their associated pseudometrics

$$\rho_X^2(i, j) = \mathbb{E}(X_i - X_j)^2 \quad \text{and} \quad \rho_Y^2(i, j) = \mathbb{E}(Y_i - Y_j)^2.$$

The Sudakov–Fernique comparison is stated in exactly this way.

Theorem 5.27 (Sudakov–Fernique) *Given a pair of zero-mean N-dimensional Gaussian vectors $(X_1, \ldots, X_N)$ and $(Y_1, \ldots, Y_N$, suppose that*

$$\mathbb{E}[(X_i - X_j)^2] \leq \mathbb{E}[(Y_i - Y_j)^2] \qquad \text{for all } (i, j) \in [N] \times [N]. \tag{5.59}$$

Then $\mathbb{E}[\max_{j=1,\ldots,N} X_j] \leq \mathbb{E}[\max_{j=1,\ldots,N} Y_j]$.

Remark: It is worth noting that the Sudakov–Fernique theorem also yields Slepian's inequality as a corollary. In particular, if the Slepian conditions (5.56a) hold, then it can be seen that the Sudakov–Fernique condition (5.59) also holds. The proof of Theorem 5.27 is more involved than that of Slepian's inequality; see the bibliographic section for references to some proofs.

5.4.3 Gaussian contraction inequality

One important consequence of the Sudakov–Fernique comparison is the *Gaussian contraction inequality*, which applies to functions $\phi_j\colon \mathbb{R} \to \mathbb{R}$ that are 1-Lipschitz, meaning that $|\phi_j(s) - \phi_j(t)| \leq |s - t|$ for all $s, t \in \mathbb{R}$, and satisfy the centering relation $\phi_j(0) = 0$. Given a vector $\theta \in \mathbb{R}^d$, we define (with a minor abuse of notation) the vector

$$\phi(\theta) := \big(\phi_1(\theta_1), \ \ \phi_2(\theta_2), \ \ \cdots, \ \ \phi_d(\theta_d)\big) \in \mathbb{R}^d.$$

Lastly, given a set $\mathbb{T} \subset \mathbb{R}^d$, we let $\phi(\mathbb{T}) = \{\phi(\theta) \mid \theta \in \mathbb{T}\}$ denote its image under the mapping ϕ. The following result shows that the Gaussian complexity of this image is never larger than the Gaussian complexity $\mathcal{G}(\mathbb{T})$ of the original set.

Proposition 5.28 (Gaussian contraction inequality) *For any set $\mathbb{T} \subseteq \mathbb{R}^d$ and any family of centered 1-Lipschitz functions $\{\phi_j,\ j = 1, \ldots, d\}$, we have*

$$\underbrace{\mathbb{E}\Big[\sup_{\theta \in \mathbb{T}} \sum_{j=1}^{d} w_j \phi_j(\theta_j)\Big]}_{\mathcal{G}(\phi(\mathbb{T}))} \leq \underbrace{\mathbb{E}\Big[\sup_{\theta \in \mathbb{T}} \sum_{j=1}^{d} w_j \theta_j\Big]}_{\mathcal{G}(\mathbb{T})}. \tag{5.60}$$

We leave the proof of this claim for the reader (see Exercise 5.12). For future reference, we also note that, with an additional factor of 2, an analogous result holds for the Rademacher complexity—namely

$$\mathcal{R}(\phi(\mathbb{T})) \leq 2\mathcal{R}(\mathbb{T}) \tag{5.61}$$

for any family of centered 1-Lipschitz functions. The proof of this result is somewhat more delicate than the Gaussian case; see the bibliographic section for further discussion.

Let us illustrate the use of the Gaussian contraction inequality (5.60) with some examples.

Example 5.29 Given a function class $\mathscr{F}$ and a collection of design points x_1^n, we have previously studied the Gaussian complexity of the set $\mathscr{F}(x_1^n) \subset \mathbb{R}^n$ defined in equation (5.30). In various statistical problems, it is often more natural to consider the Gaussian complexity of the set

$$\mathscr{F}^2(x_1^n) := \{(f^2(x_1), f^2(x_2), \ldots, f^2(x_n)) \mid f \in \mathscr{F}\} \subset \mathbb{R}^n,$$

where $f^2(x) = [f(x)]^2$ are the squared function values. The contraction inequality allows us to upper bound the Gaussian complexity of this set in terms of the original set $\mathscr{F}(x_1^n)$. In particular, suppose that the function class is b-uniformly bounded, so that $\|f\|_\infty \leq b$ for all $f \in \mathscr{F}$. We then claim that

$$\mathcal{G}(\mathscr{F}^2(x_1^n)) \leq 2b\, \mathcal{G}(\mathscr{F}(x_1^n)), \tag{5.62}$$

so that the Gaussian complexity of $\mathscr{F}^2(x_1^n)$ is not essentially larger than that of $\mathscr{F}(x_1^n)$.

In order to establish this bound, define the function $\phi_b \colon \mathbb{R} \to \mathbb{R}$ via

$$\phi_b(t) := \begin{cases} t^2/(2b) & \text{if } |t| \leq b, \\ b/2 & \text{otherwise.} \end{cases}$$

Since $|f(x_i)| \leq b$, we have $\phi_b(f(x_i)) = \frac{f^2(x_i)}{2b}$ for all $f \in \mathscr{F}$ and $i = 1, 2, \ldots, n$, and hence

$$\frac{1}{2b}\mathcal{G}(\mathscr{F}^2(x_1^n)) = \mathbb{E}\left[\sup_{f \in \mathscr{F}} \sum_{i=1}^{n} w_i \frac{f^2(x_i)}{2b}\right] = \mathbb{E}\left[\sup_{f \in \mathscr{F}} \sum_{i=1}^{n} w_i \phi_b(f(x_i))\right].$$

Moreover, it is straightforward to verify that ϕ_b is a contraction according to our definition, and hence applying Proposition 5.28 yields

$$\mathbb{E}\left[\sup_{f \in \mathscr{F}} \sum_{i=1}^{n} w_i \phi_b(f(x_i))\right] \leq \mathbb{E}\left[\sup_{f \in \mathscr{F}} \sum_{i=1}^{n} w_i f(x_i)\right] = \mathcal{G}(\mathscr{F}(x_1^n)).$$

Putting together the pieces yields the claim (5.62). ♣

5.5 Sudakov's lower bound

In previous sections, we have derived two upper bounds on the expected supremum of a sub-Gaussian process over a given set: the simple one-step discretization in Proposition 5.17, and the more refined Dudley integral bound in Theorem 5.22. In this section, we turn to the complementary question of deriving lower bounds. In contrast to the upper bounds in the preceding sections, these lower bounds are specialized to the case of *Gaussian* processes, since a general sub-Gaussian process might have different behavior than its Gaussian analog. For instance, compare the Rademacher and Gaussian complexity of the ℓ_1-ball, as discussed in Example 5.14.

This section is devoted to the exploration of a lower bound known as the *Sudakov minoration*, which is obtained by exploiting the Gaussian comparison inequalities discussed in the previous section.

Theorem 5.30 (Sudakov minoration) *Let $\{X_\theta, \theta \in \mathbb{T}\}$ be a zero-mean Gaussian process defined on the non-empty set $\mathbb{T}$. Then*

$$\mathbb{E}\Big[\sup_{\theta \in \mathbb{T}} X_\theta\Big] \geq \sup_{\delta > 0} \frac{\delta}{2} \sqrt{\log M_X(\delta\,;\mathbb{T})}, \tag{5.63}$$

where $M_X(\delta; \mathbb{T})$ is the δ-packing number of $\mathbb{T}$ in the metric $\rho_X(\theta, \widetilde{\theta}) := \sqrt{\mathbb{E}[(X_\theta - X_{\widetilde{\theta}})^2]}$.

Proof For any $\delta > 0$, let $\{\theta^1, \ldots, \theta^M\}$ be a δ-packing of $\mathbb{T}$, and consider the sequence $\{Y_i\}_{i=1}^M$ with elements $Y_i := X_{\theta^i}$. Note that by construction, we have the lower bound

$$\mathbb{E}[(Y_i - Y_j)^2] = \rho_X^2(\theta^i, \theta^j) > \delta^2 \qquad \text{for all } i \neq j.$$

Now let us define an i.i.d. sequence of Gaussian random variables $Z_i \sim \mathcal{N}(0, \delta^2/2)$ for $i = 1, \ldots, M$. Since $\mathbb{E}[(Z_i - Z_j)^2] = \delta^2$ for all $i \neq j$, the pair of random vectors Y and Z satisfy the Sudakov–Fernique condition (5.59), so that we are guaranteed that

$$\mathbb{E}\Big[\sup_{\theta \in \mathbb{T}} X_\theta\Big] \geq \mathbb{E}\Big[\max_{i=1,\ldots,M} Y_i\Big] \geq \mathbb{E}\Big[\max_{i=1,\ldots,M} Z_i\Big].$$

Since the variables $\{Z_i\}_{i=1}^M$ are zero-mean Gaussian and i.i.d., we can apply standard results on i.i.d. Gaussian maxima (viz. Exercise 2.11) to obtain the lower bound $\mathbb{E}[\max_{i=1,\ldots,M} Z_i] \geq \frac{\delta}{2}\sqrt{\log M}$, thereby completing the proof. □

Let us illustrate the Sudakov lower bound with some examples.

Example 5.31 (Gaussian complexity of ℓ_2-ball) We have shown previously that the Gaussian complexity $\mathcal{G}(\mathbb{B}_2^d)$ of the d-dimensional Euclidean ball is upper bounded as $\mathcal{G}(\mathbb{B}_2^d) \leq \sqrt{d}$. We have verified this fact both by direct calculation and through use of the upper bound in

Proposition 5.17. Here let us show how the Sudakov minoration captures the complementary lower bound. From Example 5.9, the metric entropy of the ball $\mathbb{B}_2^d$ in ℓ_2-norm is lower bounded as $\log N_2(\delta; \mathbb{B}^d) \geq d \log(1/\delta)$. Thus, by Lemma 5.5, the same lower bound applies to $\log M_2(\delta\,;\, \mathbb{B}^d)$. Therefore, the Sudakov bound (5.63) implies that

$$\mathcal{G}(\mathbb{B}_2^d) \geq \sup_{\delta>0} \left\{ \frac{\delta}{2} \sqrt{d \log(1/\delta)} \right\} \geq \frac{\sqrt{\log 4}}{8} \sqrt{d},$$

where we set $\delta = 1/4$ in order to obtain the second inequality. Thus, in this simple case, the Sudakov lower bound recovers the correct scaling as a function of $\sqrt{d}$, albeit with suboptimal control of the constant. ♣

We can also use the Sudakov minoration to upper bound the metric entropy of a set $\mathbb{T}$, assuming that we have an upper bound on its Gaussian complexity, as illustrated in the following example.

Example 5.32 (Metric entropy of ℓ_1-ball) Let us use the Sudakov minoration to upper bound the metric entropy of the ℓ_1-ball $\mathbb{B}_1^d = \{\theta \in \mathbb{R}^d \mid \sum_{i=1}^d |\theta_i| \leq 1\}$. We first observe that its Gaussian complexity can be upper bounded as

$$\mathcal{G}(\mathbb{B}_1) = \mathbb{E}\left[\sup_{\|\theta\|_1 \leq 1} \langle w, \theta \rangle \right] = \mathbb{E}[\|w\|_\infty] \leq 2 \sqrt{\log d},$$

where we have used the duality between the ℓ_1- and ℓ_∞-norms, and standard results on Gaussian maxima (see Exercise 2.11). Applying Sudakov's minoration, we conclude that the metric entropy of the d-dimensional ball $\mathbb{B}_1^d$ in the ℓ_2-norm is upper bounded as

$$\log N(\delta\,;\, \mathbb{B}_1^d, \|\cdot\|_2) \leq c(1/\delta)^2 \log d. \tag{5.64}$$

It is known that (for the most relevant range of δ) this upper bound on the metric entropy of $\mathbb{B}_1^d$ is tight up to constant factors; see the bibliographic section for further discussion. We thus see in a different way how the ℓ_1-ball is *much* smaller than the ℓ_2-ball, since its metric entropy scales logarithmically in dimension, as opposed to linearly. ♣

As another example, let us now return to some analysis of the singular values of Gaussian random matrices.

Example 5.33 (Lower bounds on maximum singular value) As a continuation of Example 5.19, let us use the Sudakov minoration to *lower* bound the maximum singular value of a standard Gaussian random matrix $\mathbf{W} \in \mathbb{R}^{n\times d}$. Recall that we can write

$$\mathbb{E}[|||\mathbf{W}|||_2] = \mathbb{E}\left[\sup_{\mathbf{\Theta} \in \mathbb{M}^{n,d}(1)} \langle\langle \mathbf{W},\, \mathbf{\Theta} \rangle\rangle \right],$$

where the set $\mathbb{M}^{n,d}(1)$ was previously defined (5.37). Consequently, in order to lower bound $\mathbb{E}[|||\mathbf{W}|||_2]$ via Sudakov minoration, it suffices to lower bound the metric entropy of $\mathbb{M}^{n,d}(1)$ in the Frobenius norm. In Exercise 5.13, we show that there is a universal constant c_1 such that

$$\log M(\delta; \mathbb{M}^{n,d}(1); |||\cdot|||_F) \geq c_1^2 (n+d) \log(1/\delta) \qquad \text{for all } \delta \in (0, \tfrac{1}{2}).$$

Setting $\delta = 1/4$, the Sudakov minoration implies that

$$\mathbb{E}[|||\mathbf{W}|||_2] \geq \frac{c_1}{8} \sqrt{(n+d)} \sqrt{\log 4} \geq c_1' (\sqrt{n} + \sqrt{d}).$$

Comparing with the upper bound from Example 5.19, we see that this lower bound has the correct scaling in the pair (n, d). ♣

5.6 Chaining and Orlicz processes

In Section 5.3.3, we introduced the idea of chaining, and showed how it can be used to obtain upper bounds on the expected supremum of a sub-Gaussian process. When the process is actually Gaussian, then classical concentration results can be used to show that the supremum is sharply concentrated around this expectation (see Exercise 5.10). For more general sub-Gaussian processes, it is useful to be able to derive similar bounds on the probability of deviations above the tail. Moreover, there are many processes that do not have sub-Gaussian tails, but rather instead are sub-exponential in nature. It is also useful to obtain bounds on the expected supremum and associated deviation bounds for such processes.

The notion of an *Orlicz norm* allows us to treat both sub-Gaussian and sub-exponential processes in a unified manner. For a given parameter $q \in [1, 2]$, consider the function $\psi_q(t) := \exp(t^q) - 1$. This function can be used to define a norm on the space of random variables as follows:

Definition 5.34 (Orlicz norm) The ψ_q-Orlicz norm of a zero-mean random variable X is given by

$$\|X\|_{\psi_q} := \inf\{\lambda > 0 \mid \mathbb{E}[\psi_q(|X|/\lambda)] \leq 1\}. \tag{5.65}$$

The Orlicz norm is infinite if there is no $\lambda \in \mathbb{R}$ for which the given expectation is finite.

Any random variable with a bounded Orlicz norm satisfies a concentration inequality specified in terms of the function ψ_q. In particular, we have

$$\mathbb{P}[|X| \geq t] \overset{\text{(i)}}{=} \mathbb{P}[\psi_q(|X|/\|X\|_{\psi_q}) \geq \psi_q(t/\|X\|_{\psi_q})] \overset{\text{(ii)}}{\leq} \frac{1}{\psi_q(t/\|X\|_{\psi_q})},$$

where the equality (i) follows because ψ_q is an increasing function, and the bound (ii) follows from Markov's inequality. In the case $q = 2$, this bound is essentially equivalent to our usual sub-Gaussian tail bound; see Exercise 2.18 for further details.

Based on the notion of the Orlicz norm, we can now define an interesting generalization of a sub-Gaussian process:

Definition 5.35 A zero-mean stochastic process $\{X_\theta, \theta \in \mathbb{T}\}$ is a ψ_q-process with respect to a metric ρ if

$$\|X_\theta - X_{\widetilde{\theta}}\|_{\psi_q} \leq \rho(\theta, \widetilde{\theta}) \qquad \text{for all } \theta, \widetilde{\theta} \in \mathbb{T}. \tag{5.66}$$

As a particular example, in this new terminology, it can be verified that the canonical Gaussian process is a ψ_2-process with respect to the (scaled) Euclidean metric $\rho(\theta, \widetilde{\theta}) = 2\,\|\theta - \widetilde{\theta}\|_2$.

We define the generalized Dudley entropy integral

$$\mathcal{J}_q(\delta; D) := \int_\delta^D \psi_q^{-1}(N(u; \mathbb{T}, \rho))\, du, \tag{5.67}$$

where ψ_q^{-1} is the inverse function of ψ_q, and $D = \sup_{\theta, \widetilde{\theta} \in \mathbb{T}} \rho(\theta, \widetilde{\theta})$ is the diameter of the set $\mathbb{T}$ under ρ. For the exponential-type functions considered here, note that we have

$$\psi_q^{-1}(u) = [\log(1 + u)]^{1/q}. \tag{5.68}$$

With this set-up, we have the following result:

Theorem 5.36 *Let $\{X_\theta, \theta \in \mathbb{T}\}$ be a ψ_q-process with respect to ρ. Then there is a universal constant c_1 such that*

$$\mathbb{P}\left[\sup_{\theta, \widetilde{\theta} \in \mathbb{T}} |X_\theta - X_{\widetilde{\theta}}| \geq c_1(\mathcal{J}_q(0; D) + t)\right] \leq 2e^{-\frac{t^q}{D^q}} \quad \textit{for all } t > 0. \tag{5.69}$$

A few comments on this result are in order. Note that the bound (5.69) involves the generalized Dudley entropy integral (5.67) for $\delta = 0$. As with our earlier statement of Dudley's entropy integral bound, there is a generalization of Theorem 5.36 that involves the truncated form, along with some discretization error. Otherwise, Theorem 5.36 should be understood as generalizing Theorem 5.22 in two ways. First, it applies to general Orlicz processes for $q \in [1, 2]$, with the sub-Gaussian setting corresponding to the special case $q = 2$. Second, it provides a tail bound on the random variable, as opposed to a bound only on its expectation. (Note that a bound on the expectation can be recovered by integrating the tail bound, in the usual way.)

Proof We begin by stating an auxiliary lemma that is of independent interest. For any measurable set A and random variable Y, let us introduce the shorthand notation $\mathbb{E}_A[Y] = \int_A Y\, d\mathbb{P}$. Note that we have $\mathbb{E}_A[Y] = \mathbb{E}[Y \mid Y \in A]\, \mathbb{P}[A]$ by construction.

Lemma 5.37 *Suppose that $Y_1, \ldots, Y_N$ are non-negative random variables such that $\|Y_i\|_{\psi_q} \leq 1$. Then for any measurable set A, we have*

$$\mathbb{E}_A[Y_i] \leq \mathbb{P}[A]\psi_q^{-1}(1/\mathbb{P}(A)) \qquad \text{for all } i = 1, 2, \ldots, N, \tag{5.70}$$

and moreover

$$\mathbb{E}_A\Big[\max_{i=1,\ldots,N} Y_i\Big] \leq \mathbb{P}[A]\psi_q^{-1}\Big(\frac{N}{\mathbb{P}(A)}\Big). \tag{5.71}$$

Proof Let us first establish the inequality (5.70). By definition, we have

$$\begin{aligned}
\mathbb{E}_A[Y] &= \mathbb{P}[A]\,\frac{1}{\mathbb{P}[A]}\,\mathbb{E}_A[\psi_q^{-1}(\psi_q(Y))] \\
&\overset{\text{(i)}}{\leq} \mathbb{P}[A]\,\psi_q^{-1}\Big(\mathbb{E}_A[\psi_q(Y)]\frac{1}{\mathbb{P}[A]}\Big) \\
&\overset{\text{(ii)}}{\leq} \mathbb{P}[A]\,\psi_q^{-1}\Big(\frac{1}{\mathbb{P}[A]}\Big),
\end{aligned}$$

where step (i) uses concavity of ψ_q^{-1} and Jensen's inequality (noting that the ratio $\frac{\mathbb{E}_A[\cdot]}{\mathbb{P}[A]}$ defines a conditional distribution); whereas step (ii) uses the fact that $\mathbb{E}_A[\psi_q(Y)] \leq \mathbb{E}[\psi_q(Y)] \leq 1$, which follows since $\psi_q(Y)$ is non-negative, and the Orlicz norm of Y is at most one, combined with the fact that ψ_q^{-1} is an increasing function.

We now prove its extension (5.71). Any measurable set A can be partitioned into a disjoint union of sets A_i, $i = 1, 2, \ldots, N$, such that $Y_i = \max_{j=1,\ldots,N} Y_j$ on A_i. Using this partition, we have

$$\begin{aligned}
\mathbb{E}_A\Big[\max_{i=1,\ldots,N} Y_i\Big] = \sum_{i=1}^N \mathbb{E}_{A_i}[Y_i] &\leq \mathbb{P}[A]\sum_{i=1}^N \frac{\mathbb{P}[A_i]}{\mathbb{P}[A]}\,\psi_q^{-1}\Big(\frac{1}{\mathbb{P}[A_i]}\Big) \\
&\leq \mathbb{P}[A]\,\psi_q^{-1}\Big(\frac{N}{\mathbb{P}[A]}\Big),
\end{aligned}$$

where the last step uses the concavity of ψ_q^{-1}, and Jensen's inequality with the weights $\mathbb{P}[A_i]/\mathbb{P}[A]$. □

In order to appreciate the relevance of this lemma for Theorem 5.36, let us use it to show that the supremum $Z := \sup_{\theta,\widetilde{\theta}\in\mathbb{T}} |X_\theta - X_{\widetilde{\theta}}|$ satisfies the inequality

$$\mathbb{E}_A[Z] \leq 8\,\mathbb{P}[A]\int_0^D \psi_q^{-1}\Big(\frac{N(u;\mathbb{T},\rho)}{\mathbb{P}[A]}\Big)\,du. \tag{5.72}$$

Choosing A to be the full probability space immediately yields an upper bound on the expected supremum—namely $\mathbb{E}[Z] \leq 8\mathcal{J}_q(D)$. On the other hand, if we choose $A = \{Z \geq t\}$, then we have

$$\mathbb{P}[Z \geq t] \overset{\text{(i)}}{\leq} \frac{1}{t}\,\mathbb{E}_A[Z] \overset{\text{(ii)}}{\leq} 8\frac{\mathbb{P}[Z \geq t]}{t}\int_0^D \psi_q^{-1}\Big(\frac{N(u;\mathbb{T},\rho)}{\mathbb{P}[Z \geq t]}\Big)\,du,$$

where step (i) follows from a version of Markov's inequality, and step (ii) follows from the bound (5.72). Canceling out a factor of $\mathbb{P}[Z \geq t]$ from both sides, and using the inequality $\psi_q^{-1}(st) \leq c(\psi_q^{-1}(s) + \psi_q^{-1}(t))$, we obtain

$$t \leq 8c\left\{ \int_0^D \psi_q^{-1}(N(u; \mathbb{T}, \rho))\, du + D\psi_q^{-1}\left(\frac{1}{\mathbb{P}[Z \geq t]}\right)\right\}.$$

Let $\delta > 0$ be arbitrary, and set $t = 8c(\mathcal{J}_q(D) + \delta)$. Some algebra then yields the inequality $\delta \leq D\psi_q^{-1}\left(\frac{1}{\mathbb{P}[Z \geq t]}\right)$, or equivalently

$$\mathbb{P}[Z \geq 8c(\mathcal{J}_q(D) + \delta)] \leq \frac{1}{\psi_q(\delta/D)},$$

as claimed.

In order to prove Theorem 5.36, it suffices to establish the bound (5.72). We do so by combining Lemma 5.37 with the chaining argument previously used to prove Theorem 5.22. Let us recall the set-up from that earlier proof: by following the one-step discretization argument, our problem was reduced to bounding the quantity $\mathbb{E}[\sup_{\theta,\widetilde{\theta}\in\mathbb{U}} |X_\theta - X_{\widetilde{\theta}}|]$, where $\mathbb{U} = \{\theta^1, \ldots, \theta^N\}$ was a δ-cover of the original set. For each $m = 1, 2, \ldots, L$, let $\mathbb{U}_m$ be a minimal $D2^{-m}$-cover of $\mathbb{U}$ in the metric ρ_X, so that at the mth step, the set $\mathbb{U}_m$ has $N_m = N_X(\epsilon_m\, ; \mathbb{U})$ elements. Similarly, define the mapping $\pi_m \colon \mathbb{U} \to \mathbb{U}_m$ via $\pi_m(\theta) = \arg\min_{\gamma\in\mathbb{U}_m} \rho_X(\theta, \gamma)$, so that $\pi_m(\theta)$ is the best approximation of $\theta \in \mathbb{U}$ from the set $\mathbb{U}_m$. Using this notation, we derived the chaining upper bound

$$\mathbb{E}_A\left[\max_{\theta,\widetilde{\theta}\in\mathbb{U}} |X_\theta - X_{\widetilde{\theta}}|\right] \leq 2\sum_{m=1}^{L} \mathbb{E}_A\left[\max_{\gamma\in\mathbb{U}_m} |X_\gamma - X_{\pi_{m-1}(\gamma)}|\right]. \tag{5.73}$$

(Previously, we had the usual expectation, as opposed to the object $\mathbb{E}_A$ used here.) For each $\gamma \in \mathbb{U}_m$, we are guaranteed that

$$\|X_\gamma - X_{\pi_{m-1}(\gamma)}\|_{\psi_q} \leq \rho_X(\gamma, \pi_{m-1}(\gamma)) \leq D2^{-(m-1)}.$$

Since $|\mathbb{U}_m| = N(D2^{-m})$, Lemma 5.37 implies that

$$\mathbb{E}_A\left[\max_{\gamma\in\mathbb{U}_m} |X_\gamma - X_{\pi_{m-1}}(\gamma)|\right] \leq \mathbb{P}[A]\, D2^{-(m-1)}\psi_q^{-1}\left(\frac{N(D2^{-m})}{\mathbb{P}(A)}\right),$$

for every measurable set A. Consequently, from the upper bound (5.73), we obtain

$$\begin{aligned}
\mathbb{E}_A\left[\max_{\theta,\widetilde{\theta}\in\mathbb{U}} |X_\theta - X_{\widetilde{\theta}}|\right] &\leq 2\mathbb{P}[A]\sum_{m=1}^{L} D2^{-(m-1)}\psi_q^{-1}\left(\frac{N(D2^{-m})}{\mathbb{P}(A)}\right) \\
&\leq c\,\mathbb{P}[A]\int_0^D \psi_q^{-1}\left(\frac{N_X(u; \mathbb{U})}{\mathbb{P}(A)}\right) du,
\end{aligned}$$

since the sum can be upper bounded by the integral. □

5.7 Bibliographic details and background

The notion of metric entropy was introduced by Kolmogorov (1956; 1958) and further developed by various authors; see the paper by Kolmogorov and Tikhomirov (1959) for an overview and some discussion of the early history. Metric entropy, along with related notions of the "sizes" of various function classes, are central objects of study in the field of approximation theory; see the books (DeVore and Lorentz, 1993; Pinkus, 1985; Carl and Stephani, 1990) for further details on approximation and operator theory. Examples 5.10 and 5.11 are discussed in depth by Kolmogorov and Tikhomirov (1959), as is the metric entropy bound for the special ellipsoid given in Example 5.12. Mitjagin (1961) proves a more general result, giving a sharp characterization of the metric entropy for any ellipsoid; see also Lorentz (1966) for related results.

The pioneering work of Dudley (1967) established the connection between the entropy integral and the behavior of Gaussian processes. The idea of chaining itself dates back to Kolmogorov and others. Upper bounds based on entropy integrals are not always the best possible. Sharp upper and lower bounds for expected Gaussian suprema can be derived by the generic chaining method of Talagrand (2000). The proof of the Orlicz-norm generalization of Dudley's entropy integral in Theorem 5.36 is based on Ledoux and Talagrand (1991).

The metric entropy of the ℓ_1-ball was discussed in Example 5.32; more generally, sharp upper and lower bounds on the entropy numbers of ℓ_q-balls for $q \in (0, 1]$ were obtained by Schütt (1984) and Kühn (2001). Raskutti et al. (2011) convert these estimates to upper and lower bounds on the metric entropy; see Lemma 2 in their paper.

Gaussian comparison inequalities have a lengthy and rich history in probability theory and geometric functional analysis (e.g., Slepian, 1962; Fernique, 1974; Gordon, 1985; Kahane, 1986; Milman and Schechtman, 1986; Gordon, 1986, 1987; Ledoux and Talagrand, 1991). A version of Slepian's inequality was first established in the paper (Slepian, 1962). Ledoux and Talagrand (1991) provide a detailed discussion of Gaussian comparison inequalities, including Slepian's inequality, the Sudakov–Fernique inequality and Gordon's inequalities. The proofs of Theorems 5.25 and 5.36 follow this development. Chatterjee (2005) provides a self-contained proof of the Sudakov–Fernique inequality, including control on the slack in the bound; see also Chernozhukov et al. (2013) for related results. Among other results, Gordon (1987) provides generalizations of Slepian's inequality and related results to elliptically contoured distribution. Section 4.2 of Ledoux and Talagrand (1991) contains a proof of the contraction inequality (5.61) for the Rademacher complexity.

The bound (5.49) on the metric entropy of a VC class is proved in Theorem 2.6.7 of van der Vaart and Wellner (1996). Exercise 5.4, adapted from this same book, works through the proof of a weaker bound.

5.8 Exercises

Exercise 5.1 (Failure of total boundedness) Let $C([0, 1], b)$ denote the class of all convex functions f defined on the unit interval such that $\|f\|_\infty \leq b$. Show that $C([0, 1], b)$ is *not* totally bounded in the sup-norm. (*Hint:* Try to construct an infinite collection of functions $\{f^j\}_{j=1}^\infty$ such that $\|f^j - f^k\|_\infty \geq 1/2$ for all $j \neq k$.)

Exercise 5.2 (Packing and covering) Prove the following relationships between packing and covering numbers:

$$M(2\delta\,; \mathbb{T}, \rho) \overset{\text{(a)}}{\leq} N(\delta\,; \mathbb{T}, \rho) \overset{\text{(b)}}{\leq} M(\delta\,; \mathbb{T}, \rho).$$

Exercise 5.3 (Packing of Boolean hypercube) Recall from Example 5.3 the binary hypercube $\mathbb{H}^d = \{0, 1\}^d$ equipped with the rescaled Hamming metric (5.1b). Prove that the packing number satisfies the bound

$$\frac{\log M(\delta; \mathbb{H}^d)}{d} \leq D(\delta/2 \,\|\, 1/2) + \frac{\log(d+1)}{d},$$

where $D(\delta/2 \,\|\, 1/2) = \frac{\delta}{2} \log \frac{\delta/2}{1/2} + (1 - \frac{\delta}{2}) \log \frac{1-\delta/2}{1/2}$ is the Kullback–Leibler divergence between the Bernoulli distributions with parameter $\delta/2$ and $1/2$. (*Hint:* You may find the result of Exercise 2.10 to be useful.)

Exercise 5.4 (From VC dimension to metric entropy) In this exercise, we explore the connection between VC dimension and metric entropy. Given a set class $\mathcal{S}$ with finite VC dimension ν, we show that the function class $\mathscr{F}_{\mathcal{S}} := \{\mathbb{I}_S, S \in \mathcal{S}\}$ of indicator functions has metric entropy at most

$$N(\delta; \mathscr{F}_{\mathcal{S}}, L^1(\mathbb{P})) \leq K(\nu) \left(\frac{3}{\delta}\right)^{2\nu}, \qquad \text{for a constant } K(\nu). \tag{5.74}$$

Let $\{\mathbb{I}_{S^1}, \ldots, \mathbb{I}_{S^N}\}$ be a maximal δ-packing in the $L^1(\mathbb{P})$-norm, so that

$$\|\mathbb{I}_{S_i} - \mathbb{I}_{S_j}\|_1 = \mathbb{E}[|\mathbb{I}_{S_i}(X) - \mathbb{I}_{S_j}(X)|] > \delta \qquad \text{for all } i \neq j.$$

By Exercise 5.2, this N is an upper bound on the δ-covering number.

(a) Suppose that we generate n samples X_i, $i = 1, \ldots, n$, drawn i.i.d. from $\mathbb{P}$. Show that the probability that every set S_i picks out a different subset of $\{X_1, \ldots, X_n\}$ is at least $1 - \binom{N}{2}(1-\delta)^n$.
(b) Using part (a), show that for $N \geq 2$ and $n = \frac{3 \log N}{\delta}$, there exists a set of n points from which $\mathcal{S}$ picks out at least N subsets, and conclude that $N \leq (\frac{3 \log N}{\delta})^\nu$.
(c) Use part (b) to show that the bound (5.74) holds with $K(\nu) := (2\nu)^{2\nu - 1}$.

Exercise 5.5 (Gaussian and Rademacher complexity) In this problem, we explore the connection between the Gaussian and Rademacher complexity of a set.

(a) Show that for any set $\mathbb{T} \subseteq \mathbb{R}^d$, the Rademacher complexity satisfies the upper bound $\mathcal{R}(\mathbb{T}) \leq \sqrt{\frac{\pi}{2}}\, \mathcal{G}(\mathbb{T})$. Give an example of a set for which this bound is met with equality.
(b) Show that $\mathcal{G}(\mathbb{T}) \leq 2\sqrt{\log d}\, \mathcal{R}(\mathbb{T})$ for any set $\mathbb{T} \subseteq \mathbb{R}^d$. Give an example for which this upper bound is tight up to the constant pre-factor. (*Hint:* In proving this bound, you may assume the Rademacher analog of the contraction inequality, namely that $\mathcal{R}(\phi(\mathbb{T})) \leq \mathcal{R}(\mathbb{T})$ for any contraction.)

Exercise 5.6 (Gaussian complexity for ℓ_q-balls) The ℓ_q-ball of unit radius is given by

$$\mathbb{B}_q^d(1) = \{\theta \in \mathbb{R}^d \mid \|\theta\|_q \leq 1\},$$

where $\|\theta\|_q = \left(\sum_{j=1}^d |\theta_j|^q\right)^{1/q}$ for $q \in [1, \infty)$ and $\|\theta\|_\infty = \max_j |\theta_j|$.

(a) For $q \in (1, \infty)$, show that there are constants c_q such that

$$\sqrt{\frac{2}{\pi}} \leq \frac{\mathcal{G}(\mathbb{B}_q^d(1))}{d^{1-\frac{1}{q}}} \leq c_q.$$

(b) Compute the Gaussian complexity $\mathcal{G}(\mathbb{B}^d_\infty(1))$ exactly.

Exercise 5.7 (Upper bounds for ℓ_0-"balls") Consider the set

$$\mathbb{T}^d(s) := \{\theta \in \mathbb{R}^d \mid \|\theta\|_0 \leq s, \ \|\theta\|_2 \leq 1\},$$

corresponding to all s-sparse vectors contained within the Euclidean unit ball. In this exercise, we prove that its Gaussian complexity is upper bounded as

$$\mathcal{G}(\mathbb{T}^d(s)) \precsim \sqrt{s \log\left(\frac{ed}{s}\right)}. \tag{5.75}$$

(a) First show that $\mathcal{G}(\mathbb{T}^d(s)) = \mathbb{E}\Big[\max_{|S|=s} \|w_S\|_2\Big]$, where $w_S \in \mathbb{R}^{|S|}$ denotes the subvector of $(w_1, \ldots, w_d)$ indexed by the subset $S \subset \{1, 2, \ldots, d\}$.
(b) Next show that

$$\mathbb{P}[\|w_S\|_2 \geq \sqrt{s} + \delta] \leq e^{-\delta^2/2}$$

for any fixed subset S of cardinality s.
(c) Use the preceding parts to establish the bound (5.75).

Exercise 5.8 (Lower bounds for ℓ_0-"balls") In Exercise 5.7, we established an upper bound on the Gaussian complexity of the set

$$\mathbb{T}^d(s) := \{\theta \in \mathbb{R}^d \mid \|\theta\|_0 \leq s, \ \|\theta\|_2 \leq 1\}.$$

The goal of this exercise to establish the matching lower bound.

(a) Derive a lower bound on the $1/\sqrt{2}$ covering number of $\mathbb{T}^d(s)$ in the Euclidean norm. (*Hint:* The Gilbert–Varshamov lemma could be useful to you).
(b) Use part (a) and a Gaussian comparison result to show that

$$\mathcal{G}(\mathbb{T}^d(s)) \succsim \sqrt{s \log\left(\frac{ed}{s}\right)}.$$

Exercise 5.9 (Gaussian complexity of ellipsoids) Recall that the space $\ell^2(\mathbb{N})$ consists of all real sequences $(\theta_j)_{j=1}^\infty$ such that $\sum_{j=1}^\infty \theta_j^2 < \infty$. Given a strictly positive sequence $(\mu_j)_{j=1}^\infty \in \ell^2(\mathbb{N})$, consider the associated ellipse

$$\mathcal{E} := \left\{(\theta_j)_{j=1}^\infty \;\middle|\; \sum_{j=1}^\infty \theta_j^2/\mu_j^2 \leq 1\right\}.$$

Ellipses of this form will play an important role in our subsequent analysis of the statistical properties of reproducing kernel Hilbert spaces.

(a) Prove that the Gaussian complexity satisfies the bounds

$$\sqrt{\frac{2}{\pi}}\left(\sum_{j=1}^\infty \mu_j^2\right)^{1/2} \leq \mathcal{G}(\mathcal{E}) \leq \left(\sum_{j=1}^\infty \mu_j^2\right)^{1/2}.$$

(*Hint:* Parts of previous problems may be helpful to you.)

(b) For a given radius $r > 0$ consider the truncated set

$$\widetilde{\mathcal{E}}(r) := \mathcal{E} \cap \left\{ (\theta_j)_{j=1}^{\infty} \;\middle|\; \sum_{j=1}^{\infty} \theta_j^2 \leq r^2 \right\}.$$

Obtain upper and lower bounds on the Gaussian complexity $\mathcal{G}(\widetilde{\mathcal{E}}(r))$ that are tight up to universal constants, independent of r and $(\mu_j)_{j=1}^{\infty}$. (*Hint:* Try to reduce the problem to an instance of (a).)

Exercise 5.10 (Concentration of Gaussian suprema) Let $\{X_\theta, \theta \in \mathbb{T}\}$ be a zero-mean Gaussian process, and define $Z = \sup_{\theta \in \mathbb{T}} X_\theta$. Prove that

$$\mathbb{P}[|Z - \mathbb{E}[Z]| \geq \delta] \leq 2e^{-\frac{\delta^2}{2\sigma^2}},$$

where $\sigma^2 := \sup_{\theta \in \mathbb{T}} \operatorname{var}(X_\theta)$ is the maximal variance of the process.

Exercise 5.11 (Details of Example 5.19) In this exercise, we work through the details of Example 5.19.

(a) Show that the maximum singular value $|||\mathbf{W}|||_2$ has the variational representation (5.38).
(b) Defining the random variable $X_\Theta = \langle\langle \mathbf{W}, \Theta \rangle\rangle$, show that the stochastic process $\{X_\Theta, \Theta \in \mathbb{M}^{n,d}(1)\}$ is zero-mean, and sub-Gaussian with respect to the Frobenius norm $|||\Theta - \Theta'|||_F$.
(c) Prove the upper bound (5.40).
(d) Prove the upper bound (5.41) on the metric entropy.

Exercise 5.12 (Gaussian contraction inequality) For each $j = 1, \ldots, d$, let $\phi_j : \mathbb{R} \to \mathbb{R}$ be a centered 1-Lipschitz function, meaning that $\phi_j(0) = 0$, and $|\phi_j(s) - \phi_j(t)| \leq |s - t|$ for all $s, t \in \mathbb{R}$. Given a set $\mathbb{T} \subseteq \mathbb{R}^d$, consider the set

$$\phi(\mathbb{T}) := \{(\phi_1(\theta_1), \phi_2(\theta_2), \ldots, \phi_d(\theta_d)) \mid \theta \in \mathbb{T}\} \subseteq \mathbb{R}^d.$$

Prove the Gaussian contraction inequality $\mathcal{G}(\phi(\mathbb{T})) \leq \mathcal{G}(\mathbb{T})$.

Exercise 5.13 (Details of Example 5.33) Recall the set $\mathbb{M}^{n,d}(1)$ from Example 5.33. Show that

$$\log M(\delta; \mathbb{M}^{n,d}(1); |||\cdot|||_F) \gtrsim (n + d)\log(1/\delta) \qquad \text{for all } \delta \in (0, 1/2).$$

Exercise 5.14 (Maximum singular value of Gaussian random matrices) In this exercise, we explore one method for obtaining tail bounds on the maximal singular value of a Gaussian random matrix $\mathbf{W} \in \mathbb{R}^{n \times d}$ with i.i.d. $N(0, 1)$ entries.

(a) To build intuition, let us begin by doing a simple simulation. Write a short computer program to generate Gaussian random matrices $\mathbf{W} \in \mathbb{R}^{n \times d}$ for $n = 1000$ and $d = \lceil \alpha n \rceil$, and to compute the maximum singular value of $\mathbf{W}/\sqrt{n}$, denoted by $\sigma_{\max}(\mathbf{W})/\sqrt{n}$. Perform $T = 20$ trials for each value of α in the set $\{0.1 + k(0.025), k = 1, \ldots, 100\}$. Plot the resulting curve of α versus the average of $\sigma_{\max}(\mathbf{W})/\sqrt{n}$.

(b) Now let's do some analysis to understand this behavior. Prove that

$$\sigma_{\max}(\mathbf{W}) = \sup_{u \in \mathbb{S}^{n-1}} \sup_{v \in \mathbb{S}^{d-1}} u^{\mathrm{T}} \mathbf{W} v,$$

where $\mathbb{S}^{d-1} = \{y \in \mathbb{R}^d \mid \|y\|_2 = 1\}$ is the d-dimensional Euclidean sphere.

(c) Observe that $Z_{u,v} := u^{\mathrm{T}} \mathbf{W} v$ defines a Gaussian process indexed by the Cartesian product $\mathbb{T} := \mathbb{S}^{n-1} \times \mathbb{S}^{d-1}$. Prove the upper bound

$$\mathbb{E}[\sigma_{\max}(\mathbf{W})] = \mathbb{E}\left[\sup_{(u,v) \in \mathbb{T}} u^{\mathrm{T}} \mathbf{W} v\right] \le \sqrt{n} + \sqrt{d}.$$

(*Hint:* For $(u, v) \in \mathbb{S}^{n-1} \times \mathbb{S}^{d-1}$, consider the zero-mean Gaussian variable $Y_{u,v} = \langle g, u \rangle + \langle h, v \rangle$, where $g \in N(0, I_{n \times n})$ and $h \sim N(0, I_{d \times d})$ are independent Gaussian random vectors. We thus obtain a second Gaussian process $\{Y_{u,v}, (u, v) \in \mathbb{S}^{n-1} \times \mathbb{S}^{d-1}\}$, and you may find it useful to compare $\{Z_{u,v}\}$ and $\{Y_{u,v}\}$.)

(d) Prove that

$$\mathbb{P}\left[\sigma_{\max}(\mathbf{W}) / \sqrt{n} \ge 1 + \sqrt{\frac{d}{n}} + t\right] \le 2e^{-\frac{nt^2}{2}}.$$

6

Random matrices and covariance estimation

Covariance matrices play a central role in statistics, and there exist a variety of methods for estimating them based on data. The problem of covariance estimation dovetails with random matrix theory, since the sample covariance is a particular type of random matrix. A classical framework allows the sample size n to tend to infinity while the matrix dimension d remains fixed; in such a setting, the behavior of the sample covariance matrix is characterized by the usual limit theory. By contrast, for high-dimensional random matrices in which the data dimension is either comparable to the sample size ($d \asymp n$), or possibly much larger than the sample size ($d \gg n$), many new phenomena arise.

High-dimensional random matrices play an important role in many branches of science, mathematics and engineering, and have been studied extensively. Part of high-dimensional theory is asymptotic in nature, such as the Wigner semicircle law and the Marčenko–Pastur law for the asymptotic distribution of the eigenvalues of a sample covariance matrix (see Chapter 1 for illustration of the latter). By contrast, this chapter is devoted to an exploration of random matrices in a non-asymptotic setting, with the goal of obtaining explicit deviation inequalities that hold for all sample sizes and matrix dimensions. Beginning with the simplest case—namely ensembles of Gaussian random matrices—we then discuss more general sub-Gaussian ensembles, and then move onwards to ensembles with milder tail conditions. Throughout our development, we bring to bear the techniques from concentration of measure, comparison inequalities and metric entropy developed previously in Chapters 2 through 5. In addition, this chapter introduces new some techniques, among them a class of matrix tail bounds developed over the past decade (see Section 6.4).

6.1 Some preliminaries

We begin by introducing notation and preliminary results used throughout this chapter, before setting up the problem of covariance estimation more precisely.

6.1.1 Notation and basic facts

Given a rectangular matrix $\mathbf{A} \in \mathbb{R}^{n \times m}$ with $n \geq m$, we write its ordered singular values as

$$\sigma_{\max}(\mathbf{A}) = \sigma_1(\mathbf{A}) \geq \sigma_2(\mathbf{A}) \geq \cdots \geq \sigma_m(\mathbf{A}) = \sigma_{\min}(\mathbf{A}) \geq 0.$$

Note that the minimum and maximum singular values have the variational characterization

$$\sigma_{\max}(\mathbf{A}) = \max_{v \in \mathbb{S}^{m-1}} \|\mathbf{A}v\|_2 \quad \text{and} \quad \sigma_{\min}(\mathbf{A}) = \min_{v \in \mathbb{S}^{m-1}} \|\mathbf{A}v\|_2, \tag{6.1}$$

where $\mathbb{S}^{d-1} := \{v \in \mathbb{R}^d \mid \|v\|_2 = 1\}$ is the Euclidean unit sphere in $\mathbb{R}^d$. Note that we have the equivalence $|||\mathbf{A}|||_2 = \sigma_{\max}(\mathbf{A})$.

Since covariance matrices are symmetric, we also focus on the set of symmetric matrices in $\mathbb{R}^d$, denoted $\mathcal{S}^{d\times d} := \{\mathbf{Q} \in \mathbb{R}^{d\times d} \mid \mathbf{Q} = \mathbf{Q}^{\mathrm{T}}\}$, as well as the subset of positive semidefinite matrices given by

$$\mathcal{S}_+^{d\times d} := \{\mathbf{Q} \in \mathcal{S}^{d\times d} \mid \mathbf{Q} \succeq 0\}. \tag{6.2}$$

From standard linear algebra, we recall the facts that any matrix $\mathbf{Q} \in \mathcal{S}^{d\times d}$ is diagonalizable via a unitary transformation, and we use $\gamma(\mathbf{Q}) \in \mathbb{R}^d$ to denote its vector of eigenvalues, ordered as

$$\gamma_{\max}(\mathbf{Q}) = \gamma_1(\mathbf{Q}) \geq \gamma_2(\mathbf{Q}) \geq \cdots \geq \gamma_d(\mathbf{Q}) = \gamma_{\min}(\mathbf{Q}).$$

Note that a matrix $\mathbf{Q}$ is positive semidefinite—written $\mathbf{Q} \succeq 0$ for short—if and only if $\gamma_{\min}(\mathbf{Q}) \geq 0$.

Our analysis frequently exploits the Rayleigh–Ritz variational characterization of the minimum and maximum eigenvalues—namely

$$\gamma_{\max}(\mathbf{Q}) = \max_{v\in\mathbb{S}^{d-1}} v^{\mathrm{T}}\mathbf{Q}v \quad \text{and} \quad \gamma_{\min}(\mathbf{Q}) = \min_{v\in\mathbb{S}^{d-1}} v^{\mathrm{T}}\mathbf{Q}v. \tag{6.3}$$

For any symmetric matrix $\mathbf{Q}$, the ℓ_2-operator norm can be written as

$$|||\mathbf{Q}|||_2 = \max\{\gamma_{\max}(\mathbf{Q}), |\gamma_{\min}(\mathbf{Q})|\}, \tag{6.4a}$$

by virtue of which it inherits the variational representation

$$|||\mathbf{Q}|||_2 := \max_{v\in\mathbb{S}^{d-1}} \left|v^{\mathrm{T}}\mathbf{Q}v\right|. \tag{6.4b}$$

Finally, given a rectangular matrix $\mathbf{A} \in \mathbb{R}^{n\times m}$ with $n \geq m$, suppose that we define the m-dimensional symmetric matrix $\mathbf{R} := \mathbf{A}^{\mathrm{T}}\mathbf{A}$. We then have the relationship

$$\gamma_j(\mathbf{R}) = (\sigma_j(\mathbf{A}))^2 \qquad \text{for } j = 1, \ldots, m.$$

6.1.2 Set-up of covariance estimation

Let us now define the problem of covariance matrix estimation. Let $\{x_1, \ldots, x_n\}$ be a collection of n independent and identically distributed samples[1] from a distribution in $\mathbb{R}^d$ with zero mean, and covariance matrix $\boldsymbol{\Sigma} = \mathrm{cov}(x_1) \in \mathcal{S}_+^{d\times d}$. A standard estimator of $\boldsymbol{\Sigma}$ is the *sample covariance matrix*

$$\widehat{\boldsymbol{\Sigma}} := \frac{1}{n}\sum_{i=1}^{n} x_i x_i^{\mathrm{T}}. \tag{6.5}$$

Since each x_i has zero mean, we are guaranteed that $\mathbb{E}[x_i x_i^T] = \boldsymbol{\Sigma}$, and hence that the random matrix $\widehat{\boldsymbol{\Sigma}}$ is an unbiased estimator of the population covariance $\boldsymbol{\Sigma}$. Consequently, the error matrix $\widehat{\boldsymbol{\Sigma}} - \boldsymbol{\Sigma}$ has mean zero, and our goal in this chapter is to obtain bounds on the error

[1] In this chapter, we use a lower case x to denote a random vector, so as to distinguish it from a random matrix.

measured in the ℓ_2-operator norm. By the variational representation (6.4b), a bound of the form $|||\widehat{\Sigma} - \Sigma|||_2 \leq \epsilon$ is equivalent to asserting that

$$\max_{v \in \mathbb{S}^{d-1}} \left| \frac{1}{n} \sum_{i=1}^{n} \langle x_i, v_i \rangle^2 - v^{\mathrm{T}} \Sigma v \right| \leq \epsilon. \tag{6.6}$$

This representation shows that controlling the deviation $|||\widehat{\Sigma}-\Sigma|||_2$ is equivalent to establishing a uniform law of large numbers for the class of functions $x \mapsto \langle x, v \rangle^2$, indexed by vectors $v \in \mathbb{S}^{d-1}$. See Chapter 4 for further discussion of such uniform laws in a general setting.

Control in the operator norm also guarantees that the eigenvalues of $\widehat{\Sigma}$ are uniformly close to those of Σ. In particular, by a corollary of Weyl's theorem (see the bibliographic section for details), we have

$$\max_{j=1,\ldots,d} |\gamma_j(\widehat{\Sigma}) - \gamma_j(\Sigma)| \leq |||\widehat{\Sigma} - \Sigma|||_2. \tag{6.7}$$

A similar type of guarantee can be made for the eigenvectors of the two matrices, but only if one has additional control on the separation between adjacent eigenvalues. See our discussion of principal component analysis in Chapter 8 for more details.

Finally, we point out the connection to the singular values of the random matrix $\mathbf{X} \in \mathbb{R}^{n \times d}$, denoted by $\{\sigma_j(\mathbf{X})\}_{j=1}^{\min\{n,d\}}$. Since the matrix $\mathbf{X}$ has the vector x_i^{T} as its ith row, we have

$$\widehat{\Sigma} = \frac{1}{n} \sum_{i=1}^{n} x_i x_i^{\mathrm{T}} = \frac{1}{n} \mathbf{X}^{\mathrm{T}} \mathbf{X},$$

and hence it follows that the eigenvalues of $\widehat{\Sigma}$ are the squares of the singular values of $\mathbf{X}/\sqrt{n}$.

6.2 Wishart matrices and their behavior

We begin by studying the behavior of singular values for random matrices with Gaussian rows. More precisely, let us suppose that each sample x_i is drawn i.i.d. from a multivariate $\mathcal{N}(0, \Sigma)$ distribution, in which case we say that the associated matrix $\mathbf{X} \in \mathbb{R}^{n \times d}$, with x_i^{T} as its ith row, is drawn from the *Σ-Gaussian ensemble.* The associated sample covariance $\widehat{\Sigma} = \frac{1}{n}\mathbf{X}^{\mathrm{T}}\mathbf{X}$ is said to follow a multivariate Wishart distribution.

Theorem 6.1 *Let $\mathbf{X} \in \mathbb{R}^{n \times d}$ be drawn according to the Σ-Gaussian ensemble. Then for all $\delta > 0$, the maximum singular value $\sigma_{\max}(\mathbf{X})$ satisfies the upper deviation inequality*

$$\mathbb{P}\left[\frac{\sigma_{\max}(\mathbf{X})}{\sqrt{n}} \geq \gamma_{\max}(\sqrt{\Sigma})(1 + \delta) + \sqrt{\frac{\mathrm{tr}(\Sigma)}{n}} \right] \leq e^{-n\delta^2/2}. \tag{6.8}$$

Moreover, for $n \geq d$, the minimum singular value $\sigma_{\min}(\mathbf{X})$ satisfies the analogous lower *deviation inequality*

$$\mathbb{P}\left[\frac{\sigma_{\min}(\mathbf{X})}{\sqrt{n}} \leq \gamma_{\min}(\sqrt{\boldsymbol{\Sigma}})\,(1-\delta) - \sqrt{\frac{\operatorname{tr}(\boldsymbol{\Sigma})}{n}}\right] \leq e^{-n\delta^2/2}. \tag{6.9}$$

Before proving this result, let us consider some illustrative examples.

Example 6.2 (Operator norm bounds for the standard Gaussian ensemble) Consider a random matrix $\mathbf{W} \in \mathbb{R}^{n\times d}$ generated with i.i.d. $\mathcal{N}(0,1)$ entries. This choice yields an instance of $\boldsymbol{\Sigma}$-Gaussian ensemble, in particular with $\boldsymbol{\Sigma} = \mathbf{I}_d$. By specializing Theorem 6.1, we conclude that for $n \geq d$, we have

$$\frac{\sigma_{\max}(\mathbf{W})}{\sqrt{n}} \leq 1 + \delta + \sqrt{\frac{d}{n}} \quad \text{and} \quad \frac{\sigma_{\min}(\mathbf{W})}{\sqrt{n}} \geq 1 - \delta - \sqrt{\frac{d}{n}}, \tag{6.10}$$

where both bounds hold with probability greater than $1 - 2e^{-n\delta^2/2}$. These bounds on the singular values of $\mathbf{W}$ imply that

$$\left|\!\left|\!\left| \frac{1}{n}\mathbf{W}^{\mathrm{T}}\mathbf{W} - \mathbf{I}_d \right|\!\right|\!\right|_2 \leq 2\epsilon + \epsilon^2, \qquad \text{where } \epsilon = \sqrt{\frac{d}{n}} + \delta, \tag{6.11}$$

with the same probability. Consequently, the sample covariance $\widehat{\boldsymbol{\Sigma}} = \frac{1}{n}\mathbf{W}^{\mathrm{T}}\mathbf{W}$ is a consistent estimate of the identity matrix $\mathbf{I}_d$ whenever $d/n \to 0$. ♣

The preceding example has interesting consequences for the problem of sparse linear regression using standard Gaussian random matrices, as in compressed sensing; in particular, see our discussion of the restricted isometry property in Chapter 7. On the other hand, from the perspective of covariance estimation, estimating the identity matrix is not especially interesting. However, a minor modification does lead to a more realistic family of problems.

Example 6.3 (Gaussian covariance estimation) Let $\mathbf{X} \in \mathbb{R}^{n\times d}$ be a random matrix from the $\boldsymbol{\Sigma}$-Gaussian ensemble. By standard properties of the multivariate Gaussian, we can write $\mathbf{X} = \mathbf{W}\sqrt{\boldsymbol{\Sigma}}$, where $\mathbf{W} \in \mathbb{R}^{n\times d}$ is a standard Gaussian random matrix, and hence

$$\left|\!\left|\!\left| \frac{1}{n}\mathbf{X}^{\mathrm{T}}\mathbf{X} - \boldsymbol{\Sigma} \right|\!\right|\!\right|_2 = \left|\!\left|\!\left| \sqrt{\boldsymbol{\Sigma}}\left(\frac{1}{n}\mathbf{W}^{\mathrm{T}}\mathbf{W} - \mathbf{I}_d\right)\sqrt{\boldsymbol{\Sigma}} \right|\!\right|\!\right|_2 \leq |\!|\!|\boldsymbol{\Sigma}|\!|\!|_2 \left|\!\left|\!\left| \frac{1}{n}\mathbf{W}^{\mathrm{T}}\mathbf{W} - \mathbf{I}_d \right|\!\right|\!\right|_2.$$

Consequently, by exploiting the bound (6.11), we are guaranteed that, for all $\delta > 0$,

$$\frac{|\!|\!|\widehat{\boldsymbol{\Sigma}} - \boldsymbol{\Sigma}|\!|\!|_2}{|\!|\!|\boldsymbol{\Sigma}|\!|\!|_2} \leq 2\sqrt{\frac{d}{n}} + 2\delta + \left(\sqrt{\frac{d}{n}} + \delta\right)^2, \tag{6.12}$$

with probability at least $1 - 2e^{-n\delta^2/2}$. Overall, we conclude that the relative error $|\!|\!|\widehat{\boldsymbol{\Sigma}} - \boldsymbol{\Sigma}|\!|\!|_2/|\!|\!|\boldsymbol{\Sigma}|\!|\!|_2$ converges to zero as long the ratio d/n converges to zero. ♣

It is interesting to consider Theorem 6.1 in application to sequences of matrices that satisfy additional structure, one being control on the eigenvalues of the covariance matrix $\boldsymbol{\Sigma}$.

Example 6.4 (Faster rates under trace constraints) Recall that $\{\gamma_j(\boldsymbol{\Sigma})\}_{j=1}^d$ denotes the ordered sequence of eigenvalues of the matrix $\boldsymbol{\Sigma}$, with $\gamma_1(\boldsymbol{\Sigma})$ being the maximum eigenvalue.

Now consider a non-zero covariance matrix $\mathbf{\Sigma}$ that satisfies a "trace constraint" of the form

$$\frac{\text{tr}(\mathbf{\Sigma})}{|||\mathbf{\Sigma}|||_2} = \frac{\sum_{j=1}^{d} \gamma_j(\mathbf{\Sigma})}{\gamma_1(\mathbf{\Sigma})} \leq C, \tag{6.13}$$

where C is some constant independent of dimension. Note that this ratio is a rough measure of the matrix rank, since inequality (6.13) always holds with $C = \text{rank}(\mathbf{\Sigma})$. Perhaps more interesting are matrices that are full-rank but that exhibit a relatively fast eigendecay, with a canonical instance being matrices that belong to the Schatten q-"balls" of matrices. For symmetric matrices, these sets take the form

$$\mathbb{B}_q(R_q) := \left\{ \mathbf{\Sigma} \in \mathcal{S}^{d\times d} \;\middle|\; \sum_{j=1}^{d} |\gamma_j(\mathbf{\Sigma})|^q \leq R_q \right\}, \tag{6.14}$$

where $q \in [0, 1]$ is a given parameter, and $R_q > 0$ is the radius. If we restrict to matrices with eigenvalues in $[-1, 1]$, these matrix families are nested: the smallest set with $q = 0$ corresponds to the case of matrices with rank at most R_0, whereas the other extreme $q = 1$ corresponds to an explicit trace constraint. Note that any non-zero matrix $\mathbf{\Sigma} \in \mathbb{B}_q(R_q)$ satisfies a bound of the form (6.13) with the parameter $C = R_q/(\gamma_1(\mathbf{\Sigma}))^q$.

For any matrix class satisfying the bound (6.13), Theorem 6.1 guarantees that, with high probability, the maximum singular value is bounded above as

$$\frac{\sigma_{\max}(\mathbf{X})}{\sqrt{n}} \leq \gamma_{\max}(\sqrt{\mathbf{\Sigma}}) \left(1 + \delta + \sqrt{\frac{C}{n}}\right). \tag{6.15}$$

By comparison to the earlier bound (6.10) for $\mathbf{\Sigma} = \mathbf{I}_d$, we conclude that the parameter C plays the role of the *effective dimension*. ♣

We now turn to the proof of Theorem 6.1.

Proof In order to simplify notation in the proof, let us introduce the convenient shorthand $\bar{\sigma}_{\max} = \gamma_{\max}(\sqrt{\mathbf{\Sigma}})$ and $\bar{\sigma}_{\min} = \gamma_{\min}(\sqrt{\mathbf{\Sigma}})$. Our proofs of both the upper and lower bounds consist of two steps: first, we use concentration inequalities (see Chapter 2) to argue that the random singular value is close to its expectation with high probability, and second, we use Gaussian comparison inequalities (see Chapter 5) to bound the expected values.

Maximum singular value: As noted previously, by standard properties of the multivariate Gaussian distribution, we can write $\mathbf{X} = \mathbf{W}\sqrt{\mathbf{\Sigma}}$, where the random matrix $\mathbf{W} \in \mathbb{R}^{n\times d}$ has i.i.d. $\mathcal{N}(0, 1)$ entries. Now let us view the mapping $\mathbf{W} \mapsto \frac{\sigma_{\max}(\mathbf{W}\sqrt{\mathbf{\Sigma}})}{\sqrt{n}}$ as a real-valued function on $\mathbb{R}^{nd}$. By the argument given in Example 2.32, this function is Lipschitz with respect to the Euclidean norm with constant at most $L = \bar{\sigma}_{\max}/\sqrt{n}$. By concentration of measure for Lipschitz functions of Gaussian random vectors (Theorem 2.26), we conclude that

$$\mathbb{P}[\sigma_{\max}(\mathbf{X}) \geq \mathbb{E}[\sigma_{\max}(\mathbf{X})] + \sqrt{n}\bar{\sigma}_{\max}\delta] \leq e^{-n\delta^2/2}.$$

Consequently, it suffices to show that

$$\mathbb{E}[\sigma_{\max}(\mathbf{X})] \leq \sqrt{n}\bar{\sigma}_{\max} + \sqrt{\text{tr}(\mathbf{\Sigma})}. \tag{6.16}$$

In order to do so, we first write $\sigma_{\max}(\mathbf{X})$ in a variational fashion, as the maximum of a suitably defined Gaussian process. By definition of the maximum singular value, we have $\sigma_{\max}(\mathbf{X}) = \max_{v' \in \mathbb{S}^{d-1}} \|\mathbf{X}v'\|_2$, where $\mathbb{S}^{d-1}$ denotes the Euclidean unit sphere in $\mathbb{R}^d$. Recalling the representation $\mathbf{X} = \mathbf{W}\sqrt{\mathbf{\Sigma}}$ and making the substitution $v = \sqrt{\mathbf{\Sigma}}\, v'$, we can write

$$\sigma_{\max}(\mathbf{X}) = \max_{v \in \mathbb{S}^{d-1}(\mathbf{\Sigma}^{-1})} \|\mathbf{W}v\|_2 = \max_{u \in \mathbb{S}^{n-1}} \max_{v \in \mathbb{S}^{d-1}(\mathbf{\Sigma}^{-1})} \underbrace{u^{\mathrm{T}}\mathbf{W}v}_{Z_{u,v}},$$

where $\mathbb{S}^{d-1}(\mathbf{\Sigma}^{-1}) := \{v \in \mathbb{R}^d \mid \|\mathbf{\Sigma}^{-\frac{1}{2}} v\|_2 = 1\}$ is an ellipse. Consequently, obtaining bounds on the maximum singular value corresponds to controlling the supremum of the zero-mean Gaussian process $\{Z_{u,v}, (u, v) \in \mathbb{T}\}$ indexed by the set $\mathbb{T} := \mathbb{S}^{n-1} \times \mathbb{S}^{d-1}(\mathbf{\Sigma}^{-1})$.

We upper bound the expected value of this supremum by constructing another Gaussian process $\{Y_{u,v}, (u, v) \in \mathbb{T}\}$ such that $\mathbb{E}[(Z_{u,v} - Z_{\widetilde{u},\widetilde{v}})^2] \leq \mathbb{E}[(Y_{u,v} - Y_{\widetilde{u},\widetilde{v}})^2]$ for all pairs (u, v) and $(\widetilde{u}, \widetilde{v})$ in $\mathbb{T}$. We can then apply the Sudakov–Fernique comparison (Theorem 5.27) to conclude that

$$\mathbb{E}[\sigma_{\max}(\mathbf{X})] = \mathbb{E}\Big[\max_{(u,v)\in\mathbb{T}} Z_{u,v}\Big] \leq \mathbb{E}\Big[\max_{(u,v)\in\mathbb{T}} Y_{u,v}\Big]. \tag{6.17}$$

Introducing the Gaussian process $Z_{u,v} := u^{\mathrm{T}}\mathbf{W}v$, let us first compute the induced pseudometric ρ_Z. Given two pairs (u, v) and $(\widetilde{u}, \widetilde{v})$, we may assume without loss of generality that $\|v\|_2 \leq \|\widetilde{v}\|_2$. (If not, we simply reverse the roles of (u, v) and $(\widetilde{u}, \widetilde{v})$ in the argument to follow.) We begin by observing that $Z_{u,v} = \langle\!\langle \mathbf{W}, uv^{\mathrm{T}} \rangle\!\rangle$, where we use $\langle\!\langle A, B \rangle\!\rangle := \sum_{j=1}^{n} \sum_{k=1}^{d} A_{jk}B_{jk}$ to denote the trace inner product. Since the matrix $\mathbf{W}$ has i.i.d. $\mathcal{N}(0, 1)$ entries, we have

$$\mathbb{E}[(Z_{u,v} - Z_{\widetilde{u},\widetilde{v}})^2] = \mathbb{E}[(\langle\!\langle \mathbf{W}, uv^{\mathrm{T}} - \widetilde{u}\widetilde{v}^{\mathrm{T}} \rangle\!\rangle)^2] = |\!|\!| uv^{\mathrm{T}} - \widetilde{u}\widetilde{v}^{\mathrm{T}} |\!|\!|_F^2.$$

Rearranging and expanding out this Frobenius norm, we find that

$$\begin{aligned}
|\!|\!| uv^{\mathrm{T}} - \widetilde{u}\widetilde{v}^{\mathrm{T}} |\!|\!|_F^2 &= |\!|\!| u(v - \widetilde{v})^{\mathrm{T}} + (u - \widetilde{u})\widetilde{v}^{\mathrm{T}} |\!|\!|_{\mathrm{F}}^2 \\
&= |\!|\!| (u - \widetilde{u})\widetilde{v}^{\mathrm{T}} |\!|\!|_{\mathrm{F}}^2 + |\!|\!| u(v - \widetilde{v})^{\mathrm{T}} |\!|\!|_{\mathrm{F}}^2 + 2\langle\!\langle u(v - \widetilde{v})^{\mathrm{T}}, (u - \widetilde{u})\widetilde{v}^{\mathrm{T}} \rangle\!\rangle \\
&\leq \|\widetilde{v}\|_2^2 \|u - \widetilde{u}\|_2^2 + \|u\|_2^2 \|v - \widetilde{v}\|_2^2 + 2(\|u\|_2^2 - \langle u, \widetilde{u} \rangle)(\langle v, \widetilde{v} \rangle - \|\widetilde{v}\|_2^2).
\end{aligned}$$

Now since $\|u\|_2 = \|\widetilde{u}\|_2 = 1$ by definition of the set $\mathbb{T}$, we have $\|u\|_2^2 - \langle u, \widetilde{u} \rangle \geq 0$. On the other hand, we have

$$|\langle v, \widetilde{v} \rangle| \overset{\text{(i)}}{\leq} \|v\|_2 \|\widetilde{v}\|_2 \overset{\text{(ii)}}{\leq} \|\widetilde{v}\|_2^2,$$

where step (i) follows from the Cauchy–Schwarz inequality, and step (ii) follows from our initial assumption that $\|v\|_2 \leq \|\widetilde{v}\|_2$. Combined with our previous bound on $\|u\|_2^2 - \langle u, \widetilde{u} \rangle$, we conclude that

$$\underbrace{(\|u\|_2^2 - \langle u, \widetilde{u} \rangle)}_{\geq 0}\ \underbrace{(\langle v, \widetilde{v} \rangle - \|\widetilde{v}\|_2^2)}_{\leq 0} \leq 0.$$

Putting together the pieces, we conclude that

$$|\!|\!| uv^{\mathrm{T}} - \widetilde{u}\widetilde{v}^{\mathrm{T}} |\!|\!|_{\mathrm{F}}^2 \leq \|\widetilde{v}\|_2^2 \|u - \widetilde{u}\|_2^2 + \|v - \widetilde{v}\|_2^2.$$

Finally, by definition of the set $\mathbb{S}^{d-1}(\mathbf{\Sigma}^{-1})$, we have $\|\widetilde{v}\|_2 \leq \bar{\sigma}_{\max} = \gamma_{\max}(\sqrt{\mathbf{\Sigma}})$, and hence

$$\mathbb{E}[(Z_{u,v} - Z_{\widetilde{u},\widetilde{v}})^2] \leq \bar{\sigma}_{\max}^2 \|u - \widetilde{u}\|_2^2 + \|v - \widetilde{v}\|_2^2.$$

Motivated by this inequality, we define the Gaussian process $Y_{u,v} := \bar{\sigma}_{\max} \langle g, u \rangle + \langle h, v \rangle$, where $g \in \mathbb{R}^n$ and $h \in \mathbb{R}^d$ are both standard Gaussian random vectors (i.e., with i.i.d. $\mathcal{N}(0,1)$ entries), and mutually independent. By construction, we have

$$\mathbb{E}[(Y_\theta - Y_{\widetilde{\theta}})^2] = \bar{\sigma}^2_{\max} \|u - \widetilde{u}\|_2^2 + \|v - \widetilde{v}\|_2^2.$$

Thus, we may apply the Sudakov–Fernique bound (6.17) to conclude that

$$\begin{aligned}
\mathbb{E}[\sigma_{\max}(\mathbf{X})] &\leq \mathbb{E}\Big[\sup_{(u,v) \in \mathbb{T}} Y_{u,v}\Big] \\
&= \bar{\sigma}_{\max}\, \mathbb{E}\Big[\sup_{u \in \mathbb{S}^{n-1}} \langle g, u \rangle\Big] + \mathbb{E}\Big[\sup_{v \in \mathbb{S}^{d-1}(\Sigma^{-1})} \langle h, v \rangle\Big] \\
&= \bar{\sigma}_{\max} \mathbb{E}[\|g\|_2] + \mathbb{E}[\|\sqrt{\Sigma} h\|_2]
\end{aligned}$$

By Jensen's inequality, we have $\mathbb{E}[\|g\|_2] \leq \sqrt{n}$, and similarly,

$$\mathbb{E}[\|\sqrt{\Sigma} h\|_2] \leq \sqrt{\mathbb{E}[h^{\mathrm{T}} \Sigma h]} = \sqrt{\mathrm{tr}(\Sigma)},$$

which establishes the claim (6.16).

The lower bound on the minimum singular value is based on a similar argument, but requires somewhat more technical work, so that we defer it to the Appendix (Section 6.6). □

6.3 Covariance matrices from sub-Gaussian ensembles

Various aspects of our development thus far have crucially exploited different properties of the Gaussian distribution, especially our use of the Gaussian comparison inequalities. In this section, we show how a somewhat different approach—namely, discretization and tail bounds—can be used to establish analogous bounds for general sub-Gaussian random matrices, albeit with poorer control of the constants.

In particular, let us assume that the random vector $x_i \in \mathbb{R}^d$ is zero-mean, and sub-Gaussian with parameter at most σ, by which we mean that, for each fixed $v \in \mathbb{S}^{d-1}$,

$$\mathbb{E}[e^{\lambda \langle v,\, x_i \rangle}] \leq e^{\frac{\lambda^2 \sigma^2}{2}} \qquad \text{for all } \lambda \in \mathbb{R}. \tag{6.18}$$

Equivalently stated, we assume that the scalar random variable $\langle v, x_i \rangle$ is zero-mean and sub-Gaussian with parameter at most σ. (See Chapter 2 for an in-depth discussion of sub-Gaussian variables.) Let us consider some examples to illustrate:

(a) Suppose that the matrix $\mathbf{X} \in \mathbb{R}^{n \times d}$ has i.i.d. entries, where each entry x_{ij} is zero-mean and sub-Gaussian with parameter $\sigma = 1$. Examples include the standard Gaussian ensemble ($x_{ij} \sim \mathcal{N}(0,1)$), the Rademacher ensemble ($x_{ij} \in \{-1, +1\}$ equiprobably), and, more generally, any zero-mean distribution supported on the interval $[-1, +1]$. In all of these cases, for any vector $v \in \mathbb{S}^{d-1}$, the random variable $\langle v, x_i \rangle$ is sub-Gaussian with parameter at most σ, using the i.i.d. assumption on the entries of $x_i \in \mathbb{R}^d$, and standard properties of sub-Gaussian variables.

(b) Now suppose that $x_i \sim \mathcal{N}(0, \Sigma)$. For any $v \in \mathbb{S}^{d-1}$, we have $\langle v, x_i \rangle \sim \mathcal{N}(0, v^T \Sigma v)$. Since $v^T \Sigma v \leq |||\Sigma|||_2$, we conclude that x_i is sub-Gaussian with parameter at most $\sigma^2 = |||\Sigma|||_2$.

When the random matrix $\mathbf{X} \in \mathbb{R}^{n \times d}$ is formed by drawing each row $x_i \in \mathbb{R}^d$ in an i.i.d. manner from a σ-sub-Gaussian distribution, then we say that $\mathbf{X}$ is a sample from a *row-wise σ-sub-Gaussian ensemble*. For any such random matrix, we have the following result:

Theorem 6.5 *There are universal constants $\{c_j\}_{j=0}^3$ such that, for any row-wise σ-sub-Gaussian random matrix $\mathbf{X} \in \mathbb{R}^{n \times d}$, the sample covariance $\widehat{\Sigma} = \frac{1}{n} \sum_{i=1}^n x_i x_i^T$ satisfies the bounds*

$$\mathbb{E}[e^{\lambda |||\widehat{\Sigma} - \Sigma|||_2}] \leq e^{c_0 \frac{\lambda^2 \sigma^4}{n} + 4d} \qquad \text{for all } |\lambda| < \frac{n}{64 e^2 \sigma^2}, \tag{6.19a}$$

and hence

$$\mathbb{P}\left[\frac{|||\widehat{\Sigma} - \Sigma|||_2}{\sigma^2} \geq c_1 \left\{ \sqrt{\frac{d}{n}} + \frac{d}{n} \right\} + \delta \right] \leq c_2 e^{-c_3 n \min\{\delta, \delta^2\}} \qquad \text{for all } \delta \geq 0. \tag{6.19b}$$

Remarks: Given the bound (6.19a) on the moment generating function of the random variable $|||\widehat{\Sigma} - \Sigma|||_2$, the tail bound (6.19b) is a straightforward consequence of the Chernoff technique (see Chapter 2). When $\Sigma = \mathbf{I}_d$ and each x_i is sub-Gaussian with parameter $\sigma = 1$, the tail bound (6.19b) implies that

$$|||\widehat{\Sigma} - \mathbf{I}_d|||_2 \precsim \sqrt{\frac{d}{n}} + \frac{d}{n}$$

with high probability. For $n \geq d$, this bound implies that the singular values of $\mathbf{X}/\sqrt{n}$ satisfy the sandwich relation

$$1 - c' \sqrt{\frac{d}{n}} \leq \frac{\sigma_{\min}(\mathbf{X})}{\sqrt{n}} \leq \frac{\sigma_{\max}(\mathbf{X})}{\sqrt{n}} \leq 1 + c' \sqrt{\frac{d}{n}}, \tag{6.20}$$

for some universal constant $c' > 1$. It is worth comparing this result to the earlier bounds (6.10), applicable to the special case of a standard Gaussian matrix. The bound (6.20) has a qualitatively similar form, except that the constant c' is larger than one.

Proof For notational convenience, we introduce the shorthand $\mathbf{Q} := \widehat{\Sigma} - \Sigma$. Recall from Section 6.1 the variational representation $|||\mathbf{Q}|||_2 = \max_{v \in \mathbb{S}^{d-1}} |\langle v, \mathbf{Q} v \rangle|$. We first reduce the supremum to a finite maximum via a discretization argument (see Chapter 5). Let $\{v^1, \ldots, v^N\}$ be a $\frac{1}{8}$-covering of the sphere $\mathbb{S}^{d-1}$ in the Euclidean norm; from Example 5.8, there exists such a covering with $N \leq 17^d$ vectors. Given any $v \in \mathbb{S}^{d-1}$, we can write $v = v^j + \Delta$ for some v^j in the cover, and an error vector Δ such that $\|\Delta\|_2 \leq \frac{1}{8}$, and hence

$$\langle v, \mathbf{Q} v \rangle = \langle v^j, \mathbf{Q} v^j \rangle + 2\langle \Delta, \mathbf{Q} v^j \rangle + \langle \Delta, \mathbf{Q} \Delta \rangle .$$

Applying the triangle inequality and the definition of operator norm yields

$$\begin{aligned}
|\langle v, \mathbf{Q}v\rangle| &\le |\langle v^j, \mathbf{Q}v^j\rangle| + 2\|\Delta\|_2 \, |||\mathbf{Q}|||_2 \, \|v^j\|_2 + |||\mathbf{Q}|||_2 \, \|\Delta\|_2^2 \\
&\le |\langle v^j, \mathbf{Q}v^j\rangle| + \tfrac{1}{4} |||\mathbf{Q}|||_2 + \tfrac{1}{64} |||\mathbf{Q}|||_2 \\
&\le |\langle v^j, \mathbf{Q}v^j\rangle| + \tfrac{1}{2} |||\mathbf{Q}|||_2.
\end{aligned}$$

Rearranging and then taking the supremum over $v \in \mathbb{S}^{d-1}$, and the associated maximum over $j \in \{1, 2, \ldots, N\}$, we obtain

$$|||\mathbf{Q}|||_2 = \max_{v \in \mathbb{S}^{d-1}} |\langle v, \mathbf{Q}v\rangle| \le 2 \max_{j=1,\ldots,N} |\langle v^j, \mathbf{Q}v^j\rangle|.$$

Consequently, we have

$$\mathbb{E}[e^{\lambda |||\mathbf{Q}|||_2}] \le \mathbb{E}\left[\exp\left(2\lambda \max_{j=1,\ldots,N} |\langle v^j, \mathbf{Q}v^j\rangle|\right)\right] \le \sum_{j=1}^{N} \{\mathbb{E}[e^{2\lambda\langle v^j, \mathbf{Q}v^j\rangle}] + \mathbb{E}[e^{-2\lambda\langle v^j, \mathbf{Q}v^j\rangle}]\}. \tag{6.21}$$

Next we claim that for any fixed unit vector $u \in \mathbb{S}^{d-1}$,

$$\mathbb{E}[e^{t\langle u, \mathbf{Q}u\rangle}] \le e^{512 \frac{t^2}{n} e^4 \sigma^4} \qquad \text{for all } |t| \le \frac{n}{32 e^2 \sigma^2}. \tag{6.22}$$

We take this bound as given for the moment, and use it to complete the theorem's proof. For each vector v^j in the covering set, we apply the bound (6.22) twice—once with $t = 2\lambda$ and once with $t = -2\lambda$. Combining the resulting bounds with inequality (6.21), we find that

$$\mathbb{E}[e^{\lambda |||\mathbf{Q}|||_2}] \le 2N e^{2048 \frac{\lambda^2}{n} e^4 \sigma^4} \le e^{c_0 \frac{\lambda^2 \sigma^4}{n} + 4d},$$

valid for all $|\lambda| < \frac{n}{64 e^2 \sigma^2}$, where the final step uses the fact that $2(17^d) \le e^{4d}$. Having established the moment generating function bound (6.19a), the tail bound (6.19b) follows as a consequence of Proposition 2.9.

Proof of the bound (6.22)*:* The only remaining detail is to prove the bound (6.22). By the definition of $\mathbf{Q}$ and the i.i.d. assumption, we have

$$\mathbb{E}[e^{t\langle u, \mathbf{Q}u\rangle}] = \prod_{i=1}^{n} \mathbb{E}[e^{\frac{t}{n}\{\langle x_i, u\rangle^2 - \langle u, \Sigma u\rangle\}}] = (\mathbb{E}[e^{\frac{t}{n}\{\langle x_1, u\rangle^2 - \langle u, \Sigma u\rangle\}}])^n. \tag{6.23}$$

Letting $\varepsilon \in \{-1, +1\}$ denote a Rademacher variable, independent of x_1, a standard symmetrization argument (see Proposition 4.11) implies that

$$\begin{aligned}
\mathbb{E}_{x_1}[e^{\frac{t}{n}\{\langle x_1, u\rangle^2 - \langle u, \Sigma u\rangle\}}] \le \mathbb{E}_{x_1,\varepsilon}[e^{\frac{2t}{n}\varepsilon\langle x_1, u\rangle^2}] &\overset{\text{(i)}}{=} \sum_{k=0}^{\infty} \frac{1}{k!}\left(\frac{2t}{n}\right)^k \mathbb{E}[\varepsilon^k \langle x_1, u\rangle^{2k}] \\
&\overset{\text{(ii)}}{=} 1 + \sum_{\ell=1}^{\infty} \frac{1}{(2\ell)!}\left(\frac{2t}{n}\right)^{2\ell} \mathbb{E}[\langle x_1, u\rangle^{4\ell}],
\end{aligned}$$

where step (i) follows by the power-series expansion of the exponential, and step (ii) follows since ε and x_1 are independent, and all odd moments of the Rademacher term vanish. By property (III) in Theorem 2.6 on equivalent characterizations of sub-Gaussian variables, we are guaranteed that

$$\mathbb{E}[\langle x_1, u\rangle^{4\ell}] \leq \frac{(4\ell)!}{2^{2\ell}(2\ell)!}(\sqrt{8}e\sigma)^{4\ell} \qquad \text{for all } \ell = 1, 2, \ldots,$$

and hence

$$\begin{aligned}\mathbb{E}_{x_1}[e^{\frac{t}{n}\{\langle x_1, u\rangle^2 - \langle u, \Sigma u\rangle\}}] &\leq 1 + \sum_{\ell=1}^{\infty} \frac{1}{(2\ell)!}\left(\frac{2t}{n}\right)^{2\ell} \frac{(4\ell)!}{2^{2\ell}(2\ell)!}(\sqrt{8}e\sigma)^{4\ell} \\ &\leq 1 + \sum_{\ell=1}^{\infty} \Big(\underbrace{\frac{16t}{n}e^2\sigma^2}_{f(t)}\Big)^{2\ell},\end{aligned}$$

where we have used the fact that $(4\ell)! \leq 2^{2\ell}[(2\ell)!]^2$. As long as $f(t) := \frac{16t}{n}e^2\sigma^2 < \frac{1}{2}$, we can write

$$1 + \sum_{\ell=1}^{\infty}[f^2(t)]^{\ell} \stackrel{\text{(i)}}{=} \frac{1}{1 - f^2(t)} \stackrel{\text{(ii)}}{\leq} \exp(2f^2(t)),$$

where step (i) follows by summing the geometric series, and step (ii) follows because $\frac{1}{1-a} \leq e^{2a}$ for all $a \in [0, \frac{1}{2}]$. Putting together the pieces and combining with our earlier bound (6.23), we have shown that $\mathbb{E}[e^{t\langle u, \mathbf{Q}u\rangle}] \leq e^{2nf^2(t)}$, valid for all $|t| < \frac{n}{32e^2\sigma^2}$, which establishes the claim (6.22). □

6.4 Bounds for general matrices

The preceding sections were devoted to bounds applicable to sample covariances under Gaussian or sub-Gaussian tail conditions. This section is devoted to developing extensions to more general tail conditions. In order to do so, it is convenient to introduce some more general methodology that applies not only to sample covariance matrices, but also to more general random matrices. The main results in this section are Theorems 6.15 and 6.17, which are (essentially) matrix-based analogs of our earlier Hoeffding and Bernstein bounds for random variables. Before proving these results, we develop some useful matrix-theoretic generalizations of ideas from Chapter 2, including various types of tail conditions, as well as decompositions for the moment generating function for independent random matrices.

6.4.1 Background on matrix analysis

We begin by introducing some additional background on matrix-valued functions. Recall the class $\mathcal{S}^{d\times d}$ of symmetric $d \times d$ matrices. Any function $f\colon \mathbb{R} \to \mathbb{R}$ can be extended to a map from the set $\mathcal{S}^{d\times d}$ to itself in the following way. Given a matrix $\mathbf{Q} \in \mathcal{S}^{d\times d}$, consider its eigendecomposition $\mathbf{Q} = \mathbf{U}^{\mathrm{T}}\mathbf{\Gamma}\mathbf{U}$. Here the matrix $\mathbf{U} \in \mathbb{R}^{d\times d}$ is a unitary matrix, satisfying the relation $\mathbf{U}^{\mathrm{T}}\mathbf{U} = \mathbf{I}_d$, whereas $\mathbf{\Gamma} := \operatorname{diag}(\gamma(\mathbf{Q}))$ is a diagonal matrix specified by the vector of eigenvalues $\gamma(\mathbf{Q}) \in \mathbb{R}^d$. Using this notation, we consider the mapping from $\mathcal{S}^{d\times d}$ to itself defined via

$$\mathbf{Q} \mapsto f(\mathbf{Q}) := \mathbf{U}^{\mathrm{T}} \operatorname{diag}(f(\gamma_1(\mathbf{Q})), \ldots, f(\gamma_d(\mathbf{Q})))\mathbf{U}.$$

In words, we apply the original function f elementwise to the vector of eigenvalues $\gamma(\mathbf{Q})$, and then rotate the resulting matrix $\text{diag}(f(\gamma(\mathbf{Q})))$ back to the original coordinate system defined by the eigenvectors of $\mathbf{Q}$. By construction, this extension of f to $\mathcal{S}^{d\times d}$ is unitarily invariant, meaning that

$$f(\mathbf{V}^{\mathrm{T}}\mathbf{Q}\mathbf{V}) = \mathbf{V}^{\mathrm{T}} f(\mathbf{Q})\mathbf{V} \qquad \text{for all unitary matrices } \mathbf{V} \in \mathbb{R}^{d\times d},$$

since it affects only the eigenvalues (but not the eigenvectors) of $\mathbf{Q}$. Moreover, the eigenvalues of $f(\mathbf{Q})$ transform in a simple way, since we have

$$\gamma(f(\mathbf{Q})) = \{f(\gamma_j(\mathbf{Q})),\ j = 1, \ldots, d\}. \tag{6.24}$$

In words, the eigenvalues of the matrix $f(\mathbf{Q})$ are simply the eigenvalues of $\mathbf{Q}$ transformed by f, a result often referred to as the *spectral mapping property*.

Two functions that play a central role in our development of matrix tail bounds are the matrix exponential and the matrix logarithm. As a particular case of our construction, the matrix exponential has the power-series expansion $e^{\mathbf{Q}} = \sum_{k=0}^{\infty} \frac{\mathbf{Q}^k}{k!}$. By the spectral mapping property, the eigenvalues of $e^{\mathbf{Q}}$ are positive, so that it is a positive definite matrix for any choice of $\mathbf{Q}$. Parts of our analysis also involve the matrix logarithm; when restricted to the cone of strictly positive definite matrices, as suffices for our purposes, the matrix logarithm corresponds to the inverse of the matrix exponential.

A function f on $\mathcal{S}^{d\times d}$ is said to be *matrix monotone* if $f(\mathbf{Q}) \preceq f(\mathbf{R})$ whenever $\mathbf{Q} \preceq \mathbf{R}$. A useful property of the logarithm is that it is a matrix monotone function, a result known as the *Löwner–Heinz theorem*. By contrast, the exponential is *not* a matrix monotone function, showing that matrix monotonicity is more complex than the usual notion of monotonicity. See Exercise 6.5 for further exploration of these properties.

Finally, a useful fact is the following: if $f\colon \mathbb{R} \to \mathbb{R}$ is any continuous and non-decreasing function in the usual sense, then for any pair of symmetric matrices such that $\mathbf{Q} \preceq \mathbf{R}$, we are guaranteed that

$$\text{tr}(f(\mathbf{Q})) \le \text{tr}(f(\mathbf{R})). \tag{6.25}$$

See the bibliographic section for further discussion of such *trace inequalities*.

6.4.2 Tail conditions for matrices

Given a symmetric random matrix $\mathbf{Q} \in \mathcal{S}^{d\times d}$, its polynomial moments, assuming that they exist, are the matrices defined by $\mathbb{E}[\mathbf{Q}^j]$. As shown in Exercise 6.6, the variance of $\mathbf{Q}$ is a positive semidefinite matrix given by $\text{var}(\mathbf{Q}) := \mathbb{E}[\mathbf{Q}^2] - (\mathbb{E}[\mathbf{Q}])^2$. The moment generating function of a random matrix $\mathbf{Q}$ is the matrix-valued mapping $\Psi_{\mathbf{Q}}\colon \mathbb{R} \to \mathcal{S}^{d\times d}$ given by

$$\Psi_{\mathbf{Q}}(\lambda) := \mathbb{E}[e^{\lambda \mathbf{Q}}] = \sum_{k=0}^{\infty} \frac{\lambda^k}{k!} \mathbb{E}[\mathbf{Q}^k]. \tag{6.26}$$

Under suitable conditions on $\mathbf{Q}$—or equivalently, suitable conditions on the polynomial moments of $\mathbf{Q}$—it is guaranteed to be finite for all λ in an interval centered at zero. In parallel with our discussion in Chapter 2, various tail conditions are based on imposing bounds on this moment generating function. We begin with the simplest case:

Definition 6.6 A zero-mean symmetric random matrix $\mathbf{Q} \in \mathcal{S}^{d\times d}$ is sub-Gaussian with matrix parameter $\mathbf{V} \in \mathcal{S}_+^{d\times d}$ if

$$\Psi_{\mathbf{Q}}(\lambda) \preceq e^{\frac{\lambda^2 \mathbf{V}}{2}} \qquad \text{for all } \lambda \in \mathbb{R}. \tag{6.27}$$

This definition is best understood by working through some simple examples.

Example 6.7 Suppose that $\mathbf{Q} = \varepsilon \mathbf{B}$ where $\varepsilon \in \{-1, +1\}$ is a Rademacher variable, and $\mathbf{B} \in \mathcal{S}^{d\times d}$ is a fixed matrix. Random matrices of this form frequently arise as the result of symmetrization arguments, as discussed at more length in the sequel. Note that we have $\mathbb{E}[\mathbf{Q}^{2k+1}] = 0$ and $\mathbb{E}[\mathbf{Q}^{2k}] = \mathbf{B}^{2k}$ for all $k = 1, 2, \ldots$, and hence

$$\mathbb{E}[e^{\lambda \mathbf{Q}}] = \sum_{k=0}^{\infty} \frac{\lambda^{2k}}{(2k)!} \mathbf{B}^{2k} \preceq \sum_{k=1}^{\infty} \frac{1}{k!} \left(\frac{\lambda^2 \mathbf{B}^2}{2} \right)^k = e^{\frac{\lambda^2 \mathbf{B}^2}{2}},$$

showing that the sub-Gaussian condition (6.27) holds with $\mathbf{V} = \mathbf{B}^2 = \text{var}(\mathbf{Q})$. ♣

As we show in Exercise 6.7, more generally, a random matrix of the form $\mathbf{Q} = g\mathbf{B}$, where $g \in \mathbb{R}$ is a σ-sub-Gaussian variable with distribution symmetric around zero, satisfies the condition (6.27) with matrix parameter $\mathbf{V} = \sigma^2 \mathbf{B}^2$.

Example 6.8 As an extension of the previous example, consider a random matrix of the form $\mathbf{Q} = \varepsilon \mathbf{C}$, where ε is a Rademacher variable as before, and $\mathbf{C}$ is now a random matrix, independent of ε with its spectral norm bounded as $|||\mathbf{C}|||_2 \le b$. First fixing $\mathbf{C}$ and taking expectations over the Rademacher variable, the previous example yields $\mathbb{E}_\varepsilon[e^{\lambda \varepsilon \mathbf{C}}] \preceq e^{\frac{\lambda^2}{2} \mathbf{C}^2}$. Since $|||\mathbf{C}|||_2 \le b$, we have $e^{\frac{\lambda^2}{2}\mathbf{C}^2} \preceq e^{\frac{\lambda^2}{2} b^2 \mathbf{I}_d}$, and hence

$$\Psi_{\mathbf{Q}}(\lambda) \preceq e^{\frac{\lambda^2}{2} b^2 \mathbf{I}_d} \qquad \text{for all } \lambda \in \mathbb{R},$$

showing that $\mathbf{Q}$ is sub-Gaussian with matrix parameter $\mathbf{V} = b^2 \mathbf{I}_d$. ♣

In parallel with our treatment of scalar random variables in Chapter 2, it is natural to consider various weakenings of the sub-Gaussian requirement.

Definition 6.9 (Sub-exponential random matrices) A zero-mean random matrix is sub-exponential with parameters $(\mathbf{V}, \alpha)$ if

$$\Psi_{\mathbf{Q}}(\lambda) \preceq e^{\frac{\lambda^2 \mathbf{V}}{2}} \qquad \text{for all } |\lambda| < \tfrac{1}{\alpha}. \tag{6.28}$$

Thus, any sub-Gaussian random matrix is also sub-exponential with parameters $(\mathbf{V}, 0)$. However, there also exist sub-exponential random matrices that are not sub-Gaussian. One example is the zero-mean random matrix $\mathbf{M} = \varepsilon g^2 \mathbf{B}$, where $\varepsilon \in \{-1, +1\}$ is a Rademacher variable, the variable $g \sim \mathcal{N}(0,1)$ is independent of ε, and $\mathbf{B}$ is a fixed symmetric matrix.

The Bernstein condition for random matrices provides one useful way of certifying the sub-exponential condition:

Definition 6.10 (Bernstein's condition for matrices) A zero-mean symmetric random matrix $\mathbf{Q}$ satisfies a Bernstein condition with parameter $b > 0$ if

$$\mathbb{E}[\mathbf{Q}^j] \preceq \tfrac{1}{2} j!\, b^{j-2} \operatorname{var}(\mathbf{Q}) \qquad \text{for } j = 3, 4, \ldots. \tag{6.29}$$

We note that (a stronger form of) Bernstein's condition holds whenever the matrix $\mathbf{Q}$ has a bounded operator norm—say $|||\mathbf{Q}|||_2 \leq b$ almost surely. In this case, it can be shown (see Exercise 6.9) that

$$\mathbb{E}[\mathbf{Q}^j] \preceq b^{j-2} \operatorname{var}(\mathbf{Q}) \qquad \text{for all } j = 3, 4, \ldots. \tag{6.30}$$

Exercise 6.11 gives an example of a random matrix with unbounded operator norm for which Bernstein's condition holds.

The following lemma shows how the general Bernstein condition (6.29) implies the sub-exponential condition. More generally, the argument given here provides an explicit bound on the moment generating function:

Lemma 6.11 *For any symmetric zero-mean random matrix satisfying the Bernstein condition* (6.29), *we have*

$$\Psi_{\mathbf{Q}}(\lambda) \preceq \exp\left(\frac{\lambda^2 \operatorname{var}(\mathbf{Q})}{2(1 - b|\lambda|)}\right) \qquad \textit{for all } |\lambda| < \frac{1}{b}. \tag{6.31}$$

Proof Since $\mathbb{E}[\mathbf{Q}] = 0$, applying the definition of the matrix exponential for a suitably small $\lambda \in \mathbb{R}$ yields

$$\begin{aligned}
\mathbb{E}[e^{\lambda \mathbf{Q}}] &= \mathbf{I}_d + \frac{\lambda^2 \operatorname{var}(\mathbf{Q})}{2} + \sum_{j=3}^{\infty} \frac{\lambda^j \mathbb{E}[\mathbf{Q}^j]}{j!} \\
&\overset{\text{(i)}}{\preceq} \mathbf{I}_d + \frac{\lambda^2 \operatorname{var}(\mathbf{Q})}{2} \left\{ \sum_{j=0}^{\infty} |\lambda|^j b^j \right\} \\
&\overset{\text{(ii)}}{=} \mathbf{I}_d + \frac{\lambda^2 \operatorname{var}(\mathbf{Q})}{2(1 - b|\lambda|)} \\
&\overset{\text{(iii)}}{\preceq} \exp\left(\frac{\lambda^2 \operatorname{var}(\mathbf{Q})}{2(1 - b|\lambda|)}\right),
\end{aligned}$$

where step (i) applies the Bernstein condition, step (ii) is valid for any $|\lambda| < 1/b$, a choice for which the geometric series is summable, and step (iii) follows from the matrix inequality

$\mathbf{I}_d + \mathbf{A} \preceq e^{\mathbf{A}}$, which is valid for any symmetric matrix $\mathbf{A}$. (See Exercise 6.4 for more discussion of this last property.) □

6.4.3 Matrix Chernoff approach and independent decompositions

The Chernoff approach to tail bounds, as discussed in Chapter 2, is based on controlling the moment generating function of a random variable. In this section, we begin by showing that the trace of the matrix moment generating function (6.26) plays a similar role in bounding the operator norm of random matrices.

Lemma 6.12 (Matrix Chernoff technique) *Let* $\mathbf{Q}$ *be a zero-mean symmetric random matrix whose moment generating function* $\Psi_{\mathbf{Q}}$ *exists in an open interval* $(-a, a)$. *Then for any* $\delta > 0$, *we have*

$$\mathbb{P}[\gamma_{\max}(\mathbf{Q}) \geq \delta] \leq \operatorname{tr}(\Psi_{\mathbf{Q}}(\lambda))e^{-\lambda\delta} \qquad \text{for all } \lambda \in [0, a), \tag{6.32}$$

where $\operatorname{tr}(\cdot)$ *denotes the trace operator on matrices. Similarly, we have*

$$\mathbb{P}[|||\mathbf{Q}|||_2 \geq \delta] \leq 2\operatorname{tr}(\Psi_{\mathbf{Q}}(\lambda))e^{-\lambda\delta} \qquad \text{for all } \lambda \in [0, a). \tag{6.33}$$

Proof For each $\lambda \in [0, a)$, we have

$$\mathbb{P}[\gamma_{\max}(\mathbf{Q}) \geq \delta] = \mathbb{P}[e^{\gamma_{\max}(\lambda\mathbf{Q})} \geq e^{\lambda\delta}] \stackrel{\text{(i)}}{=} \mathbb{P}[\gamma_{\max}(e^{\lambda\mathbf{Q}}) \geq e^{\lambda\delta}], \tag{6.34}$$

where step (i) uses the functional calculus relating the eigenvalues of $\lambda\mathbf{Q}$ to those of $e^{\lambda\mathbf{Q}}$. Applying Markov's inequality yields

$$\mathbb{P}[\gamma_{\max}(e^{\lambda\mathbf{Q}}) \geq e^{\lambda\delta}] \leq \mathbb{E}[\gamma_{\max}(e^{\lambda\mathbf{Q}})]e^{-\lambda\delta} \stackrel{\text{(i)}}{\leq} \mathbb{E}[\operatorname{tr}(e^{\lambda\mathbf{Q}})]e^{-\lambda\delta}. \tag{6.35}$$

Here inequality (i) uses the upper bound $\gamma_{\max}(e^{\lambda\mathbf{Q}}) \leq \operatorname{tr}(e^{\lambda\mathbf{Q}})$, which holds since $e^{\lambda\mathbf{Q}}$ is positive definite. Finally, since trace and expectation commute, we have

$$\mathbb{E}[\operatorname{tr}(e^{\lambda\mathbf{Q}})] = \operatorname{tr}(\mathbb{E}[e^{\lambda\mathbf{Q}}]) = \operatorname{tr}(\Psi_{\mathbf{Q}}(\lambda)).$$

Note that the same argument can be applied to bound the event $\gamma_{\max}(-\mathbf{Q}) \geq \delta$, or equivalently the event $\gamma_{\min}(\mathbf{Q}) \leq -\delta$. Since $|||\mathbf{Q}|||_2 = \max\{\gamma_{\max}(\mathbf{Q}), |\gamma_{\min}(\mathbf{Q})|\}$, the tail bound on the operator norm (6.33) follows. □

An important property of independent random variables is that the moment generating function of their sum can be decomposed as the product of the individual moment generating functions. For random matrices, this type of decomposition is no longer guaranteed to hold with equality, essentially because matrix products need not commute. However, for independent random matrices, it is nonetheless possible to establish an upper bound in terms of the trace of the product of moment generating functions, as we now show.

Lemma 6.13 *Let $\mathbf{Q}_1, \ldots, \mathbf{Q}_n$ be independent symmetric random matrices whose moment generating functions exist for all $\lambda \in I$, and define the sum $\mathbf{S}_n := \sum_{i=1}^n \mathbf{Q}_i$. Then*

$$\mathrm{tr}(\Psi_{\mathbf{S}_n}(\lambda)) \leq \mathrm{tr}\Big(e^{\sum_{i=1}^n \log \Psi_{\mathbf{Q}_i}(\lambda)}\Big) \quad \textit{for all } \lambda \in I. \tag{6.36}$$

Remark: In conjunction with Lemma 6.12, this lemma provides an avenue for obtaining tail bounds on the operator norm of sums of independent random matrices. In particular, if we apply the upper bound (6.33) to the random matrix $\mathbf{S}_n/n$, we find that

$$\mathbb{P}\left[\left|\left|\left|\frac{1}{n}\sum_{i=1}^n \mathbf{Q}_i\right|\right|\right|_2 \geq \delta\right] \leq 2\,\mathrm{tr}\Big(e^{\sum_{i=1}^n \log \Psi_{\mathbf{Q}_i}(\lambda)}\Big) e^{-\lambda n\delta} \qquad \text{for all } \lambda \in [0, a). \tag{6.37}$$

Proof In order to prove this lemma, we require the following result due to Lieb (1973): for any fixed matrix $\mathbf{H} \in \mathcal{S}^{d\times d}$, the function $f\colon \mathcal{S}_+^{d\times d} \to \mathbb{R}$ given by

$$f(\mathbf{A}) := \mathrm{tr}(e^{\mathbf{H}+\log(\mathbf{A})})$$

is concave. Introducing the shorthand notation $G(\lambda) := \mathrm{tr}(\Psi_{\mathbf{S}_n}(\lambda))$, we note that, by linearity of trace and expectation, we have

$$G(\lambda) = \mathrm{tr}\Big(\mathbb{E}[e^{\lambda\mathbf{S}_{n-1}+\log\exp(\lambda\mathbf{Q}_n)}]\Big) = \mathbb{E}_{\mathbf{S}_{n-1}}\mathbb{E}_{\mathbf{Q}_n}[\mathrm{tr}(e^{\lambda\mathbf{S}_{n-1}+\log\exp(\lambda\mathbf{Q}_n)})].$$

Using concavity of the function f with $\mathbf{H} = \lambda\mathbf{S}_{n-1}$ and $\mathbf{A} = e^{\lambda\mathbf{Q}_n}$, Jensen's inequality implies that

$$\mathbb{E}_{\mathbf{Q}_n}[\mathrm{tr}(e^{\lambda\mathbf{S}_{n-1}+\log\exp(\lambda\mathbf{Q}_n)})] \leq \mathrm{tr}(e^{\lambda\mathbf{S}_{n-1}+\log\mathbb{E}_{\mathbf{Q}_n}\exp(\lambda\mathbf{Q}_n)}),$$

so that we have shown that $G(\lambda) \leq \mathbb{E}_{\mathbf{S}_{n-1}}[\mathrm{tr}(e^{\lambda\mathbf{S}_{n-1}+\log\Psi_{\mathbf{Q}_n}(\lambda)})]$.

We now recurse this argument, in particular peeling off the term involving $\mathbf{Q}_{n-1}$, so that we have

$$G(\lambda) \leq \mathbb{E}_{\mathbf{S}_{n-2}}\mathbb{E}_{\mathbf{Q}_{n-1}}[\,\mathrm{tr}(e^{\lambda\mathbf{S}_{n-2}+\log\Psi_{\mathbf{Q}_n}(\lambda)+\log\exp(\lambda\mathbf{Q}_{n-1})})].$$

We again exploit the concavity of the function f, this time with the choices $\mathbf{H} = \lambda\mathbf{S}_{n-2} + \log\Psi_{\mathbf{Q}_n}(\lambda)$ and $\mathbf{A} = e^{\lambda\mathbf{Q}_{n-1}}$, thereby finding that

$$G(\lambda) \leq \mathbb{E}_{\mathbf{S}_{n-2}}[\,\mathrm{tr}(e^{\lambda\mathbf{S}_{n-2}+\log\Psi_{\mathbf{Q}_{n-1}}(\lambda)+\log\Psi_{\mathbf{Q}_n}(\lambda)})].$$

Continuing in this manner completes the proof of the claim. □

In many cases, our goal is to bound the maximum eigenvalue (or operator norm) of sums of centered random matrices of the form $\mathbf{Q}_i = \mathbf{A}_i - \mathbb{E}[\mathbf{A}_i]$. In this and other settings, it is often convenient to perform an additional symmetrization step, so that we can deal instead with matrices $\widetilde{\mathbf{Q}}_i$ that are guaranteed to have distribution symmetric around zero (meaning that $\widetilde{\mathbf{Q}}_i$ and $-\widetilde{\mathbf{Q}}_i$ follow the same distribution).

Example 6.14 (Rademacher symmetrization for random matrices) Let $\{\mathbf{A}_i\}_{i=1}^n$ be a sequence of independent symmetric random matrices, and suppose that our goal is to bound the maximum eigenvalue of the matrix sum $\sum_{i=1}^n(\mathbf{A}_i - \mathbb{E}[\mathbf{A}_i])$. Since the maximum eigenvalue can be represented as the supremum of an empirical process, the symmetrization techniques from Chapter 4 can be used to reduce the problem to one involving the new matrices

$\widetilde{\mathbf{Q}}_i = \varepsilon_i \mathbf{A}_i$, where ε_i is an independent Rademacher variable. Let us now work through this reduction. By Markov's inequality, we have

$$\mathbb{P}\left[\gamma_{\max}\left(\sum_{i=1}^n \{\mathbf{A}_i - \mathbb{E}[\mathbf{A}_i]\}\right) \geq \delta\right] \leq \mathbb{E}[e^{\lambda \gamma_{\max}(\sum_{i=1}^n \{\mathbf{A}_i - \mathbb{E}[\mathbf{A}_i]\})}] e^{-\lambda\delta}.$$

By the variational representation of the maximum eigenvalue, we have

$$\begin{aligned}
\mathbb{E}[e^{\lambda \gamma_{\max}(\sum_{i=1}^n \{\mathbf{A}_i - \mathbb{E}[\mathbf{A}_i]\})}] &= \mathbb{E}\left[\exp\left(\lambda \sup_{\|u\|_2=1} \left\langle u, \left(\sum_{i=1}^n (\mathbf{A}_i - \mathbb{E}[\mathbf{A}_i])\right) u\right\rangle\right)\right] \\
&\overset{\text{(i)}}{\leq} \mathbb{E}\left[\exp\left(2\lambda \sup_{\|u\|_2=1} \left\langle u, \left(\sum_{i=1}^n \varepsilon_i \mathbf{A}_i\right) u\right\rangle\right)\right] \\
&= \mathbb{E}[e^{2\lambda \gamma_{\max}(\sum_{i=1}^n \varepsilon_i \mathbf{A}_i)}] \\
&\overset{\text{(ii)}}{=} \mathbb{E}[\gamma_{\max}(e^{2\lambda \sum_{i=1}^n \varepsilon_i \mathbf{A}_i})],
\end{aligned}$$

where inequality (i) makes use of the symmetrization inequality from Proposition 4.11(b) with $\Phi(t) = e^{\lambda t}$, and step (ii) uses the spectral mapping property (6.24). Continuing on, we have

$$\mathbb{E}[\gamma_{\max}(e^{2\lambda \sum_{i=1}^n \varepsilon_i \mathbf{A}_i})] \leq \operatorname{tr}\left(\mathbb{E}[e^{2\lambda \sum_{i=1}^n \varepsilon_i \mathbf{A}_i}]\right) \leq \operatorname{tr}\left(e^{\sum_{i=1}^n \log \Psi_{\widetilde{\mathbf{Q}}_i}(2\lambda)}\right),$$

where the final step follows from applying Lemma 6.13 to the symmetrized matrices $\widetilde{\mathbf{Q}}_i = \varepsilon_i \mathbf{A}_i$. Consequently, apart from the factor of 2, we may assume without loss of generality when bounding maximum eigenvalues that our matrices have a distribution symmetric around zero. ♣

6.4.4 Upper tail bounds for random matrices

We now have collected the ingredients necessary for stating and proving various tail bounds for the deviations of sums of zero-mean independent random matrices.

Sub-Gaussian case

We begin with a tail bound for sub-Gaussian random matrices. It provides an approximate analog of the Hoeffding-type tail bound for random variables (Proposition 2.5).

Theorem 6.15 (Hoeffding bound for random matrices) *Let $\{\mathbf{Q}_i\}_{i=1}^n$ be a sequence of zero-mean independent symmetric random matrices that satisfy the sub-Gaussian condition with parameters $\{\mathbf{V}_i\}_{i=1}^n$. Then for all $\delta > 0$, we have the upper tail bound*

$$\mathbb{P}\left[\left\|\left\|\frac{1}{n}\sum_{i=1}^n \mathbf{Q}_i\right\|\right\|_2 \geq \delta\right] \leq 2\operatorname{rank}\left(\sum_{i=1}^n \mathbf{V}_i\right) e^{-\frac{n\delta^2}{2\sigma^2}} \leq 2d e^{-\frac{n\delta^2}{2\sigma^2}}, \tag{6.38}$$

where $\sigma^2 = |||\frac{1}{n}\sum_{i=1}^n \mathbf{V}_i|||_2$.

Proof We first prove the claim in the case when $\mathbf{V} := \sum_{i=1}^n \mathbf{V}_i$ is full-rank, and then show how to prove the general case. From Lemma 6.13, it suffices to upper bound $\operatorname{tr}\left(e^{\sum_{i=1}^n \log \Psi_{\mathbf{Q}_i}(\lambda)}\right)$. From Definition 6.6, the assumed sub-Gaussianity, and the monotonicity of the matrix logarithm, we have

$$\sum_{i=1}^n \log \Psi_{\mathbf{Q}_i}(\lambda) \preceq \frac{\lambda^2}{2} \sum_{i=1}^n \mathbf{V}_i,$$

where we have used the fact that the logarithm is matrix monotone. Now since the exponential is an increasing function, the trace bound (6.25) implies that

$$\operatorname{tr}\left(e^{\sum_{i=1}^n \log \Psi_{\mathbf{Q}_i}(\lambda)}\right) \leq \operatorname{tr}\left(e^{\frac{\lambda^2}{2} \sum_{i=1}^n \mathbf{V}_i}\right).$$

This upper bound, when combined with the matrix Chernoff bound (6.37), yields

$$\mathbb{P}\left[\left|\!\left|\!\left| \frac{1}{n} \sum_{i=1}^n \mathbf{Q}_i \right|\!\right|\!\right|_2 \geq \delta\right] \leq 2 \operatorname{tr}\left(e^{\frac{\lambda^2}{2} \sum_{i=1}^n \mathbf{V}_i}\right) e^{-\lambda n \delta}.$$

For any d-dimensional symmetric matrix $\mathbf{R}$, we have $\operatorname{tr}(e^{\mathbf{R}}) \leq d e^{|||\mathbf{R}|||_2}$. Applying this inequality to the matrix $\mathbf{R} = \frac{\lambda^2}{2} \sum_{i=1}^n \mathbf{V}_i$, for which we have $|||\mathbf{R}|||_2 = \frac{\lambda^2}{2} n\sigma^2$, yields the bound

$$\mathbb{P}\left[\left|\!\left|\!\left| \frac{1}{n} \sum_{i=1}^n \mathbf{Q}_i \right|\!\right|\!\right|_2 \geq \delta\right] \leq 2d e^{\frac{\lambda^2}{2} n\sigma^2 - \lambda n \delta}.$$

This upper bound holds for all $\lambda \geq 0$ and setting $\lambda = \delta/\sigma^2$ yields the claim.

Now suppose that the matrix $\mathbf{V} := \sum_{i=1}^n \mathbf{V}_i$ is not full-rank, say of rank $r < d$. In this case, an eigendecomposition yields $\mathbf{V} = \mathbf{U}\mathbf{D}\mathbf{U}^{\mathrm{T}}$, where $\mathbf{U} \in \mathbb{R}^{d \times r}$ has orthonormal columns. Introducing the shorthand $\mathbf{Q} := \sum_{i=1}^n \mathbf{Q}_i$, the r-dimensional matrix $\widetilde{\mathbf{Q}} = \mathbf{U}^{\mathrm{T}}\mathbf{Q}\mathbf{U}$ then captures all randomness in $\mathbf{Q}$, and in particular we have $|||\widetilde{\mathbf{Q}}|||_2 = |||\mathbf{Q}|||_2$. We can thus apply the same argument to bound $|||\widetilde{\mathbf{Q}}|||_2$, leading to a pre-factor of r instead of d. □

An important fact is that inequality (6.38) also implies an analogous bound for general independent but potentially non-symmetric and/or non-square matrices, with d replaced by $(d_1 + d_2)$. More specifically, a problem involving general zero-mean random matrices $\mathbf{A}_i \in \mathbb{R}^{d_1 \times d_2}$ can be transformed to a symmetric version by defining the $(d_1 + d_2)$-dimensional square matrices

$$\mathbf{Q}_i := \begin{bmatrix} \mathbf{0}_{d_1 \times d_1} & \mathbf{A}_i \\ \mathbf{A}_i^{\mathrm{T}} & \mathbf{0}_{d_2 \times d_2} \end{bmatrix}, \tag{6.39}$$

and imposing some form of moment generating function bound—for instance, the sub-Gaussian condition (6.27)—on the symmetric matrices $\mathbf{Q}_i$. See Exercise 6.10 for further details.

A significant feature of the tail bound (6.38) is the appearance of either the rank or the dimension d in front of the exponent. In certain cases, this dimension-dependent factor is superfluous, and leads to sub-optimal bounds. However, it cannot be avoided in general. The following example illustrates these two extremes.

Example 6.16 (Looseness/sharpness of Theorem 6.15) For simplicity, let us consider examples with $n = d$. For each $i = 1, 2, \ldots, d$, let $\mathbf{E}_i \in \mathcal{S}^{d\times d}$ denote the diagonal matrix with 1 in position (i, i), and 0s elsewhere. Define $\mathbf{Q}_i = y_i \mathbf{E}_i$, where $\{y_i\}_{i=1}^n$ is an i.i.d. sequence of 1-sub-Gaussian variables. Two specific cases to keep in mind are Rademacher variables $\{\varepsilon_i\}_{i=1}^n$, and $\mathcal{N}(0, 1)$ variables $\{g_i\}_{i=1}^n$.

For any such choice of sub-Gaussian variables, a calculation similar to that of Example 6.7 shows that each $\mathbf{Q}_i$ satisfies the sub-Gaussian bound (6.27) with $\mathbf{V}_i = \mathbf{E}_i$, and hence $\sigma^2 = |||\frac{1}{d}\sum_{i=1}^d \mathbf{V}_i|||_2 = 1/d$. Consequently, an application of Theorem 6.15 yields the tail bound

$$\mathbb{P}\left[\left|\left|\left|\frac{1}{d}\sum_{i=1}^d \mathbf{Q}_i\right|\right|\right|_2 \geq \delta\right] \leq 2de^{-\frac{d^2\delta^2}{2}} \quad \text{for all } \delta > 0, \tag{6.40}$$

which implies that $|||\frac{1}{d}\sum_{j=1}^d \mathbf{Q}_j|||_2 \precsim \frac{\sqrt{2\log(2d)}}{d}$ with high probability. On the other hand, an explicit calculation shows that

$$\left|\left|\left|\frac{1}{d}\sum_{i=1}^n \mathbf{Q}_i\right|\right|\right|_2 = \max_{i=1,\ldots,d} \frac{|y_i|}{d}. \tag{6.41}$$

Comparing the exact result (6.41) with the bound (6.40) yields a range of behavior. At one extreme, for i.i.d. Rademacher variables $y_i = \varepsilon_i \in \{-1, +1\}$, we have $|||\frac{1}{d}\sum_{i=1}^n \mathbf{Q}_i|||_2 = 1/d$, showing that the bound (6.40) is off by the order $\sqrt{\log d}$. On the other hand, for i.i.d. Gaussian variables $y_i = g_i \sim \mathcal{N}(0, 1)$, we have

$$\left|\left|\left|\frac{1}{d}\sum_{i=1}^d \mathbf{Q}_i\right|\right|\right|_2 = \max_{i=1,\ldots,d} \frac{|g_i|}{d} \simeq \frac{\sqrt{2\log d}}{d},$$

using the fact that the maximum of d i.i.d. $\mathcal{N}(0, 1)$ variables scales as $\sqrt{2\log d}$. Consequently, Theorem 6.15 cannot be improved for this class of random matrices. ♣

Bernstein-type bounds for random matrices

We now turn to bounds on random matrices that satisfy sub-exponential tail conditions, in particular of the Bernstein form (6.29).

Theorem 6.17 (Bernstein bound for random matrices) *Let $\{\mathbf{Q}_i\}_{i=1}^n\}$ be a sequence of independent, zero-mean, symmetric random matrices that satisfy the Bernstein condition (6.29) with parameter $b > 0$. Then for all $\delta \geq 0$, the operator norm satisfies the tail bound*

$$\mathbb{P}\left[\frac{1}{n}\left|\left|\left|\sum_{i=1}^n \mathbf{Q}_i\right|\right|\right|_2 \geq \delta\right] \leq 2\,\mathrm{rank}\left(\sum_{i=1}^n \mathrm{var}(\mathbf{Q}_i)\right)\exp\left\{-\frac{n\delta^2}{2(\sigma^2 + b\delta)}\right\}, \tag{6.42}$$

where $\sigma^2 := \frac{1}{n}|||\sum_{j=1}^n \mathrm{var}(\mathbf{Q}_j)|||_2$.

Proof By Lemma 6.13, we have $\operatorname{tr}(\Psi_{\mathbf{S}_n}(\lambda)) \leq \operatorname{tr}\big(e^{\sum_{i=1}^n \log \Psi_{\mathbf{Q}_i}(\lambda)}\big)$. By Lemma 6.11, the Bernstein condition combined with matrix monotonicity of the logarithm yields the bound $\log \Psi_{\mathbf{Q}_i}(\lambda) \preceq \frac{\lambda^2 \operatorname{var}(\mathbf{Q}_i)}{1-b|\lambda|}$ for any $|\lambda| < \frac{1}{b}$. Putting together the pieces yields

$$\operatorname{tr}\big(e^{\sum_{i=1}^n \log \Psi_{\mathbf{Q}_i}(\lambda)}\big) \leq \operatorname{tr}\left(\exp\left(\frac{\lambda^2 \sum_{i=1}^n \operatorname{var}(\mathbf{Q}_i)}{1-b|\lambda|}\right)\right) \leq \operatorname{rank}\left(\sum_{i=1}^n \operatorname{var}(\mathbf{Q}_i)\right) e^{\frac{n\lambda^2\sigma^2}{1-b|\lambda|}},$$

where the final inequality follows from the same argument as the proof of Theorem 6.15. Combined with the upper bound (6.37), we find that

$$\mathbb{P}\left[\left\|\frac{1}{n}\sum_{i=1}^n \mathbf{Q}_i\right\|_2 \geq \delta\right] \leq 2\operatorname{rank}\left(\sum_{i=1}^n \operatorname{var}(\mathbf{Q}_i)\right) e^{\frac{n\sigma^2\lambda^2}{1-b|\lambda|} - \lambda n \delta},$$

valid for all $\lambda \in [0, 1/b)$. Setting $\lambda = \frac{\delta}{\sigma^2 + b\delta} \in (0, \frac{1}{b})$ and simplifying yields the claim (6.42). □

Remarks: Note that the tail bound (6.42) is of the sub-exponential type, with two regimes of behavior depending on the relative sizes of the parameters σ^2 and b. Thus, it is a natural generalization of the classical Bernstein bound for scalar random variables. As with Theorem 6.15, Theorem 6.17 can also be generalized to non-symmetric (and potentially non-square) matrices $\{\mathbf{A}_i\}_{i=1}^n$ by introducing the sequence of $\{\mathbf{Q}_i\}_{i=1}^n$ symmetric matrices defined in equation (6.39), and imposing the Bernstein condition on it. As one special case, if $|||\mathbf{A}_i|||_2 \leq b$ almost surely, then it can be verified that the matrices $\{\mathbf{Q}_i\}_{i=1}^n$ satisfy the Bernstein condition with b and the quantity

$$\sigma^2 := \max\left\{\left\|\frac{1}{n}\sum_{i=1}^n \mathbb{E}[\mathbf{A}_i\mathbf{A}_i^{\mathrm{T}}]\right\|_2, \left\|\frac{1}{n}\sum_{i=1}^n \mathbb{E}[\mathbf{A}_i^{\mathrm{T}}\mathbf{A}_i]\right\|_2\right\}. \tag{6.43}$$

We provide an instance of this type of transformation in Example 6.18 to follow.

The problem of matrix completion provides an interesting class of examples in which Theorem 6.17 can be fruitfully applied. See Chapter 10 for a detailed description of the underlying problem, which motivates the following discussion.

Example 6.18 (Tail bounds in matrix completion) Consider an i.i.d. sequence of matrices of the form $\mathbf{A}_i = \xi_i \mathbf{X}_i \in \mathbb{R}^{d\times d}$, where ξ_i is a zero-mean sub-exponential variable that satisfies the Bernstein condition with parameter b and variance ν^2, and $\mathbf{X}_i$ is a random "mask matrix", independent from ξ_i, with a single entry equal to d in a position chosen uniformly at random from all d^2 entries, and all remaining entries equal to zero. By construction, for any fixed matrix $\mathbf{\Theta} \in \mathbb{R}^{d\times d}$, we have $\mathbb{E}[\langle\langle \mathbf{A}_i, \mathbf{\Theta}\rangle\rangle^2] = \nu^2 |||\mathbf{\Theta}|||_{\mathrm{F}}^2$—a property that plays an important role in our later analysis of matrix completion.

As noted in Example 6.14, apart from constant factors, there is no loss of generality in assuming that the random matrices $\mathbf{A}_i$ have distributions that are symmetric around zero; in this particular, this symmetry condition is equivalent to requiring that the scalar random variables ξ_i and $-\xi_i$ follow the same distribution. Moreover, as defined, the matrices $\mathbf{A}_i$ are not symmetric (meaning that $\mathbf{A}_i \neq \mathbf{A}_i^{\mathrm{T}}$), but as discussed following Theorem 6.17, we can bound the operator norm $|||\frac{1}{n}\sum_{i=1}^{n}\mathbf{A}_i|||_2$ in terms of the operator norm $|||\frac{1}{n}\sum_{i=1}^{n}\mathbf{Q}_i|||_2$, where the symmetrized version $\mathbf{Q}_i \in \mathbb{R}^{2d\times 2d}$ was defined in equation (6.39).

By the independence between ξ_i and $\mathbf{A}_i$ and the symmetric distribution of ξ_i, we have $\mathbb{E}[\mathbf{Q}_i^{2m+1}] = 0$ for all $m = 0, 1, 2, \ldots$. Turning to the even moments, suppose that entry (a, b) is the only non-zero in the mask matrix $\mathbf{X}_i$. We then have

$$\mathbf{Q}_i^{2m} = (\xi_i)^{2m} d^{2m} \begin{bmatrix} \mathbf{D}_a & 0 \\ 0 & \mathbf{D}_b \end{bmatrix} \quad \text{for all } m = 1, 2, \ldots, \tag{6.44}$$

where $\mathbf{D}_a \in \mathbb{R}^{d\times d}$ is the diagonal matrix with a single 1 in entry (a, a), with $\mathbf{D}_b$ defined analogously. By the Bernstein condition, we have $\mathbb{E}[\xi_i^{2m}] \le \frac{1}{2}(2m)!b^{2m-2}\nu^2$ for all $m = 1, 2, \ldots$.

On the other hand, $\mathbb{E}[\mathbf{D}_a] = \frac{1}{d}\mathbf{I}_d$ since the probability of choosing a in the first coordinate is $1/d$. We thus see that $\mathrm{var}(\mathbf{Q}_i) = \nu^2 d\mathbf{I}_{2d}$. Putting together the pieces, we have shown that

$$\mathbb{E}[\mathbf{Q}_i^{2m}] \le \frac{1}{2}(2m)!b^{2m-2}\nu^2 d^{2m}\frac{1}{d}\mathbf{I}_{2d} = \frac{1}{2}(2m)!(bd)^{2m-2}\,\mathrm{var}(\mathbf{Q}_i),$$

showing that $\mathbf{Q}_i$ satisfies the Bernstein condition with parameters bd and

$$\sigma^2 := \left|\left|\left|\frac{1}{n}\sum_{i=1}^{n}\mathrm{var}(\mathbf{Q}_i)\right|\right|\right|_2 \le \nu^2 d.$$

Consequently, Theorem 6.17 implies that

$$\mathbb{P}\left[\left|\left|\left|\frac{1}{n}\sum_{i=1}^{n}\mathbf{A}_i\right|\right|\right|_2 \ge \delta\right] \le 4de^{-\frac{n\delta^2}{2d(\nu^2+b\delta)}}. \tag{6.45}$$

♣

In certain cases, it is possible to sharpen the dimension dependence of Theorem 6.17—in particular, by replacing the rank-based pre-factor, which can be as large as d, by a quantity that is potentially much smaller. We illustrate one instance of such a sharpened result in the following example.

Example 6.19 (Bernstein bounds with sharpened dimension dependence) Consider a sequence of independent zero-mean random matrices $\mathbf{Q}_i$ bounded as $|||\mathbf{Q}_i|||_2 \le 1$ almost surely, and suppose that our goal is to upper bound the maximum eigenvalue $\gamma_{\max}(\mathbf{S}_n)$ of the sum $\mathbf{S}_n := \sum_{i=1}^{n}\mathbf{Q}_i$. Defining the function $\phi(\lambda) := e^{\lambda} - \lambda - 1$, we note that it is monotonically increasing on the positive real line. Consequently, as verified in Exercise 6.12, for any pair $\delta > 0$, we have

$$\mathbb{P}[\gamma_{\max}(\mathbf{S}_n) \ge \delta] \le \inf_{\lambda>0}\frac{\mathrm{tr}(\mathbb{E}[\phi(\lambda\mathbf{S}_n)])}{\phi(\lambda\delta)}. \tag{6.46}$$

Moreover, using the fact that $|||\mathbf{Q}_i|||_2 \le 1$, the same exercise shows that

$$\log\Psi_{\mathbf{Q}_i}(\lambda) \le \phi(\lambda)\,\mathrm{var}(\mathbf{Q}_i) \tag{6.47a}$$

and

$$\mathrm{tr}(\mathbb{E}[\phi(\lambda\mathbf{S}_n)]) \le \frac{\mathrm{tr}(\bar{\mathbf{V}})}{|||\bar{\mathbf{V}}|||_2}\, e^{\phi(\lambda)|||\bar{\mathbf{V}}|||_2}, \tag{6.47b}$$

where $\bar{\mathbf{V}} := \sum_{i=1}^n \operatorname{var}(\mathbf{Q}_i)$. Combined with the initial bound (6.46), we conclude that

$$\mathbb{P}[\gamma_{\max}(\mathbf{S}_n) \geq \delta] \leq \frac{\operatorname{tr}(\bar{\mathbf{V}})}{|||\bar{\mathbf{V}}|||_2} \inf_{\lambda>0} \left\{ \frac{e^{\phi(\lambda)|||\bar{\mathbf{V}}|||_2}}{\phi(\lambda\delta)} \right\}. \tag{6.48}$$

The significance of this bound is the appearance of the trace ratio $\frac{\operatorname{tr}(\bar{\mathbf{V}})}{|||\bar{\mathbf{V}}|||_2}$ as a pre-factor, as opposed to the quantity $\operatorname{rank}(\bar{\mathbf{V}}) \leq d$ that arose in our previous derivation. Note that we always have $\frac{\operatorname{tr}(\bar{\mathbf{V}})}{|||\bar{\mathbf{V}}|||_2} \leq \operatorname{rank}(\bar{\mathbf{V}})$, and in certain cases, the trace ratio can be substantially smaller than the rank. See Exercise 6.13 for one such case. ♣

6.4.5 Consequences for covariance matrices

We conclude with a useful corollary of Theorem 6.17 for the estimation of covariance matrices.

Corollary 6.20 *Let $x_1, \ldots, x_n$ be i.i.d. zero-mean random vectors with covariance $\mathbf{\Sigma}$ such that $\|x_j\|_2 \leq \sqrt{b}$ almost surely. Then for all $\delta > 0$, the sample covariance matrix $\widehat{\mathbf{\Sigma}} = \frac{1}{n}\sum_{i=1}^n x_i x_i^{\mathrm{T}}$ satisfies*

$$\mathbb{P}[|||\widehat{\mathbf{\Sigma}} - \mathbf{\Sigma}|||_2 \geq \delta] \leq 2d \exp\left(-\frac{n\delta^2}{2b(|||\mathbf{\Sigma}|||_2 + \delta)}\right). \tag{6.49}$$

Proof We apply Theorem 6.17 to the zero-mean random matrices $\mathbf{Q}_i := x_i x_i^{\mathrm{T}} - \mathbf{\Sigma}$. These matrices have controlled operator norm: indeed, by the triangle inequality, we have

$$|||\mathbf{Q}_i|||_2 \leq \|x_i\|_2^2 + |||\mathbf{\Sigma}|||_2 \leq b + |||\mathbf{\Sigma}|||_2.$$

Since $\mathbf{\Sigma} = \mathbb{E}[x_i x_i^{\mathrm{T}}]$, we have $|||\mathbf{\Sigma}|||_2 = \max_{v \in \mathbb{S}^{d-1}} \mathbb{E}[\langle v,\, x_i\rangle^2] \leq b$, and hence $|||\mathbf{Q}_i|||_2 \leq 2b$. Turning to the variance of $\mathbf{Q}_i$, we have

$$\operatorname{var}(\mathbf{Q}_i) = \mathbb{E}[(x_i x_i^{\mathrm{T}})^2] - \mathbf{\Sigma}^2 \preceq \mathbb{E}[\|x_i\|_2^2\, x_i x_i^{\mathrm{T}}] \preceq b\mathbf{\Sigma},$$

so that $|||\operatorname{var}(\mathbf{Q}_i)|||_2 \leq b|||\mathbf{\Sigma}|||_2$. Substituting into the tail bound (6.42) yields the claim. □

Let us illustrate some consequences of this corollary with some examples.

Example 6.21 (Random vectors uniform on a sphere) Suppose that the random vectors x_i are chosen uniformly from the sphere $\mathbb{S}^{d-1}(\sqrt{d})$, so that $\|x_i\|_2 = \sqrt{d}$ for all $i = 1, \ldots, n$. By construction, we have $\mathbb{E}[x_i x_i^{\mathrm{T}}] = \mathbf{\Sigma} = \mathbf{I}_d$, and hence $|||\mathbf{\Sigma}|||_2 = 1$. Applying Corollary 6.20 yields

$$\mathbb{P}[|||\widehat{\mathbf{\Sigma}} - \mathbf{I}_d|||_2 \geq \delta] \leq 2de^{-\frac{n\delta^2}{2d+2d\delta}} \qquad \text{for all } \delta \geq 0. \tag{6.50}$$

This bound implies that

$$|||\widehat{\mathbf{\Sigma}} - \mathbf{I}_d|||_2 \precsim \sqrt{\frac{d \log d}{n}} + \frac{d \log d}{n} \tag{6.51}$$

with high probability, so that the sample covariance is a consistent estimate as long as $\frac{d \log d}{n} \to 0$. This result is close to optimal, with only the extra logarithmic factor being superfluous in this particular case. It can be removed, for instance, by noting that x_i is a sub-Gaussian random vector, and then applying Theorem 6.5. ♣

Example 6.22 ("Spiked" random vectors) We now consider an ensemble of random vectors that are rather different than the previous example, but still satisfy the same bound. In particular, consider a random vector of the form $x_i = \sqrt{d} e_{a(i)}$, where $a(i)$ is an index chosen uniformly at random from $\{1, \ldots, d\}$, and $e_{a(i)} \in \mathbb{R}^d$ is the canonical basis vector with 1 in position $a(i)$. As before, we have $\|x_i\|_2 = \sqrt{d}$, and $\mathbb{E}[x_i x_i^{\mathrm{T}}] = \mathbf{I}_d$ so that the tail bound (6.50) also applies to this ensemble. An interesting fact is that, for this particular ensemble, the bound (6.51) is sharp, meaning it cannot be improved beyond constant factors. ♣

6.5 Bounds for structured covariance matrices

In the preceding sections, our primary focus has been estimation of general unstructured covariance matrices via the sample covariance. When a covariance matrix is equipped with additional structure, faster rates of estimation are possible using different estimators than the sample covariance matrix. In this section, we explore the faster rates that are achievable for sparse and/or graph-structured matrices.

In the simplest setting, the covariance matrix is known to be sparse, and the positions of the non-zero entries are known. In such settings, it is natural to consider matrix estimators that are non-zero only in these known positions. For instance, if we are given *a priori* knowledge that the covariance matrix is diagonal, then it would be natural to use the estimate $\widehat{\mathbf{D}} := \mathrm{diag}\{\widehat{\Sigma}_{11}, \widehat{\Sigma}_{22}, \ldots, \widehat{\Sigma}_{dd}\}$, corresponding to the diagonal entries of the sample covariance matrix $\widehat{\mathbf{\Sigma}}$. As we explore in Exercise 6.15, the performance of this estimator can be substantially better: in particular, for sub-Gaussian variables, it achieves an estimation error of the order $\sqrt{\frac{\log d}{n}}$, as opposed to the order $\sqrt{\frac{d}{n}}$ rates in the unstructured setting. Similar statements apply to other forms of known sparsity.

6.5.1 Unknown sparsity and thresholding

More generally, suppose that the covariance matrix $\mathbf{\Sigma}$ is known to be relatively sparse, but that the positions of the non-zero entries are no longer known. It is then natural to consider estimators based on thresholding. Given a parameter $\lambda > 0$, the *hard-thresholding operator* is given by

$$T_\lambda(u) := u \, \mathbb{I}[|u| > \lambda] = \begin{cases} u & \text{if } |u| > \lambda, \\ 0 & \text{otherwise.} \end{cases} \tag{6.52}$$

With a minor abuse of notation, for a matrix $\mathbf{M}$, we write $T_\lambda(\mathbf{M})$ for the matrix obtained by applying the thresholding operator to each element of $\mathbf{M}$. In this section, we study the performance of the estimator $T_{\lambda_n}(\widehat{\mathbf{\Sigma}})$, where the parameter $\lambda_n > 0$ is suitably chosen as a function of the sample size n and matrix dimension d.

The sparsity of the covariance matrix can be measured in various ways. Its zero pattern is captured by the adjacency matrix $\mathbf{A} \in \mathbb{R}^{d \times d}$ with entries $A_{j\ell} = \mathbb{I}[\Sigma_{j\ell} \neq 0]$. This adjacency matrix defines the edge structure of an undirected graph G on the vertices $\{1, 2, \ldots, d\}$, with edge (j, ℓ) included in the graph if and only if $\Sigma_{j\ell} \neq 0$, along with the self-edges (j, j) for each of the diagonal entries. The operator norm $|||\mathbf{A}|||_2$ of the adjacency matrix provides a natural measure of sparsity. In particular, it can be verified that $|||\mathbf{A}|||_2 \leq d$, with equality holding when G is fully connected, meaning that $\mathbf{\Sigma}$ has no zero entries. More generally, as shown in Exercise 6.2, we have $|||\mathbf{A}|||_2 \leq s$ whenever $\mathbf{\Sigma}$ has at most s non-zero entries per row, or equivalently when the graph G has maximum degree at most $s - 1$. The following result provides a guarantee for the thresholded sample covariance matrix that involves the graph adjacency matrix $\mathbf{A}$ defined by $\mathbf{\Sigma}$.

Theorem 6.23 (Thresholding-based covariance estimation) *Let $\{x_i\}_{i=1}^n$ be an i.i.d. sequence of zero-mean random vectors with covariance matrix $\mathbf{\Sigma}$, and suppose that each component x_{ij} is sub-Gaussian with parameter at most σ. If $n > \log d$, then for any $\delta > 0$, the thresholded sample covariance matrix $T_{\lambda_n}(\widehat{\mathbf{\Sigma}})$ with $\lambda_n/\sigma^2 = 8\sqrt{\frac{\log d}{n}} + \delta$ satisfies*

$$\mathbb{P}\big[|||T_{\lambda_n}(\widehat{\mathbf{\Sigma}}) - \mathbf{\Sigma}|||_2 \geq 2|||\mathbf{A}|||_2 \lambda_n\big] \leq 8e^{-\frac{n}{16}\min\{\delta, \delta^2\}}. \tag{6.53}$$

Underlying the proof of Theorem 6.23 is the following (deterministic) result: for any choice of λ_n such that $\|\widehat{\mathbf{\Sigma}} - \mathbf{\Sigma}\|_{\max} \leq \lambda_n$, we are guaranteed that

$$|||T_{\lambda_n}(\widehat{\mathbf{\Sigma}}) - \mathbf{\Sigma}|||_2 \leq 2|||\mathbf{A}|||_2 \lambda_n. \tag{6.54}$$

The proof of this intermediate claim is straightforward. First, for any index pair (j, ℓ) such that $\Sigma_{j\ell} = 0$, the bound $\|\widehat{\mathbf{\Sigma}} - \mathbf{\Sigma}\|_{\max} \leq \lambda_n$ guarantees that $|\widehat{\Sigma}_{j\ell}| \leq \lambda_n$, and hence that $T_{\lambda_n}(\widehat{\Sigma}_{j\ell}) = 0$ by definition of the thresholding operator. On the other hand, for any pair (j, ℓ) for which $\Sigma_{j\ell} \neq 0$, we have

$$|T_{\lambda_n}(\widehat{\Sigma}_{j\ell}) - \Sigma_{j\ell}| \overset{\text{(i)}}{\leq} |T_{\lambda_n}(\widehat{\Sigma}_{j\ell}) - \widehat{\Sigma}_{j\ell}| + |\widehat{\Sigma}_{j\ell} - \Sigma_{j\ell}| \overset{\text{(ii)}}{\leq} 2\lambda_n,$$

where step (i) follows from the triangle inequality, and step (ii) follows from the fact that $|T_{\lambda_n}(\widehat{\Sigma}_{j\ell}) - \widehat{\Sigma}_{j\ell}| \leq \lambda_n$, and a second application of the assumption $\|\widehat{\mathbf{\Sigma}} - \mathbf{\Sigma}\|_{\max} \leq \lambda_n$. Consequently, we have shown that the matrix $\mathbf{B} := |T_{\lambda_n}(\widehat{\mathbf{\Sigma}}) - \mathbf{\Sigma}|$ satisfies the elementwise inequality $\mathbf{B} \leq 2\lambda_n \mathbf{A}$. Since both $\mathbf{B}$ and $\mathbf{A}$ have non-negative entries, we are guaranteed that $|||\mathbf{B}|||_2 \leq 2\lambda_n |||\mathbf{A}|||_2$, and hence that $|||T_{\lambda_n}(\widehat{\mathbf{\Sigma}}) - \mathbf{\Sigma}|||_2 \leq 2\lambda_n |||\mathbf{A}|||_2$ as claimed. (See Exercise 6.3 for the details of these last steps.)

Theorem 6.23 has a number of interesting corollaries for particular classes of covariance matrices.

Corollary 6.24 *Suppose that, in addition to the conditions of Theorem 6.23, the covariance matrix $\mathbf{\Sigma}$ has at most s non-zero entries per row. Then with $\lambda_n/\sigma^2 = 8\sqrt{\frac{\log d}{n}} + \delta$ for some $\delta > 0$, we have*

$$\mathbb{P}[|||T_{\lambda_n}(\widehat{\mathbf{\Sigma}}) - \mathbf{\Sigma}|||_2 \geq 2s\lambda_n] \leq 8e^{-\frac{n}{16}\min\{\delta,\delta^2\}}. \tag{6.55}$$

In order to establish these claims from Theorem 6.23, it suffices to show that $|||\mathbf{A}|||_2 \leq s$. Since $\mathbf{A}$ has at most s ones per row (with the remaining entries equal to zero), this claim follows from the result of Exercise 6.2.

Example 6.25 (Sparsity and adjacency matrices) In certain ways, the bound (6.55) is more appealing than the bound (6.53), since it is based on a local quantity—namely, the maximum degree of the graph defined by the covariance matrix, as opposed to the spectral norm $|||\mathbf{A}|||_2$. In certain cases, these two bounds coincide. As an example, consider any graph with maximum degree $s-1$ that contains an s-clique (i.e., a subset of s nodes that are all joined by edges). As we explore in Exercise 6.16, for any such graph, we have $|||\mathbf{A}|||_2 = s$, so that the two bounds are equivalent.

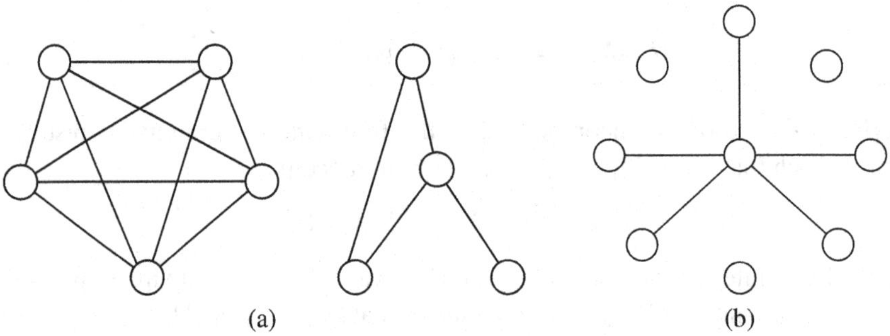

Figure 6.1 (a) An instance of a graph on $d = 9$ nodes containing an $s = 5$ clique. For this class of graphs, the bounds (6.53) and (6.55) coincide. (b) A hub-and-spoke graph on $d = 9$ nodes with maximum degree $s = 5$. For this class of graphs, the bounds differ by a factor of $\sqrt{s}$.

However, in general, the bound (6.53) can be substantially sharper than the bound (6.55). As an example, consider a hub-and-spoke graph, in which one central node known as the hub is connected to s of the remaining $d-1$ nodes, as illustrated in Figure 6.1(b). For such a graph, we have $|||\mathbf{A}|||_2 = 1 + \sqrt{s-1}$, so that in this case Theorem 6.23 guarantees that

$$|||T_{\lambda_n}(\widehat{\mathbf{\Sigma}}) - \mathbf{\Sigma}|||_2 \precsim \sqrt{\frac{s\log d}{n}},$$

with high probability, a bound that is sharper by a factor of order $\sqrt{s}$ compared to the bound (6.55) from Corollary 6.24. ♣

We now turn to the proof of the remainder of Theorem 6.23. Based on the reasoning leading to equation (6.54), it suffices to establish a high-probability bound on the elementwise infinity norm of the error matrix $\widehat{\mathbf{\Delta}} := \widehat{\mathbf{\Sigma}} - \mathbf{\Sigma}$.

Lemma 6.26 *Under the conditions of Theorem 6.23, we have*

$$\mathbb{P}[\|\widehat{\mathbf{\Delta}}\|_{\max}/\sigma^2 \geq t] \leq 8e^{-\frac{n}{16}\min\{t, t^2\}+2\log d} \qquad \textit{for all } t > 0. \tag{6.56}$$

Setting $t = \lambda_n/\sigma^2 = 8\sqrt{\frac{\log d}{n}} + \delta$ in the bound (6.56) yields

$$\mathbb{P}[\|\widehat{\mathbf{\Delta}}\|_{\max} \geq \lambda_n] \leq 8e^{-\frac{n}{16}\min\{\delta, \delta^2\}},$$

where we have used the fact that $n > \log d$ by assumption.

It remains to prove Lemma 6.26. Note that the rescaled vector x_i/σ is sub-Gaussian with parameter at most 1. Consequently, we may assume without loss of generality that $\sigma = 1$, and then rescale at the end. First considering a diagonal entry, the result of Exercise 6.15(a) guarantees that there are universal positive constants c_1, c_2 such that

$$\mathbb{P}[|\widehat{\Delta}_{jj}| \geq c_1\delta] \leq 2e^{-c_2 n\delta^2} \qquad \text{for all } \delta \in (0, 1). \tag{6.57}$$

Turning to the non-diagonal entries, for any $j \neq \ell$, we have

$$2\widehat{\Delta}_{j\ell} = \frac{2}{n}\sum_{i=1}^{n} x_{ij}x_{i\ell} - 2\Sigma_{j\ell} = \frac{1}{n}\sum_{i=1}^{n}(x_{ij} + x_{i\ell})^2 - (\Sigma_{jj} + \Sigma_{\ell\ell} + 2\Sigma_{j\ell}) - \widehat{\Delta}_{jj} - \widehat{\Delta}_{\ell\ell}.$$

Since x_{ij} and $x_{i\ell}$ are both zero-mean and sub-Gaussian with parameter σ, the sum $x_{ij} + x_{i\ell}$ is zero-mean and sub-Gaussian with parameter at most $2\sqrt{2}\sigma$ (see Exercise 2.13(c)). Consequently, there are universal constants c_2, c_3 such that for all $\delta \in (0, 1)$, we have

$$\mathbb{P}\left[\left|\frac{1}{n}\sum_{i=1}^{n}(x_{ij} + x_{i\ell})^2 - (\Sigma_{jj} + \Sigma_{\ell\ell} + 2\Sigma_{j\ell})\right| \geq c_3\delta\right] \leq 2e^{-c_2 n\delta^2},$$

and hence, combining with our earlier diagonal bound (6.57), we obtain the tail bound $\mathbb{P}[|\widehat{\Delta}_{j\ell}| \geq c_1'\delta] \leq 6e^{-c_2 n\delta^2}$. Finally, combining this bound with the earlier inequality (6.57) and then taking a union bound over all d^2 entries of the matrix yields the stated claim (6.56).

6.5.2 Approximate sparsity

Given a covariance matrix $\mathbf{\Sigma}$ with no entries that are exactly zero, the bounds of Theorem 6.23 are very poor. In particular, for a completely dense matrix, the associated adjacency matrix $\mathbf{A}$ is simply the all-ones matrix, so that $|||\mathbf{A}|||_2 = d$. Intuitively, one might expect that these bounds could be improved if $\mathbf{\Sigma}$ had a large number of non-zero entries, but many of them were "near zero".

Recall that one way in which to measure the sparsity of $\mathbf{\Sigma}$ is in terms of the maximum number of non-zero entries per row. A generalization of this idea is to measure the ℓ_q-"norm" of each row. More specifically, given a parameter $q \in [0, 1]$ and a radius R_q, we impose the constraint

$$\max_{j=1,\ldots,d} \sum_{\ell=1}^{d} |\Sigma_{j\ell}|^q \leq R_q. \tag{6.58}$$

(See Figure 7.1 in Chapter 7 for an illustration of these types of sets.) In the special case $q = 0$, this constraint is equivalent to requiring that each row of $\mathbf{\Sigma}$ have at most R_0 non-zero entries. For intermediate values $q \in (0, 1]$, it allows for many non-zero entries but requires that their absolute magnitudes (if ordered from largest to smallest) drop off relatively quickly.

Theorem 6.27 (Covariance estimation under ℓ_q-sparsity) *Suppose that the covariance matrix $\mathbf{\Sigma}$ satisfies the ℓ_q-sparsity constraint* (6.58). *Then for any λ_n such that $\|\widehat{\mathbf{\Sigma}} - \mathbf{\Sigma}\|_{\max} \leq \lambda_n/2$, we are guaranteed that*

$$|\!|\!| T_{\lambda_n}(\widehat{\mathbf{\Sigma}}) - \mathbf{\Sigma} |\!|\!|_2 \leq 4 R_q \lambda_n{}^{1-q}. \tag{6.59a}$$

Consequently, if the sample covariance is formed using i.i.d. samples $\{x_i\}_{i=1}^n$ that are zero-mean with sub-Gaussian parameter at most σ, then with $\lambda_n/\sigma^2 = 8\sqrt{\frac{\log d}{n}} + \delta$, we have

$$\mathbb{P}[|\!|\!| T_{\lambda_n}(\widehat{\mathbf{\Sigma}}) - \mathbf{\Sigma} |\!|\!|_2 \geq 4 R_q \lambda_n{}^{1-q}] \leq 8 e^{-\frac{n}{16}\min\{\delta, \delta^2\}} \qquad \textit{for all } \delta > 0. \tag{6.59b}$$

Proof Given the deterministic claim (6.59a), the probabilistic bound (6.59b) follows from standard tail bounds on sub-exponential variables. The deterministic claim is based on the assumption that $\|\widehat{\mathbf{\Sigma}} - \mathbf{\Sigma}\|_{\max} \leq \lambda_n/2$. By the result of Exercise 6.2, the operator norm can be upper bounded as

$$|\!|\!| T_{\lambda_n}(\widehat{\mathbf{\Sigma}}) - \mathbf{\Sigma} |\!|\!|_2 \leq \max_{j=1,\ldots,d} \sum_{\ell=1}^{d} |T_{\lambda_n}(\widehat{\Sigma}_{j\ell}) - \Sigma_{j\ell}|.$$

Fixing an index $j \in \{1, 2, \ldots, d\}$, define the set $S_j(\lambda_n/2) = \big\{\ell \in \{1, \ldots, d\} \mid |\Sigma_{j\ell}| > \lambda_n/2\big\}$. For any index $\ell \in S_j(\lambda_n/2)$, we have

$$|T_{\lambda_n}(\widehat{\Sigma}_{j\ell}) - \Sigma_{j\ell}| \leq |T_{\lambda_n}(\widehat{\Sigma}_{j\ell}) - \widehat{\Sigma}_{j\ell}| + |\widehat{\Sigma}_{j\ell} - \Sigma_{j\ell}| \leq \frac{3}{2}\lambda_n.$$

On the other hand, for any index $\ell \notin S_j(\lambda_n/2)$, we have $T_{\lambda_n}(\widehat{\Sigma}_{j\ell}) = 0$, by definition of the thresholding operator, and hence

$$|T_{\lambda_n}(\widehat{\Sigma}_{j\ell}) - \Sigma_{j\ell}| = |\Sigma_{j\ell}|.$$

Putting together the pieces, we have

$$\sum_{\ell=1}^{d} |T_{\lambda_n}(\widehat{\Sigma}_{j\ell}) - \Sigma_{j\ell}| = \sum_{\ell \in S_j(\lambda_n)} |T_{\lambda_n}(\widehat{\Sigma}_{j\ell}) - \Sigma_{j\ell}| + \sum_{\ell \notin S_j(\lambda_n)} |T_{\lambda_n}(\widehat{\Sigma}_{j\ell}) - \Sigma_{j\ell}|$$
$$\leq |S_j(\lambda_n/2)| \frac{3}{2}\lambda_n + \sum_{\ell \notin S_j(\lambda_n)} |\Sigma_{j\ell}|. \tag{6.60}$$

Now we have

$$\sum_{\ell \notin S_j(\lambda_n/2)} |\Sigma_{j\ell}| = \frac{\lambda_n}{2} \sum_{\ell \notin S_j(\lambda_n/2)} \frac{|\Sigma_{j\ell}|}{\lambda_n/2} \overset{\text{(i)}}{\leq} \frac{\lambda_n}{2} \sum_{\ell \notin S_j(\lambda_n/2)} \left(\frac{|\Sigma_{j\ell}|}{\lambda_n/2}\right)^q \overset{\text{(ii)}}{\leq} \lambda_n{}^{1-q} R_q,$$

where step (i) follows since $|\Sigma_{j\ell}| \leq \lambda_n/2$ for all $\ell \notin S_j(\lambda_n/2)$ and $q \in [0,1]$, and step (ii) follows by the assumption (6.58). On the other hand, we have

$$R_q \geq \sum_{\ell=1}^{d} |\Sigma_{j\ell}|^q \geq |S_j(\lambda_n/2)| \left(\frac{\lambda_n}{2}\right)^q,$$

whence $|S_j(\lambda_n/2)| \leq 2^q R_q \lambda_n^{-q}$. Combining these ingredients with the inequality (6.60), we find that

$$\sum_{\ell=1}^{d} |T_{\lambda_n}(\widehat{\Sigma}_{j\ell}) - \Sigma_{j\ell}| \leq 2^q R_q \lambda_n^{1-q} \frac{3}{2} + R_q \lambda_n^{1-q} \leq 4 R_q \lambda_n^{1-q}.$$

Since this same argument holds for each index $j = 1, \ldots, d$, the claim (6.59a) follows. □

6.6 Appendix: Proof of Theorem 6.1

It remains to prove the lower bound (6.9) on the minimal singular value. In order to do so, we follow an argument similar to that used to upper bound the maximal singular value. Throughout this proof, we assume that $\mathbf{\Sigma}$ is strictly positive definite (and hence invertible); otherwise, its minimal singular value is zero, and the claimed lower bound is vacuous. We begin by lower bounding the expectation using a Gaussian comparison principle due to Gordon (1985). By definition, the minimum singular value has the variational representation $\sigma_{\min}(\mathbf{X}) = \min_{v' \in \mathbb{S}^{d-1}} \|\mathbf{X}v'\|_2$. Let us reformulate this representation slightly for later theoretical convenience. Recalling the shorthand notation $\bar{\sigma}_{\min} = \sigma_{\min}(\sqrt{\mathbf{\Sigma}})$, we define the radius $R = 1/\bar{\sigma}_{\min}$, and then consider the set

$$\mathcal{V}(R) := \{z \in \mathbb{R}^d \mid \|\sqrt{\mathbf{\Sigma}}z\|_2 = 1,\ \|z\|_2 \leq R\}. \tag{6.61}$$

We claim that it suffices to show that a lower bound of the form

$$\min_{z \in \mathcal{V}(R)} \frac{\|\mathbf{X}z\|_2}{\sqrt{n}} \geq 1 - \delta - R\sqrt{\frac{\operatorname{tr}(\mathbf{\Sigma})}{n}} \tag{6.62}$$

holds with probability at least $1 - e^{-n\delta^2/2}$. Indeed, suppose that inequality (6.62) holds. Then for any $v' \in \mathbb{S}^{d-1}$, we can define the rescaled vector $z := \frac{v'}{\|\sqrt{\mathbf{\Sigma}}v'\|_2}$. By construction, we have

$$\|\sqrt{\mathbf{\Sigma}}z\|_2 = 1 \quad \text{and} \quad \|z\|_2 = \frac{1}{\|\sqrt{\mathbf{\Sigma}}v'\|_2} \leq \frac{1}{\sigma_{\min}(\sqrt{\mathbf{\Sigma}})} = R,$$

so that $z \in \mathcal{V}(R)$. We now observe that

$$\frac{\|\mathbf{X}v'\|_2}{\sqrt{n}} = \|\sqrt{\boldsymbol{\Sigma}}v'\|_2 \frac{\|\mathbf{X}z\|_2}{\sqrt{n}} \geq \sigma_{\min}(\sqrt{\boldsymbol{\Sigma}}) \min_{z \in \mathcal{V}(R)} \frac{\|\mathbf{X}z\|_2}{\sqrt{n}}.$$

Since this bound holds for all $v' \in \mathbb{S}^{d-1}$, we can take the minimum on the left-hand side, thereby obtaining

$$\begin{aligned}
\min_{v' \in \mathbb{S}^{d-1}} \frac{\|\mathbf{X}v'\|_2}{\sqrt{n}} &\geq \bar{\sigma}_{\min} \min_{z \in \mathcal{V}(R)} \frac{\|\mathbf{X}z\|_2}{\sqrt{n}} \\
&\overset{\text{(i)}}{\geq} \bar{\sigma}_{\min}\left\{1 - R\sqrt{\frac{\operatorname{tr}(\boldsymbol{\Sigma})}{n}} - \delta\right\} \\
&= (1-\delta)\bar{\sigma}_{\min} - R\sqrt{\frac{\operatorname{tr}(\boldsymbol{\Sigma})}{n}},
\end{aligned}$$

where step (i) follows from the bound (6.62).

It remains to prove the lower bound (6.62). We begin by showing concentration of the random variable $\min_{v \in \mathcal{V}(R)} \|\mathbf{X}v\|_2/\sqrt{n}$ around its expected value. Since the matrix $\mathbf{X} \in \mathbb{R}^{n \times d}$ has i.i.d. rows, each drawn from the $\mathcal{N}(0, \boldsymbol{\Sigma})$ distribution, we can write $\mathbf{X} = \mathbf{W}\sqrt{\boldsymbol{\Sigma}}$, where the random matrix $\mathbf{W}$ is standard Gaussian. Using the fact that $\|\sqrt{\boldsymbol{\Sigma}}v\|_2 = 1$ for all $v \in \mathcal{V}(R)$, it follows that the function $\mathbf{W} \mapsto \min_{v \in \mathcal{V}(R)} \frac{\|\mathbf{W}\sqrt{\boldsymbol{\Sigma}}v\|_2}{\sqrt{n}}$ is Lipschitz with parameter $L = 1/\sqrt{n}$. Applying Theorem 2.26, we conclude that

$$\min_{v \in \mathcal{V}(R)} \frac{\|\mathbf{X}v\|_2}{\sqrt{n}} \geq \mathbb{E}\left[\min_{v \in \mathcal{V}(R)} \frac{\|\mathbf{X}v\|_2}{\sqrt{n}}\right] - \delta$$

with probability at least $1 - e^{-n\delta^2/2}$.

Consequently, the proof will be complete if we can show that

$$\mathbb{E}\left[\min_{v \in \mathcal{V}(R)} \frac{\|\mathbf{X}v\|_2}{\sqrt{n}}\right] \geq 1 - R\sqrt{\frac{\operatorname{tr}(\boldsymbol{\Sigma})}{n}}. \tag{6.63}$$

In order to do so, we make use of an extension of the Sudakov–Fernique inequality, known as Gordon's inequality, which we now state. Let $\{Z_{u,v}\}$ and $\{Y_{u,v}\}$ be a pair of zero-mean Gaussian processes indexed by a non-empty index set $\mathbb{T} = U \times V$. Suppose that

$$\mathbb{E}[(Z_{u,v} - Z_{\widetilde{u},\widetilde{v}})^2] \leq \mathbb{E}[(Y_{u,v} - Y_{\widetilde{u},\widetilde{v}})^2] \qquad \text{for all pairs } (u,v) \text{ and } (\widetilde{u},\widetilde{v}) \in \mathbb{T}, \tag{6.64}$$

and moreover that this inequality holds with *equality* whenever $v = \widetilde{v}$. Under these conditions, Gordon's inequality guarantees that

$$\mathbb{E}\left[\max_{v \in V} \min_{u \in U} Z_{u,v}\right] \leq \mathbb{E}\left[\max_{v \in V} \min_{u \in U} Y_{u,v}\right]. \tag{6.65}$$

In order to exploit this result, we first observe that

$$-\min_{z \in \mathcal{V}(R)} \|\mathbf{X}z\|_2 = \max_{z \in \mathcal{V}(R)} \{-\|\mathbf{X}z\|_2\} = \max_{z \in \mathcal{V}(R)} \min_{u \in \mathbb{S}^{n-1}} u^{\mathrm{T}}\mathbf{X}z.$$

As before, if we introduce the standard Gaussian random matrix $\mathbf{W} \in \mathbb{R}^{n \times d}$, then for any

$z \in \mathcal{V}(R)$, we can write $u^{\mathrm{T}}\mathbf{X}z = u^{\mathrm{T}}\mathbf{W}v$, where $v := \sqrt{\mathbf{\Sigma}}z$. Whenever $z \in \mathcal{V}(R)$, then the vector v must belong to the set $\mathcal{V}'(R) := \{v \in \mathbb{S}^{d-1} \mid \|\mathbf{\Sigma}^{-\frac{1}{2}}v\|_2 \leq R\}$, and we have shown that

$$\min_{z\in\mathcal{V}(R)} \|\mathbf{X}z\|_2 = \max_{v\in\mathcal{V}'(R)} \min_{u\in\mathbb{S}^{n-1}} \underbrace{u^{\mathrm{T}}\mathbf{W}v}_{Z_{u,v}}.$$

Let (u, v) and $(\widetilde{u}, \widetilde{v})$ be any two members of the Cartesian product space $\mathbb{S}^{n-1} \times \mathcal{V}'(R)$. Since $\|u\|_2 = \|\widetilde{u}\|_2 = \|v\|_2 = \|\widetilde{v}\|_2 = 1$, following the same argument as in bounding the maximal singular value shows that

$$\rho_Z^2((u, v), (\widetilde{u}, \widetilde{v})) \leq \|u - \widetilde{u}\|_2^2 + \|v - \widetilde{v}\|_2^2, \tag{6.66}$$

with equality holding when $v = \widetilde{v}$. Consequently, if we define the Gaussian process $Y_{u,v} := \langle g, u\rangle + \langle h, v\rangle$, where $g \in \mathbb{R}^n$ and $h \in \mathbb{R}^d$ are standard Gaussian vectors and mutually independent, then we have

$$\rho_Y^2((u, v), (\widetilde{u}, \widetilde{v})) = \|u - \widetilde{u}\|_2^2 + \|v - \widetilde{v}\|_2^2,$$

so that the Sudakov–Fernique increment condition (6.64) holds. In addition, for a pair such that $v = \widetilde{v}$, equality holds in the upper bound (6.66), which guarantees that $\rho_Z((u, v), (\widetilde{u}, v)) = \rho_Y((u, v), (\widetilde{u}, v))$. Consequently, we may apply Gordon's inequality (6.65) to conclude that

$$\begin{aligned}
\mathbb{E}\Big[-\min_{z\in\mathcal{V}(R)} \|\mathbf{X}z\|_2\Big] &\leq \mathbb{E}\Big[\max_{v\in\mathcal{V}'(R)} \min_{u\in\mathbb{S}^{n-1}} Y_{u,v}\Big] \\
&= \mathbb{E}\Big[\min_{u\in\mathbb{S}^{n-1}} \langle g, u\rangle\Big] + \mathbb{E}\Big[\max_{v\in\mathcal{V}'(R)} \langle h, v\rangle\Big] \\
&\leq -\mathbb{E}[\|g\|_2] + \mathbb{E}[\|\sqrt{\mathbf{\Sigma}}h\|_2]R,
\end{aligned}$$

where we have used the upper bound $|\langle h, v\rangle| = |\langle \sqrt{\mathbf{\Sigma}}h, \mathbf{\Sigma}^{-\frac{1}{2}}v\rangle| \leq \|\sqrt{\mathbf{\Sigma}}h\|_2 R$, by definition of the set $\mathcal{V}'(R)$.

We now claim that

$$\frac{\mathbb{E}[\|\sqrt{\mathbf{\Sigma}}h\|_2]}{\sqrt{\mathrm{tr}(\mathbf{\Sigma})}} \leq \frac{\mathbb{E}[\|h\|_2]}{\sqrt{d}}. \tag{6.67}$$

Indeed, by the rotation invariance of the Gaussian distribution, we may assume that $\mathbf{\Sigma}$ is diagonal, with non-negative entries $\{\gamma_j\}_{j=1}^d$, and the claim is equivalent to showing that the function $F(\gamma) := \mathbb{E}[(\sum_{j=1}^d \gamma_j h_j^2)^{1/2}]$ achieves its maximum over the probability simplex at the uniform vector (i.e., with all entries $\gamma_j = 1/d$). Since F is continuous and the probability simplex is compact, the maximum is achieved. By the rotation invariance of the Gaussian, the function F is also permutation invariant—i.e., $F(\gamma) = F(\Pi(\gamma))$ for all permutation matrices Π. Since F is also concave, the maximum must be achieved at $\gamma_j = 1/d$, which establishes the inequality (6.67).

Recalling that $R = 1/\bar{\sigma}_{\min}$, we then have

$$-\mathbb{E}[\|g\|_2] + R\,\mathbb{E}[\|\sqrt{\Sigma}h\|_2] \le -\mathbb{E}[\|g\|_2] + \frac{\sqrt{\operatorname{tr}(\Sigma)}}{\bar{\sigma}_{\min}} \frac{\mathbb{E}[\|h\|_2]}{\sqrt{d}}$$
$$= \underbrace{\{-\mathbb{E}[\|g\|_2] + \mathbb{E}[\|h\|_2]\}}_{T_1} + \underbrace{\left\{\sqrt{\frac{\operatorname{tr}(\Sigma)}{\bar{\sigma}^2_{\min} d}} - 1\right\} \mathbb{E}[\|h\|_2]}_{T_2}.$$

By Jensen's inequality, we have $\mathbb{E}\|h\|_2 \le \sqrt{\mathbb{E}\|h\|_2^2} = \sqrt{d}$. Since $\frac{\operatorname{tr}(\Sigma)}{\bar{\sigma}^2_{\min} d} \ge 1$, we conclude that $T_2 \le \left\{\sqrt{\frac{\operatorname{tr}(\Sigma)}{\bar{\sigma}^2_{\min} d}} - 1\right\} \sqrt{d}$. On the other hand, a direct calculation, using our assumption that $n \ge d$, shows that $T_1 \le -\sqrt{n} + \sqrt{d}$. Combining the pieces, we conclude that

$$\mathbb{E}\left[-\min_{z \in \mathcal{V}(R)} \|\mathbf{X}z\|_2\right] \le -\sqrt{n} + \sqrt{d} + \left\{\sqrt{\frac{\operatorname{tr}(\Sigma)}{\bar{\sigma}^2_{\min} d}} - 1\right\} \sqrt{d}$$
$$= -\sqrt{n} + \frac{\sqrt{\operatorname{tr}(\Sigma)}}{\bar{\sigma}_{\min}},$$

which establishes the initial claim (6.62), thereby completing the proof.

6.7 Bibliographic details and background

The two-volume series by Horn and Johnson (1985; 1991) is a standard reference on linear algebra. A statement of Weyl's theorem and its corollaries can be found in section 4.3 of the first volume (Horn and Johnson, 1985). The monograph by Bhatia (1997) is more advanced in nature, taking a functional-analytic perspective, and includes discussion of Lidskii's theorem (see section III.4). The notes by Carlen (2009) contain further background on trace inequalities, such as inequality (6.25).

Some classical papers on asymptotic random matrix theory include those by Wigner (1955; 1958), Marčenko and Pastur (1967), Pastur (1972), Wachter (1978) and Geman (1980). Mehta (1991) provides an overview of asymptotic random matrix theory, primarily from the physicist's perspective, whereas the book by Bai and Silverstein (2010) takes a more statistical perspective. The lecture notes of Vershynin (2011) focus on the non-asymptotic aspects of random matrix theory, as partially covered here. Davidson and Szarek (2001) describe the use of Sudakov–Fernique (Slepian) and Gordon inequalities in bounding expectations of random matrices; see also the earlier papers by Gordon (1985; 1986; 1987) and Szarek (1991). The results in Davidson and Szarek (2001) are for the special case of the standard Gaussian ensemble ($\Sigma = \mathbf{I}_d$), but the underlying arguments are easily extended to the general case, as given here.

The proof of Theorem 6.5 is based on the lecture notes of Vershynin (2011). The underlying discretization argument is classical, used extensively in early work on random constructions in Banach space geometry (e.g., see the book by Pisier (1989) and references therein). Note that this discretization argument is the one-step version of the more sophisticated chaining methods described in Chapter 5.

Bounds on the expected operator norm of a random matrix follow a class of results known as non-commutative Bernstein inequalities, as derived initially by Rudelson (1999). Ahlswede and Winter (2002) developed techniques for matrix tail bounds based on controlling the matrix moment generating function, and exploiting the Golden–Thompson inequality. Other authors, among them Oliveira (2010), Gross (2011) and Recht (2011), developed various extensions and refinements of the original Ahlswede–Winter approach. Tropp (2010) introduced the idea of controlling the matrix generating function directly, and developed the argument that underlies Lemma 6.13. Controlling the moment generating function in this way leads to tail bounds involving the variance parameter $\sigma^2 := \frac{1}{n}|||\sum_{i=1}^n \mathrm{var}(\mathbf{Q}_i)|||_2$ as opposed to the potentially larger quantity $\tilde{\sigma}^2 := \frac{1}{n}\sum_{i=1}^n |||\mathrm{var}(\mathbf{Q}_i)|||_2$ that follows from the original Ahlswede–Winter argument. By the triangle inequality for the operator norm, we have $\sigma^2 \leq \tilde{\sigma}^2$, and the latter quantity can be substantially larger. Independent work by Oliveira (2010) also derived bounds involving the variance parameter σ^2, using a related technique that sharpened the original Ahlswede–Winter approach. Tropp (2010) also provides various extensions of the basic Bernstein bound, among them results for matrix martingales as opposed to the independent random matrices considered here. Mackey et al. (2014) show how to derive matrix concentration bounds with sharp constants using the method of exchangeable pairs introduced by Chatterjee (2007). Matrix tail bounds with refined forms of dimension dependence have been developed by various authors (Minsker, 2011; Hsu et al., 2012a); the specific sharpening sketched out in Example 6.19 and Exercise 6.12 is due to Minsker (2011).

For covariance estimation, Adamczak et al. (2010) provide sharp results on the deviation $|||\widehat{\mathbf{\Sigma}} - \mathbf{\Sigma}|||_2$ for distributions with sub-exponential tails. These results remove the superfluous logarithmic factor that arises from an application of Corollary 6.20 to a sub-exponential ensemble. Srivastava and Vershynin (2013) give related results under very weak moment conditions. For thresholded sample covariances, the first high-dimensional analyses were undertaken in independent work by Bickel and Levina (2008a) and El Karoui (2008). Bickel and Levina studied the problem under sub-Gaussian tail conditions, and introduced the row-wise sparsity model, defined in terms of the maximum ℓ_q-"norm" taken over the rows. By contrast, El Karoui imposed a milder set of moment conditions, and measured sparsity in terms of the growth rates of path lengths in the graph; this approach is essentially equivalent to controlling the operator norm $|||\mathbf{A}|||_2$ of the adjacency matrix, as in Theorem 6.23. The star graph is an interesting example that illustrates the difference between the row-wise sparsity model, and the operator norm approach.

An alternative model for covariance matrices is a banded decay model, in which entries decay according to their distance from the diagonal. Bickel and Levina (2008b) introduced this model in the covariance setting, and proposed a certain kind of tapering estimator. Cai et al. (2010) analyzed the minimax-optimal rates associated with this class of covariance matrices, and provided a modified estimator that achieves these optimal rates.

6.8 Exercises

Exercise 6.1 (Bounds on eigenvalues) Given two symmetric matrices $\mathbf{A}$ and $\mathbf{B}$, show directly, without citing any other theorems, that

$$|\gamma_{\max}(\mathbf{A}) - \gamma_{\max}(\mathbf{B})| \leq |||\mathbf{A} - \mathbf{B}|||_2 \quad \text{and} \quad |\gamma_{\min}(\mathbf{A}) - \gamma_{\min}(\mathbf{B})| \leq |||\mathbf{A} - \mathbf{B}|||_2.$$

Exercise 6.2 (Relations between matrix operator norms) For a rectangular matrix $\mathbf{A}$ with real entries and a scalar $q \in [1, \infty]$, the $(\ell_q \to \ell_q)$-operator norms are given by

$$|||\mathbf{A}|||_q = \sup_{\|x\|_q=1} \|\mathbf{A}x\|_q.$$

(a) Derive explicit expressions for the operator norms $|||\mathbf{A}|||_2$, $|||\mathbf{A}|||_1$ and $|||\mathbf{A}|||_\infty$ in terms of elements and/or singular values of $\mathbf{A}$.
(b) Prove that $|||\mathbf{AB}|||_q \leq |||\mathbf{A}|||_q |||\mathbf{B}|||_q$ for any size-compatible matrices $\mathbf{A}$ and $\mathbf{B}$.
(c) For a square matrix $\mathbf{A}$, prove that $|||\mathbf{A}|||_2^2 \leq |||\mathbf{A}|||_1 |||\mathbf{A}|||_\infty$. What happens when $\mathbf{A}$ is symmetric?

Exercise 6.3 (Non-negative matrices and operator norms) Given two d-dimensional symmetric matrices $\mathbf{A}$ and $\mathbf{B}$, suppose that $0 \leq \mathbf{A} \leq \mathbf{B}$ in an elementwise sense (i.e., $0 \leq A_{j\ell} \leq B_{j\ell}$ for all $j, \ell = 1, \ldots, d$.)

(a) Show that $0 \leq \mathbf{A}^m \leq \mathbf{B}^m$ for all integers $m = 1, 2, \ldots$.
(b) Use part (a) to show that $|||\mathbf{A}|||_2 \leq |||\mathbf{B}|||_2$.
(c) Use a similar argument to show that $|||\mathbf{C}|||_2 \leq |||\,|\mathbf{C}|\,|||_2$ for any symmetric matrix $\mathbf{C}$, where $|\mathbf{C}|$ denotes the absolute value function applied elementwise.

Exercise 6.4 (Inequality for matrix exponential) Let $\mathbf{A} \in \mathcal{S}^{d\times d}$ be any symmetric matrix. Show that $\mathbf{I}_d + \mathbf{A} \preceq e^{\mathbf{A}}$. (*Hint:* First prove the statement for a diagonal matrix $\mathbf{A}$, and then show how to reduce to the diagonal case.)

Exercise 6.5 (Matrix monotone functions) A function $f\colon \mathcal{S}_+^{d\times d} \to \mathcal{S}_+^{d\times d}$ on the space of symmetric positive semidefinite matrices is said to be *matrix monotone* if

$$f(\mathbf{A}) \preceq f(\mathbf{B}) \qquad \text{whenever } \mathbf{A} \preceq \mathbf{B}.$$

Here $\preceq$ denotes the positive semidefinite ordering on $\mathcal{S}_+^{d\times d}$.

(a) Show by counterexample that the function $f(\mathbf{A}) = \mathbf{A}^2$ is *not* matrix monotone. (*Hint:* Note that $(\mathbf{A}+t\mathbf{C})^2 = \mathbf{A}^2+t^2\mathbf{C}^2+t(\mathbf{AC}+\mathbf{CA})$, and search for a pair of positive semidefinite matrices such that $\mathbf{AC} + \mathbf{CA}$ has a negative eigenvalue.)
(b) Show by counterexample that the matrix exponential function $f(\mathbf{A}) = e^{\mathbf{A}}$ is *not* matrix monotone. (*Hint:* Part (a) could be useful.)
(c) Show that the matrix logarithm function $f(\mathbf{A}) = \log \mathbf{A}$ is matrix monotone on the cone of strictly positive definite matrices. (*Hint:* You may use the fact that $g(\mathbf{A}) = \mathbf{A}^p$ is matrix monotone for all $p \in [0, 1]$.)

Exercise 6.6 (Variance and positive semidefiniteness) Recall that the variance of a symmetric random matrix $\mathbf{Q}$ is given by $\text{var}(\mathbf{Q}) = \mathbb{E}[\mathbf{Q}^2] - (\mathbb{E}[\mathbf{Q}])^2$. Show that $\text{var}(\mathbf{Q}) \succeq 0$.

Exercise 6.7 (Sub-Gaussian random matrices) Consider the random matrix $\mathbf{Q} = g\mathbf{B}$, where $g \in \mathbb{R}$ is a zero-mean σ-sub-Gaussian variable.

(a) Assume that g has a distribution symmetric around zero, and $\mathbf{B} \in \mathcal{S}^{d\times d}$ is a deterministic matrix. Show that $\mathbf{Q}$ is sub-Gaussian with matrix parameter $\mathbf{V} = c^2\sigma^2\mathbf{B}^2$, for some universal constant c.

(b) Now assume that $\mathbf{B} \in \mathcal{S}^{d\times d}$ is random and independent of g, with $|||\mathbf{B}|||_2 \le b$ almost surely. Prove that $\mathbf{Q}$ is sub-Gaussian with matrix parameter given by $\mathbf{V} = c^2\sigma^2 b^2 \mathbf{I}_d$.

Exercise 6.8 (Sub-Gaussian matrices and mean bounds) Consider a sequence of independent, zero-mean random matrices $\{\mathbf{Q}_i\}_{i=1}^n$ in $\mathcal{S}^{d\times d}$, each sub-Gaussian with matrix parameter $\mathbf{V}_i$. In this exercise, we provide bounds on the expected value of eigenvalues and operator norm of $\mathbf{S}_n = \frac{1}{n}\sum_{i=1}^n \mathbf{Q}_i$.

(a) Show that $\mathbb{E}[\gamma_{\max}(\mathbf{S}_n)] \le \sqrt{\frac{2\sigma^2 \log d}{n}}$, where $\sigma^2 = |||\frac{1}{n}\sum_{i=1}^n \mathbf{V}_i|||_2$.
(*Hint:* Start by showing that $\mathbb{E}[e^{\lambda\gamma_{\max}(\mathbf{S}_n)}] \le d e^{\frac{\lambda^2\sigma^2}{2n}}$.)

(b) Show that

$$\mathbb{E}\left[\left|\left|\left|\frac{1}{n}\sum_{i=1}^n \mathbf{Q}_i\right|\right|\right|_2\right] \le \sqrt{\frac{2\sigma^2\log(2d)}{n}}. \tag{6.68}$$

Exercise 6.9 (Bounded matrices and Bernstein condition) Let $\mathbf{Q} \in \mathcal{S}^{d\times d}$ be an arbitrary symmetric matrix.

(a) Show that the bound $|||\mathbf{Q}|||_2 \le b$ implies that $\mathbf{Q}^{j-2} \preceq b^{j-2}\mathbf{I}_d$.
(b) Show that the positive semidefinite order is preserved under left–right multiplication, meaning that if $\mathbf{A} \preceq \mathbf{B}$, then we also have $\mathbf{QAQ} \preceq \mathbf{QBQ}$ for any matrix $\mathbf{Q} \in \mathcal{S}^{d\times d}$.
(c) Use parts (a) and (b) to prove the inequality (6.30).

Exercise 6.10 (Tail bounds for non-symmetric matrices) In this exercise, we prove that a version of the tail bound (6.42) holds for general independent zero-mean matrices $\{\mathbf{A}_i\}_{i=1}^n$ that are almost surely bounded as $|||\mathbf{A}_i|||_2 \le b$, as long as we adopt the new definition (6.43) of σ^2.

(a) Given a general matrix $\mathbf{A}_i \in \mathbb{R}^{d_1\times d_2}$, define a symmetric matrix of dimension $(d_1 + d_2)$ via

$$\mathbf{Q}_i := \begin{bmatrix} \mathbf{0}_{d_1\times d_2} & \mathbf{A}_i \\ \mathbf{A}_i^T & \mathbf{0}_{d_2\times d_1} \end{bmatrix}.$$

Prove that $|||\mathbf{Q}_i|||_2 = |||\mathbf{A}_i|||_2$.
(b) Prove that $|||\frac{1}{n}\sum_{i=1}^n \mathrm{var}(\mathbf{Q}_i)|||_2 \le \sigma^2$ where σ^2 is defined in equation (6.43).
(c) Conclude that

$$\mathbb{P}\left[\left|\left|\left|\sum_{i=1}^n \mathbf{A}_i\right|\right|\right|_2 \ge n\delta\right] \le 2(d_1 + d_2)e^{-\frac{n\delta^2}{2(\sigma^2+b\delta)}}. \tag{6.69}$$

Exercise 6.11 (Unbounded matrices and Bernstein bounds) Consider an independent sequence of random matrices $\{\mathbf{A}_i\}_{i=1}^n$ in $\mathbb{R}^{d_1\times d_2}$, each of the form $\mathbf{A}_i = g_i\mathbf{B}_i$, where $g_i \in \mathbb{R}$ is a zero-mean scalar random variable, and $\mathbf{B}_i$ is an independent random matrix. Suppose that $\mathbb{E}[g_i^j] \le \frac{j!}{2} b_1^{j-2}\sigma^2$ for $j = 2, 3, \ldots$, and that $|||\mathbf{B}_i|||_2 \le b_2$ almost surely.

(a) For any $\delta > 0$, show that

$$\mathbb{P}\left[\left\|\left\|\frac{1}{n}\sum_{i=1}^{n}\mathbf{A}_i\right\|\right\|_2 \geq \delta\right] \leq (d_1 + d_2)e^{-\frac{n\delta^2}{2(\sigma^2 b_2^2 + b_1 b_2 \delta)}}.$$

(*Hint:* The result of Exercise 6.10(a) could be useful.)

(b) Show that

$$\mathbb{E}\left[\left\|\left\|\frac{1}{n}\sum_{i=1}^{n}\mathbf{A}_i\right\|\right\|_2\right] \leq \frac{2\sigma b_2}{\sqrt{n}}\left\{\sqrt{\log(d_1 + d_2)} + \sqrt{\pi}\right\} + \frac{4b_1 b_2}{n}\{\log(d_1 + d_2) + 1\}.$$

(*Hint:* The result of Exercise 2.8 could be useful.)

Exercise 6.12 (Sharpened matrix Bernstein inequality) In this exercise, we work through various steps of the calculation sketched in Example 6.19.

(a) Prove the bound (6.46).
(b) Show that for any symmetric zero-mean random matrix $\mathbf{Q}$ such that $|||\mathbf{Q}|||_2 \leq 1$ almost surely, the moment generating function is bounded as

$$\log \Psi_{\mathbf{Q}}(\lambda) \leq \underbrace{(e^{\lambda} - \lambda - 1)}_{\phi(\lambda)} \operatorname{var}(\mathbf{Q}).$$

(c) Prove the upper bound (6.47b).

Exercise 6.13 (Bernstein's inequality for vectors) In this exercise, we consider the problem of obtaining a Bernstein-type bound on random variable $\| \sum_{i=1}^{n} x_i \|_2$, where $\{x_i\}_{i=1}^{n}$ is an i.i.d. sequence of zero-mean random vectors such that $\|x_i\|_2 \leq 1$ almost surely, and $\operatorname{cov}(x_i) = \mathbf{\Sigma}$. In order to do so, we consider applying either Theorem 6.17 or the bound (6.48) to the $(d + 1)$-dimensional symmetric matrices

$$\mathbf{Q}_i := \begin{bmatrix} 0 & x_i^{\mathrm{T}} \\ x_i & \mathbf{0}_d \end{bmatrix}.$$

Define the matrix $\mathbf{V}_n = \sum_{i=1}^{n} \operatorname{var}(\mathbf{Q}_i)$.

(a) Show that the best bound obtainable from Theorem 6.17 will have a pre-factor of the form $\operatorname{rank}(\mathbf{\Sigma}) + 1$, which can be as large as $d + 1$.
(b) By way of contrast, show that the bound (6.48) yields a dimension-independent pre-factor of 2.

Exercise 6.14 (Random packings) The goal of this exercise is to prove that there exists a collection of vectors $\mathcal{P} = \{\theta^1, \ldots, \theta^M\}$ belonging to the sphere $\mathbb{S}^{d-1}$ such that:

(a) the set $\mathcal{P}$ forms a $1/2$-packing in the Euclidean norm;
(b) the set $\mathcal{P}$ has cardinality $M \geq e^{c_0 d}$ for some universal constant c_0;
(c) the inequality $|||\frac{1}{M}\sum_{j=1}^{M}(\theta^j \otimes \theta^j)|||_2 \leq \frac{2}{d}$ holds.

(*Note:* You may assume that d is larger than some universal constant so as to avoid annoying subcases.)

Exercise 6.15 (Estimation of diagonal covariances) Let $\{x_i\}_{i=1}^n$ be an i.i.d. sequence of d-dimensional vectors, drawn from a zero-mean distribution with diagonal covariance matrix $\mathbf{\Sigma} = \mathbf{D}$. Consider the estimate $\widehat{\mathbf{D}} = \mathrm{diag}(\widehat{\mathbf{\Sigma}})$, where $\widehat{\mathbf{\Sigma}}$ is the usual sample covariance matrix.

(a) When each vector x_i is sub-Gaussian with parameter at most σ, show that there are universal positive constants c_j such that

$$\mathbb{P}\left[|||\widehat{\mathbf{D}} - \mathbf{D}|||_2/\sigma^2 \geq c_0\sqrt{\frac{\log d}{n}} + \delta\right] \leq c_1 e^{-c_2 n \min\{\delta, \delta^2\}}, \qquad \text{for all } \delta > 0.$$

(b) Instead of a sub-Gaussian tail condition, suppose that for some even integer $m \geq 2$, there is a universal constant K_m such that

$$\underbrace{\mathbb{E}[(x_{ij}^2 - \Sigma_{jj})^m]}_{\|x_{ij}^2 - \Sigma_{jj}\|_m^m} \leq K_m \qquad \text{for each } i = 1, \ldots, n \text{ and } j = 1, \ldots, d.$$

Show that

$$\mathbb{P}\left[|||\widehat{\mathbf{D}} - \mathbf{D}|||_2 \geq 4\delta\sqrt{\frac{d^{2/m}}{n}}\right] \leq K'_m\left(\frac{1}{2\delta}\right)^m \qquad \text{for all } \delta > 0,$$

where K'_m is another universal constant.

Hint: You may find Rosenthal's inequality useful: given zero-mean independent random variables Z_i such that $\|Z_i\|_m < +\infty$, there is a universal constant C_m such that

$$\left\|\sum_{i=1}^n Z_i\right\|_m \leq C_m\left\{\left(\sum_{i=1}^n \mathbb{E}[Z_i^2]\right)^{1/2} + \left(\sum_{i=1}^n \mathbb{E}[|Z_i|^m]\right)^{1/m}\right\}.$$

Exercise 6.16 (Graphs and adjacency matrices) Let G be a graph with maximum degree $s - 1$ that contains an s-clique. Letting $\mathbf{A}$ denote its adjacency matrix (defined with ones on the diagonal), show that $|||\mathbf{A}|||_2 = s$.

7

Sparse linear models in high dimensions

The linear model is one of the most widely used in statistics, and has a history dating back to the work of Gauss on least-squares prediction. In its low-dimensional instantiation, in which the number of predictors d is substantially less than the sample size n, the associated theory is classical. By contrast, our aim in this chapter is to develop theory that is applicable to the high-dimensional regime, meaning that it allows for scalings such that $d \asymp n$, or even $d \gg n$. As one might intuitively expect, if the model lacks any additional structure, then there is no hope of obtaining consistent estimators when the ratio d/n stays bounded away from zero.[1] For this reason, when working in settings in which $d > n$, it is necessary to impose additional structure on the unknown regression vector $\theta^* \in \mathbb{R}^d$, and this chapter focuses on different types of sparse models.

7.1 Problem formulation and applications

Let $\theta^* \in \mathbb{R}^d$ be an unknown vector, referred to as the regression vector. Suppose that we observe a vector $y \in \mathbb{R}^n$ and a matrix $\mathbf{X} \in \mathbb{R}^{n\times d}$ that are linked via the standard linear model

$$y = \mathbf{X}\theta^* + w, \tag{7.1}$$

where $w \in \mathbb{R}^n$ is a vector of noise variables. This model can also be written in a scalarized form: for each index $i = 1, 2, \ldots, n$, we have $y_i = \langle x_i, \theta^* \rangle + w_i$, where $x_i^{\mathrm{T}} \in \mathbb{R}^d$ is the ith row of $\mathbf{X}$, and y_i and w_i are (respectively) the ith entries of the vectors y and w. The quantity $\langle x_i, \theta^* \rangle := \sum_{j=1}^d x_{ij}\theta^*_j$ denotes the usual Euclidean inner product between the vector $x_i \in \mathbb{R}^d$ of predictors (or covariates), and the regression vector $\theta^* \in \mathbb{R}^d$. Thus, each response y_i is a noisy version of a linear combination of d covariates.

The focus of this chapter is settings in which the sample size n is smaller than the number of predictors d. In this case, it can also be of interest in certain applications to consider a *noiseless linear model*, meaning the special case of equation (7.1) with $w = 0$. When $n < d$, the equations $y = \mathbf{X}\theta^*$ define an underdetermined linear system, and the goal is to understand the structure of its sparse solutions.

7.1.1 Different sparsity models

At the same time, when $d > n$, it is impossible to obtain any meaningful estimates of θ^* unless the model is equipped with some form of low-dimensional structure. One of the

[1] Indeed, this intuition will be formalized as a theorem in Chapter 15 using information-theoretic methods.

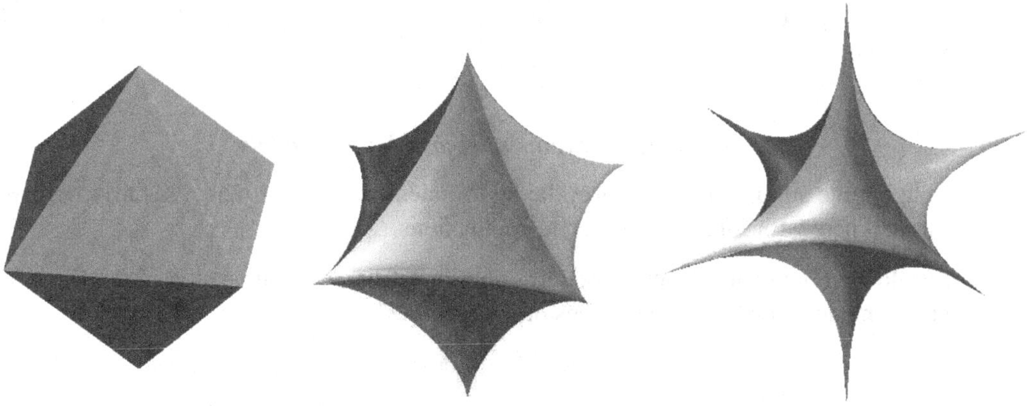

Figure 7.1 Illustrations of the ℓ_q-"balls" for different choices of the parameter $q \in (0, 1]$. (a) For $q = 1$, the set $\mathbb{B}_1(R_q)$ corresponds to the usual ℓ_1-ball shown here. (b) For $q = 0.75$, the ball is a non-convex set obtained by collapsing the faces of the ℓ_1-ball towards the origin. (c) For $q = 0.5$, the set becomes more "spiky", and it collapses into the hard sparsity constraint as $q \to 0^+$. As shown in Exercise 7.2(a), for all $q \in (0, 1]$, the set $\mathbb{B}_q(1)$ is star-shaped around the origin.

simplest kinds of structure in a linear model is a *hard sparsity* assumption, meaning that the set

$$S(\theta^*) := \{j \in \{1, 2, \ldots, d\} \mid \theta^*_j \neq 0\}, \tag{7.2}$$

known as the *support set* of θ^*, has cardinality $s := |S(\theta^*)|$ substantially smaller than d. Assuming that the model is exactly supported on s coefficients may be overly restrictive, in which case it is also useful to consider various relaxations of hard sparsity, which leads to the notion of weak sparsity. Roughly speaking, a vector θ^* is *weakly sparse* if it can be closely approximated by a sparse vector.

There are different ways in which to formalize such an idea, one way being via the ℓ_q-"norms". For a parameter $q \in [0, 1]$ and radius $R_q > 0$, consider the set

$$\mathbb{B}_q(R_q) = \left\{\theta \in \mathbb{R}^d \;\middle|\; \sum_{j=1}^d |\theta_j|^q \leq R_q\right\}. \tag{7.3}$$

It is known as the ℓ_q-ball of radius R_q. As illustrated in Figure 7.1, for $q \in [0, 1)$, it is not a ball in the strict sense of the word, since it is a non-convex set. In the special case $q = 0$, any vector $\theta^* \in \mathbb{B}_0(R_0)$ can have at most $s = R_0$ non-zero entries. More generally, for values of q in $(0, 1]$, membership in the set $\mathbb{B}_q(R_q)$ has different interpretations. One of them involves how quickly the ordered coefficients

$$\underbrace{|\theta^*_{(1)}|}_{\max_{j=1,2,\ldots,d} |\theta^*_j|} \geq |\theta^*_{(2)}| \geq \cdots \geq |\theta^*_{(d-1)}| \geq \underbrace{|\theta^*_{(d)}|}_{\min_{j=1,2,\ldots,d} |\theta^*_j|} \tag{7.4}$$

decay. More precisely, as we explore in Exercise 7.2, if these ordered coefficients satisfy the bound $|\theta^*_{(j)}| \le C j^{-\alpha}$ for a suitable exponent α, then θ^* belongs to $\mathbb{B}_q(R_q)$ for a radius R_q depending on (C, α).

7.1.2 Applications of sparse linear models

Although quite simple in appearance, the high-dimensional linear model is fairly rich. We illustrate it here with some examples and applications.

Example 7.1 (Gaussian sequence model) In a finite-dimensional version of the Gaussian sequence model, we make observations of the form

$$y_i = \sqrt{n}\theta^*_i + w_i, \qquad \text{for } i = 1, 2, \ldots, n, \tag{7.5}$$

where $w_i \sim \mathcal{N}(0, \sigma^2)$ are i.i.d. noise variables. This model is a special case of the general linear regression model (7.1) with $n = d$, and a design matrix $\mathbf{X} = \sqrt{n}\mathbf{I}_n$. It is a truly high-dimensional model, since the sample size n is equal to the number of parameters d. Although it appears simple on the surface, it is a surprisingly rich model: indeed, many problems in nonparametric estimation, among them regression and density estimation, can be reduced to an "equivalent" instance of the Gaussian sequence model, in the sense that the optimal rates for estimation are the same under both models. For nonparametric regression, when the function f belongs to a certain type of function class (known as a Besov space), then the vector of its wavelet coefficients belongs to a certain type of ℓ_q-ball with $q \in (0, 1)$, so that the estimation problem corresponds to a version of the Gaussian sequence problem with an ℓ_q-sparsity constraint. Various methods for estimation, such as wavelet thresholding, exploit this type of approximate sparsity. See the bibliographic section for additional references on this connection. ♣

Example 7.2 (Signal denoising in orthonormal bases) Sparsity plays an important role in signal processing, both for compression and for denoising of signals. In abstract terms, a signal can be represented as a vector $\beta^* \in \mathbb{R}^d$. Depending on the application, the signal length d could represent the number of pixels in an image, or the number of discrete samples of a time series. In a denoising problem, one makes noisy observations of the form $\widetilde{y} = \beta^* + \widetilde{w}$, where the vector $\widetilde{w}$ corresponds to some kind of additive noise. Based on the observation vector $\widetilde{y} \in \mathbb{R}^d$, the goal is to "denoise" the signal, meaning to reconstruct β^* as accurately as possible. In a compression problem, the goal is to produce a representation of β^*, either exact or approximate, that can be stored more compactly than its original representation.

Many classes of signals exhibit sparsity when transformed into an appropriate basis, such as a wavelet basis. This sparsity can be exploited both for compression and for denoising. In abstract terms, any such transform can be represented as an orthonormal matrix $\boldsymbol{\Psi} \in \mathbb{R}^{d \times d}$, constructed so that $\theta^* := \boldsymbol{\Psi}^{\mathrm{T}}\beta^* \in \mathbb{R}^d$ corresponds to the vector of transform coefficients. If the vector θ^* is known to be sparse, then it can be compressed by retaining only some number $s < d$ of its coefficients, say the largest s in absolute value. Of course, if θ^* were exactly sparse, then this representation would be exact. It is more realistic to assume that θ^* satisfies some form of approximate sparsity, and, as we explore in Exercise 7.2, such conditions can be used to provide guarantees on the accuracy of the reconstruction.

Returning to the denoising problem, in the transformed space, the observation model takes the form $y = \theta^* + w$, where $y := \boldsymbol{\Psi}^{\mathrm{T}}\widetilde{y}$ and $w := \boldsymbol{\Psi}^{\mathrm{T}}\widetilde{w}$ are the transformed observation and noise vector, respectively. When the observation noise is assumed to be i.i.d. Gaussian (and hence invariant under orthogonal transformation), then both the original and the transformed observations are instances of the Gaussian sequence model from Example 7.1, both with $n = d$.

If the vector θ^* is known to be sparse, then it is natural to consider estimators based on thresholding. In particular, for a threshold $\lambda > 0$ to be chosen, the hard-thresholded estimate of θ^* is defined as

$$[H_\lambda(y)]_i = \begin{cases} y_i & \text{if } |y_i| \geq \lambda, \\ 0 & \text{otherwise.} \end{cases} \tag{7.6a}$$

Closely related is the soft-thresholded estimate given by

$$[T_\lambda(y)]_i = \begin{cases} \operatorname{sign}(y_i)(|y_i| - \lambda) & \text{if } |y_i| \geq \lambda, \\ 0 & \text{otherwise.} \end{cases} \tag{7.6b}$$

As we explore in Exercise 7.1, each of these estimators have interpretations as minimizing the quadratic cost function $\theta \mapsto \|y - \theta\|_2^2$ subject to ℓ_0- and ℓ_1-constraints, respectively. ♣

Example 7.3 (Lifting and nonlinear functions) Despite its superficial appearance as representing purely linear functions, augmenting the set of predictors allows for nonlinear models to be represented by the standard equation (7.1). As an example, let us consider polynomial functions in a scalar variable $t \in \mathbb{R}$ of degree k, say of the form

$$f_\theta(t) = \theta_1 + \theta_2 t + \cdots + \theta_{k+1} t^k.$$

Suppose that we observe n samples of the form $\{(y_i, t_i)\}_{i=1}^n$, where each pair is linked via the observation model $y_i = f_\theta(t_i) + w_i$. This problem can be converted into an instance of the linear regression model by using the sample points $(t_1, \ldots, t_n)$ to define the $n \times (k+1)$ matrix

$$\mathbf{X} = \begin{bmatrix} 1 & t_1 & t_1^2 & \cdots & t_1^k \\ 1 & t_2 & t_2^2 & \cdots & t_2^k \\ \vdots & \vdots & \vdots & \ddots & \vdots \\ 1 & t_n & t_n^2 & \cdots & t_n^k \end{bmatrix}.$$

When expressed in this lifted space, the polynomial functions are linear in θ, and so we can write the observations $\{(y_i, t_i)\}_{i=1}^n$ in the standard vector form $y = \mathbf{X}\theta + w$.

This lifting procedure is not limited to polynomial functions. The more general setting is to consider functions that are linear combinations of some set of basis functions—say of the form

$$f_\theta(t) = \sum_{j=1}^{b} \theta_j \phi_j(t),$$

where $\{\phi_1, \ldots, \phi_b\}$ are some known functions. Given n observation pairs (y_i, t_i), this model can also be reduced to the form $y = \mathbf{X}\theta + w$, where the design matrix $\mathbf{X} \in \mathbb{R}^{n \times d}$ has entries $X_{ij} = \phi_j(t_i)$.

Although the preceding discussion has focused on univariate functions, the same ideas apply to multivariate functions, say in D dimensions. Returning to the case of polynomial functions, we note that there are $\binom{D}{k}$ possible multinomials of degree k in dimension D. This leads to the model dimension growing exponentially as D^k, so that sparsity assumptions become essential in order to produce manageable classes of models. ♣

Example 7.4 (Signal compression in overcomplete bases) We now return to an extension of the signal processing problem introduced in Example 7.2. As we observed previously, many classes of signals exhibit sparsity when represented in an appropriate basis, such as a wavelet basis, and this sparsity can be exploited for both compression and denoising purposes. Given a signal $y \in \mathbb{R}^n$, classical approaches to signal denoising and compression are based on orthogonal transformations, where the basis functions are represented by the columns of an orthonormal matrix $\mathbf{\Psi} \in \mathbb{R}^{n\times n}$. However, it can be useful to consider an *overcomplete* set of basis functions, represented by the columns of a matrix $\mathbf{X} \in \mathbb{R}^{n\times d}$ with $d > n$. Within this framework, signal compression can be performed by finding a vector $\theta \in \mathbb{R}^d$ such that $y = \mathbf{X}\theta$. Since $\mathbf{X}$ has rank n, we can always find a solution with at most n non-zero coordinates, but the hope is to find a solution $\theta^* \in \mathbb{R}^d$ with $\|\theta^*\|_0 = s \ll n$ non-zeros.

Problems involving ℓ_0-constraints are computationally intractable, so that it is natural to consider relaxations. As we will discuss at more length later in the chapter, the ℓ_1-relaxation has proven very successful. In particular, one seeks a sparse solution by solving the convex program

$$\widehat{\theta} \in \arg\min_{\theta\in\mathbb{R}^d} \underbrace{\sum_{j=1}^{d} |\theta_j|}_{\|\theta\|_1} \qquad \text{such that } y = \mathbf{X}\theta.$$

Later sections of the chapter will provide theory under which the solution to this ℓ_1-relaxation is equivalent to the original ℓ_0-problem. ♣

Example 7.5 (Compressed sensing) Compressed sensing is based on the combination of ℓ_1-relaxation with the random projection method, which was previously described in Example 2.12 from Chapter 2. It is motivated by the inherent wastefulness of the classical approach to exploiting sparsity for signal compression. As previously described in Example 7.2, given a signal $\beta^* \in \mathbb{R}^d$, the standard approach is first to compute the full vector $\theta^* = \mathbf{\Psi}^{\mathrm{T}}\beta^* \in \mathbb{R}^d$ of transform coefficients, and then to *discard* all but the top s coefficients. Is there a more direct way of estimating β^*, without pre-computing the full vector θ^* of its transform coefficients?

The compressed sensing approach is to take $n \ll d$ random projections of the original signal $\beta^* \in \mathbb{R}^d$, each of the form $y_i = \langle x_i, \beta^*\rangle := \sum_{j=1}^d x_{ij}\beta^*_j$, where $x_i \in \mathbb{R}^d$ is a random vector. Various choices are possible, including the standard Gaussian ensemble ($x_{ij} \sim \mathcal{N}(0,1)$, i.i.d.), or the Rademacher ensemble ($x_{ij} \in \{-1,+1\}$, i.i.d.). Let $\mathbf{X} \in \mathbb{R}^{n\times d}$ be a measurement matrix with x_i^{T} as its ith row and $y \in \mathbb{R}^n$ be the concatenated set of random projections. In matrix–vector notation, the problem of exact reconstruction amounts to finding a solution $\beta \in \mathbb{R}^d$ of the underdetermined linear system $\mathbf{X}\beta = \mathbf{X}\beta^*$ such that $\mathbf{\Psi}^{\mathrm{T}}\beta$ is as sparse as possible. Recalling that $y = \mathbf{X}\beta^*$, the standard ℓ_1-relaxation of this problem takes the form

$\min_{\beta\in\mathbb{R}^d} \|\mathbf{\Psi}^{\mathrm{T}}\beta\|_1$ such that $y = \mathbf{X}\beta$, or equivalently, in the transform domain,

$$\min_{\theta\in\mathbb{R}^d} \|\theta\|_1 \qquad \text{such that } y = \widetilde{\mathbf{X}}\theta, \tag{7.7}$$

where $\widetilde{\mathbf{X}} := \mathbf{X}\mathbf{\Psi}$. In asserting this equivalence, we have used the orthogonality relation $\mathbf{\Psi}\mathbf{\Psi}^{\mathrm{T}} = \mathbf{I}_d$. This is another instance of the basis pursuit linear program (LP) with a random design matrix $\widetilde{\mathbf{X}}$.

Compressed sensing is a popular approach to recovering sparse signals, with a number of applications. Later in the chapter, we will develop theory that guarantees the success of ℓ_1-relaxation for the random design matrices that arise from taking random projections. ♣

Example 7.6 (Selection of Gaussian graphical models) Any zero-mean Gaussian random vector $(Z_1, \ldots, Z_d)$ with a non-degenerate covariance matrix has a density of the form

$$p_{\mathbf{\Theta}^*}(z_1, \ldots, z_d) = \frac{1}{\sqrt{(2\pi)^d \det((\mathbf{\Theta}^*)^{-1})}} \exp(-\tfrac{1}{2} z^{\mathrm{T}}\mathbf{\Theta}^* z),$$

where $\mathbf{\Theta}^* \in \mathbb{R}^{d\times d}$ is the inverse covariance matrix, also known as the precision matrix. For many interesting models, the precision matrix is sparse, with relatively few non-zero entries. The problem of Gaussian graphical model selection, as discussed at more length in Chapter 11, is to infer the non-zero entries in the matrix $\mathbf{\Theta}^*$.

This problem can be reduced to an instance of sparse linear regression as follows. For a given index $s \in V := \{1, 2, \ldots, d\}$, suppose that we are interested in recovering its neighborhood, meaning the subset $\mathcal{N}(s) := \{t \in V \mid \Theta^*_{st} \neq 0\}$. In order to do so, imagine performing a linear regression of the variable Z_s on the $(d-1)$-dimensional vector $Z_{\backslash\{s\}} := \{Z_t,\ t \in V \setminus \{s\}\}$. As we explore in Exercise 11.3 in Chapter 11, we can write

$$\underbrace{Z_s}_{\text{response } y} = \langle \underbrace{Z_{\backslash\{s\}}}_{\text{predictors}}, \theta^* \rangle + w_s,$$

where w_s is a zero-mean Gaussian variable, independent of the vector $Z_{\backslash\{s\}}$. Moreover, the vector $\theta^* \in \mathbb{R}^{d-1}$ has the same sparsity pattern as the sth off-diagonal row $(\mathbf{\Theta}^*_{st},\ t \in V \setminus \{s\})$ of the precision matrix. ♣

7.2 Recovery in the noiseless setting

In order to build intuition, we begin by focusing on the simplest case in which the observations are perfect or noiseless. More concretely, we wish to find a solution θ to the linear system $y = \mathbf{X}\theta$, where $y \in \mathbb{R}^n$ and $\mathbf{X} \in \mathbb{R}^{n\times d}$ are given. When $d > n$, this is an *underdetermined* set of linear equations, so that there is a whole subspace of solutions. But what if we are told that there is a sparse solution? In this case, we know that there is some vector $\theta^* \in \mathbb{R}^d$ with at most $s \ll d$ non-zero entries such that $y = \mathbf{X}\theta^*$. Our goal is to find this sparse solution to the linear system. This noiseless problem has applications in signal representation and compression, as discussed in Examples 7.4 and 7.5.

7.2.1 ℓ_1-based relaxation

This problem can be cast as a (non-convex) optimization problem involving the ℓ_0-"norm". Let us define

$$\|\theta\|_0 := \sum_{j=1}^{d} \mathbb{I}[\theta_j \neq 0],$$

where the function $t \mapsto \mathbb{I}[t \neq 0]$ is equal to one if $t \neq 0$, and zero otherwise. Strictly speaking, this is not a norm, but it serves to count the number of non-zero entries in the vector $\theta \in \mathbb{R}^d$. We now consider the optimization problem

$$\min_{\theta \in \mathbb{R}^d} \|\theta\|_0 \qquad \text{such that } \mathbf{X}\theta = y. \tag{7.8}$$

If we could solve this problem, then we would obtain a solution to the linear equations that has the fewest number of non-zero entries.

But how to solve the problem (7.8)? Although the constraint set is simply a subspace, the cost function is non-differentiable and non-convex. The most direct approach would be to search exhaustively over subsets of the columns of $\mathbf{X}$. In particular, for each subset $S \subset \{1, \ldots, d\}$, we could form the matrix $\mathbf{X}_S \in \mathbb{R}^{n \times |S|}$ consisting of the columns of $\mathbf{X}$ indexed by S, and then examine the linear system $y = \mathbf{X}_S \theta$ to see whether or not it had a solution $\theta \in \mathbb{R}^{|S|}$. If we iterated over subsets in increasing cardinality, then the first solution found would be the sparsest solution. Let's now consider the associated computational cost. If the sparsest solution contained s non-zero entries, then we would have to search over at least $\sum_{j=1}^{s-1} \binom{d}{j}$ subsets before finding it. But the number of such subsets grows exponentially in s, so the procedure would not be computationally feasible for anything except toy problems.

Given the computational difficulties associated with ℓ_0-minimization, a natural strategy is to replace the troublesome ℓ_0-objective by the nearest convex member of the ℓ_q-family, namely the ℓ_1-norm. This is an instance of a *convex relaxation*, in which a non-convex optimization problem is approximated by a convex program. In this setting, doing so leads to the optimization problem

$$\min_{\theta \in \mathbb{R}^d} \|\theta\|_1 \qquad \text{such that } \mathbf{X}\theta = y. \tag{7.9}$$

Unlike the ℓ_0-version, this is now a convex program, since the constraint set is a subspace (hence convex), and the cost function is piecewise linear and thus convex as well. More precisely, the problem (7.9) is a linear program, since any piecewise linear convex cost can always be reformulated as the maximum of a collection of linear functions. We refer to the optimization problem (7.9) as the *basis pursuit linear program*, after Chen, Donoho and Saunders (1998).

7.2.2 Exact recovery and restricted nullspace

We now turn to an interesting theoretical question: when is solving the basis pursuit program (7.9) equivalent to solving the original ℓ_0-problem (7.8)? More concretely, let us suppose that there is a vector $\theta^* \in \mathbb{R}^d$ such that $y = \mathbf{X}\theta^*$, and moreover, the vector θ^* has support

$S \subset \{1, 2, \ldots, d\}$, meaning that $\theta^*_j = 0$ for all $j \in S^c$ (where S^c denotes the complement of S).

Intuitively, the success of basis pursuit should depend on how the nullspace of $\mathbf{X}$ is related to this support, as well as the geometry of the ℓ_1-ball. To make this concrete, recall that the nullspace of $\mathbf{X}$ is given by $\text{null}(\mathbf{X}) := \{\Delta \in \mathbb{R}^d \mid \mathbf{X}\Delta = 0\}$. Since $\mathbf{X}\theta^* = y$ by assumption, any vector of the form $\theta^* + \Delta$ for some $\Delta \in \text{null}(\mathbf{X})$ is feasible for the basis pursuit program. Now let us consider the *tangent cone* of the ℓ_1-ball at θ^*, given by

$$\mathbb{T}(\theta^*) = \{\Delta \in \mathbb{R}^d \mid \|\theta^* + t\Delta\|_1 \leq \|\theta^*\|_1 \text{ for some } t > 0\}. \tag{7.10}$$

As illustrated in Figure 7.2, this set captures the set of all directions relative to θ^* along which the ℓ_1-norm remains constant or decreases. As noted earlier, the set $\theta^* + \text{null}(\mathbf{X})$, drawn with a solid line in Figure 7.2, corresponds to the set of all vectors that are feasible for the basis pursuit LP. Consequently, if θ^* is the unique optimal solution of the basis pursuit LP, then it must be the case that the intersection of the nullspace $\text{null}(\mathbf{X})$ with this tangent cone contains only the zero vector. This favorable case is shown in Figure 7.2(a), whereas Figure 7.2(b) shows the non-favorable case, in which θ^* need not be optimal.

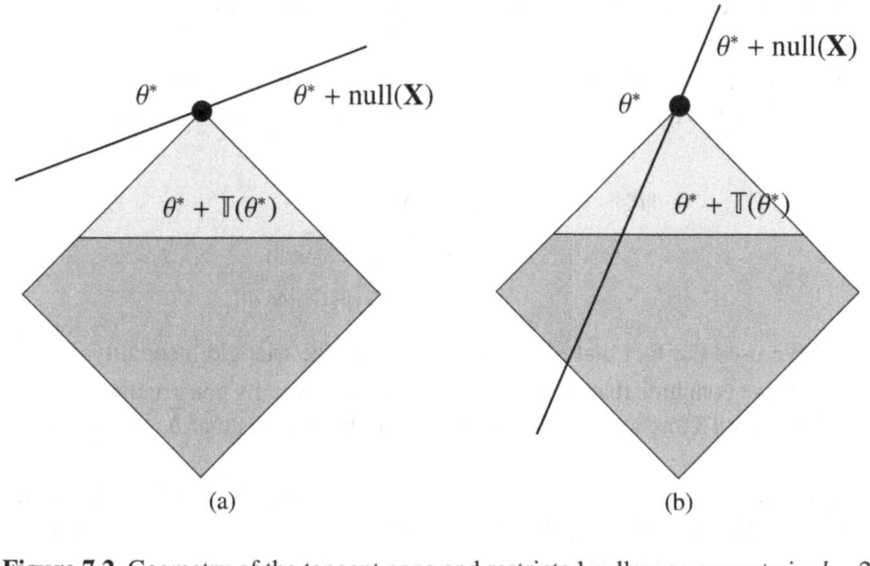

Figure 7.2 Geometry of the tangent cone and restricted nullspace property in $d = 2$ dimensions. (a) The favorable case in which the set $\theta^* + \text{null}(\mathbf{X})$ intersects the tangent cone only at θ^*. (b) The unfavorable setting in which the set $\theta^* + \text{null}(\mathbf{X})$ passes directly through the tangent cone.

This intuition leads to a condition on $\mathbf{X}$ known as the *restricted nullspace property*. Let us define the subset

$$\mathbb{C}(S) = \{\Delta \in \mathbb{R}^d \mid \|\Delta_{S^c}\|_1 \leq \|\Delta_S\|_1\},$$

corresponding to the cone of vectors whose ℓ_1-norm off the support is dominated by the ℓ_1-norm on the support. The following definition links the nullspace of a matrix $\mathbf{X}$ to this set:

Definition 7.7 The matrix $\mathbf{X}$ satisfies the *restricted nullspace property* with respect to S if $\mathbb{C}(S) \cap \text{null}(\mathbf{X}) = \{0\}$.

As shown in the proof of Theorem 7.8 to follow, the difference set $\mathbb{C}(S)$ provides an alternative way of capturing the behavior of the tangent cone $\mathbb{T}(\theta^*)$, one that is independent of θ^*. In particular, the proof establishes that, for any S-sparse vector θ^*, the tangent cone $\mathbb{T}(\theta^*)$ is contained within $\mathbb{C}(S)$, and conversely, that $\mathbb{C}(S)$ is contained in the union of such tangent cones. More precisely, the restricted nullspace property is equivalent to the success of the basis pursuit LP in the following sense:

Theorem 7.8 *The following two properties are equivalent:*

(a) *For any vector $\theta^* \in \mathbb{R}^d$ with support S, the basis pursuit program (7.9) applied with $y = \mathbf{X}\theta^*$ has unique solution $\widehat{\theta} = \theta^*$.*
(b) *The matrix $\mathbf{X}$ satisfies the restricted nullspace property with respect to S.*

Proof We first show that (b) $\Rightarrow$ (a). Since both $\widehat{\theta}$ and θ^* are feasible for the basis pursuit program, and since $\widehat{\theta}$ is optimal, we have $\|\widehat{\theta}\|_1 \leq \|\theta^*\|_1$. Defining the error vector $\widehat{\Delta} := \widehat{\theta} - \theta^*$, we have

$$\begin{aligned} \|\theta^*_S\|_1 = \|\theta^*\|_1 &\geq \|\theta^* + \widehat{\Delta}\|_1 \\ &= \|\theta^*_S + \widehat{\Delta}_S\|_1 + \|\widehat{\Delta}_{S^c}\|_1 \\ &\geq \|\theta^*_S\|_1 - \|\widehat{\Delta}_S\|_1 + \|\widehat{\Delta}_{S^c}\|_1, \end{aligned}$$

where we have used the fact that $\theta^*_{S^c} = 0$, and applied the triangle inequality. Rearranging this inequality, we conclude that the error $\widehat{\Delta} \in \mathbb{C}(S)$. However, by construction, we also have $\mathbf{X}\widehat{\Delta} = 0$, so $\widehat{\Delta} \in \text{null}(\mathbf{X})$ as well. By our assumption, this implies that $\widehat{\Delta} = 0$, or equivalently that $\widehat{\theta} = \theta^*$.

In order to establish the implication (a) $\Rightarrow$ (b), it suffices to show that, if the ℓ_1-relaxation succeeds for all S-sparse vectors, then the set $\text{null}(\mathbf{X}) \setminus \{0\}$ has no intersection with $\mathbb{C}(S)$. For a given vector $\theta^* \in \text{null}(\mathbf{X}) \setminus \{0\}$, consider the basis pursuit problem

$$\min_{\beta \in \mathbb{R}^d} \|\beta\|_1 \quad \text{such that } \mathbf{X}\beta = \mathbf{X}\begin{bmatrix} \theta^*_S \\ 0 \end{bmatrix}. \tag{7.11}$$

By assumption, the unique optimal solution will be $\widehat{\beta} = [\theta^*_S \quad 0]^{\mathrm{T}}$. Since $\mathbf{X}\theta^* = 0$ by assumption, the vector $[0 \quad -\theta^*_{S^c}]^{\mathrm{T}}$ is also feasible for the problem, and, by uniqueness, we must have $\|\theta^*_S\|_1 < \|\theta^*_{S^c}\|_1$, implying that $\theta^* \notin \mathbb{C}(S)$ as claimed. □

7.2.3 Sufficient conditions for restricted nullspace

In order for Theorem 7.8 to be a useful result in practice, one requires a certificate that the restricted nullspace property holds. The earliest sufficient conditions were based on the incoherence parameter of the design matrix, namely the quantity

$$\delta_{\mathrm{PW}}(\mathbf{X}) := \max_{j,k=1,\ldots,d} \left| \frac{|\langle X_j, X_k \rangle|}{n} - \mathbb{I}[j = k] \right|, \tag{7.12}$$

where X_j denotes the jth column of $\mathbf{X}$, and $\mathbb{I}[j = k]$ denotes the $\{0, 1\}$-valued indicator for the event $\{j = k\}$. Here we have chosen to rescale matrix columns by $1/\sqrt{n}$, as it makes results for random designs more readily interpretable.

The following result shows that a small pairwise incoherence is sufficient to guarantee a uniform version of the restricted nullspace property.

Proposition 7.9 *If the pairwise incoherence satisfies the bound*

$$\delta_{\mathrm{PW}}(\mathbf{X}) \leq \frac{1}{3s}, \tag{7.13}$$

then the restricted nullspace property holds for all subsets S of cardinality at most s.

We guide the reader through the steps involved in the proof of this claim in Exercise 7.3.

A related but more sophisticated sufficient condition is the restricted isometry property (RIP). It can be understood as a natural generalization of the pairwise incoherence condition, based on looking at conditioning of larger subsets of columns.

Definition 7.10 (Restricted isometry property) For a given integer $s \in \{1, \ldots, d\}$, we say that $\mathbf{X} \in \mathbb{R}^{n \times d}$ satisfies a restricted isometry property of order s with constant $\delta_s(\mathbf{X}) > 0$ if

$$\left|\!\left|\!\left| \frac{\mathbf{X}_S^{\mathrm{T}} \mathbf{X}_S}{n} - \mathbf{I}_s \right|\!\right|\!\right|_2 \leq \delta_s(\mathbf{X}) \qquad \text{for all subsets } S \text{ of size at most } s. \tag{7.14}$$

In this definition, we recall that $|||\cdot|||_2$ denotes the ℓ_2-operator norm of a matrix, corresponding to its maximum singular value. For $s = 1$, the RIP condition implies that the rescaled columns of $\mathbf{X}$ are near-unit-norm—that is, we are guaranteed that $\frac{\|X_j\|_2^2}{n} \in [1 - \delta_1, 1 + \delta_1]$ for all $j = 1, 2, \ldots, d$. For $s = 2$, the RIP constant δ_2 is very closely related to the pairwise incoherence parameter $\delta_{\mathrm{PW}}(\mathbf{X})$. This connection is most apparent when the matrix $\mathbf{X}/\sqrt{n}$ has unit-norm columns, in which case, for any pair of columns $\{j, k\}$, we have

$$\frac{\mathbf{X}_{\{j,k\}}^{\mathrm{T}} \mathbf{X}_{\{j,k\}}}{n} - \begin{bmatrix} 1 & 0 \\ 0 & 1 \end{bmatrix} = \begin{bmatrix} \frac{\|X_j\|_2^2}{n} - 1 & \frac{\langle X_j, X_k \rangle}{n} \\ \frac{\langle X_j, X_k \rangle}{n} & \frac{\|X_k\|_2^2}{n} - 1 \end{bmatrix} \stackrel{\text{(i)}}{=} \begin{bmatrix} 0 & \frac{\langle X_j, X_k \rangle}{n} \\ \frac{\langle X_j, X_k \rangle}{n} & 0 \end{bmatrix},$$

where the final equality (i) uses the column normalization condition. Consequently, we find

that

$$\delta_2(\mathbf{X}) = \left|\left|\left|\frac{\mathbf{X}_{\{j,k\}}^{\mathrm{T}}\mathbf{X}_{\{j,k\}}}{n} - \mathbf{I}_2\right|\right|\right|_2 = \max_{j\neq k}\left|\frac{\langle X_j, X_k\rangle}{n}\right| = \delta_{\mathrm{PW}}(\mathbf{X}),$$

where the final step again uses the column normalization condition. More generally, as we show in Exercise 7.4, for any matrix $\mathbf{X}$ and sparsity level $s \in \{2,\ldots,d\}$, we have the sandwich relation

$$\delta_{\mathrm{PW}}(\mathbf{X}) \overset{\text{(i)}}{\leq} \delta_s(\mathbf{X}) \overset{\text{(ii)}}{\leq} s\delta_{\mathrm{PW}}(\mathbf{X}), \tag{7.15}$$

and neither bound can be improved in general. (We also show that there exist matrices for which $\delta_s(\mathbf{X}) = \sqrt{s}\,\delta_{\mathrm{PW}}(\mathbf{X})$.) Although RIP imposes constraints on much larger submatrices than pairwise incoherence, the magnitude of the constraints required to guarantee the uniform restricted nullspace property can be milder.

The following result shows that suitable control on the RIP constants implies that the restricted nullspace property holds:

Proposition 7.11 *If the RIP constant of order $2s$ is bounded as $\delta_{2s}(\mathbf{X}) < 1/3$, then the uniform restricted nullspace property holds for any subset S of cardinality $|S| \leq s$.*

Proof Let $\theta \in \mathrm{null}(\mathbf{X})$ be an arbitrary non-zero member of the nullspace. For any subset A, we let $\theta_A \in \mathbb{R}^{|A|}$ denote the subvector of elements indexed by A, and we define the vector $\widetilde{\theta}_A \in \mathbb{R}^d$ with elements

$$\widetilde{\theta}_j = \begin{cases} \theta_j & \text{if } j \in A,\\ 0 & \text{otherwise.}\end{cases}$$

We frequently use the fact that $\|\widetilde{\theta}_A\| = \|\theta_A\|$ for any elementwise separable norm, such as the ℓ_1- or ℓ_2-norms.

Let S be the subset of $\{1,2,\ldots,d\}$ corresponding to the s entries of θ that are largest in absolute value. It suffices to show that $\|\theta_{S^c}\|_1 > \|\theta_S\|_1$ for this subset. Let us write $S^c = \bigcup_{j\geq 1} S_j$, where S_1 is the subset of indices given by the s largest values of $\widetilde{\theta}_{S^c}$; the subset S_2 is the largest s in the subset $S^c \setminus S_1$, and the final subset may contain fewer than s entries. Using this notation, we have the decomposition $\theta = \widetilde{\theta}_S + \sum_{k\geq 1}\widetilde{\theta}_{S_k}$.

The RIP property guarantees that $\|\widetilde{\theta}_S\|_2^2 \leq \frac{1}{1-\delta_{2s}}\left\|\frac{1}{\sqrt{n}}\mathbf{X}\widetilde{\theta}_S\right\|_2^2$. Moreover, since $\theta \in \mathrm{null}(\mathbf{X})$, we have $\mathbf{X}\widetilde{\theta}_S = -\sum_{j\geq 1}\mathbf{X}\widetilde{\theta}_{S_j}$, and hence

$$\|\widetilde{\theta}_{S_0}\|_2^2 \leq \frac{1}{1-\delta_{2s}}\left|\sum_{j\geq 1}\frac{\langle \mathbf{X}\widetilde{\theta}_{S_0}, \mathbf{X}\widetilde{\theta}_{S_j}\rangle}{n}\right| \overset{\text{(i)}}{=} \frac{1}{1-\delta_{2s}}\left|\sum_{j\geq 1}\widetilde{\theta}_{S_0}\left[\frac{\mathbf{X}^{\mathrm{T}}\mathbf{X}}{n} - \mathbf{I}_d\right]\widetilde{\theta}_{S_j}\right|,$$

where equality (i) uses the fact that $\langle\widetilde{\theta}_S, \widetilde{\theta}_{S_j}\rangle = 0$.

By the RIP property, for each $j \geq 1$, the $\ell_2 \to \ell_2$ operator norm satisfies the bound

$|||n^{-1}\mathbf{X}_{S_0 \cup S_j}^{\mathrm{T}} \mathbf{X}_{S_0 \cup S_j} - \mathbf{I}_{2s}|||_2 \leq \delta_{2s}$, and hence we have

$$\|\widetilde{\theta}_{S_0}\|_2 \leq \frac{\delta_{2s}}{1-\delta_{2s}} \sum_{j \geq 1} \|\widetilde{\theta}_{S_j}\|_2, \tag{7.16}$$

where we have canceled out a factor of $\|\widetilde{\theta}_{S_0}\|_2$ from each side. Finally, by construction of the sets S_j, for each $j \geq 1$, we have $\|\widetilde{\theta}_{S_j}\|_\infty \leq \frac{1}{s} \|\widetilde{\theta}_{S_{j-1}}\|_1$, which implies that $\|\widetilde{\theta}_{S_j}\|_2 \leq \frac{1}{\sqrt{s}} \|\widetilde{\theta}_{S_{j-1}}\|_1$. Applying these upper bounds to the inequality (7.16), we obtain

$$\|\widetilde{\theta}_{S_0}\|_1 \leq \sqrt{s}\, \|\widetilde{\theta}_{S_0}\|_2 \leq \frac{\delta_{2s}}{1-\delta_{2s}} \Big\{ \|\widetilde{\theta}_{S_0}\|_1 + \sum_{j \geq 1} \|\widetilde{\theta}_{S_j}\|_1 \Big\},$$

or equivalently $\|\widetilde{\theta}_{S_0}\|_1 \leq \frac{\delta_{2s}}{1-\delta_{2s}} \{ \|\widetilde{\theta}_{S_0}\|_1 + \|\widetilde{\theta}_{S^c}\|_1 \}$. Some simple algebra verifies that this inequality implies that $\|\widetilde{\theta}_{S_0}\|_1 < \|\widetilde{\theta}_{S^c}\|_1$ as long as $\delta_{2s} < 1/3$. □

Like the pairwise incoherence constant, control on the RIP constants is a sufficient condition for the basis pursuit LP to succeed. A major advantage of the RIP approach is that for various classes of random design matrices, of particular interest in compressed sensing (see Example 7.5), it can be used to guarantee exactness of basis pursuit using a sample size n that is much smaller than that guaranteed by pairwise incoherence. As we explore in Exercise 7.7, for sub-Gaussian random matrices with i.i.d. elements, the pairwise incoherence is bounded by $\frac{1}{3s}$ with high probability as long as $n \gtrsim s^2 \log d$. By contrast, this same exercise also shows that the RIP constants for certain classes of random design matrices $\mathbf{X}$ are well controlled as long as $n \gtrsim s \log(ed/s)$. Consequently, the RIP approach overcomes the "quadratic barrier"—namely, the requirement that the sample size n scales quadratically in the sparsity s, as in the pairwise incoherence approach.

It should be noted that, unlike the restricted nullspace property, neither the pairwise incoherence condition nor the RIP condition are necessary conditions. Indeed, the basis pursuit LP succeeds for many classes of matrices for which both pairwise incoherence and RIP conditions are violated. For example, consider a random matrix $\mathbf{X} \in \mathbb{R}^{n \times d}$ with i.i.d. rows $X_i \sim \mathcal{N}(0, \mathbf{\Sigma})$. Letting $\mathbb{1} \in \mathbb{R}^d$ denote the all-ones vector, consider the family of covariance matrices

$$\mathbf{\Sigma} := (1-\mu)\mathbf{I}_d + \mu \mathbb{1}\mathbb{1}^{\mathrm{T}}, \tag{7.17}$$

for a parameter $\mu \in [0, 1)$. In Exercise 7.8, we show that, for any fixed $\mu \in (0, 1)$, the pairwise incoherence bound (7.13) is violated with high probability for large s, and moreover that the condition number of any $2s$-sized subset grows at the rate $\mu\sqrt{s}$ with high probability, so that the RIP constants will (with high probability) grow unboundedly as $s \to +\infty$ for any fixed $\mu \in (0, 1)$. Nonetheless, for any $\mu \in [0, 1)$, the basis pursuit LP relaxation still succeeds with high probability with sample size $n \gtrsim s \log(ed/s)$, as illustrated in Figure 7.4. Later in the chapter, we provide a result on random matrices that allows for direct verification of the restricted nullspace property for various families, including (among others) the family (7.17). See Theorem 7.16 and the associated discussion for further details.

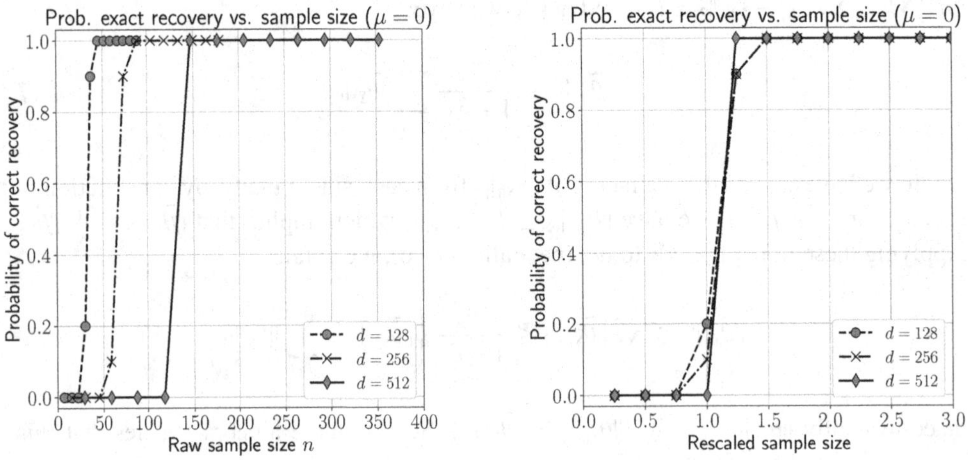

(a) (b)

Figure 7.3 (a) Probability of basis pursuit success versus the raw sample size n for random design matrices drawn with i.i.d. $\mathcal{N}(0,1)$ entries. Each curve corresponds to a different problem size $d \in \{128, 256, 512\}$ with sparsity $s = \lceil 0.1d \rceil$. (b) The same results replotted versus the rescaled sample size $n/(s \log(ed/s))$. The curves exhibit a phase transition at the same value of this rescaled sample size.

7.3 Estimation in noisy settings

Let us now turn to the noisy setting, in which we observe the vector–matrix pair $(y, \mathbf{X}) \in \mathbb{R}^n \times \mathbb{R}^{n \times d}$ linked by the observation model $y = \mathbf{X}\theta^* + w$. The new ingredient here is the noise vector $w \in \mathbb{R}^n$. A natural extension of the basis pursuit program is based on minimizing a weighted combination of the data-fidelity term $\|y - \mathbf{X}\theta\|_2^2$ with the ℓ_1-norm penalty, say of the form

$$\widehat{\theta} \in \arg\min_{\theta \in \mathbb{R}^d} \left\{ \frac{1}{2n} \|y - \mathbf{X}\theta\|_2^2 + \lambda_n \|\theta\|_1 \right\}. \tag{7.18}$$

Here $\lambda_n > 0$ is a *regularization parameter* to be chosen by the user. Following Tibshirani (1996), we refer to it as the *Lasso program*.

Alternatively, one can consider different constrained forms of the Lasso, that is either

$$\min_{\theta \in \mathbb{R}^d} \left\{ \frac{1}{2n} \|y - \mathbf{X}\theta\|_2^2 \right\} \quad \text{such that } \|\theta\|_1 \leq R \tag{7.19}$$

for some radius $R > 0$, or

$$\min_{\theta \in \mathbb{R}^d} \|\theta\|_1 \quad \text{such that } \frac{1}{2n} \|y - \mathbf{X}\theta\|_2^2 \leq b^2 \tag{7.20}$$

for some noise tolerance $b > 0$. The constrained version (7.20) is referred to as *relaxed basis pursuit* by Chen et al. (1998). By Lagrangian duality theory, all three families of convex programs are equivalent. More precisely, for any choice of radius $R > 0$ in the constrained variant (7.19), there is a regularization parameter $\lambda \geq 0$ such that solving the Lagrangian

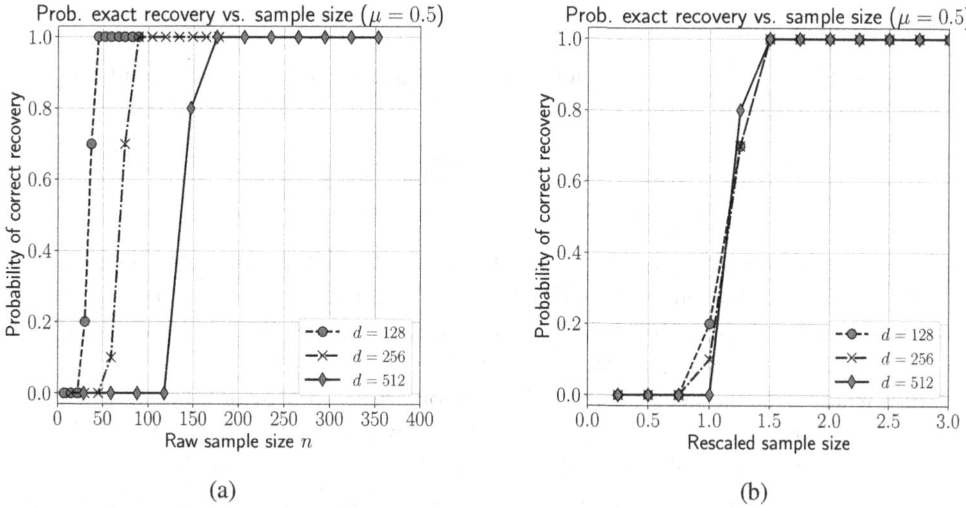

Figure 7.4 (a) Probability of basis pursuit success versus the raw sample size n for random design matrices drawn with i.i.d. rows $X_i \sim \mathcal{N}(0, \boldsymbol{\Sigma})$, where $\mu = 0.5$ in the model (7.17). Each curve corresponds to a different problem size $d \in \{128, 256, 512\}$ with sparsity $s = \lceil 0.1d \rceil$. (b) The same results replotted versus the rescaled sample size $n/(s \log(ed/s))$. The curves exhibit a phase transition at the same value of this rescaled sample size.

version (7.18) is equivalent to solving the constrained version (7.19). Similar statements apply to choices of $b > 0$ in the constrained variant (7.20).

7.3.1 Restricted eigenvalue condition

In the noisy setting, we can no longer expect to achieve perfect recovery. Instead, we focus on bounding the ℓ_2-error $\|\widehat{\theta} - \theta^*\|_2$ between a Lasso solution $\widehat{\theta}$ and the unknown regression vector θ^*. In the presence of noise, we require a condition that is closely related to but slightly stronger than the restricted nullspace property—namely, that the restricted eigenvalues of the matrix $\frac{\mathbf{X}^{\mathrm{T}}\mathbf{X}}{n}$ are lower bounded over a cone. In particular, for a constant $\alpha \geq 1$, let us define the set

$$\mathbb{C}_\alpha(S) := \{\Delta \in \mathbb{R}^d \mid \|\Delta_{S^c}\|_1 \leq \alpha \|\Delta_S\|_1\}. \tag{7.21}$$

This definition generalizes the set $\mathbb{C}(S)$ used in our definition of the restricted nullspace property, which corresponds to the special case $\alpha = 1$.

Definition 7.12 The matrix $\mathbf{X}$ satisfies the *restricted eigenvalue* (RE) condition over S with parameters (κ, α) if

$$\frac{1}{n}\|\mathbf{X}\Delta\|_2^2 \geq \kappa\|\Delta\|_2^2 \qquad \text{for all } \Delta \in \mathbb{C}_\alpha(S). \tag{7.22}$$

Note that the RE condition is a strengthening of the restricted nullspace property. In particular, if the RE condition holds with parameters $(\kappa, 1)$ for any $\kappa > 0$, then the restricted nullspace property holds. Moreover, we will prove that under the RE condition, the error $\|\widehat{\theta} - \theta^*\|_2$ in the Lasso solution is well controlled.

From where does the need for the RE condition arise? To provide some intuition, let us consider the constrained version (7.19) of the Lasso, with radius $R = \|\theta^*\|_1$. With this setting, the true parameter vector θ^* is feasible for the problem. By definition, the Lasso estimate $\widehat{\theta}$ minimizes the quadratic cost function $\mathcal{L}_n(\theta) = \frac{1}{2n}\|y - \mathbf{X}\theta\|_2^2$ over the ℓ_1-ball of radius R. As the amount of data increases, we expect that θ^* should become a near-minimizer of the same cost function, so that $\mathcal{L}_n(\widehat{\theta}) \approx \mathcal{L}_n(\theta^*)$. But when does closeness in cost imply that the error vector $\Delta := \widehat{\theta} - \theta^*$ is small? As illustrated in Figure 7.5, the link between the cost difference $\delta\mathcal{L}_n := \mathcal{L}_n(\theta^*) - \mathcal{L}_n(\widehat{\theta})$ and the error $\Delta = \widehat{\theta} - \theta^*$ is controlled by the curvature of the cost function. In the favorable setting of Figure 7.5(a), the cost has a high curvature around its optimum $\widehat{\theta}$, so that a small excess loss $\delta\mathcal{L}_n$ implies that the error vector Δ is small. This curvature no longer holds for the cost function in Figure 7.5(b), for which it is possible that $\delta\mathcal{L}_n$ could be small while the error Δ is relatively large.

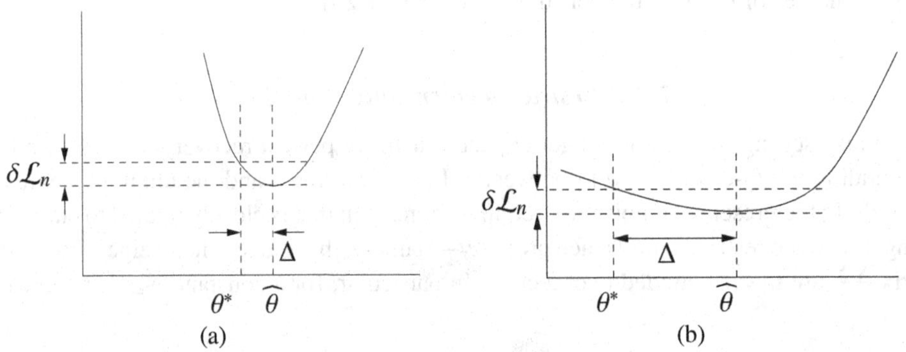

Figure 7.5 Illustration of the connection between curvature (strong convexity) of the cost function, and estimation error. (a) In a favorable setting, the cost function is sharply curved around its minimizer $\widehat{\theta}$, so that a small change $\delta\mathcal{L}_n := \mathcal{L}_n(\theta^*) - \mathcal{L}_n(\widehat{\theta})$ in the cost implies that the error vector $\Delta = \widehat{\theta} - \theta^*$ is not too large. (b) In an unfavorable setting, the cost is very flat, so that a small cost difference $\delta\mathcal{L}_n$ need not imply small error.

Figure 7.5 illustrates a one-dimensional function, in which case the curvature can be captured by a scalar. For a function in d dimensions, the curvature of a cost function is captured by the structure of its Hessian matrix $\nabla^2\mathcal{L}_n(\theta)$, which is a symmetric positive semidefinite

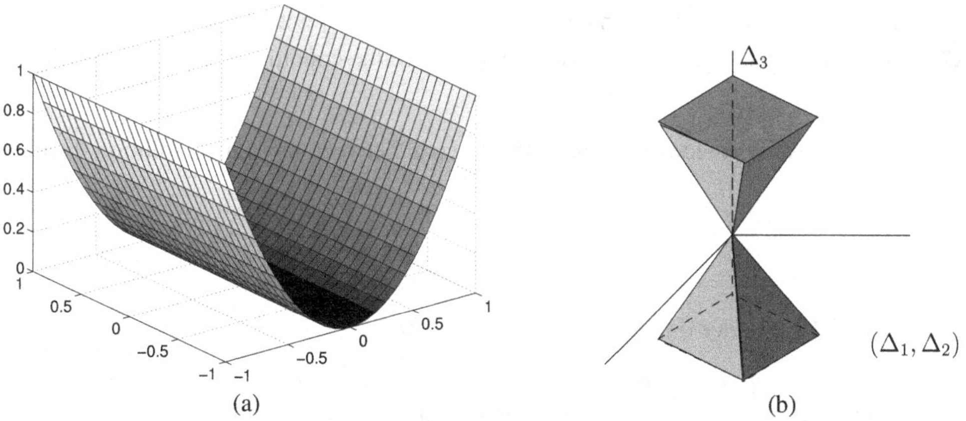

Figure 7.6 (a) A convex cost function in high-dimensional settings (with $d \gg n$) cannot be strongly convex; rather, it will be curved in some directions but flat in others. (b) The Lasso error $\widehat{\Delta}$ must lie in the restricted subset $\mathbb{C}_\alpha(S)$ of $\mathbb{R}^d$. For this reason, it is only necessary that the cost function be curved in certain directions of space.

matrix. In the special case of the quadratic cost function that underlies the Lasso, the Hessian is easily calculated as

$$\nabla^2 \mathcal{L}_n(\theta) = \frac{1}{n}\mathbf{X}^{\mathrm{T}}\mathbf{X}. \tag{7.23}$$

If we could guarantee that the eigenvalues of this matrix were uniformly bounded away from zero, say that

$$\frac{\|\mathbf{X}\Delta\|_2^2}{n} \geq \kappa \|\Delta\|_2^2 > 0 \qquad \text{for all } \Delta \in \mathbb{R}^d \setminus \{0\}, \tag{7.24}$$

then we would be assured of having curvature in all directions.

In the high-dimensional setting with $d > n$, this Hessian is a $d \times d$ matrix with rank at most n, so that it is impossible to guarantee that it has a positive curvature in all directions. Rather, the quadratic cost function always has the form illustrated in Figure 7.6(a): although it may be curved in some directions, there is always a $(d - n)$-dimensional subspace of directions in which it is completely flat! Consequently, the uniform lower bound (7.24) is never satisfied. For this reason, we need to relax the stringency of the uniform curvature condition, and require that it holds only for a subset $\mathbb{C}_\alpha(S)$ of vectors, as illustrated in Figure 7.6(b). If we can be assured that the subset $\mathbb{C}_\alpha(S)$ is well aligned with the curved directions of the Hessian, then a small difference in the cost function will translate into bounds on the difference between $\widehat{\theta}$ and θ^*.

7.3.2 Bounds on ℓ_2-error for hard sparse models

With this intuition in place, we now state a result that provides a bound on the error $\|\widehat{\theta} - \theta^*\|_2$ in the case of a "hard sparse" vector θ^*. In particular, let us impose the following conditions:

(A1) The vector θ^* is supported on a subset $S \subseteq \{1, 2, \ldots, d\}$ with $|S| = s$.

(A2) The design matrix satisfies the restricted eigenvalue condition (7.22) over S with parameters $(\kappa, 3)$.

The following result provides bounds on the ℓ_2-error between any Lasso solution $\widehat{\theta}$ and the true vector θ^*.

Theorem 7.13 *Under assumptions (A1) and (A2):*

(a) *Any solution of the Lagrangian Lasso* (7.18) *with regularization parameter lower bounded as $\lambda_n \geq 2\left\|\frac{\mathbf{X}^{\mathrm{T}}w}{n}\right\|_\infty$ satisfies the bound*

$$\|\widehat{\theta} - \theta^*\|_2 \leq \frac{3}{\kappa}\sqrt{s}\lambda_n. \tag{7.25a}$$

(b) *Any solution of the constrained Lasso* (7.19) *with $R = \|\theta^*\|_1$ satisfies the bound*

$$\|\widehat{\theta} - \theta^*\|_2 \leq \frac{4}{\kappa}\sqrt{s}\left\|\frac{\mathbf{X}^{\mathrm{T}}w}{n}\right\|_\infty. \tag{7.25b}$$

(c) *Any solution of the relaxed basis pursuit program* (7.20) *with $b^2 \geq \frac{\|w\|_2^2}{2n}$ satisfies the bound*

$$\|\widehat{\theta} - \theta^*\|_2 \leq \frac{4}{\kappa}\sqrt{s}\left\|\frac{\mathbf{X}^{\mathrm{T}}w}{n}\right\|_\infty + \frac{2}{\sqrt{\kappa}}\sqrt{b^2 - \frac{\|w\|_2^2}{2n}}. \tag{7.25c}$$

In addition, all three solutions satisfy the ℓ_1-bound $\|\widehat{\theta} - \theta^*\|_1 \leq 4\sqrt{s}\|\widehat{\theta} - \theta^*\|_2$.

In order to develop intuition for these claims, we first discuss them at a high level, and then illustrate them with some concrete examples. First, it is important to note that these results are deterministic, and apply to any set of linear regression equations. As stated, however, the results involve unknown quantities stated in terms of w and/or θ^*. Obtaining results for specific statistical models—as determined by assumptions on the noise vector w and/or the design matrix—involves bounding or approximating these quantities. Based on our earlier discussion of the role of strong convexity, it is natural that all three upper bounds are inversely proportional to the restricted eigenvalue constant $\kappa > 0$. Their scaling with $\sqrt{s}$ is also natural, since we are trying to estimate the unknown regression vector with s unknown entries. The remaining terms in the bound involve the unknown noise vector, either via the quantity $\|\frac{\mathbf{X}^{\mathrm{T}}w}{n}\|_\infty$ in parts (a), (b) and (c), or additionally via $\frac{\|w\|_2^2}{n}$ in part (c).

Let us illustrate some concrete consequences of Theorem 7.13 for some linear regression models that are commonly used and studied.

Example 7.14 (Classical linear Gaussian model) We begin with the classical linear Gaussian model from statistics, for which the noise vector $w \in \mathbb{R}^n$ has i.i.d. $\mathcal{N}(0, \sigma^2)$ entries. Let us consider the case of deterministic design, meaning that the matrix $\mathbf{X} \in \mathbb{R}^{n \times d}$ is fixed.

Suppose that $\mathbf{X}$ satisfies the RE condition (7.22) and that it is C-column normalized, meaning that $\max_{j=1,\ldots,d} \frac{\|X_j\|_2}{\sqrt{n}} \leq C$, where $X_j \in \mathbb{R}^n$ denotes the jth column of $\mathbf{X}$. With this set-up, the random variable $\left\|\frac{\mathbf{X}^{\mathrm{T}}w}{n}\right\|_\infty$ corresponds to the absolute maximum of d zero-mean Gaussian variables, each with variance at most $\frac{C^2\sigma^2}{n}$. Consequently, from standard Gaussian tail bounds (Exercise 2.12), we have

$$\mathbb{P}\left[\left\|\frac{\mathbf{X}^{\mathrm{T}}w}{n}\right\|_\infty \geq C\sigma\left(\sqrt{\frac{2\log d}{n}} + \delta\right)\right] \leq 2e^{-\frac{n\delta^2}{2}} \qquad \text{for all } \delta > 0.$$

Consequently, if we set $\lambda_n = 2C\sigma\left(\sqrt{\frac{2\log d}{n}} + \delta\right)$, then Theorem 7.13(a) implies that any optimal solution of the Lagrangian Lasso (7.18) satisfies the bound

$$\|\widehat{\theta} - \theta^*\|_2 \leq \frac{6C\sigma}{\kappa}\sqrt{s}\left\{\sqrt{\frac{2\log d}{n}} + \delta\right\} \tag{7.26}$$

with probability at least $1 - 2e^{-\frac{n\delta^2}{2}}$. Similarly, Theorem 7.13(b) implies that any optimal solution of the constrained Lasso (7.19) satisfies the bound

$$\|\widehat{\theta} - \theta^*\|_2 \leq \frac{4C\sigma}{\kappa}\sqrt{s}\left\{\sqrt{\frac{2\log d}{n}} + \delta\right\} \tag{7.27}$$

with the same probability. Apart from constant factors, these two bounds are equivalent. Perhaps the most significant difference is that the constrained Lasso (7.19) assumes exact knowledge of the ℓ_1-norm $\|\theta^*\|_1$, whereas the Lagrangian Lasso only requires knowledge of the noise variance σ^2. In practice, it is relatively straightforward to estimate the noise variance, whereas the ℓ_1-norm is a more delicate object.

Turning to Theorem 7.13(c), given the Gaussian noise vector w, the rescaled variable $\frac{\|w\|_2^2}{\sigma^2 n}$ is χ^2 with n degrees of freedom. From Example 2.11, we have

$$\mathbb{P}\left[\left|\frac{\|w\|_2^2}{n} - \sigma^2\right| \geq \sigma^2\delta\right] \leq 2e^{-n\delta^2/8} \qquad \text{for all } \delta \in (0,1).$$

Consequently, Theorem 7.13(c) implies that any optimal solution of the relaxed basis pursuit program (7.20) with $b^2 = \frac{\sigma^2}{2}(1+\delta)$ satisfies the bound

$$\|\widehat{\theta} - \theta^*\|_2 \leq \frac{8C\sigma}{\kappa}\sqrt{s}\left\{\sqrt{\frac{2\log d}{n}} + \delta\right\} + \frac{2\sigma}{\sqrt{\kappa}}\sqrt{\delta} \qquad \text{for all } \delta \in (0,1),$$

with probability at least $1 - 4e^{-\frac{n\delta^2}{8}}$. ♣

Example 7.15 (Compressed sensing) In the domain of compressed sensing, the design matrix $\mathbf{X}$ can be chosen by the user, and one standard choice is the standard Gaussian matrix with i.i.d. $\mathcal{N}(0,1)$ entries. Suppose that the noise vector $w \in \mathbb{R}^n$ is deterministic, say with bounded entries ($\|w\|_\infty \leq \sigma$). Under these assumptions, each variable $X_j^{\mathrm{T}}w/\sqrt{n}$ is a zero-mean Gaussian with variance at most σ^2. Thus, by following the same argument as in the preceding example, we conclude that the Lasso estimates will again satisfy the bounds (7.26) and (7.27), this time with $C = 1$. Similarly, if we set $b^2 = \frac{\sigma^2}{2}$, then the relaxed basis pursuit program (7.19) will satisfy the bound

$$\|\widehat{\theta} - \theta^*\|_2 \leq \frac{8\sigma}{\kappa} \sqrt{s} \left\{ \sqrt{\frac{2 \log d}{n}} + \delta \right\} + \frac{2\sigma}{\sqrt{\kappa}}$$

with probability at least $1 - 2e^{-\frac{n\delta^2}{2}}$. ♣

With these examples in hand, we now turn to the proof of Theorem 7.13.

Proof (b) We begin by proving the error bound (7.25b) for the constrained Lasso (7.19). Given the choice $R = \|\theta^*\|_1$, the target vector θ^* is feasible. Since $\widehat{\theta}$ is optimal, we have the inequality $\frac{1}{2n}\|y - \mathbf{X}\widehat{\theta}\|_2^2 \leq \frac{1}{2n}\|y - \mathbf{X}\theta^*\|_2^2$. Defining the error vector $\widehat{\Delta} := \widehat{\theta} - \theta^*$ and performing some algebra yields the *basic inequality*

$$\frac{\|\mathbf{X}\widehat{\Delta}\|_2^2}{n} \leq \frac{2w^{\mathrm{T}}\mathbf{X}\widehat{\Delta}}{n}. \tag{7.28}$$

Applying Hölder's inequality to the right-hand side yields $\frac{\|\mathbf{X}\widehat{\Delta}\|_2^2}{n} \leq 2\left\|\frac{\mathbf{X}^{\mathrm{T}}w}{n}\right\|_\infty \|\widehat{\Delta}\|_1$. As shown in the proof of Theorem 7.8, whenever $\|\widehat{\theta}\|_1 \leq \|\theta^*\|_1$ for an S-sparse vector, the error $\widehat{\Delta}$ belongs to the cone $\mathbb{C}_1(S)$, whence

$$\|\widehat{\Delta}\|_1 = \|\widehat{\Delta}_S\|_1 + \|\widehat{\Delta}_{S^c}\|_1 \leq 2\|\widehat{\Delta}_S\|_1 \leq 2\sqrt{s}\,\|\widehat{\Delta}\|_2.$$

Since $\mathbb{C}_1(S)$ is a subset of $\mathbb{C}_3(S)$, we may apply the restricted eigenvalue condition (7.22) to the left-hand side of the inequality (7.28), thereby obtaining $\frac{\|\mathbf{X}\widehat{\Delta}\|_2^2}{n} \geq \kappa\|\widehat{\Delta}\|_2^2$. Putting together the pieces yields the claimed bound.

(c) Next we prove the error bound (7.25c) for the relaxed basis pursuit (RBP) program. Note that $\frac{1}{2n}\|y - \mathbf{X}\theta^*\|_2^2 = \frac{\|w\|_2^2}{2n} \leq b^2$, where the inequality follows by our assumed choice of b. Thus, the target vector θ^* is feasible, and since $\widehat{\theta}$ is optimal, we have $\|\widehat{\theta}\|_1 \leq \|\theta^*\|_1$. As previously reasoned, the error vector $\widehat{\Delta} = \widehat{\theta} - \theta^*$ must then belong to the cone $\mathbb{C}_1(S)$. Now by the feasibility of $\widehat{\theta}$, we have

$$\frac{1}{2n}\|y - \mathbf{X}\widehat{\theta}\|_2^2 \leq b^2 = \frac{1}{2n}\|y - \mathbf{X}\theta^*\|_2^2 + \left(b^2 - \frac{\|w\|_2^2}{2n}\right).$$

Rearranging yields the modified basic inequality

$$\frac{\|\mathbf{X}\widehat{\Delta}\|_2^2}{n} \leq 2\frac{w^{\mathrm{T}}\mathbf{X}\widehat{\Delta}}{n} + 2\left(b^2 - \frac{\|w\|_2^2}{2n}\right).$$

Applying the same argument as in part (b)—namely, the RE condition to the left-hand side and the cone inequality to the right-hand side—we obtain

$$\kappa\|\widehat{\Delta}\|_2^2 \leq 4\sqrt{s}\,\|\widehat{\Delta}\|_2 \left\|\frac{\mathbf{X}^{\mathrm{T}}w}{n}\right\|_\infty + 2\left(b^2 - \frac{\|w\|_2^2}{2n}\right),$$

which implies that $\|\widehat{\Delta}\|_2 \leq \frac{8}{\kappa}\sqrt{s}\left\|\frac{\mathbf{X}^{\mathrm{T}}w}{n}\right\|_\infty + \frac{2}{\sqrt{\kappa}}\sqrt{b^2 - \frac{\|w\|_2^2}{2n}}$, as claimed.

(a) Finally, we prove the bound (7.25a) for the Lagrangian Lasso (7.18). Our first step is to show that, under the condition $\lambda_n \geq 2\|\frac{\mathbf{X}^{\mathrm{T}}w}{n}\|_\infty$, the error vector $\widehat{\Delta}$ belongs to $\mathbb{C}_3(S)$. To establish this intermediate claim, let us define the Lagrangian $L(\theta; \lambda_n) = \frac{1}{2n}\|y - \mathbf{X}\theta\|_2^2 + \lambda_n\|\theta\|_1$. Since $\widehat{\theta}$ is optimal, we have

$$L(\widehat{\theta}; \lambda_n) \leq L(\theta^*; \lambda_n) = \frac{1}{2n}\|w\|_2^2 + \lambda_n\|\theta^*\|_1.$$

Rearranging yields the *Lagrangian basic inequality*

$$0 \leq \frac{1}{2n}\|\mathbf{X}\widehat{\Delta}\|_2^2 \leq \frac{w^{\mathrm{T}}\mathbf{X}\widehat{\Delta}}{n} + \lambda_n\{\|\theta^*\|_1 - \|\widehat{\theta}\|_1\}. \tag{7.29}$$

Now since θ^* is S-sparse, we can write

$$\|\theta^*\|_1 - \|\widehat{\theta}\|_1 = \|\theta^*_S\|_1 - \|\theta^*_S + \widehat{\Delta}_S\|_1 - \|\widehat{\Delta}_{S^c}\|_1.$$

Substituting into the basic inequality (7.29) yields

$$\begin{aligned} 0 \leq \frac{1}{n}\|\mathbf{X}\widehat{\Delta}\|_2^2 &\leq 2\frac{w^{\mathrm{T}}\mathbf{X}\widehat{\Delta}}{n} + 2\lambda_n\{\|\theta^*_S\|_1 - \|\theta^*_S + \widehat{\Delta}_S\|_1 - \|\widehat{\Delta}_{S^c}\|_1\} \\ &\overset{\text{(i)}}{\leq} 2\,\|\mathbf{X}^{\mathrm{T}}w/n\|_\infty\,\|\widehat{\Delta}\|_1 + 2\lambda_n\{\|\widehat{\Delta}_S\|_1 - \|\widehat{\Delta}_{S^c}\|_1\} \\ &\overset{\text{(ii)}}{\leq} \lambda_n\{3\,\|\widehat{\Delta}_S\|_1 - \|\widehat{\Delta}_{S^c}\|_1\}, \end{aligned} \tag{7.30}$$

where step (i) follows from a combination of Hölder's inequality and the triangle inequality, whereas step (ii) follows from the choice of λ_n. Inequality (7.30) shows that $\widehat{\Delta} \in \mathbb{C}_3(S)$, so that the RE condition may be applied. Doing so, we obtain $\kappa\|\widehat{\Delta}\|_2^2 \leq 3\lambda_n\sqrt{s}\,\|\widehat{\Delta}\|_2$, which implies the claim (7.25a). □

7.3.3 Restricted nullspace and eigenvalues for random designs

Theorem 7.13 is based on assuming that the design matrix $\mathbf{X}$ satisfies the restricted eigenvalue (RE) condition (7.22). In practice, it is difficult to verify that a given design matrix $\mathbf{X}$ satisfies this condition. Indeed, developing methods to "certify" design matrices in this way is one line of on-going research. However, it is possible to give high-probability results in the case of random design matrices. As discussed previously, pairwise incoherence and RIP conditions are one way in which to certify the restricted nullspace and eigenvalue properties, and are well suited to isotropic designs (in which the population covariance matrix of the rows X_i is the identity). Many other random design matrices encountered in practice do not have such an isotropic structure, so that it is desirable to have alternative direct verifications of the restricted nullspace property.

The following theorem provides a result along these lines. It involves the maximum diagonal entry $\rho^2(\mathbf{\Sigma})$ of a covariance matrix $\mathbf{\Sigma}$.

Theorem 7.16 *Consider a random matrix* $\mathbf{X} \in \mathbb{R}^{n \times d}$*, in which each row* $x_i \in \mathbb{R}^d$ *is drawn i.i.d. from a* $\mathcal{N}(0, \Sigma)$ *distribution. Then there are universal positive constants* $c_1 < 1 < c_2$ *such that*

$$\frac{\|\mathbf{X}\theta\|_2^2}{n} \geq c_1 \|\sqrt{\Sigma}\,\theta\|_2^2 - c_2 \rho^2(\Sigma) \frac{\log d}{n} \|\theta\|_1^2 \qquad \text{for all } \theta \in \mathbb{R}^d \tag{7.31}$$

with probability at least $1 - \frac{e^{-n/32}}{1-e^{-n/32}}$.

Remark: The proof of this result is provided in the Appendix (Section 7.6). It makes use of techniques discussed in other chapters, including the Gordon–Slepian inequalities (Chapters 5 and 6) and concentration of measure for Gaussian functions (Chapter 2). Concretely, we show that the bound (7.31) holds with $c_1 = \frac{1}{8}$ and $c_2 = 50$, but sharper constants can be obtained with a more careful argument. It can be shown (Exercise 7.11) that a lower bound of the form (7.31) implies that an RE condition (and hence a restricted nullspace condition) holds over $\mathbb{C}_3(S)$, uniformly over all subsets of cardinality $|S| \leq \frac{c_1}{32 c_2} \frac{\gamma_{\min}(\Sigma)}{\rho^2(\Sigma)} \frac{n}{\log d}$.

Theorem 7.16 can be used to establish restricted nullspace and eigenvalue conditions for various matrix ensembles that do not satisfy incoherence or RIP conditions. Let us consider a few examples to illustrate.

Example 7.17 (Geometric decay) Consider a covariance matrix with the Toeplitz structure $\mathbf{\Sigma}_{ij} = \nu^{|i-j|}$ for some parameter $\nu \in [0, 1)$. This type of geometrically decaying covariance structure arises naturally from autoregressive processes, where the parameter ν allows for tuning of the memory in the process. By classical results on eigenvalues of Toeplitz matrices, we have $\gamma_{\min}(\mathbf{\Sigma}) \geq (1 - \nu)^2 > 0$ and $\rho^2(\mathbf{\Sigma}) = 1$, independently of the dimension d. Consequently, Theorem 7.16 implies that, with high probability, the sample covariance matrix $\widehat{\mathbf{\Sigma}} = \frac{\mathbf{X}^T\mathbf{X}}{n}$ obtained by sampling from this distribution will satisfy the RE condition for all subsets S of cardinality at most $|S| \leq \frac{c_1}{32c_2}(1 - \nu)^2 \frac{n}{\log d}$. This provides an example of a matrix family with substantial correlation between covariates for which the RE property still holds. ♣

We now consider a matrix family with an even higher amount of dependence among the covariates.

Example 7.18 (Spiked identity model) Recall from our earlier discussion the spiked identity family (7.17) of covariance matrices. This family of covariance matrices is parameterized by a scalar $\mu \in [0, 1)$, and we have $\gamma_{\min}(\mathbf{\Sigma}) = 1 - \mu$ and $\rho^2(\mathbf{\Sigma}) = 1$, again independent of the dimension. Consequently, Theorem 7.16 implies that, with high probability, the sample covariance based on i.i.d. draws from this ensemble satisfies the restricted eigenvalue and restricted nullspace conditions uniformly over all subsets of cardinality at most $|S| \leq \frac{c_1}{32c_2}(1 - \mu)\frac{n}{\log d}$.

However, for any $\mu \neq 0$, the spiked identity matrix is very poorly conditioned, and also has poorly conditioned submatrices. This fact implies that both the pairwise incoherence and restricted isometry property will be violated with high probability, regardless of how large the sample size is taken. To see this, for an arbitrary subset S of size s, consider the associated $s \times s$ submatrix of $\mathbf{\Sigma}$, which we denote by $\mathbf{\Sigma}_{SS}$. The maximal eigenvalue of $\mathbf{\Sigma}_{SS}$

scales as $1 + \mu(s-1)$, which diverges as s increases for any fixed $\mu > 0$. As we explore in Exercise 7.8, this fact implies that both pairwise incoherence and RIP will be violated with high probability. ♣

When a bound of the form (7.31) holds, it is also possible to prove a more general result on the Lasso error, known as an *oracle inequality*. This result holds without any assumptions whatsoever on the underlying regression vector $\theta^* \in \mathbb{R}^d$, and it actually yields a family of upper bounds with a tunable parameter to be optimized. The flexibility in tuning this parameter is akin to that of an oracle, which would have access to the ordered coefficients of θ^*. In order to minimize notational clutter, we introduce the convenient shorthand notation $\bar{\kappa} := \gamma_{\min}(\mathbf{\Sigma})$.

Theorem 7.19 (Lasso oracle inequality) *Under the condition (7.31), consider the Lagrangian Lasso (7.18) with regularization parameter $\lambda_n \geq 2\|\mathbf{X}^{\mathrm{T}} w/n\|_\infty$. For any $\theta^* \in \mathbb{R}^d$, any optimal solution $\widehat{\theta}$ satisfies the bound*

$$\|\widehat{\theta} - \theta^*\|_2^2 \leq \underbrace{\frac{144}{c_1^2} \frac{\lambda_n^2}{\bar{\kappa}^2} |S|}_{\text{estimation error}} + \underbrace{\frac{16}{c_1} \frac{\lambda_n}{\bar{\kappa}} \|\theta^*_{S^c}\|_1 + \frac{32 c_2}{c_1} \frac{\rho^2(\mathbf{\Sigma})}{\bar{\kappa}} \frac{\log d}{n} \|\theta^*_{S^c}\|_1^2}_{\text{approximation error}}, \tag{7.32}$$

valid for any subset S with cardinality $|S| \leq \frac{c_1}{64 c_2} \frac{\bar{\kappa}}{\rho^2(\mathbf{\Sigma})} \frac{n}{\log d}$.

Note that inequality (7.32) actually provides a family of upper bounds, one for each valid choice of the subset S. The optimal choice of S is based on trading off the two sources of error. The first term grows linearly with the cardinality $|S|$, and corresponds to the error associated with estimating a total of $|S|$ unknown coefficients. The second term corresponds to approximation error, and depends on the unknown regression vector via the tail sum $\|\theta^*_{S^c}\|_1 = \sum_{j \notin S} |\theta^*_j|$. An optimal bound is obtained by choosing S to balance these two terms. We illustrate an application of this type of trade-off in Exercise 7.12.

Proof Throughout the proof, we use ρ^2 as a shorthand for $\rho^2(\mathbf{\Sigma})$. Recall the argument leading to the bound (7.30). For a general vector $\theta^* \in \mathbb{R}^d$, the same argument applies with any subset S except that additional terms involving $\|\theta^*_{S^c}\|_1$ must be tracked. Doing so yields that

$$0 \leq \frac{1}{2n} \|\mathbf{X}\hat{\Delta}\|_2^2 \leq \frac{\lambda_n}{2} \{3\|\hat{\Delta}_S\|_1 - \|\hat{\Delta}_{S^c}\|_1 + 2\|\theta^*_{S^c}\|_1\}. \tag{7.33}$$

This inequality implies that the error vector $\hat{\Delta}$ satisfies the constraint

$$\|\hat{\Delta}\|_1^2 \leq (4\|\hat{\Delta}_S\|_1 + 2\|\theta^*_{S^c}\|_1)^2 \leq 32\,|S|\,\|\hat{\Delta}\|_2^2 + 8\|\theta^*_{S^c}\|_1^2. \tag{7.34}$$

Combined with the bound (7.31), we find that

$$\frac{\|\mathbf{X}\hat{\Delta}\|_2^2}{n} \geq \left\{c_1\bar{\kappa} - 32c_2\rho^2|S|\frac{\log d}{n}\right\}\|\hat{\Delta}\|_2^2 - 8c_2\rho^2\frac{\log d}{n}\|\theta^*_{S^c}\|_1^2$$
$$\geq c_1\frac{\bar{\kappa}}{2}\|\hat{\Delta}\|_2^2 - 8c_2\rho^2\frac{\log d}{n}\|\theta^*_{S^c}\|_1^2, \tag{7.35}$$

where the final inequality uses the condition $32c_2\rho^2|S|\frac{\log d}{n} \leq c_1\frac{\bar{\kappa}}{2}$. We split the remainder of the analysis into two cases.

Case 1: First suppose that $c_1\frac{\bar{\kappa}}{4}\|\hat{\Delta}\|_2^2 \geq 8c_2\rho^2\frac{\log d}{n}\|\theta^*_{S^c}\|_1^2$. Combining the bounds (7.35) and (7.33) yields

$$c_1\frac{\bar{\kappa}}{4}\|\widehat{\Delta}\|_2^2 \leq \frac{\lambda_n}{2}\{3\sqrt{|S|}\,\|\widehat{\Delta}\|_2 + 2\|\theta^*_{S^c}\|_1\}. \tag{7.36}$$

This bound involves a quadratic form in $\|\widehat{\Delta}\|_2$; computing the zeros of this quadratic form, we find that

$$\|\widehat{\Delta}\|_2^2 \leq \frac{144\lambda_n^2}{c_1^2\bar{\kappa}^2}|S| + \frac{16\lambda_n\|\theta^*_{S^c}\|_1}{c_1\bar{\kappa}}.$$

Case 2: Otherwise, we must have $c_1\frac{\bar{\kappa}}{4}\|\hat{\Delta}\|_2^2 < 8c_2\rho^2\frac{\log d}{n}\|\theta^*_{S^c}\|_1^2$.

Taking into account both cases, we combine this bound with the earlier inequality (7.36), thereby obtaining the claim (7.32). □

7.4 Bounds on prediction error

In the previous analysis, we have focused exclusively on the problem of parameter recovery, either in noiseless or noisy settings. In other applications, the actual value of the regression vector θ^* may not be of primary interest; rather, we might be interested in finding a good predictor, meaning a vector $\widehat{\theta} \in \mathbb{R}^d$ such that the *mean-squared prediction error*

$$\frac{\|\mathbf{X}(\widehat{\theta} - \theta^*)\|_2^2}{n} = \frac{1}{n}\sum_{i=1}^{n}(\langle x_i, \widehat{\theta} - \theta^*\rangle)^2 \tag{7.37}$$

is small. To understand why the quantity (7.37) is a measure of prediction error, suppose that we estimate $\widehat{\theta}$ on the basis of the response vector $y = \mathbf{X}\theta^* + w$. Suppose that we then receive a "fresh" vector of responses, say $\widetilde{y} = \mathbf{X}\theta^* + \widetilde{w}$, where $\widetilde{w} \in \mathbb{R}^n$ is a noise vector, with i.i.d. zero-mean entries with variance σ^2. We can then measure the quality of our vector $\widehat{\theta}$ by how well it predicts the vector $\widetilde{y}$ in terms of squared error, taking averages over instantiations of the noise vector $\widetilde{w}$. Following some algebra, we find that

$$\frac{1}{n}\mathbb{E}[\|\widetilde{y} - \mathbf{X}\widehat{\theta}\|_2^2] = \frac{1}{n}\|\mathbf{X}(\widehat{\theta} - \theta^*)\|_2^2 + \sigma^2,$$

so that apart from the constant additive factor of σ^2, the quantity (7.37) measures how well we can predict a new vector of responses, with the design matrix held fixed.

It is important to note that, at least in general, the problem of finding a good predictor should be easier than estimating θ^* well in ℓ_2-norm. Indeed, the prediction problem does not require that θ^* even be identifiable: unlike in parameter recovery, the problem can still be solved if two columns of the design matrix $\mathbf{X}$ are identical.

Theorem 7.20 (Prediction error bounds) *Consider the Lagrangian Lasso (7.18) with a strictly positive regularization parameter* $\lambda_n \geq 2\|\frac{\mathbf{X}^T w}{n}\|_\infty$.

(a) *Any optimal solution* $\widehat{\theta}$ *satisfies the bound*

$$\frac{\|\mathbf{X}(\widehat{\theta} - \theta^*)\|_2^2}{n} \leq 12\|\theta^*\|_1 \lambda_n. \tag{7.38}$$

(b) *If* θ^* *is supported on a subset* S *of cardinality* s, *and the design matrix satisfies the* $(\kappa; 3)$*-RE condition over* S, *then any optimal solution satisfies the bound*

$$\frac{\|\mathbf{X}(\widehat{\theta} - \theta^*)\|_2^2}{n} \leq \frac{9}{\kappa} s \lambda_n^2. \tag{7.39}$$

Remarks: As previously discussed in Example 7.14, when the noise vector w has i.i.d. zero-mean σ-sub-Gaussian entries and the design matrix is C-column normalized, the choice $\lambda_n = 2C\sigma(\sqrt{\frac{2\log d}{n}} + \delta)$ is valid with probability at least $1 - 2e^{-\frac{n\delta^2}{2}}$. In this case, Theorem 7.20(a) implies the upper bound

$$\frac{\|\mathbf{X}(\widehat{\theta} - \theta^*)\|_2^2}{n} \leq 24\,\|\theta^*\|_1 C\sigma\left(\sqrt{\frac{2\log d}{n}} + \delta\right) \tag{7.40}$$

with the same high probability. For this bound, the requirements on the design matrix are extremely mild—only the column normalization condition $\max_{j=1,\ldots,d} \frac{\|X_j\|_2}{\sqrt{n}} \leq C$. Thus, the matrix $\mathbf{X}$ could have many identical columns, and this would have no effect on the prediction error. In fact, when the only constraint on θ^* is the ℓ_1-norm bound $\|\theta^*\|_1 \leq R$, then the bound (7.40) is unimprovable—see the bibliographic section for further discussion.

On the other hand, when θ^* is s-sparse and in addition, the design matrix satisfies an RE condition, then Theorem 7.20(b) guarantees the bound

$$\frac{\|\mathbf{X}(\widehat{\theta} - \theta^*)\|_2^2}{n} \leq \frac{72}{\kappa} C^2\sigma^2\left(\frac{2s\log d}{n} + s\delta^2\right) \tag{7.41}$$

with the same high probability. This error bound can be significantly smaller than the $\sqrt{\frac{\log d}{n}}$ error bound (7.40) guaranteed under weaker assumptions. For this reason, the bounds (7.38) and (7.39) are often referred to as the *slow rates* and *fast rates*, respectively, for prediction error. It is natural to question whether or not the RE condition is needed for achieving the fast rate (7.39); see the bibliography section for discussion of some subtleties surrounding this issue.

Proof Throughout the proof, we adopt the usual notation $\widehat{\Delta} = \widehat{\theta} - \theta^*$ for the error vector.

(a) We first show that $\|\widehat{\Delta}\|_1 \leq 4\|\theta^*\|_1$ under the stated conditions. From the Lagrangian basic inequality (7.29), we have

$$0 \leq \frac{1}{2n}\|\mathbf{X}\widehat{\Delta}\|_2^2 \leq \frac{w^{\mathrm{T}}\mathbf{X}\widehat{\Delta}}{n} + \lambda_n\{\|\theta^*\|_1 - \|\widehat{\theta}\|_1\}. \tag{7.42}$$

By Hölder's inequality and our choice of λ_n, we have

$$\left|\frac{w^{\mathrm{T}}\mathbf{X}\widehat{\Delta}}{n}\right| \leq \left\|\frac{\mathbf{X}^{\mathrm{T}}w}{n}\right\|_\infty \|\widehat{\Delta}\|_1 \leq \frac{\lambda_n}{2}\{\|\theta^*\|_1 + \|\widehat{\theta}\|_1\},$$

where the final step also uses the triangle inequality. Putting together the pieces yields

$$0 \leq \frac{\lambda_n}{2}\{\|\theta^*\|_1 + \|\widehat{\theta}\|_1\} + \lambda_n\{\|\theta^*\|_1 - \|\widehat{\theta}\|_1\},$$

which (for $\lambda_n > 0$) implies that $\|\widehat{\theta}\|_1 \leq 3\|\theta^*\|_1$. Consequently, a final application of the triangle inequality yields $\|\widehat{\Delta}\|_1 \leq \|\theta^*\|_1 + \|\widehat{\theta}\|_1 \leq 4\|\theta^*\|_1$, as claimed.

We can now complete the proof. Returning to our earlier inequality (7.42), we have

$$\frac{\|\mathbf{X}\widehat{\Delta}\|_2^2}{2n} \leq \frac{\lambda_n}{2}\|\widehat{\Delta}\|_1 + \lambda_n\{\|\theta^*\|_1 - \|\theta^* + \widehat{\Delta}\|_1\} \overset{\text{(i)}}{\leq} \frac{3\lambda_n}{2}\|\widehat{\Delta}\|_1,$$

where step (i) is based on the triangle inequality bound $\|\theta^* + \widehat{\Delta}\|_1 \geq \|\theta^*\|_1 - \|\widehat{\Delta}\|_1$. Combined with the upper bound $\|\widehat{\Delta}\|_1 \leq 4\|\theta^*\|_1$, the proof is complete.

(b) In this case, the same argument as in the proof of Theorem 7.13(a) leads to the basic inequality

$$\frac{\|\mathbf{X}\widehat{\Delta}\|_2^2}{n} \leq 3\lambda_n\|\widehat{\Delta}_S\|_1 \leq 3\lambda_n\sqrt{s}\|\widehat{\Delta}\|_2.$$

Similarly, the proof of Theorem 7.13(a) shows that the error vector $\widehat{\Delta}$ belongs to $\mathbb{C}_3(S)$, whence the $(\kappa; 3)$-RE condition can be applied, this time to the right-hand side of the basic inequality. Doing so yields $\|\widehat{\Delta}\|_2^2 \leq \frac{1}{\kappa}\frac{\|\mathbf{X}\widehat{\Delta}\|_2^2}{n}$, and hence that $\frac{\|\mathbf{X}\widehat{\Delta}\|_2}{\sqrt{n}} \leq \frac{3}{\sqrt{\kappa}}\sqrt{s}\lambda_n$, as claimed. □

7.5 Variable or subset selection

Thus far, we have focused on results that guarantee that either the ℓ_2-error or the prediction error of the Lasso is small. In other settings, we are interested in a somewhat more refined question, namely whether or not a Lasso estimate $\widehat{\theta}$ has non-zero entries in the same positions as the true regression vector θ^*. More precisely, suppose that the true regression vector θ^* is s-sparse, meaning that it is supported on a subset $S(\theta^*)$ of cardinality $s = |S(\theta^*)|$. In such a setting, a natural goal is to correctly identify the subset $S(\theta^*)$ of relevant variables. In terms of the Lasso, we ask the following question: given an optimal Lasso solution $\widehat{\theta}$, when is its support set—denoted by $S(\widehat{\theta})$—exactly equal to the true support $S(\theta^*)$? We refer to this property as *variable selection consistency*.

Note that it is possible for the ℓ_2-error $\|\widehat{\theta} - \theta^*\|_2$ to be quite small even if $\widehat{\theta}$ and θ^* have different supports, as long as $\widehat{\theta}$ is non-zero for all "suitably large" entries of θ^*, and not too large in positions where θ^* is zero. On the other hand, as we discuss in the sequel, given an estimate $\widehat{\theta}$ that correctly recovers the support of θ^*, we can estimate θ^* very well (in ℓ_2-norm, or other metrics) simply by performing an ordinary least-squares regression restricted to this subset.

7.5.1 Variable selection consistency for the Lasso

We begin by addressing the issue of variable selection in the context of deterministic design matrices $\mathbf{X}$. (Such a result can be extended to random design matrices, albeit with additional effort.) It turns out that variable selection requires some assumptions that are related to but distinct from the restricted eigenvalue condition (7.22). In particular, consider the following conditions:

(A3) *Lower eigenvalue:* The smallest eigenvalue of the sample covariance submatrix indexed by S is bounded below:

$$\gamma_{\min}\left(\frac{\mathbf{X}_S^{\mathrm{T}}\mathbf{X}_S}{n}\right) \geq c_{\min} > 0. \tag{7.43a}$$

(A4) *Mutual incoherence:* There exists some $\alpha \in [0, 1)$ such that

$$\max_{j \in S^c} \|(\mathbf{X}_S^{\mathrm{T}}\mathbf{X}_S)^{-1}\mathbf{X}_S^{\mathrm{T}}X_j\|_1 \leq \alpha. \tag{7.43b}$$

To provide some intuition, the first condition (A3) is very mild: in fact, it would be required in order to ensure that the model is identifiable, *even if* the support set S were known *a priori*. In particular, the submatrix $\mathbf{X}_S \in \mathbb{R}^{n \times s}$ corresponds to the subset of covariates that are in the support set, so that if assumption (A3) were violated, then the submatrix $\mathbf{X}_S$ would have a non-trivial nullspace, leading to a non-identifiable model. Assumption (A4) is a more subtle condition. In order to gain intuition, suppose that we tried to predict the column vector X_j using a linear combination of the columns of $\mathbf{X}_S$. The best weight vector $\widehat{\omega} \in \mathbb{R}^{|S|}$ is given by

$$\widehat{\omega} = \arg\min_{\omega \in \mathbb{R}^{|S|}} \|X_j - \mathbf{X}_S\omega\|_2^2 = (\mathbf{X}_S^{\mathrm{T}}\mathbf{X}_S)^{-1}\mathbf{X}_S^{\mathrm{T}}X_j,$$

and the mutual incoherence condition is a bound on $\|\widehat{\omega}\|_1$. In the ideal case, if the column space of $\mathbf{X}_S$ were orthogonal to X_j, then the optimal weight vector $\widehat{\omega}$ would be identically zero. In general, we cannot expect this orthogonality to hold, but the mutual incoherence condition (A4) imposes a type of approximate orthogonality.

With this set-up, the following result applies to the Lagrangian Lasso (7.18) when applied to an instance of the linear observation model such that the true parameter θ^* is supported on a subset S with cardinality s. In order to state the result, we introduce the convenient shorthand $\Pi_{S^\perp}(\mathbf{X}) = \mathbf{I}_n - \mathbf{X}_S(\mathbf{X}_S^{\mathrm{T}}\mathbf{X}_S)^{-1}\mathbf{X}_S^{\mathrm{T}}$, a type of orthogonal projection matrix.

Theorem 7.21 *Consider an S-sparse linear regression model for which the design matrix satisfies conditions (A3) and (A4). Then for any choice of regularization parameter such that*

$$\lambda_n \geq \frac{2}{1-\alpha} \left\| \mathbf{X}_{S^c}^{\mathrm{T}} \Pi_{S^\perp}(\mathbf{X}) \frac{w}{n} \right\|_\infty, \tag{7.44}$$

the Lagrangian Lasso (7.18) *has the following properties:*

(a) *Uniqueness: There is a unique optimal solution $\widehat{\theta}$.*
(b) *No false inclusion: This solution has its support set $\widehat{S}$ contained within the true support set S.*
(c) *ℓ_∞-bounds: The error $\widehat{\theta} - \theta^*$ satisfies*

$$\|\widehat{\theta}_S - \theta^*_S\|_\infty \leq \underbrace{\left\| \left(\frac{\mathbf{X}_S^{\mathrm{T}} \mathbf{X}_S}{n} \right)^{-1} \mathbf{X}_S^{\mathrm{T}} \frac{w}{n} \right\|_\infty + \left|\left|\left| \left(\frac{\mathbf{X}_S^{\mathrm{T}} \mathbf{X}_S}{n} \right)^{-1} \right|\right|\right|_\infty \lambda_n}_{B(\lambda_n;\mathbf{X})}, \tag{7.45}$$

where $|||A|||_\infty = \max_{i=1,\ldots,s} \sum_j |A_{ij}|$ is the matrix ℓ_∞-norm.
(d) *No false exclusion: The Lasso includes all indices $i \in S$ such that $|\theta^*_i| > B(\lambda_n; \mathbf{X})$, and hence is variable selection consistent if $\min_{i \in S} |\theta^*_i| > B(\lambda_n; \mathbf{X})$.*

Before proving this result, let us try to interpret its main claims. First, the uniqueness claim in part (a) is not trivial in the high-dimensional setting, because, as discussed previously, although the Lasso objective is convex, it can never be strictly convex when $d > n$. Based on the uniqueness claim, we can talk unambiguously about the support of the Lasso estimate $\widehat{\theta}$. Part (b) guarantees that the Lasso does not falsely include variables that are not in the support of θ^*, or equivalently that $\widehat{\theta}_{S^c} = 0$, whereas part (d) is a consequence of the sup-norm bound from part (c): as long as the minimum value of $|\theta^*_i|$ over indices $i \in S$ is not too small, then the Lasso is variable selection consistent in the full sense.

As with our earlier result (Theorem 7.13) on ℓ_2-error bounds, Theorem 7.21 is a deterministic result that applies to any set of linear regression equations. It implies more concrete results when we make specific assumptions about the noise vector w, as we show here.

Corollary 7.22 *Consider the S-sparse linear model based on a noise vector w with zero-mean i.i.d. σ-sub-Gaussian entries, and a deterministic design matrix $\mathbf{X}$ that satisfies assumptions (A3) and (A4), as well as the C-column normalization condition ($\max_{j=1,\ldots,d} \|X_j\|_2/\sqrt{n} \leq C$). Suppose that we solve the Lagrangian Lasso* (7.18) *with regularization parameter*

$$\lambda_n = \frac{2C\sigma}{1-\alpha} \left\{ \sqrt{\frac{2\log(d-s)}{n}} + \delta \right\} \tag{7.46}$$

for some $\delta > 0$. Then the optimal solution $\widehat{\theta}$ is unique with its support contained within S, and satisfies the ℓ_∞-error bound

$$\|\widehat{\theta}_S - \theta^*_S\|_\infty \le \frac{\sigma}{\sqrt{c_{\min}}}\left\{\sqrt{\frac{2\log s}{n}} + \delta\right\} + \left|\!\left|\!\left|\left(\frac{\mathbf{X}_S^{\mathrm{T}}\mathbf{X}_S}{n}\right)^{-1}\right|\!\right|\!\right|_\infty \lambda_n, \tag{7.47}$$

all with probability at least $1 - 4e^{-\frac{n\delta^2}{2}}$.

Proof We first verify that the given choice (7.46) of regularization parameter satisfies the bound (7.44) with high probability. It suffices to bound the maximum absolute value of the random variables

$$Z_j := X_j^{\mathrm{T}} \underbrace{[\mathbf{I}_n - \mathbf{X}_S(\mathbf{X}_S^{\mathrm{T}}\mathbf{X}_S)^{-1}\mathbf{X}_S^{\mathrm{T}}]}_{\Pi_{S^\perp}(\mathbf{X})}\left(\frac{w}{n}\right) \qquad \text{for } j \in S^c.$$

Since $\Pi_{S^\perp}(\mathbf{X})$ is an orthogonal projection matrix, we have

$$\|\Pi_{S^\perp}(\mathbf{X})X_j\|_2 \le \|X_j\|_2 \overset{\text{(i)}}{\le} C\sqrt{n},$$

where inequality (i) follows from the column normalization assumption. Therefore, each variable Z_j is sub-Gaussian with parameter at most $C^2\sigma^2/n$. From standard sub-Gaussian tail bounds (Chapter 2), we have

$$\mathbb{P}\left[\max_{j\in S^c} |Z_j| \ge t\right] \le 2(d-s)e^{-\frac{nt^2}{2C^2\sigma^2}},$$

from which we see that our choice (7.46) of λ_n ensures that the bound (7.44) holds with the claimed probability.

The only remaining step is to simplify the ℓ_∞-bound (7.45). The second term in this bound is a deterministic quantity, so we focus on bounding the first term. For each $i = 1, \ldots, s$, consider the random variable $\widetilde{Z}_i := e_i^{\mathrm{T}}(\frac{1}{n}\mathbf{X}_S^{\mathrm{T}}\mathbf{X}_S)^{-1}\mathbf{X}_S^{\mathrm{T}}w/n$. Since the elements of the vector w are i.i.d. σ-sub-Gaussian, the variable $\widetilde{Z}_i$ is zero-mean and sub-Gaussian with parameter at most

$$\frac{\sigma^2}{n}\left|\!\left|\!\left|\left(\frac{1}{n}\mathbf{X}_S^{\mathrm{T}}\mathbf{X}_S\right)^{-1}\right|\!\right|\!\right|_2 \le \frac{\sigma^2}{c_{\min} n},$$

where we have used the eigenvalue condition (7.43a). Consequently, for any $\delta > 0$, we have $\mathbb{P}\left[\max_{i=1,\ldots,s} |\widetilde{Z}_i| > \frac{\sigma}{\sqrt{c_{\min}}}\left\{\sqrt{\frac{2\log s}{n}} + \delta\right\}\right] \le 2e^{-\frac{n\delta^2}{2}}$, from which the claim follows. □

Corollary 7.22 applies to linear models with a fixed matrix $\mathbf{X}$ of covariates. An analogous result—albeit with a more involved proof—can be proved for Gaussian random covariate matrices. Doing so involves showing that a random matrix $\mathbf{X}$ drawn from the $\mathbf{\Sigma}$-Gaussian ensemble, with rows sampled i.i.d. from a $\mathcal{N}(0, \mathbf{\Sigma})$ distribution, satisfies the α-incoherence condition with high probability (whenever the population matrix $\mathbf{\Sigma}$ satisfies this condition, and the sample size n is sufficiently large). We work through a version of this result in Exercise 7.19, showing that the incoherence condition holds with high probability with $n \gtrsim s\log(d-s)$ samples. Figure 7.7 shows that this theoretical prediction is actually sharp, in that the Lasso undergoes a phase transition as a function of the control parameter $\frac{n}{s\log(d-s)}$. See the bibliographic section for further discussion of this phenomenon.

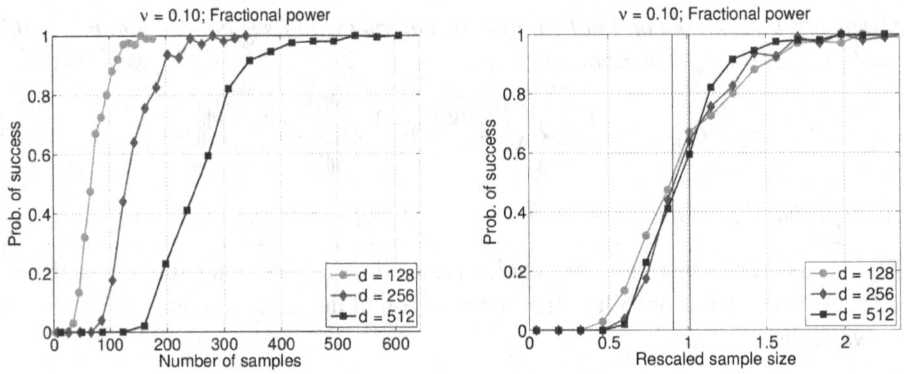

(a) (b)

Figure 7.7 Thresholds for correct variable selection using the Lasso. (a) Probability of correct variable selection $\mathbb{P}[\widehat{S} = S]$ versus the raw sample size n for three different problem sizes $d \in \{128, 256, 512\}$ and square-root sparsity $s = \lceil \sqrt{d} \rceil$. Each point corresponds to the average of 20 random trials, using a random covariate matrix drawn from the Toeplitz ensemble of Example 7.17 with $\nu = 0.1$. Note that larger problems require more samples before the Lasso is able to recover the correct support. (b) The same simulation results replotted versus the rescaled sample size $\frac{n}{s \log(d-s)}$. Notice how all three curves are now well aligned, and show a threshold behavior, consistent with theoretical predictions.

7.5.2 *Proof of Theorem 7.21*

We begin by developing the necessary and sufficient conditions for optimality in the Lasso. A minor complication arises because the ℓ_1-norm is not differentiable, due to its sharp point at the origin. Instead, we need to work in terms of the subdifferential of the ℓ_1-norm. Given a convex function $f\colon \mathbb{R}^d \to \mathbb{R}$, we say that $z \in \mathbb{R}^d$ is a subgradient of f at θ, denoted by $z \in \partial f(\theta)$, if we have

$$f(\theta + \Delta) \geq f(\theta) + \langle z, \Delta \rangle \qquad \text{for all } \Delta \in \mathbb{R}^d.$$

When $f(\theta) = \|\theta\|_1$, it can be seen that $z \in \partial \|\theta\|_1$ if and only if $z_j = \text{sign}(\theta_j)$ for all $j = 1, 2, \ldots, d$. Here we allow $\text{sign}(0)$ to be any number in the interval $[-1, 1]$. In application to the Lagrangian Lasso program (7.18), we say that a pair $(\widehat{\theta}, \widehat{z}) \in \mathbb{R}^d \times \mathbb{R}^d$ is *primal–dual optimal* if $\widehat{\theta}$ is a minimizer and $\widehat{z} \in \partial \|\widehat{\theta}\|_1$. Any such pair must satisfy the zero-subgradient condition

$$\frac{1}{n}\mathbf{X}^{\mathrm{T}}(\mathbf{X}\widehat{\theta} - y) + \lambda_n \widehat{z} = 0, \tag{7.48}$$

which is the analog of a zero-gradient condition in the non-differentiable setting.

Our proof of Theorem 7.21 is based on a constructive procedure, known as a *primal–dual witness method*, which constructs a pair $(\widehat{\theta}, \widehat{z})$ satisfying the zero-subgradient condition (7.48), and such that $\widehat{\theta}$ has the correct signed support. When this procedure succeeds, the constructed pair is primal–dual optimal, and acts as a witness for the fact that the Lasso has a unique optimal solution with the correct signed support. In more detail, the procedure consists of the following steps:

Primal–dual witness (PDW) construction:

1 Set $\widehat{\theta}_{S^c} = 0$.

2 Determine $(\widehat{\theta}_S, \widehat{z}_S) \in \mathbb{R}^s \times \mathbb{R}^s$ by solving the *oracle subproblem*

$$\widehat{\theta}_S \in \arg\min_{\theta_S \in \mathbb{R}^s} \Big\{ \underbrace{\frac{1}{2n}\|y - \mathbf{X}_S\theta_S\|_2^2 + \lambda_n\|\theta_S\|_1}_{=: f(\theta_S)} \Big\}, \tag{7.49}$$

and then choosing $\widehat{z}_S \in \partial\|\widehat{\theta}_S\|_1$ such that $\nabla f(\theta_S)\big|_{\theta_S = \widehat{\theta}_S} + \lambda_n \widehat{z}_S = 0$.

3 Solve for $\widehat{z}_{S^c} \in \mathbb{R}^{d-s}$ via the zero-subgradient equation (7.48), and check whether or not the *strict dual feasibility* condition $\|\widehat{z}_{S^c}\|_\infty < 1$ holds.

Note that the vector $\widehat{\theta}_{S^c} \in \mathbb{R}^{d-s}$ is determined in step 1, whereas the remaining three subvectors are determined in steps 2 and 3. By construction, the subvectors $\widehat{\theta}_S$, $\widehat{z}_S$ and $\widehat{z}_{S^c}$ satisfy the zero-subgradient condition (7.48). By using the fact that $\widehat{\theta}_{S^c} = \theta^*_{S^c} = 0$ and writing out this condition in block matrix form, we obtain

$$\frac{1}{n}\begin{bmatrix} \mathbf{X}_S^{\mathrm{T}}\mathbf{X}_S & \mathbf{X}_S^{\mathrm{T}}\mathbf{X}_{S^c} \\ \mathbf{X}_{S^c}^{\mathrm{T}}\mathbf{X}_S & \mathbf{X}_{S^c}^{\mathrm{T}}\mathbf{X}_{S^c} \end{bmatrix} \begin{bmatrix} \widehat{\theta}_S - \theta^*_S \\ 0 \end{bmatrix} - \frac{1}{n}\begin{bmatrix} \mathbf{X}_S^{\mathrm{T}} w \\ \mathbf{X}_{S^c}^{\mathrm{T}} w \end{bmatrix} + \lambda_n \begin{bmatrix} \widehat{z}_S \\ \widehat{z}_{S^c} \end{bmatrix} = \begin{bmatrix} 0 \\ 0 \end{bmatrix}. \tag{7.50}$$

We say that the PDW construction succeeds if the vector $\widehat{z}_{S^c}$ constructed in step 3 satisfies the strict dual feasibility condition. The following result shows that this success acts as a witness for the Lasso:

Lemma 7.23 *If the lower eigenvalue condition (A3) holds, then success of the PDW construction implies that the vector $(\widehat{\theta}_S, 0) \in \mathbb{R}^d$ is the unique optimal solution of the Lasso.*

Proof When the PDW construction succeeds, then $\widehat{\theta} = (\widehat{\theta}_S, 0)$ is an optimal solution with associated subgradient vector $\widehat{z} \in \mathbb{R}^d$ satisfying $\|\widehat{z}_{S^c}\|_\infty < 1$, and $\langle \widehat{z}, \widehat{\theta}\rangle = \|\widehat{\theta}\|_1$. Now let $\widetilde{\theta}$ be any other optimal solution. If we introduce the shorthand notation $F(\theta) = \frac{1}{2n}\|y - \mathbf{X}\theta\|_2^2$, then we are guaranteed that $F(\widehat{\theta}) + \lambda_n\langle \widehat{z}, \widehat{\theta}\rangle = F(\widetilde{\theta}) + \lambda_n\|\widetilde{\theta}\|_1$, and hence

$$F(\widehat{\theta}) - \lambda_n\langle \widehat{z}, \widetilde{\theta} - \widehat{\theta}\rangle = F(\widetilde{\theta}) + \lambda_n(\|\widetilde{\theta}\|_1 - \langle \widehat{z}, \widetilde{\theta}\rangle).$$

But by the zero-subgradient conditions (7.48), we have $\lambda_n \widehat{z} = -\nabla F(\widehat{\theta})$, which implies that

$$F(\widehat{\theta}) + \langle \nabla F(\widehat{\theta}), \widetilde{\theta} - \widehat{\theta}\rangle - F(\widetilde{\theta}) = \lambda_n(\|\widetilde{\theta}\|_1 - \langle \widehat{z}, \widetilde{\theta}\rangle).$$

By convexity of F, the left-hand side is negative, which implies that $\|\widetilde{\theta}\|_1 \leq \langle \widehat{z}, \widetilde{\theta}\rangle$. But since we also have $\langle \widehat{z}, \widetilde{\theta}\rangle \leq \|\widehat{z}\|_\infty\|\widetilde{\theta}\|_1$, we must have $\|\widetilde{\theta}\|_1 = \langle \widehat{z}, \widetilde{\theta}\rangle$. Since $\|\widehat{z}_{S^c}\|_\infty < 1$, this equality can only occur if $\widetilde{\theta}_j = 0$ for all $j \in S^c$.

Thus, all optimal solutions are supported only on S, and hence can be obtained by solving the oracle subproblem (7.49). Given the lower eigenvalue condition (A3), this subproblem is strictly convex, and so has a unique minimizer. □

Thus, in order to prove Theorem 7.21(a) and (b), it suffices to show that the vector $\widehat{z}_{S^c} \in \mathbb{R}^{d-s}$ constructed in step 3 satisfies the strict dual feasibility condition. Using the zero-subgradient conditions (7.50), we can solve for the vector $\widehat{z}_{S^c} \in \mathbb{R}^{d-s}$, thereby finding that

$$\widehat{z}_{S^c} = -\frac{1}{\lambda_n n} \mathbf{X}_{S^c}^{\mathrm{T}} \mathbf{X}_S (\widehat{\theta}_S - \theta_S^*) + \mathbf{X}_{S^c}^{\mathrm{T}} \left(\frac{w}{\lambda_n n} \right). \tag{7.51}$$

Similarly, using the assumed invertibility of $\mathbf{X}_S^{\mathrm{T}} \mathbf{X}_S$ in order to solve for the difference $\widehat{\theta}_S - \theta_S^*$ yields

$$\widehat{\theta}_S - \theta_S^* = (\mathbf{X}_S^{\mathrm{T}} \mathbf{X}_S)^{-1} \mathbf{X}_S^{\mathrm{T}} w - \lambda_n n (\mathbf{X}_S^{\mathrm{T}} \mathbf{X}_S)^{-1} \widehat{z}_S. \tag{7.52}$$

Substituting this expression back into equation (7.51) and simplifying yields

$$\widehat{z}_{S^c} = \underbrace{\mathbf{X}_{S^c}^{\mathrm{T}} \mathbf{X}_S (\mathbf{X}_S^{\mathrm{T}} \mathbf{X}_S)^{-1} \widehat{z}_S}_{\mu} + \underbrace{\mathbf{X}_{S^c}^{\mathrm{T}} [I - \mathbf{X}_S (\mathbf{X}_S^{\mathrm{T}} \mathbf{X}_S)^{-1} \mathbf{X}_S^{\mathrm{T}}] \left(\frac{w}{\lambda_n n} \right)}_{V_{S^c}}. \tag{7.53}$$

By the triangle inequality, we have $\|\widehat{z}_{S^c}\|_\infty \leq \|\mu\|_\infty + \|V_{S^c}\|_\infty$. By the mutual incoherence condition (7.43b), we have $\|\mu\|_\infty \leq \alpha$. By our choice (7.44) of regularization parameter, we have $\|V_{S^c}\|_\infty \leq \frac{1}{2}(1-\alpha)$. Putting together the pieces, we conclude that $\|\widehat{z}_{S^c}\|_\infty \leq \frac{1}{2}(1+\alpha) < 1$, which establishes the strict dual feasibility condition.

It remains to establish a bound on the ℓ_∞-norm of the error $\widehat{\theta}_S - \theta_S^*$. From equation (7.52) and the triangle inequality, we have

$$\|\widehat{\theta}_S - \theta^*{}_S\|_\infty \leq \left\| \left(\frac{\mathbf{X}_S^{\mathrm{T}} \mathbf{X}_S}{n} \right)^{-1} \mathbf{X}_S^{\mathrm{T}} \frac{w}{n} \right\|_\infty + \left|\left|\left| \left(\frac{\mathbf{X}_S^{\mathrm{T}} \mathbf{X}_S}{n} \right)^{-1} \right|\right|\right|_\infty \lambda_n, \tag{7.54}$$

which completes the proof.

7.6 Appendix: Proof of Theorem 7.16

By a rescaling argument, it suffices to restrict attention to vectors belonging to the ellipse $\mathbb{S}^{d-1}(\mathbf{\Sigma}) = \{\theta \in \mathbb{R}^d \mid \|\sqrt{\mathbf{\Sigma}}\,\theta\|_2 = 1\}$. Define the function $g(t) := 2\rho(\mathbf{\Sigma}) \sqrt{\frac{\log d}{n}}\, t$, and the associated "bad" event

$$\mathcal{E} := \left\{ \mathbf{X} \in \mathbb{R}^{n \times d} \;\middle|\; \inf_{\theta \in \mathbb{S}^{d-1}(\mathbf{\Sigma})} \frac{\|\mathbf{X}\theta\|_2}{\sqrt{n}} \leq \frac{1}{4} - 2g(\|\theta\|_1) \right\}. \tag{7.55}$$

We first claim that on the complementary set $\mathcal{E}^c$, the lower bound (7.31) holds. Let $\theta \in \mathbb{S}^{d-1}(\mathbf{\Sigma})$ be arbitrary. Defining $a = \frac{1}{4}$, $b = 2g(\|\theta\|_1)$ and $c = \frac{\|\mathbf{X}\theta\|_2}{\sqrt{n}}$, we have $c \geq \max\{a - b, 0\}$ on the event $\mathcal{E}^c$. We claim that this lower bound implies that $c^2 \geq (1-\delta)^2 a^2 - \frac{1}{\delta^2} b^2$ for any $\delta \in (0, 1)$. Indeed, if $\frac{b}{\delta} \geq a$, then the claimed lower bound is trivial. Otherwise, we may assume that $b \leq \delta a$, in which case the bound $c \geq a - b$ implies that $c \geq (1-\delta)a$, and hence that $c^2 \geq (1-\delta)^2 a^2$. Setting $(1-\delta)^2 = \frac{1}{2}$ then yields the claim. Thus, the remainder of our proof is devoted to upper bounding $\mathbb{P}[\mathcal{E}]$.

For a pair (r_ℓ, r_u) of radii such that $0 \leq r_\ell < r_u$, define the sets

$$\mathbb{K}(r_\ell, r_u) := \{\theta \in \mathbb{S}^{d-1}(\mathbf{\Sigma}) \mid g(\|\theta\|_1) \in [r_\ell, r_u]\}, \tag{7.56a}$$

along with the events

$$\mathcal{A}(r_\ell, r_u) := \left\{ \inf_{\theta \in \mathbb{K}(r_\ell, r_u)} \frac{\|\mathbf{X}\theta\|_2}{\sqrt{n}} \leq \frac{1}{2} - 2r_u \right\}. \tag{7.56b}$$

Given these objects, the following lemma is the central technical result in the proof:

Lemma 7.24 *For any pair of radii $0 \leq r_\ell < r_u$, we have*

$$\mathbb{P}[\mathcal{A}(r_\ell, r_u)] \leq e^{-\frac{n}{32}} e^{-\frac{n}{2} r_u^2}. \tag{7.57a}$$

Moreover, for $\mu = 1/4$, we have

$$\mathcal{E} \subseteq \mathcal{A}(0, \mu) \cup \left(\bigcup_{\ell=1}^{\infty} \mathcal{A}(2^{\ell-1}\mu, 2^\ell \mu) \right). \tag{7.57b}$$

Based on this lemma, the remainder of the proof is straightforward. From the inclusion (7.57b) and the union bound, we have

$$\mathbb{P}[\mathcal{E}] \leq \mathbb{P}[\mathcal{A}(0, \mu)] + \sum_{\ell=1}^{\infty} \mathbb{P}[\mathcal{A}(2^{\ell-1}\mu, 2^\ell \mu)] \leq e^{-\frac{n}{32}} \left\{ \sum_{\ell=0}^{\infty} e^{-\frac{n}{2} 2^{2\ell} \mu^2} \right\}.$$

Since $\mu = 1/4$ and $2^{2\ell} \geq 2\ell$, we have

$$\mathbb{P}[\mathcal{E}] \leq e^{-\frac{n}{32}} \sum_{\ell=0}^{\infty} e^{-\frac{n}{2} 2^{2\ell} \mu^2} \leq e^{-\frac{n}{32}} \sum_{\ell=0}^{\infty} (e^{-n\mu^2})^\ell \leq \frac{e^{-\frac{n}{32}}}{1 - e^{-\frac{n}{32}}}.$$

It remains to prove the lemma.

Proof of Lemma 7.24: We begin with the inclusion (7.57b). Let $\theta \in \mathbb{S}^{d-1}(\mathbf{\Sigma})$ be a vector that certifies the event $\mathcal{E}$; then it must belong either to the set $\mathbb{K}(0, \mu)$ or to a set $\mathbb{K}(2^{\ell-1}\mu, 2^\ell \mu)$ for some $\ell = 1, 2, \ldots$.

Case 1: First suppose that $\theta \in \mathbb{K}(0, \mu)$, so that $g(\|\theta\|_1) \leq \mu = 1/4$. Since θ certifies the event $\mathcal{E}$, we have

$$\frac{\|\mathbf{X}\theta\|_2}{\sqrt{n}} \leq \frac{1}{4} - 2g(\|\theta\|_1) \leq \frac{1}{4} = \frac{1}{2} - \mu,$$

showing that event $\mathcal{A}(0, \mu)$ must happen.

Case 2: Otherwise, we must have $\theta \in \mathbb{K}(2^{\ell-1}\mu, 2^\ell \mu)$ for some $\ell = 1, 2, \ldots$, and moreover

$$\frac{\|\mathbf{X}\theta\|_2}{\sqrt{n}} \leq \frac{1}{4} - 2g(\|\theta\|_1) \leq \frac{1}{2} - 2(2^{\ell-1}\mu) \leq \frac{1}{2} - 2^\ell \mu,$$

which shows that the event $\mathcal{A}(2^{\ell-1}\mu, 2^{\ell}\mu)$ must happen.

We now establish the tail bound (7.57a). It is equivalent to upper bound the random variable $T(r_\ell, r_u) := -\inf_{\theta \in \mathbb{K}(r_\ell, r_u)} \frac{\|\mathbf{X}\theta\|_2}{\sqrt{n}}$. By the variational representation of the ℓ_2-norm, we have

$$T(r_\ell, r_u) = -\inf_{\theta \in \mathbb{K}(r_\ell, r_u)} \sup_{u \in \mathbb{S}^{n-1}} \frac{\langle u, \mathbf{X}\theta \rangle}{\sqrt{n}} = \sup_{\theta \in \mathbb{K}(r_\ell, r_u)} \inf_{u \in \mathbb{S}^{n-1}} \frac{\langle u, \mathbf{X}\theta \rangle}{\sqrt{n}}.$$

Consequently, if we write $\mathbf{X} = \mathbf{W}\sqrt{\mathbf{\Sigma}}$, where $\mathbf{W} \in \mathbb{R}^{n \times d}$ is a standard Gaussian matrix and define the transformed vector $v = \sqrt{\mathbf{\Sigma}}\, \theta$, then

$$-\inf_{\theta \in \mathbb{K}(r_\ell, r_u)} \frac{\|\mathbf{X}\theta\|_2}{\sqrt{n}} = \sup_{v \in \widetilde{\mathbb{K}}(r_\ell, r)} \inf_{u \in \mathbb{S}^{n-1}} \underbrace{\frac{\langle u, \mathbf{W}v \rangle}{\sqrt{n}}}_{Z_{u,v}}, \tag{7.58}$$

where $\widetilde{\mathbb{K}}(r_\ell, r_u) = \{v \in \mathbb{R}^d \mid \|v\|_2 = 1,\ g(\mathbf{\Sigma}^{-\frac{1}{2}} v) \in [r_\ell, r_u]\}$.

Since (u, v) range over a subset of $\mathbb{S}^{n-1} \times \mathbb{S}^{d-1}$, each variable $Z_{u,v}$ is zero-mean Gaussian with variance n^{-1}. Furthermore, the Gaussian comparison principle due to Gordon, previously used in the proof of Theorem 6.1, may be applied. More precisely, we may compare the Gaussian process $\{Z_{u,v}\}$ to the zero-mean Gaussian process with elements

$$Y_{u,v} := \frac{\langle g, u \rangle}{\sqrt{n}} + \frac{\langle h, v \rangle}{\sqrt{n}}, \qquad \text{where } g \in \mathbb{R}^n,\ h \in \mathbb{R}^d \text{ have i.i.d. } \mathcal{N}(0, 1) \text{ entries.}$$

Applying Gordon's inequality (6.65), we find that

$$\begin{aligned}
\mathbb{E}[T(r_\ell, r_u)] = \mathbb{E}\Big[\sup_{v \in \widetilde{\mathbb{K}}(r_\ell, r_u)} \inf_{u \in \mathbb{S}^{n-1}} Z_{u,v}\Big] &\le \mathbb{E}\Big[\sup_{v \in \widetilde{\mathbb{K}}(r_\ell, r_u)} \inf_{u \in \mathbb{S}^{n-1}} Y_{u,v}\Big] \\
&= \mathbb{E}\Big[\sup_{v \in \widetilde{\mathbb{K}}(r_\ell, r_u)} \frac{\langle h, v \rangle}{\sqrt{n}}\Big] + \mathbb{E}\Big[\inf_{u \in \mathbb{S}^{n-1}} \frac{\langle g, u \rangle}{\sqrt{n}}\Big] \\
&= \mathbb{E}\Big[\sup_{\theta \in \mathbb{K}(r_\ell, r_u)} \frac{\big\langle \sqrt{\mathbf{\Sigma}}\, h, \theta \big\rangle}{\sqrt{n}}\Big] - \mathbb{E}\Big[\frac{\|g\|_2}{\sqrt{n}}\Big].
\end{aligned}$$

On one hand, we have $\mathbb{E}[\|g\|_2] \ge \sqrt{n}\sqrt{\frac{2}{\pi}}$. On the other hand, applying Hölder's inequality yields

$$\mathbb{E}\Big[\sup_{\theta \in \mathbb{K}(r_\ell, r_u)} \frac{\langle \sqrt{\mathbf{\Sigma}}\, h, \theta \rangle}{\sqrt{n}}\Big] \le \mathbb{E}\Big[\sup_{\theta \in \mathbb{K}(r_\ell, r_u)} \|\theta\|_1 \frac{\|\sqrt{\mathbf{\Sigma}}\, h\|_\infty}{\sqrt{n}}\Big] \overset{\text{(i)}}{\le} r_u,$$

where step (i) follows since $\mathbb{E}[\frac{\|\sqrt{\mathbf{\Sigma}}h\|_\infty}{\sqrt{n}}] \le 2\rho(\mathbf{\Sigma})\sqrt{\frac{\log d}{n}}$ and $\sup_{\theta \in \mathbb{K}(r_\ell, r_u)} \|\theta\|_1 \le \frac{r_u}{2\rho(\mathbf{\Sigma})\sqrt{(\log d)/n}}$, by the definition (7.56a) of $\mathbb{K}$. Putting together the pieces, we have shown that

$$\mathbb{E}[T(r_\ell, r_u)] \le -\sqrt{\frac{2}{\pi}} + r_u. \tag{7.59}$$

From the representation (7.58), we see that the random variable $\sqrt{n}\, T(r_\ell, r_u)$ is a 1-Lipschitz function of the standard Gaussian matrix $\mathbf{W}$, so that Theorem 2.26 implies the upper tail bound $\mathbb{P}[T(r_\ell, r_u) \geq \mathbb{E}[T(r_\ell, r_u)] + \delta] \leq e^{-n\delta^2/2}$ for all $\delta > 0$. Define the constant $C = \sqrt{\frac{2}{\pi}} - \frac{1}{2} \geq \frac{1}{4}$. Setting $\delta = C + r_u$ and using our upper bound on the mean (7.59) yields

$$\mathbb{P}[T(r_\ell, r_u) \geq -\tfrac{1}{2} + 2r_u] \leq e^{-\frac{n}{2}C^2} e^{-\frac{n}{2}r_u^2} \leq e^{-\frac{n}{32}} e^{-\frac{n}{2}r_u^2},$$

as claimed.

7.7 Bibliographic details and background

The Gaussian sequence model discussed briefly in Example 7.1 has been the subject of intensive study. Among other reasons, it is of interest because many nonparametric estimation problems can be "reduced" to equivalent versions in the (infinite-dimensional) normal sequence model. The book by Johnstone (2015) provides a comprehensive introduction; see also the references therein. Donoho and Johnstone (1994) derive sharp upper and lower bounds on the minimax risk in ℓ_p-norm for a vector belonging to an ℓ_q-ball, $q \in [0, 1]$, for the case of the Gaussian sequence model. The problem of bounding the in-sample prediction error for nonparametric least squares, as studied in Chapter 13, can also be understood as a special case of the Gaussian sequence model.

The use of ℓ_1-regularization for ill-posed inverse problems has a lengthy history, with early work in geophysics (e.g., Levy and Fullagar, 1981; Oldenburg et al., 1983; Santosa and Symes, 1986); see Donoho and Stark (1989) for further discussion. Alliney and Ruzinsky (1994) studied various algorithmic issues associated with ℓ_1-regularization, which soon became the subject of more intensive study in statistics and applied mathematics following the seminal papers of Chen, Donoho and Saunders (1998) on the basis pursuit program (7.9), and Tibshirani (1996) on the Lasso (7.18). Other authors have also studied various forms of non-convex regularization for enforcing sparsity; for instance, see the papers (Fan and Li, 2001; Zou and Li, 2008; Fan and Lv, 2011; Zhang, 2012; Zhang and Zhang, 2012; Loh and Wainwright, 2013; Fan et al., 2014) and references therein.

Early work on the basis pursuit linear program (7.9) focused on the problem of representing a signal in a pair of bases, in which n is the signal length, and $p = 2n$ indexes the union of the two bases of $\mathbb{R}^n$. The incoherence condition arose from this line of work (e.g., Donoho and Huo, 2001; Elad and Bruckstein, 2002); the necessary and sufficient conditions that constitute the restricted nullspace property seem to have been isolated for the first time by Feuer and Nemirovski (2003). However, the terminology and precise definition of restricted nullspace used here was given by Cohen et al. (2008).

Juditsky and Nemirovsky (2000), Nemirovski (2000) and Greenshtein and Ritov (2004) were early authors to provide some high-dimensional guarantees for estimators based on ℓ_1-regularization, in particular in the context of function aggregation problems. Candès and Tao (2005) and Donoho (2006a; 2006b) analyzed the basis pursuit method for the case of random Gaussian or unitary matrices, and showed that it can succeed with $n \gtrsim s\log(ed/s)$ samples. Donoho and Tanner (2008) provided a sharp analysis of this threshold phenomenon in the noiseless case, with connections to the structure of random polytopes. The restricted isometry property was introduced by Candès and Tao (2005; 2007). They also proposed the Dantzig selector, an alternative ℓ_1-based relaxation closely related to the Lasso, and proved bounds on noisy recovery for ensembles that satisfy the RIP condition. Bickel et al. (2009) introduced the weaker restricted eigenvalue (RE) condition, slightly different than but essentially equivalent to the version stated here, and provided a unified way to derive ℓ_2-error and prediction error bounds for both the Lasso and the Dantzig selector. Exercises 7.13 and 7.14 show how to derive ℓ_∞-bounds on the Lasso error by using ℓ_∞-analogs of the ℓ_2-restricted eigenvalues; see Ye and Zhang (2010) for bounds on the Lasso and Dantzig errors using these and other types of restricted eigenvalues. Van de Geer and Bühlmann (2009) provide a comprehensive overview of different types of RE conditions, and the relationships among them; see also their book (Bühlmann and van de Geer, 2011).

The proof of Theorem 7.13(a) is inspired by the proof technique of Bickel et al. (2009); see also the material in Chapter 9, and the paper by Negahban et al. (2012) for a general viewpoint on regularized M-estimators. There are many variants and extensions of the basic Lasso, including the square-root Lasso (Belloni et al., 2011), the elastic net (Zou and Hastie, 2005), the fused Lasso (Tibshirani et al., 2005), the adaptive Lasso (Zou, 2006; Huang et al., 2008) and the group Lasso (Yuan and Lin, 2006). See Exercise 7.17 in this chapter for discussion of the square-root Lasso, and Chapter 9 for discussion of some of these other extensions.

Theorem 7.16 was proved by Raskutti et al. (2010). Rudelson and Zhou (2013) prove an analogous result for more general ensembles of sub-Gaussian random matrices; this analysis requires substantially different techniques, since Gaussian comparison results are no longer available. Both of these results apply to a very broad class of random matrices; for instance, it is even possible to sample the rows of the random matrix $\mathbf{X} \in \mathbb{R}^{n \times d}$ from a distribution with a degenerate covariance matrix, and/or with its maximum eigenvalue diverging with the problem size, and these results can still be applied to show that a (lower) restricted eigenvalue condition holds with high probability. Exercise 7.10 is based on results of Loh and Wainwright (2012).

Exercise 7.12 explores the ℓ_2-error rates achievable by the Lasso for vectors that belong to an ℓ_q-ball. These results are known to be minimax-optimal, as can be shown using information-theoretic techniques for lower bounding the minimax rate. See Chapter 15 for details on techniques for proving lower bounds, and the papers (Ye and Zhang, 2010; Raskutti et al., 2011) for specific lower bounds in the context of sparse linear regression.

The slow rate and fast rates for prediction—that is, the bounds in equations (7.40) and (7.41) respectively—have been derived in various papers (e.g., Bunea et al., 2007; Candès and Tao, 2007; Bickel et al., 2009). It is natural to wonder whether the restricted eigenvalue conditions, which control correlation between the columns of the design matrix, should be required for achieving the fast rate. From a fundamental point of view, such conditions are not necessary: an ℓ_0-based estimator, one that performs an exhaustive search over all $\binom{d}{s}$ subsets of size s, can achieve the fast rate with only a column normalization condition on the design matrix (Bunea et al., 2007; Raskutti et al., 2011); see Example 13.16 for an explicit derivation of the fast bound for this method. It can be shown that the Lasso itself is sub-optimal: a number of authors (Foygel and Srebro, 2011; Dalalyan et al., 2014) have given design matrices $\mathbf{X}$ and 2-sparse vectors for which the Lasso squared prediction error is lower bounded as $1/\sqrt{n}$. Zhang et al. (2017) construct a harder design matrix for which the ℓ_0-based method can achieve the fast rate, but for which a broad class of M-estimators, one that includes the Lasso as well as estimators based on non-convex regularizers, has prediction error lower bounded as $1/\sqrt{n}$. If, in addition, we restrict attention to methods that are required to output an s-sparse estimator, then Zhang et al. (2014) show that, under a standard conjecture in complexity theory, no polynomial-time algorithm can achieve the fast rate (7.41) without the lower RE condition.

Irrepresentable conditions for variable selection consistency were introduced independently by Fuchs (2004) and Tropp (2006) in signal processing, and by Meinshausen and Buhlmann (2006) and Zhao and Yu (2006) in statistics. The primal–dual witness proof of Theorem 7.21 follows the argument of Wainwright (2009b); see also this paper for extensions to general random Gaussian designs. The proof of Lemma 7.23 was suggested by Caramanis (personal communication, 2010). The primal–dual witness method that underlies the proof of Theorem 7.21 has been applied in a variety of other settings, including analysis of group Lasso (Obozinski et al., 2011; Wang et al., 2015) and related relaxations (Jalali et al., 2010; Negahban and Wainwright, 2011b), graphical Lasso (Ravikumar et al., 2011), methods for Gaussian graph selection with hidden variables (Chandrasekaran et al., 2012b), and variable selection in nonparametric models (Xu et al., 2014). Lee et al. (2013) describe a general framework for deriving consistency results using the primal–dual witness method.

The results in this chapter were based on theoretically derived choices of the regularization parameter λ_n, all of which involved the (unknown) standard deviation σ of the additive noise. One way to circumvent this difficulty is by using the square-root Lasso estimator (Belloni et al., 2011), for which the optimal choice of regularization parameter does not depend on σ. See Exercise 7.17 for a description and analysis of this estimator.

7.8 Exercises

Exercise 7.1 (Optimization and threshold estimators)

(a) Show that the hard-thresholding estimator (7.6a) corresponds to the optimal solution $\widehat{\theta}$ of the non-convex program

$$\min_{\theta \in \mathbb{R}^n} \left\{ \frac{1}{2} \|y - \theta\|_2^2 + \frac{1}{2} \lambda^2 \|\theta\|_0 \right\}.$$

(b) Show that the soft-thresholding estimator (7.6b) corresponds to the optimal solution $\widehat{\theta}$ of the ℓ_1-regularized quadratic program

$$\min_{\theta \in \mathbb{R}^n} \left\{ \frac{1}{2} \|y - \theta\|_2^2 + \lambda \|\theta\|_1 \right\}.$$

Exercise 7.2 (Properties of ℓ_q-balls) For a given $q \in (0, 1]$, recall the (strong) ℓ_q-ball

$$\mathbb{B}_q(R_q) := \left\{ \theta \in \mathbb{R}^d \,\middle|\, \sum_{j=1}^{d} |\theta_j|^q \leq R_q \right\}. \tag{7.60}$$

The weak ℓ_q-ball with parameters (C, α) is defined as

$$\mathbb{B}_{w(\alpha)}(C) := \{\theta \in \mathbb{R}^d \mid |\theta|_{(j)} \leq C j^{-\alpha} \text{ for } j = 1, \ldots, d\}. \tag{7.61}$$

Here $|\theta|_{(j)}$ denote the order statistics of θ^* in absolute value, ordered from largest to smallest (so that $|\theta|_{(1)} = \max_{j=1,2,\ldots,d} |\theta_j|$ and $|\theta|_{(d)} = \min_{j=1,2,\ldots,d} |\theta_j|$.)

(a) Show that the set $\mathbb{B}_q(R_q)$ is star-shaped around the origin. (A set $C \subseteq \mathbb{R}^d$ is star-shaped around the origin if $\theta \in C \Rightarrow t\theta \in C$ for all $t \in [0, 1]$.)

(b) For any $\alpha > 1/q$, show that there is a radius R_q depending on (C, α) such that $\mathbb{B}_{w(\alpha)}(C) \subseteq \mathbb{B}_q(R_q)$. This inclusion underlies the terminology "strong" and "weak", respectively.

(c) For a given integer $s \in \{1, 2, \ldots, d\}$, the best s-term approximation to a vector $\theta^* \in \mathbb{R}^d$ is given by

$$\Pi_s(\theta^*) := \arg \min_{\|\theta\|_0 \leq s} \|\theta - \theta^*\|_2^2. \tag{7.62}$$

Give a closed-form expression for $\Pi_s(\theta^*)$.

(d) When $\theta^* \in \mathbb{B}_q(R_q)$ for some $q \in (0, 1]$, show that the best s-term approximation satisfies

$$\|\Pi_s(\theta^*) - \theta^*\|_2^2 \leq (R_q)^{2/q} \left(\frac{1}{s}\right)^{\frac{2}{q}-1}. \tag{7.63}$$

Exercise 7.3 (Pairwise incoherence) Given a matrix $\mathbf{X} \in \mathbb{R}^{n \times d}$, suppose that it has pairwise incoherence (7.12) upper bounded as $\delta_{\mathrm{PW}}(\mathbf{X}) < \frac{\gamma}{s}$.

(a) Let $S \subset \{1, 2, \ldots, d\}$ be any subset of size s. Show that there is a function $\gamma \mapsto c(\gamma)$ such that $\gamma_{\min}\left(\frac{\mathbf{X}_S^T \mathbf{X}_S}{n}\right) \geq c(\gamma) > 0$, as long as γ is sufficiently small.

(b) Prove that $\mathbf{X}$ satisfies the restricted nullspace property with respect to S as long as $\gamma < 1/3$. (Do this from first principles, without using any results on restricted isometry.)

Exercise 7.4 (RIP and pairwise incoherence) In this exercise, we explore the relation between the pairwise incoherence and RIP constants.

(a) Prove the sandwich relation (7.15) for the pairwise incoherence and RIP constants. Give a matrix for which inequality (i) is tight, and another matrix for which inequality (ii) is tight.

(b) Construct a matrix such that $\delta_s(\mathbf{X}) = \sqrt{s}\,\delta_{\mathrm{PW}}(\mathbf{X})$.

Exercise 7.5 (ℓ_2-RE $\Rightarrow$ ℓ_1-RE) Let $S \subset \{1, 2, \ldots, d\}$ be a subset of cardinality s. A matrix $\mathbf{X} \in \mathbb{R}^{n \times d}$ satisfies an ℓ_1-RE condition over S with parameters (γ_1, α_1) if

$$\frac{\|\mathbf{X}\theta\|_2^2}{n} \geq \gamma_1 \frac{\|\theta\|_1^2}{s} \qquad \text{for all } \theta \in \mathbb{C}(S; \alpha_1).$$

Show that any matrix satisfying the ℓ_2-RE condition (7.22) with parameters (γ_2, α_2) satisfies the ℓ_1-RE condition with parameters $\gamma_1 = \frac{\gamma_2}{(1+\alpha_2^2)}$ and $\alpha_1 = \alpha_2$.

Exercise 7.6 (Weighted ℓ_1-norms) In many applications, one has additional information about the relative scalings of different predictors, so that it is natural to use a weighted ℓ_1-norm, of the form $\|\theta\|_{\nu(1)} := \sum_{j=1}^d \omega_j |\theta_j|$, where $\omega \in \mathbb{R}^d$ is a vector of strictly positive weights. In the case of noiseless observations, this leads to the weighted basis pursuit LP

$$\min_{\theta \in \mathbb{R}^d} \|\theta\|_{\nu(1)} \qquad \text{such that } \mathbf{X}\theta = y.$$

(a) State and prove necessary and sufficient conditions on $\mathbf{X}$ for the weighted basis pursuit LP to (uniquely) recover all k-sparse vectors θ^*.

(b) Suppose that θ^* is supported on a subset S of cardinality s, and the weight vector ω satisfies

$$\omega_j = \begin{cases} 1 & \text{if } j \in S, \\ t & \text{otherwise}, \end{cases}$$

for some $t \geq 1$. State and prove a sufficient condition for recovery in terms of $c_{\min} = \gamma_{\min}(\mathbf{X}_S^{\mathrm{T}}\mathbf{X}_S/n)$, the pairwise incoherence $\delta_{\mathrm{PW}}(\mathbf{X})$ and the scalar t. How do the conditions on $\mathbf{X}$ behave as $t \to +\infty$?

Exercise 7.7 (Pairwise incoherence and RIP for isotropic ensembles) Consider a random matrix $\mathbf{X} \in \mathbb{R}^{n\times d}$ with i.i.d. $\mathcal{N}(0,1)$ entries.

(a) For a given $s \in \{1, 2, \ldots, d\}$, suppose that $n \gtrsim s^2 \log d$. Show that the pairwise incoherence satisfies the bound $\delta_{\mathrm{PW}}(\mathbf{X}) < \frac{1}{3s}$ with high probability.
(b) Now suppose that $n \gtrsim s \log\left(\frac{es}{d}\right)$. Show that the RIP constant satisfies the bound $\delta_{2s} < 1/3$ with high probability.

Exercise 7.8 (Violations of pairwise incoherence and RIP) Recall the ensemble of spiked identity covariance matrices from Example 7.18 with a constant $\mu > 0$, and consider an arbitrary sparsity level $s \in \{1, 2, \ldots, d\}$.

(a) Violation of pairwise incoherence: show that

$$\mathbb{P}[\delta_{\mathrm{PW}}(\mathbf{X}) > \mu - 3\delta] \geq 1 - 6e^{-n\delta^2/8} \qquad \text{for all } \delta \in (0, 1/\sqrt{2}).$$

Consequently, a pairwise incoherence condition cannot hold unless $\mu \ll \frac{1}{s}$.
(b) Violation of RIP: Show that

$$\mathbb{P}[\delta_{2s}(\mathbf{X}) \geq (1 + (\sqrt{2s} - 1)\mu)\delta] \geq 1 - e^{-n\delta^2/8} \qquad \text{for all } \delta \in (0, 1).$$

Consequently, a RIP condition cannot hold unless $\mu \ll \frac{1}{\sqrt{s}}$.

Exercise 7.9 (Relations between ℓ_0 and ℓ_1 constraints) For an integer $k \in \{1, \ldots, d\}$, consider the following two subsets:

$$\mathbb{L}_0(k) := \mathbb{B}_2(1) \cap \mathbb{B}_0(k) = \{\theta \in \mathbb{R}^d \mid \|\theta\|_2 \leq 1 \text{ and } \|\theta\|_0 \leq k\},$$
$$\mathbb{L}_1(k) := \mathbb{B}_2(1) \cap \mathbb{B}_1(\sqrt{k}) = \{\theta \in \mathbb{R}^d \mid \|\theta\|_2 \leq 1 \text{ and } \|\theta\|_1 \leq \sqrt{k}\}.$$

For any set $\mathbb{L}$, let $\overline{\mathrm{conv}}(\mathbb{L})$ denote the closure of its convex hull.

(a) Prove that $\overline{\mathrm{conv}}(\mathbb{L}_0(k)) \subseteq \mathbb{L}_1(k)$.
(b) Prove that $\mathbb{L}_1(k) \subseteq 2\,\overline{\mathrm{conv}}(\mathbb{L}_0(k))$.

(*Hint:* For part (b), you may find it useful to consider the support functions of the two sets.)

Exercise 7.10 (Sufficient conditions for RE) Consider an arbitrary symmetric matrix $\mathbf{\Gamma}$ for which there is a scalar $\delta > 0$ such that

$$|\theta^{\mathrm{T}}\mathbf{\Gamma}\theta| \leq \delta \qquad \text{for all } \theta \in \mathbb{L}_0(2s),$$

where the set $\mathbb{L}_0$ was defined in Exercise 7.9.

(a) Show that

$$|\theta^{\mathrm{T}}\mathbf{\Gamma}\theta| \le \begin{cases} 12\delta\|\theta\|_2^2 & \text{for all vectors such that } \|\theta\|_1 \le \sqrt{s}\,\|\theta\|_2, \\ \frac{12\delta}{s}\|\theta\|_1^2 & \text{otherwise.} \end{cases}$$

(*Hint:* Part (b) of Exercise 7.9 could be useful.)

(b) Use part (a) to show that RIP implies the RE condition.

(c) Give an example of a matrix family that violates RIP for which part (a) can be used to guarantee the RE condition.

Exercise 7.11 (Weaker sufficient conditions for RE) Consider a covariance matrix $\mathbf{\Sigma}$ with minimum eigenvalue $\gamma_{\min}(\mathbf{\Sigma}) > 0$ and maximum variance $\rho^2(\mathbf{\Sigma})$.

(a) Show that the lower bound (7.31) implies that the RE condition (7.22) holds with parameter $\kappa = \frac{c_1}{2}\gamma_{\min}(\mathbf{\Sigma})$ over $\mathbb{C}_\alpha(S)$, uniformly for all subsets S of cardinality at most $|S| \le \frac{c_1}{2c_2}\frac{\gamma_{\min}(\mathbf{\Sigma})}{\rho^2(\mathbf{\Sigma})}(1+\alpha)^{-2}\frac{n}{\log d}$.

(b) Give a sequence of covariance matrices $\{\mathbf{\Sigma}^{(d)}\}$ for which $\gamma_{\max}(\mathbf{\Sigma}^{(d)})$ diverges, but part (a) can still be used to guarantee the RE condition.

Exercise 7.12 (Estimation over ℓ_q-"balls") In this problem, we consider linear regression with a vector $\theta^* \in \mathbb{B}_q(R_q)$ for some radius $R_q \ge 1$ and parameter $q \in (0, 1]$ under the following conditions: (a) the design matrix $\mathbf{X}$ satisfies the lower bound (7.31) and uniformly bounded columns ($\|X_j\|_2/\sqrt{n} \le 1$ for all $j = 1, \ldots, d$); (b) the noise vector $w \in \mathbb{R}^n$ has i.i.d. zero-mean entries that are sub-Gaussian with parameter σ.

Using Theorem 7.19 and under an appropriate lower bound on the sample size n in terms of (d, R_q, σ, q), show that there are universal constants (c_0, c_1, c_2) such that, with probability $1 - c_1 e^{-c_2 \log d}$, any Lasso solution $\widehat{\theta}$ satisfies the bound

$$\|\widehat{\theta} - \theta^*\|_2^2 \le c_0 R_q \left(\frac{\sigma^2 \log d}{n}\right)^{1-\frac{q}{2}}.$$

(*Note:* The universal constants can depend on quantities related to $\mathbf{\Sigma}$, as in the bound (7.31).)

Exercise 7.13 (ℓ_∞-bounds for the Lasso) Consider the sparse linear regression model $y = \mathbf{X}\theta^* + w$, where $w \sim \mathcal{N}(0, \sigma^2 I_{n\times n})$ and $\theta^* \in \mathbb{R}^d$ is supported on a subset S. Suppose that the sample covariance matrix $\widehat{\mathbf{\Sigma}} = \frac{1}{n}\mathbf{X}^{\mathrm{T}}\mathbf{X}$ has its diagonal entries uniformly upper bounded by one, and that for some parameter $\gamma > 0$, it also satisfies an ℓ_∞-curvature condition of the form

$$\|\widehat{\mathbf{\Sigma}}\Delta\|_\infty \ge \gamma\|\Delta\|_\infty \qquad \text{for all } \Delta \in \mathbb{C}_3(S). \tag{7.64}$$

Show that with the regularization parameter $\lambda_n = 4\sigma\sqrt{\frac{\log d}{n}}$, any Lasso solution satisfies the ℓ_∞-bound

$$\|\widehat{\theta} - \theta^*\|_\infty \le \frac{6\sigma}{\gamma}\sqrt{\frac{\log d}{n}}$$

with high probability.

Exercise 7.14 (Verifying ℓ_∞-curvature conditions) This problem is a continuation of Exercise 7.13. Suppose that we form a random design matrix $\mathbf{X} \in \mathbb{R}^{n \times d}$ with rows drawn i.i.d. from a $\mathcal{N}(0, \mathbf{\Sigma})$ distribution, and moreover that

$$\|\mathbf{\Sigma}\Delta\|_\infty \geq \gamma\|\Delta\|_\infty \qquad \text{for all vectors } \Delta \in \mathbb{C}_3(S).$$

Show that, with high probability, the sample covariance $\widehat{\mathbf{\Sigma}} := \frac{1}{n}\mathbf{X}^T\mathbf{X}$ satisfies this same property with $\gamma/2$ as long as $n \gtrsim s^2 \log d$.

Exercise 7.15 (Sharper bounds for Lasso) Let $\mathbf{X} \in \mathbb{R}^{n \times d}$ be a fixed design matrix such that $\frac{|||\mathbf{X}_S|||_2}{\sqrt{n}} \leq C$ for all subsets S of cardinality at most s. In this exercise, we show that, with high probability, any solution of the constrained Lasso (7.19) with $R = \|\theta^*\|_1$ satisfies the bound

$$\|\widehat{\theta} - \theta^*\|_2 \precsim \frac{\sigma}{\kappa}\sqrt{\frac{s \log(ed/s)}{n}}, \tag{7.65}$$

where $s = \|\theta^*\|_0$. Note that this bound provides an improvement for linear sparsity (i.e., whenever $s = \alpha d$ for some constant $\alpha \in (0, 1)$).

(a) Define the random variable

$$Z := \sup_{\Delta \in \mathbb{R}^d} \left| \left\langle \Delta, \frac{1}{n}\mathbf{X}^{\mathrm{T}} w \right\rangle \right| \qquad \text{such that } \|\Delta\|_2 \leq 1 \text{ and } \|\Delta\|_1 \leq \sqrt{s}, \tag{7.66}$$

where $w \sim \mathcal{N}(0, \sigma^2 I)$. Show that

$$\mathbb{P}\left[\frac{Z}{C\sigma} \geq c_1 \sqrt{\frac{s \log(ed/s)}{n}} + \delta\right] \leq c_2 e^{-c_3 n \delta^2}$$

for universal constants (c_1, c_2, c_3). (*Hint:* The result of Exercise 7.9 may be useful here.)

(b) Use part (a) and results from the chapter to show that if $\mathbf{X}$ satisfies an RE condition, then any optimal Lasso solution $\widehat{\theta}$ satisfies the bound (7.65) with probability $1 - c_2' e^{-c_3' s \log\left(\frac{ed}{s}\right)}$.

Exercise 7.16 (Analysis of weighted Lasso) In this exercise, we analyze the weighted Lasso estimator

$$\widehat{\theta} \in \arg\min_{\theta \in \mathbb{R}^d} \left\{ \frac{1}{2n}\|y - \mathbf{X}\theta\|_2^2 + \lambda_n \|\theta\|_{\nu(1)} \right\},$$

where $\|\theta\|_{\nu(1)} := \sum_{j=1}^d \nu_j |\theta_j|$ denotes the *weighted ℓ_1-norm* defined by a positive weight vector $\nu \in \mathbb{R}^d$. Define $C_j = \frac{\|X_j\|_2}{\sqrt{n}}$, where $X_j \in \mathbb{R}^n$ denotes the jth column of the design matrix, and let $\widehat{\Delta} = \widehat{\theta} - \theta^*$ be the error vector associated with an optimal solution $\widehat{\theta}$.

(a) Suppose that we choose a regularization parameter $\lambda_n \geq 2 \max_{j=1,\ldots,d} \frac{|\langle X_j, w\rangle|}{n \nu_j}$. Show that the vector $\widehat{\Delta}$ belongs to the modified cone set

$$\mathbb{C}_3(S; \nu) := \{\Delta \in \mathbb{R}^d \mid \|\Delta_{S^c}\|_{\nu(1)} \leq 3\|\Delta_S\|_{\nu(1)}\}. \tag{7.67}$$

(b) Assuming that $\mathbf{X}$ satisfies a κ-RE condition over $\mathbb{C}_\nu(S; 3)$, show that

$$\|\widehat{\theta} - \theta^*\|_2 \leq \frac{6}{\kappa}\lambda_n \sqrt{\sum_{j \in S} \nu_j^2}.$$

(c) For a general design matrix, the rescaled column norms $C_j = \|X_j\|_2/\sqrt{n}$ may vary widely. Give a choice of weights for which the weighted Lasso error bound is superior to the ordinary Lasso bound. (*Hint:* You should be able to show an improvement by a factor of $\frac{\max_{j\in S} C_j}{\max_{j=1,\ldots,d} C_j}$.)

Exercise 7.17 (Analysis of square-root Lasso) The square-root Lasso is given by

$$\widehat{\theta} \in \arg\min_{\theta\in\mathbb{R}^d} \left\{ \frac{1}{\sqrt{n}} \|y - \mathbf{X}\theta\|_2 + \gamma_n \|\theta\|_1 \right\}.$$

(a) Suppose that the regularization parameter γ_n is varied over the interval $(0, \infty)$. Show that the resulting set of solutions coincides with those of the Lagrangian Lasso as λ_n is varied.

(b) Show that any square-root Lasso estimate $\widehat{\theta}$ satisfies the equality

$$\frac{\frac{1}{n}\mathbf{X}^{\mathrm{T}}(\mathbf{X}\widehat{\theta} - y)}{\frac{1}{\sqrt{n}}\|y - \mathbf{X}\widehat{\theta}\|_2} + \gamma_n \widehat{z} = 0,$$

where $\widehat{z} \in \mathbb{R}^d$ belongs to the subdifferential of the ℓ_1-norm at $\widehat{\theta}$.

(c) Suppose $y = \mathbf{X}\theta^* + w$ where the unknown regression vector θ^* is S-sparse. Use part (b) to establish that the error $\widehat{\Delta} = \widehat{\theta} - \theta^*$ satisfies the basic inequality

$$\frac{1}{n}\|\mathbf{X}\widehat{\Delta}\|_2^2 \le \left\langle \widehat{\Delta}, \frac{1}{n}\mathbf{X}^{\mathrm{T}} w \right\rangle + \gamma_n \frac{\|y - \mathbf{X}\widehat{\theta}\|_2}{\sqrt{n}} \{\|\widehat{\Delta}_S\|_1 - \|\widehat{\Delta}_{S^c}\|_1\}.$$

(d) Suppose that $\gamma_n \ge 2\frac{\|\mathbf{X}^{\mathrm{T}} w\|_\infty}{\sqrt{n}\|w\|_2}$. Show that the error vector satisfies the cone constraint $\|\widehat{\Delta}_{S^c}\|_1 \le 3\|\widehat{\Delta}_S\|_1$.

(e) Suppose in addition that $\mathbf{X}$ satisfies an RE condition over the set $\mathbb{C}_3(S)$. Show that there is a universal constant c such that

$$\|\widehat{\theta} - \theta^*\|_2 \le c\frac{\|w\|_2}{\sqrt{n}} \gamma_n \sqrt{s}.$$

Exercise 7.18 (From pairwise incoherence to irrepresentable condition) Consider a matrix $\mathbf{X} \in \mathbb{R}^{n\times d}$ whose pairwise incoherence (7.12) satisfies the bound $\delta_{\mathrm{PW}}(\mathbf{X}) < \frac{1}{2s}$. Show that the irrepresentable condition (7.43b) holds for any subset S of cardinality at most s.

Exercise 7.19 (Irrepresentable condition for random designs) Let $\mathbf{X} \in \mathbb{R}^{n\times d}$ be a random matrix with rows $\{x_i\}_{i=1}^n$ sampled i.i.d. according to a $\mathcal{N}(0, \mathbf{\Sigma})$ distribution. Suppose that the diagonal entries of $\mathbf{\Sigma}$ are at most 1, and that it satisfies the irrepresentable condition with parameter $\alpha \in [0, 1)$—that is,

$$\max_{j\in S^c} \|\mathbf{\Sigma}_{jS}(\mathbf{\Sigma}_{SS})^{-1}\|_1 \le \alpha < 1.$$

Let $z \in \mathbb{R}^s$ be a random vector that depends only on the submatrix $\mathbf{X}_S$.

(a) Show that, for each $j \in S^c$,

$$|X_j^{\mathrm{T}}\mathbf{X}_S(\mathbf{X}_S^{\mathrm{T}}\mathbf{X}_S)^{-1}z| \le \alpha + |W_j^{\mathrm{T}}\mathbf{X}_S(\mathbf{X}_S^{\mathrm{T}}\mathbf{X}_S)^{-1}z|,$$

where $W_j \in \mathbb{R}^n$ is a Gaussian random vector, independent of $\mathbf{X}_S$.

(b) Use part (a) and random matrix/vector tail bounds to show that

$$\max_{j \in S^c} |X_j^{\mathrm{T}} \mathbf{X}_S (\mathbf{X}_S^{\mathrm{T}} \mathbf{X}_S)^{-1} z| \leq \alpha' := \tfrac{1}{2}(1+\alpha),$$

with probability at least $1 - 4e^{-c \log d}$, as long as $n > \frac{16}{(1-\alpha)\sqrt{c_{\min}}} s \log(d-s)$, where $c_{\min} = \gamma_{\min}(\mathbf{\Sigma}_{SS})$.

Exercise 7.20 (Analysis of ℓ_0-regularization) Consider a design matrix $\mathbf{X} \in \mathbb{R}^{n \times d}$ satisfying the ℓ_0-based upper/lower RE condition

$$\gamma_\ell \|\Delta\|_2^2 \leq \frac{\|\mathbf{X}\Delta\|_2^2}{n} \leq \gamma_u \|\Delta\|_2^2 \qquad \text{for all } \|\Delta\|_0 \leq 2s. \tag{7.68}$$

Suppose that we observe noisy samples $y = \mathbf{X}\theta^* + w$ for some s-sparse vector θ^*, where the noise vector has i.i.d. $\mathcal{N}(0, \sigma^2)$ entries. In this exercise, we analyze an estimator based on the ℓ_0-constrained quadratic program

$$\min_{\theta \in \mathbb{R}^d} \left\{ \frac{1}{2n} \|y - \mathbf{X}\theta\|_2^2 \right\} \qquad \text{such that } \|\theta\|_0 \leq s. \tag{7.69}$$

(a) Show that the non-convex program (7.69) has a unique optimal solution $\widehat{\theta} \in \mathbb{R}^d$.
(b) Using the "basic inequality" proof technique, show that

$$\|\widehat{\theta} - \theta^*\|_2^2 \precsim \frac{\sigma^2 \gamma_u}{\gamma_\ell^2} \frac{s \log(ed/s)}{n}$$

with probability at least $1 - c_1 e^{-c_2 s \log(ed/s)}$. (*Hint:* The result of Exercise 5.7 could be useful to you.)

8

Principal component analysis in high dimensions

Principal component analysis (PCA) is a standard technique for exploratory data analysis and dimension reduction. It is based on seeking the maximal variance components of a distribution, or equivalently, a low-dimensional subspace that captures the majority of the variance. Given a finite collection of samples, the empirical form of principal component analysis involves computing some subset of the top eigenvectors of the sample covariance matrix. Of interest is when these eigenvectors provide a good approximation to the subspace spanned by the top eigenvectors of the population covariance matrix. In this chapter, we study these issues in a high-dimensional and non-asymptotic framework, both for classical unstructured forms of PCA as well as for more modern structured variants.

8.1 Principal components and dimension reduction

Let $\mathcal{S}_+^{d\times d}$ denote the space of d-dimensional positive semidefinite matrices, and denote the d-dimensional unit sphere by $\mathbb{S}^{d-1} = \{v \in \mathbb{R}^d \mid \|v\|_2 = 1\}$. Consider a d-dimensional random vector X, say with a zero-mean vector and covariance matrix $\mathbf{\Sigma} \in \mathcal{S}_+^{d\times d}$. We use

$$\gamma_1(\mathbf{\Sigma}) \geq \gamma_2(\mathbf{\Sigma}) \geq \cdots \geq \gamma_d(\mathbf{\Sigma}) \geq 0$$

to denote the ordered eigenvalues of the covariance matrix. In its simplest instantiation, principal component analysis asks: along what unit-norm vector $v \in \mathbb{S}^{d-1}$ is the variance of the random variable $\langle v, X\rangle$ maximized? This direction is known as the first principal component at the population level, assumed here for the sake of discussion to be unique. In analytical terms, we have

$$v^* = \arg\max_{v\in\mathbb{S}^{d-1}} \operatorname{var}(\langle v, X\rangle) = \arg\max_{v\in\mathbb{S}^{d-1}} \mathbb{E}[\,\langle v, X\rangle^2\,] = \arg\max_{v\in\mathbb{S}^{d-1}} \langle v, \mathbf{\Sigma} v\rangle, \tag{8.1}$$

so that by definition, the first principal component is the maximum eigenvector of the covariance matrix $\mathbf{\Sigma}$. More generally, we can define the top r principal components at the population level by seeking an orthonormal matrix $\mathbf{V} \in \mathbb{R}^{d\times r}$, formed with unit-norm and orthogonal columns $\{v_1, \ldots, v_r\}$, that maximizes the quantity

$$\mathbb{E}\|\mathbf{V}^{\mathrm{T}}X\|_2^2 = \sum_{j=1}^{r} \mathbb{E}[\langle v_j, X\rangle^2]. \tag{8.2}$$

As we explore in Exercise 8.4, these principal components are simply the top r eigenvectors of the population covariance matrix $\mathbf{\Sigma}$.

In practice, however, we do not know the covariance matrix, but rather only have access to a finite collection of samples, say $\{x_i\}_{i=1}^n$, each drawn according to $\mathbb{P}$. Based on these samples (and using the zero-mean assumption), we can form the sample covariance matrix $\widehat{\Sigma} = \frac{1}{n}\sum_{i=1}^n x_i x_i^{\mathrm{T}}$. The empirical version of PCA is based on the "plug-in" principle, namely replacing the unknown population covariance $\mathbf{\Sigma}$ with this empirical version $\widehat{\mathbf{\Sigma}}$. For instance, the empirical analog of the first principal component (8.1) is given by the optimization problem

$$\widehat{v} = \arg\max_{v \in \mathbb{S}^{d-1}} \left\langle v,\, \widehat{\mathbf{\Sigma}} v \right\rangle. \tag{8.3}$$

Consequently, from the statistical point of view, we need to understand in what sense the maximizers of these empirically defined problems provide good approximations to their population analogs. Alternatively phrased, we need to determine how the eigenstructures of the population and sample covariance matrices are related.

8.1.1 Interpretations and uses of PCA

Before turning to the analysis of PCA, let us consider some of its interpretations and applications.

Example 8.1 (PCA as matrix approximation) Principal component analysis can be interpreted in terms of low-rank approximation. In particular, given some unitarily invariant[1] matrix norm $|||\cdot|||$, consider the problem of finding the best rank-r approximation to a given matrix $\mathbf{\Sigma}$—that is,

$$\mathbf{Z}^* = \arg\min_{\mathrm{rank}(\mathbf{Z}) \leq r} \left\{ |||\mathbf{\Sigma} - \mathbf{Z}|||^2 \right\}. \tag{8.4}$$

In this interpretation, the matrix $\mathbf{\Sigma}$ need only be symmetric, not necessarily positive semidefinite as it must be when it is a covariance matrix. A classical result known as the *Eckart–Young–Mirsky theorem* guarantees that an optimal solution $\mathbf{Z}^*$ exists, and takes the form of a truncated eigendecomposition, specified in terms of the top r eigenvectors of the matrix $\mathbf{\Sigma}$. More precisely, recall that the symmetric matrix $\mathbf{\Sigma}$ has an orthonormal basis of eigenvectors, say $\{v_1, \ldots, v_d\}$, associated with its ordered eigenvalues $\{\gamma_j(\mathbf{\Sigma})\}_{j=1}^d$. In terms of this notation, the optimal rank-r approximation takes the form

$$\mathbf{Z}^* = \sum_{j=1}^{r} \gamma_j(\mathbf{\Sigma})\,(v_j \otimes v_j), \tag{8.5}$$

where $v_j \otimes v_j := v_j v_j^{\mathrm{T}}$ is the rank-one outer product. For the Frobenius matrix norm $|||\mathbf{M}|||_{\mathrm{F}} = \sqrt{\sum_{j,k=1}^d M_{jk}^2}$, the error in the optimal approximation is given by

$$|||\mathbf{Z}^* - \mathbf{\Sigma}|||_{\mathrm{F}}^2 = \sum_{j=r+1}^{d} \gamma_j^2(\mathbf{\Sigma}). \tag{8.6}$$

Figure 8.1 provides an illustration of the matrix approximation view of PCA. We first

[1] For a symmetric matrix $\mathbf{M}$, a matrix norm is unitarily invariant if $|||\mathbf{M}||| = |||\mathbf{V}^{\mathrm{T}}\mathbf{M}\mathbf{V}|||$ for any orthonormal matrix $\mathbf{V}$. See Exercise 8.2 for further discussion.

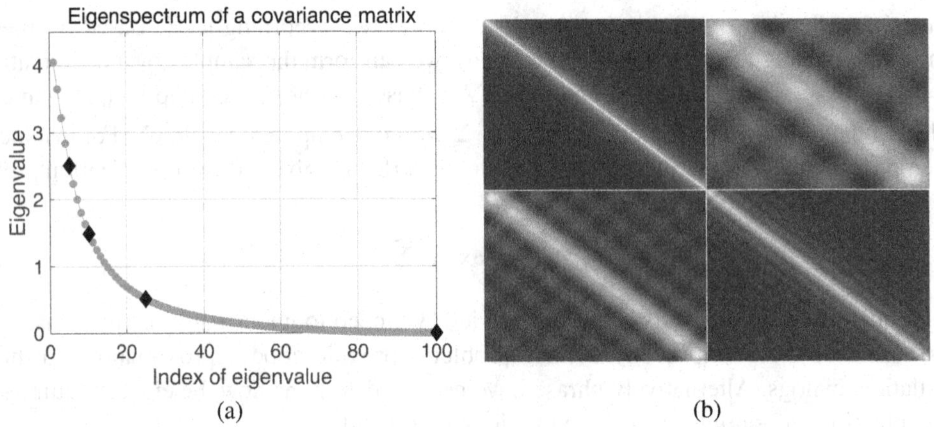

Figure 8.1 Illustration of PCA for low-rank matrix approximation. (a) Eigenspectrum of a matrix $\mathbf{\Sigma} \in \mathcal{S}_+^{100\times 100}$ generated as described in the text. Note the extremely rapid decay of the sorted eigenspectrum. Dark diamonds mark the rank cutoffs $r \in \{5, 10, 25, 100\}$, the first three of which define three approximations to the whole matrix ($r = 100$.) (b) Top left: original matrix. Top right: approximation based on $r = 5$ components. Bottom left: approximation based on $r = 10$ components. Bottom right: approximation based on $r = 25$ components.

generated the Toeplitz matrix $\mathbf{T} \in \mathcal{S}_+^{d\times d}$ with entries $T_{jk} = e^{-\alpha\sqrt{|j-k|}}$ with $\alpha = 0.95$, and then formed the recentered matrix $\mathbf{\Sigma} := \mathbf{T} - \gamma_{\min}(\mathbf{T})\mathbf{I}_d$. Figure 8.1(a) shows the eigenspectrum of the matrix $\mathbf{\Sigma}$: note that the rapid decay of the eigenvalues that renders it amenable to an accurate low-rank approximation. The top left image in Figure 8.1(b) corresponds to the original matrix $\mathbf{\Sigma}$, whereas the remaining images illustrate approximations with increasing rank ($r = 5$ in top right, $r = 10$ in bottom left and $r = 25$ in bottom right). Although the defects in approximations with rank $r = 5$ or $r = 10$ are readily apparent, the approximation with rank $r = 25$ seems reasonable. ♣

Example 8.2 (PCA for data compression) Principal component analysis can also be interpreted as a linear form of data compression. Given a zero-mean random vector $X \in \mathbb{R}^d$, a simple way in which to compress it is via projection to a lower-dimensional subspace $\mathbb{V}$—say via a projection operator of the form $\Pi_{\mathbb{V}}(X)$. For a fixed dimension r, how do we choose the subspace $\mathbb{V}$? Consider the criterion that chooses $\mathbb{V}$ by minimizing the mean-squared error

$$\mathbb{E}[\|X - \Pi_{\mathbb{V}}(X)\|_2^2].$$

This optimal subspace need not be unique in general, but will be when there is a gap between the eigenvalues $\gamma_r(\mathbf{\Sigma})$ and $\gamma_{r+1}(\mathbf{\Sigma})$. In this case, the optimal subspace $\mathbb{V}^*$ is spanned by the top r eigenvectors of the covariance matrix $\mathbf{\Sigma} = \text{cov}(X)$. In particular, the projection operator $\Pi_{\mathbb{V}^*}$ can be written as $\Pi_{\mathbb{V}^*}(x) = \mathbf{V}_r\mathbf{V}_r^{\mathrm{T}} x$, where $\mathbf{V}_r \in \mathbb{R}^{d\times r}$ is an orthonormal matrix with the top r eigenvectors $\{v_1, \ldots, v_r\}$ as its columns. Using this optimal projection, the minimal

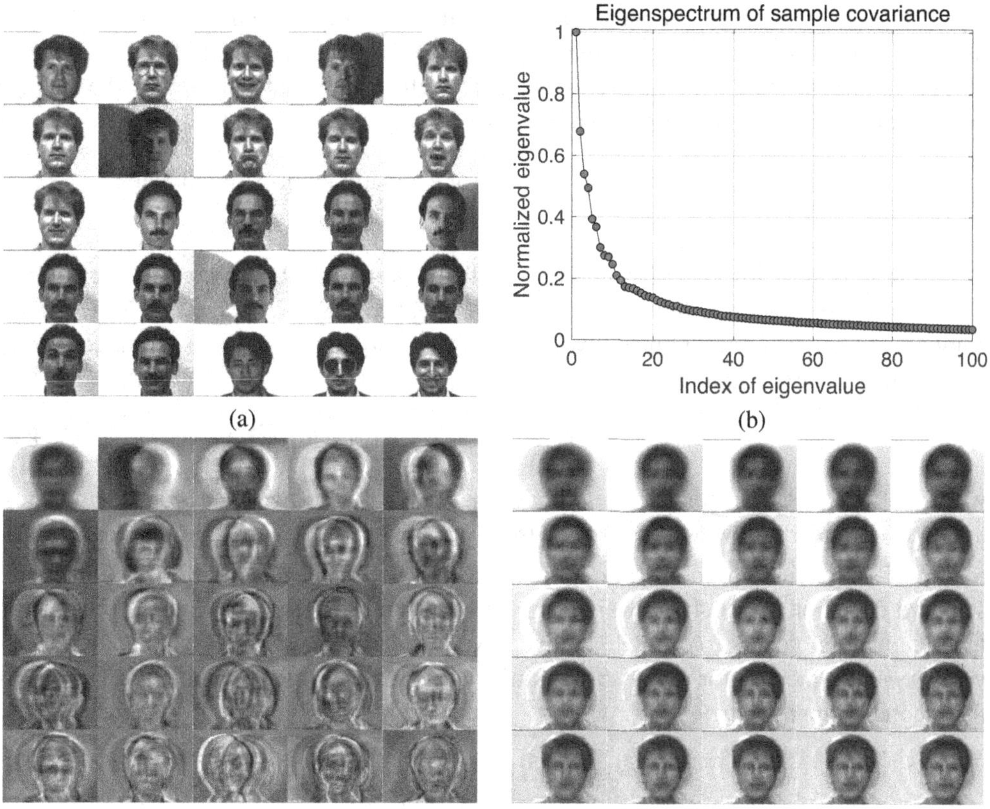

(a) (b) (c) (d)

Figure 8.2 (a) Samples of face images from the Yale Face Database. (b) First 100 eigenvalues of the sample covariance matrix. (c) First 25 eigenfaces computed from the sample covariance matrix. (d) Reconstructions based on the first 25 eigenfaces plus the average face.

reconstruction error based on a rank-r projection is given by

$$\mathbb{E}[\|X - \Pi_{\mathbb{V}^*}(X)\|_2^2] = \sum_{j=r+1}^{d} \gamma_j^2(\mathbf{\Sigma}), \tag{8.7}$$

where $\{\gamma_j(\mathbf{\Sigma})\}_{j=1}^d$ are the ordered eigenvalues of $\mathbf{\Sigma}$. See Exercise 8.4 for further exploration of these and other properties.

The problem of face analysis provides an interesting illustration of PCA for data compression. Consider a large database of face images, such as those illustrated in Figure 8.2(a). Taken from the Yale Face Database, each image is gray-scale with dimensions 243×320. By vectorizing each image, we obtain a vector x in $d = 243 \times 320 = 77\,760$ dimensions. We compute the average image $\bar{x} = \frac{1}{n}\sum_{i=1}^n x_i$ and the sample covariance matrix $\widehat{\mathbf{\Sigma}} = \frac{1}{n-1}\sum_{i=1}^n (x_i - \bar{x})(x_i - \bar{x})^{\mathrm{T}}$ based on $n = 165$ samples. Figure 8.2(b) shows the relatively fast decay of the first 100 eigenvalues of this sample covariance matrix. Figure 8.2(c) shows the average face (top left image) along with the first 24 "eigenfaces", meaning the top 25 eigenvectors of the sample covariance matrix, each converted back to a 243×320 image. Finally, for a particular sample, Figure 8.2(d) shows a sequence of reconstructions of a given face, starting with the average face (top left image), and followed by the average face in conjunction with principal components 1 through 24. ♣

Principal component analysis can also be used for estimation in mixture models.

Example 8.3 (PCA for Gaussian mixture models) Let $\phi(\cdot;\mu,\mathbf{\Sigma})$ denote the density of a Gaussian random vector with mean vector $\mu \in \mathbb{R}^d$ and covariance matrix $\mathbf{\Sigma} \in \mathcal{S}_+^{d\times d}$. A two-component Gaussian mixture model with isotropic covariance structure is a random vector $X \in \mathbb{R}^d$ drawn according to the density

$$f(x;\theta) = \alpha\,\phi(x;-\theta^*,\sigma^2\mathbf{I}_d) + (1-\alpha)\,\phi(x;\theta^*,\sigma^2\mathbf{I}_d), \tag{8.8}$$

where $\theta^* \in \mathbb{R}^d$ is a vector parameterizing the means of the two Gaussian components, $\alpha \in (0,1)$ is a mixture weight and $\sigma > 0$ is a dispersion term. Figure 8.3 provides an illustration of such a mixture model in $d = 2$ dimensions, with mean vector $\theta^* = \begin{bmatrix}0.6 & -0.6\end{bmatrix}^{\mathrm{T}}$, standard deviation $\sigma = 0.4$ and weight $\alpha = 0.4$. Given samples $\{x_i\}_{i=1}^n$ drawn from such a model, a natural goal is to estimate the mean vector θ^*. Principal component analysis provides a natural method for doing so. In particular, a straightforward calculation yields that the second-moment matrix

$$\mathbf{\Gamma} := \mathbb{E}[X \otimes X] = \theta^* \otimes \theta^* + \sigma^2\mathbf{I}_d,$$

where $X \otimes X := XX^{\mathrm{T}}$ is the $d \times d$ rank-one outer product matrix. Thus, we see that θ^* is proportional to the maximal eigenvector of $\mathbf{\Gamma}$. Consequently, a reasonable estimator $\widehat{\theta}$ is

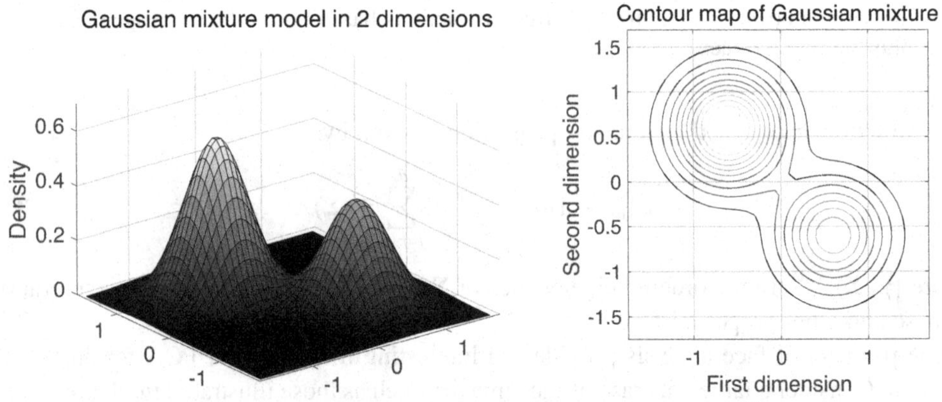

(a) (b)

Figure 8.3 Use of PCA for Gaussian mixture models. (a) Density function of a two-component Gaussian mixture (8.8) with mean vector $\theta^* = [0.6 \quad -0.6]^{\mathrm{T}}$, standard deviation $\sigma = 0.4$ and weight $\alpha = 0.4$. (b) Contour plots of the density function, which provide intuition as to why PCA should be useful in recovering the mean vector θ^*.

given by the maximal eigenvector of the sample second moment[2] matrix $\widehat{\Gamma} = \frac{1}{n}\sum_{i=1}^n x_i x_i^{\mathrm{T}}$. We study the properties of this estimator in Exercise 8.6. ♣

8.1.2 Perturbations of eigenvalues and eigenspaces

Thus far, we have seen that the eigenvectors of population and sample covariance matrices are interesting objects with a range of uses. In practice, PCA is always applied to the sample covariance matrix, and the central question of interest is how well the sample-based eigenvectors approximate those of the population covariance.

Before addressing this question, let us make a brief detour into matrix perturbation theory. Let us consider the following general question: given a symmetric matrix $\mathbf{R}$, how does its eigenstructure relate to the perturbed matrix $\mathbf{Q} = \mathbf{R}+\mathbf{P}$? Here $\mathbf{P}$ is another symmetric matrix, playing the role of the perturbation. It turns out that the eigenvalues of $\mathbf{Q}$ and $\mathbf{R}$ are related in a straightforward manner. Understanding how the eigenspaces change, however, requires some more care.

Let us begin with changes in the eigenvalues. From the standard variational definition of the maximum eigenvalue, we have

$$\gamma_1(\mathbf{Q}) = \max_{v\in\mathbb{S}^{d-1}} \langle v, (\mathbf{R}+\mathbf{P})v\rangle \leq \max_{v\in\mathbb{S}^{d-1}} \langle v, \mathbf{R}v\rangle + \max_{v\in\mathbb{S}^{d-1}} \langle v, \mathbf{P}v\rangle \leq \gamma_1(\mathbf{R}) + |||\mathbf{P}|||_2.$$

Since the same argument holds with the roles of $\mathbf{Q}$ and $\mathbf{R}$ reversed, we conclude that $|\gamma_1(\mathbf{Q}) - \gamma_1(\mathbf{R})| \leq |||\mathbf{Q}-\mathbf{R}|||_2$. Thus, the maximum eigenvalues of $\mathbf{Q}$ and $\mathbf{R}$ can differ by at most the operator norm of their difference. More generally, we have

$$\max_{j=1,\ldots,d} \big|\gamma_j(\mathbf{Q}) - \gamma_j(\mathbf{R})\big| \leq |||\mathbf{Q}-\mathbf{R}|||_2. \tag{8.9}$$

This bound is a consequence of a more general result known as *Weyl's inequality*; we work through its proof in Exercise 8.3.

Although eigenvalues are generically stable, the same does not hold for eigenvectors and eigenspaces, unless further conditions are imposed. The following example provides an illustration of such instability:

Example 8.4 (Sensitivity of eigenvectors) For a parameter $\epsilon \in [0,1]$, consider the family of symmetric matrices

$$\mathbf{Q}_\epsilon := \begin{bmatrix} 1 & \epsilon \\ \epsilon & 1.01 \end{bmatrix} = \underbrace{\begin{bmatrix} 1 & 0 \\ 0 & 1.01 \end{bmatrix}}_{\mathbf{Q}_0} + \epsilon \underbrace{\begin{bmatrix} 0 & 1 \\ 1 & 0 \end{bmatrix}}_{\mathbf{P}}. \tag{8.10}$$

By construction, the matrix $\mathbf{Q}_\epsilon$ is a perturbation of a diagonal matrix $\mathbf{Q}_0$ by an ϵ-multiple of the fixed matrix $\mathbf{P}$. Since $|||\mathbf{P}|||_2 = 1$, the magnitude of the perturbation is directly controlled by ϵ. On one hand, the eigenvalues remain stable to this perturbation: in terms of the shorthand $a = 1.01$, we have $\gamma(\mathbf{Q}_0) = \{1, a\}$ and

$$\gamma(\mathbf{Q}_\epsilon) = \Big\{\tfrac{1}{2}\big[(a+1) + \sqrt{(a-1)^2 + 4\epsilon^2}\big],\ \ \tfrac{1}{2}\big[(a+1) - \sqrt{(a-1)^2 + 4\epsilon^2}\big]\Big\}.$$

[2] This second-moment matrix coincides with the usual covariance matrix for the special case of an equally weighted mixture pair with $\alpha = 0.5$.

Thus, we find that

$$\max_{j=1,2} |\gamma_j(\mathbf{Q}_0) - \gamma_j(\mathbf{Q}_\epsilon)| = \frac{1}{2}\left|(a-1) - \sqrt{(a-1)^2 + 4\epsilon^2}\right| \leq \epsilon,$$

which confirms the validity of Weyl's inequality (8.9) in this particular case.

On the other hand, the maximal eigenvector of $\mathbf{Q}_\epsilon$ is very different from that of $\mathbf{Q}_0$, even for relatively small values of ϵ. For $\epsilon = 0$, the matrix $\mathbf{Q}_0$ has the unique maximal eigenvector $v_0 = [0 \quad 1]^{\mathrm{T}}$. However, if we set $\epsilon = 0.01$, a numerical calculation shows that the maximal eigenvector of $\mathbf{Q}_\epsilon$ is $v_\epsilon \approx [0.53 \quad 0.85]^{\mathrm{T}}$. Note that $\|v - v_\epsilon\|_2 \gg \epsilon$, showing that eigenvectors can be extremely sensitive to perturbations. ♣

What is the underlying problem? The issue is that, while $\mathbf{Q}_0$ has a unique maximal eigenvector, the gap between the largest eigenvalue $\gamma_1(\mathbf{Q}_0) = 1.01$ and the second largest eigenvalue $\gamma_2(\mathbf{Q}_0) = 1$ is very small. Consequently, even small perturbations of the matrix lead to "mixing" between the spaces spanned by the top and second largest eigenvectors. On the other hand, if this eigengap can be bounded away from zero, then it turns out that we can guarantee stability of the eigenvectors. We now turn to this type of theory.

8.2 Bounds for generic eigenvectors

We begin our exploration of eigenvector bounds with the generic case, in which no additional structure is imposed on the eigenvectors. In later sections, we turn to structured variants of eigenvector estimation.

8.2.1 A general deterministic result

Consider a symmetric positive semidefinite matrix $\mathbf{\Sigma}$ with eigenvalues ordered as

$$\gamma_1(\mathbf{\Sigma}) \geq \gamma_2(\mathbf{\Sigma}) \geq \gamma_3(\mathbf{\Sigma}) \geq \cdots \geq \gamma_d(\mathbf{\Sigma}) \geq 0.$$

Let $\theta^* \in \mathbb{R}^d$ denote its maximal eigenvector, assumed to be unique. Now consider a perturbed version $\widehat{\mathbf{\Sigma}} = \mathbf{\Sigma} + \mathbf{P}$ of the original matrix. As suggested by our notation, in the context of PCA, the original matrix corresponds to the population covariance matrix, whereas the perturbed matrix corresponds to the sample covariance. However, at least for the time being, our theory should be viewed as general.

As should be expected based on Example 8.4, any theory relating the maximum eigenvectors of $\mathbf{\Sigma}$ and $\widehat{\mathbf{\Sigma}}$ should involve the *eigengap* $\nu := \gamma_1(\mathbf{\Sigma}) - \gamma_2(\mathbf{\Sigma})$, assumed to be strictly positive. In addition, the following result involves the transformed perturbation matrix

$$\widetilde{\mathbf{P}} := \mathbf{U}^{\mathrm{T}}\mathbf{P}\mathbf{U} = \begin{bmatrix} \tilde{p}_{11} & \tilde{p}^{\mathrm{T}} \\ \tilde{p} & \widetilde{\mathbf{P}}_{22} \end{bmatrix}, \tag{8.11}$$

where $\tilde{p}_{11} \in \mathbb{R}$, $\tilde{p} \in \mathbb{R}^{d-1}$ and $\widetilde{\mathbf{P}}_{22} \in \mathbb{R}^{(d-1)\times(d-1)}$. Here $\mathbf{U}$ is an orthonormal matrix with the eigenvectors of $\mathbf{\Sigma}$ as its columns.

Theorem 8.5 *Consider a positive semidefinite matrix $\mathbf{\Sigma}$ with maximum eigenvector $\theta^* \in \mathbb{S}^{d-1}$ and eigengap $\nu = \gamma_1(\mathbf{\Sigma}) - \gamma_2(\mathbf{\Sigma}) > 0$. Given any matrix $\mathbf{P} \in \mathcal{S}^{d\times d}$ such that $|||\mathbf{P}|||_2 < \nu/2$, the perturbed matrix $\widehat{\mathbf{\Sigma}} := \mathbf{\Sigma} + \mathbf{P}$ has a unique maximal eigenvector $\widehat{\theta}$ satisfying the bound*

$$\|\widehat{\theta} - \theta^*\|_2 \leq \frac{2\|\tilde{p}\|_2}{\nu - 2|||\mathbf{P}|||_2}. \tag{8.12}$$

In general, this bound is sharp in the sense that there are problems for which the requirement $|||\mathbf{P}|||_2 < \nu/2$ cannot be loosened. As an example, suppose that $\mathbf{\Sigma} = \mathrm{diag}\{2, 1\}$ so that $\nu = 2 - 1 = 1$. Given $\mathbf{P} = \mathrm{diag}\{-\frac{1}{2}, +\frac{1}{2}\}$, the perturbed matrix $\widehat{\mathbf{\Sigma}} = \mathbf{\Sigma} + \mathbf{P} = \frac{3}{2}\mathbf{I}_2$ no longer has a unique maximal eigenvector. Note that this counterexample lies just at the boundary of our requirement, since $|||\mathbf{P}|||_2 = \frac{1}{2} = \frac{\nu}{2}$.

Proof Our proof is variational in nature, based on the optimization problems that characterize the maximal eigenvectors of the matrices $\mathbf{\Sigma}$ and $\widehat{\mathbf{\Sigma}}$, respectively. Define the error vector $\widehat{\Delta} = \widehat{\theta} - \theta^*$, and the function

$$\Psi(\Delta; \mathbf{P}) := \langle \Delta, \mathbf{P}\Delta \rangle + 2\langle \Delta, \mathbf{P}\theta^* \rangle. \tag{8.13}$$

In parallel to our analysis of sparse linear regression from Chapter 7, the first step in our analysis is to prove the *basic inequality for PCA*. For future reference, we state this inequality in a slightly more general form than required for the current proof. In particular, given any subset $C \subseteq \mathbb{S}^{d-1}$, let θ^* and $\widehat{\theta}$ maximize the quadratic objectives

$$\max_{\theta \in C} \langle \theta, \mathbf{\Sigma}\theta \rangle \quad \text{and} \quad \max_{\theta \in C} \langle \theta, \widehat{\mathbf{\Sigma}}\theta \rangle, \tag{8.14}$$

respectively. The current proof involves the choice $C = \mathbb{S}^{d-1}$.

It is convenient to bound the distance between $\widehat{\theta}$ and θ^* in terms of the inner product $\varrho := \langle \widehat{\theta}, \theta^* \rangle$. Due to the sign ambiguity in eigenvector estimation, we may assume without loss of generality that $\widehat{\theta}$ is chosen such that $\varrho \in [0, 1]$.

Lemma 8.6 (PCA basic inequality) *Given a matrix $\mathbf{\Sigma}$ with eigengap $\nu > 0$, the error $\widehat{\Delta} = \widehat{\theta} - \theta^*$ is bounded as*

$$\nu\left(1 - \left\langle \widehat{\theta}, \theta^* \right\rangle^2\right) \leq \left|\Psi(\widehat{\Delta}; \mathbf{P})\right|. \tag{8.15}$$

Taking this inequality as given for the moment, the remainder of the proof is straightforward. Recall the transformation $\widetilde{\mathbf{P}} = \mathbf{U}^{\mathrm{T}}\mathbf{P}\mathbf{U}$, or equivalently $\mathbf{P} = \mathbf{U}\widetilde{\mathbf{P}}\mathbf{U}^{\mathrm{T}}$. Substituting this expression into equation (8.13) yields

$$\Psi(\widehat{\Delta}; \mathbf{P}) = \left\langle \mathbf{U}^{\mathrm{T}}\widehat{\Delta}, \widetilde{\mathbf{P}}\mathbf{U}^{\mathrm{T}}\widehat{\Delta} \right\rangle + 2\left\langle \mathbf{U}^{\mathrm{T}}\widehat{\Delta}, \widetilde{\mathbf{P}}\,\mathbf{U}^{\mathrm{T}}\theta^* \right\rangle. \tag{8.16}$$

In terms of the inner product $\varrho = \left\langle \widehat{\theta}, \theta^* \right\rangle$, we may write $\widehat{\theta} = \varrho\,\theta^* + \sqrt{1 - \varrho^2}\, z$, where $z \in \mathbb{R}^d$ is a vector orthogonal to θ^*. Since the matrix $\mathbf{U}$ is orthonormal with its first column given by

θ^*, we have $\mathbf{U}^T\theta^* = e_1$. Letting $\mathbf{U}_2 \in \mathbb{R}^{d\times(d-1)}$ denote the submatrix formed by the remaining $d-1$ eigenvectors and defining the vector $\widetilde{z} = U_2^T z \in \mathbb{R}^{d-1}$, we can write

$$\mathbf{U}^T\widehat{\Delta} = \begin{bmatrix}(\varrho - 1) & (1-\varrho^2)^{\frac{1}{2}}\,\widetilde{z}\end{bmatrix}^T.$$

Substituting these relations into equation (8.16) yields that

$$\begin{aligned}\Psi(\widehat{\Delta}; \mathbf{P}) &= (\varrho-1)^2\tilde{p}_{11} + 2(\varrho-1)\sqrt{1-\varrho^2}\,\langle\widetilde{z},\, \tilde{p}\rangle + (1-\varrho^2)\left\langle\widetilde{z},\, \widetilde{\mathbf{P}}_{22}\widetilde{z}\right\rangle \\ &\quad + 2(\varrho-1)\tilde{p}_{11} + 2\sqrt{1-\varrho^2}\,\langle\widetilde{z},\, \tilde{p}\rangle \\ &= (\varrho^2-1)\tilde{p}_{11} + 2\varrho\sqrt{1-\varrho^2}\,\langle\widetilde{z},\, \tilde{p}\rangle + (1-\varrho^2)\left\langle\widetilde{z},\, \widetilde{\mathbf{P}}_{22}\widetilde{z}\right\rangle.\end{aligned}$$

Putting together the pieces, since $\|\widetilde{z}\|_2 \leq 1$ and $|\tilde{p}_{11}| \leq |||\widetilde{\mathbf{P}}|||_2$, we have

$$|\Psi(\widehat{\Delta}; \mathbf{P})| \leq 2(1-\varrho^2)|||\widetilde{\mathbf{P}}|||_2 + 2\varrho\sqrt{1-\varrho^2}\|\tilde{p}\|_2.$$

Combined with the basic inequality (8.15), we find that

$$\nu(1-\varrho^2) \leq 2(1-\varrho^2)|||\mathbf{P}|||_2 + 2\varrho\sqrt{1-\varrho^2}\|\tilde{p}\|_2.$$

Whenever $\nu > 2|||\mathbf{P}|||_2$, this inequality implies that $\sqrt{1-\varrho^2} \leq \frac{2\varrho\|\tilde{p}\|_2}{\nu - 2|||\mathbf{P}|||_2}$. Noting that $\|\widehat{\Delta}\|_2 = \sqrt{2(1-\varrho)}$, we thus conclude that

$$\|\widehat{\Delta}\|_2 \leq \frac{\sqrt{2}\varrho}{\sqrt{1+\varrho}}\left(\frac{2\|\tilde{p}\|_2}{\nu - 2|||\mathbf{P}|||_2}\right) \leq \frac{2\|\tilde{p}\|_2}{\nu - 2|||\mathbf{P}|||_2},$$

where the final step follows since $2\varrho^2 \leq 1 + \varrho$ for all $\varrho \in [0, 1]$.

Let us now return to prove the PCA basic inequality (8.15).

Proof of Lemma 8.6: Since $\widehat{\theta}$ and θ^* are optimal and feasible, respectively, for the programs (8.14), we are guaranteed that $\left\langle\theta^*,\, \widehat{\mathbf{\Sigma}}\,\theta^*\right\rangle \leq \left\langle\widehat{\theta},\, \widehat{\mathbf{\Sigma}}\widehat{\theta}\right\rangle$. Defining the matrix perturbation $\mathbf{P} = \widehat{\mathbf{\Sigma}} - \mathbf{\Sigma}$, we have

$$\langle\langle \mathbf{\Sigma},\ \theta^*\otimes\theta^* - \widehat{\theta}\otimes\widehat{\theta}\rangle\rangle \leq -\langle\langle \mathbf{P},\ \theta^*\otimes\theta^* - \widehat{\theta}\otimes\widehat{\theta}\rangle\rangle,$$

where $\langle\langle \mathbf{A},\ \mathbf{B}\rangle\rangle$ is the trace inner product, and $a\otimes a = aa^T$ denotes the rank-one outer product. Following some simple algebra, the right-hand side is seen to be equal to $-\Psi(\widehat{\Delta}; \mathbf{P})$. The final step is to show that

$$\langle\langle \mathbf{\Sigma},\ \theta^*\otimes\theta^* - \widehat{\theta}\otimes\widehat{\theta}\rangle\rangle \geq \frac{\nu}{2}\|\widehat{\Delta}\|_2^2. \tag{8.17}$$

Recall the representation $\widehat{\theta} = \varrho\theta^* + (\sqrt{1-\varrho^2})\,z$, where the vector $z \in \mathbb{R}^d$ is orthogonal to θ^*, and $\varrho \in [0, 1]$. Using the shorthand notation $\gamma_j \equiv \gamma_j(\mathbf{\Sigma})$ for $j = 1, 2$, define the matrix $\mathbf{\Gamma} = \mathbf{\Sigma} - \gamma_1(\theta^*\otimes\theta^*)$, and note that $\mathbf{\Gamma}\theta^* = 0$ and $|||\mathbf{\Gamma}|||_2 \leq \gamma_2$ by construction. Consequently, we can write

$$\begin{aligned}\langle\langle \mathbf{\Sigma},\ \theta^*\otimes\theta^* - \widehat{\theta}\otimes\widehat{\theta}\rangle\rangle &= \gamma_1\langle\langle \theta^*\otimes\theta^*,\ \theta^*\otimes\theta^* - \widehat{\theta}\otimes\widehat{\theta}\rangle\rangle + \langle\langle \mathbf{\Gamma},\ \theta^*\otimes\theta^* - \widehat{\theta}\otimes\widehat{\theta}\rangle\rangle \\ &= (1-\varrho^2)\{\gamma_1 - \langle\langle \mathbf{\Gamma},\ z\otimes z\rangle\rangle\}.\end{aligned}$$

Since $|||\mathbf{\Gamma}|||_2 \leq \gamma_2$, we have $|\langle\langle \mathbf{\Gamma},\ z \otimes z \rangle\rangle| \leq \gamma_2$. Putting together the pieces, we have shown that

$$\langle\langle \mathbf{\Sigma},\ \theta^* \otimes \theta^* - \widehat{\theta} \otimes \widehat{\theta} \rangle\rangle \geq (1 - \varrho^2)\{\gamma_1 - \gamma_2\} = (1 - \varrho^2)\,\nu,$$

from which the claim (8.15) follows. □

8.2.2 Consequences for a spiked ensemble

Theorem 8.5 applies to any form of matrix perturbation. In the context of principal component analysis, this perturbation takes a very specific form—namely, as the difference between the sample and population covariance matrices. More concretely, suppose that we have drawn n i.i.d. samples $\{x_i\}_{i=1}^n$ from a zero-mean random vector with covariance $\mathbf{\Sigma}$. Principal component analysis is then based on the eigenstructure of the sample covariance matrix $\widehat{\mathbf{\Sigma}} = \frac{1}{n}\sum_{i=1}^n x_i x_i^{\mathrm{T}}$, and the goal is to draw conclusions about the eigenstructure of the population matrix.

In order to bring sharper focus to this issue, let us study how PCA behaves for a very simple class of covariance matrices, known as spiked covariance matrices. A sample $x_i \in \mathbb{R}^d$ from the *spiked covariance ensemble* takes the form

$$x_i \stackrel{\mathrm{d}}{=} \sqrt{\nu}\,\xi_i\,\theta^* + w_i, \tag{8.18}$$

where $\xi_i \in \mathbb{R}$ is a zero-mean random variable with unit variance, and $w_i \in \mathbb{R}^d$ is a random vector independent of ξ_i, with zero mean and covariance matrix $\mathbf{I}_d$. Overall, the random vector x_i has zero mean, and a covariance matrix of the form

$$\mathbf{\Sigma} := \nu\,\theta^*\,(\theta^*)^{\mathrm{T}} + \mathbf{I}_d. \tag{8.19}$$

By construction, for any $\nu > 0$, the vector θ^* is the unique maximal eigenvector of $\mathbf{\Sigma}$ with eigenvalue $\gamma_1(\mathbf{\Sigma}) = \nu + 1$. All other eigenvalues of $\mathbf{\Sigma}$ are located at 1, so that we have an eigengap $\gamma_1(\mathbf{\Sigma}) - \gamma_2(\mathbf{\Sigma}) = \nu$.

In the following result, we say that the vector $x_i \in \mathbb{R}^d$ has sub-Gaussian tails if both ξ_i and w_i are sub-Gaussian with parameter at most one.

Corollary 8.7 *Given i.i.d. samples $\{x_i\}_{i=1}^n$ from the spiked covariance ensemble* (8.18) *with sub-Gaussian tails, suppose that $n > d$ and $\sqrt{\frac{\nu+1}{\nu^2}}\sqrt{\frac{d}{n}} \leq \frac{1}{128}$. Then, with probability at least $1 - c_1 e^{-c_2 n \min\{\sqrt{\nu}\delta,\, \nu\delta^2\}}$, there is a unique maximal eigenvector $\widehat{\theta}$ of the sample covariance matrix $\widehat{\mathbf{\Sigma}} = \frac{1}{n}\sum_{i=1}^n x_i x_i^{\mathrm{T}}$ such that*

$$\|\widehat{\theta} - \theta^*\|_2 \leq c_0 \sqrt{\frac{\nu+1}{\nu^2}}\sqrt{\frac{d}{n}} + \delta. \tag{8.20}$$

Figure 8.4 shows the results of simulations that confirm the qualitative scaling predicted by Corollary 8.7. In each case, we drew $n = 500$ samples from a spiked covariance matrix with the signal-to-noise parameter ν ranging over the interval [0.75, 5]. We then computed the ℓ_2-distance $\|\widehat{\theta} - \theta^*\|_2$ between the maximal eigenvectors of the sample and population

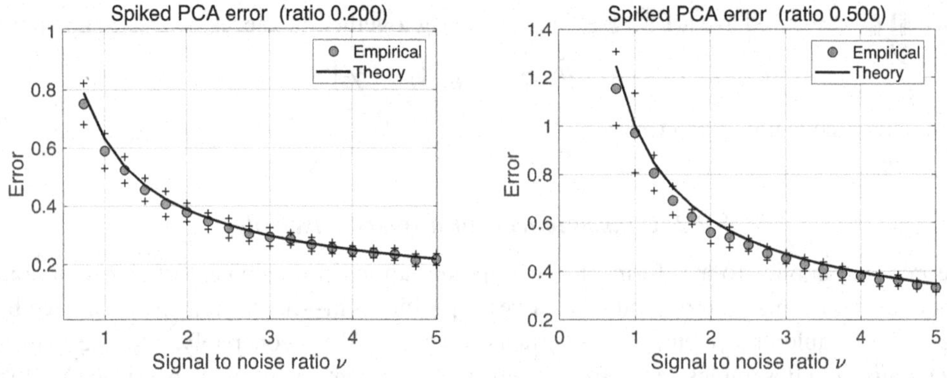

(a) (b)

Figure 8.4 Plots of the error $\|\widehat{\theta} - \theta^*\|_2$ versus the signal-to-noise ratio, as measured by the eigengap ν. Both plots are based on a sample size $n = 500$. Dots show the average of 100 trials, along with the standard errors (crosses). The full curve shows the theoretical bound $\sqrt{\frac{\nu+1}{\nu^2}}\sqrt{\frac{d}{n}}$. (a) Dimension $d = 100$. (b) Dimension $d = 250$.

covariances, respectively, performing $T = 100$ trials for each setting of ν. The circles in Figure 8.4 show the empirical means, along with standard errors in crosses, whereas the solid curve corresponds to the theoretical prediction $\sqrt{\frac{\nu+1}{\nu^2}}\sqrt{\frac{d}{n}}$. Note that Corollary 8.7 predicts this scaling, but with a looser leading constant ($c_0 > 1$). As shown by Figure 8.4, Corollary 8.7 accurately captures the scaling behavior of the error as a function of the signal-to-noise ratio.

Proof Let $\mathbf{P} = \widehat{\mathbf{\Sigma}} - \mathbf{\Sigma}$ be the difference between the sample and population covariance matrices. In order to apply Theorem 8.5, we need to upper bound the quantities $|||\mathbf{P}|||_2$ and $\|\tilde{p}\|_2$. Defining the random vector $\bar{w} := \frac{1}{n}\sum_{i=1}^n \xi_i w_i$, the perturbation matrix $\mathbf{P}$ can be decomposed as

$$\mathbf{P} = \underbrace{\nu\left(\frac{1}{n}\sum_{i=1}^n \xi_i^2 - 1\right)\theta^*(\theta^*)^{\mathrm{T}}}_{\mathbf{P}_1} + \underbrace{\sqrt{\nu}\left(\bar{w}(\theta^*)^{\mathrm{T}} + \theta^*\bar{w}^{\mathrm{T}}\right)}_{\mathbf{P}_2} + \underbrace{\left(\frac{1}{n}\sum_{i=1}^n w_i w_i^{\mathrm{T}} - \mathbf{I}_d\right)}_{\mathbf{P}_3}. \tag{8.21}$$

Since $\|\theta^*\|_2 = 1$, the operator norm of $\mathbf{P}$ can be bounded as

$$|||\mathbf{P}|||_2 \leq \nu\left|\frac{1}{n}\sum_{i=1}^n \xi_i^2 - 1\right| + 2\sqrt{\nu}\|\bar{w}\|_2 + \left|\left|\left|\frac{1}{n}\sum_{i=1}^n w_i w_i^{\mathrm{T}} - \mathbf{I}_d\right|\right|\right|_2. \tag{8.22a}$$

Let us derive a similar upper bound on $\|\tilde{p}\|_2$ using the decomposition (8.11). Since θ^* is the unique maximal eigenvector of $\mathbf{\Sigma}$, it forms the first column of the matrix $\mathbf{U}$. Let $\mathbf{U}_2 \in \mathbb{R}^{d\times(d-1)}$ denote the matrix formed of the remaining $(d-1)$ columns. With this notation, we have $\tilde{p} = \mathbf{U}_2^{\mathrm{T}}\mathbf{P}\theta^*$. Using the decomposition (8.21) of the perturbation matrix and the fact that $\mathbf{U}_2^{\mathrm{T}}\theta^* = 0$, we find that $\tilde{p} = \sqrt{\nu}\,\mathbf{U}_2^{\mathrm{T}}\bar{w} + \frac{1}{n}\sum_{i=1}^n \mathbf{U}_2^{\mathrm{T}} w_i \,\langle w_i, \theta^*\rangle$. Since $\mathbf{U}_2$ has orthonormal columns, we have $\|\mathbf{U}_2^{\mathrm{T}}\bar{w}\|_2 \leq \|\bar{w}\|_2$ and also

$$\|\sum_{i=1}^{n} \mathbf{U}_2^{\mathrm{T}} w_i \langle w_i, \theta^* \rangle \|_2 = \sup_{\|v\|_2=1} \left|(\mathbf{U}_2 v)^{\mathrm{T}}(\sum_{i=1}^{n} w_i w_i^{\mathrm{T}} - \mathbf{I}_d)\theta^*\right| \leq |||\frac{1}{n}\sum_{i=1}^{n} w_i w_i^{\mathrm{T}} - \mathbf{I}_d|||_2.$$

Putting together the pieces, we have shown that

$$\|\tilde{p}\|_2 \leq \sqrt{\nu}\,\|\bar{w}\|_2 + |||\frac{1}{n}\sum_{i=1}^{n} w_i w_i^{\mathrm{T}} - \mathbf{I}_d|||_2. \tag{8.22b}$$

The following lemma allows us to control the quantities appearing the bounds (8.22a) and (8.22b):

Lemma 8.8 *Under the conditions of Corollary 8.7, we have*

$$\mathbb{P}[|\frac{1}{n}\sum_{i=1}^{n} \xi_i^2 - 1| \geq \delta_1] \leq 2e^{-c_2 n \min\{\delta_1, \delta_1^2\}}, \tag{8.23a}$$

$$\mathbb{P}[\|\bar{w}\|_2 \geq 2\sqrt{\frac{d}{n}} + \delta_2] \leq 2e^{-c_2 n \min\{\delta_2, \delta_2^2\}} \tag{8.23b}$$

and

$$\mathbb{P}\left[|||\frac{1}{n}\sum_{i=1}^{n} w_i w_i^{\mathrm{T}} - \mathbf{I}_d|||_2 \geq c_3\sqrt{\frac{d}{n}} + \delta_3\right] \leq 2e^{-c_2 n \min\{\delta_3, \delta_3^2\}}. \tag{8.23c}$$

We leave the proof of this claim as an exercise, since it is straightforward application of results and techniques from previous chapters. For future reference, we define

$$\phi(\delta_1, \delta_2, \delta_3) := 2e^{-c_2 n \min\{\delta_1, \delta_1^2\}} + 2e^{-c_2 n \min\{\delta_2, \delta_2^2\}} + 2e^{-c_2 n \min\{\delta_3, \delta_3^2\}}, \tag{8.24}$$

corresponding to the probability with which at least one of the bounds in Lemma 8.8 is violated.

In order to apply Theorem 8.5, we need to first show that $|||\mathbf{P}|||_2 < \frac{\nu}{4}$ with high probability. Beginning with the inequality (8.22a) and applying Lemma 8.8 with $\delta_1 = \frac{1}{16}$, $\delta_2 = \frac{\delta}{4\sqrt{\nu}}$ and $\delta_3 = \delta/16 \in (0, 1)$, we have

$$|||\mathbf{P}|||_2 \leq \frac{\nu}{16} + 8(\sqrt{\nu} + 1)\sqrt{\frac{d}{n}} + \delta \leq \frac{\nu}{16} + 16(\sqrt{\nu + 1})\sqrt{\frac{d}{n}} + \delta$$

with probability at least $1 - \phi(\frac{1}{4}, \frac{\delta}{3\sqrt{\nu}}, \frac{\delta}{16})$. Consequently, as long as $\sqrt{\frac{\nu+1}{\nu^2}}\sqrt{\frac{d}{n}} \leq \frac{1}{128}$, we have

$$|||\mathbf{P}|||_2 \leq \frac{3}{16}\nu + \delta < \frac{\nu}{4}, \qquad \text{for all } \delta \in (0, \tfrac{\nu}{16}).$$

It remains to bound $\|\tilde{p}\|_2$. Applying Lemma 8.8 to the inequality (8.22b) with the previously specified choices of $(\delta_1, \delta_2, \delta_3)$, we have

$$\|\tilde{p}\|_2 \leq 2(\sqrt{\nu} + 1)\sqrt{\frac{d}{n}} + \delta \leq 4\sqrt{\nu + 1}\sqrt{\frac{d}{n}} + \delta$$

with probability at least $1-\phi(\frac{1}{4}, \frac{\delta}{3\sqrt{\nu}}, \frac{\delta}{16})$. We have shown that conditions of Theorem 8.5 are satisfied, so that the claim (8.20) follows as a consequence of the bound (8.12). □

8.3 Sparse principal component analysis

Note that Corollary 8.7 requires that the sample size n be larger than the dimension d in order for ordinary PCA to perform well. One might wonder whether this requirement is fundamental: does PCA still perform well in the high-dimensional regime $n < d$?

The answer to this question turns out to be a dramatic “no”. As discussed at more length in the bibliography section, for any fixed signal-to-noise ratio, if the ratio d/n stays suitably bounded away from zero, then the eigenvectors of the sample covariance in a spiked covariance model become *asymptotically orthogonal* to their population analogs. Thus, the classical PCA estimate is no better than ignoring the data, and drawing a vector uniformly at random from the Euclidean sphere. Given this total failure of classical PCA, a next question to ask is whether the eigenvectors might be estimated consistently using a method more sophisticated than PCA. This question also has a negative answer: as we discuss in Chapter 15, for the standard spiked model (8.18), it can be shown via the framework of minimax theory that *no method* can produce consistent estimators of the population eigenvectors when d/n stays bounded away from zero. See Example 15.19 in Chapter 15 for the details of this minimax lower bound.

In practice, however, it is often reasonable to impose structure on eigenvectors, and this structure can be exploited to develop effective estimators even when $n < d$. Perhaps the simplest such structure is that of sparsity in the eigenvectors, which allows for both effective estimation in high-dimensional settings, as well as increased interpretability. Accordingly, this section is devoted to the sparse version of principal component analysis.

Let us illustrate the idea of sparse eigenanalysis by revisiting the eigenfaces from Example 8.2.

Example 8.9 (Sparse eigenfaces) We used the images from the Yale Face Database to set up a PCA problem in $d = 77\,760$ dimensions. In this example, we used an iterative method to approximate sparse eigenvectors with at most $s = \lfloor 0.25d \rfloor = 19\,440$ non-zero coefficients. In particular, we applied a thresholded version of the matrix power method for computing sparse eigenvalues and eigenvectors. (See Exercise 8.5 for exploration of the standard matrix power method.)

Figure 8.5(a) shows the average face (top left image), along with approximations to the first 24 sparse eigenfaces. Each sparse eigenface was restricted to have at most 25% of its pixels non-zero, corresponding to a savings of a factor of 4 in storage. Note that the sparse eigenfaces are more localized than their PCA analogs from Figure 8.2. Figure 8.5(b) shows reconstruction using the average face in conjunction with the first 100 sparse eigenfaces, which require equivalent storage (in terms of pixel values) to the first 25 regular eigenfaces. ♣

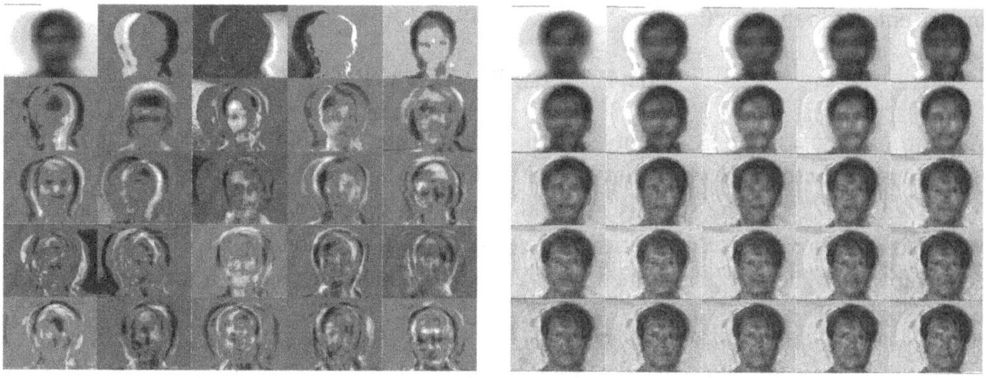

(a) (b)

Figure 8.5 Illustration of sparse eigenanalysis for the Yale Face Database. (a) Average face (top left image), and approximations to the first 24 sparse eigenfaces, obtained by a greedy iterative thresholding procedure applied to the eigenvalue power method. Eigenfaces were restricted to have at most 25% of their pixels non-zero, corresponding to a 1/4 reduction in storage. (b) Reconstruction based on sparse eigenfaces.

8.3.1 A general deterministic result

We now turn to the question of how to estimate a maximal eigenvector that is known *a priori* to be sparse. A natural approach is to augment the quadratic objective function underlying classical PCA with an additional sparsity constraint or penalty. More concretely, we analyze both the constrained problem

$$\widehat{\theta} \in \arg\max_{\|\theta\|_2=1} \left\{ \langle \theta, \widehat{\mathbf{\Sigma}}\, \theta \rangle \right\} \qquad \text{such that } \|\theta\|_1 \leq R, \tag{8.25a}$$

as well as the penalized variant

$$\widehat{\theta} \in \arg\max_{\|\theta\|_2=1} \left\{ \langle \theta, \widehat{\mathbf{\Sigma}}\, \theta \rangle - \lambda_n \|\theta\|_1 \right\} \qquad \text{such that } \|\theta\|_1 \leq \left(\tfrac{n}{\log d}\right)^{1/4}. \tag{8.25b}$$

In our analysis of the constrained version (8.25a), we set $R = \|\theta^*\|_1$. The advantage of the penalized variant (8.25b) is that the regularization parameter λ_n can be chosen without knowledge of the true eigenvector θ^*. In both formulations, the matrix $\widehat{\mathbf{\Sigma}}$ represents some type of approximation to the population covariance matrix $\mathbf{\Sigma}$, with the sample covariance being a canonical example. Note that neither estimator is convex, since they involve maximization of a positive semidefinite quadratic form. Nonetheless, it is instructive to analyze them in order to understand the statistical behavior of sparse PCA, and in the exercises, we describe some relaxations of these non-convex programs.

Naturally, the proximity of $\widehat{\theta}$ to the maximum eigenvector θ^* of $\mathbf{\Sigma}$ depends on the perturbation matrix $\mathbf{P} := \widehat{\mathbf{\Sigma}} - \mathbf{\Sigma}$. How to measure the effect of the perturbation? As will become clear, much of our analysis of ordinary PCA can be modified in a relatively straightforward way so as to obtain results for the sparse version. In particular, a central object in our analysis of ordinary PCA was the basic inequality stated in Lemma 8.6: it shows that the perturbation matrix enters via the function

$$\Psi(\Delta; \mathbf{P}) := \langle \Delta, \mathbf{P}\Delta \rangle + 2 \langle \Delta, \mathbf{P}\theta^* \rangle.$$

As with our analysis of PCA, our general deterministic theorem for sparse PCA involves imposing a form of uniform control on $\Psi(\Delta; \mathbf{P})$ as Δ ranges over all vectors of the form $\theta - \theta^*$ with $\theta \in \mathbb{S}^{d-1}$. The sparsity constraint enters in the form of this uniform bound that we assume. More precisely, letting $\varphi_\nu(n, d)$ and $\psi_\nu(n, d)$ be non-negative functions of the eigengap ν, sample size and dimension, we assume that there exists a universal constant $c_0 > 0$ such that

$$\sup_{\substack{\Delta = \theta - \theta^* \\ \|\theta\|_2 = 1}} \left|\Psi(\Delta; \mathbf{P})\right| \leq c_0\, \nu \|\Delta\|_2^2 + \varphi_\nu(n, d)\|\Delta\|_1 + \psi_\nu^2(n, d)\|\Delta\|_1^2. \tag{8.26}$$

As a concrete example, for a sparse version of the spiked PCA ensemble (8.18) with sub-Gaussian tails, this condition is satisfied with high probability with $\varphi_\nu^2(n, d) \asymp (\nu + 1)\frac{\log d}{n}$ and $\psi_\nu^2(n, d) \asymp \frac{1}{\nu} \frac{\log d}{n}$. This fact will be established in the proof of Corollary 8.12 to follow.

Theorem 8.10 *Given a matrix $\mathbf{\Sigma}$ with a unique, unit-norm, s-sparse maximal eigenvector θ^* with eigengap ν, let $\widehat{\mathbf{\Sigma}}$ be any symmetric matrix satisfying the uniform deviation condition* (8.26) *with constant $c_0 < \frac{1}{6}$, and $16 s\, \psi_\nu^2(n, d) \leq c_0 \nu$.*

(a) *For any optimal solution $\widehat{\theta}$ to the constrained program* (8.25a) *with $R = \|\theta^*\|_1$,*

$$\min\left\{\|\widehat{\theta} - \theta^*\|_2, \|\widehat{\theta} + \theta^*\|_2\right\} \leq \frac{8}{\nu(1 - 4c_0)} \sqrt{s}\, \varphi_\nu(n, d). \tag{8.27}$$

(b) *Consider the penalized program* (8.25b) *with the regularization parameter lower bounded as $\lambda_n \geq 4\left(\frac{n}{\log d}\right)^{1/4} \psi_\nu^2(n, d) + 2\varphi_\nu(n, d)$. Then any optimal solution $\widehat{\theta}$ satisfies the bound*

$$\min\left\{\|\widehat{\theta} - \theta^*\|_2, \|\widehat{\theta} + \theta^*\|_2\right\} \leq \frac{2\left(\frac{\lambda_n}{\varphi_\nu(n,d)} + 4\right)}{\nu(1 - 4c_0)} \sqrt{s}\, \varphi_\nu(n, d). \tag{8.28}$$

Proof We begin by analyzing the constrained estimator, and then describe the modifications necessary for the regularized version.

Argument for constrained estimator: Note that $\|\widehat{\theta}\|_1 \leq R = \|\theta^*\|_1$ by construction of the estimator, and moreover $\theta^*_{S^c} = 0$ by assumption. By splitting the ℓ_1-norm into two components, indexed by S and S^c, respectively, it can be shown[3] that the error $\hat{\Delta} = \widehat{\theta} - \theta^*$ satisfies the inequality $\|\hat{\Delta}_{S^c}\|_1 \leq \|\hat{\Delta}_S\|_1$. So as to simplify our treatment of the regularized estimator, let us proceed by assuming only the weaker inequality $\|\hat{\Delta}_{S^c}\|_1 \leq 3\|\hat{\Delta}_S\|_1$, which implies that $\|\hat{\Delta}\|_1 \leq 4\sqrt{s}\|\hat{\Delta}\|_2$. Combining this inequality with the uniform bound (8.26) on Ψ, we find

[3] We leave this calculation as an exercise for the reader: helpful details can be found in Chapter 7.

that

$$\left|\Psi(\hat{\Delta};\mathbf{P})\right| \leq c_0\,\nu\,\|\hat{\Delta}\|_2^2 + 4\,\sqrt{s}\,\varphi_\nu(n,d)\|\hat{\Delta}\|_2 + 16\,s\,\psi_\nu^2(n,d)\|\hat{\Delta}\|_2^2. \tag{8.29}$$

Substituting back into the basic inequality (8.15) and performing some algebra yields

$$\underbrace{\nu\,\left\{\tfrac{1}{2} - c_0 - 16\frac{s}{\nu}\,\psi_\nu^2(n,d)\right\}}_{\kappa}\|\hat{\Delta}\|_2^2 \leq 4\,\sqrt{s}\,\varphi_\nu(n,d)\,\|\hat{\Delta}\|_2.$$

Note that our assumptions imply that $\kappa > \frac{1}{2}(1 - 4c_0) > 0$, so that the bound (8.27) follows after canceling a term $\|\widehat{\Delta}\|_2$ and rearranging.

Argument for regularized estimator: We now turn to the regularized estimator (8.25b). With the addition of the regularizer, the basic inequality (8.15) now takes the slightly modified form

$$\frac{\nu}{2}\|\hat{\Delta}\|_2^2 - |\Psi(\hat{\Delta};\mathbf{P})| \leq \lambda_n\{\|\theta^*\|_1 - \|\widehat{\theta}\|_1\} \leq \lambda_n\{\|\hat{\Delta}_S\|_1 - \|\hat{\Delta}_{S^c}\|_1\}, \tag{8.30}$$

where the second inequality follows by the S-sparsity of θ^* and the triangle inequality (see Chapter 7 for details).

We claim that the error vector $\widehat{\Delta}$ still satisfies a form of the cone inequality. Let us state this claim as a separate lemma.

Lemma 8.11 *Under the conditions of Theorem 8.10, the error vector* $\widehat{\Delta} = \widehat{\theta} - \theta^*$ *satisfies the cone inequality*

$$\|\widehat{\Delta}_{S^c}\|_1 \leq 3\|\widehat{\Delta}_S\|_1 \quad \text{and hence} \quad \|\widehat{\Delta}\|_1 \leq 4\sqrt{s}\|\widehat{\Delta}\|_2. \tag{8.31}$$

Taking this lemma as given, let us complete the proof of the theorem. Given Lemma 8.11, the previously derived upper bound (8.29) on $|\Psi(\hat{\Delta};\mathbf{P})|$ is also applicable to the regularized estimator. Substituting this bound into our basic inequality, we find that

$$\underbrace{\nu\left\{\tfrac{1}{2} - c_0 - \frac{16}{\nu}s\,\psi_\nu^2(n,d)\right\}}_{\kappa}\|\hat{\Delta}\|_2^2 \leq \sqrt{s}\left(\lambda_n + 4\,\varphi_\nu(n,d)\right)\|\hat{\Delta}\|_2.$$

Our assumptions imply that $\kappa \geq \frac{1}{2}(1 - 4c_0) > 0$, from which claim (8.28) follows.

It remains to prove Lemma 8.11. Combining the uniform bound with the basic inequality (8.30)

$$0 \leq \underbrace{\nu\left(\tfrac{1}{2} - c_0\right)}_{>0}\|\Delta\|_2^2 \leq \varphi_\nu(n,d)\|\Delta\|_1 + \psi_\nu^2(n,d)\|\Delta\|_1^2 + \lambda_n\left\{\|\widehat{\Delta}_S\|_1 - \|\widehat{\Delta}_{S^c}\|_1\right\}.$$

Introducing the shorthand $R = \left(\frac{n}{\log d}\right)^{1/4}$, the feasibility of $\widehat{\theta}$ and θ^* implies that $\|\widehat{\Delta}\|_1 \leq 2R$, and hence

$$0 \le \underbrace{\Big\{\varphi_\nu(n,d) + 2R\psi_\nu^2(n,d)\Big\}}_{\le \frac{\lambda_n}{2}} \|\hat{\Delta}\|_1 + \lambda_n\Big\{\|\hat{\Delta}_S\|_1 - \|\hat{\Delta}_{S^c}\|_1\Big\}$$

$$\le \lambda_n\{\tfrac{3}{2}\|\hat{\Delta}_S\|_1 - \tfrac{1}{2}\|\hat{\Delta}_{S^c}\|_1\},$$

and rearranging yields the claim. □

8.3.2 Consequences for the spiked model with sparsity

Theorem 8.10 is a general deterministic guarantee that applies to any matrix with a sparse maximal eigenvector. In order to obtain more concrete results in a particular case, let us return to the spiked covariance model previously introduced in equation (8.18), and analyze a sparse variant of it. More precisely, consider a random vector $x_i \in \mathbb{R}^d$ generated from the usual spiked ensemble—namely, as $x_i \overset{d}{=} \sqrt{\nu}\,\xi_i\theta^* + w_i$, where $\theta^* \in \mathbb{S}^{d-1}$ is an s-sparse vector, corresponding to the maximal eigenvector of $\mathbf{\Sigma} = \mathrm{cov}(x_i)$. As before, we assume that both the random variable ξ_i and the random vector $w_i \in \mathbb{R}^d$ are independent, each sub-Gaussian with parameter 1, in which case we say that the random vector $x_i \in \mathbb{R}^d$ has sub-Gaussian tails.

Corollary 8.12 *Consider n i.i.d. samples $\{x_i\}_{i=1}^n$ from an s-sparse spiked covariance matrix with eigengap $\nu > 0$ and suppose that $\frac{s \log d}{n} \le c \min\{1, \frac{\nu^2}{\nu+1}\}$ for a sufficiently small constant $c > 0$. Then for any $\delta \in (0,1)$, any optimal solution $\widehat{\theta}$ to the constrained program (8.25a) with $R = \|\theta^*\|_1$, or to the penalized program (8.25b) with $\lambda_n = c_3\sqrt{\nu+1}\Big\{\sqrt{\frac{\log d}{n}} + \delta\Big\}$, satisfies the bound*

$$\min\Big\{\|\widehat{\theta} - \theta^*\|_2, \|\widehat{\theta} + \theta^*\|_2\Big\} \le c_4\sqrt{\frac{\nu+1}{\nu^2}}\left\{\sqrt{\frac{s\log d}{n}} + \delta\right\} \quad \text{for all } \delta \in (0,1) \tag{8.32}$$

with probability at least $1 - c_1 e^{-c_2(n/s)\min\{\delta^2, \nu^2, \nu\}}$.

Proof Letting $\mathbf{P} = \widehat{\mathbf{\Sigma}} - \mathbf{\Sigma}$ be the deviation between the sample and population covariance matrices, our goal is to show that $\Psi(\cdot, \mathbf{P})$ satisfies the uniform deviation condition (8.26). In particular, we claim that, uniformly over $\Delta \in \mathbb{R}^d$, we have

$$\big|\Psi(\Delta; \mathbf{P})\big| \le \underbrace{\frac{1}{8}}_{c_0}\nu\|\Delta\|_2^2 + \underbrace{16\sqrt{\nu+1}\left\{\sqrt{\frac{\log d}{n}} + \delta\right\}}_{\varphi_\nu(n,d)}\|\Delta\|_1 + \underbrace{\frac{c_3'}{\nu}\frac{\log d}{n}}_{\psi_\nu^2(n,d)}\|\Delta\|_1^2, \tag{8.33}$$

with probability at least $1 - c_1 e^{-c_2 n\min\{\delta^2, \nu^2\}}$. Here (c_1, c_2, c_3') are universal constants. Taking this intermediate claim as given, let us verify that the bound (8.32) follows as a consequence of Theorem 8.10. We have

$$\frac{9s\psi_\nu^2(n,d)}{c_0} = \frac{72c_3'}{\nu}\frac{s\log d}{n} \le \nu\left\{72c_3'\frac{\nu+1}{\nu^2}\frac{s\log d}{n}\right\} \le \nu,$$

using the assumed upper bound on the ratio $\frac{s\log d}{n}$ for a sufficiently small constant c. Consequently, the bound for the constrained estimator follows from Theorem 8.10. For the penalized estimator, there are a few other conditions to be verified: let us first check that $\|\theta^*\|_1 \le \nu\sqrt{\frac{n}{\log d}}$. Since θ^* is s-sparse with $\|\theta^*\|_2 = 1$, it suffices to have $\sqrt{s} \le \nu\sqrt{\frac{n}{\log d}}$, or equivalently $\frac{1}{\nu^2}\frac{s\log d}{n} \le 1$, which follows from our assumptions. Finally, we need to check that λ_n satisfies the lower bound requirement in Theorem 8.10. We have

$$\begin{aligned}
4R\,\psi_\nu^2(n,d) + 2\varphi_\nu(n,d) &\le 4\nu\sqrt{\frac{n}{\log d}}\,\frac{c_3'}{\nu}\frac{\log d}{n} + 24\sqrt{\nu+1}\left\{\sqrt{\frac{\log d}{n}} + \delta\right\} \\
&\le \underbrace{c_3\sqrt{\nu+1}\left\{\sqrt{\frac{\log d}{n}} + \delta\right\}}_{\lambda_n}
\end{aligned}$$

as required.

It remains to prove the uniform bound (8.33). Recall the decomposition $\mathbf{P} = \sum_{j=1}^3 \mathbf{P}_j$ given in equation (8.21). By linearity of the function Ψ in its second argument, this decomposition implies that $\Psi(\Delta;\mathbf{P}) = \sum_{j=1}^3 \Psi(\Delta;\mathbf{P}_j)$. We control each of these terms in turn.

Control of first component: Lemma 8.8 guarantees that $\left|\frac{1}{n}\sum_{i=1}^n \xi_i^2 - 1\right| \le \frac{1}{16}$ with probability at least $1 - 2e^{-cn}$. Conditioned on this bound, for any vector of the form $\Delta = \theta - \theta^*$ with $\theta \in \mathbb{S}^{d-1}$, we have

$$|\Psi(\Delta;\mathbf{P}_1)| \le \frac{\nu}{16}\langle\Delta,\,\theta^*\rangle^2 = \frac{\nu}{16}\big(1 - \langle\theta^*,\,\theta\rangle\big)^2 \le \frac{\nu}{32}\|\Delta\|_2^2, \tag{8.34}$$

where we have used the fact that $2\big(1 - \langle\theta^*,\,\theta\rangle\big)^2 \le 2\big(1 - \langle\theta^*,\,\theta\rangle\big) = \|\Delta\|_2^2$.

Control of second component: We have

$$\begin{aligned}
|\Psi(\Delta;\mathbf{P}_2)| &\le 2\sqrt{\nu}\Big\{\langle\Delta,\,\bar{w}\rangle\langle\Delta,\,\theta^*\rangle + \langle\bar{w},\,\Delta\rangle + \langle\theta^*,\,\bar{w}\rangle\langle\Delta,\,\theta^*\rangle\Big\} \\
&\le 4\sqrt{\nu}\|\Delta\|_1\|\bar{w}\|_\infty + 2\sqrt{\nu}|\langle\theta^*,\,\bar{w}\rangle|\frac{\|\Delta\|_2^2}{2}.
\end{aligned} \tag{8.35}$$

The following lemma provides control on the two terms in this upper bound:

Lemma 8.13 *Under the conditions of Corollary 8.12, we have*

$$\mathbb{P}\left[\|\bar{w}\|_\infty \geq 2\sqrt{\frac{\log d}{n}} + \delta\right] \leq c_1 e^{-c_2 n\delta^2} \quad \textit{for all } \delta \in (0,1), \textit{ and} \tag{8.36a}$$

$$\mathbb{P}\left[|\langle \theta^*, \bar{w}\rangle| \geq \frac{\sqrt{\nu}}{32}\right] \leq c_1 e^{-c_2 n\nu}. \tag{8.36b}$$

We leave the proof of these bounds as an exercise for the reader, since they follow from standard results in Chapter 2. Combining Lemma 8.13 with the bound (8.35) yields

$$|\Psi(\Delta; \mathbf{P}_2)| \leq \frac{\nu}{32}\|\Delta\|_2^2 + 8\sqrt{\nu+1}\left\{\sqrt{\frac{\log d}{n}} + \delta\right\}\|\Delta\|_1. \tag{8.37}$$

Control of third term: Recalling that $\mathbf{P}_3 = \frac{1}{n}\mathbf{W}^{\mathrm{T}}\mathbf{W} - \mathbf{I}_d$, we have

$$|\Psi(\Delta; \mathbf{P}_3)| \leq \left|\langle \Delta, \mathbf{P}_3\Delta\rangle\right| + 2\|\mathbf{P}_3\theta^*\|_\infty\|\Delta\|_1. \tag{8.38}$$

Our final lemma controls the two terms in this bound:

Lemma 8.14 *Under the conditions of Corollary 8.12, for all $\delta \in (0,1)$, we have*

$$\|\mathbf{P}_3\theta^*\|_\infty \leq 2\sqrt{\frac{\log d}{n}} + \delta \tag{8.39a}$$

and

$$\sup_{\Delta\in\mathbb{R}^d}\left|\langle \Delta, \mathbf{P}_3\Delta\rangle\right| \leq \frac{\nu}{16}\|\Delta\|_2^2 + \frac{c_3'}{\nu}\frac{\log d}{n}\|\Delta\|_1^2, \tag{8.39b}$$

where both inequalities hold with probability greater than $1 - c_1 e^{-c_2 n \min\{\nu, \nu^2, \delta^2\}}$.

Combining this lemma with our earlier inequality (8.38) yields the bound

$$|\Psi(\Delta; \mathbf{P}_3)| \leq \frac{\nu}{16}\|\Delta\|_2^2 + 8\left\{\sqrt{\frac{\log d}{n}} + \delta\right\}\|\Delta\|_1 + \frac{c_3'}{\nu}\frac{\log d}{n}\|\Delta\|_1^2. \tag{8.40}$$

Finally, combining the bounds (8.34), (8.37) and (8.40) yields the claim (8.33).

The only remaining detail is the proof of Lemma 8.14. The proof of the tail bound (8.39a) is a simple exercise, using the sub-exponential tail bounds from Chapter 2. The proof of the bound (8.39b) requires more involved argument, one that makes use of both Exercise 7.10 and our previous results on estimation of sample covariances from Chapter 6.

For a constant $\xi > 0$ to be chosen, consider the positive integer $k := \lceil \xi\nu^2 \frac{n}{\log d}\rceil$, and the collection of submatrices $\{(\mathbf{P}_3)_{SS}, |S| = k\}$. Given a parameter $\alpha \in (0, 1)$ to be chosen, a combination of the union bound and Theorem 6.5 imply that there are universal constants c_1 and c_2 such that

$$\mathbb{P}\left[\max_{|S|=k} |||(\mathbf{P}_3)_{SS}|||_2 \geq c_1 \sqrt{\frac{k}{n}} + \alpha\nu\right] \leq 2e^{-c_2 n\alpha^2\nu^2 + \log\binom{d}{k}}.$$

Since $\log\binom{d}{k} \leq 2k\log(d) \leq 4\xi\nu^2 n$, this probability is at most $e^{-c_2 n\nu^2(\alpha^2 - 4\xi)} = e^{-c_2 n\nu^2\alpha^2/2}$, as long as we set $\xi = \alpha^2/8$. The result of Exercise 7.10 then implies that

$$\left|\langle\Delta, \mathbf{P}_3\Delta\rangle\right| \leq 27c_1'\alpha\nu\left\{\|\Delta\|_2^2 + \frac{8}{\alpha^2\nu^2}\frac{\log d}{n}\|\Delta\|_1^2\right\} \qquad \text{for all } \Delta \in \mathbb{R}^d,$$

with the previously stated probability. Setting $\alpha = \frac{1}{(16\times 27)c_1'}$ yields the claim (8.39b) with $c_3' = (2\alpha^2)^{-1}$. □

8.4 Bibliographic details and background

Further details on PCA and its applications can be found in books by Anderson (1984) (cf. chapter 11), Jollife (2004) and Muirhead (2008). See the two-volume set by Horn and Johnson (1985; 1991) for background on matrix analysis, as well as the book by Bhatia (1997) for a general operator-theoretic viewpoint. The book by Stewart and Sun (1980) is more specifically focused on matrix perturbation theory, whereas Stewart (1971) provides perturbation theory in the more general setting of closed linear operators.

Johnstone (2001) introduced the spiked covariance model (8.18), and investigated the high-dimensional asymptotics of its eigenstructure; see also the papers by Baik and Silverstein (2006) and Paul (2007) for high-dimensional asymptotics. Johnstone and Lu (2009) introduced the sparse variant of the spiked ensemble, and proved consistency results for a simple estimator based on thresholding the diagonal entries of the sample covariance matrix. Amini and Wainwright (2009) provided a more refined analysis of this same estimator, as well as of a semidefinite programming (SDP) relaxation proposed by d'Asprémont et al. (2007). See Exercise 8.8 for the derivation of this latter SDP relaxation. The non-convex estimator (8.25a) was first proposed by Joliffe et al. (2003), and called the `SCOTLASS` criterion; Witten et al. (2009) derive an alternating algorithm for finding a local optimum of this criterion. Other authors, including Ma (2010; 2013) and Yuan and Zhang (2013), have studied iterative algorithms for sparse PCA based on combining the power method with soft or hard thresholding.

Minimax lower bounds for estimating principal components in various types of spiked ensembles can be derived using techniques discussed in Chapter 15. These lower bounds show that the upper bounds obtained in Corollaries 8.7 and 8.12 for ordinary and sparse PCA, respectively, are essentially optimal. See Birnbaum et al. (2012) and Vu and Lei (2012) for lower bounds on the ℓ_2-norm error in sparse PCA. Amini and Wainwright (2009) derived lower bounds for the problem of variable selection in sparse PCA. Some of these lower bounds are covered in this book: in particular, see Example 15.19 for minimax lower bounds on ℓ_2-error in ordinary PCA, Example 15.20 for lower bounds on variable selection in sparse PCA, and Exercise 15.16 for ℓ_2-error lower bounds on sparse PCA. Berthet and Rigollet (2013) derived certain hardness results for the problem of sparse PCA detection, based on relating it to the (conjectured) average-case hardness of the planted k-clique problem in Erdős–Rényi random graphs. Ma and Wu (2013) developed a related but distinct reduction, one which applies to a Gaussian detection problem over a family of sparse-plus-low-rank matrices. See also the papers (Wang et al., 2014; Cai et al., 2015; Gao et al., 2015) for related results using the conjectured hardness of the k-clique problem.

8.5 Exercises

Exercise 8.1 (Courant–Fischer variational representation) For a given integer $j \in \{2, \ldots, d\}$, let $\mathcal{V}_{j-1}$ denote the collection of all subspaces of dimension $j-1$. For any symmetric matrix $\mathbf{Q}$, show that the jth largest eigenvalue is given by

$$\gamma_j(\mathbf{Q}) = \min_{\mathbb{V} \in \mathcal{V}_{j-1}} \max_{u \in \mathbb{V}^\perp \cap \mathbb{S}^{d-1}} \langle u, \mathbf{Q}u \rangle, \tag{8.41}$$

where $\mathbb{V}^\perp$ denotes the orthogonal subspace to $\mathbb{V}$.

Exercise 8.2 (Unitarily invariant matrix norms) For positive integers $d_1 \leq d_2$, a matrix norm on $\mathbb{R}^{d_1 \times d_2}$ is *unitarily invariant* if $|||\mathbf{M}||| = |||\mathbf{VMU}|||$ for all orthonormal matrices $\mathbf{V} \in \mathbb{R}^{d_1 \times d_1}$ and $\mathbf{U} \in \mathbb{R}^{d_2 \times d_2}$.

(a) Which of the following matrix norms are unitarily invariant?
   (i) The Frobenium norm $|||\mathbf{M}|||_{\mathrm{F}}$.
   (ii) The nuclear norm $|||\mathbf{M}|||_{\mathrm{nuc}}$.
   (iii) The ℓ_2-operator norm $|||\mathbf{M}|||_2 = \sup_{\|u\|_2=1} \|\mathbf{M}u\|_2$.
   (iv) The ℓ_∞-operator norm $|||\mathbf{M}|||_\infty = \sup_{\|u\|_\infty=1} \|\mathbf{M}u\|_\infty$.

(b) Let ρ be a norm on $\mathbb{R}^{d_1}$ that is invariant to permutations and sign changes—that is

$$\rho(x_1, \ldots, x_{d_1}) = \rho(z_1 x_{\pi(1)}, \ldots, z_{d_1} x_{\pi(d_1)})$$

for all binary strings $z \in \{-1, 1\}^{d_1}$ and permutations π on $\{1, \ldots, d_1\}$. Such a function is known as a *symmetric gauge function*. Letting $\{\sigma_j(\mathbf{M})\}_{j=1}^{d_1}$ denote the singular values of $\mathbf{M}$, show that

$$|||\mathbf{M}|||_\rho := \rho(\underbrace{\sigma_1(\mathbf{M}), \ldots, \sigma_{d_1}(\mathbf{M})}_{\sigma(\mathbf{M}) \in \mathbb{R}^{d_1}})$$

defines a matrix norm. (*Hint*: For any pair of $d_1 \times d_2$ matrices $\mathbf{M}$ and $\mathbf{N}$, we have $\mathrm{trace}(\mathbf{N}^{\mathrm{T}}\mathbf{M}) \leq \langle \sigma(\mathbf{N}), \sigma(\mathbf{M}) \rangle$, where $\sigma(\mathbf{M})$ denotes the ordered vector of singular values.)

(c) Show that all matrix norms in the family from part (b) are unitarily invariant.

Exercise 8.3 (Weyl's inequality) Prove Weyl's inequality (8.9). (*Hint:* Exercise 8.1 may be useful.)

Exercise 8.4 (Variational characterization of eigenvectors) Show that the orthogonal matrix $\mathbf{V} \in \mathbb{R}^{d \times r}$ maximizing the criterion (8.2) has columns formed by the top r eigenvectors of $\mathbf{\Sigma} = \mathrm{cov}(X)$.

Exercise 8.5 (Matrix power method) Let $\mathbf{Q} \in \mathcal{S}^{d\times d}$ be a strictly positive definite symmetric matrix with a unique maximal eigenvector θ^*. Given some non-zero initial vector $\theta^0 \in \mathbb{R}^d$, consider the sequence $\{\theta^t\}_{t=0}^{\infty}$,

$$\theta^{t+1} = \frac{\mathbf{Q}\theta^t}{\|\mathbf{Q}\theta^t\|_2}. \tag{8.42}$$

(a) Prove that there is a large set of initial vectors θ^0 for which the sequence $\{\theta^t\}_{t=0}^{\infty}$ converges to θ^*.
(b) Give a "bad" initialization for which this convergence does not take place.
(c) Based on part (b), specify a procedure to compute the second largest eigenvector, assuming it is also unique.

Exercise 8.6 (PCA for Gaussian mixture models) Consider an instance of the Gaussian mixture model from Example 8.3 with equal mixture weights ($\alpha = 0.5$) and unit-norm mean vector ($\|\theta^*\|_2 = 1$), and suppose that we implement the PCA-based estimator $\widehat{\theta}$ for the mean vector θ^*.

(a) Prove that if the sample size is lower bounded as $n > c_1\sigma^2(1 + +\sigma^2)d$ for a sufficiently large constant c_1, this estimator satisfies a bound of the form

$$\|\widehat{\theta} - \theta^*\|_2 \leq c_2\sigma\sqrt{1+\sigma^2}\sqrt{\frac{d}{n}}$$

with high probability.
(b) Explain how to use your estimator to build a classification rule—that is, a mapping $x \mapsto \psi(x) \in \{-1, +1\}$, where the binary labels code whether sample x has mean $-\theta^*$ or $+\theta^*$.
(c) Does your method still work if the shared covariance matrix is *not* a multiple of the identity?

Exercise 8.7 (PCA for retrieval from absolute values) Suppose that our goal is to estimate an unknown vector $\theta^* \in \mathbb{R}^d$ based on n i.i.d. samples $\{(x_i, y_i)\}_{i=1}^n$ of the form $y_i = \left|\langle x_i, \theta^*\rangle\right|$, where $x_i \sim \mathcal{N}(0, \mathbf{I}_d)$. This model is a real-valued idealization of the problem of phase retrieval, to be discussed at more length in Chapter 10. Suggest a PCA-based method for estimating θ^* that is consistent in the limit of infinite data. (*Hint:* Using the pair (x, y), try to construct a random matrix $\mathbf{Z}$ such that $\mathbb{E}[\mathbf{Z}] = \sqrt{\frac{2}{\pi}}(\theta^* \otimes \theta^* + \mathbf{I}_d)$.)

Exercise 8.8 (Semidefinite relaxation of sparse PCA) Recall the non-convex problem (8.25a), also known as the SCOTLASS estimator. In this exercise, we derive a convex relaxation of the objective, due to d'Aspremont et al. (2007).

(a) Show that the non-convex problem (8.25a) is equivalent to the optimization problem

$$\max_{\theta \in \mathcal{S}_+^{d\times d}} \operatorname{trace}(\widehat{\mathbf{\Sigma}}\mathbf{\Theta}) \qquad \text{such that } \operatorname{trace}(\mathbf{\Theta}) = 1,\ \textstyle\sum_{j,k=1}^d |\mathbf{\Theta}_{jk}| \leq R^2 \text{ and } \operatorname{rank}(\mathbf{\Theta}) = 1,$$

where $\mathcal{S}_+^{d\times d}$ denotes the cone of symmetric, positive semidefinite matrices.

(b) Dropping the rank constraint yields the convex program

$$\max_{\theta \in \mathcal{S}_+^{d\times d}} \operatorname{trace}(\widehat{\boldsymbol{\Sigma}}\boldsymbol{\Theta}) \qquad \text{such that } \operatorname{trace}(\boldsymbol{\Theta}) = 1 \text{ and } \textstyle\sum_{j,k=1}^{d} |\Theta_{jk}| \leq R^2.$$

What happens when its optimum is achieved at a rank-one matrix?

Exercise 8.9 (Primal–dual witness for sparse PCA) The SDP relaxation from Exercise 8.8(b) can be written in the equivalent Lagrangian form

$$\max_{\substack{\boldsymbol{\Theta}\in\mathcal{S}_+^{d\times d}\\ \operatorname{trace}(\boldsymbol{\Theta})=1}} \left\{ \operatorname{trace}(\widehat{\boldsymbol{\Sigma}}\boldsymbol{\Theta}) - \lambda_n \sum_{j,k=1}^{d} |\Theta_{jk}| \right\}. \tag{8.43}$$

Suppose that there exists a vector $\widehat{\theta} \in \mathbb{R}^d$ and a matrix $\widehat{\mathbf{U}} \in \mathbb{R}^{d\times d}$ such that

$$\widehat{U}_{jk} = \begin{cases} \operatorname{sign}(\widehat{\theta}_j\widehat{\theta}_k) & \text{if } \widehat{\theta}_j\widehat{\theta}_k \neq 0, \\ \in [-1, 1] & \text{otherwise,} \end{cases}$$

and moreover such that $\widehat{\theta}$ is a maximal eigenvector of the matrix $\widehat{\boldsymbol{\Sigma}} - \lambda_n\widehat{\mathbf{U}}$. Prove that the rank-one matrix $\widehat{\boldsymbol{\Theta}} = \widehat{\theta} \otimes \widehat{\theta}$ is an optimal solution to the SDP relaxation (8.43).

9

Decomposability and restricted strong convexity

In Chapter 7, we studied the class of sparse linear models, and the associated use of ℓ_1-regularization. The basis pursuit and Lasso programs are special cases of a more general family of estimators, based on combining a cost function with a regularizer. Minimizing such an objective function yields an estimation method known as an *M-estimator*. The goal of this chapter is to study this more general family of regularized M-estimators, and to develop techniques for bounding the associated estimation error for high-dimensional problems. Two properties are essential to obtaining consistent estimators in high dimensions: decomposability of the regularizer, and a certain type of lower restricted curvature condition on the cost function.

9.1 A general regularized M-estimator

Our starting point is an indexed family of probability distributions $\{\mathbb{P}_\theta, \theta \in \Omega\}$, where θ represents some type of "parameter" to be estimated. As we discuss in the sequel, the space Ω of possible parameters can take various forms, including subsets of vectors, matrices, or—in the nonparametric setting to be discussed in Chapters 13 and 14—subsets of regression or density functions. Suppose that we observe a collection of n samples $Z_1^n = (Z_1, \ldots, Z_n)$, where each sample Z_i takes values in some space $\mathcal{Z}$, and is drawn independently according to some distribution $\mathbb{P}$. In the simplest setting, known as the well-specified case, the distribution $\mathbb{P}$ is a member of our parameterized family—say $\mathbb{P} = \mathbb{P}_{\theta^*}$—and our goal is to estimate the unknown parameter θ^*. However, our set-up will also allow for mis-specified models, in which case the target parameter θ^* is defined as the minimizer of the population cost function—in particular, see equation (9.2) below.

The first ingredient of a general M-estimator is a cost function $\mathcal{L}_n : \Omega \times \mathcal{Z}^n \to \mathbb{R}$, where the value $\mathcal{L}_n(\theta; Z_1^n)$ provides a measure of the fit of parameter θ to the data Z_1^n. Its expectation defines the *population cost function*—namely the quantity

$$\bar{\mathcal{L}}(\theta) := \mathbb{E}[\mathcal{L}_n(\theta; Z_1^n)]. \tag{9.1}$$

Implicit in this definition is that the expectation does not depend on the sample size n, a condition which holds in many settings (with appropriate scalings). For instance, it is often the case that the cost function has an additive decomposition of the form $\mathcal{L}_n(\theta; Z_1^n) = \frac{1}{n}\sum_{i=1}^n \mathcal{L}(\theta; Z_i)$, where $\mathcal{L} : \Omega \times \mathcal{Z} \to \mathbb{R}$ is the cost defined for a single sample. Of course, any likelihood-based cost function decomposes in this way when the samples are drawn in an independent and identically distributed manner, but such cost functions can also be useful for dependent data.

Next we define the *target parameter* as the minimum of the population cost function

$$\theta^* = \arg\min_{\theta\in\Omega} \bar{\mathcal{L}}(\theta). \tag{9.2}$$

In many settings—in particular, when $\mathcal{L}_n$ is the negative log-likelihood of the data—this minimum is achieved at an interior point of Ω, in which case θ^* must satisfy the zero-gradient equation $\nabla\bar{\mathcal{L}}(\theta^*) = 0$. However, we do not assume this condition in our general analysis.

With this set-up, our goal is to estimate θ^* on the basis of the observed samples $Z_1^n = \{Z_1, \ldots, Z_n\}$. In order to do so, we combine the empirical cost function with a regularizer or penalty function $\Phi\colon \Omega \to \mathbb{R}$. As will be clarified momentarily, the purpose of this regularizer is to enforce a certain type of structure expected in θ^*. Our overall estimator is based on solving the optimization problem

$$\widehat{\theta} \in \arg\min_{\theta\in\Omega} \{\mathcal{L}_n(\theta; Z_1^n) + \lambda_n\Phi(\theta)\}, \tag{9.3}$$

where $\lambda_n > 0$ is a user-defined regularization weight. The estimator (9.3) is known as an *M-estimator*, where the "M" stands for minimization (or maximization).

Remark: An important remark on notation is needed before proceeding. From here onwards, we will frequently adopt $\mathcal{L}_n(\theta)$ as a shorthand for $\mathcal{L}_n(\theta; Z_1^n)$, remembering that the subscript n reflects implicitly the dependence on the underlying samples. We also adopt the same notation for the derivatives of the empirical cost function.

Let us illustrate this set-up with some examples.

Example 9.1 (Linear regression and Lasso) We begin with the problem of linear regression previously studied in Chapter 7. In this case, each sample takes the form $Z_i = (x_i, y_i)$, where $x_i \in \mathbb{R}^d$ is a covariate vector, and $y_i \in \mathbb{R}$ is a response variable. In the simplest case, we assume that the data are generated exactly from a linear model, so that $y_i = \langle x_i, \theta^*\rangle + w_i$, where w_i is some type of stochastic noise variable, assumed to be independent of x_i. The least-squares estimator is based on the quadratic cost function

$$\mathcal{L}_n(\theta) = \frac{1}{n}\sum_{i=1}^{n}\frac{1}{2}\big(y_i - \langle x_i, \theta\rangle\big)^2 = \frac{1}{2n}\|y - \mathbf{X}\theta\|_2^2,$$

where we recall from Chapter 7 our usual notation for the vector $y \in \mathbb{R}^n$ of response variables and design matrix $\mathbf{X} \in \mathbb{R}^{n\times d}$. When the response–covariate pairs (y_i, x_i) are drawn from a linear model with regression vector θ^*, then the population cost function takes the form

$$\mathbb{E}_{x,y}[\tfrac{1}{2}(y - \langle x, \theta\rangle)^2] = \frac{1}{2}(\theta - \theta^*)^{\mathrm{T}}\mathbf{\Sigma}(\theta - \theta^*) + \frac{1}{2}\sigma^2 = \frac{1}{2}\|\sqrt{\mathbf{\Sigma}}\,(\theta - \theta^*)\|_2^2 + \frac{1}{2}\sigma^2,$$

where $\mathbf{\Sigma} := \operatorname{cov}(x_1)$ and $\sigma^2 := \operatorname{var}(w_1)$. Even when the samples are not drawn from a linear model, we can still define θ^* as a minimizer of the population cost function $\theta \mapsto \mathbb{E}_{x,y}[(y - \langle x, \theta\rangle)^2]$. In this case, the linear function $x \mapsto \langle x, \theta^*\rangle$ provides the best linear approximation of the regression function $x \mapsto \mathbb{E}[y \mid x]$.

As discussed in Chapter 7, there are many cases in which the target regression vector θ^*

is expected to be sparse, and in such settings, a good choice of regularizer Φ is the ℓ_1-norm $\Phi(\theta) = \sum_{j=1}^{d} |\theta_j|$. In conjunction with the least-squares loss, we obtain the Lasso estimator

$$\widehat{\theta} \in \arg\min_{\theta \in \mathbb{R}^d} \left\{ \frac{1}{2n} \sum_{i=1}^{n} (y_i - \langle x_i, \theta \rangle)^2 + \lambda_n \sum_{j=1}^{d} |\theta_j| \right\} \tag{9.4}$$

as a special case of the general estimator (9.3). See Chapter 7 for an in-depth analysis of this particular M-estimator. ♣

As our first extension of the basic Lasso (9.4), we now consider a more general family of regression problems.

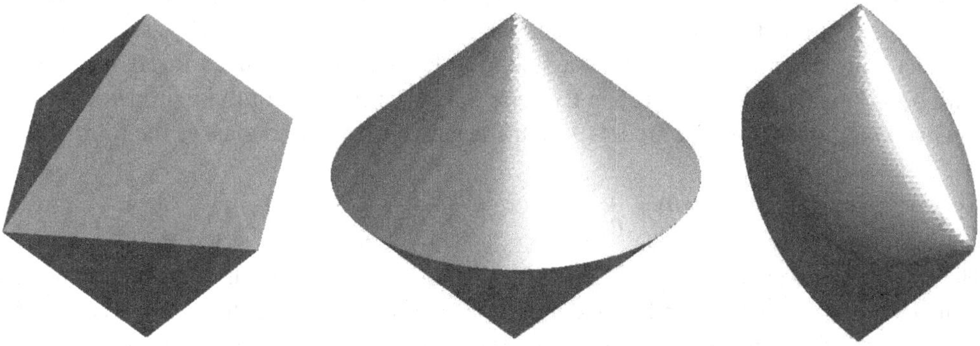

(a) (b) (c)

Figure 9.1 Illustration of unit balls of different norms in $\mathbb{R}^3$. (a) The ℓ_1-ball generated by $\Phi(\theta) = \sum_{j=1}^{3} |\theta_j|$. (b) The group Lasso ball generated by $\Phi(\theta) = \sqrt{\theta_1^2 + \theta_2^2} + |\theta_3|$. (c) A group Lasso ball with overlapping groups, generated by $\Phi(\theta) = \sqrt{\theta_1^2 + \theta_2^2} + \sqrt{\theta_1^2 + \theta_3^2}$.

Example 9.2 (Generalized linear models and ℓ_1-regularization) We again consider samples of the form $Z_i = (x_i, y_i)$ where $x_i \in \mathbb{R}^d$ is a vector of covariates, but now the response variable y_i is allowed to take values in an arbitrary space $\mathcal{Y}$. The previous example of linear regression corresponds to the case $\mathcal{Y} = \mathbb{R}$. A different example is the problem of binary classification, in which the response y_i represents a class label belonging to $\mathcal{Y} = \{0, 1\}$. For applications that involve responses that take on non-negative integer values—for instance, photon counts in imaging applications—the choice $\mathcal{Y} = \{0, 1, 2, \ldots\}$ is appropriate.

The family of *generalized linear models*, or GLMs for short, provides a unified approach to these different types of regression problems. Any GLM is based on modeling the conditional distribution of the response $y \in \mathcal{Y}$ given the covariate $x \in \mathbb{R}^d$ in an exponential family form, namely as

$$\mathbb{P}_{\theta^*}(y \mid x) = h_\sigma(y) \exp\left\{ \frac{y \langle x, \theta^* \rangle - \psi(\langle x, \theta^* \rangle)}{c(\sigma)} \right\}, \tag{9.5}$$

where $c(\sigma)$ is a scale parameter, and the function $\psi\colon \mathbb{R} \to \mathbb{R}$ is the partition function of the underlying exponential family.

Many standard models are special cases of the generalized linear family (9.5). First, consider the standard linear model $y = \langle x, \theta^* \rangle + w$, where $w \sim \mathcal{N}(0, \sigma^2)$. Setting $c(\sigma) = \sigma^2$ and $\psi(t) = t^2/2$, the conditional distribution (9.5) corresponds to that of a $\mathcal{N}(\langle x, \theta^* \rangle, \sigma^2)$ variate, as required. Similarly, in the logistic model for binary classification, we assume that the log-odds ratio is given by $\langle x, \theta^* \rangle$—that is,

$$\log \frac{\mathbb{P}_{\theta^*}(y = 1 \mid x)}{\mathbb{P}_{\theta^*}(y = 0 \mid x)} = \langle x, \theta^* \rangle. \tag{9.6}$$

This assumption again leads to a special case of the generalized linear model (9.5), this time with $c(\sigma) \equiv 1$ and $\psi(t) = \log(1 + \exp(t))$. As a final example, when the response $y \in \{0, 1, 2, \ldots\}$ represents some type of count, it can be appropriate to model y as conditionally Poisson with mean $\mu = e^{\langle x, \theta^* \rangle}$. This assumption leads to a generalized linear model (9.5) with $\psi(t) = \exp(t)$ and $c(\sigma) \equiv 1$. See Exercise 9.3 for verification of these properties.

Given n samples from the model (9.5), the negative log-likelihood takes the form

$$\mathcal{L}_n(\theta) = \frac{1}{n} \sum_{i=1}^{n} \psi(\langle x_i, \theta \rangle) - \left\langle \frac{1}{n} \sum_{i=1}^{n} y_i x_i, \theta \right\rangle. \tag{9.7}$$

Here we have rescaled the log-likelihood by $1/n$ for later convenience, and also dropped the scale factor $c(\sigma)$, since it is independent of θ. When the true regression vector θ^* is expected to be sparse, then it is again reasonable to use the ℓ_1-norm as a regularizer, and combining with the cost function (9.7) leads to the *generalized linear Lasso*

$$\widehat{\theta} \in \arg\min_{\theta \in \mathbb{R}^d} \left\{ \frac{1}{n} \sum_{i=1}^{n} \psi(\langle x_i, \theta \rangle) - \left\langle \frac{1}{n} \sum_{i=1}^{n} y_i x_i, \theta \right\rangle + \lambda_n \|\theta\|_1 \right\}. \tag{9.8}$$

When $\psi(t) = t^2/2$, this objective function is equivalent to the standard Lasso, apart from the constant term $\frac{1}{2n} \sum_{i=1}^{n} y_i^2$ that has no effect on $\widehat{\theta}$. ♣

Thus far, we have discussed only the ℓ_1-norm. There are various extensions of the ℓ_1-norm that are based on some type of grouping of the coefficients.

Example 9.3 (Group Lasso) Let $\mathcal{G} = \{g_1, \ldots, g_T\}$ be a disjoint partition of the index set $\{1, \ldots, d\}$—that is, each group g_j is a subset of the index set, disjoint from every other group, and the union of all T groups covers the full index set. See panel (a) in Figure 9.3 for an example of a collection of overlapping groups.

For a given vector $\theta \in \mathbb{R}^d$, we let θ_g denote the d-dimensional vector with components equal to θ on indices within g, and zero in all other positions. For a given base norm $\|\cdot\|$, we then define the *group Lasso norm*

$$\Phi(\theta) := \sum_{g \in \mathcal{G}} \|\theta_g\|. \tag{9.9}$$

The standard form of the group Lasso uses the ℓ_2-norm as the base norm, so that we obtain a block ℓ_1/ℓ_2-norm—namely, the ℓ_1-norm of the ℓ_2-norms within each group. See Figure 9.1(b) for an illustration of the norm (9.9) with the blocks $g_1 = \{1, 2\}$ and $g_2 = \{3\}$. The block ℓ_1/ℓ_∞-version of the group Lasso has also been studied extensively. Apart from the basic group Lasso (9.9), another variant involves associating a positive weight ω_g with each group. ♣

In the preceding example, the groups were non-overlapping. The same regularizer (9.9) can also be used in the case of overlapping groups; it remains a norm as long as the groups cover the space. For instance, Figure 9.1(c) shows the unit ball generated by the overlapping groups $g_1 = \{1, 2\}$ and $g_2 = \{1, 3\}$ in $\mathbb{R}^3$. However, the standard group Lasso (9.9) with overlapping groups has a property that can be undesirable. Recall that the motivation for group-structured penalties is to estimate parameter vectors whose support lies within a union of a (relatively small) subset of groups. However, when used as a regularizer in an M-estimator, the standard group Lasso (9.9) with overlapping groups typically leads to solutions with support contained in the *complement* of a union of groups. For instance, in the example shown in Figure 9.1(c) with groups $g_1 = \{1, 2\}$ and $g_2 = \{1, 3\}$, apart from the all-zero solution that has empty support set, or a solution with the complete support $\{1, 2, 3\}$, the penalty encourages solutions with supports equal to either $g_1^c = \{3\}$ or $g_2^c = \{2\}$.

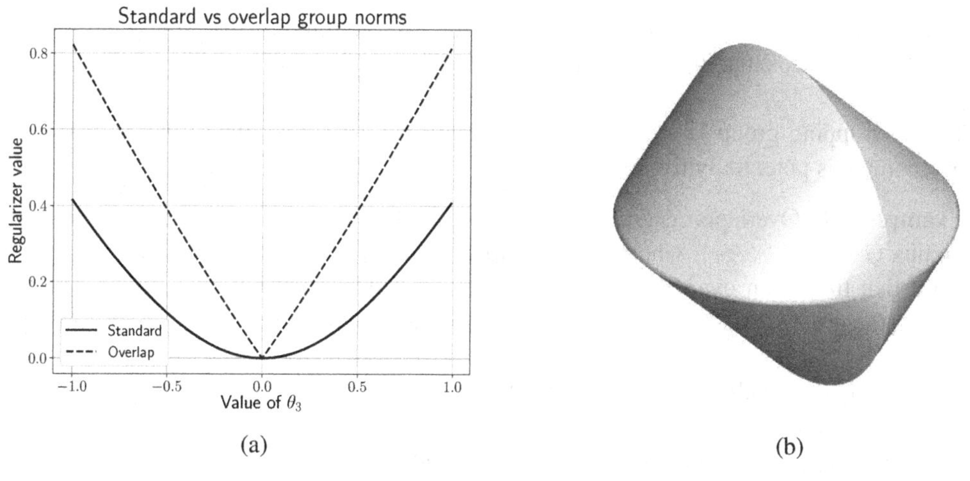

Figure 9.2 (a) Plots of the residual penalty $f(\theta_3) = \Phi(1, 1, \theta_3) - \Phi(1, 1, 0)$ for the standard group Lasso (9.9) with a solid line and overlap group Lasso (9.10) with a dashed line, in the case of the groups $g_1 = \{1, 2\}$ and $g_2 = \{1, 3\}$. (b) Plot of the unit ball of the overlapping group Lasso norm (9.10) for the same groups as in panel (a).

Why is this the case? In the example given above, consider a vector $\theta \in \mathbb{R}^3$ such that θ_1, a variable shared by both groups, is active. For concreteness, say that $\theta_1 = \theta_2 = 1$, and consider the residual penalty $f(\theta_3) := \Phi(1, 1, \theta_3) - \Phi(1, 1, 0)$ on the third coefficient. It takes the form

$$f(\theta_3) = \|(1, 1)\|_2 + \|(1, \theta_3)\|_2 - \|(1, 1)\|_2 - \|(1, 0)\|_2 = \sqrt{1 + \theta_3^2} - 1.$$

As shown by the solid curve in Figure 9.2(a), the function f is differentiable at $\theta_3 = 0$. Indeed, since $f'(\theta_3)\big|_{\theta_3=0} = 0$, this penalty *does not* encourage sparsity of the third coefficient.

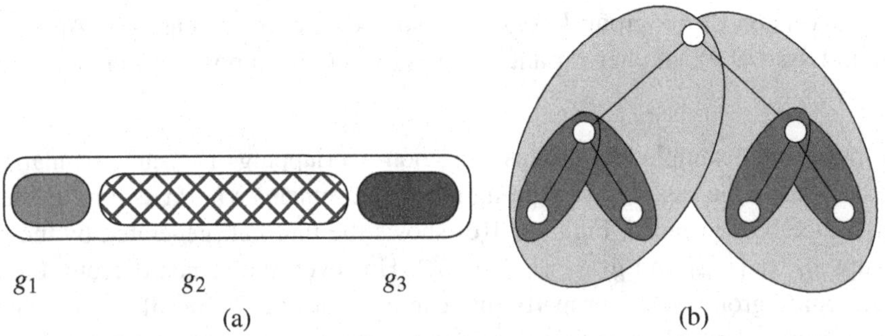

Figure 9.3 (a) Group Lasso penalty with non-overlapping groups. The groups $\{g_1, g_2, g_3\}$ form a disjoint partition of the index set $\{1, 2, \ldots, d\}$. (b) A total of $d = 7$ variables are associated with the vertices of a binary tree, and sub-trees are used to define a set of overlapping groups. Such overlapping group structures arise naturally in multiscale signal analysis.

A similar argument applies with the roles of θ_2 and θ_3 reversed. Consequently, if the shared first variable is active in an optimal solution, it is usually the case that the second and third variables will also be active, leading to a fully dense solution. See the bibliographic discussion for references that discuss this phenomenon in greater detail.

The overlapping group Lasso is a closely related but different penalty that is designed to overcome this potentially troublesome issue.

Example 9.4 (Overlapping group Lasso) As in Example 9.3, consider a collection of groups $\mathcal{G} = \{g_1, \ldots, g_T\}$, where each group is a subset of the index set $\{1, \ldots, d\}$. We require that the union over all groups covers the full index set, but we allow for overlaps among the groups. See panel (b) in Figure 9.3 for an example of a collection of overlapping groups.

When there actually is overlap, any vector θ has many possible group representations, meaning collections $\{w_g,\ g \in \mathcal{G}\}$ such that $\sum_{g \in \mathcal{G}} w_g = \theta$. The *overlap group norm* is based on minimizing over all such representations, as follows:

$$\Phi_{\mathrm{over}}(\theta) := \inf_{\substack{\theta = \sum_{g \in \mathcal{G}} w_g \\ w_g,\ g \in \mathcal{G}}} \left\{ \sum_{g \in \mathcal{G}} \|w_g\| \right\}. \tag{9.10}$$

As we verify in Exercise 9.1, the variational representation (9.10) defines a valid norm on $\mathbb{R}^d$. Of course, when the groups are non-overlapping, this definition reduces to the previous one (9.9). Figure 9.2(b) shows the overlapping group norm (9.10) in the special case of the groups $g_1 = \{1, 2\}$ and $g_2 = \{1, 3\}$. Notice how it differs from the standard group Lasso (9.9) with the same choice of groups, as shown in Figure 9.1(c). ♣

When used as a regularizer in the general M-estimator (9.3), the overlapping group Lasso (9.10) tends to induce solution vectors with their support contained within a union of the groups. To understand this issue, let us return to the group set $g_1 = \{1, 2\}$ and $g_2 = \{1, 3\}$, and suppose once again that the first two variables are active, say $\theta_1 = \theta_2 = 1$. The residual penalty on θ_3 then takes the form

$$f_{\text{over}}(\theta_3) := \Phi_{\text{over}}(1, 1, \theta_3) - \Phi_{\text{over}}(1, 1, 0) = \inf_{\alpha \in \mathbb{R}} \{\|(\alpha, 1)\|_2 + \|(1 - \alpha, \theta_3)\|_2\} - \sqrt{2}.$$

It can be shown that this function behaves like the ℓ_1-norm around the origin, so that it tends to encourage sparsity in θ_3. See Figure 9.2(b) for an illustration.

Up to this point, we have considered vector estimation problems, in which the parameter space Ω is some subspace of $\mathbb{R}^d$. We now turn to various types of matrix estimation problems, in which the parameter space is some subset of $\mathbb{R}^{d_1 \times d_2}$, the space of all $(d_1 \times d_2)$-dimensional matrices. Of course, any such problem can be viewed as a vector estimation problem, simply by transforming the matrix to a $D = d_1 d_2$ vector. However, it is often more natural to retain the matrix structure of the problem. Let us consider some examples.

Example 9.5 (Estimation of Gaussian graphical models) Any zero-mean Gaussian random vector with a strictly positive definite covariance matrix $\mathbf{\Sigma} > 0$ has a density of the form

$$\mathbb{P}(x_1, \ldots, x_d; \mathbf{\Theta}^*) \propto \sqrt{\det(\mathbf{\Theta}^*)}\, e^{-\frac{1}{2} x^{\mathrm{T}} \mathbf{\Theta}^* x}, \tag{9.11}$$

where $\mathbf{\Theta}^* = (\mathbf{\Sigma})^{-1}$ is the inverse covariance matrix, also known as the precision matrix. In many cases, the components of the random vector $X = (X_1, \ldots, X_d)$ satisfy various types of conditional independence relationships: for instance, it might be the case that X_j is conditionally independent of X_k given the other variables $X_{\setminus\{j,k\}}$. In the Gaussian case, it is a consequence of the Hammersley–Clifford theorem that this conditional independence statement holds if and only if the precision matrix $\mathbf{\Theta}^*$ has a zero in position (j, k). Thus, conditional independence is directly captured by the sparsity of the precision matrix. See Chapter 11 for further details on this relationship between conditional independence, and the structure of $\mathbf{\Theta}^*$.

Given a Gaussian model that satisfies many conditional independence relationships, the precision matrix will be sparse, in which case it is natural to use the elementwise ℓ_1-norm $\Phi(\mathbf{\Theta}) = \sum_{j \neq k} |\Theta_{jk}|$ as a regularizer. Here we have chosen not to regularize the diagonal entries, since they all must be non-zero so as to ensure strict positive definiteness. Combining this form of ℓ_1-regularization with the Gaussian log-likelihood leads to the estimator

$$\widehat{\mathbf{\Theta}} \in \arg\min_{\mathbf{\Theta} \in \mathcal{S}^{d \times d}} \left\{ \langle\langle \mathbf{\Theta}, \widehat{\mathbf{\Sigma}} \rangle\rangle - \log\det \mathbf{\Theta} + \lambda_n \sum_{j \neq k} |\Theta_{jk}| \right\}, \tag{9.12}$$

where $\widehat{\mathbf{\Sigma}} = \frac{1}{n} \sum_{i=1}^n x_i x_i^{\mathrm{T}}$ is the sample covariance matrix. This combination corresponds to another special case of the general estimator (9.3), known as the *graphical Lasso*, which we analyze in Chapter 11. ♣

The problem of multivariate regression is a natural extension of a standard regression problem, which involves scalar response variables, to the vector-valued setting.

Example 9.6 (Multivariate regression) In a multivariate regression problem, we observe samples of the form $(z_i, y_i) \in \mathbb{R}^p \times \mathbb{R}^T$, and our goal is to use the vector of features z_i to predict the vector of responses $y_i \in \mathbb{R}^T$. Let $\mathbf{Y} \in \mathbb{R}^{n\times T}$ and $\mathbf{Z} \in \mathbb{R}^{n\times p}$ be matrices with y_i and z_i, respectively, as their ith row. In the simplest case, we assume that the response matrix $\mathbf{Y}$ and covariate matrix $\mathbf{Z}$ are linked via the linear model

$$\mathbf{Y} = \mathbf{Z}\mathbf{\Theta}^* + \mathbf{W}, \tag{9.13}$$

where $\mathbf{\Theta}^* \in \mathbb{R}^{p\times T}$ is a matrix of regression coefficients, and $\mathbf{W} \in \mathbb{R}^{n\times T}$ is a stochastic noise matrix. See Figure 9.4 for an illustration.

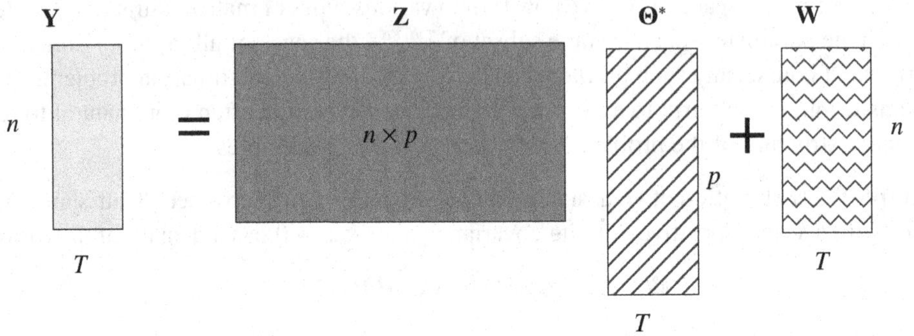

Figure 9.4 Illustration of the multivariate linear regression model: a data set of n observations consists of a matrix $\mathbf{Y} \in \mathbb{R}^{n\times T}$ of multivariate responses, and a matrix $\mathbf{Z} \in \mathbb{R}^{n\times p}$ of covariates, in this case shared across the tasks. Our goal is to estimate the matrix $\mathbf{\Theta}^* \in \mathbb{R}^{p\times T}$ of regression coefficients.

One way in which to view the model (9.13) is as a collection of T different p-dimensional regression problems of the form

$$Y_{\cdot,t} = \mathbf{Z}\Theta^*_{\cdot,t} + W_{\cdot,t}, \qquad \text{for } t = 1, \ldots, T,$$

where $Y_{\cdot,t} \in \mathbb{R}^n$, $\Theta^*_{\cdot,t} \in \mathbb{R}^p$ and $W_{\cdot,t} \in \mathbb{R}^n$ are the tth columns of the matrices $\mathbf{Y}$, $\mathbf{\Theta}^*$ and $\mathbf{W}$, respectively. One could then estimate each column $\Theta^*_{\cdot,t}$ separately by solving a standard univariate regression problem.

However, many applications lead to interactions between the different columns of $\mathbf{\Theta}^*$, which motivates solving the univariate regression problems in a joint manner. For instance, it is often the case that there is a subset of features—that is, a subset of the rows of $\mathbf{\Theta}^*$—that are relevant for prediction in all T regression problems. For estimating such a row-sparse matrix, a natural regularizer is the row-wise $(2,1)$-norm $\Phi(\mathbf{\Theta}) := \sum_{j=1}^p \|\Theta_{j,\cdot}\|_2$, where $\Theta_{j,\cdot} \in \mathbb{R}^T$ denotes the jth row of the matrix $\mathbf{\Theta} \in \mathbb{R}^{p\times T}$. Note that this regularizer is a special case of the general group penalty (9.9). Combining this regularizer with the least-squares cost, we obtain

$$\widehat{\mathbf{\Theta}} \in \arg\min_{\mathbf{\Theta}\in\mathbb{R}^{p\times T}} \left\{ \frac{1}{2n} |||\mathbf{Y} - \mathbf{Z}\mathbf{\Theta}|||_F^2 + \lambda_n \sum_{j=1}^p \|\Theta_{j,\cdot}\|_2 \right\}. \tag{9.14}$$

This estimator is often referred to as the *multivariate group Lasso*, for obvious reasons. The underlying optimization problem is an instance of a second-order cone problem (SOCP), and can be solved efficiently by a variety of algorithms; see the bibliography section for further discussion. ♣

Other types of structure are also possible in multivariate regression problems, and lead to different types of regularization.

Example 9.7 (Overlapping group Lasso and multivariate regression) There is an interesting extension of the row-sparse model from Example 9.6, one which leads to an instance of the overlapping group Lasso (9.10). The row-sparse model assumes that there is a relatively small subset of predictors, each of which is active in *all* of the T tasks. A more flexible model allows for the possibility of a subset of predictors that are shared among all tasks, coupled with a subset of predictors that appear in only one (or relatively few) tasks. This type of structure can be modeled by decomposing the regression matrix $\mathbf{\Theta}^*$ as the sum of a row-sparse matrix $\mathbf{\Omega}^*$ along with an elementwise-sparse matrix $\mathbf{\Gamma}^*$. If we impose a group $\ell_{1,2}$-norm on the row-sparse component and an ordinary ℓ_1-norm on the element-sparse component, then we are led to the estimator

$$(\widehat{\mathbf{\Omega}}, \widehat{\mathbf{\Gamma}}) \in \arg \min_{\mathbf{\Omega}, \mathbf{\Gamma} \in \mathbb{R}^{d \times T}} \left\{ \frac{1}{2n} |||\mathbf{Y} - \mathbf{Z}(\mathbf{\Omega} + \mathbf{\Gamma})|||_{\mathrm{F}}^2 + \lambda_n \sum_{j=1}^{d} \|\mathbf{\Omega}_{j,\cdot}\|_2 + \mu_n \|\mathbf{\Gamma}\|_1 \right\}, \tag{9.15}$$

where $\lambda_n, \mu_n > 0$ are regularization parameters to be chosen. Any solution to this optimization problem defines an estimate of the full regression matrix via $\widehat{\mathbf{\Theta}} = \widehat{\mathbf{\Omega}} + \widehat{\mathbf{\Gamma}}$.

We have defined the estimator (9.15) as an optimization problem over the matrix pair $(\mathbf{\Omega}, \mathbf{\Gamma})$, using a separate regularizer for each matrix component. Alternatively, we can formulate it as a direct estimator for $\widehat{\mathbf{\Theta}}$. In particular, by making the substitution $\mathbf{\Theta} = \mathbf{\Omega} + \mathbf{\Gamma}$, and minimizing over both $\mathbf{\Theta}$ and the pair $(\mathbf{\Omega}, \mathbf{\Gamma})$ subject to this linear constraint, we obtain the equivalent formulation

$$\widehat{\mathbf{\Theta}} \in \arg \min_{\mathbf{\Theta} \in \mathbb{R}^{d \times T}} \left\{ \frac{1}{2n} |||\mathbf{Y} - \mathbf{Z}\mathbf{\Theta}|||_{\mathrm{F}}^2 + \lambda_n \underbrace{\left\{ \inf_{\mathbf{\Omega} + \mathbf{\Gamma} = \mathbf{\Theta}} \|\mathbf{\Omega}\|_{1,2} + \omega_n \|\mathbf{\Gamma}\|_1 \right\}}_{\Phi_{\text{over}}(\mathbf{\Theta})} \right\}, \tag{9.16}$$

where $\omega_n = \frac{\mu_n}{\lambda_n}$. In this direct formulation, we see that the assumed decomposition leads to an interesting form of the overlapping group norm. We return to study the estimator (9.16) in Section 9.7. ♣

In other applications of multivariate regression, one might imagine that the individual regression vectors—that is, the columns $\Theta^*_{\cdot,t} \in \mathbb{R}^p$—all lie within some low-dimensional subspace, corresponding to some hidden meta-features, so that it has relatively low rank. Many other problems, to be discussed in more detail in Chapter 10, also lead to estimation problems that involve rank constraints. In such settings, the ideal approach would be to impose an explicit rank constraint within our estimation procedure. Unfortunately, when viewed as function on the space of $d_1 \times d_2$ matrices, the rank function is non-convex, so that this approach is not computationally feasible. Accordingly, we are motivated to study convex relaxations of rank constraints.

Example 9.8 (Nuclear norm as a relaxation of rank) The *nuclear norm* provides a natural relaxation of the rank of a matrix, one which is analogous to the ℓ_1-norm as a relaxation of the cardinality of a vector. In order to define the nuclear norm, we first recall the *singular value decomposition*, or SVD for short, of a matrix $\Theta \in \mathbb{R}^{d_1 \times d_2}$. Letting $d' = \min\{d_1, d_2\}$, the SVD takes the form

$$\mathbf{\Theta} = \mathbf{U}\mathbf{D}\mathbf{V}^{\mathrm{T}}, \tag{9.17}$$

where $\mathbf{U} \in \mathbb{R}^{d_1 \times d'}$ and $\mathbf{V} \in \mathbb{R}^{d_2 \times d'}$ are orthonormal matrices (meaning that $\mathbf{U}^{\mathrm{T}}\mathbf{U} = \mathbf{V}^{\mathrm{T}}\mathbf{V} = \mathbf{I}_{d'}$). The matrix $\mathbf{D} \in \mathbb{R}^{d' \times d'}$ is diagonal with its entries corresponding to the singular values of $\mathbf{\Theta}$, denoted by

$$\sigma_1(\mathbf{\Theta}) \geq \sigma_2(\mathbf{\Theta}) \geq \sigma_3(\mathbf{\Theta}) \geq \cdots \geq \sigma_{d'}(\mathbf{\Theta}) \geq 0. \tag{9.18}$$

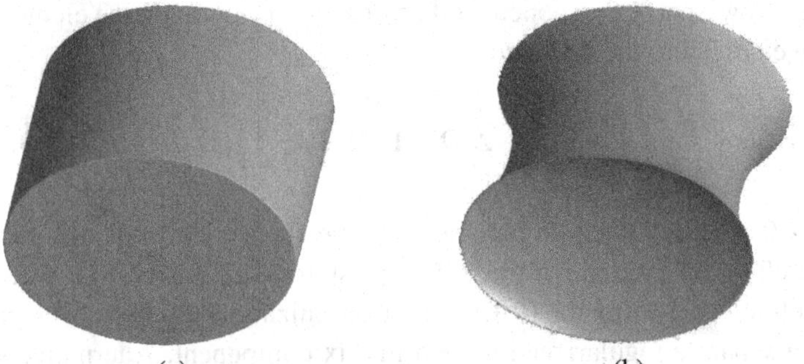

(a) (b)

Figure 9.5 Illustration of the nuclear norm ball as a relaxation of a rank constraint. (a) Set of all matrices of the form $\mathbf{\Theta} = \begin{bmatrix} \theta_1 & \theta_2 \\ \theta_2 & \theta_3 \end{bmatrix}$ such that $|||\mathbf{\Theta}|||_{\text{nuc}} \leq 1$. This is a projection of the unit ball of the nuclear norm ball onto the space of symmetric matrices. (b) For a parameter $q > 0$, the ℓ_q-"ball" of matrices is defined by $\mathbb{B}_q(1) = \{\mathbf{\Theta} \in \mathbb{R}^{2\times 2} \mid \sum_{j=1}^{2} \sigma_j(\mathbf{\Theta})^q \leq 1\}$. For all $q \in [0, 1)$, this is a non-convex set, and it is equivalent to the set of all rank-one matrices for $q = 0$.

Observe that the number of strictly positive singular values specifies the rank—that is, we have $\text{rank}(\mathbf{\Theta}) = \sum_{j=1}^{d'} \mathbb{I}[\sigma_j(\mathbf{\Theta}) > 0]$. This observation, though not practically useful on its own, suggests a natural convex relaxation of a rank constraint, namely the *nuclear norm*

$$|||\mathbf{\Theta}|||_{\text{nuc}} = \sum_{j=1}^{d'} \sigma_j(\mathbf{\Theta}), \tag{9.19}$$

corresponding to the ℓ_1-norm of the singular values.[1] As shown in Figure 9.5(a), the nuclear norm provides a convex relaxation of the set of low-rank matrices. ♣

There are a variety of other statistical models—in addition to multivariate regression—in which rank constraints play a role, and the nuclear norm relaxation is useful for many of them. These problems are discussed in detail in Chapter 10 to follow.

[1] No absolute value is necessary, since singular values are non-negative by definition.

9.2 Decomposable regularizers and their utility

Having considered a general family of M-estimators (9.3) and illustrated it with various examples, we now turn to the development of techniques for bounding the estimation error $\widehat{\theta} - \theta^*$. The first ingredient in our analysis is a property of the regularizer known as decomposability. It is a geometric property, based on how the regularizer behaves over certain pairs of subspaces. The ℓ_1-norm is the canonical example of a decomposable norm, but various other norms also share this property. Decomposability implies that any optimum $\widehat{\theta}$ to the M-estimator (9.3) belongs to a very special set, as shown in Proposition 9.13.

From here onwards, we assume that the set Ω is endowed with an inner product $\langle\cdot, \cdot\rangle$, and we use $\|\cdot\|$ to denote the norm induced by this inner product. The standard examples to keep in mind are

- the space $\mathbb{R}^d$ with the usual Euclidean inner product, or more generally with a weighted Euclidean inner product, and
- the space $\mathbb{R}^{d_1 \times d_2}$ equipped with the trace inner product (10.1).

Given a vector $\theta \in \Omega$ and a subspace $\mathbb{S}$ of Ω, we use $\theta_{\mathbb{S}}$ to denote the projection of θ onto $\mathbb{S}$. More precisely, we have

$$\theta_{\mathbb{S}} := \arg\min_{\widetilde{\theta} \in \mathbb{S}} \|\widetilde{\theta} - \theta\|^2. \tag{9.20}$$

These projections play an important role in the sequel; see Exercise 9.2 for some examples.

9.2.1 Definition and some examples

The notion of a decomposable regularizer is defined in terms of a pair of subspaces $\mathbb{M} \subseteq \overline{\mathbb{M}}$ of $\mathbb{R}^d$. The role of the *model subspace* $\mathbb{M}$ is to capture the constraints specified by the model; for instance, as illustrated in the examples to follow, it might be the subspace of vectors with a particular support or a subspace of low-rank matrices. The orthogonal complement of the space $\overline{\mathbb{M}}$, namely the set

$$\overline{\mathbb{M}}^\perp := \left\{v \in \mathbb{R}^d \mid \langle u, v\rangle = 0 \quad \text{for all } u \in \overline{\mathbb{M}}\right\}, \tag{9.21}$$

is referred to as the *perturbation subspace*, representing deviations away from the model subspace $\mathbb{M}$. In the ideal case, we have $\overline{\mathbb{M}}^\perp = \mathbb{M}^\perp$, but the definition allows for the possibility that $\overline{\mathbb{M}}$ is strictly larger than $\mathbb{M}$, so that $\overline{\mathbb{M}}^\perp$ is strictly smaller than $\mathbb{M}^\perp$. This generality is needed for treating the case of low-rank matrices and nuclear norm, as discussed in Chapter 10.

Definition 9.9 Given a pair of subspaces $\mathbb{M} \subseteq \overline{\mathbb{M}}$, a norm-based regularizer Φ is *decomposable* with respect to $(\mathbb{M}, \overline{\mathbb{M}}^\perp)$ if

$$\Phi(\alpha + \beta) = \Phi(\alpha) + \Phi(\beta) \qquad \text{for all } \alpha \in \mathbb{M} \text{ and } \beta \in \overline{\mathbb{M}}^\perp. \tag{9.22}$$

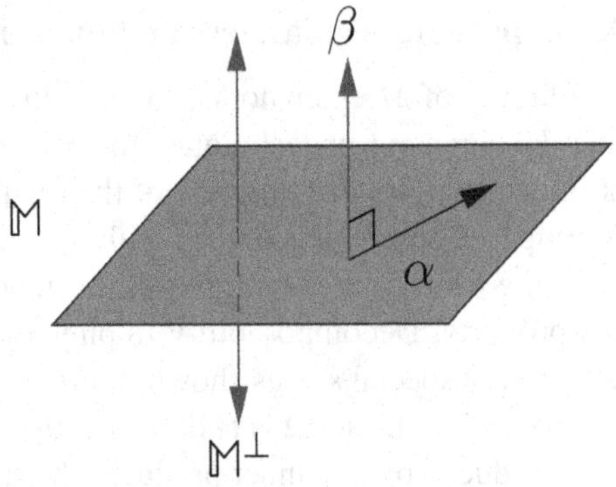

Figure 9.6 In the ideal case, decomposability is defined in terms of a subspace pair $(\mathbb{M}, \mathbb{M}^\perp)$. For any $\alpha \in \mathbb{M}$ and $\beta \in \mathbb{M}^\perp$, the regularizer should decompose as $\Phi(\alpha + \beta) = \Phi(\alpha) + \Phi(\beta)$.

See Figure 9.6 for the geometry of this definition. In order to build some intuition, let us consider the ideal case $\mathbb{M} = \overline{\mathbb{M}}$, so that the decomposition (9.22) holds for all pairs $(\alpha, \beta) \in \mathbb{M} \times \mathbb{M}^\perp$. For any given pair (α, β) of this form, the vector $\alpha + \beta$ can be interpreted as perturbation of the model vector α away from the subspace $\mathbb{M}$, and it is desirable that the regularizer penalize such deviations as much as possible. By the triangle inequality for a norm, we always have $\Phi(\alpha + \beta) \leq \Phi(\alpha) + \Phi(\beta)$, so that the decomposability condition (9.22) holds if and only if the triangle inequality is tight for all pairs $(\alpha, \beta) \in (\mathbb{M}, \mathbb{M}^\perp)$. It is exactly in this setting that the regularizer penalizes deviations away from the model subspace $\mathbb{M}$ as much as possible.

Let us consider some illustrative examples:

Example 9.10 (Decomposability and sparse vectors) We begin with the ℓ_1-norm, which is the canonical example of a decomposable regularizer. Let S be a given subset of the index set $\{1, \ldots, d\}$ and S^c be its complement. We then define the model subspace

$$\mathbb{M} \equiv \mathbb{M}(S) := \{\theta \in \mathbb{R}^d \mid \theta_j = 0 \quad \text{for all } j \in S^c\}, \tag{9.23}$$

corresponding to the set of all vectors that are supported on S. Observe that

$$\mathbb{M}^\perp(S) = \{\theta \in \mathbb{R}^d \mid \theta_j = 0 \quad \text{for all } j \in S\}.$$

With these definitions, it is then easily seen that for any pair of vectors $\alpha \in \mathbb{M}(S)$ and $\beta \in \mathbb{M}^\perp(S)$, we have

$$\|\alpha + \beta\|_1 = \|\alpha\|_1 + \|\beta\|_1,$$

showing that the ℓ_1-norm is decomposable with respect to the pair $(\mathbb{M}(S), \mathbb{M}^\perp(S))$. ♣

Example 9.11 (Decomposability and group sparse norms) We now turn to the notion of decomposability for the group Lasso norm (9.9). In this case, the subspaces are defined in terms of subsets of groups. More precisely, given any subset $S_{\mathcal{G}} \subset \mathcal{G}$ of the group index set, consider the set

$$\mathbb{M}(S_{\mathcal{G}}) := \{\theta \in \Omega \mid \theta_g = 0 \quad \text{for all } g \notin S_{\mathcal{G}}\}, \tag{9.24}$$

corresponding to the subspace of vectors supported only on groups indexed by $S_{\mathcal{G}}$. Note that the orthogonal subspace is given by $\mathbb{M}^{\perp}(S_{\mathcal{G}}) = \{\theta \in \Omega \mid \theta_g = 0 \text{ for all } g \in S_{\mathcal{G}}\}$. Letting $\alpha \in \mathbb{M}(S_{\mathcal{G}})$ and $\beta \in \mathbb{M}^{\perp}(S_{\mathcal{G}})$ be arbitrary, we have

$$\Phi(\alpha + \beta) = \sum_{g \in S_{\mathcal{G}}} \|\alpha_g\| + \sum_{g \in S_{\mathcal{G}}^c} \|\beta_g\| = \Phi(\alpha) + \Phi(\beta),$$

thus showing that the group norm is decomposable with respect to the pair $(\mathbb{M}(S_{\mathcal{G}}), \mathbb{M}^{\perp}(S_{\mathcal{G}}))$. ♣

In the preceding example, we considered the case of non-overlapping groups. It is natural to ask whether the same decomposability—that is, with respect to the pair $(\mathbb{M}(S_{\mathcal{G}}), \mathbb{M}^{\perp}(S_{\mathcal{G}}))$—continues to hold for the ordinary group Lasso $\|\theta\|_{\mathcal{G}} = \sum_{g \in \mathcal{G}} \|\theta_g\|$ when the groups are allowed to be overlapping. A little thought shows that this is not the case in general: for instance, in the case $\theta \in \mathbb{R}^4$, consider the overlapping groups $g_1 = \{1, 2\}$, $g_2 = \{2, 3\}$ and $g_3 = \{3, 4\}$. If we let $S_{\mathcal{G}} = \{g_1\}$, then

$$\mathbb{M}^{\perp}(S_{\mathcal{G}}) = \{\theta \in \mathbb{R}^4 \mid \theta_1 = \theta_2 = 0\}.$$

The vector $\alpha = \begin{bmatrix} 0 & 1 & 0 & 0 \end{bmatrix}$ belongs to $\mathbb{M}(S_{\mathcal{G}})$, and the vector $\beta = \begin{bmatrix} 0 & 0 & 1 & 0 \end{bmatrix}$ belongs to $\mathbb{M}^{\perp}(S_{\mathcal{G}})$. In the case of the group ℓ_1/ℓ_2-norm $\|\theta\|_{\mathcal{G},2} = \sum_{g \in \mathcal{G}} \|\theta_g\|_2$, we have $\|\alpha + \beta\|_{\mathcal{G},2} = 1 + \sqrt{2} + 1$, but

$$\|\alpha\|_{\mathcal{G},2} + \|\beta\|_{\mathcal{G},2} = 1 + 1 + 1 + 1 = 4 > 2 + \sqrt{2}, \tag{9.25}$$

showing that decomposability is violated. However, this issue can be addressed by a different choice of subspace pair, one that makes use of the additional freedom provided by allowing for $\overline{\mathbb{M}} \supsetneq \mathbb{M}$. We illustrate this procedure in the following:

Example 9.12 (Decomposability of ordinary group Lasso with overlapping groups) As before, let $S_{\mathcal{G}}$ be a subset of the group index set $\mathcal{G}$, and define the subspace $\mathbb{M}(S_{\mathcal{G}})$. We then define the augmented group set

$$\widetilde{S}_{\mathcal{G}} := \Big\{g \in \mathcal{G} \mid g \cap \bigcup_{h \in S_{\mathcal{G}}} h \neq \emptyset\Big\}, \tag{9.26}$$

corresponding to the set of groups with non-empty intersection with some group in $S_{\mathcal{G}}$. Note that in the case of non-overlapping groups, we have $\widetilde{S}_{\mathcal{G}} = S_{\mathcal{G}}$, whereas $\widetilde{S}_{\mathcal{G}} \supseteq S_{\mathcal{G}}$ in the more general case of overlapping groups. This augmented set defines the subspace $\overline{\mathbb{M}} := \mathbb{M}(\widetilde{S}_{\mathcal{G}}) \supseteq \mathbb{M}(S_{\mathcal{G}})$, and we claim that the overlapping group norm is decomposable with respect to the pair $(\mathbb{M}(S_{\mathcal{G}}), \mathbb{M}^{\perp}(\widetilde{S}_{\mathcal{G}}))$.

Indeed, let α and β be arbitrary members of $\mathbb{M}(S_{\mathcal{G}})$ and $\mathbb{M}^{\perp}(\widetilde{S}_{\mathcal{G}})$, respectively. Note that any element of $\mathbb{M}^{\perp}(\widetilde{S}_{\mathcal{G}})$ can have support only on the subset $\bigcup_{h \notin \widetilde{S}_{\mathcal{G}}} h$; at the same time, this subset has no overlap with $\bigcup_{g \in S_{\mathcal{G}}} g$, and any element of $\mathbb{M}(S_{\mathcal{G}})$ is supported on this latter subset. As a consequence of these properties, we have

$$\|\alpha + \beta\|_{\mathcal{G}} = \sum_{g \in \mathcal{G}} (\alpha + \beta)_g = \sum_{g \in \widetilde{S}_{\mathcal{G}}} \alpha_g + \sum_{g \notin \widetilde{S}_{\mathcal{G}}} \beta_g = \|\alpha\|_{\mathcal{G}} + \|\beta\|_{\mathcal{G}},$$

as claimed. ♣

It is worthwhile observing how our earlier counterexample (9.25) is excluded by the construction given in Example 9.12. With the groups $g_1 = \{1, 2\}$, $g_2 = \{2, 3\}$ and $g_3 = \{3, 4\}$, combined with the subset $S_{\mathcal{G}} = \{g_1\}$, we have $\widetilde{S}_{\mathcal{G}} = \{g_1, g_2\}$. The vector $\beta = \begin{bmatrix} 0 & 0 & 1 & 0 \end{bmatrix}$ belongs to the subspace

$$\mathbb{M}^\perp(S_{\mathcal{G}}) = \{\theta \in \mathbb{R}^d \mid \theta_1 = \theta_2 = 0\},$$

but it does *not* belong to the smaller subspace

$$\mathbb{M}^\perp(\widetilde{S}_{\mathcal{G}}) = \{\theta \in \mathbb{R}^4 \mid \theta_1 = \theta_2 = \theta_3 = 0\}.$$

Consequently, it does not violate the decomposability property. However, note that there is a statistical price to be paid by enlarging to the augmented set $\mathbb{M}(\widetilde{S}_{\mathcal{G}})$: as our later results demonstrate, the statistical estimation error scales as a function of the size of this set.

As discussed previously, many problems involve estimating low-rank matrices, in which context the nuclear norm (9.19) plays an important role. In Chapter 10, we show how the nuclear norm is decomposable with respect to appropriately chosen subspaces. Unlike our previous examples (in which $\mathbb{M} = \overline{\mathbb{M}}$), in this case we need to use the full flexibility of our definition, and choose $\overline{\mathbb{M}}$ to be a strict superset of $\mathbb{M}$.

Finally, it is worth noting that sums of decomposable regularizers over disjoint sets of parameters remain decomposable: that is, if Φ_1 and Φ_2 are decomposable with respect to subspaces over Ω_1 and Ω_2 respectively, then the sum $\Phi_1 + \Phi_2$ remains decomposable with respect to the same subspaces extended to the Cartesian product space $\Omega_1 \times \Omega_2$. For instance, this property is useful for the matrix decomposition problems discussed in Chapter 10, which involve a pair of matrices $\mathbf{\Lambda}$ and $\mathbf{\Gamma}$, and the associated regularizers $\Phi_1(\mathbf{\Lambda}) = |||\mathbf{\Lambda}|||_{\text{nuc}}$ and $\Phi_2(\mathbf{\Gamma}) = \|\mathbf{\Gamma}\|_1$.

9.2.2 A key consequence of decomposability

Why is decomposability important in the context of M-estimation? Ultimately, our goal is to provide bounds on the error vector $\widehat{\Delta} := \widehat{\theta} - \theta^*$ between any global optimum of the optimization problem (9.3) and the unknown parameter θ^*. In this section, we show that decomposability—in conjunction with a suitable choice for the regularization weight λ_n—ensures that the error $\widehat{\Delta}$ must lie in a very restricted set.

In order to specify a "suitable" choice of regularization parameter λ_n, we need to define the notion of the dual norm associated with our regularizer. Given any norm $\Phi \colon \mathbb{R}^d \to \mathbb{R}$, its dual norm is defined in a variational manner as

$$\Phi^*(v) := \sup_{\Phi(u) \leq 1} \langle u, v \rangle . \tag{9.27}$$

Regularizer Φ	Dual norm Φ^*
ℓ_1-norm $\quad \Phi(u) = \sum_{j=1}^{d} \|u_j\|$	ℓ_∞-norm $\quad \Phi^*(v) = \|v\|_\infty = \max_{j=1,\ldots,d} \|v_j\|$
Group ℓ_1/ℓ_p-norm $\quad \Phi(u) = \sum_{g\in\mathcal{G}} \|u_g\|_p$ Non-overlapping groups	Group ℓ_∞/ℓ_q-norm $\quad \Phi^*(v) = \max_{g\in\mathcal{G}} \|v_g\|_q$ $\frac{1}{p} + \frac{1}{q} = 1$
Nuclear norm $\quad \Phi(\mathbf{M}) = \sum_{j=1}^{d} \sigma_j(\mathbf{M})$	ℓ_2-operator norm $\quad \Phi^*(\mathbf{N}) = \max_{j=1,\ldots,d} \sigma_j(\mathbf{N})$ $d = \min\{d_1, d_2\}$
Overlap group norm $\Phi(u) = \inf_{u=\sum_{g\in\mathcal{G}} w_g} \|w_g\|_p$	Overlap dual norm $\Phi^*(v) = \max_{g\in\mathcal{G}} \|v_g\|_q$
Sparse-low-rank decomposition norm $\Phi_\omega(\mathbf{M}) = \inf_{\mathbf{M}=\mathbf{A}+\mathbf{B}} \{\|\mathbf{A}\|_1 + \omega \vert\vert\vert \mathbf{B} \vert\vert\vert_{\text{nuc}}\}$	Weighted max. norm $\Phi^*(\mathbf{N}) = \max\{\|\mathbf{N}\|_{\max}, \omega^{-1} \vert\vert\vert \mathbf{N} \vert\vert\vert_2\}$

Table 9.1 *Primal and dual pairs of regularizers in various cases. See Exercises 9.4 and 9.5 for verification of some of these correspondences.*

Table 9.1 gives some examples of various dual norm pairs.

Our choice of regularization parameter is specified in terms of the random vector $\nabla\mathcal{L}_n(\theta^*)$ —the gradient of the empirical cost evaluated at θ^*, also referred to as the *score function*. Under mild regularity conditions, we have $\mathbb{E}[\nabla\mathcal{L}_n(\theta^*)] = \nabla\bar{\mathcal{L}}(\theta^*)$. Consequently, when the target parameter θ^* lies in the interior of the parameter space Ω, by the optimality conditions for the minimization (9.2), the random vector $\nabla\mathcal{L}_n(\theta^*)$ has zero mean. Under ideal circumstances, we expect that the score function will not be too large, and we measure its fluctuations in terms of the dual norm, thereby defining the "good event"

$$\mathbb{G}(\lambda_n) := \left\{\Phi^*(\nabla\mathcal{L}_n(\theta^*)) \leq \frac{\lambda_n}{2}\right\}. \tag{9.28}$$

With this set-up, we are now ready for the statement of the main technical result of this section. The reader should recall the definition of the subspace projection operator (9.20).

Proposition 9.13 *Let $\mathcal{L}_n : \Omega \to \mathbb{R}$ be a convex function, let the regularizer $\Phi : \Omega \to [0, \infty)$ be a norm, and consider a subspace pair $(\mathbb{M}, \bar{\mathbb{M}}^\perp)$ over which Φ is decomposable. Then conditioned on the event $\mathbb{G}(\lambda_n)$, the error $\widehat{\Delta} = \widehat{\theta} - \theta^*$ belongs to the set*

$$\mathbb{C}_{\theta^*}(\mathbb{M}, \bar{\mathbb{M}}^\perp) := \{\Delta \in \Omega \mid \Phi(\Delta_{\bar{\mathbb{M}}^\perp}) \leq 3\Phi(\Delta_{\bar{\mathbb{M}}}) + 4\Phi(\theta^*_{\mathbb{M}^\perp})\}. \tag{9.29}$$

When the subspaces $(\mathbb{M}, \bar{\mathbb{M}}^\perp)$ and parameter θ^* are clear from the context, we adopt the shorthand notation $\mathbb{C}$. Figure 9.7 provides an illustration of the geometric structure of the set $\mathbb{C}$. To understand its significance, let us consider the special case when $\theta^* \in \mathbb{M}$, so that $\theta^*_{\mathbb{M}^\perp} = 0$. In this case, membership of $\widehat{\Delta}$ in $\mathbb{C}$ implies that $\Phi(\widehat{\Delta}_{\bar{\mathbb{M}}^\perp}) \leq 3\Phi(\widehat{\Delta}_{\bar{\mathbb{M}}})$, and hence that

$$\Phi(\widehat{\Delta}) = \Phi(\widehat{\Delta}_{\bar{\mathbb{M}}} + \widehat{\Delta}_{\bar{\mathbb{M}}^\perp}) \leq \Phi(\widehat{\Delta}_{\bar{\mathbb{M}}}) + \Phi(\widehat{\Delta}_{\bar{\mathbb{M}}^\perp}) \leq 4\Phi(\widehat{\Delta}_{\bar{\mathbb{M}}}). \tag{9.30}$$

Consequently, when measured in the norm defined by the regularizer, the vector $\widehat{\Delta}$ is only a constant factor larger than the projected quantity $\widehat{\Delta}_{\bar{\mathbb{M}}}$. Whenever the subspace $\bar{\mathbb{M}}$ is relatively small, this inequality provides significant control on $\widehat{\Delta}$.

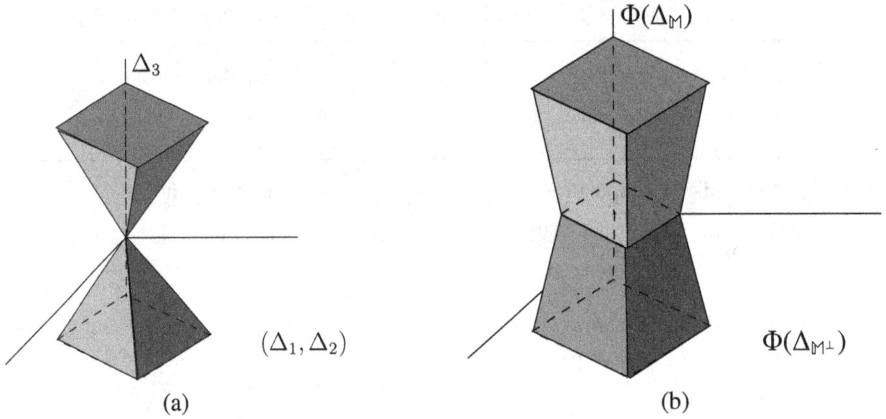

Figure 9.7 Illustration of the set $\mathbb{C}_{\theta^*}(\mathbb{M}, \bar{\mathbb{M}}^\perp)$ in the special case $\Delta = (\Delta_1, \Delta_2, \Delta_3) \in \mathbb{R}^3$ and regularizer $\Phi(\Delta) = \|\Delta\|_1$, relevant for sparse vectors (Example 9.1). This picture shows the case $S = \{3\}$, so that the model subspace is $\mathbb{M}(S) = \{\Delta \in \mathbb{R}^3 \mid \Delta_1 = \Delta_2 = 0\}$, and its orthogonal complement is given by $\mathbb{M}^\perp(S) = \{\Delta \in \mathbb{R}^3 \mid \Delta_3 = 0\}$. (a) In the special case when $\theta^*_1 = \theta^*_2 = 0$, so that $\theta^* \in \mathbb{M}$, the set $\mathbb{C}(\mathbb{M}, \mathbb{M}^\perp)$ is a cone, with no dependence on θ^*. (b) When θ^* does not belong to $\mathbb{M}$, the set $\mathbb{C}(\mathbb{M}, \mathbb{M}^\perp)$ is enlarged in the coordinates (Δ_1, Δ_2) that span $\mathbb{M}^\perp$. It is no longer a cone, but is still a star-shaped set.

We now turn to the proof of the proposition:

Proof Our argument is based on the function $\mathcal{F} : \Omega \to \mathbb{R}$ given by

$$\mathcal{F}(\Delta) := \mathcal{L}_n(\theta^* + \Delta) - \mathcal{L}_n(\theta^*) + \lambda_n\{\Phi(\theta^* + \Delta) - \Phi(\theta^*)\}. \tag{9.31}$$

By construction, we have $\mathcal{F}(0) = 0$, and so the optimality of $\widehat{\theta}$ implies that the error vector $\widehat{\Delta} = \widehat{\theta} - \theta^*$ must satisfy the condition $\mathcal{F}(\widehat{\Delta}) \leq 0$, corresponding to a basic inequality in this general setting. Our goal is to exploit this fact in order to establish the inclusion (9.29). In order to do so, we require control on the two separate pieces of $\mathcal{F}$, as summarized in the following:

Lemma 9.14 (Deviation inequalities) *For any decomposable regularizer and parameters θ^* and Δ, we have*

$$\Phi(\theta^* + \Delta) - \Phi(\theta^*) \geq \Phi(\Delta_{\bar{\mathbb{M}}^\perp}) - \Phi(\Delta_{\bar{\mathbb{M}}}) - 2\Phi(\theta^*_{\mathbb{M}^\perp}). \tag{9.32}$$

Moreover, for any convex function $\mathcal{L}_n$, conditioned on the event $\mathbb{G}(\lambda_n)$, we have

$$\mathcal{L}_n(\theta^* + \Delta) - \mathcal{L}_n(\theta^*) \geq -\frac{\lambda_n}{2}\left[\Phi(\Delta_{\bar{\mathbb{M}}}) + \Phi(\Delta_{\bar{\mathbb{M}}^\perp})\right]. \tag{9.33}$$

Given this lemma, the claim of Proposition 9.13 follows immediately. Indeed, combining the two lower bounds (9.32) and (9.33), we obtain

$$\begin{aligned} 0 \geq \mathcal{F}(\widehat{\Delta}) &\geq \lambda_n\left\{\Phi(\Delta_{\bar{\mathbb{M}}^\perp}) - \Phi(\Delta_{\bar{\mathbb{M}}}) - 2\Phi(\theta^*_{\mathbb{M}^\perp})\right\} - \frac{\lambda_n}{2}\left\{\Phi(\Delta_{\bar{\mathbb{M}}}) + \Phi(\Delta_{\bar{\mathbb{M}}^\perp})\right\} \\ &= \frac{\lambda_n}{2}\left\{\Phi(\Delta_{\bar{\mathbb{M}}^\perp}) - 3\Phi(\Delta_{\bar{\mathbb{M}}}) - 4\Phi(\theta^*_{\mathbb{M}^\perp})\right\}, \end{aligned}$$

from which the claim follows.

Thus, it remains to prove Lemma 9.14, and here we exploit decomposability of the regularizer. Since $\Phi(\theta^* + \Delta) = \Phi\left(\theta^*_{\mathbb{M}} + \theta^*_{\mathbb{M}^\perp} + \Delta_{\bar{\mathbb{M}}} + \Delta_{\bar{\mathbb{M}}^\perp}\right)$, applying the triangle inequality yields

$$\Phi(\theta^* + \Delta) \geq \Phi(\theta^*_{\mathbb{M}} + \Delta_{\bar{\mathbb{M}}^\perp}) - \Phi(\theta^*_{\mathbb{M}^\perp} + \Delta_{\bar{\mathbb{M}}}) \geq \Phi(\theta^*_{\mathbb{M}} + \Delta_{\bar{\mathbb{M}}^\perp}) - \Phi(\theta^*_{\mathbb{M}^\perp}) - \Phi(\Delta_{\bar{\mathbb{M}}}).$$

By decomposability applied to $\theta^*_{\mathbb{M}}$ and $\Delta_{\bar{\mathbb{M}}^\perp}$, we have $\Phi\left(\theta^*_{\mathbb{M}} + \Delta_{\bar{\mathbb{M}}^\perp}\right) = \Phi\left(\theta^*_{\mathbb{M}}\right) + \Phi(\Delta_{\bar{\mathbb{M}}^\perp})$, so that

$$\Phi(\theta^* + \Delta) \geq \Phi(\theta^*_{\mathbb{M}}) + \Phi(\Delta_{\bar{\mathbb{M}}^\perp}) - \Phi(\theta^*_{\mathbb{M}^\perp}) - \Phi(\Delta_{\bar{\mathbb{M}}}). \tag{9.34}$$

Similarly, by the triangle inequality, we have $\Phi(\theta^*) \leq \Phi\left(\theta^*_{\mathbb{M}}\right) + \Phi\left(\theta^*_{\mathbb{M}^\perp}\right)$. Combining this inequality with the bound (9.34), we obtain

$$\begin{aligned} \Phi(\theta^* + \Delta) - \Phi(\theta^*) &\geq \Phi(\theta^*_{\mathbb{M}}) + \Phi(\Delta_{\bar{\mathbb{M}}^\perp}) - \Phi(\theta^*_{\mathbb{M}^\perp}) - \Phi(\Delta_{\bar{\mathbb{M}}}) - \{\Phi(\theta^*_{\mathbb{M}}) + \Phi(\theta^*_{\mathbb{M}^\perp})\} \\ &= \Phi(\Delta_{\bar{\mathbb{M}}^\perp}) - \Phi(\Delta_{\bar{\mathbb{M}}}) - 2\Phi(\theta^*_{\mathbb{M}^\perp}), \end{aligned}$$

which yields the claim (9.32).

Turning to the cost difference, using the convexity of the cost function $\mathcal{L}_n$, we have

$$\mathcal{L}_n(\theta^* + \Delta) - \mathcal{L}_n(\theta^*) \geq \langle\nabla\mathcal{L}_n(\theta^*), \Delta\rangle \geq -|\langle\nabla\mathcal{L}_n(\theta^*), \Delta\rangle|.$$

Applying the Hölder inequality with the regularizer and its dual (see Exercise 9.7), we have

$$|\langle\nabla\mathcal{L}_n(\theta^*), \Delta\rangle| \leq \Phi^*(\nabla\mathcal{L}_n(\theta^*))\,\Phi(\Delta) \leq \frac{\lambda_n}{2}\left[\Phi(\Delta_{\bar{\mathbb{M}}}) + \Phi(\Delta_{\bar{\mathbb{M}}^\perp})\right],$$

where the final step uses the triangle inequality, and the assumed bound $\lambda_n \geq 2\Phi^*(\nabla\mathcal{L}_n(\theta^*))$. Putting together the pieces yields the claimed bound (9.33). This completes the proof of Lemma 9.14, and hence the proof of the proposition. □

9.3 Restricted curvature conditions

We now turn to the second component of a general framework, which concerns the curvature of the cost function. Before discussing the general high-dimensional setting, let us recall the classical role of curvature in maximum likelihood estimation, where it enters via the Fisher information matrix. Under i.i.d. sampling, the principle of maximum likelihood is equivalent to minimizing the cost function

$$\mathcal{L}_n(\theta) := -\frac{1}{n}\sum_{i=1}^{n} \log \mathbb{P}_\theta(z_i). \tag{9.35}$$

The Hessian of this cost function $\nabla^2\mathcal{L}_n(\theta)$ is the sample version of the Fisher information matrix; as the sample size n increases to infinity with d fixed, it converges in a pointwise sense to the population *Fisher information* $\nabla^2\overline{\mathcal{L}}(\theta)$. Recall that the population cost function $\overline{\mathcal{L}}$ was defined previously in equation (9.1). The Fisher information matrix evaluated at θ^* provides a lower bound on the accuracy of any statistical estimator via the Cramér–Rao bound. As a second derivative, the Fisher information matrix $\nabla^2\overline{\mathcal{L}}(\theta^*)$ captures the curvature of the cost function around the point θ^*.

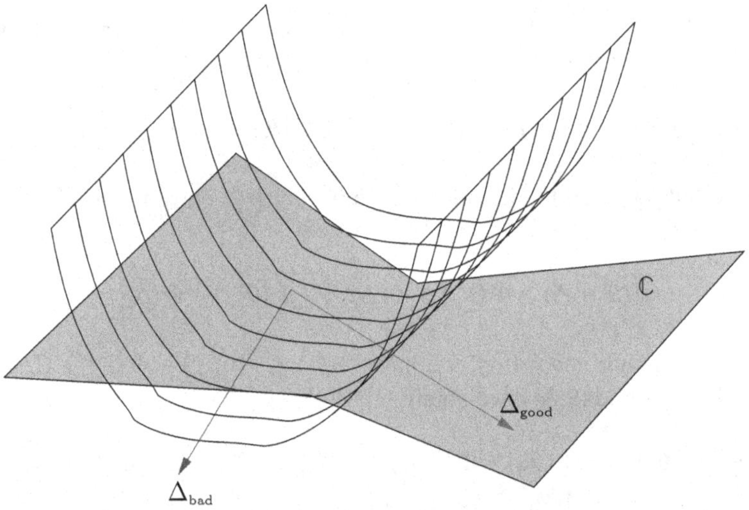

Figure 9.8 Illustration of the cost function $\theta \mapsto \mathcal{L}_n(\theta; Z_1^n)$. In the high-dimensional setting ($d > n$), although it may be curved in certain directions (e.g., Δ_{good}), there are $d - n$ directions in which it is flat up to second order (e.g., Δ_{bad}).

In the high-dimensional setting, the story becomes a little more complicated. In particular, whenever $n < d$, then the sample Fisher information matrix $\nabla^2\mathcal{L}_n(\theta^*)$ is rank-degenerate. Geometrically, this rank degeneracy implies that the cost function takes the form shown in Figure 9.8: while curved upwards in certain directions, there are $d - n$ directions in which it is flat up to second order. Consequently, the high-dimensional setting precludes any type of uniform lower bound on the curvature, and we can only hope to obtain some form of *restricted curvature*. There are several ways in which to develop such notions, and we describe two in the sections to follow, the first based on lower bounding the error in the first-order

Taylor-series expansion, and the second by directly lower bounding the curvature of the gradient mapping.

9.3.1 Restricted strong convexity

We begin by describing the notion of restricted strong convexity, which is defined by the Taylor-series expansion. Given any differentiable cost function, we can use the gradient to form the first-order Taylor approximation, which then defines the *first-order Taylor-series error*

$$\mathcal{E}_n(\Delta) := \mathcal{L}_n(\theta^* + \Delta) - \mathcal{L}_n(\theta^*) - \langle \nabla \mathcal{L}_n(\theta^*), \Delta \rangle. \tag{9.36}$$

Whenever the function $\theta \mapsto \mathcal{L}_n(\theta)$ is convex, this error term is always guaranteed to be non-negative.[2] Strong convexity requires that this lower bound holds with a quadratic slack: in particular, for a given norm $\|\cdot\|$, the cost function is locally *κ-strongly convex* at θ^* if the first-order Taylor error is lower bounded as

$$\mathcal{E}_n(\Delta) \geq \frac{\kappa}{2}\|\Delta\|^2 \tag{9.37}$$

for all Δ in a neighborhood of the origin. As previously discussed, this notion of strong convexity cannot hold for a generic high-dimensional problem. But for decomposable regularizers, we have seen (Proposition 9.13) that the error vector must belong to a very special set, and we use this fact to define the notion of restricted strong convexity.

Definition 9.15 For a given norm $\|\cdot\|$ and regularizer $\Phi(\cdot)$, the cost function satisfies a *restricted strong convexity* (RSC) condition with radius $R > 0$, curvature $\kappa > 0$ and tolerance τ_n^2 if

$$\mathcal{E}_n(\Delta) \geq \frac{\kappa}{2}\|\Delta\|^2 - \tau_n^2\, \Phi^2(\Delta) \qquad \text{for all } \Delta \in \mathbb{B}(R). \tag{9.38}$$

To clarify a few aspects of this definition, the set $\mathbb{B}(R)$ is the unit ball defined by the given norm $\|\cdot\|$. In our applications of RSC, the norm $\|\cdot\|$ will be derived from an inner product on the space Ω. Standard cases include the usual Euclidean norm on $\mathbb{R}^d$, and the Frobenius norm on the matrix space $\mathbb{R}^{d_1 \times d_2}$. Various types of weighted quadratic norms also fall within this general class.

Note that, if we set the tolerance term $\tau_n^2 = 0$, then the RSC condition (9.38) is equivalent to asserting that $\mathcal{L}_n$ is locally strongly convex in a neighborhood of θ^* with coefficient κ. As previously discussed, such a strong convexity condition cannot hold in the high-dimensional setting. However, given our goal of proving error bounds on M-estimators, we are not interested in all directions, but rather only the directions in which the error vector $\widehat{\Delta} = \widehat{\theta} - \theta^*$ can lie. For decomposable regularizers, Proposition 9.13 guarantees that the error vector must lie in the very special "cone-like" sets $\mathbb{C}_{\theta^*}(\mathbb{M}, \overline{\mathbb{M}}^\perp)$. Even with a strictly positive tolerance $\tau_n^2 > 0$, an RSC condition of the form (9.38) can be used to guarantee a lower curvature over

[2] Indeed, for differentiable functions, this property may be viewed as an equivalent definition of convexity.

this restricted set, as long as the sample size is sufficiently large. We formalize this intuition after considering a few concrete instances of Definition 9.15.

Example 9.16 (Restricted eigenvalues for least-squares cost) In this example, we show how the restricted eigenvalue conditions (see Definition 7.12 in Chapter 7) correspond to a special case of restricted strong convexity. For the least-squares objective $\mathcal{L}_n(\theta) = \frac{1}{2n}\|y - \mathbf{X}\theta\|_2^2$, an easy calculation yields that the first-order Taylor error is given by $\mathcal{E}_n(\Delta) = \frac{\|\mathbf{X}\Delta\|_2^2}{2n}$. A restricted strong convexity condition with the ℓ_1-norm then takes the form

$$\frac{\|\mathbf{X}\Delta\|_2^2}{2n} \geq \frac{\kappa}{2}\|\Delta\|_2^2 - \tau_n^2\|\Delta\|_1^2 \qquad \text{for all } \Delta \in \mathbb{R}^d. \tag{9.39}$$

For various types of sub-Gaussian matrices, bounds of this form hold with high probability for the choice $\tau_n^2 \asymp \frac{\log d}{n}$. Theorem 7.16 in Chapter 7 provides one instance of such a result.

As a side remark, this example shows that the least-squares objective is special in two ways: the first-order Taylor error is independent of θ^* and, moreover, it is a positively homogeneous function of degree two—that is, $\mathcal{E}_n(t\Delta) = t^2\mathcal{E}_n(\Delta)$ for all $t \in \mathbb{R}$. The former property implies that we need not be concerned about uniformity in θ^*, whereas the latter implies that it is not necessary to localize Δ to a ball $\mathbb{B}(R)$. ♣

Later in Section 9.8, we provide more general results, showing that a broader class of cost functions satisfy a restricted strong convexity condition of the type (9.39). Let us consider one example here:

Example 9.17 (RSC for generalized linear models) Recall the family of generalized linear models from Example 9.2, and the cost function (9.7) defined by the negative log-likelihood. Suppose that we draw n i.i.d. samples, in which the covariates $\{x_i\}_{i=1}^n$ are drawn from a zero-mean sub-Gaussian distribution with non-degenerate covariance matrix $\mathbf{\Sigma}$. As a consequence of a result to follow (Theorem 9.36), the Taylor-series error of various GLM log-likelihoods satisfies a lower bound of the form

$$\mathcal{E}_n(\Delta) \geq \frac{\kappa}{2}\|\Delta\|_2^2 - c_1\frac{\log d}{n}\|\Delta\|_1^2 \qquad \text{for all } \|\Delta\|_2 \leq 1 \tag{9.40}$$

with probability greater than $1 - c_2\exp(-c_3 n)$.

Theorem 9.36 actually provides a more general guarantee in terms of the quantity

$$\mu_n(\Phi^*) := \mathbb{E}_{x,\varepsilon}\left[\Phi^*\left(\frac{1}{n}\sum_{i=1}^n \varepsilon_i x_i\right)\right], \tag{9.41}$$

where Φ^* denotes the dual norm, and $\{\varepsilon_i\}_{i=1}^n$ is a sequence of i.i.d. Rademacher variables. With this notation, we have

$$\mathcal{E}_n(\Delta) \geq \frac{\kappa}{2}\|\Delta\|_2^2 - c_1\,\mu_n^2(\Phi^*)\,\Phi^2(\Delta) \qquad \text{for all } \|\Delta\|_2 \leq 1 \tag{9.42}$$

with probability greater than $1 - c_2\exp(-c_3 n)$. This result is a generalization of our previous bound (9.40), since $\mu_n(\Phi^*) \precsim \sqrt{\frac{\log d}{n}}$ in the case of ℓ_1-regularization.

In Exercise 9.8, we bound the quantity (9.41) for various norms. For group Lasso with group set $\mathcal{G}$ and maximum group size m, we show that

$$\mu_n(\Phi^*) \precsim \sqrt{\frac{m}{n}} + \sqrt{\frac{\log |\mathcal{G}|}{n}}, \tag{9.43a}$$

whereas for the nuclear norm for $d_1 \times d_2$ matrices, we show that

$$\mu_n(\Phi^*) \precsim \sqrt{\frac{d_1}{n}} + \sqrt{\frac{d_2}{n}}. \tag{9.43b}$$

We also show how these results, in conjunction with the lower bound (9.42), imply suitable forms of restricted convexity as long as the sample size is sufficiently large. ♣

We conclude this section with the definition of one last geometric parameter that plays an important role. As we have just seen, in the context of ℓ_1-regularization and the RE condition, the cone constraint is very useful; in particular, it implies that $\|\Delta\|_1 \leq 4\sqrt{s}\|\Delta\|_2$, a bound used repeatedly in Chapter 7. Returning to the general setting, we need to study how to translate between $\Phi(\Delta_{\mathbb{M}})$ and $\|\Delta_{\mathbb{M}}\|$ for an arbitrary decomposable regularizer and error norm.

Definition 9.18 (Subspace Lipschitz constant) For any subspace $\mathbb{S}$ of $\mathbb{R}^d$, the *subspace Lipschitz constant* with respect to the pair $(\Phi, \|\cdot\|)$ is given by

$$\Psi(\mathbb{S}) := \sup_{u \in \mathbb{S}\setminus\{0\}} \frac{\Phi(u)}{\|u\|}. \tag{9.44}$$

To clarify our terminology, this quantity is the Lipschitz constant of the regularizer with respect to the error norm, but as restricted to the subspace $\mathbb{S}$. It corresponds to the worst-case price of translating between the Φ- and $\|\cdot\|$-norms for any vector in $\mathbb{S}$.

To illustrate its use, let us consider it in the special case when $\theta^* \in \mathbb{M}$. Then for any $\Delta \in \mathbb{C}_{\theta^*}(\mathbb{M}, \bar{\mathbb{M}}^\perp)$, we have

$$\Phi(\Delta) \overset{\text{(i)}}{\leq} \Phi(\Delta_{\bar{\mathbb{M}}}) + \Phi(\Delta_{\bar{\mathbb{M}}^\perp}) \overset{\text{(ii)}}{\leq} 4\Phi(\Delta_{\bar{\mathbb{M}}}) \overset{\text{(iii)}}{\leq} 4\,\Psi(\bar{\mathbb{M}})\|\Delta\|, \tag{9.45}$$

where step (i) follows from the triangle inequality, step (ii) from membership in $\mathbb{C}(\mathbb{M}, \mathbb{M}^\perp)$, and step (iii) from the definition of $\Psi(\bar{\mathbb{M}})$.

As a simple example, if $\mathbb{M}$ is a subspace of s-sparse vectors, then with regularizer $\Phi(u) = \|u\|_1$ and error norm $\|u\| = \|u\|_2$, we have $\Psi(\mathbb{M}) = \sqrt{s}$. In this way, we see that inequality (9.45) is a generalization of the familiar inequality $\|\Delta\|_2 \leq 4\sqrt{s}\|\Delta\|_1$ in the context of sparse vectors. The subspace Lipschitz constant appears explicitly in the main results, and also arises in establishing restricted strong convexity.

9.4 Some general theorems

Thus far, we have discussed the notion of decomposable regularizers, and some related notions of restricted curvature for the cost function. In this section, we state and prove some results on the estimation error, namely, the quantity $\widehat{\theta} - \theta^*$, where $\widehat{\theta}$ denotes any optimum of the regularized M-estimator (9.3).

9.4.1 Guarantees under restricted strong convexity

We begin by stating and proving a general result that holds under the restricted strong convexity condition given in Section 9.3.1. Let us summarize the assumptions that we impose throughout this section:

(A1) The cost function is convex, and satisfies the local RSC condition (9.38) with curvature κ, radius R and tolerance τ_n^2 with respect to an inner-product induced norm $\|\cdot\|$.
(A2) There is a pair of subspaces $\mathbb{M} \subseteq \overline{\mathbb{M}}$ such that the regularizer decomposes over $(\mathbb{M}, \overline{\mathbb{M}}^\perp)$.

We state the result as a deterministic claim, but conditioned on the "good" event

$$\mathbb{G}(\lambda_n) := \left\{\Phi^*(\nabla \mathcal{L}_n(\theta^*)) \leq \frac{\lambda_n}{2}\right\}. \tag{9.46}$$

Our bound involves the quantity

$$\varepsilon_n^2(\overline{\mathbb{M}}, \mathbb{M}^\perp) := \underbrace{9\,\frac{\lambda_n^2}{\kappa^2}\,\Psi^2(\overline{\mathbb{M}})}_{\text{estimation error}} + \underbrace{\frac{8}{\kappa}\left\{\lambda_n \Phi(\theta^*_{\mathbb{M}^\perp}) + 16\tau_n^2 \Phi^2(\theta^*_{\mathbb{M}^\perp})\right\}}_{\text{approximation error}}, \tag{9.47}$$

which depends on the choice of our subspace pair $(\overline{\mathbb{M}}, \mathbb{M}^\perp)$.

Theorem 9.19 (Bounds for general models) *Under conditions (A1) and (A2), consider the regularized M-estimator* (9.3) *conditioned on the event* $\mathbb{G}(\lambda_n)$,

(a) *Any optimal solution satisfies the bound*

$$\Phi(\widehat{\theta} - \theta^*) \leq 4\left\{\Psi(\overline{\mathbb{M}})\,\|\widehat{\theta} - \theta^*\| + \Phi(\theta^*_{\mathbb{M}^\perp})\right\}. \tag{9.48a}$$

(b) *For any subspace pair* $(\overline{\mathbb{M}}, \mathbb{M}^\perp)$ *such that* $\tau_n^2 \Psi^2(\overline{\mathbb{M}}) \leq \frac{\kappa}{64}$ *and* $\varepsilon_n(\overline{\mathbb{M}}, \mathbb{M}^\perp) \leq R$, *we have*

$$\|\widehat{\theta} - \theta^*\|^2 \leq \varepsilon_n^2(\overline{\mathbb{M}}, \mathbb{M}^\perp). \tag{9.48b}$$

It should be noted that Theorem 9.19 is actually a deterministic result. Probabilistic conditions enter in certifying that the RSC condition holds with high probability (see Section 9.8), and in verifying that, for a concrete choice of regularization parameter, the *dual norm bound* $\lambda_n \geq 2\Phi^*(\nabla \mathcal{L}_n(\theta^*))$ defining the event $\mathbb{G}(\lambda_n)$ holds with high probability. The dual norm bound cannot be explicitly verified, since it presumes knowledge of θ^*, but it suffices to give choices of λ_n for which it holds with high probability. We illustrate such choices in various examples to follow.

Equations (9.48a) and (9.48b) actually specify a family of upper bounds, one for each subspace pair $(\overline{\mathbb{M}}, \overline{\mathbb{M}}^\perp)$ over which the regularizer Φ decomposes. The optimal choice of these subspaces serves to trade off the estimation and approximation error terms in the bound. The upper bound (9.48b) corresponds to an *oracle inequality*, since it applies to any parameter θ^*, and gives a family of upper bounds involving two sources of error. The term labeled "estimation error" represents the statistical cost of estimating a parameter belong to the subspace $\mathbb{M} \subseteq \bar{\mathbb{M}}$; naturally, it increases as $\mathbb{M}$ grows. The second quantity represents "approximation error" incurred by estimating only within the subspace $\mathbb{M}$, and it shrinks as $\mathbb{M}$ is increased. Thus, the optimal bound is obtained by choosing the model subspace to balance these two types of error. We illustrate such choices in various examples to follow.

In the special case that the target parameter θ^* is contained within a subspace $\mathbb{M}$, Theorem 9.19 has the following corollary:

Corollary 9.20 *Suppose that, in addition to the conditions of Theorem 9.19, the optimal parameter θ^* belongs to $\mathbb{M}$. Then any optimal solution $\widehat{\theta}$ to the optimization problem* (9.3) *satisfies the bounds*

$$\Phi(\widehat{\theta} - \theta^*) \leq 6\frac{\lambda_n}{\kappa}\Psi^2(\bar{\mathbb{M}}), \tag{9.49a}$$

$$\|\widehat{\theta} - \theta^*\|^2 \leq 9\,\frac{\lambda_n^2}{\kappa^2}\Psi^2(\bar{\mathbb{M}}). \tag{9.49b}$$

This corollary can be applied directly to obtain concrete estimation error bounds for many problems, as we illustrate in the sequel.

We now turn to the proof of Theorem 9.19.

Proof We begin by proving part (a). Letting $\widehat{\Delta} = \widehat{\theta} - \theta^*$ be the error, by the triangle inequality, we have

$$\begin{aligned}
\Phi(\widehat{\Delta}) &\leq \Phi(\widehat{\Delta}_{\bar{\mathbb{M}}}) + \Phi(\widehat{\Delta}_{\bar{\mathbb{M}}^\perp}) \\
&\overset{\text{(i)}}{\leq} \Phi(\widehat{\Delta}_{\bar{\mathbb{M}}}) + \left\{3\Phi(\widehat{\Delta}_{\bar{\mathbb{M}}}) + 4\Phi(\theta^*_{\mathbb{M}^\perp})\right\} \\
&\overset{\text{(ii)}}{\leq} 4\left\{\Psi(\bar{\mathbb{M}})\,\|\widehat{\theta} - \theta^*\| + \Phi(\theta^*_{\mathbb{M}^\perp})\right\},
\end{aligned}$$

where inequality (i) follows from Proposition 9.13 under event $\mathbb{G}(\lambda_n)$ and inequality (ii) follows from the definition of the optimal subspace constant.

Turning to the proof of part (b), in order to simplify notation, we adopt the shorthand $\mathbb{C}$ for the set $\mathbb{C}_{\theta^*}(\mathbb{M}, \bar{\mathbb{M}}^\perp)$. Letting $\delta \in (0, R]$ be a given error radius to be chosen, the following lemma shows that it suffices to control the sign of the function $\mathcal{F}$ from equation (9.31) over the set $\mathbb{K}(\delta) := \mathbb{C} \cap \{\|\Delta\| = \delta\}$.

Lemma 9.21 *If $\mathcal{F}(\Delta) > 0$ for all vectors $\Delta \in \mathbb{K}(\delta)$, then $\|\widehat{\Delta}\| \le \delta$.*

Proof We prove the contrapositive statement: in particular, we show that if for some optimal solution $\widehat{\theta}$, the associated error vector $\widehat{\Delta} = \widehat{\theta} - \theta^*$ satisfies the inequality $\|\widehat{\Delta}\| > \delta$, then there must be some vector $\widetilde{\Delta} \in \mathbb{K}(\delta)$ such that $\mathcal{F}(\widetilde{\Delta}) \le 0$. If $\|\widehat{\Delta}\| > \delta$, then since $\mathbb{C}$ is star-shaped around the origin (see the Appendix, Section 9.9), the line joining $\widehat{\Delta}$ to 0 must intersect the set $\mathbb{K}(\delta)$ at some intermediate point of the form $t^*\widehat{\Delta}$ for some $t^* \in [0, 1]$. See Figure 9.9 for an illustration.

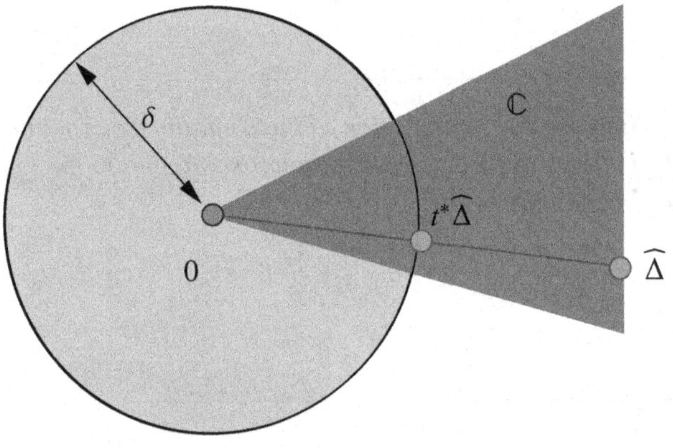

Figure 9.9 Geometry of the proof of Lemma 9.21. When $\|\widehat{\Delta}\| > \delta$ and the set $\mathbb{C}$ is star-shaped around the origin, any line joining $\widehat{\Delta}$ and the origin 0 must intersect the set $\mathbb{K}(\delta) = \{\|\Delta\| = \delta\} \cap \mathbb{C}$ at some intermediate point of the form $t^*\widehat{\Delta}$ for some $t^* \in [0, 1]$.

Since the cost function $\mathcal{L}_n$ and regularizer Φ are convex, the function $\mathcal{F}$ is also convex for any non-negative choice of the regularization parameter. Given the convexity of $\mathcal{F}$, we can apply Jensen's inequality so as to obtain

$$\mathcal{F}(t^*\widehat{\Delta}) = \mathcal{F}(t^*\widehat{\Delta} + (1 - t^*)\,0) \le t^*\,\mathcal{F}(\widehat{\Delta}) + (1 - t^*)\mathcal{F}(0) \overset{\text{(i)}}{=} t^*\mathcal{F}(\widehat{\Delta}),$$

where equality (i) uses the fact that $\mathcal{F}(0) = 0$ by construction. But since $\widehat{\Delta}$ is optimal, we must have $\mathcal{F}(\widehat{\Delta}) \le 0$, and hence $\mathcal{F}(t^*\Delta) \le 0$ as well. Thus, we have constructed a vector $\widetilde{\Delta} = t^*\Delta$ with the claimed properties, thereby establishing the claim in the lemma. □

We now return to the proof of Theorem 9.19. Fix some radius $\delta \in (0, R]$, whose value will be specified later in the proof (see equation (9.53)). On the basis of Lemma 9.21, the proof of Theorem 9.19 will be complete if we can establish a lower bound on the function value

$\mathcal{F}(\Delta)$ for all vectors $\Delta \in \mathbb{K}(\delta)$. For an arbitrary vector $\Delta \in \mathbb{K}(\delta)$, we have

$$\begin{aligned}\mathcal{F}(\Delta) &= \mathcal{L}_n(\theta^* + \Delta) - \mathcal{L}_n(\theta^*) + \lambda_n\{\Phi(\theta^* + \Delta) - \Phi(\theta^*)\}\\ &\overset{\text{(i)}}{\geq} \langle \nabla \mathcal{L}_n(\theta^*),\, \Delta\rangle + \frac{\kappa}{2}\|\Delta\|^2 - \tau_n^2\Phi^2(\Delta) + \lambda_n\{\Phi(\theta^* + \Delta) - \Phi(\theta^*)\} \qquad (9.50)\\ &\overset{\text{(ii)}}{\geq} \langle \nabla \mathcal{L}_n(\theta^*),\, \Delta\rangle + \frac{\kappa}{2}\|\Delta\|^2 - \tau_n^2\Phi^2(\Delta) + \lambda_n\{\Phi(\Delta_{\bar{\mathbb{M}}^\perp}) - \Phi(\Delta_{\bar{\mathbb{M}}}) - 2\Phi(\theta^*_{\mathbb{M}^\perp})\},\end{aligned}$$

where inequality (i) follows from the RSC condition, and inequality (ii) follows from the bound (9.32).

By applying Hölder's inequality with the regularizer Φ and its dual Φ^*, we find that

$$|\langle \nabla \mathcal{L}_n(\theta^*),\, \Delta\rangle| \leq \Phi^*(\nabla \mathcal{L}_n(\theta^*))\, \Phi(\Delta).$$

Under the event $\mathbb{G}(\lambda_n)$, the regularization parameter is lower bounded as $\lambda_n \geq 2\Phi^*(\nabla \mathcal{L}_n(\theta^*))$, which implies that $|\langle \nabla \mathcal{L}_n(\theta^*),\, \Delta\rangle| \leq \frac{\lambda_n}{2}\Phi(\Delta)$. Consequently, we have

$$\mathcal{F}(\Delta) \geq \frac{\kappa}{2}\|\Delta\|^2 - \tau_n^2\Phi^2(\Delta) + \lambda_n\Big\{\Phi(\Delta_{\bar{\mathbb{M}}^\perp}) - \Phi(\Delta_{\bar{\mathbb{M}}}) - 2\Phi(\theta^*_{\mathbb{M}^\perp})\Big\} - \frac{\lambda_n}{2}\Phi(\Delta).$$

The triangle inequality implies that

$$\Phi(\Delta) = \Phi(\Delta_{\bar{\mathbb{M}}^\perp} + \Delta_{\bar{\mathbb{M}}}) \leq \Phi(\Delta_{\bar{\mathbb{M}}^\perp}) + \Phi(\Delta_{\bar{\mathbb{M}}}),$$

and hence, following some algebra, we find that

$$\begin{aligned}\mathcal{F}(\Delta) &\geq \frac{\kappa}{2}\|\Delta\|^2 - \tau_n^2\Phi^2(\Delta) + \lambda_n\Big\{\frac{1}{2}\Phi(\Delta_{\bar{\mathbb{M}}^\perp}) - \frac{3}{2}\Phi(\Delta_{\bar{\mathbb{M}}}) - 2\Phi(\theta^*_{\mathbb{M}^\perp})\Big\}\\ &\geq \frac{\kappa}{2}\|\Delta\|^2 - \tau_n^2\Phi^2(\Delta) - \frac{\lambda_n}{2}\Big\{3\Phi(\Delta_{\bar{\mathbb{M}}}) + 4\Phi(\theta^*_{\mathbb{M}^\perp})\Big\}. \qquad (9.51)\end{aligned}$$

Now definition (9.44) of the subspace Lipschitz constant implies that $\Phi(\Delta_{\bar{\mathbb{M}}}) \leq \Psi(\bar{\mathbb{M}})\|\Delta_{\bar{\mathbb{M}}}\|$. Since the projection $\Delta \mapsto \Delta_{\bar{\mathbb{M}}}$ is defined in terms of the norm $\|\cdot\|$, it is non-expansive. Since $0 \in \bar{\mathbb{M}}$, we have

$$\|\Delta_{\bar{\mathbb{M}}}\| = \|\Pi_{\bar{\mathbb{M}}}(\Delta) - \Pi_{\bar{\mathbb{M}}}(0)\| \overset{\text{(i)}}{\leq} \|\Delta - 0\| = \|\Delta\|,$$

where inequality (i) uses non-expansiveness of the projection. Combining with the earlier bound, we conclude that $\Phi(\Delta_{\bar{\mathbb{M}}}) \leq \Psi(\bar{\mathbb{M}})\|\Delta\|$.

Similarly, for any $\Delta \in \mathbb{C}$, we have

$$\begin{aligned}\Phi^2(\Delta) &\leq \Big\{4\Phi(\Delta_{\bar{\mathbb{M}}}) + 4\Phi(\theta^*_{\mathbb{M}^\perp})\Big\}^2 \leq 32\Phi^2(\Delta_{\bar{\mathbb{M}}}) + 32\Phi^2(\theta^*_{\mathbb{M}^\perp})\\ &\leq 32\Psi^2(\bar{\mathbb{M}})\,\|\Delta\|^2 + 32\Phi^2(\theta^*_{\mathbb{M}^\perp}). \qquad (9.52)\end{aligned}$$

Substituting into the lower bound (9.51), we obtain the inequality

$$\begin{aligned}\mathcal{F}(\Delta) &\geq \Big\{\frac{\kappa}{2} - 32\tau_n^2\Psi^2(\bar{\mathbb{M}})\Big\}\,\|\Delta\|^2 - 32\tau_n^2\Phi^2(\theta^*_{\mathbb{M}^\perp}) - \frac{\lambda_n}{2}\Big\{3\Psi(\bar{\mathbb{M}})\,\|\Delta\| + 4\Phi(\theta^*_{\mathbb{M}^\perp})\Big\}\\ &\overset{\text{(ii)}}{\geq} \frac{\kappa}{4}\|\Delta\|^2 - \frac{3\lambda_n}{2}\Psi(\bar{\mathbb{M}})\,\|\Delta\| - 32\tau_n^2\Phi^2(\theta^*_{\mathbb{M}^\perp}) - 2\lambda_n\Phi(\theta^*_{\mathbb{M}^\perp}),\end{aligned}$$

where step (ii) uses the assumed bound $\tau_n^2\Psi^2(\bar{\mathbb{M}}) < \frac{\kappa}{64}$.

The right-hand side of this inequality is a strictly positive definite quadratic form in $\|\Delta\|$, and so will be positive for $\|\Delta\|$ sufficiently large. In particular, some algebra shows that this is the case as long as

$$\|\Delta\|^2 \geq \varepsilon_n^2(\overline{\mathbb{M}}, \mathbb{M}^\perp) := 9\,\frac{\lambda_n^2}{\kappa^2}\,\Psi^2(\overline{\mathbb{M}}) + \frac{8}{\kappa}\Big\{\lambda_n \Phi(\theta^*_{\mathbb{M}^\perp}) + 16\tau_n^2 \Phi^2(\theta^*_{\mathbb{M}^\perp})\Big\}. \tag{9.53}$$

This argument is valid as long as $\varepsilon_n \leq R$, as assumed in the statement. □

9.4.2 Bounds under Φ^*-curvature

We now turn to an alternative form of restricted curvature, one which involves a lower bound on the gradient of the cost function. In order to motivate the definition to follow, note that an alternative way of characterizing strong convexity of a differentiable cost function is via the behavior of its gradient. More precisely, a differentiable function $\mathcal{L}_n$ is locally κ-strongly convex at θ^*, in the sense of the earlier definition (9.37), if and only if

$$\langle \nabla\mathcal{L}_n(\theta^* + \Delta)) - \nabla\mathcal{L}_n(\theta^*),\, \Delta\rangle \geq \kappa\|\Delta\|^2 \tag{9.54}$$

for all Δ in some ball around zero. See Exercise 9.9 for verification of the equivalence between the property (9.54) and the earlier definition (9.37). When the underlying norm $\|\cdot\|$ is the ℓ_2-norm, then the condition (9.54), combined with the Cauchy–Schwarz inequality, implies that

$$\|\nabla\mathcal{L}_n(\theta^* + \Delta) - \nabla\mathcal{L}_n(\theta^*)\|_2 \geq \kappa\|\Delta\|_2.$$

This implication suggests that it could be useful to consider alternative notions of curvature based on different choices of the norm. Here we consider such a notion based on the dual norm Φ^*:

Definition 9.22 The cost function satisfies a Φ^*-*norm curvature condition* with curvature κ, tolerance τ_n and radius R if

$$\Phi^*\Big(\nabla\mathcal{L}_n(\theta^* + \Delta) - \nabla\mathcal{L}_n(\theta^*)\Big) \geq \kappa\,\Phi^*(\Delta) - \tau_n\Phi(\Delta) \tag{9.55}$$

for all $\Delta \in \mathbb{B}_{\Phi^*}(R) := \{\theta \in \Omega \mid \Phi^*(\theta) \leq R\}$.

As with restricted strong convexity, this definition is most easily understood in application to the classical case of least-squares cost and ℓ_1-regularization:

Example 9.23 (Restricted curvature for least-squares cost) For the least-squares cost function, we have $\nabla\mathcal{L}_n(\theta) = \frac{1}{n}\mathbf{X}^T\mathbf{X}(\theta - \theta^*) = \widehat{\Sigma}\,(\theta - \theta^*)$, where $\widehat{\Sigma} = \frac{1}{n}\mathbf{X}^T\mathbf{X}$ is the sample covariance matrix. For the ℓ_1-norm as the regularizer Φ, the dual norm Φ^* is the ℓ_∞-norm, so that the restricted curvature condition (9.55) is equivalent to the lower bound

$$\big\|\widehat{\Sigma}\Delta\big\|_\infty \geq \kappa\|\Delta\|_\infty - \tau_n\|\Delta\|_1 \qquad \text{for all } \Delta \in \mathbb{R}^d. \tag{9.56}$$

In this particular example, localization to the ball $\mathbb{B}_\infty(R)$ is actually unnecessary, since the lower bound is invariant to rescaling of Δ. The bound (9.56) is very closely related to what are known as *ℓ_∞-restricted eigenvalues* of the sample covariance matrix $\widehat{\Sigma}$. More precisely, such conditions involve lower bounds of the form

$$\left\|\widehat{\Sigma}\Delta\right\|_\infty \geq \kappa' \, \|\Delta\|_\infty \qquad \text{for all } \Delta \in \mathbb{C}(S;\alpha), \tag{9.57}$$

where $\mathbb{C}(S;\alpha) := \{\Delta \in \mathbb{R}^d \mid \|\Delta_{S^c}\|_1 \leq \alpha\|\Delta_S\|_1\}$, and (κ', α) are given positive constants. In Exercise 9.11, we show that a bound of the form (9.56) implies a form of the ℓ_∞-RE condition (9.57) as long as $n \gtrsim |S|^2 \log d$. Moreover, as we show in Exercise 7.13, such an ℓ_∞-RE condition can be used to derive bounds on the ℓ_∞-error of the Lasso.

Finally, as with ℓ_2-restricted eigenvalue conditions (recall Example 9.16), a lower bound of the form (9.56) holds with high probability with constant κ and tolerance $\tau_n \asymp \sqrt{\frac{\log d}{n}}$ for various types of random design matrices, Exercise 7.14 provides details on one such result. ♣

With this definition in place, we are ready to state the assumptions underlying the main result of this section:

(A1′) The cost satisfies the Φ^*-curvature condition (9.55) with parameters $(\kappa, \tau_n; R)$.

(A2) The regularizer is decomposable with respect to the subspace pair $(\mathbb{M}, \overline{\mathbb{M}}^\perp)$ with $\mathbb{M} \subseteq \overline{\mathbb{M}}$.

Under these conditions, we have the following:

Theorem 9.24 *Given a target parameter $\theta^* \in \mathbb{M}$, consider the regularized M-estimator (9.3) under conditions (A1′) and (A2), and suppose that $\tau_n \Psi^2(\overline{\mathbb{M}}) < \frac{\kappa}{32}$. Conditioned on the event $\mathbb{G}(\lambda_n) \cap \{\Phi^*(\widehat{\theta} - \theta^*) \leq R\}$, any optimal solution $\widehat{\theta}$ satisfies the bound*

$$\Phi^*(\widehat{\theta} - \theta^*) \leq 3\frac{\lambda_n}{\kappa}. \tag{9.58}$$

Like Theorem 9.19, this claim is deterministic given the stated conditioning. Probabilistic claims enter in certifying that the "good" event $\mathbb{G}(\lambda_n)$ holds with high probability with a specified choice of λ_n. Moreover, except for the special case of least squares, we need to use related results (such as those in Theorem 9.19) to certify that $\Phi^*(\widehat{\theta} - \theta^*) \leq R$, before applying this result.

Proof The proof is relatively straightforward given our development thus far. By standard optimality conditions for a convex program, for any optimum $\widehat{\theta}$, there must exist a subgradient vector $\widehat{z} \in \partial\Phi(\widehat{\theta})$ such that $\nabla\mathcal{L}_n(\widehat{\theta}) + \lambda_n\widehat{z} = 0$. Introducing the error vector $\widehat{\Delta} := \widehat{\theta} - \theta^*$, some algebra yields

$$\nabla\mathcal{L}_n(\theta^* + \widehat{\Delta}) - \nabla\mathcal{L}_n(\theta^*) = -\nabla\mathcal{L}_n(\theta^*) - \lambda_n\widehat{z}.$$

Taking the Φ^*-norm of both sides and applying the triangle inequality yields

$$\Phi^*(\nabla\mathcal{L}_n(\theta^* + \Delta) - \nabla\mathcal{L}_n(\theta^*)) \leq \Phi^*(\nabla\mathcal{L}_n(\theta^*)) + \lambda_n\Phi^*(\widehat{z}).$$

On one hand, on the event $\mathbb{G}(\lambda_n)$, we have that $\Phi^*(\nabla\mathcal{L}_n(\theta^*)) \leq \lambda_n/2$, whereas, on the other hand, Exercise 9.6 implies that $\Phi^*(\widetilde{z}) \leq 1$. Putting together the pieces, we find that $\Phi^*(\nabla\mathcal{L}_n(\theta^* + \Delta) - \nabla\mathcal{L}_n(\theta^*)) \leq \frac{3\lambda_n}{2}$. Finally, applying the curvature condition (9.55), we obtain

$$\kappa\,\Phi^*(\widehat{\Delta}) \leq \frac{3}{2}\lambda_n + \tau_n\Phi(\widehat{\Delta}). \tag{9.59}$$

It remains to bound $\Phi(\widehat{\Delta})$ in terms of the dual norm $\Phi^*(\widehat{\Delta})$. Since this result is useful in other contexts, we state it as a separate lemma here:

Lemma 9.25 *If $\theta^* \in \mathbb{M}$, then*

$$\Phi(\Delta) \leq 16\Psi^2(\overline{\mathbb{M}})\,\Phi^*(\Delta) \qquad \textit{for any } \Delta \in \mathbb{C}_{\theta^*}(\mathbb{M}, \overline{\mathbb{M}}^\perp). \tag{9.60}$$

Before returning to prove this lemma, we use it to complete the proof of the theorem. On the event $\mathbb{G}(\lambda_n)$, Proposition 9.13 may be applied to guarantee that $\widehat{\Delta} \in \mathbb{C}_{\theta^*}(\mathbb{M}, \overline{\mathbb{M}}^\perp)$. Consequently, the bound (9.60) applies to $\widehat{\Delta}$. Substituting into the earlier bound (9.59), we find that $(\kappa - 16\Psi^2(\mathbb{M})\tau_n)\,\Phi^*(\widehat{\Delta}) \leq \frac{3}{2}\lambda_n$, from which the claim follows by the assumption that $\Psi^2(\mathbb{M})\tau_n \leq \frac{\kappa}{32}$.

We now return to prove Lemma 9.25. From our earlier calculation (9.45), whenever $\theta^* \in \mathbb{M}$ and $\Delta \in \mathbb{C}_{\theta^*}(\mathbb{M}, \overline{\mathbb{M}}^\perp)$, then $\Phi(\Delta) \leq 4\Psi(\overline{\mathbb{M}})\,\|\Delta\|$. Moreover, by Hölder's inequality, we have

$$\|\Delta\|^2 \leq \Phi(\Delta)\,\Phi^*(\Delta) \leq 4\Psi(\overline{\mathbb{M}})\|\Delta\|\,\Phi^*(\Delta),$$

whence $\|\Delta\| \leq 4\Psi(\overline{\mathbb{M}})\Phi^*(\Delta)$. Putting together the pieces, we have

$$\Phi(\Delta) \leq 4\Psi(\overline{\mathbb{M}})\|\Delta\| \leq 16\Psi^2(\overline{\mathbb{M}})\,\Phi^*(\Delta),$$

as claimed. This completes the proof of the lemma, and hence of the theorem. □

Thus far, we have derived two general bounds on the error $\widehat{\theta} - \theta^*$ associated with optima of the M-estimator (9.3). In the remaining sections, we specialize these general results to particular classes of statistical models.

9.5 Bounds for sparse vector regression

We now turn to some consequences of our general theory for the problem of sparse regression. In developing the theory for the full class of generalized linear models, this section provides an alternative and more general complement to our discussion of the sparse linear model in Chapter 7.

9.5.1 Generalized linear models with sparsity

All results in the following two sections are applicable to samples the form $\{(x_i, y_i)\}_{i=1}^n$ where:

(G1) The covariates are C-column normalized: $\max_{j=1,\ldots,d} \sqrt{\frac{\sum_{j=1}^{d} x_{ij}^2}{n}} \leq C$.

(G2) Conditionally on x_i, each response y_i is drawn i.i.d. according to a conditional distribution of the form

$$\mathbb{P}_{\theta^*}(y \mid x) \propto \exp\left\{\frac{y\,\langle x,\,\theta^*\rangle - \psi(\langle x,\,\theta^*\rangle)}{c(\sigma)}\right\},$$

where the partition function ψ has a bounded second derivative ($\|\psi''\|_\infty \leq B^2$).

We analyze the ℓ_1-regularized version of the GLM log-likelihood estimator, namely

$$\widehat{\theta} \in \arg\min_{\theta\in\mathbb{R}^d}\Big\{\underbrace{\frac{1}{n}\sum_{i=1}^{n}\{\psi(\langle x_i,\,\theta\rangle) - y_i\,\langle x_i,\,\theta\rangle\}}_{\mathcal{L}_n(\theta)} + \lambda_n\|\theta\|_1\Big\}. \tag{9.61}$$

For short, we refer to this M-estimator as the *GLM Lasso*. Note that the usual linear model description $y_i = \langle x_i,\,\theta^*\rangle + w_i$ with $w_i \sim \mathcal{N}(0,\sigma^2)$ falls into this class with $B = 1$, in which the case the estimator (9.61) is equivalent to the ordinary Lasso. It also includes as special cases the problems of logistic regression and multinomial regression, but excludes the case of Poisson regression, due to the boundedness condition (G2).

9.5.2 Bounds under restricted strong convexity

We begin by proving bounds when the Taylor-series error around θ^* associated with the negative log-likelihood (9.61) satisfies the RSC condition

$$\mathcal{E}_n(\Delta) \geq \frac{\kappa}{2}\|\Delta\|_2^2 - c_1\frac{\log d}{n}\|\Delta\|_1^2 \qquad \text{for all } \|\Delta\|_2 \leq 1. \tag{9.62}$$

As discussed in Example 9.17, when the covariates $\{x_i\}_{i=1}^n$ are drawn from a zero-mean sub-Gaussian distribution, a bound of this form holds with high probability for any GLM.

The following result applies to any solution $\widehat{\theta}$ of the GLM Lasso (9.61) with regularization parameter $\lambda_n = 4\,B\,C\left\{\sqrt{\frac{\log d}{n}} + \delta\right\}$ for some $\delta \in (0, 1)$.

Corollary 9.26 *Consider a GLM satisfying conditions (G1) and (G2), the RSC condition (9.62), and suppose the true regression vector θ^* is supported on a subset S of cardinality s. Given a sample size n large enough to ensure that $s\left\{\lambda_n^2 + \frac{\log d}{n}\right\} < \min\left\{\frac{4\kappa^2}{9}, \frac{\kappa}{64c_1}\right\}$, any GLM Lasso solution $\widehat{\theta}$ satisfies the bounds*

$$\|\widehat{\theta} - \theta^*\|_2^2 \leq \frac{9}{4}\frac{s\lambda_n^2}{\kappa^2} \quad \textit{and} \quad \|\widehat{\theta} - \theta^*\|_1 \leq \frac{6}{\kappa}\,s\,\lambda_n, \tag{9.63}$$

both with probability at least $1 - 2\,e^{-2n\delta^2}$.

We have already proved results of this form in Chapter 7 for the special case of the linear model; the proof here illustrates the application of our more general techniques.

Proof Both results follow via an application of Corollary 9.20 with the subspaces

$$\mathbb{M}(S) = \bar{\mathbb{M}}(S) = \{\theta \in \mathbb{R}^d \mid \theta_j = 0 \quad \text{for all } j \notin S\}.$$

With this choice, note that we have $\Psi^2(\mathbb{M}) = s$; moreover, the assumed RSC condition (9.62) is a special case of our general definition with $\tau_n^2 = c_1 \frac{\log d}{n}$. In order to apply Corollary 9.20, we need to ensure that $\tau_n^2 \Psi^2(\mathbb{M}) < \frac{\kappa}{64}$, and since the local RSC holds over a ball with radius $R = 1$, we also need to ensure that $\frac{9}{4}\frac{\Psi^2(\mathbb{M})\lambda_n^2}{\kappa^2} < 1$. Both of these conditions are guaranteed by our assumed lower bound on the sample size.

The only remaining step is to verify that the good event $\mathbb{G}(\lambda_n)$ holds with the probability stated in Corollary 9.26. Given the form (9.61) of the GLM log-likelihood, we can write the score function as the i.i.d. sum $\nabla \mathcal{L}_n(\theta^*) = \frac{1}{n}\sum_{i=1}^n V_i$, where $V_i \in \mathbb{R}^d$ is a zero-mean random vector with components

$$V_{ij} = \{\psi'(\langle x_i, \theta^* \rangle) - y_i\}\, x_{ij}.$$

Let us upper bound the moment generating function of these variables. For any $t \in \mathbb{R}$, we have

$$\begin{aligned}\log \mathbb{E}[e^{-tV_{ij}}] &= \log \mathbb{E}[e^{ty_i x_{ij}}] - t x_{ij}\psi'\big(\langle x_i, \theta^* \rangle\big)\\ &= \psi\big(t x_{ij} + \langle x_i, \theta^* \rangle\big) - \psi\big(\langle x_i, \theta^* \rangle\big) - t x_{ij}\psi'\big(\langle x_i, \theta^* \rangle\big).\end{aligned}$$

By a Taylor-series expansion, there is some intermediate $\tilde{t}$ such that

$$\log \mathbb{E}[e^{-tV_{ij}}] = \frac{1}{2} t^2 x_{ij}^2 \psi''\big(\tilde{t} x_{ij} + \langle x_i, \theta^* \rangle\big) \le \frac{B^2 t^2 x_{ij}^2}{2},$$

where the final inequality follows from the boundedness condition (G2). Using independence of the samples, we have

$$\frac{1}{n} \log \mathbb{E}\left[e^{-t\sum_{i=1}^n V_{ij}}\right] \le \frac{t^2 B^2}{2}\left(\frac{1}{n}\sum_{i=1}^n x_{ij}^2\right) \le \frac{t^2 B^2 C^2}{2},$$

where the final step uses the column normalization (G1) on the columns of the design matrix $\mathbf{X}$. Since this bound holds for any $t \in \mathbb{R}$, we have shown that each element of the score function $\nabla \mathcal{L}_n(\theta^*) \in \mathbb{R}^d$ is zero-mean and sub-Gaussian with parameter at most $BC/\sqrt{n}$. Thus, sub-Gaussian tail bounds combined with the union bound guarantee that

$$\mathbb{P}\big[\|\nabla \mathcal{L}_n(\theta^*)\|_\infty \ge t\big] \le 2\exp\left(-\frac{nt^2}{2B^2C^2} + \log d\right).$$

Setting $t = 2BC\left\{\sqrt{\frac{\log d}{n}} + \delta\right\}$ completes the proof. □

9.5.3 Bounds under ℓ_∞-curvature conditions

The preceding results were devoted to error bounds in terms of quadratic-type norms, such as the Euclidean vector and Frobenius matrix norms. On the other hand, Theorem 9.24 provides bounds in terms of the dual norm Φ^*—that is, in terms of the ℓ_∞-norm in the case of ℓ_1-regularization. We now turn to exploration of such bounds in the case of generalized linear models. As we discuss, ℓ_∞-bounds also lead to bounds in terms of the ℓ_2- and ℓ_1-norms, so that the resulting guarantees are in some sense stronger.

Recall that Theorem 9.24 is based on a restricted curvature condition (9.55). In the earlier Example 9.23, we discussed the specialization of this condition to the least-squares cost, and in Exercise 9.14, we work through the proof of an analogous result for generalized linear models with bounded cumulant generating functions ($\|\psi''\|_\infty \leq B$). More precisely, when the population cost satisfies an ℓ_∞-curvature condition over the ball $\mathbb{B}_2(R)$, and the covariates are i.i.d. and sub-Gaussian with parameter C, then the GLM log-likelihood $\mathcal{L}_n$ from equation (9.61) satisfies a bound of the form

$$\|\nabla \mathcal{L}_n(\theta^* + \Delta) - \nabla \mathcal{L}_n(\theta^*)\|_\infty \geq \kappa \|\Delta\|_\infty - \frac{c_0}{32} \sqrt{\frac{\log d}{n}} \|\Delta\|_1, \tag{9.64}$$

uniformly over $\mathbb{B}_\infty(1)$. Here is c_0 is a constant that depends only on the parameters (B, C).

Corollary 9.27 *In addition to the conditions of Corollary 9.26, suppose that the ℓ_∞-curvature condition (9.64) holds, and that the sample size is lower bounded as $n > c_0^2 s^2 \log d$. Then any optimal solution $\widehat{\theta}$ to the GLM Lasso (9.61) with regularization parameter $\lambda_n = 2\,BC\left(\sqrt{\frac{\log d}{n}} + \delta\right)$ satisfies*

$$\|\widehat{\theta} - \theta^*\|_\infty \leq 3\,\frac{\lambda_n}{\kappa} \tag{9.65}$$

with probability at least $1 - 2\,e^{-2n\delta^2}$.

Proof We prove this corollary by applying Theorem 9.24 with the familiar subspaces

$$\overline{\mathbb{M}}(S) = \mathbb{M}(S) = \{\theta \in \mathbb{R}^d \mid \theta_{S^c} = 0\},$$

for which we have $\Psi^2(\overline{\mathbb{M}}(S)) = s$. By assumption (9.64), the ℓ_∞-curvature condition holds with tolerance $\tau_n = \frac{c_0}{32}\sqrt{\frac{\log d}{n}}$, so that the condition $\tau_n \Psi^2(\mathbb{M}) < \frac{\kappa}{32}$ is equivalent to the lower bound $n > c_0^2\, s^2 \log d$ on the sample size.

Since we have assumed the conditions of Corollary 9.26, we are guaranteed that the error vector $\widehat{\Delta} = \widehat{\theta} - \theta^*$ satisfies the bound $\|\widehat{\Delta}\|_\infty \leq \|\widehat{\Delta}\|_2 \leq 1$ with high probability. This localization allows us to apply the local ℓ_∞-curvature condition to the error vector $\widehat{\Delta} = \widehat{\theta} - \theta^*$.

Finally, as shown in the proof of Corollary 9.26, if we choose the regularization parameter $\lambda_n = 2BC\left\{\sqrt{\frac{\log d}{n}} + \delta\right\}$, then the event $\mathbb{G}(\lambda_n)$ holds with probability at least $1 - e^{-2n\delta^2}$. We have thus verified that all the conditions needed to apply Theorem 9.24 are satisfied. □

The ℓ_∞-bound (9.65) is a stronger guarantee than our earlier bounds in terms of the ℓ_1- and ℓ_2-norms. For instance, under additional conditions on the smallest non-zero absolute values of θ^*, the ℓ_∞-bound (9.65) can be used to construct an estimator that has variable selection guarantees, which may not be possible with bounds in other norms. Moreover, as we explore in Exercise 9.13, when combined with other properties of the error vector, Corollary 9.27 implies bounds on the ℓ_1- and ℓ_2-norm errors that are analogous to those in Corollary 9.26.

9.6 Bounds for group-structured sparsity

We now turn to the consequences of Theorem 9.19 for estimators based on the group Lasso penalty with non-overlapping groups, previously discussed in Example 9.3. For concreteness, we focus on the ℓ_2-version of the group Lasso penalty $\|\theta\|_{\mathcal{G},2} = \sum_{g\in\mathcal{G}} \|\theta_g\|_2$. As discussed in Example 9.6, one motivation for the group Lasso penalty are multivariate regression problems, in which the regression coefficients are assumed to appear on–off in a groupwise manner. The linear multivariate regression problem from Example 9.6 is the simplest example. In this section, we analyze the extension to generalized linear models. Accordingly, let us consider the *group GLM Lasso*

$$\widehat{\theta} \in \arg\min_{\theta\in\mathbb{R}^d} \left\{ \frac{1}{n} \sum_{i=1}^{n} \{\psi(\langle\theta,\, x_i\rangle) - y_i \langle\theta,\, x_i\rangle\} + \lambda_n \sum_{g\in\mathcal{G}} \|\theta_g\|_2 \right\}, \tag{9.66}$$

a family of estimators that includes the least-squares version of the group Lasso (9.14) as a particular case.

As with our previous corollaries, we assume that the samples $\{(x_i, y_i)\}_{i=1}^n$ are drawn i.i.d. from a generalized linear model (GLM) satisfying condition (G2). Letting $\mathbf{X}_g \in \mathbb{R}^{n\times|g|}$ denote the submatrix indexed by g, we also impose the following variant of condition (G1) on the design:

(G1′) The covariates satisfy the group normalization condition $\max_{g\in\mathcal{G}} \frac{|||\mathbf{X}_g|||_2}{\sqrt{n}} \leq C$.

Moreover, we assume an RSC condition of the form

$$\mathcal{E}_n(\Delta) \geq \kappa\|\Delta\|_2^2 - c_1 \left\{ \frac{m}{n} + \frac{\log|\mathcal{G}|}{n} \right\} \|\Delta\|_{\mathcal{G},2}^2 \qquad \text{for all } \|\Delta\|_2 \leq 1, \tag{9.67}$$

where m denotes the maximum size over all groups. As shown in Example 9.17 and Theorem 9.36, a lower bound of this form holds with high probability when the covariates $\{x_i\}_{i=1}^n$ are drawn i.i.d. from a zero-mean sub-Gaussian distribution. Our bound applies to any solution $\widehat{\theta}$ to the group GLM Lasso (9.66) based on a regularization parameter

$$\lambda_n = 4BC\left\{ \sqrt{\frac{m}{n}} + \sqrt{\frac{\log|\mathcal{G}|}{n}} + \delta \right\} \qquad \text{for some } \delta \in (0, 1).$$

Corollary 9.28 *Given n i.i.d. samples from a GLM satisfying conditions (G1′), (G2), the RSC condition* (9.67), *suppose that the true regression vector θ^* has group support $S_{\mathcal{G}}$. As long as $|S_{\mathcal{G}}|\left\{\lambda_n^2 + \frac{m}{n} + \frac{\log|\mathcal{G}|}{n}\right\} < \min\left\{\frac{4\kappa^2}{9}, \frac{\kappa}{64c_1}\right\}$, the estimate $\widehat{\theta}$ satisfies the bound*

$$\|\widehat{\theta} - \theta^*\|_2^2 \leq \frac{9}{4} \frac{|S_{\mathcal{G}}|\lambda_n^2}{\kappa^2} \tag{9.68}$$

with probability at least $1 - 2e^{-2n\delta^2}$.

In order to gain some intuition for this corollary, it is worthwhile to consider some special cases. The ordinary Lasso is a special case of the group Lasso, in which there are $|\mathcal{G}| = d$ groups, each of size $m = 1$. In this case, if we use the regularization parameter $\lambda_n = 8BC\sqrt{\frac{\log d}{n}}$, the bound (9.68) implies that

$$\|\widehat{\theta} - \theta^*\|_2 \precsim \frac{BC}{\kappa}\sqrt{\frac{|S_{\mathcal{G}}| \log d}{n}},$$

showing that Corollary 9.28 is a natural generalization of Corollary 9.26.

The problem of multivariate regression provides a more substantive example of the potential gains of using the group Lasso. Throughout this example, we take the regularization parameter $\lambda_n = 8BC\left\{\sqrt{\frac{m}{n}} + \sqrt{\frac{\log d}{n}}\right\}$ as given.

Example 9.29 (Faster rates for multivariate regression) As previously discussed in Example 9.6, the problem of multivariate regression is based on the linear observation model $\mathbf{Y} = \mathbf{Z}\mathbf{\Theta}^* + \mathbf{W}$, where $\mathbf{\Theta}^* \in \mathbb{R}^{p\times T}$ is a matrix of regression coefficients, $\mathbf{Y} \in \mathbb{R}^{n\times T}$ is a matrix of observations, and $\mathbf{W} \in \mathbb{R}^{n\times T}$ is a noise matrix. A natural group structure is defined by the rows of the regression matrix $\mathbf{\Theta}^*$, so that we have a total of p groups each of size T.

A naive approach would be to ignore the group sparsity, and simply apply the elementwise ℓ_1-norm as a regularizer to the matrix $\mathbf{\Theta}$. This set-up corresponds to a Lasso problem with $d = pT$ coefficients and elementwise sparsity $T|S_{\mathcal{G}}|$, so that Corollary 9.26 would guarantee an estimation error bound of the form

$$|||\widehat{\mathbf{\Theta}} - \mathbf{\Theta}^*|||_F \precsim \sqrt{\frac{|S_{\mathcal{G}}|\,T\,\log(pT)}{n}}. \tag{9.69a}$$

By contrast, if we used the group Lasso estimator, which does explicitly model the grouping in the sparsity, then Corollary 9.28 would guarantee an error of the form

$$|||\widehat{\mathbf{\Theta}} - \mathbf{\Theta}^*|||_F \precsim \sqrt{\frac{|S_{\mathcal{G}}|\,T}{n}} + \sqrt{\frac{|S_{\mathcal{G}}| \log p}{n}}. \tag{9.69b}$$

For $T > 1$, it can be seen that this error bound is always better than the Lasso error bound (9.69a), showing that the group Lasso is a better estimator when $\mathbf{\Theta}^*$ has a sparse group structure. In Chapter 15, we will develop techniques that can be used to show that the rate (9.69b) is the best possible for any estimator. Indeed, the two components in this rate have a very concrete interpretation: the first corresponds to the error associated with estimating $|S_{\mathcal{G}}|T$ parameters, assuming that the group structure is known. For $|S_{\mathcal{G}}| \ll p$, the second term is proportional to $\log\binom{p}{|S_{\mathcal{G}}|}$, and corresponds to the search complexity associated with finding the subset of $|S_{\mathcal{G}}|$ rows out of p that contain non-zero coefficients. ♣

We now turn to the proof of Corollary 9.28.

Proof We apply Corollary 9.20 using the model subspace $\mathbb{M}(S_{\mathcal{G}})$ defined in equation (9.24). From Definition 9.18 of the subspace constant with $\Phi(\theta) = \|\theta\|_{\mathcal{G},2}$, we have

$$\Psi(\mathbb{M}(S_{\mathcal{G}})) := \sup_{\theta \in \mathbb{M}(S_{\mathcal{G}})\setminus\{0\}} \frac{\sum_{g\in\mathcal{G}} \|\theta_g\|_2}{\|\theta\|_2} = \sqrt{|S_{\mathcal{G}}|}.$$

The assumed RSC condition (9.62) is a special case of our general definition with the tolerance parameter $\tau_n^2 = c_1 \left\{\frac{m}{n} + \frac{\log|\mathcal{G}|}{n}\right\}$ and radius $R = 1$. In order to apply Corollary 9.20, we need to ensure that $\tau_n^2 \Psi^2(\mathbb{M}) < \frac{\kappa}{64}$, and since the local RSC holds over a ball with radius $R = 1$, we also need to ensure that $\frac{9}{4}\frac{\Psi^2(\mathbb{M})\lambda_n^2}{\kappa^2} < 1$. Both of these conditions are guaranteed by our assumed lower bound on the sample size.

It remains to verify that, given the specified choice of regularization parameter λ_n, the event $\mathbb{G}(\lambda_n)$ holds with high probability.

Verifying the event $\mathbb{G}(\lambda_n)$: Using the form of the dual norm given in Table 9.1, we have $\Phi^*(\nabla\mathcal{L}_n(\theta^*)) = \max_{g\in\mathcal{G}} \|(\nabla\mathcal{L}_n(\theta^*))_g\|_2$. Based on the form of the GLM log-likelihood, we have $\nabla\mathcal{L}_n(\theta^*) = \frac{1}{n}\sum_{i=1}^n V_i$ where the random vector $V_i \in \mathbb{R}^d$ has components

$$V_{ij} = \left\{\psi'(\langle x_i, \theta^*\rangle) - y_i\right\} x_{ij}.$$

For each group g, we let $V_{i,g} \in \mathbb{R}^{|g|}$ denote the subvector indexed by elements of g. With this notation, we then have

$$\|(\nabla\mathcal{L}_n(\theta^*))_g\|_2 = \|\frac{1}{n}\sum_{i=1}^n V_{i,g}\|_2 = \sup_{u\in\mathbb{S}^{|g|-1}} \left\langle u, \frac{1}{n}\sum_{i=1}^n V_{i,g}\right\rangle,$$

where $\mathbb{S}^{|g|-1}$ is the Euclidean sphere in $\mathbb{R}^{|g|}$. From Example 5.8, we can find a $1/2$-covering of $\mathbb{S}^{|g|-1}$ in the Euclidean norm—say $\{u^1, \ldots, u^N\}$—with cardinality at most $N \leq 5^{|g|}$. By the standard discretization arguments from Chapter 5, we have

$$\|(\nabla\mathcal{L}_n(\theta^*))_g\|_2 \leq 2 \max_{j=1,\ldots,N} \left\langle u^j, \frac{1}{n}\sum_{i=1}^n V_{i,g}\right\rangle.$$

Using the same proof as Corollary 9.26, the random variable $\left\langle u^j, \frac{1}{n}\sum_{i=1}^n V_{i,g}\right\rangle$ is sub-Gaussian with parameter at most

$$\frac{B}{\sqrt{n}}\sqrt{\frac{1}{n}\sum_{i=1}^n \left\langle u^j, x_{i,g}\right\rangle^2} \leq \frac{BC}{\sqrt{n}},$$

where the inequality follows from condition (G1′). Consequently, from the union bound and standard sub-Gaussian tail bounds, we have

$$\mathbb{P}\left[\|(\nabla\mathcal{L}_n(\theta^*))_g\|_2 \geq 2t\right] \leq 2\exp\left(-\frac{nt^2}{2B^2C^2} + |g|\log 5\right).$$

Taking the union over all $|\mathcal{G}|$ groups yields

$$\mathbb{P}\left[\max_{g\in\mathcal{G}} \|(\nabla\mathcal{L}_n(\theta^*))_g\|_2 \geq 2t\right] \leq 2\exp\left(-\frac{nt^2}{2B^2C^2} + m\log 5 + \log|\mathcal{G}|\right),$$

where we have used the maximum group size m as an upper bound on each group size $|g|$. Setting $t^2 = \lambda_n^2$ yields the result. □

9.7 Bounds for overlapping decomposition-based norms

In this section, we turn to the analysis of the more "exotic" overlapping group Lasso norm, as previously introduced in Example 9.4. In order to motivate this estimator, let us return to the problem of multivariate regression.

Example 9.30 (Matrix decomposition in multivariate regression) Recall the problem of linear multivariate regression from Example 9.6: it is based on the linear observation model $\mathbf{Y} = \mathbf{Z}\mathbf{\Theta}^* + \mathbf{W}$, where $\mathbf{\Theta}^* \in \mathbb{R}^{p \times T}$ is an unknown matrix of regression coefficients. As discussed previously, the ordinary group Lasso is often applied in this setting, using the rows of the regression matrix to define the underlying set of groups. When the true regression matrix $\mathbf{\Theta}^*$ is actually row-sparse, then we can expect the group Lasso to yield a more accurate estimate than the usual elementwise Lasso: compare the bounds (9.69a) and (9.69b).

However, now suppose that we apply the group Lasso estimator to a problem for which the true regression matrix $\mathbf{\Theta}^*$ *violates* the row-sparsity assumption: concretely, let us suppose that $\mathbf{\Theta}^*$ has s total non-zero entries, each contained within a row of its own. In this setting, Corollary 9.28 guarantees a bound of the order

$$|||\widehat{\mathbf{\Theta}} - \mathbf{\Theta}^*|||_F \precsim \sqrt{\frac{sT}{n}} + \sqrt{\frac{s \log p}{n}}. \tag{9.70}$$

However, if we were to apply the ordinary elementwise Lasso to this problem, then Corollary 9.26 would guarantee a bound of the form

$$|||\widehat{\mathbf{\Theta}} - \mathbf{\Theta}^*|||_F \precsim \sqrt{\frac{s \log(pT)}{n}}. \tag{9.71}$$

This error bound is always smaller than the group Lasso bound (9.70), and substantially so for large T. Consequently, the ordinary group Lasso has the undesirable feature of being less statistically efficient than the ordinary Lasso in certain settings, despite its higher computational cost.

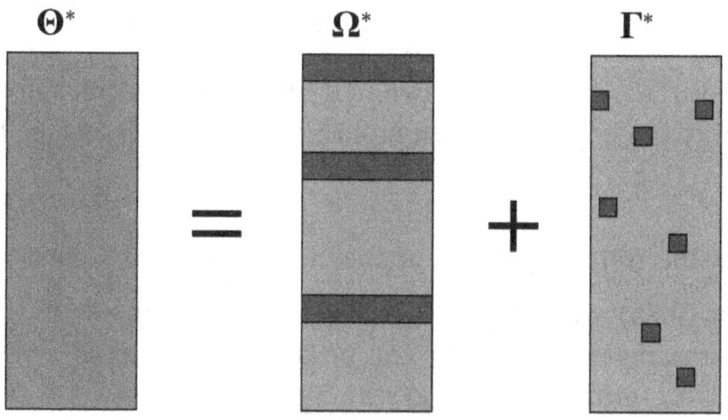

Figure 9.10 Illustration of the matrix decomposition norm (9.72) for the group Lasso applied to the matrix rows, combined with the elementwise ℓ_1-norm. The norm is defined by minimizing over all additive decompositions of $\mathbf{\Theta}^*$ as the sum of a row-sparse matrix $\mathbf{\Omega}^*$ with an elementwise-sparse matrix $\mathbf{\Gamma}^*$.

How do we remedy this issue? What would be desirable is an *adaptive estimator*, one that achieves the ordinary Lasso rate (9.71) when the sparsity structure is elementwise, and the group Lasso rate (9.70) when the sparsity is row-wise. To this end, let us consider decomposing the regression matrix $\mathbf{\Theta}^*$ as a sum $\mathbf{\Omega}^* + \mathbf{\Gamma}^*$, where $\mathbf{\Omega}^*$ is a row-sparse matrix and $\mathbf{\Gamma}^*$ is elementwise sparse, as shown in Figure 9.10. Minimizing a weighted combination of the group Lasso and ℓ_1-norms over all such decompositions yields the norm

$$\Phi_\omega(\mathbf{\Theta}) = \inf_{\mathbf{\Omega}+\mathbf{\Gamma}=\mathbf{\Theta}} \left\{ \|\mathbf{\Gamma}\|_1 + \omega \sum_{j=1}^{p} \|\mathbf{\Omega}_j\|_2 \right\}, \tag{9.72}$$

which is a special case of the overlapping group Lasso (9.10). Our analysis to follow will show that an M-estimator based on such a regularizer exhibits the desired adaptivity. ♣

Let us return to the general setting, in which we view the parameter $\theta \in \mathbb{R}^d$ as a vector,[3] and consider the more general ℓ_1-plus-group overlap norm

$$\Phi_\omega(\theta) := \inf_{\alpha+\beta=\theta} \left\{ \|\alpha\|_1 + \omega \|\beta\|_{\mathcal{G},2} \right\}, \tag{9.73}$$

where $\mathcal{G}$ is a set of disjoint groups, each of size at most m. The overlap norm (9.72) is a special case, where the groups are specified by the rows of the underlying matrix. For reasons to become clear in the proof, we use the weight

$$\omega := \frac{\sqrt{m} + \sqrt{\log |\mathcal{G}|}}{\sqrt{\log d}}. \tag{9.74}$$

With this set-up, the following result applies to the *adaptive group GLM Lasso*,

$$\widehat{\theta} \in \arg\min_{\theta \in \mathbb{R}^d} \Big\{ \underbrace{\frac{1}{n} \sum_{i=1}^{n} \{\psi(\langle \theta, x_i \rangle) - \langle \theta, x_i y_i \rangle\}}_{\mathcal{L}_n(\theta)} + \lambda_n \Phi_\omega(\theta) \Big\}, \tag{9.75}$$

for which the Taylor-series error satisfies the RSC condition

$$\mathcal{E}_n(\Delta) \geq \frac{\kappa}{2} \|\Delta\|_2^2 - c_1 \frac{\log d}{n} \Phi_\omega^2(\Delta) \qquad \text{for all } \|\Delta\|_2 \leq 1. \tag{9.76}$$

Again, when the covariates $\{x_i\}_{i=1}^n$ are drawn i.i.d. from a zero-mean sub-Gaussian distribution, a bound of this form holds with high probability for any GLM (see Example 9.17 and Exercise 9.8).

With this set-up, the following result applies to any optimal solution $\widehat{\theta}$ of the adaptive group GLM Lasso (9.75) with $\lambda_n = 4BC(\sqrt{\frac{\log d}{n}} + \delta)$ for some $\delta \in (0, 1)$. Moreover, it supposes that the true regression vector can be decomposed as $\theta^* = \alpha^* + \beta^*$, where α is S_{elt}-sparse, and β^* is $S_{\mathcal{G}}$-group-sparse, and with $S_{\mathcal{G}}$ disjoint from S_{elt}.

[3] The problem of multivariate regression can be thought of as a particular case of the vector model with vector dimension $d = pT$, via the transformation $\mathbf{\Theta} \mapsto \text{vec}(\mathbf{\Theta}) \in \mathbb{R}^{pT}$.

Corollary 9.31 *Given n i.i.d. samples from a GLM satisfying conditions (G1′) and (G2), suppose that the RSC condition* (9.76) *with curvature* $\kappa > 0$ *holds, and that* $\left\{\sqrt{|S_{\text{elt}}|} + \omega\sqrt{|S_{\mathcal{G}}|}\right\}^2 \left\{\lambda_n^2 + \frac{\log d}{n}\right\} < \min\left\{\frac{\kappa^2}{36}, \frac{\kappa}{64 c_1}\right\}$. *Then the adaptive group GLM Lasso estimate* $\widehat{\theta}$ *satisfies the bounds*

$$\|\widehat{\theta} - \theta^*\|_2^2 \leq \frac{36\lambda_n^2}{\kappa^2}\left\{\sqrt{|S_{\text{elt}}|} + \omega\sqrt{|S_{\mathcal{G}}|}\right\}^2 \tag{9.77}$$

with probability at least $1 - 3e^{-8n\delta^2}$.

Remark: The most important feature of the bound (9.77) is its adaptivity to the elementwise versus group sparsity. This adaptivity stems from the fact that the choices of $S_{\mathcal{G}}$ and S_{elt} can be optimized so as to obtain the tightest possible bound, depending on the structure of the regression vector θ^*. To be concrete, consider the bound with the choice $\lambda_n = 8BC\sqrt{\frac{\log d}{n}}$. At one extreme, suppose that the true regression vector $\theta^* \in \mathbb{R}^d$ is purely elementwise sparse, in that each group contains at most one non-zero entry. In this case, we can apply the bound (9.77) with $S_{\mathcal{G}} = \emptyset$, leading to

$$\|\widehat{\theta} - \theta^*\|_2^2 \precsim \frac{B^2 C^2}{\kappa^2} \frac{s \log d}{n},$$

where $s = |S_{\text{elt}}|$ denotes the sparsity of θ^*. We thus recover our previous Lasso bound from Corollary 9.26 in this special case. At the other extreme, consider a vector that is "purely" group-sparse, in the sense that it has some subset of active groups $S_{\mathcal{G}}$, but no isolated sparse entries. The bound (9.77) with $S_{\text{elt}} = \emptyset$ then yields

$$\|\widehat{\theta} - \theta^*\|_2^2 \precsim \frac{B^2 C^2}{\kappa^2}\left\{\frac{m|S_{\mathcal{G}}|}{n} + \frac{|S_{\mathcal{G}}| \log d}{n}\right\},$$

so that, in this special case, the decomposition method obtains the group Lasso rate from Corollary 9.28.

Let us now prove the corollary:

Proof In this case, we work through the details carefully, as the decomposability of the overlap norm needs some care. Recall the function $\mathcal{F}$ from equation (9.31), and let $\widehat{\Delta} = \widehat{\theta} - \theta^*$. Our proof is based on showing that any vector of the form $\Delta = t\widehat{\Delta}$ for some $t \in [0, 1]$ satisfies the bounds

$$\Phi_\omega(\Delta) \leq 4\left\{\sqrt{|S_{\text{elt}}|} + \omega\sqrt{|S_{\mathcal{G}}|}\right\}\|\Delta\|_2 \tag{9.78a}$$

and

$$\mathcal{F}(\Delta) \geq \frac{\kappa}{2}\|\Delta\|_2^2 - c_1\frac{\log d}{n}\Phi_\omega^2(\Delta) - \frac{3\lambda_n}{2}\left\{\sqrt{|S_{\text{elt}}|} + \omega\sqrt{|S_{\mathcal{G}}|}\right\}\|\Delta\|_2. \tag{9.78b}$$

Let us take these bounds as given for the moment, and then return to prove them. Substituting the bound (9.78a) into inequality (9.78b) and rearranging yields

$$\mathcal{F}(\Delta) \geq \|\Delta\|_2\left\{\kappa'\|\Delta\|_2 - \frac{3\lambda_n}{2}\left(\sqrt{|S_{\text{elt}}|} + \omega\sqrt{|S_{\mathcal{G}}|}\right)\right\},$$

where $\kappa' := \frac{\kappa}{2} - 16c_1 \frac{\log d}{n}(\sqrt{|S_{\text{elt}}|} + \omega\sqrt{|S_{\mathcal{G}}|})^2$. Under the stated bound on the sample size n, we have $\kappa' \geq \frac{\kappa}{4}$, so that $\mathcal{F}$ is non-negative whenever

$$\|\Delta\|_2 \geq \frac{6\lambda_n}{\kappa}\left(\sqrt{|S_{\text{elt}}|} + \omega\sqrt{|S_{\mathcal{G}}|}\right).$$

Finally, following through the remainder of the proof of Theorem 9.19 yields the claimed bound (9.77).

Let us now return to prove the bounds (9.78a) and (9.78b). To begin, a straightforward calculation shows that the dual norm is given by

$$\Phi^*_{\omega}(v) = \max\left\{\|v\|_\infty, \frac{1}{\omega}\max_{g\in\mathcal{G}}\|v_g\|_2\right\}.$$

Consequently, the event $\mathbb{G}(\lambda_n) := \{\Phi^*_{\omega}(\nabla\mathcal{L}_n(\theta^*)) \leq \frac{\lambda_n}{2}\}$ is equivalent to

$$\|\nabla\mathcal{L}_n(\theta^*)\|_\infty \leq \frac{\lambda_n}{2} \quad\text{and}\quad \max_{g\in\mathcal{G}}\|(\nabla\mathcal{L}_n(\theta^*))_g\|_2 \leq \frac{\lambda_n\omega}{2}. \tag{9.79}$$

We assume that these conditions hold for the moment, returning to verify them at the end of the proof.

Define $\Delta = t\widehat{\Delta}$ for some $t \in [0,1]$. Fix some decomposition $\theta^* = \alpha^* + \beta^*$, where α^* is S_{elt}-sparse and β^* is $S_{\mathcal{G}}$-group-sparse, and note that

$$\Phi_\omega(\theta^*) \leq \|\alpha^*\|_1 + \omega\|\beta^*\|_{\mathcal{G},2}.$$

Similarly, let us write $\Delta = \Delta_\alpha + \Delta_\beta$ for some pair such that

$$\Phi_\omega(\theta^* + \Delta) = \|\Delta_\alpha\|_1 + \omega\|\Delta_\beta\|_{\mathcal{G},2}.$$

Proof of inequality (9.78a): Define the function

$$\mathcal{F}(\Delta) := \mathcal{L}_n(\theta^* + \Delta) - \mathcal{L}_n(\theta^*) + \lambda_n\{\Phi_\omega(\theta^* + \Delta) - \Phi_\omega(\theta^*)\}.$$

Consider a vector of the form $\Delta = t\widehat{\Delta}$ for some scalar $t \in [0,1]$. Noting that $\mathcal{F}$ is convex and minimized at $\widehat{\Delta}$, we have

$$\mathcal{F}(\Delta) = \mathcal{F}(t\widehat{\Delta} + (1-t)0) \leq t\mathcal{F}(\widehat{\Delta}) + (1-t)\mathcal{F}(0) \leq \mathcal{F}(0).$$

Recalling that $\mathcal{E}_n(\Delta) = \mathcal{L}_n(\theta^* + \Delta) - \mathcal{L}_n(\theta^*) - \langle\nabla\mathcal{L}_n(\theta^*), \Delta\rangle$, some algebra then leads to the inequality

$$\begin{aligned}
\mathcal{E}_n(\Delta) &\leq \left|\langle\nabla\mathcal{L}_n(\theta^*), \Delta\rangle\right| - \lambda_n\{\|\alpha^* + \Delta_\alpha\|_1 - \|\alpha^*\|_1\} - \lambda_n\omega\left\{\|\beta^* + \Delta_\beta\|_{\mathcal{G},2} - \|\beta^*\|_{\mathcal{G},2}\right\}\\
&\overset{\text{(i)}}{\leq} \left|\langle\nabla\mathcal{L}_n(\theta^*), \Delta\rangle\right| + \lambda_n\left\{\|(\Delta_\alpha)_{S_{\text{elt}}}\|_1 - \|(\Delta_\alpha)_{S^c_{\text{elt}}}\|_1\right\} + \lambda_n\omega\left\{\|(\Delta_\beta)_{S_{\mathcal{G}}}\|_{\mathcal{G},2} - \|(\Delta_\beta)_{S_{\mathcal{G}}}\|_{\mathcal{G},2}\right\}\\
&\overset{\text{(ii)}}{\leq} \frac{\lambda_n}{2}\left\{3\|(\Delta_\alpha)_{S_{\text{elt}}}\|_1 - \|(\Delta_\alpha)_{S^c_{\text{elt}}}\|_1\right\} + \frac{\lambda_n\omega}{2}\left\{\|(\Delta_\beta)_{S_{\mathcal{G}}}\|_{\mathcal{G},2} - \|(\Delta_\beta)_{S_{\mathcal{G}}}\|_{\mathcal{G},2}\right\}.
\end{aligned}$$

Here step (i) follows by decomposability of the ℓ_1 and the group norm, and step (ii) follows by using the inequalities (9.79). Since $\mathcal{E}_n(\Delta) \geq 0$ by convexity, rearranging yields

$$\begin{aligned}\|\Delta_\alpha\|_1 + \omega\|\Delta_\beta\|_{\mathcal{G},2} &\leq 4\Big\{\|(\Delta_\alpha)_{S_{\mathrm{elt}}}\|_1 + \omega\|(\Delta_\beta)_{S_{\mathcal{G}}}\|_{\mathcal{G},2}\Big\} \\ &\overset{\text{(iii)}}{\leq} 4\Big\{\sqrt{|S_{\mathrm{elt}}|}\,\|(\Delta_\alpha)_{S_{\mathrm{elt}}}\|_2 + \omega\sqrt{|S_{\mathcal{G}}|}\,\|(\Delta_\beta)_{S_{\mathcal{G}}}\|_2\Big\} \\ &\leq 4\Big\{\sqrt{|S_{\mathrm{elt}}|} + \omega\sqrt{|S_{\mathcal{G}}|}\Big\}\Big\{\|(\Delta_\alpha)_{S_{\mathrm{elt}}}\|_2 + \|(\Delta_\beta)_{S_{\mathcal{G}}}\|_2\Big\},\end{aligned} \tag{9.80}$$

where step (iii) follows from the subspace constants for the two decomposable norms. The overall vector Δ has the decomposition $\Delta = (\Delta_\alpha)_{S_{\mathrm{elt}}} + (\Delta_\beta)_{S_{\mathcal{G}}} + \Delta_T$, where T is the complement of the indices in S_{elt} and $S_{\mathcal{G}}$. Noting that all three sets are disjoint by construction, we have

$$\|(\Delta_\alpha)_{S_{\mathrm{elt}}}\|_2 + \|(\Delta_\beta)_{S_{\mathcal{G}}}\|_2 = \|(\Delta_\alpha)_{S_{\mathrm{elt}}} + (\Delta_\beta)_{S_{\mathcal{G}}}\|_2 \leq \|\Delta\|_2.$$

Combining with inequality (9.80) completes the proof of the bound (9.78a).

Proof of inequality (9.78b)*:* From the proof of Theorem 9.19, recall the lower bound (9.50). This inequality, combined with the RSC condition, guarantees that the function value $\mathcal{F}(\Delta)$ is at least

$$\begin{aligned}&\frac{\kappa}{2}\|\Delta\|_2^2 - c_1\frac{\log d}{n}\Phi_\omega^2(\Delta) - \big|\langle\nabla\mathcal{L}_n(\theta^*),\, \Delta\rangle\big| \\ &\quad + \lambda_n\{\|\alpha^* + \Delta_\alpha\|_1 - \|\alpha^*\|_1\} + \lambda_n\omega\{\|\beta^* + \Delta_\beta\|_{\mathcal{G},2} - \|\beta^*\|_{\mathcal{G},2}\}.\end{aligned}$$

Again, applying the dual norm bounds (9.79) and exploiting decomposability leads to the lower bound (9.78b).

Verifying inequalities (9.79)*:* The only remaining detail is to verify that the conditions (9.79) defining the event $\mathbb{G}(\lambda_n)$. From the proof of Corollary 9.26, we have

$$\mathbb{P}[\|\nabla\mathcal{L}_n(\theta^*)\|_\infty \geq t] \leq d\, e^{-\frac{nt^2}{2B^2C^2}}.$$

Similarly, from the proof of Corollary 9.28, we have

$$\mathbb{P}\Big[\frac{1}{\omega}\max_{g\in\mathcal{G}}\|(\nabla\mathcal{L}_n(\theta^*))_g\|_2 \geq 2t\Big] \leq 2\exp\left(-\frac{n\omega^2t^2}{2B^2C^2} + m\log 5 + \log|\mathcal{G}|\right).$$

Setting $t = 4BC\left\{\sqrt{\frac{\log d}{n}} + \delta\right\}$ and performing some algebra yields the claimed lower bound $\mathbb{P}[\mathbb{G}(\lambda_n)] \geq 1 - 3e^{-8n\delta^2}$. □

9.8 Techniques for proving restricted strong convexity

All of the previous results rely on the empirical cost function satisfying some form of restricted curvature condition. In this section, we turn to a deeper investigation of the conditions under which restricted strong convexity conditions, as previously formalized in Definition 9.15, are satisfied.

Before proceeding, let us set up some notation. Given a collection of samples $Z_1^n = \{Z_i\}_{i=1}^n$, we write the empirical cost as $\mathcal{L}_n(\theta) = \frac{1}{n}\sum_{i=1}^n \mathcal{L}(\theta; Z_i)$, where $\mathcal{L}$ is the loss applied to a single sample. We can then define the error in the first-order Taylor expansion of $\mathcal{L}$ for sample Z_i, namely

$$\mathcal{E}(\Delta\,; Z_i) := \mathcal{L}(\theta^* + \Delta; Z_i) - \mathcal{L}(\theta^*; Z_i) - \langle \nabla \mathcal{L}(\theta^*; Z_i),\, \Delta \rangle .$$

By construction, we have $\mathcal{E}_n(\Delta) = \frac{1}{n}\sum_{i=1}^n \mathcal{E}(\Delta\,; Z_i)$. Given the population cost function $\bar{\mathcal{L}}(\theta) := \mathbb{E}[\mathcal{L}_n(\theta; Z_1^n)]$, a local form of strong convexity can be defined in terms of its Taylor-series error

$$\bar{\mathcal{E}}(\Delta) := \bar{\mathcal{L}}(\theta^* + \Delta) - \bar{\mathcal{L}}(\theta^*) - \left\langle \nabla \bar{\mathcal{L}}(\theta^*),\, \Delta \right\rangle . \tag{9.81}$$

We say that the population cost is (locally) *κ-strongly convex* around the minimizer θ^* if there exists a radius $R > 0$ such that

$$\bar{\mathcal{E}}(\Delta) \geq \kappa \|\Delta\|_2^2 \qquad \text{for all } \Delta \in \mathbb{B}_2(R) := \{\Delta \in \Omega \mid \|\theta\|_2 \leq R\}. \tag{9.82}$$

We wish to see when this type of curvature condition is inherited by the sample-based error $\mathcal{E}_n(\Delta)$. At a high level, then, our goal is clear: in order to establish a form of restricted strong convexity (RSC), we need to derive some type of uniform law of large numbers (see Chapter 4) for the zero-mean stochastic process

$$\left\{ \mathcal{E}_n(\Delta) - \bar{\mathcal{E}}(\Delta),\ \Delta \in \mathbb{S} \right\}, \tag{9.83}$$

where $\mathbb{S}$ is a suitably chosen subset of $\mathbb{B}_2(R)$.

Example 9.32 (Least squares) To gain intuition in a specific example, recall the quadratic cost function $\mathcal{L}(\theta; y_i, x_i) = \frac{1}{2}(y - \langle \theta,\, x_i\rangle)^2$ that underlies least-squares regression. In this case, we have $\mathcal{E}(\Delta\,; x_i, y_i) = \frac{1}{2}\langle \Delta,\, x_i\rangle^2$, and hence

$$\mathcal{E}_n(\Delta) = \frac{1}{2n}\sum_{i=1}^n \langle \Delta,\, x_i\rangle^2 = \frac{1}{2n}\|\mathbf{X}\Delta\|_2^2,$$

where $\mathbf{X} \in \mathbb{R}^{n\times d}$ is the usual design matrix. Denoting $\mathbf{\Sigma} = \mathrm{cov}(x)$, we find that

$$\bar{\mathcal{E}}(\Delta) = \mathbb{E}[\mathcal{E}_n(\Delta)] = \tfrac{1}{2}\Delta^{\mathrm{T}}\mathbf{\Sigma}\Delta.$$

Thus, our specific goal in this case is to establish a uniform law for the family of random variables

$$\left\{ \frac{1}{2}\Delta^{\mathrm{T}}\Big(\frac{\mathbf{X}^{\mathrm{T}}\mathbf{X}}{n} - \mathbf{\Sigma}\Big)\Delta,\ \Delta \in \mathbb{S} \right\}. \tag{9.84}$$

When $\mathbb{S} = \mathbb{B}_2(1)$, the supremum over this family is equal to the operator norm $|\!|\!|\frac{\mathbf{X}^{\mathrm{T}}\mathbf{X}}{n} - \mathbf{\Sigma}|\!|\!|_2$, a quantity that we studied in Chapter 6. When $\mathbb{S}$ involves an additional ℓ_1-constraint, then a uniform law over this family amounts to establishing a restricted eigenvalue condition, as studied in Chapter 7. ♣

9.8.1 Lipschitz cost functions and Rademacher complexity

This section is devoted to showing how the problem of establishing RSC for Lipschitz cost functions can be reduced to controlling a version of the Rademacher complexity. As the reader might expect, the symmetrization and contraction techniques from Chapter 4 turn out to be useful.

We say that $\mathcal{L}$ is locally *L-Lipschitz* over the ball $\mathbb{B}_2(R)$ if for each sample $Z = (x, y)$

$$\left|\mathcal{L}(\theta; Z) - \mathcal{L}(\widetilde{\theta}; Z)\right| \leq L \left|\langle\theta,\ x\rangle - \left\langle\widetilde{\theta},\ x\right\rangle\right| \qquad \text{for all } \theta, \widetilde{\theta} \in \mathbb{B}_2(R). \tag{9.85}$$

Let us illustrate this definition with an example.

Example 9.33 (Cost functions for binary classification) The class of Lipschitz cost functions includes various objective functions for binary classification, in which the goal is to use the covariates $x \in \mathbb{R}^d$ to predict an underlying class label $y \in \{-1, 1\}$. The simplest approach is based on a linear classification rule: given a weight vector $\theta \in \mathbb{R}^d$, the sign of the inner product $\langle\theta,\ x\rangle$ is used to make decisions. If we disregard computational issues, the most natural cost function is the 0–1 cost $\mathbb{I}[y\,\langle\theta,\ x\rangle < 0]$, which assigns a penalty of 1 if the decision is incorrect, and returns 0 otherwise. (Note that $y\,\langle\theta,\ x\rangle < 0$ if and only if $\operatorname{sign}(\langle\theta,\ x\rangle) \neq y$.)

For instance, the *logistic cost* takes the form

$$\mathcal{L}(\theta; (x, y)) := \log(1 + e^{\langle\theta,\ x\rangle}) - y\,\langle\theta,\ x\rangle, \tag{9.86}$$

and it is straightforward to verify that this cost function satisfies the Lipschitz condition with $L = 2$. Similarly, the support vector machine approach to classification is based on the *hinge cost*

$$\mathcal{L}(\theta; (x, y)) := \max\{0, 1 - y\,\langle\theta,\ x\rangle\} \equiv (1 - y\,\langle\theta,\ x\rangle)_+, \tag{9.87}$$

which is Lipschitz with parameter $L = 1$. Note that the least-squares cost function $\mathcal{L}(\theta; (x, y)) = \frac{1}{2}(y - \langle\theta,\ x\rangle)^2$ is *not* Lipschitz unless additional boundedness conditions are imposed. A similar observation applies to the exponential cost function $\mathcal{L}(\theta; (x, y)) = e^{-y\langle\theta,\ x\rangle}$. ♣

In this section, we prove that Lipschitz functions with regression-type data $z = (x, y)$ satisfy a certain form of restricted strong convexity, depending on the tail fluctuations of the covariates. The result itself involves a complexity measure associated with the norm ball of the regularizer Φ. More precisely, letting $\{\varepsilon_i\}_{i=1}^n$ be an i.i.d. sequence of Rademacher variables, we define the symmetrized random vector $\bar{x}_n = \frac{1}{n}\sum_{i=1}^n \varepsilon_i x_i$, and the random variable

$$\Phi^*(\bar{x}_n) := \sup_{\Phi(\theta)\leq 1} \left\langle\theta,\ \frac{1}{n}\sum_{i=1}^n \varepsilon_i x_i\right\rangle. \tag{9.88}$$

When $x_i \sim \mathcal{N}(0, \mathbf{I}_d)$, the mean $\mathbb{E}[\Phi^*(\bar{x}_n)]$ is proportional to the Gaussian complexity of the unit ball $\{\theta \in \mathbb{R}^d \mid \Phi(\theta) \leq 1\}$. (See Chapter 5 for an in-depth discussion of the Gaussian complexity and its properties.) More generally, the quantity (9.88) reflects the size of the Φ-unit ball with respect to the fluctuations of the covariates.

The following theorem applies to any norm Φ that dominates the Euclidean norm, in the sense that $\Phi(\Delta) \geq \|\Delta\|_2$ uniformly. For a pair of radii $0 < R_\ell < R_u$, it guarantees a form of restricted strong convexity over the "donut" set

$$\mathbb{B}_2(R_\ell, R_u) := \{\Delta \in \mathbb{R}^d \mid R_\ell \leq \|\Delta\|_2 \leq R_u\}. \tag{9.89}$$

The high-probability statement is stated in terms of the random variable $\Phi^*(\bar{x}_n)$, as well as the quantity $M_n(\Phi; R) := 4\log\left(\frac{R_u}{R_\ell}\right)\,\log\sup_{\theta\neq 0}\left(\frac{\Phi(\theta)}{\|\theta\|_2}\right)$, which arises for technical reasons.

Theorem 9.34 *Suppose that the cost function $\mathcal{L}$ is L-Lipschitz (9.85), and the population cost $\bar{\mathcal{L}}$ is locally κ-strongly convex (9.82) over the ball $\mathbb{B}_2(R_u)$. Then for any $\delta > 0$, the first-order Taylor error $\mathcal{E}_n$ satisfies*

$$\left|\mathcal{E}_n(\Delta) - \bar{\mathcal{E}}(\Delta)\right| \leq 16L\,\Phi(\Delta)\,\delta \qquad \text{for all } \Delta \in \mathbb{B}_2(R_\ell, R_u) \tag{9.90}$$

with probability at least $1 - M_n(\Phi; R)\,\inf_{\lambda>0}\mathbb{E}[e^{\lambda(\Phi^*(\bar{x}_n)-\delta)}]$.

For Lipschitz functions, this theorem reduces the question of establishing RSC to that of controlling the random variable $\Phi^*(\bar{x}_n)$. Let us consider a few examples to illustrate the consequences of Theorem 9.34.

Example 9.35 (Lipschitz costs and group Lasso) Consider the group Lasso norm $\Phi(\theta) = \sum_{g\in\mathcal{G}}\|\theta_g\|_2$, where we take groups of equal size m for simplicity. Suppose that the covariates $\{x_i\}_{i=1}^n$ are drawn i.i.d. as $\mathcal{N}(0, \Sigma)$ vectors, and let $\sigma^2 = |||\Sigma|||_2$. In this case, we show that for any L-Lipschitz cost function, the inequality

$$\left|\mathcal{E}_n(\Delta) - \bar{\mathcal{E}}(\Delta)\right| \leq 16L\sigma\left\{\sqrt{\frac{m}{n}} + \sqrt{\frac{2\log|\mathcal{G}|}{n}} + \delta\right\}\sum_{g\in\mathcal{G}}\|\Delta_g\|_2$$

holds uniformly for all $\Delta \in \mathbb{B}_2(\frac{1}{d}, 1)$ with probability at least $1 - 4\log^2(d)\,e^{-\frac{n\delta^2}{2}}$.

In order to establish this claim, we begin by noting that $\Phi^*(\bar{x}_n) = \max_{g\in\mathcal{G}}\|(\bar{x}_n)_g\|_2$ from Table 9.1. Consequently, we have

$$\mathbb{E}[e^{\lambda\Phi^*(\bar{x}_n)}] \leq \sum_{g\in\mathcal{G}}\mathbb{E}\left[e^{\lambda(\|(\bar{x}_n)_g\|_2)}\right] = \sum_{g\in\mathcal{G}}\mathbb{E}\left[e^{\lambda(\|(\bar{x}_n)_g\|_2 - \mathbb{E}[\|(\bar{x}_n)_g\|_2])}\right]\,e^{\lambda\mathbb{E}[\|(\bar{x}_n)_g\|_2]}.$$

By Theorem 2.26, the random variable $\|(\bar{x}_n)_g\|_2$ has sub-Gaussian concentration around its mean with parameter $\sigma/\sqrt{n}$, whence $\mathbb{E}[e^{\lambda(\|(\bar{x}_n)_g\|_2 - \mathbb{E}[\|(\bar{x}_n)_g\|_2])}] \leq e^{\frac{\lambda^2\sigma^2}{2n}}$. By Jensen's inequality, we have

$$\mathbb{E}[\|(\bar{x}_n)_g\|_2] \leq \sqrt{\mathbb{E}[\|(\bar{x}_n)_g\|_2^2]} \leq \sigma\sqrt{\frac{m}{n}},$$

using the fact that $\sigma^2 = |||\Sigma|||_2$. Putting together the pieces, we have shown that

$$\inf_{\lambda>0}\log\mathbb{E}[e^{\lambda(\Phi^*(\bar{x}_n)-(\epsilon+\sigma\sqrt{\frac{m}{n}}))}] \leq \log|\mathcal{G}| + \inf_{\lambda>0}\left\{\frac{\lambda^2\sigma^2}{2n} - \lambda\epsilon\right\} = \log|\mathcal{G}| - \frac{n\epsilon^2}{2\sigma^2}.$$

With the choices $R_u = 1$ and $R_\ell = \frac{1}{d}$, we have

$$M_n(\Phi; R) = 4\log(d)\,\log|\mathcal{G}| \leq 4\log^2(d),$$

since $|\mathcal{G}| \leq d$. Thus, setting $\epsilon = 2\sigma\left\{\sqrt{\frac{\log|\mathcal{G}|}{n}} + \epsilon\right\}$ and applying Theorem 9.34 yields the stated claim. ♣

In Chapter 10, we discuss some consequences of Theorem 9.34 for estimating low-rank matrices. Let us now turn to its proof.

Proof Recall that

$$\mathcal{E}(\Delta; z_i) := \mathcal{L}(\theta^* + \Delta; z_i) - \mathcal{L}(\theta^*; z_i) - \langle \nabla \mathcal{L}(\theta^*; z_i), \Delta \rangle$$

denotes the Taylor-series error associated with a single sample $z_i = (x_i, y_i)$.

Showing the Taylor error is Lipschitz: We first show that $\mathcal{E}$ is a $2L$-Lipschitz function in $\langle \Delta, x_i \rangle$. To establish this claim, note if that we let $\frac{\partial \mathcal{L}}{\partial u}$ denote the derivative of $\mathcal{L}$ with respect to $u = \langle \theta, x \rangle$, then the Lipschitz condition implies that $\|\frac{\partial \mathcal{L}}{\partial u}\|_\infty \leq L$. Consequently, by the chain rule, for any sample $z_i \in \mathcal{Z}$ and parameters $\Delta, \widetilde{\Delta} \in \mathbb{R}^d$, we have

$$\left|\langle \nabla \mathcal{L}(\theta^*; Z_i), \Delta - \widetilde{\Delta} \rangle\right| \leq \left|\frac{\partial \mathcal{L}}{\partial u}(\theta^*; Z_i)\right| \left|\langle \Delta, x_i \rangle - \langle \widetilde{\Delta}, x_i \rangle\right| \leq L \left|\langle \Delta, x_i \rangle - \langle \widetilde{\Delta}, x_i \rangle\right|. \tag{9.91}$$

Putting together the pieces, for any pair $\Delta, \widetilde{\Delta}$, we have

$$\begin{aligned} \left|\mathcal{E}(\Delta; Z_i) - \mathcal{E}(\widetilde{\Delta}; Z_i)\right| &\leq \left|\mathcal{L}(\theta^* + \Delta; Z_i) - \mathcal{L}(\theta^* + \widetilde{\Delta}; Z_i)\right| + \left|\langle \nabla \mathcal{L}(\theta^*; Z_i), \Delta - \widetilde{\Delta} \rangle\right| \\ &\leq 2L |\langle \Delta, x_i \rangle - \langle \widetilde{\Delta}, x_i \rangle|, \end{aligned} \tag{9.92}$$

where the second inequality applies our Lipschitz assumption, and the gradient bound (9.91). Thus, the Taylor error is a $2L$-Lipschitz function in $\langle \Delta, x_i \rangle$.

Tail bound for fixed radii: Next we control the difference $|\mathcal{E}_n(\Delta) - \bar{\mathcal{E}}(\Delta)|$ uniformly over certain sets defined by fixed radii. More precisely, for positive quantities (r_1, r_2), define the set

$$\mathbb{C}(r_1, r_2) := \mathbb{B}_2(r_2) \cap \{\Phi(\Delta) \leq r_1 \|\Delta\|_2\},$$

and the random variable $A_n(r_1, r_2) := \frac{1}{4 r_1 r_2 L} \sup_{\Delta \in \mathbb{C}(r_1, r_2)} \left|\mathcal{E}_n(\Delta) - \bar{\mathcal{E}}(\Delta)\right|$. The choice of radii can be implicitly understood, so that we adopt the shorthand A_n.

Our goal is to control the probability of the event $\{A_n \geq \delta\}$, and we do so by controlling the moment generating function. By our assumptions, the Taylor error has the additive decomposition $\mathcal{E}_n(\Delta) = \frac{1}{n} \sum_{i=}^{n} \mathcal{E}(\Delta; Z_i)$. Thus, letting $\{\varepsilon_i\}_{i=1}^n$ denote an i.i.d. Rademacher sequence, applying the symmetrization upper bound from Proposition 4.11(b) yields

$$\mathbb{E}[e^{\lambda A_n}] \leq \mathbb{E}_{Z, \varepsilon}\left[\exp\left(2\lambda \sup_{\Delta \in \mathbb{C}(r_1, r_2)} \left|\frac{1}{4 L r_1 r_2} \frac{1}{n} \sum_{i=1}^n \varepsilon_i \, \mathcal{E}(\Delta; Z_i)\right|\right)\right].$$

Now we have

$$\mathbb{E}[e^{\lambda A_n}] \overset{\text{(i)}}{\leq} \mathbb{E}\left[\exp\left(\frac{\lambda}{r_1 r_2} \sup_{\Delta \in \mathbb{C}(r_1, r_2)} \left|\frac{1}{n} \sum_{i=1}^n \varepsilon_i \langle \Delta, x_i \rangle\right|\right)\right] \overset{\text{(ii)}}{\leq} \mathbb{E}\left[\exp\left\{\lambda \, \Phi^*\left(\frac{1}{n} \sum_{i=1}^n \varepsilon_i x_i\right)\right\}\right],$$

where step (i) uses the Lipschitz property (9.92) and the Ledoux–Talagrand contraction inequality (5.61), whereas step (ii) follows from applying Hölder's inequality to the regularizer and its dual (see Exercise 9.7), and uses the fact that $\Phi^*(\Delta) \leq r_1 r_2$ for any vector

$\Delta \in \mathbb{C}(r_1, r_2)$. Adding and subtracting the scalar $\delta > 0$ then yields

$$\log \mathbb{E}[e^{\lambda(A_n - \delta)}] \le -\lambda\delta + \log \mathbb{E}\left[\exp\left\{\lambda\, \Phi^*\Big(\frac{1}{n}\sum_{i=1}^n \varepsilon_i x_i\Big)\right\}\right],$$

and consequently, by Markov's inequality,

$$\mathbb{P}[A_n(r_1, r_2) \ge \delta] \le \inf_{\lambda > 0} \mathbb{E}\left[\exp\left(\lambda\{\Phi^*(\bar{x}_n) - \delta\}\right)\right]. \tag{9.93}$$

Extension to uniform radii via peeling: This bound (9.93) applies to fixed choice of quantities (r_1, r_2), whereas the claim of Theorem 9.34 applies to possibly random choices—namely, $\frac{\Phi(\Delta)}{\|\Delta\|_2}$ and $\|\Delta\|_2$, respectively, where Δ might be chosen in a way dependent on the data. In order to extend the bound to all choices, we make use of a peeling argument.

Let $\mathcal{E}$ be the event that the bound (9.90) is violated. For positive integers (k, ℓ), define the sets

$$\mathbb{S}_{k,\ell} := \left\{\Delta \in \mathbb{R}^d \mid 2^{k-1} \le \frac{\Phi(\Delta)}{\|\Delta\|_2} \le 2^k \text{ and } 2^{\ell-1} R_\ell \le \|\Delta\|_2 \le 2^\ell R_\ell\right\}.$$

By construction, any vector that can possibly violate the bound (9.90) is contained in the union $\bigcup_{k=1}^{N_1} \bigcup_{\ell=1}^{N_2} \mathbb{S}_{k,\ell}$, where $N_1 := \lceil \log \sup_{\theta \neq 0} \frac{\Phi(\theta)}{\|\theta\|} \rceil$ and $N_2 := \lceil \log \frac{R_u}{R_\ell} \rceil$. Suppose that the bound (9.90) is violated by some $\widehat{\Delta} \in \mathbb{S}_{k,\ell}$. In this case, we have

$$\left|\mathcal{E}_n(\widehat{\Delta}) - \bar{\mathcal{E}}(\widehat{\Delta})\right| \ge 16L \frac{\Phi(\widehat{\Delta})}{\|\widehat{\Delta}\|_2} \|\widehat{\Delta}\|_2\, \delta \ge 16L 2^{k-1} 2^{\ell-1} R_\ell\, \delta = 4L 2^k 2^\ell R_\ell\, \delta,$$

which implies that $A_n(2^k, 2^\ell R_\ell) \ge \delta$. Consequently, we have shown that

$$\mathbb{P}[\mathcal{E}] \le \sum_{k=1}^{N_1} \sum_{\ell=1}^{N_2} \mathbb{P}[A_n(2^k, 2^\ell R_\ell) \ge \delta] \le N_1 N_2 \inf_{\lambda > 0} \mathbb{E}[e^{\lambda(\Phi^*(\bar{x}_n) - \delta)}],$$

where the final step follows by the union bound, and the tail bound (9.93). Given the upper bound $N_1 N_2 \le 4 \log(\sup_{\theta \neq 0} \frac{\Phi(\theta)}{\|\theta\|}) \log(\frac{R_u}{R_\ell}) = M_n(\Phi; R)$, the claim follows. □

9.8.2 A one-sided bound via truncation

In the previous section, we actually derived two-sided bounds on the difference between the empirical $\mathcal{E}_n$ and population $\bar{\mathcal{E}}$ form of the Taylor-series error. The resulting upper bounds on $\mathcal{E}_n$ guarantee a form of *restricted smoothness*, one which is useful in proving fast convergence rates of optimization algorithms. (See the bibliographic section for further details.) However, for proving bounds on the estimation error, as has been our focus in this chapter, it is only *restricted strong convexity*—that is, the lower bound on the Taylor-series error—that is required.

In this section, we show how a truncation argument can be used to derive restricted strong convexity for generalized linear models. Letting $\{\varepsilon_i\}_{i=1}^n$ denote an i.i.d. sequence of Rademacher variables, we define a complexity measure involving the dual norm Φ^*—namely

$$\mu_n(\Phi^*) := \mathbb{E}_{x,\varepsilon}\left[\Phi^*\left(\frac{1}{n}\sum_{i=1}^n \varepsilon_i x_i\right)\right] = \mathbb{E}\left[\sup_{\Phi(\Delta)\le 1} \frac{1}{n}\sum_{i=1}^n \varepsilon_i \langle \Delta,\, x_i\rangle\right].$$

This is simply the Rademacher complexity of the linear function class $x \mapsto \langle \Delta, x\rangle$ as Δ ranges over the unit ball of the norm Φ.

Our theory applies to covariates $\{x_i\}_{i=1}^n$ drawn i.i.d. from a zero-mean distribution such that, for some positive constants (α, β), we have

$$\mathbb{E}\left[\langle \Delta,\, x\rangle^2\right] \ge \alpha \quad \text{and} \quad \mathbb{E}\left[\langle \Delta,\, x\rangle^4\right] \le \beta \qquad \text{for all vectors } \Delta \in \mathbb{R}^d \text{ with } \|\Delta\|_2 = 1. \tag{9.94}$$

Theorem 9.36 *Consider any generalized linear model with covariates drawn from a zero-mean distribution satisfying the condition* (9.94). *Then the Taylor-series error* $\mathcal{E}_n$ *in the log-likelihood is lower bounded as*

$$\mathcal{E}_n(\Delta) \ge \frac{\kappa}{2}\|\Delta\|_2^2 - c_0\, \mu_n^2(\Phi^*)\, \Phi^2(\Delta) \qquad \textit{for all } \Delta \in \mathbb{R}^d \textit{ with } \|\Delta\|_2 \le 1 \tag{9.95}$$

with probability at least $1 - c_1 e^{-c_2 n}$.

In this statement, the constants (κ, c_0, c_1, c_2) can depend on the GLM, the fixed vector θ^* and (α, β), but are independent of dimension, sample size, and regularizer.

Proof Using a standard formula for the remainder in the Taylor series, we have

$$\mathcal{E}_n(\Delta) = \frac{1}{n}\sum_{i=1}^n \psi''\big(\langle \theta^*,\, x_i\rangle + t\langle \Delta,\, x_i\rangle\big)\langle \Delta,\, x_i\rangle^2,$$

for some scalar $t \in [0, 1]$. We proceed via a truncation argument. Fix some vector $\Delta \in \mathbb{R}^d$ with Euclidean norm $\|\Delta\|_2 = \delta \in (0, 1]$, and set $\tau = K\delta$ for a constant $K > 0$ to be chosen. Since the function $\varphi_\tau(u) = u^2\mathbb{I}[|u| \le 2\tau]$ lower bounds the quadratic and ψ'' is positive, we have

$$\mathcal{E}_n(\Delta) \ge \frac{1}{n}\sum_{i=1}^n \psi''\big(\langle \theta^*,\, x_i\rangle + t\langle \Delta,\, x_i\rangle\big)\, \varphi_\tau(\langle \Delta,\, x_i\rangle)\, \mathbb{I}[|\langle \theta^*,\, x_i\rangle| \le T], \tag{9.96}$$

where T is a second truncation parameter to be chosen. Since φ_τ vanishes outside the interval $[-2\tau, 2\tau]$ and $\tau \le K$, for any positive term in this sum, the absolute value $|\langle \theta^*,\, x_i\rangle + t\langle \Delta,\, x_i\rangle|$ is at most $T + 2K$, and hence

$$\mathcal{E}_n(\Delta) \ge \gamma\, \frac{1}{n}\sum_{i=1}^n \varphi_\tau(\langle \Delta,\, x_i\rangle)\; \mathbb{I}[|\langle \theta^*,\, x_i\rangle| \le T] \qquad \text{where } \gamma := \min_{|u|\le T+2K} \psi''(u).$$

Based on this lower bound, it suffices to show that for all $\delta \in (0, 1]$ and for $\Delta \in \mathbb{R}^d$ with $\|\Delta\|_2 = \delta$, we have

$$\frac{1}{n}\sum_{i=1}^n \varphi_{\tau(\delta)}(\langle \Delta,\, x_i\rangle)\; \mathbb{I}[|\langle \theta^*,\, x_i\rangle| \le T] \ge c_3\delta^2 - c_4\mu_n(\Phi^*)\Phi(\Delta)\,\delta. \tag{9.97}$$

When this bound holds, then inequality (9.95) holds with constants (κ, c_0) depending on (c_3, c_4, γ). Moreover, we claim that the problem can be reducing to proving the bound (9.97) for $\delta = 1$. Indeed, given any vector with Euclidean norm $\|\Delta\|_2 = \delta > 0$, we can apply the bound (9.97) to the rescaled unit-norm vector Δ/δ to obtain

$$\frac{1}{n}\sum_{i=1}^{n} \varphi_{\tau(1)}(\langle \Delta/\delta,\, x_i\rangle)\, \mathbb{I}[|\langle \theta^*,\, x_i\rangle| \leq T] \geq c_3\left\{1 - c_4\mu_n(\Phi^*)\frac{\Phi(\Delta)}{\delta}\right\},$$

where $\tau(1) = K$, and $\tau(\delta) = K\delta$. Noting that $\varphi_{\tau(1)}(u/\delta) = (1/\delta)^2\varphi_{\tau(\delta)}(u)$, the claim follows by multiplying both sides by δ^2. Thus, the remainder of our proof is devoted to proving (9.97) with $\delta = 1$. In fact, in order to make use of a contraction argument for Lipschitz functions, it is convenient to define a new truncation function

$$\widetilde{\varphi}_\tau(u) = u^2\, \mathbb{I}[|u| \leq \tau] + (u - 2\tau)^2\, \mathbb{I}[\tau < u \leq 2\tau] + (u + 2\tau)^2\, \mathbb{I}[-2\tau \leq u < -\tau].$$

Note that it is Lipschitz with parameter 2τ. Since $\widetilde{\varphi}_\tau$ lower bounds φ_τ, it suffices to show that for all unit-norm vectors Δ, we have

$$\frac{1}{n}\sum_{i=1}^{n} \widetilde{\varphi}_\tau\big(\langle \Delta,\, x_i\rangle\big)\, \mathbb{I}[|\langle \theta^*,\, x_i\rangle| \leq T] \geq c_3 - c_4\mu_n(\Phi^*)\Phi(\Delta). \tag{9.98}$$

For a given radius $r \geq 1$, define the random variable

$$Z_n(r) := \sup_{\substack{\|\Delta\|_2 = 1\\ \Phi(\Delta) \leq r}} \left|\frac{1}{n}\sum_{i=1}^{n} \widetilde{\varphi}_\tau(\langle \Delta,\, x_i\rangle)\, \mathbb{I}[|\langle \theta^*,\, x_i\rangle| \leq T] - \mathbb{E}[\widetilde{\varphi}_\tau(\langle \Delta,\, x\rangle)\mathbb{I}[|\langle \theta^*,\, x\rangle| \leq T]]\right|.$$

Suppose that we can prove that

$$\mathbb{E}\Big[\widetilde{\varphi}_\tau(\langle \Delta,\, x\rangle)\mathbb{I}[|\langle \theta^*,\, x\rangle| \leq T]\Big] \geq \frac{3}{4}\alpha \tag{9.99a}$$

and

$$\mathbb{P}\Big[Z_n(r) > \alpha/2 + c_4 r\mu_n(\Phi^*)\Big] \leq \exp\left(-c_2\frac{nr^2\mu_n^2(\Phi^*)}{\sigma^2} - c_2 n\right). \tag{9.99b}$$

The bound (9.98) with $c_3 = \alpha/4$ then follows for all vectors with unit Euclidean norm and $\Phi(\Delta) \leq r$. Accordingly, we prove the bounds (9.99a) and (9.99b) here for a fixed radius r. A peeling argument can be used to extend it to all radii, as in the proof of Theorem 9.34, with the probability still upper bounded by $c_1 e^{-c_2 n}$.

Proof of the expectation bound (9.99a)*:* We claim that it suffices to show that

$$\mathbb{E}\Big[\widetilde{\varphi}_\tau(\langle \Delta,\, x\rangle)\Big] \overset{\text{(i)}}{\geq} \frac{7}{8}\alpha, \quad \text{and} \quad \mathbb{E}\Big[\widetilde{\varphi}_\tau(\langle \Delta,\, x\rangle)\, \mathbb{I}[|\langle \theta^*,\, x\rangle| > T]\Big] \overset{\text{(ii)}}{\leq} \frac{1}{8}\alpha.$$

Indeed, if these two inequalities hold, then we have

$$\begin{aligned}\mathbb{E}[\widetilde{\varphi}_\tau(\langle \Delta,\, x\rangle)\mathbb{I}[|\langle \theta^*,\, x\rangle| \leq T]] &= \mathbb{E}[\widetilde{\varphi}_\tau(\langle \Delta,\, x\rangle)] - \mathbb{E}[\widetilde{\varphi}_\tau(\langle \Delta,\, x\rangle)\mathbb{I}[|\langle \theta^*,\, x\rangle| > T]]\\ &\geq \left\{\frac{7}{8} - \frac{1}{8}\right\}\alpha = \frac{3}{4}\alpha.\end{aligned}$$

We now prove inequalities (i) and (ii). Beginning with inequality (i), we have

$$\mathbb{E}[\widetilde{\varphi}_\tau(\langle \Delta,\, x\rangle)] \geq \mathbb{E}\Big[\langle \Delta,\, x\rangle^2\, \mathbb{I}[|\langle \Delta,\, x\rangle| \leq \tau]\Big] = \mathbb{E}[\langle \Delta,\, x\rangle^2] - \mathbb{E}\Big[\langle \Delta,\, x\rangle^2\, \mathbb{I}[|\langle \Delta,\, x\rangle| > \tau]\Big]$$
$$\geq \alpha - \mathbb{E}\Big[\langle \Delta,\, x\rangle^2\, \mathbb{I}[|\langle \Delta,\, x\rangle| > \tau]\Big],$$

so that it suffices to show that the last term is at most $\alpha/8$. By the condition (9.94) and Markov's inequality, we have

$$\mathbb{P}[|\langle \Delta,\, x\rangle| > \tau] \leq \frac{\mathbb{E}[\langle \Delta,\, x\rangle^4]}{\tau^4} \leq \frac{\beta}{\tau^4}$$

and

$$\mathbb{E}[\langle \Delta,\, x\rangle^4] \leq \beta.$$

Recalling that $\tau = K$ when $\delta = 1$, applying the Cauchy–Schwarz inequality yields

$$\mathbb{E}\Big[\langle \Delta,\, x\rangle^2\, \mathbb{I}[|\langle \Delta,\, x\rangle| > \tau]\Big] \leq \sqrt{\mathbb{E}[\langle \Delta,\, x\rangle^4]}\, \sqrt{\mathbb{P}[|\langle \Delta,\, x\rangle| > \tau]} \leq \frac{\beta}{K^2},$$

so that setting $K^2 = 8\beta/\alpha$ guarantees an upper bound of $\alpha/8$, which in turn implies inequality (i) by our earlier reasoning.

Turning to inequality (ii), since

$$\widetilde{\varphi}_\tau(\langle \Delta,\, x\rangle) \leq \langle \Delta,\, x\rangle^2 \quad \text{and} \quad \mathbb{P}[|\langle \theta^*,\, x\rangle| \geq T] \leq \frac{\beta\|\theta^*\|_2^4}{T^4},$$

the Cauchy–Schwarz inequality implies that

$$\mathbb{E}[\widetilde{\varphi}_\tau(\langle \Delta,\, x\rangle)\mathbb{I}[|\langle \theta^*,\, x\rangle| > T]] \leq \frac{\beta\|\theta^*\|_2^2}{T^2}.$$

Thus, setting $T^2 = 8\beta\|\theta^*\|_2^2/\alpha$ guarantees inequality (ii).

Proof of the tail bound (9.99b)*:* By our choice $\tau = K$, the empirical process defining $Z_n(r)$ is based on functions bounded in absolute value by K^2. Thus, the functional Hoeffding inequality (Theorem 3.26) implies that

$$\mathbb{P}[Z_n(r) \geq \mathbb{E}[Z_n(r)] + r\mu_n(\Phi^*) + \alpha/2] \leq e^{-c_2 n r^2 \mu_n^2(\Phi^*) - c_2 n}.$$

As for the expectation, letting $\{\varepsilon_i\}_{i=1}^n$ denote an i.i.d. sequence of Rademacher variables, the usual symmetrization argument (Proposition 4.11) implies that

$$\mathbb{E}[Z_n(r)] \leq 2 \sup \mathbb{E}_{x,\varepsilon}\Bigg[\sup_{\substack{\|\Delta\|_2=1\\ \Phi(\Delta)\leq r}} \Big|\frac{1}{n}\sum_{i=1}^n \varepsilon_i \widetilde{\varphi}_\tau(\langle \Delta,\, x_i\rangle)\, \mathbb{I}[|\langle \theta^*,\, x_i\rangle| \leq T]\Big|\Bigg].$$

Since $\mathbb{I}[|\langle \theta^*,\, x_i\rangle| \leq T] \leq 1$ and $\widetilde{\varphi}_\tau$ is Lipschitz with parameter $2K$, the contraction principle yields

$$\mathbb{E}[Z_n(r)] \leq 8K\, \mathbb{E}_{x,\varepsilon}\Bigg[\sup_{\substack{\|\Delta\|_2=1\\ \Phi(\Delta)\leq r}} \Big|\frac{1}{n}\sum_{i=1}^n \varepsilon_i \langle \Delta,\, x_i\rangle\Big|\Bigg] \leq 8Kr\, \mathbb{E}\Big[\Phi^*\Big(\frac{1}{n}\sum_{i=1}^n \varepsilon_i x_i\Big)\Big],$$

where the final step follows by applying Hölder's inequality using Φ and its dual Φ^*. □

9.9 Appendix: Star-shaped property

Recall the set $\mathbb{C}$ previously defined in Proposition 9.13. In this appendix, we prove that $\mathbb{C}$ is star-shaped around the origin, meaning that if $\Delta \in \mathbb{C}$, then $t\Delta \in \mathbb{C}$ for all $t \in [0, 1]$. This property is immediate whenever $\theta^* \in \mathbb{M}$, since $\mathbb{C}$ is then a cone, as illustrated in Figure 9.7(a). Now consider the general case, when $\theta^* \notin \mathbb{M}$. We first observe that for any $t \in (0, 1]$,

$$\Pi_{\bar{\mathbb{M}}}(t\Delta) = \arg\min_{\theta \in \bar{\mathbb{M}}} \|t\Delta - \theta\| = t \arg\min_{\theta \in \bar{\mathbb{M}}} \left\|\Delta - \frac{\theta}{t}\right\| = t\,\Pi_{\bar{\mathbb{M}}}(\Delta),$$

using the fact that θ/t also belongs to the subspace $\bar{\mathbb{M}}$. A similar argument can be used to establish the equality $\Pi_{\bar{\mathbb{M}}^\perp}(t\Delta) = t\Pi_{\bar{\mathbb{M}}^\perp}(\Delta)$. Consequently, for all $\Delta \in \mathbb{C}$, we have

$$\begin{aligned}\Phi(\Pi_{\bar{\mathbb{M}}^\perp}(t\Delta)) = \Phi(t\Pi_{\bar{\mathbb{M}}^\perp}(\Delta)) &\overset{\text{(i)}}{=} t\,\Phi(\Pi_{\bar{\mathbb{M}}^\perp}(\Delta)) \\ &\overset{\text{(ii)}}{\leq} t\,\{3\,\Phi(\Pi_{\bar{\mathbb{M}}}(\Delta)) + 4\Phi(\theta^*_{\mathbb{M}^\perp})\},\end{aligned}$$

where step (i) uses the fact that any norm is positive homogeneous,[4] and step (ii) uses the inclusion $\Delta \in \mathbb{C}$. We now observe that $3\,t\,\Phi(\Pi_{\bar{\mathbb{M}}}(\Delta)) = 3\,\Phi(\Pi_{\bar{\mathbb{M}}}(t\Delta))$, and moreover, since $t \in (0, 1]$, we have $4t\,\Phi(\theta^*_{\mathbb{M}^\perp}) \leq 4\Phi(\theta^*_{\mathbb{M}^\perp})$. Putting together the pieces, we find that

$$\Phi(\Pi_{\bar{\mathbb{M}}^\perp}(t\Delta)) \leq 3\,\Phi(\Pi_{\bar{\mathbb{M}}}(t\Delta)) + 4\,t\Phi(\theta^*_{\mathbb{M}^\perp}) \leq 3\,\Phi(\Pi_{\bar{\mathbb{M}}}(t\Delta)) + 4\Phi(\theta^*_{\mathbb{M}^\perp}),$$

showing that $t\Delta \in \mathbb{C}$ for all $t \in (0, 1]$, as claimed.

9.10 Bibliographic details and background

The definitions of decomposable regularizers and restricted strong convexity were introduced by Negahban et al. (2012), who first proved a version of Theorem 9.19. Restricted strong convexity is the natural generalization of a restricted eigenvalue to the setting of general (potentially non-quadratic) cost functions, and general decomposable regularizers. A version of Theorem 9.36 was proved in the technical report (Negahban et al., 2010) for the ℓ_1-norm; note that this result allows for the second derivative ψ'' to be unbounded, as in the Poisson case. The class of decomposable regularizers includes the atomic norms studied by Chandrasekaran et al. (2012a), whereas van de Geer (2014) introduced a generalization known as weakly decomposable regularizers.

The argument used in the proof of Theorem 9.19 exploits ideas from Ortega and Rheinboldt (2000) as well as Rothman et al. (2008), who first derived Frobenius norm error bounds on the graphical Lasso (9.12). See Chapter 11 for a more detailed discussion of the graphical Lasso, and related problems concerning graphical models. The choice of regularizer defining the "good" event $\mathbb{G}(\lambda_n)$ in Proposition 9.13 is known as the *dual norm bound*. It is a cleanly stated and generally applicable choice, sharp for many (but not all) problems. See Exercise 7.15 as well as Chapter 13 for a discussion of instances in which it can be improved. These types of dual-based quantities also arise in analyses of exact recovery based on random projections; see the papers by Mendelson et al. (2007) and Chandrasekaran et al. (2012a) for geometric perspectives of this type.

The ℓ_1/ℓ_2 group Lasso norm from Example 9.3 was introduced by Yuan and Lin (2006);

[4] Explicitly, for any norm and non-negative scalar t, we have $\|tx\| = t\|x\|$.

see also Kim et al. (2006). As a convex program, it is a special case of second-order cone program (SOCP), for which there are various efficient algorithms (Bach et al., 2012; Boyd and Vandenberghe, 2004). Turlach et al. (2005) studied the ℓ_1/ℓ_∞ version of the group Lasso norm. Several groups (Zhao et al., 2009; Baraniuk et al., 2010) have proposed unifying frameworks that include these group-structured norms as particular cases. See Bach et al. (2012) for discussion of algorithmic issues associated with optimization involving group sparse penalties. Jacob et al. (2009) introduced the overlapping group Lasso norm discussed in Example 9.4, and provide detailed discussion of why the standard group Lasso norm with overlap fails to select unions of groups. A number of authors have investigated the statistical benefits of the group Lasso versus the ordinary Lasso when the underlying regression vector is group-sparse; for instance, Obozinski et al. (2011) study the problem of variable selection, whereas the papers (Baraniuk et al., 2010; Huang and Zhang, 2010; Lounici et al., 2011) provide guarantees on the estimation error. Negahban and Wainwright (2011a) study the variable selection properties of ℓ_1/ℓ_∞-regularization for multivariate regression, and show that, while it can be more statistically efficient than ℓ_1-regularization with complete shared overlap, this gain is surprisingly non-robust: it is very easy to construct examples in which it is outperformed by the ordinary Lasso. Motivated by this deficiency, Jalali et al. (2010) study a decomposition-based estimator, in which the multivariate regression matrix is decomposed as the sum of an elementwise-sparse and row-sparse matrix (as in Section 9.7), and show that it adapts in the optimal way. The adaptive guarantee given in Corollary 9.31 is of a similar flavor, but as applied to the estimation error as opposed to variable selection.

Convex relaxations based on nuclear norm introduced in Example 9.8 have been the focus of considerable research; see Chapter 10 for an in-depth discussion.

The Φ^*-norm restricted curvature conditions discussed in Section 9.3 are a generalization of the notion of ℓ_∞-restricted eigenvalues (van de Geer and Bühlmann, 2009; Ye and Zhang, 2010; Bühlmann and van de Geer, 2011). See Exercises 7.13, 7.14 and 9.11 for some analysis of these ℓ_∞-RE conditions for the usual Lasso, and Exercise 9.14 for some analysis for Lipschitz cost functions. Section 10.2.3 provides various applications of this condition to nuclear norm regularization.

9.11 Exercises

Exercise 9.1 (Overlapping group Lasso) Show that the overlap group Lasso, as defined by the variational representation (9.10), is a valid norm.

Exercise 9.2 (Subspace projection operator) Recall the definition (9.20) of the subspace projection operator. Compute an explicit form for the following subspaces:

(a) For a fixed subset $S \subseteq \{1, 2, \ldots, d\}$, the subspace of vectors

$$\mathbb{M}(S) := \{\theta \in \mathbb{R}^d \mid \theta_j = 0 \quad \text{for all } j \notin S\}.$$

(b) For a given pair of r-dimensional subspaces $\mathbb{U}$ and $\mathbb{V}$, the subspace of matrices

$$\mathbb{M}(\mathbb{U}, \mathbb{V}) := \{\mathbf{\Theta} \in \mathbb{R}^{d \times d} \mid \text{rowspan}(\mathbf{\Theta}) \subseteq \mathbb{U},\ \text{colspan}(\mathbf{\Theta}) \subseteq \mathbb{V}\},$$

where rowspan($\mathbf{\Theta}$) and colspan($\mathbf{\Theta}$) denote the row and column spans of $\mathbf{\Theta}$.

Exercise 9.3 (Generalized linear models) This exercise treats various cases of the generalized linear model.

(a) Suppose that we observe samples of the form $y = \langle x, \theta \rangle + w$, where $w \sim N(0, \sigma^2)$. Show that the conditional distribution of y given x is of the form (9.5) with $c(\sigma) = \sigma^2$ and $\psi(t) = t^2/2$.
(b) Suppose that y is (conditionally) Poisson with mean $\lambda = e^{\langle x, \theta \rangle}$. Show that this is a special case of the log-linear model (9.5) with $c(\sigma) \equiv 1$ and $\psi(t) = e^t$.

Exercise 9.4 (Dual norms) In this exercise, we study various forms of dual norms.

(a) Show that the dual norm of the ℓ_1-norm is the ℓ_∞-norm.
(b) Consider the general group Lasso norm

$$\Phi(u) = \|u\|_{1,\mathcal{G}(p)} = \sum_{g \in \mathcal{G}} \|u_g\|_p,$$

where $p \in [1, \infty]$ is arbitrary, and the groups are non-overlapping. Show that its dual norm takes the form

$$\Phi^*(v) = \|v\|_{\infty,\mathcal{G}(q)} = \max_{g \in \mathcal{G}} \|v_g\|_q,$$

where $q = \frac{p}{p-1}$ is the conjugate exponent to p.
(c) Show that the dual norm of the nuclear norm is the ℓ_2-operator norm

$$\Phi^*(\mathbf{N}) = |||\mathbf{N}|||_2 := \sup_{\|z\|_2=1} \|\mathbf{N}z\|_2.$$

(*Hint:* Try to reduce the problem to a version of part (a).)

Exercise 9.5 (Overlapping group norm and duality) Let $p \in [1, \infty]$, and recall the overlapping group norm (9.10).

(a) Show that it has the equivalent representation

$$\Phi(u) = \max_{v \in \mathbb{R}^d} \langle v, u \rangle \quad \text{such that } \|v_g\|_q \le 1 \text{ for all } g \in \mathcal{G},$$

where $q = \frac{p}{p-1}$ is the dual exponent.
(b) Use part (a) to show that its dual norm is given by

$$\Phi^*(v) = \max_{g \in \mathcal{G}} \|v_g\|_q.$$

Exercise 9.6 (Boundedness of subgradients in the dual norm) Let $\Phi : \mathbb{R}^d \to \mathbb{R}$ be a norm, and $\theta \in \mathbb{R}^d$ be arbitrary. For any $z \in \partial\Phi(\theta)$, show that $\Phi^*(z) \le 1$.

Exercise 9.7 (Hölder's inequality) Let $\Phi : \mathbb{R}^d \to \mathbb{R}_+$ be a norm, and let $\Phi^* : \mathbb{R}^d \to \mathbb{R}_+$ be its dual norm.

(a) Show that $\left|\langle u, v \rangle\right| \le \Phi(u)\, \Phi^*(v)$ for all $u, v \in \mathbb{R}^d$.

(b) Use part (a) to prove Hölder's inequality for ℓ_p-norms, namely

$$\left|\langle u,\, v\rangle\right| \leq \|u\|_p\, \|v\|_q,$$

where the exponents (p, q) satisfy the conjugate relation $1/p + 1/q = 1$.

(c) Let $\mathbf{Q} \succ 0$ be a positive definite symmetric matrix. Use part (a) to show that

$$\left|\langle u,\, v\rangle\right| \leq \sqrt{u^{\mathrm{T}}\mathbf{Q}u}\, \sqrt{v^{\mathrm{T}}\mathbf{Q}^{-1}v} \quad \text{for all } u, v \in \mathbb{R}^d.$$

Exercise 9.8 (Complexity parameters) This exercise concerns the complexity parameter $\mu_n(\Phi^*)$ previously defined in equation (9.41). Suppose throughout that the covariates $\{x_i\}_{i=1}^n$ are drawn i.i.d., each sub-Gaussian with parameter σ.

(a) Consider the group Lasso norm (9.9) with group set $\mathcal{G}$ and maximum group size m. Show that

$$\mu_n(\Phi^*) \precsim \sigma\sqrt{\frac{m}{n}} + \sigma\sqrt{\frac{\log|\mathcal{G}|}{n}}.$$

(b) For the nuclear norm on the space of $d_1 \times d_2$ matrices, show that

$$\mu_n(\Phi^*) \precsim \sigma\sqrt{\frac{d_1}{n}} + \sigma\sqrt{\frac{d_2}{n}}.$$

Exercise 9.9 (Equivalent forms of strong convexity) Suppose that a differentiable function $f : \mathbb{R}^d \to \mathbb{R}$ is κ-strongly convex in the sense that

$$f(y) \geq f(x) + \langle\nabla f(x),\, y - x\rangle + \frac{\kappa}{2}\|y - x\|_2^2 \qquad \text{for all } x, y \in \mathbb{R}^d. \tag{9.100a}$$

Show that

$$\langle\nabla f(y) - \nabla f(x),\, y - x\rangle \geq \kappa\|y - x\|_2^2 \qquad \text{for all } x, y \in \mathbb{R}^d. \tag{9.100b}$$

Exercise 9.10 (Implications of local strong convexity) Suppose that $f : \mathbb{R}^d \to \mathbb{R}$ is a twice differentiable, convex function that is *locally* κ-strongly convex around x, in the sense that the lower bound (9.100a) holds for all vectors z in the ball $\mathbb{B}_2(x) := \{z \in \mathbb{R}^d \mid \|z - x\|_2 \leq 1\}$. Show that

$$\langle\nabla f(y) - \nabla f(x),\, y - x\rangle \geq \kappa\|y - x\|_2 \quad \text{for all } y \in \mathbb{R}^d \backslash \mathbb{B}_2(x).$$

Exercise 9.11 (ℓ_∞-curvature and RE conditions) In this exercise, we explore the link between the ℓ_∞-curvature condition (9.56) and the ℓ_∞-RE condition (9.57). Suppose that the bound (9.56) holds with $\tau_n = c_1\sqrt{\frac{\log d}{n}}$. Show that the bound (9.57) holds with $\kappa' = \frac{\kappa}{2}$ as long as $n > c_2|S|^2 \log d$ with $c_2 = \frac{4c_1^2(1+\alpha)^4}{\kappa^2}$.

Exercise 9.12 (ℓ_1-regularization and soft thresholding) Given observations from the linear model $y = \mathbf{X}\theta^* + w$, consider the M-estimator

$$\widehat{\theta} = \arg\min_{\theta\in\mathbb{R}^d}\left\{\frac{1}{2}\|\theta\|_2^2 - \left\langle\theta,\, \frac{1}{n}\mathbf{X}^{\mathrm{T}}y\right\rangle + \lambda_n\|\theta\|_1\right\}.$$

(a) Show that the optimal solution is always unique, and given by $\widehat{\theta} = T_{\lambda_n}(\frac{1}{n}\mathbf{X}^{\mathrm{T}}y)$, where the soft-thresholding operator T_{λ_n} was previously defined (7.6b).

(b) Now suppose that θ^* is s-sparse. Show that if

$$\lambda_n \geq 2\Big\{\Big\|\Big(\frac{\mathbf{X}^{\mathrm{T}}\mathbf{X}}{n} - \mathbf{I}_d\Big)\theta^*\Big\|_\infty + \Big\|\frac{\mathbf{X}^{\mathrm{T}}w}{n}\Big\|_\infty\Big\},$$

then the optimal solution satisfies the bound $\|\widehat{\theta} - \theta^*\|_2 \leq \frac{3}{2}\sqrt{s}\lambda_n$.

(c) Now suppose that the covariates $\{x_i\}_{i=1}^n$ are drawn i.i.d. from a zero-mean ν-sub-Gaussian ensemble with covariance $\mathrm{cov}(x_i) = \mathbf{I}_d$, and the noise vector w is bounded as $\|w\|_2 \leq b\sqrt{n}$ for some $b > 0$. Show that with an appropriate choice of λ_n, we have

$$\|\widehat{\theta} - \theta^*\|_2 \leq 3\nu\,(\nu\|\theta^*\|_2 + b)\,\sqrt{s}\,\Big\{\sqrt{\frac{\log d}{n}} + \delta\Big\}$$

with probability at least $1 - 4e^{-\frac{n\delta^2}{8}}$ for all $\delta \in (0, 1)$.

Exercise 9.13 (From ℓ_∞ to $\{\ell_1, \ell_2\}$-bounds) In the setting of Corollary 9.27, show that any optimal solution $\widehat{\theta}$ that satisfies the ℓ_∞-bound (9.65) also satisfies the following ℓ_1- and ℓ_2-error bounds

$$\|\widehat{\theta} - \theta^*\|_1 \leq \frac{24\sigma}{\kappa} s\sqrt{\frac{\log d}{n}} \quad \text{and} \quad \|\widehat{\theta} - \theta^*\|_2 \leq \frac{12\sigma}{\kappa}\sqrt{\frac{s\log d}{n}}.$$

(*Hint:* Proposition 9.13 is relevant here.)

Exercise 9.14 (ℓ_∞-curvature for Lipschitz cost functions) In the setting of regression-type data $z = (x, y) \in \mathcal{X} \times \mathcal{Y}$, consider a cost function whose gradient is elementwise L-Lipschitz: i.e., for any sample z and pair $\theta, \widetilde{\theta}$, the jth partial derivative satisfies

$$\Big|\frac{\partial\mathcal{L}(\theta; z_i)}{\theta_j} - \frac{\partial\mathcal{L}(\widetilde{\theta}; z_i)}{\theta_j}\Big| \leq L\big|x_{ij}\big\langle x_i,\, \theta - \widetilde{\theta}\big\rangle\big|. \tag{9.101}$$

The goal of this exercise is to show that such a function satisfies an ℓ_∞-curvature condition similar to equation (9.64), as required for applying Corollary 9.27.

(a) Show that for any GLM whose cumulant function has a uniformly bounded second derivative ($\|\psi''\|_\infty \leq B^2$), the elementwise Lipschitz condition (9.101) is satisfied with $L = \frac{B^2}{2}$.

(b) For a given radius $r > 0$ and ratio $\rho > 0$, define the set

$$\mathbb{T}(R; \rho) := \Big\{\Delta \in \mathbb{R}^d \;\Big|\; \frac{\|\Delta\|_1}{\|\Delta\|_\infty} \leq \rho, \text{ and } \|\Delta\|_\infty \leq r\Big\},$$

and consider the random vector $V \in \mathbb{R}^d$ with elements

$$V_j := \frac{1}{4Lr\rho}\sup_{\Delta\in\mathbb{T}(r;\rho)}\Big|\frac{1}{n}\sum_{i=1}^n f_j(\Delta; z_i)\Big|, \quad \text{for } j = 1, \ldots, d,$$

where, for each fixed vector Δ,

$$f_j(\Delta; z_i) := \Big\{\frac{\partial\mathcal{L}(\theta^* + \Delta; z_i)}{\theta_j} - \frac{\partial\mathcal{L}(\theta^*; z_i)}{\theta_j}\Big\} - \Big\{\frac{\partial\bar{\mathcal{L}}(\theta^* + \Delta)}{\theta_j} - \frac{\partial\bar{\mathcal{L}}(\theta^*)}{\theta_j}\Big\}$$

is a zero-mean random variable. For each $\lambda > 0$, show that

$$\mathbb{E}_x[e^{\lambda \|V\|_\infty}] \le d\, \mathbb{E}_{x,\varepsilon}\left[\exp\left(\lambda \left\| \frac{1}{n}\sum_{i=1}^{n} \varepsilon_i x_i x_i^{\mathrm{T}} \right\|_\infty\right)\right].$$

(c) Suppose that the covariates $\{x_i\}_{i=1}^n$ are sampled independently, with each x_{ij} following a zero-mean σ-sub-Gaussian distribution. Show that for all $t \in (0, \sigma^2)$,

$$\mathbb{P}[\|V\|_\infty \ge t] \le 2d^2 e^{-\frac{nt^2}{2\sigma^4}}.$$

(d) Suppose that the population function $\bar{\mathcal{L}}$ satisfies the ℓ_∞- curvature condition

$$\|\nabla \bar{\mathcal{L}}(\theta^* + \Delta) - \nabla \bar{\mathcal{L}}(\theta^*)\|_\infty \ge \kappa \|\Delta\|_\infty \quad \text{for all } \Delta \in \mathbb{T}(r; \rho).$$

Use this condition and the preceding parts to show that

$$\|\nabla \mathcal{L}_n(\theta^* + \Delta) - \nabla \mathcal{L}_n(\theta^*)\|_\infty \ge \kappa \|\Delta\|_\infty - 16\, L \sigma^2 \sqrt{\frac{\log d}{n}} \rho\, r \qquad \text{for all } \Delta \in \mathbb{T}(r; \rho)$$

with probability at least $1 - e^{-4 \log d}$.

10

Matrix estimation with rank constraints

In Chapter 8, we discussed the problem of principal component analysis, which can be understood as a particular type of low-rank estimation problem. In this chapter, we turn to other classes of matrix problems involving rank and other related constraints. We show how the general theory of Chapter 9 can be brought to bear in a direct way so as to obtain theoretical guarantees for estimators based on nuclear norm regularization, as well as various extensions thereof, including methods for additive matrix decomposition.

10.1 Matrix regression and applications

In previous chapters, we have studied various forms of vector-based regression, including standard linear regression (Chapter 7) and extensions based on generalized linear models (Chapter 9). As suggested by its name, matrix regression is the natural generalization of such vector-based problems to the matrix setting. The analog of the Euclidean inner product on the matrix space $\mathbb{R}^{d_1 \times d_2}$ is the *trace inner product*

$$\langle\langle \mathbf{A}, \mathbf{B} \rangle\rangle := \operatorname{trace}(\mathbf{A}^T \mathbf{B}) = \sum_{j_1=1}^{d_1} \sum_{j_2=1}^{d_2} A_{j_1 j_2} B_{j_1 j_2}. \tag{10.1}$$

This inner product induces the *Frobenius norm* $|||\mathbf{A}|||_F = \sqrt{\sum_{j_1=1}^{d_1} \sum_{j_2=1}^{d_2} (A_{j_1 j_2})^2}$, which is simply the Euclidean norm on a vectorized version of the matrix.

In a matrix regression model, each observation takes the form $\mathbf{Z}_i = (\mathbf{X}_i, y_i)$, where $\mathbf{X}_i \in \mathbb{R}^{d_1 \times d_2}$ is a matrix of covariates, and $y_i \in \mathbb{R}$ is a response variable. As usual, the simplest case is the linear model, in which the response–covariate pair are linked via the equation

$$y_i = \langle\langle \mathbf{X}_i, \mathbf{\Theta}^* \rangle\rangle + w_i, \tag{10.2}$$

where w_i is some type of noise variable. We can also write this observation model in a more compact form by defining the *observation operator* $\mathfrak{X}_n \colon \mathbb{R}^{d_1 \times d_2} \to \mathbb{R}^n$ with elements $[\mathfrak{X}_n(\mathbf{\Theta})]_i = \langle\langle \mathbf{X}_i, \mathbf{\Theta} \rangle\rangle$, and then writing

$$y = \mathfrak{X}_n(\mathbf{\Theta}^*) + w, \tag{10.3}$$

where $y \in \mathbb{R}^n$ and $w \in \mathbb{R}^n$ are the vectors of response and noise variables, respectively. The adjoint of the observation operator, denoted $\mathfrak{X}_n^*$, is the linear mapping from $\mathbb{R}^n$ to $\mathbb{R}^{d_1 \times d_2}$ given by $u \mapsto \sum_{i=1}^n u_i \mathbf{X}_i$. Note that the operator $\mathfrak{X}_n$ is the natural generalization of the design matrix $\mathbf{X}$, viewed as a mapping from $\mathbb{R}^d$ to $\mathbb{R}^n$ in the usual setting of vector regression.

As illustrated by the examples to follow, there are many applications in which the regression matrix $\mathbf{\Theta}^*$ is either low-rank, or well approximated by a low-rank matrix. Thus, if we were to disregard computational costs, an appropriate estimator would be a rank-penalized form of least squares. However, including a rank penalty makes this a non-convex form of least squares so that—apart from certain special cases—it is computationally difficult to solve. This obstacle motivates replacing the rank penalty with the nuclear norm, which leads to the convex program

$$\widehat{\mathbf{\Theta}} \in \arg \min_{\mathbf{\Theta} \in \mathbb{R}^{d_1 \times d_2}} \left\{ \frac{1}{2n} \|y - \mathfrak{X}_n(\mathbf{\Theta})\|_2^2 + \lambda_n |||\mathbf{\Theta}|||_{\text{nuc}} \right\}. \tag{10.4}$$

Recall that the nuclear norm of $\mathbf{\Theta}$ is given by the sum of its singular values—namely,

$$|||\mathbf{\Theta}|||_{\text{nuc}} = \sum_{j=1}^{d'} \sigma_j(\mathbf{\Theta}), \quad \text{where } d' = \min\{d_1, d_2\}. \tag{10.5}$$

See Example 9.8 for our earlier discussion of this matrix norm.

Let us illustrate these definitions with some examples, beginning with the problem of multivariate regression.

Example 10.1 (Multivariate regression as matrix regression) As previously introduced in Example 9.6, the multivariate regression observation model can be written as $\mathbf{Y} = \mathbf{Z}\mathbf{\Theta}^* + \mathbf{W}$, where $\mathbf{Z} \in \mathbb{R}^{n \times p}$ is the regression matrix, and $\mathbf{Y} \in \mathbb{R}^{n \times T}$ is the matrix of responses. The tth column $\Theta^*_{\bullet,t}$ of the $(p \times T)$-dimensional regression matrix $\mathbf{\Theta}^*$ can be thought of as an ordinary regression vector for the tth component of the response. In many applications, these vectors lie on or close to a low-dimensional subspace, which means that the matrix $\mathbf{\Theta}^*$ is low-rank, or well approximated by a low-rank matrix. A direct way of estimating $\mathbf{\Theta}^*$ would be via *reduced rank regression*, in which one minimizes the usual least-squares cost $|||\mathbf{Y} - \mathbf{Z}\mathbf{\Theta}|||_{\text{F}}^2$ while imposing a rank constraint directly on the regression matrix $\mathbf{\Theta}$. Even though this problem is non-convex due to the rank constraint, it is easily solvable in this special case; see the bibliographic section and Exercise 10.1 for further details. However, this ease of solution is very fragile and will no longer hold if other constraints, in addition to a bounded rank, are added. In such cases, it can be useful to apply nuclear norm regularization in order to impose a "soft" rank constraint.

Multivariate regression can be recast as a form of the matrix regression model (10.2) with $N = nT$ observations in total. For each $j = 1, \ldots, n$ and $\ell = 1, \ldots, T$, let $\mathbf{E}_{j\ell}$ be an $n \times T$ mask matrix, with zeros everywhere except for a one in position (j, ℓ). If we then define the matrix $\mathbf{X}_{j\ell} := \mathbf{Z}^{\mathrm{T}}\mathbf{E}_{j\ell} \in \mathbb{R}^{p \times T}$, the multivariate regression model is based on the $N = nT$ observations $(\mathbf{X}_{j\ell}, y_{j\ell})$, each of the form

$$y_{j\ell} = \langle\langle \mathbf{X}_{j\ell}, \mathbf{\Theta}^* \rangle\rangle + W_{j\ell}, \qquad \text{for } j = 1, \ldots, n \text{ and } \ell = 1, \ldots, T.$$

Consequently, multivariate regression can be analyzed via the general theory that we develop for matrix regression problems. ♣

Another example of matrix regression is the problem of matrix completion.

Example 10.2 (Low-rank matrix completion) Matrix completion refers to the problem of estimating an unknown matrix $\mathbf{\Theta}^* \in \mathbb{R}^{d_1 \times d_2}$ based on (noisy) observations of a subset of its entries. Of course, this problem is ill-posed unless further structure is imposed, and so there are various types of matrix completion problems, depending on this underlying structure. One possibility is that the unknown matrix has a low-rank, or more generally can be well approximated by a low-rank matrix.

As one motivating application, let us consider the "Netflix problem", in which the rows of $\mathbf{\Theta}^*$ correspond to people, and columns correspond to movies. Matrix entry $\Theta^*_{a,b}$ represents the rating assigned by person a (say "Alice") to a given movie b that she has seen. In this setting, the goal of matrix completion is to make recommendations to Alice—that is, to suggest other movies that she has not yet seen but would be to likely to rate highly. Given the large corpus of movies stored by Netflix, most entries of the matrix $\mathbf{\Theta}^*$ are unobserved, since any given individual can only watch a limited number of movies over his/her lifetime. Consequently, this problem is ill-posed without further structure. See Figure 10.1(a) for an illustration of this observation model. Empirically, if one computes the singular values of recommender matrices, such as those that arise in the Netflix problem, the singular value spectrum tends to exhibit a fairly rapid decay—although the matrix itself is not exactly low-rank, it can be well-approximated by a matrix of low rank. This phenomenon is illustrated for a portion of the Jester joke data set (Goldberg et al., 2001), in Figure 10.1(b).

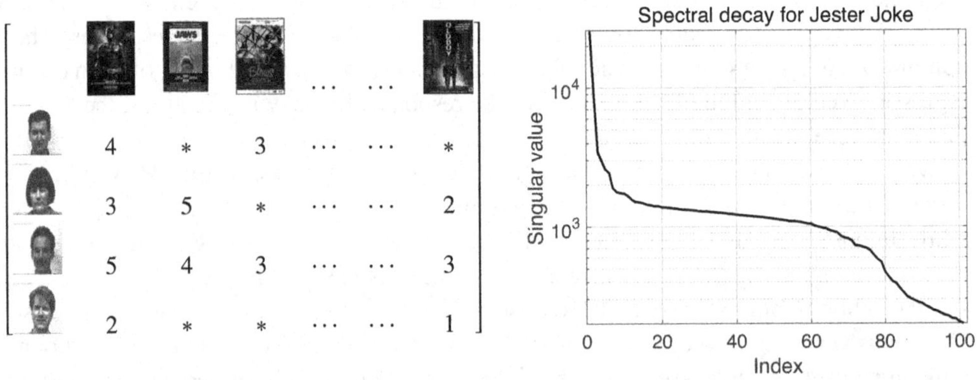

(a) (b)

Figure 10.1 (a) Illustration of the Netflix problem. Each user (rows of the matrix) rates a subset of movies (columns of the matrix) on a scale of 1 to 5. All remaining entries of the matrix are unobserved (marked with $*$), and the goal of matrix completion is to fill in these missing entries. (b) Plot of the singular values for a portion of the Jester joke data set (Goldberg et al., 2001), corresponding to ratings of jokes on a scale of $[-10, 10]$, and available at `http://eigentaste.berkeley.edu/`. Although the matrix is not exactly low-rank, it can be well approximated by a low-rank matrix.

In this setting, various observation models are possible, with the simplest being that we are given noiseless observations of a subset of the entries of $\mathbf{\Theta}^*$. A slightly more realistic model allows for noisiness—for instance, in the linear case, we might assume that

$$\widetilde{y}_i = \Theta_{a(i),b(i)} + \frac{w_i}{\sqrt{d_1 d_2}}, \tag{10.6}$$

where w_i is some form[1] of observation noise, and $(a(i), b(i))$ are the row and column indices of the ith observation.

How to reformulate the observations as an instance of matrix regression? For sample index i, define the mask matrix $\mathbf{X}_i \in \mathbb{R}^{d_1 \times d_2}$, which is zero everywhere *except* for position $(a(i), b(i))$, where it takes the value $\sqrt{d_1 d_2}$. Then by defining the rescaled observation $y_i := \sqrt{d_1 d_2}\, \widetilde{y}_i$, the observation model can be written in the trace regression form as

$$y_i = \langle\langle \mathbf{X}_i, \, \mathbf{\Theta}^* \rangle\rangle + w_i. \tag{10.7}$$

We analyze this form of matrix completion in the sequel.

Often, matrices might take on discrete values, such as for yes/no votes coded in the set $\{-1, 1\}$, or ratings belonging to some subset of the positive integers (e.g., $\{1, \ldots, 5\}$), in which case a generalized version of the basic linear model (10.6) would be appropriate. For instance, in order to model binary-valued responses $y \in \{-1, 1\}$, it could be appropriate to use the logistic model

$$\mathbb{P}(y_i \mid \mathbf{X}_i, \mathbf{\Theta}^*) = \frac{e^{y \langle\langle \mathbf{X}_i, \, \mathbf{\Theta}^* \rangle\rangle}}{1 + e^{y_i \langle\langle \mathbf{X}_i, \, \mathbf{\Theta}^* \rangle\rangle}}. \tag{10.8}$$

In this context, the parameter $\Theta^*_{a,b}$ is proportional to the log-odds ratio for whether user a likes (or dislikes) item b. ♣

We now turn to the matrix analog of the compressed sensing observation model, originally discussed in Chapter 7 for vectors. It is another special case of the matrix regression problem.

Example 10.3 (Compressed sensing for low-rank matrices) Working with the linear observation model (10.3), suppose that the design matrices $\mathbf{X}_i \in \mathbb{R}^{d_1 \times d_2}$ are drawn i.i.d from a random Gaussian ensemble. In the simplest of settings, the design matrix is chosen from the standard Gaussian ensemble, meaning that each of its $D = d_1 d_2$ entries is an i.i.d. draw from the $\mathcal{N}(0, 1)$ distribution. In this case, the random operator $\mathfrak{X}_n$ provides n random projections of the unknown matrix $\mathbf{\Theta}^*$—namely

$$y_i = \langle\langle \mathbf{X}_i, \, \mathbf{\Theta}^* \rangle\rangle \qquad \text{for } i = 1, \ldots, n. \tag{10.9}$$

In this noiseless setting, it is natural to ask how many such observations suffice to recover $\mathbf{\Theta}^*$ exactly. We address this question in Corollary 10.9 to follow in the sequel. ♣

The problem of signal phase retrieval leads to a variant of the low-rank compressed sensing problem:

Example 10.4 (Phase retrieval) Let $\theta^* \in \mathbb{R}^d$ be an unknown vector, and suppose that we make measurements of the form $\widetilde{y}_i = \left|\langle x_i, \, \theta^* \rangle\right|$ where $x_i \sim \mathcal{N}(0, \mathbf{I}_d)$ is a standard normal vector. This set-up is a real-valued idealization of the problem of phase retrieval in image

[1] Our choice of normalization by $1/\sqrt{d_1 d_2}$ is for later theoretical convenience, as clarified in the sequel—see equation (10.36).

processing, in which we observe the magnitude of complex inner products, and want the retrieve the phase of the associated complex vector. In this idealized setting, the "phase" can take only two possible values, namely the possible signs of $\langle x_i, \theta^* \rangle$.

A standard semidefinite relaxation is based on lifting the observation model to the space of matrices. Taking squares on both sides yields the equivalent observation model

$$\widetilde{y}_i^2 = (\langle x_i, \theta^* \rangle)^2 = \langle\langle x_i \otimes x_i, \theta^* \otimes \theta^* \rangle\rangle \qquad \text{for } i = 1, \ldots, n,$$

where $\theta^* \otimes \theta^* = \theta^*(\theta^*)^T$ is the rank-one outer product. By defining the scalar observation $y_i := \widetilde{y}_i^2$, as well as the matrices $\mathbf{X}_i := x_i \otimes x_i$ and $\mathbf{\Theta}^* := \theta^* \otimes \theta^*$, we obtain an equivalent version of the noiseless phase retrieval problem—namely, to find a rank-one solution to the set of matrix-linear equations $y_i = \langle\langle \mathbf{X}_i, \mathbf{\Theta}^* \rangle\rangle$ for $i = 1, \ldots, n$. This problem is nonconvex, but by relaxing the rank constraint to a nuclear norm constraint, we obtain a tractable semidefinite program (see equation (10.29) to follow).

Overall, the phase retrieval problem is a variant of the compressed sensing problem from Example 10.3, in which the random design matrices $\mathbf{X}_i$ are no longer Gaussian, but rather the outer product $x_i \otimes x_i$ of two Gaussian vectors. In Corollary 10.13 to follow, we show that the solution of the semidefinite relaxation coincides with the rank-constrained problem with high probability given $n \gtrsim d$ observations. ♣

Matrix estimation problems also arise in modeling of time series, where the goal is to describe the dynamics of an underlying process.

Example 10.5 (Time-series and vector autoregressive processes) A vector autoregressive (VAR) process in d dimensions consists of a sequence of d-dimensional random vectors $\{z^t\}_{t=1}^N$ that are generated by first choosing the random vector $z^1 \in \mathbb{R}^d$ according to some initial distribution, and then recursively setting

$$z^{t+1} = \mathbf{\Theta}^* z^t + w^t, \qquad \text{for } t = 1, 2, \ldots, N-1. \tag{10.10}$$

Here the sequence of d-dimensional random vectors $\{w^t\}_{t=1}^{N-1}$ forms the driving noise of the process; we model them as i.i.d. zero-mean random vectors with covariance $\mathbf{\Gamma} \succ 0$. Of interest to us is the matrix $\mathbf{\Theta}^* \in \mathbb{R}^{d \times d}$ that controls the dependence between successive samples of the process. Assuming that w^t is independent of z^t for each t, the covariance matrix $\mathbf{\Sigma}^t = \text{cov}(z^t)$ of the process evolves according to the recursion $\mathbf{\Sigma}^{t+1} := \mathbf{\Theta}^* \mathbf{\Sigma}^t (\mathbf{\Theta}^*)^T + \mathbf{\Gamma}$. Whenever $|||\mathbf{\Theta}^*|||_2 < 1$, it can be shown that the process is stable, meaning that the eigenvalues of $\mathbf{\Sigma}^t$ stay bounded independently of t, and the sequence $\{\mathbf{\Sigma}^t\}_{t=1}^\infty$ converges to a well-defined limiting object. (See Exercise 10.2.)

Our goal is to estimate the system parameters, namely the d-dimensional matrices $\mathbf{\Theta}^*$ and $\mathbf{\Gamma}$. When the noise covariance $\mathbf{\Gamma}$ is known and strictly positive definite, one possible estimator for $\mathbf{\Theta}^*$ is based on a sum of quadratic losses over successive samples—namely,

$$\mathcal{L}_n(\mathbf{\Theta}) = \frac{1}{2N} \sum_{t=1}^{N-1} \|z^{t+1} - \mathbf{\Theta} z^t\|_{\mathbf{\Gamma}^{-1}}^2, \tag{10.11}$$

where $\|a\|_{\mathbf{\Gamma}^{-1}} := \sqrt{\langle a, \mathbf{\Gamma}^{-1} a \rangle}$ is the quadratic norm defined by $\mathbf{\Gamma}$. When the driving noise w^t is zero-mean Gaussian with covariance $\mathbf{\Gamma}$, then this cost function is equivalent to the negative log-likelihood, disregarding terms not depending on $\mathbf{\Theta}^*$.

In many applications, among them subspace tracking and biomedical signal processing, the system matrix $\mathbf{\Theta}^*$ can be modeled as being low-rank, or well approximated by a low-rank matrix. In this case, the nuclear norm is again an appropriate choice of regularizer, and when combined with the loss function (10.11), we obtain another form of semidefinite program to solve.

Although different on the surface, this VAR observation model can be reformulated as a particular instance of the matrix regression model (10.2), in particular one with $n = d(N-1)$ observations in total. At each time $t = 2, \ldots, N$, we receive a total of d observations. Letting $e_j \in \mathbb{R}^d$ denote the canonical basis vector with a single one in position j, the jth observation in the block has the form

$$z_j^t = \langle e_j, z^t \rangle = \langle e_j, \mathbf{\Theta}^* z^{t-1} \rangle + w_j^{t-1} = \langle\!\langle e_j \otimes z^{t-1}, \mathbf{\Theta}^* \rangle\!\rangle + w_j^{t-1},$$

so that in the matrix regression observation model (10.2), we have $y_i = (z_t)_j$ and $\mathbf{X}_i = e_j \otimes z^{t-1}$ when i indexes the sample (t, j). ♣

10.2 Analysis of nuclear norm regularization

Having motivated problems of low-rank matrix regression, we now turn to the development and analysis of M-estimators based on nuclear norm regularization. Our goal is to bring to bear the general theory from Chapter 9. This general theory requires specification of certain subspaces over which the regularizer decomposes, as well as restricted strong convexity conditions related to these subspaces. This section is devoted to the development of these two ingredients in the special case of nuclear norm (10.5).

10.2.1 Decomposability and subspaces

We begin by developing appropriate choices of decomposable subspaces for the nuclear norm. For any given matrix $\mathbf{\Theta} \in \mathbb{R}^{d_1 \times d_2}$, we let rowspan$(\mathbf{\Theta}) \subseteq \mathbb{R}^{d_2}$ and colspan$(\mathbf{\Theta}) \subseteq \mathbb{R}^{d_1}$ denote its row space and column space, respectively. For a given positive integer $r \leq d' := \min\{d_1, d_2\}$, let $\mathbb{U}$ and $\mathbb{V}$ denote r-dimensional subspaces of vectors. We can then define the two subspaces of matrices

$$\mathbb{M}(\mathbb{U}, \mathbb{V}) := \{\mathbf{\Theta} \in \mathbb{R}^{d_1 \times d_2} \mid \text{rowspan}(\mathbf{\Theta}) \subseteq \mathbb{V}, \ \ \text{colspan}(\mathbf{\Theta}) \subseteq \mathbb{U}\} \tag{10.12a}$$

and

$$\bar{\mathbb{M}}^{\perp}(\mathbb{U}, \mathbb{V}) := \{\mathbf{\Theta} \in \mathbb{R}^{d_1 \times d_2} \mid \text{rowspan}(\mathbf{\Theta}) \subseteq \mathbb{V}^{\perp}, \ \ \text{colspan}(\mathbf{\Theta}) \subseteq \mathbb{U}^{\perp}\}. \tag{10.12b}$$

Here $\mathbb{U}^{\perp}$ and $\mathbb{V}^{\perp}$ denote the subspaces orthogonal to $\mathbb{U}$ and $\mathbb{V}$, respectively. When the subspaces $(\mathbb{U}, \mathbb{V})$ are clear from the context, we omit them so as to simplify notation. From the definition (10.12a), any matrix in the model space $\mathbb{M}$ has rank at most r. On the other hand, equation (10.12b) defines the subspace $\bar{\mathbb{M}}(\mathbb{U}, \mathbb{V})$ implicitly, via taking the orthogonal complement. We show momentarily that unlike other regularizers considered in Chapter 9, this definition implies that $\bar{\mathbb{M}}(\mathbb{U}, \mathbb{V})$ is a *strict superset* of $\mathbb{M}(\mathbb{U}, \mathbb{V})$.

To provide some intuition for the definition (10.12), it is helpful to consider an explicit matrix-based representation of the subspaces. Recalling that $d' = \min\{d_1, d_2\}$, let $\mathbf{U} \in \mathbb{R}^{d_1 \times d'}$

and $\mathbf{V} \in \mathbb{R}^{d_2 \times d'}$ be a pair of orthonormal matrices. These matrices can be used to define r-dimensional spaces: namely, let $\mathbb{U}$ be the span of the first r columns of $\mathbf{U}$, and similarly, let $\mathbb{V}$ be the span of the first r columns of $\mathbf{V}$. In practice, these subspaces correspond (respectively) to the spaces spanned by the top r left and right singular vectors of the target matrix $\mathbf{\Theta}^*$.

With these choices, any pair of matrices $\mathbf{A} \in \mathbb{M}(\mathbb{U}, \mathbb{V})$ and $\mathbf{B} \in \bar{\mathbb{M}}^\perp(\mathbb{U}, \mathbb{V})$ can be represented in the form

$$\mathbf{A} = \mathbf{U} \begin{bmatrix} \mathbf{\Gamma}_{11} & \mathbf{0}_{r \times (d'-r)} \\ \mathbf{0}_{(d'-r) \times r} & \mathbf{0}_{(d'-r) \times (d'-r)} \end{bmatrix} \mathbf{V}^{\mathrm{T}} \quad \text{and} \quad \mathbf{B} = \mathbf{U} \begin{bmatrix} \mathbf{0}_{r \times r} & \mathbf{0}_{r \times (d'-r)} \\ \mathbf{0}_{(d'-r) \times r} & \mathbf{\Gamma}_{22} \end{bmatrix} \mathbf{V}^{\mathrm{T}}, \tag{10.13}$$

where $\mathbf{\Gamma}_{11} \in \mathbb{R}^{r \times r}$ and $\mathbf{\Gamma}_{22} \in \mathbb{R}^{(d'-r) \times (d'-r)}$ are arbitrary matrices. Thus, we see that $\mathbb{M}$ corresponds to the subspace of matrices with non-zero left and right singular vectors contained within the span of first r columns of $\mathbf{U}$ and $\mathbf{V}$, respectively.

On the other hand, the set $\bar{\mathbb{M}}^\perp$ corresponds to the subspace of matrices with non-zero left and right singular vectors associated with the remaining $d' - r$ columns of $\mathbf{U}$ and $\mathbf{V}$. Since the trace inner product defines orthogonality, any member $\bar{\mathbf{A}}$ of $\bar{\mathbb{M}}(\mathbb{U}, \mathbb{V})$ must take the form

$$\bar{\mathbf{A}} = \mathbf{U} \begin{bmatrix} \bar{\mathbf{\Gamma}}_{11} & \bar{\mathbf{\Gamma}}_{12} \\ \bar{\mathbf{\Gamma}}_{21} & \mathbf{0} \end{bmatrix} \mathbf{V}^{\mathrm{T}}, \tag{10.14}$$

where all three matrices $\bar{\mathbf{\Gamma}}_{11} \in \mathbb{R}^{r \times r}$, $\bar{\mathbf{\Gamma}}_{12} \in \mathbb{R}^{r \times (d'-r)}$ and $\bar{\mathbf{\Gamma}}_{21} \in \mathbb{R}^{(d'-r) \times r}$ are arbitrary. In this way, we see explicitly that $\bar{\mathbb{M}}$ is a strict superset of $\mathbb{M}$ whenever $r < d'$. An important fact, however, is that $\bar{\mathbb{M}}$ is not substantially larger than $\mathbb{M}$. Whereas any matrix in $\mathbb{M}$ has rank at most r, the representation (10.14) shows that any matrix in $\bar{\mathbb{M}}$ has rank at most $2r$.

The preceding discussion also demonstrates the decomposability of the nuclear norm. Using the representation (10.13), for an arbitrary pair of matrices $\mathbf{A} \in \mathbb{M}$ and $\mathbf{B} \in \bar{\mathbb{M}}^\perp$, we have

$$\begin{aligned} |\!|\!|\mathbf{A} + \mathbf{B}|\!|\!|_{\mathrm{nuc}} &\overset{\text{(i)}}{=} \left|\!\left|\!\left| \begin{bmatrix} \mathbf{\Gamma}_{11} & \mathbf{0} \\ \mathbf{0} & \mathbf{0} \end{bmatrix} + \begin{bmatrix} \mathbf{0} & \mathbf{0} \\ \mathbf{0} & \mathbf{\Gamma}_{22} \end{bmatrix} \right|\!\right|\!\right|_{\mathrm{nuc}} = \left|\!\left|\!\left| \begin{bmatrix} \mathbf{\Gamma}_{11} & \mathbf{0} \\ \mathbf{0} & \mathbf{0} \end{bmatrix} \right|\!\right|\!\right|_{\mathrm{nuc}} + \left|\!\left|\!\left| \begin{bmatrix} \mathbf{0} & \mathbf{0} \\ \mathbf{0} & \mathbf{\Gamma}_{22} \end{bmatrix} \right|\!\right|\!\right|_{\mathrm{nuc}} \\ &\overset{\text{(ii)}}{=} |\!|\!|\mathbf{A}|\!|\!|_{\mathrm{nuc}} + |\!|\!|\mathbf{B}|\!|\!|_{\mathrm{nuc}}, \end{aligned}$$

where steps (i) and (ii) use the invariance of the nuclear norm to orthogonal transformations corresponding to multiplication by the matrices $\mathbf{U}$ or $\mathbf{V}$, respectively.

When the target matrix $\mathbf{\Theta}^*$ is of rank r, then the "best" choice of the model subspace (10.12a) is clear. In particular, the low-rank condition on $\mathbf{\Theta}^*$ means that it can be factored in the form $\mathbf{\Theta}^* = \mathbf{U}\mathbf{D}\mathbf{V}^{\mathrm{T}}$, where the diagonal matrix $\mathbf{D} \in \mathbb{R}^{d' \times d'}$ has the r non-zero singular values of $\mathbf{\Theta}^*$ in its first r diagonal entries. The matrices $\mathbf{U} \in \mathbb{R}^{d_1 \times d'}$ and $\mathbf{V} \in \mathbb{R}^{d_2 \times d'}$ are orthonormal, with their first r columns corresponding to the left and right singular vectors, respectively, of $\mathbf{\Theta}^*$. More generally, even when $\mathbf{\Theta}^*$ is not exactly of rank r, matrix subspaces of this form are useful: we simply choose the first r columns of $\mathbf{U}$ and $\mathbf{V}$ to index the singular vectors associated with the largest singular values of $\mathbf{\Theta}^*$, a subspace that we denote by $\mathbb{M}(\mathbb{U}^r, \mathbb{V}^r)$.

With these details in place, let us state for future reference a consequence of Proposition 9.13 for M-estimators involving the nuclear norm. Consider an M-estimator of the form

$$\widehat{\mathbf{\Theta}} \arg \min_{\mathbf{\Theta} \in \mathbb{R}^{d_1 \times d_2}} \{\mathcal{L}_n(\mathbf{\Theta}) + \lambda_n |\!|\!|\mathbf{\Theta}|\!|\!|_{\mathrm{nuc}}\},$$

where $\mathcal{L}_n$ is some convex and differentiable cost function. Then for any choice of regularization parameter $\lambda_n \geq 2|||\nabla \mathcal{L}_n(\Theta^*)|||_2$, the error matrix $\widehat{\Delta} = \widehat{\Theta} - \Theta^*$ must satisfy the cone-like constraint

$$|||\widehat{\Delta}_{\bar{\mathbb{M}}^\perp}|||_{\text{nuc}} \leq 3|||\widehat{\Delta}_{\bar{\mathbb{M}}}|||_{\text{nuc}} + 4|||\Theta^*_{\mathbb{M}^\perp}|||_{\text{nuc}}, \tag{10.15}$$

where $\mathbb{M} = \mathbb{M}(\mathbb{U}^r, \mathbb{V}^r)$ and $\bar{\mathbb{M}} = \bar{\mathbb{M}}(\mathbb{U}^r, \mathbb{V}^r)$. Here the reader should recall that $\widehat{\Delta}_{\bar{\mathbb{M}}}$ denotes the projection of the matrix $\widehat{\Delta}$ onto the subspace $\bar{\mathbb{M}}$, with the other terms defined similarly.

10.2.2 Restricted strong convexity and error bounds

We begin our exploration of nuclear norm regularization in the simplest setting, namely when it is coupled with a least-squares objective function. More specifically, given observations $(y, \mathfrak{X}_n)$ from the matrix regression model (10.3), consider the estimator

$$\widehat{\Theta} \in \arg \min_{\Theta \in \mathbb{R}^{d_1 \times d_2}} \left\{ \frac{1}{2n} \|y - \mathfrak{X}_n(\Theta)\|_2^2 + \lambda_n |||\Theta|||_{\text{nuc}} \right\}, \tag{10.16}$$

where $\lambda_n > 0$ is a user-defined regularization parameter. As discussed in the previous section, the nuclear norm is a decomposable regularizer and the least-squares cost is convex, and so given a suitable choice of λ_n, the error matrix $\widehat{\Delta} := \widehat{\Theta} - \Theta^*$ must satisfy the cone-like constraint (10.15).

The second ingredient of the general theory from Chapter 9 is restricted strong convexity of the loss function. For this least-squares cost, restricted strong convexity amounts to lower bounding the quadratic form $\frac{\|\mathfrak{X}_n(\Delta)\|_2^2}{2n}$. In the sequel, we show the random operator $\mathfrak{X}_n$ satisfies a uniform lower bound of the form

$$\frac{\|\mathfrak{X}_n(\Delta)\|_2^2}{2n} \geq \frac{\kappa}{2} |||\Delta|||_F^2 - c_0 \frac{(d_1 + d_2)}{n} |||\Delta|||_{\text{nuc}}^2, \qquad \text{for all } \Delta \in \mathbb{R}^{d_1 \times d_2}, \tag{10.17}$$

with high probability. Here the quantity $\kappa > 0$ is a *curvature constant*, and c_0 is another universal constant of secondary importance. In the notation of Chapter 9, this lower bound implies a form of restricted strong convexity—in particular, see Definition 9.15—with curvature κ and tolerance $\tau_n^2 = c_0 \frac{(d_1+d_2)}{n}$. We then have the following corollary of Theorem 9.19:

Proposition 10.6 *Suppose that the observation operator $\mathfrak{X}_n$ satisfies the restricted strong convexity condition* (10.17) *with parameter $\kappa > 0$. Then conditioned on the event $\mathbb{G}(\lambda_n) = \{|||\frac{1}{n}\sum_{i=1}^n w_i \mathbf{X}_i|||_2 \leq \frac{\lambda_n}{2}\}$, any optimal solution to nuclear norm regularized least squares* (10.16) *satisfies the bound*

$$|||\widehat{\Theta} - \Theta^*|||_F^2 \leq \frac{9}{2}\frac{\lambda_n^2}{\kappa^2} r + \frac{1}{\kappa}\left\{2\lambda_n \sum_{j=r+1}^{d'} \sigma_j(\Theta^*) + \frac{32 c_0 (d_1 + d_2)}{n} \left[\sum_{j=r+1}^{d'} \sigma_j(\Theta^*)\right]^2\right\}, \tag{10.18}$$

valid for any $r \in \{1, \ldots, d'\}$ such that $r \leq \frac{\kappa n}{128\, c_0\, (d_1+d_2)}$.

Remark: As with Theorem 9.19, the result of Proposition 10.6 is a type of *oracle inequality*: it applies to any matrix Θ^*, and involves a natural splitting into estimation and approximation error, parameterized by the choice of r. Note that the choice of r can be optimized so as to obtain the tightest possible bound.

The bound (10.18) takes a simpler form in special cases. For instance, suppose that $\text{rank}(\Theta^*) < d'$ and moreover that $n > 128 \frac{c_0}{\kappa} \text{rank}(\Theta^*)(d_1 + d_2)$. We then may apply the bound (10.18) with $r = \text{rank}(\Theta^*)$. Since $\sum_{j=r+1}^{d'} \sigma_j(\Theta^*) = 0$, Proposition 10.6 implies the upper bound

$$|||\widehat{\Theta} - \Theta^*|||_F^2 \leq \frac{9}{2} \frac{\lambda_n^2}{\kappa^2} \text{rank}(\Theta^*). \tag{10.19}$$

We make frequent use of this simpler bound in the sequel.

Proof For each $r \in \{1, \ldots, d'\}$, let $(\mathbb{U}^r, \mathbb{V}^r)$ be the subspaces spanned by the top r left and right singular vectors of Θ^*, and recall the subspaces $\mathbb{M}(\mathbb{U}^r, \mathbb{V}^r)$ and $\bar{\mathbb{M}}^\perp(\mathbb{U}^r, \mathbb{V}^r)$ previously defined in (10.12). As shown previously, the nuclear norm is decomposable with respect to any such subspace pair. In general, the "good" event from Chapter 9 is given by $\mathbb{G}(\lambda_n) = \{\Phi^*(\nabla \mathcal{L}_n(\Theta^*)) \leq \frac{\lambda_n}{2}\}$. From Table 9.1, the dual norm to the nuclear norm is the ℓ_2-operator norm. For the least-squares cost function, we have $\nabla \mathcal{L}_n(\Theta^*) = \frac{1}{n} \sum_{i=1}^n w_i \mathbf{X}_i$, so that the statement of Proposition 10.6 involves the specialization of this event to the nuclear norm and least-squares cost.

The assumption (10.17) is a form of restricted strong convexity with tolerance parameter $\tau_n^2 = c_0 \frac{d_1+d_2}{n}$. It only remains to verify the condition $\tau_n^2 \Psi^2(\bar{\mathbb{M}}) \leq \frac{\kappa}{64}$. The representation (10.14) reveals that any matrix $\Theta \in \bar{\mathbb{M}}(\mathbb{U}^r, \mathbb{V}^r)$ has rank at most $2r$, and hence

$$\Psi(\bar{\mathbb{M}}(\mathbb{U}^r, \mathbb{V}^r)) := \sup_{\Theta \in \bar{\mathbb{M}}(\mathbb{U}^r, \mathbb{V}^r) \setminus \{0\}} \frac{|||\Theta|||_{\text{nuc}}}{|||\Theta|||_F} \leq \sqrt{2r}.$$

Consequently, the final condition of Theorem 9.19 holds whenever the target rank r is bounded as in the statement of Proposition 10.6, which completes the proof. □

10.2.3 Bounds under operator norm curvature

In Chapter 9, we also proved a general result—namely, Theorem 9.24—that, for a given regularizer Φ, provides a bound on the estimation error in terms of the dual norm Φ^*. Recall from Table 9.1 that the dual to the nuclear norm is the ℓ_2-operator norm or spectral norm. For the least-squares cost function, the gradient is given by

$$\nabla \mathcal{L}_n(\Theta) = \frac{1}{n} \sum_{i=1}^n \mathbf{X}_i^{\mathrm{T}} \big(y_i - \langle\langle \mathbf{X}_i, \Theta \rangle\rangle\big) = \frac{1}{n} \mathfrak{X}_n^* \big(y - \mathfrak{X}_n(\Theta)\big),$$

where $\mathfrak{X}_n^* \colon \mathbb{R}^n \to \mathbb{R}^{d_1 \times d_2}$ is the adjoint operator. Consequently, in this particular case, the Φ^*-curvature condition from Definition 9.22 takes the form

$$|||\frac{1}{n} \mathfrak{X}_n^* \mathfrak{X}_n(\Delta)|||_2 \geq \kappa |||\Delta|||_2 - \tau_n |||\Delta|||_{\text{nuc}} \qquad \text{for all } \Delta \in \mathbb{R}^{d_1 \times d_2}, \tag{10.20}$$

where $\kappa > 0$ is the curvature parameter, and $\tau_n \geq 0$ is the tolerance parameter.

Proposition 10.7 *Suppose that the observation operator $\mathfrak{X}_n$ satisfies the curvature condition* (10.20) *with parameter $\kappa > 0$, and consider a matrix Θ^* with* $\operatorname{rank}(\Theta^*) < \frac{\kappa}{64\tau_n}$. *Then, conditioned on the event $\mathbb{G}(\lambda_n) = \{|||\frac{1}{n}\mathfrak{X}_n^*(w)|||_2 \leq \frac{\lambda_n}{2}\}$, any optimal solution to the M-estimator* (10.16) *satisfies the bound*

$$|||\widehat{\Theta} - \Theta^*|||_2 \leq 3\sqrt{2}\,\frac{\lambda_n}{\kappa}. \tag{10.21}$$

Remark: Note that this bound is smaller by a factor of $\sqrt{r}$ than the Frobenius norm bound (10.19) that follows from Proposition 10.6. Such a scaling is to be expected, since the Frobenius norm of a rank-r matrix is at most $\sqrt{r}$ times larger than its operator norm. The operator norm bound (10.21) is, in some sense, stronger than the earlier Frobenius norm bound. More specifically, in conjunction with the cone-like inequality (10.15), inequality (10.21) implies a bound of the form (10.19). See Exercise 10.5 for verification of these properties.

Proof In order to apply Theorem 9.24, the only remaining condition to verify is the inequality $\tau_n \Psi^2(\overline{\mathbb{M}}) < \frac{\kappa}{32}$. We have previously calculated that $\Psi^2(\overline{\mathbb{M}}) \leq 2r$, so that the stated upper bound on r ensures that this inequality holds. □

10.3 Matrix compressed sensing

Thus far, we have derived some general results on least squares with nuclear norm regularization, which apply to any model that satisfies the restricted convexity or curvature conditions. We now turn to consequences of these general results for more specific observation models that arise in particular applications. Let us begin this exploration by studying compressed sensing for low-rank matrices, as introduced previously in Example 10.3. There we discussed the standard Gaussian observation model, in which the observation matrices $\mathbf{X}_i \in \mathbb{R}^{d_1 \times d_2}$ are drawn i.i.d., with all entries of each observation matrix drawn i.i.d. from the standard Gaussian $\mathcal{N}(0, 1)$ distribution. More generally, one might draw random observation matrices $\mathbf{X}_i$ with dependent entries, for instance with $\operatorname{vec}(\mathbf{X}_i) \sim \mathcal{N}(0, \mathbf{\Sigma})$, where $\mathbf{\Sigma} \in \mathbb{R}^{(d_1 d_2) \times (d_1 d_2)}$ is the covariance matrix. In this case, we say that $\mathbf{X}_i$ is drawn from the *$\mathbf{\Sigma}$-Gaussian ensemble*.

In order to apply Proposition 10.6 to this ensemble, our first step is to establish a form of restricted strong convexity. The following result provides a high-probability lower bound on the Hessian of the least-squares cost for this ensemble. It involves the quantity

$$\rho^2(\mathbf{\Sigma}) := \sup_{\|u\|_2 = \|v\|_2 = 1} \operatorname{var}(\langle\langle \mathbf{X},\, uv^{\mathrm{T}} \rangle\rangle).$$

Note that $\rho^2(\mathbf{I}_d) = 1$ for the special case of the identity ensemble.

Theorem 10.8 *Given n i.i.d. draws $\{\mathbf{X}_i\}_{i=1}^n$ of random matrices from the $\mathbf{\Sigma}$-Gaussian ensemble, there are positive constants $c_1 < 1 < c_2$ such that*

$$\frac{\|\mathfrak{X}_n(\mathbf{\Delta})\|_2^2}{n} \geq c_1 \|\sqrt{\mathbf{\Sigma}}\,\mathrm{vec}(\mathbf{\Delta})\|_2^2 - c_2 \rho^2(\mathbf{\Sigma})\left\{\frac{d_1 + d_2}{n}\right\} |\!|\!|\mathbf{\Delta}|\!|\!|_{\mathrm{nuc}}^2 \quad \forall\, \mathbf{\Delta} \in \mathbb{R}^{d_1 \times d_2} \tag{10.22}$$

with probability at least $1 - \frac{e^{-\frac{n}{32}}}{1-e^{-\frac{n}{32}}}$.

This result can be understood as a variant of Theorem 7.16, which established a similar result for the case of sparse vectors and the ℓ_1-norm. As with this earlier theorem, Theorem 10.8 can be proved using the Gordon–Slepian comparison lemma for Gaussian processes. In Exercise 10.6, we work through a proof of a slightly simpler form of the bound.

Theorem 10.8 has an immediate corollary for the *noiseless* observation model, in which we observe $(y_i, \mathbf{X}_i)$ pairs linked by the linear equation $y_i = \langle\langle \mathbf{X}_i,\ \mathbf{\Theta}^* \rangle\rangle$. In this setting, the natural analog of the basis pursuit program from Chapter 7 is the following convex program:

$$\min_{\mathbf{\Theta} \in \mathbb{R}^{d_1 \times d_2}} |\!|\!|\mathbf{\Theta}|\!|\!|_{\mathrm{nuc}} \quad \text{such that } \langle\langle \mathbf{X}_i,\ \mathbf{\Theta} \rangle\rangle = y_i \text{ for all } i = 1, \ldots, n. \tag{10.23}$$

That is, we search over the space of matrices that match the observations perfectly to find the solution with minimal nuclear norm. As with the estimator (10.16), it can be reformulated as an instance of semidefinite program.

Corollary 10.9 *Given $n > 16 \frac{c_2}{c_1} \frac{\rho^2(\mathbf{\Sigma})}{\gamma_{\min}(\mathbf{\Sigma})} r\,(d_1 + d_2)$ i.i.d. samples from the $\mathbf{\Sigma}$-ensemble, the estimator* (10.23) *recovers the rank-r matrix $\mathbf{\Theta}^*$ exactly—i.e., it has a unique solution $\widehat{\mathbf{\Theta}} = \mathbf{\Theta}^*$—with probability at least* $1 - \frac{e^{-\frac{n}{32}}}{1-e^{-\frac{n}{32}}}$.

The requirement that the sample size n is larger than $r\,(d_1 + d_2)$ is intuitively reasonable, as can be seen by counting the degrees of freedom required to specify a rank-r matrix of size $d_1 \times d_2$. Roughly speaking, we need r numbers to specify its singular values, and rd_1 and rd_2 numbers to specify its left and right singular vectors.[2] Putting together the pieces, we conclude that the matrix has of the order $r(d_1 + d_2)$ degrees of freedom, consistent with the corollary. Let us now turn to its proof.

Proof Since $\widehat{\mathbf{\Theta}}$ and $\mathbf{\Theta}^*$ are optimal and feasible, respectively, for the program (10.23), we have $|\!|\!|\widehat{\mathbf{\Theta}}|\!|\!|_{\mathrm{nuc}} \leq |\!|\!|\mathbf{\Theta}^*|\!|\!|_{\mathrm{nuc}} = |\!|\!|\mathbf{\Theta}^*_{\mathbb{M}}|\!|\!|_{\mathrm{nuc}}$. Introducing the error matrix $\widehat{\mathbf{\Delta}} = \widehat{\mathbf{\Theta}} - \mathbf{\Theta}^*$, we have

$$|\!|\!|\widehat{\mathbf{\Theta}}|\!|\!|_{\mathrm{nuc}} = |\!|\!|\mathbf{\Theta}^* + \widehat{\mathbf{\Delta}}|\!|\!|_{\mathrm{nuc}} = |\!|\!|\mathbf{\Theta}^*_{\mathbb{M}} + \widehat{\mathbf{\Delta}}_{\bar{\mathbb{M}}^\perp} + \widehat{\mathbf{\Delta}}_{\bar{\mathbb{M}}}|\!|\!|_{\mathrm{nuc}} \overset{\text{(i)}}{\geq} |\!|\!|\mathbf{\Theta}^*_{\mathbb{M}} + \widehat{\mathbf{\Delta}}_{\bar{\mathbb{M}}^\perp}|\!|\!|_{\mathrm{nuc}} - |\!|\!|\widehat{\mathbf{\Delta}}_{\bar{\mathbb{M}}}|\!|\!|_{\mathrm{nuc}}$$

by the triangle inequality. Applying decomposability this yields $|\!|\!|\mathbf{\Theta}^*_{\mathbb{M}} + \widehat{\mathbf{\Delta}}_{\bar{\mathbb{M}}^\perp}|\!|\!|_{\mathrm{nuc}} =$

[2] The orthonormality constraints for the singular vectors reduce the degrees of freedom, so we have just given an upper bound here.

$|||\Theta^*_{\bar{\mathbb{M}}}|||_{\text{nuc}} + |||\widehat{\Delta}_{\bar{\mathbb{M}}^\perp}|||_{\text{nuc}}$. Combining the pieces, we find that $|||\widehat{\Delta}_{\bar{\mathbb{M}}^\perp}|||_{\text{nuc}} \leq |||\widehat{\Delta}_{\bar{\mathbb{M}}}|||_{\text{nuc}}$. From the representation (10.14), any matrix in $\bar{\mathbb{M}}$ has rank at most $2r$, whence

$$|||\widehat{\Delta}|||_{\text{nuc}} \leq 2|||\widehat{\Delta}_{\bar{\mathbb{M}}}|||_{\text{nuc}} \leq 2\sqrt{2r}\,|||\widehat{\Delta}|||_{\text{F}}. \tag{10.24}$$

Now let us condition on the event that the lower bound (10.22) holds. When applied to $\widehat{\Delta}$, and coupled with the inequality (10.24), we find that

$$\frac{\|\mathfrak{X}_n(\widehat{\Delta})\|_2^2}{n} \geq \left\{c_1\gamma_{\min}(\Sigma) - 8\,c_2\,\rho^2(\Sigma)\frac{r\,(d_1+d_2)}{n}\right\}|||\widehat{\Delta}|||_{\text{F}}^2 \geq \frac{c_1}{2}\gamma_{\min}(\Sigma)\,|||\widehat{\Delta}|||_{\text{F}}^2,$$

where the final inequality follows by applying the given lower bound on n, and performing some algebra. But since both $\widehat{\Theta}$ and Θ^* are feasible for the convex program (10.23), we have shown that $0 = \frac{\|\mathfrak{X}_n(\widehat{\Delta})\|_2^2}{n} \geq \frac{c_1}{2}\,\gamma_{\min}(\Sigma)\,|||\widehat{\Delta}|||_{\text{F}}^2$, which implies that $\widehat{\Delta} = 0$ as claimed. □

Theorem 10.8 can also be used to establish bounds for the least-squares estimator (10.16), based on noisy observations of the form $y_i = \langle\langle \mathbf{X}_i,\ \Theta^*\rangle\rangle + w_i$. Here we state and prove a result that is applicable to matrices of rank at most r.

Corollary 10.10 *Consider* $n > 64\frac{c_2}{c_1}\frac{\rho^2(\Sigma)}{\gamma_{\min}(\Sigma)}\,r\,(d_1+d_2)$ *i.i.d. samples* $(y_i, \mathbf{X}_i)$ *from the linear matrix regression model, where each* $\mathbf{X}_i$ *is drawn from the* Σ*-Gaussian ensemble. Then any optimal solution to the program* (10.16) *with* $\lambda_n = 10\,\sigma\,\rho(\Sigma)(\sqrt{\frac{d_1+d_2}{n}} + \delta)$ *satisfies the bound*

$$|||\widehat{\Theta} - \Theta^*|||_{\text{F}}^2 \leq 125\frac{\sigma^2\rho^2(\Sigma)}{c_1^2\,\gamma_{\min}^2(\Sigma)}\left\{\frac{r(d_1+d_2)}{n} + r\delta^2\right\} \tag{10.25}$$

with probability at least $1 - 2e^{-2n\delta^2}$.

Figure 10.2 provides plots of the behavior predicted by Corollary 10.10. We generated these plots by simulating matrix regression problems with design matrices $\mathbf{X}_i$ chosen from the standard Gaussian ensemble, and then solved the convex program (10.16) with the choice of λ_n given in Corollary 10.10, and matrices of size $d \times d$, where $d^2 \in \{400, 1600, 6400\}$ and rank $r = \lceil\sqrt{d}\rceil$. In Figure 10.2(a), we plot the Frobenius norm error $|||\widehat{\Theta} - \Theta^*|||_{\text{F}}$, averaged over $T = 10$ trials, versus the raw sample size n. Each of these error plots tends to zero as the sample size increases, showing the classical consistency of the method. However, the curves shift to the right as the matrix dimension d (and hence the rank r) is increased, showing the effect of dimensionality. Assuming that the scaling of Corollary 10.10 is sharp, it predicts that, if we plot the same Frobenius errors versus the *rescaled sample size* $\frac{n}{rd}$, then all three curves should be relatively well aligned. These rescaled curves are shown in Figure 10.2(b): consistent with the prediction of Corollary 10.10, they are now all relatively well aligned, independently of the dimension and rank, consistent with the prediction.

Let us now turn to the proof of Corollary 10.10.

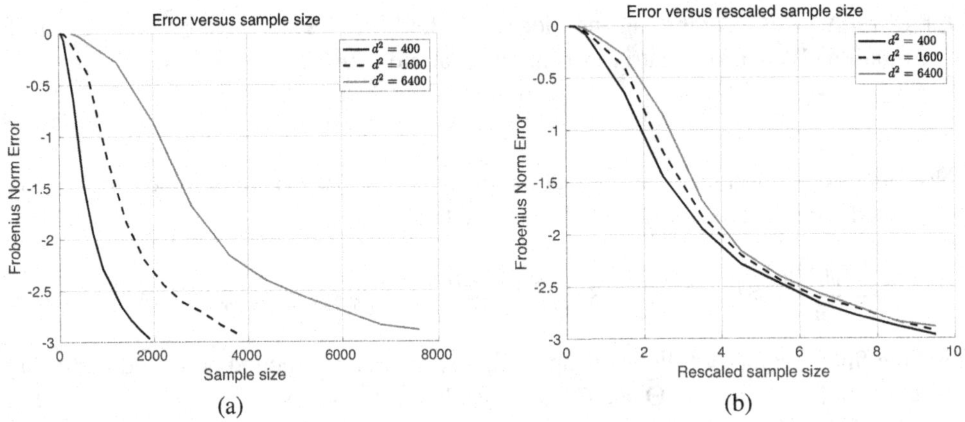

Figure 10.2 Plots of the Frobenius norm error $|||\widehat{\Theta} - \Theta^*|||_F$ for the nuclear norm regularized least-squares estimator (10.16) with design matrices $\mathbf{X}_i$ drawn from the standard Gaussian ensemble. (a) Plots of Frobenius norm error versus sample size n for three different matrix sizes $d \in \{40, 80, 160\}$ and rank $r = \lceil \sqrt{d} \rceil$. (b) Same error measurements now plotted against the rescaled sample size $\frac{n}{rd}$. As predicted by the theory, all three curves are now relatively well-aligned.

Proof We prove the bound (10.25) via an application of Proposition 10.6, in particular in the form of the bound (10.19). Theorem 10.8 shows that the RSC condition holds with $\kappa = c_1$ and $c_0 = \frac{c_2 \rho^2(\Sigma)}{2}$, so that the stated lower bound on the sample size ensures that Proposition 10.6 can be applied with $r = \text{rank}(\Theta^*)$.

It remains to verify that the event $\mathbb{G}(\lambda_n) = \{|||\frac{1}{n}\sum_{i=1}^n w_i \mathbf{X}_i|||_2 \leq \frac{\lambda_n}{2}\}$ holds with high probability. Introduce the shorthand $\mathbf{Q} = \frac{1}{n}\sum_{i=1}^n w_i \mathbf{X}_i$, and define the event $\mathcal{E} = \left\{\frac{\|w\|_2^2}{n} \leq 2\sigma^2\right\}$. We then have

$$\mathbb{P}\left[|||\mathbf{Q}|||_2 \geq \frac{\lambda_n}{2}\right] \leq \mathbb{P}[\mathcal{E}^c] + \mathbb{P}\left[|||\mathbf{Q}|||_2 \geq \frac{\lambda_n}{2} \mid \mathcal{E}\right].$$

Since the noise variables $\{w_i\}_{i=1}^n$ are i.i.d., each zero-mean and sub-Gaussian with parameter σ, we have $\mathbb{P}[\mathcal{E}^c] \leq e^{-n/8}$. It remains to upper bound the second term, which uses conditioning on $\mathcal{E}$.

Let $\{u^1, \ldots, u^M\}$ and $\{v^1, \ldots, v^N\}$ be $1/4$-covers in Euclidean norm of the spheres $\mathbb{S}^{d_1-1}$ and $\mathbb{S}^{d_2-1}$, respectively. By Lemma 5.7, we can find such covers with $M \leq 9^{d_1}$ and $N \leq 9^{d_2}$ elements respectively. For any $v \in \mathbb{S}^{d_2-1}$, we can write $v = v^\ell + \Delta$ for some vector Δ with ℓ_2 at most $1/4$, and hence

$$|||\mathbf{Q}|||_2 = \sup_{v \in \mathbb{S}^{d_2-1}} \|\mathbf{Q}v\|_2 \leq \tfrac{1}{4}|||\mathbf{Q}|||_2 + \max_{\ell=1,\ldots,N} \|\mathbf{Q}v^\ell\|_2.$$

A similar argument involving the cover of $\mathbb{S}^{d_1-1}$ yields $\|\mathbf{Q}v^\ell\|_2 \leq \frac{1}{4}|||\mathbf{Q}|||_2 + \max_{j=1,\ldots,M} \langle u^j, \mathbf{Q}v^\ell \rangle$. Thus, we have established that

$$|||\mathbf{Q}|||_2 \leq 2 \max_{j=1,\ldots,M} \max_{\ell=1,\ldots,N} |Z^{j,\ell}| \qquad \text{where } Z^{j,\ell} = \langle u^j, \mathbf{Q}\, v^\ell \rangle.$$

Fix some index pair (j, ℓ): we can then write $Z^{j,\ell} = \frac{1}{n}\sum_{i=1}^n w_i Y_i^{j,\ell}$ where $Y_i^{j,\ell} = \langle u^j, \mathbf{X}_i v^\ell \rangle$. Note that each variable $Y_i^{j,\ell}$ is zero-mean Gaussian with variance at most $\rho^2(\mathbf{\Sigma})$. Consequently, the variable $Z^{j,\ell}$ is zero-mean Gaussian with variance at most $\frac{2\sigma^2\rho^2(\mathbf{\Sigma})}{n}$, where we have used the conditioning on event $\mathcal{E}$. Putting together the pieces, we conclude that

$$\begin{aligned} \mathbb{P}\left[|||\frac{1}{n}\sum_{i=1}^n w_i \mathbf{X}_i|||_2 \geq \frac{\lambda_n}{2} \mid \mathcal{E} \right] &\leq \sum_{j=1}^M \sum_{\ell=1}^N \mathbb{P}\left[|Z^{j,\ell}| \geq \frac{\lambda_n}{4} \right] \\ &\leq 2\, e^{-\frac{n\lambda_n^2}{32\sigma^2\rho^2(\mathbf{\Sigma})} + \log M + \log N} \\ &\leq 2\, e^{-\frac{n\lambda_n^2}{32\sigma^2\rho^2(\mathbf{\Sigma})} + (d_1+d_2)\log 9}. \end{aligned}$$

Setting $\lambda_n = 10\sigma\rho(\mathbf{\Sigma})\left(\sqrt{\frac{(d_1+d_2)}{n}} + \delta\right)$, we find that $\mathbb{P}\left[|||\frac{1}{n}\sum_{i=1}^n w_i \mathbf{X}_i|||_2 \geq \frac{\lambda_n}{2} \right] \leq 2e^{-2n\delta^2}$ as claimed. □

Corollary 10.10 is stated for matrices that are exactly low-rank. However, Proposition 10.6 can also be used to derive error bounds for matrices that are not exactly low-rank, but rather *well approximated* by a low-rank matrix. For instance, suppose that $\mathbf{\Theta}^*$ belongs to the ℓ_q-"ball" of matrices given by

$$\mathbb{B}_q(R_q) := \left\{ \mathbf{\Theta} \in \mathbb{R}^{d_1 \times d_2} \mid \sum_{j=1}^d (\sigma_j(\mathbf{\Theta}))^q \leq R_q \right\}, \tag{10.26}$$

where $q \in [0, 1]$ is a parameter, and R_q is the radius. Note that this is simply the set of matrices whose vector of singular values belongs to the usual ℓ_q-ball for vectors. See Figure 9.5 for an illustration.

When the unknown matrix $\mathbf{\Theta}^*$ belongs to $\mathbb{B}_q(R_q)$, Proposition 10.6 can be used to show that the estimator (10.16) satisfies an error bound of the form

$$|||\widehat{\mathbf{\Theta}} - \mathbf{\Theta}^*|||_{\mathrm{F}}^2 \precsim R_q \left(\frac{\sigma^2 (d_1 + d_2)}{n} \right)^{1-\frac{q}{2}} \tag{10.27}$$

with high probability. Note that this bound generalizes Corollary 10.10, since in the special case $q = 0$, the set $\mathbb{B}_0(r)$ corresponds to the set of matrices with rank at most r. See Exercise 10.7 for more details.

As another extension, one can move beyond the setting of least squares, and consider more general non-quadratic cost functions. As an initial example, still in the context of matrix regression with samples $z = (\mathbf{X}, y)$, let us consider a cost function that satisfies a local L-Lipschitz condition of the form

$$\left|\mathcal{L}(\mathbf{\Theta}; z) - \mathcal{L}(\widetilde{\mathbf{\Theta}}; z)\right| \leq L \left|\langle\langle \mathbf{\Theta}, \mathbf{X} \rangle\rangle - \langle\langle \widetilde{\mathbf{\Theta}}, \mathbf{X} \rangle\rangle\right| \qquad \text{for all } \mathbf{\Theta}, \widetilde{\mathbf{\Theta}} \in \mathbb{B}_F(R).$$

For instance, if the response variables y were binary-valued, with the conditional distribution of the logistic form, as described in Example 9.2, then the log-likelihood would satisfy this condition with $L = 2$ (see Example 9.33). Similarly, in classification problems based on matrix-valued observations, the hinge loss that underlies the support vector machine would also satisfy this condition. In the following example, we show how Theorem 9.34 can be used to establish restricted strong convexity with respect to the nuclear norm for such Lipschitz losses.

Example 10.11 (Lipschitz losses and nuclear norm) As a generalization of Corollary 10.10, suppose that the $d_1 \times d_2$ design matrices $\{\mathbf{X}_i\}_{i=1}^n$ are generated i.i.d. from a ν-sub-Gaussian ensemble, by which we mean that, for each pair of unit-norm vectors (u, v), the random variable $\langle u,\, \mathbf{X}_i\, v\rangle$ is zero-mean and ν-sub-Gaussian. Note that the $\mathbf{\Sigma}$-Gaussian ensemble is a special case with $\nu = \rho(\mathbf{\Sigma})$.

Now recall that

$$\mathcal{E}_n(\mathbf{\Delta}) := \mathcal{L}_n(\mathbf{\Theta}^* + \mathbf{\Delta}) - \mathcal{L}_n(\mathbf{\Theta}^*) - \langle\!\langle \nabla \mathcal{L}_n(\mathbf{\Theta}^*),\, \mathbf{\Delta}\rangle\!\rangle$$

denotes the error in the first-order Taylor-series expansion of the empirical cost function, whereas $\bar{\mathcal{E}}(\mathbf{\Delta})$ denotes the analogous quantity for the population cost function. We claim that for any $\delta > 0$, any cost function that is L-Lipschitz over the ball $\mathbb{B}_F(1)$ satisfies the bound

$$\left|\mathcal{E}_n(\mathbf{\Delta}) - \bar{\mathcal{E}}(\mathbf{\Delta})\right| \le 16 L\, \nu\, |||\mathbf{\Delta}|||_{\text{nuc}} \left\{12\sqrt{\frac{d_1 + d_2}{n}} + \epsilon\right\} \qquad \text{for all } \mathbf{\Delta} \in \mathbb{B}_F(1/d, 1) \tag{10.28}$$

with probability at least $1 - 4(\log d)^2\, e^{-\frac{n\epsilon^2}{12}}$.

In order to establish the bound (10.28), we need to verify the conditions of Theorem 9.34. For a matrix $\mathbf{\Theta} \in \mathbb{R}^{d \times d}$, recall that we use $\{\sigma_j(\mathbf{\Theta})\}_{j=1}^d$ to denote its singular values. The dual to the nuclear norm is the ℓ_2-operator norm $|||\mathbf{\Theta}|||_2 = \max_{j=1,\ldots,d} \sigma_j(\mathbf{\Theta})$. Based on Theorem 9.34, we need to study the deviations of the random variable $|||\frac{1}{n}\sum_{i=1}^n \varepsilon_i \mathbf{X}_i|||_2$, where $\{\varepsilon_i\}_{i=1}^n$ is an i.i.d. sequence of Rademacher variables. Since the random matrices $\{\mathbf{X}_i\}_{i=1}^n$ are i.i.d., this random variable has the same distribution as $|||\mathbf{V}|||_2$, where $\mathbf{V}$ is a $\nu/\sqrt{n}$-sub-Gaussian random matrix. By the same discretization argument used in the proof of Corollary 10.10, for each $\lambda > 0$, we have $\mathbb{E}[e^{\lambda |||\mathbf{V}|||_2}] \le \sum_{j=1}^M \sum_{\ell=1}^N \mathbb{E}[e^{2\lambda Z^{j,\ell}}]$, where $M \le 9^{d_1}$ and $N \le 9^{d_2}$, and each random variable $Z^{j,\ell}$ is sub-Gaussian with parameter at most $\sqrt{2}\nu/\sqrt{n}$. Consequently, for any $\delta > 0$,

$$\inf_{\lambda > 0} \mathbb{E}[e^{\lambda(|||\mathbf{V}|||_2 - \delta)}] \le M N \inf_{\lambda > 0} e^{\frac{8\nu^2\lambda}{n} - \lambda\delta} = e^{-\frac{n\delta^2}{16\nu^2} + 9(d_1 + d_2)}.$$

Setting $\delta^2 = 144\nu^2 \frac{d_1 + d_2}{n} + \nu^2\epsilon^2$ yields the claim (10.28). ♣

10.4 Bounds for phase retrieval

We now return to the problem of phase retrieval. In the idealized model previously introduced in Example 10.4, we make n observations of the form $\tilde{y}_i = |\langle x_i,\, \theta^*\rangle|$, where the observation vector $x_i \sim N(0, \mathbf{I}_d)$ are drawn independently. A standard lifting procedure leads to the semidefinite relaxation

$$\widehat{\mathbf{\Theta}} \in \arg\min_{\mathbf{\Theta} \in \mathcal{S}_+^{d \times d}} \operatorname{trace}(\mathbf{\Theta}) \quad \text{such that} \quad \tilde{y}_i^2 = \langle\!\langle \mathbf{\Theta},\, x_i \otimes x_i\rangle\!\rangle \quad \text{for all } i = 1, \ldots, n. \tag{10.29}$$

This optimization problem is known as a semidefinite program (SDP), since it involves optimizing over the cone $\mathcal{S}_+^{d \times d}$ of positive semidefinite matrices. By construction, the rank-one matrix $\mathbf{\Theta}^* = \theta^* \otimes \theta^*$ is feasible for the optimization problem (10.29), and our goal is to understand when it is the unique optimal solution. Equivalently, our goal is to show that the error matrix $\widehat{\boldsymbol{\Delta}} = \widehat{\boldsymbol{\Theta}} - \boldsymbol{\Theta}^*$ is equal to zero.

Defining the new response variables $y_i = \widetilde{y}_i^2$ and observation matrices $\mathbf{X}_i := x_i \otimes x_i$, the constraints in the SDP (10.29) can be written in the equivalent trace inner product form $y_i = \langle\langle \mathbf{X}_i, \, \Theta \rangle\rangle$. Since both $\widehat{\boldsymbol{\Theta}}$ and $\boldsymbol{\Theta}^*$ are feasible and hence must satisfy these constraints, we see that the error matrix $\widehat{\boldsymbol{\Delta}}$ must belong to the nullspace of the linear operator $\mathfrak{X}_n \colon \mathbb{R}^{d \times d} \to \mathbb{R}^n$ with components $[\mathfrak{X}_n(\boldsymbol{\Theta})]_i = \langle\langle \mathbf{X}_i, \, \boldsymbol{\Theta} \rangle\rangle$. The following theorem shows that this random operator satisfies a version of the restricted nullspace property (recall Chapter 7):

Theorem 10.12 (Restricted nullspace/eigenvalues for phase retrieval) *For each $i = 1, \ldots, n$, consider random matrices of the form $\mathbf{X}_i = x_i \otimes x_i$ for i.i.d. $\mathcal{N}(0, \mathbf{I}_d)$ vectors. Then there are universal constants (c_0, c_1, c_2) such that for any $\rho > 0$, a sample size $n > c_0 \rho d$ suffices to ensure that*

$$\frac{1}{n} \sum_{i=1}^n \langle\langle \mathbf{X}_i, \, \boldsymbol{\Theta} \rangle\rangle^2 \geq \frac{1}{2} |||\boldsymbol{\Theta}|||_F^2 \quad \text{for all matrices such that } |||\boldsymbol{\Theta}|||_F^2 \leq \rho |||\boldsymbol{\Theta}|||_{\text{nuc}}^2 \tag{10.30}$$

with probability at least $1 - c_1 e^{-c_2 n}$.

Note that the lower bound (10.30) implies that there are no matrices in the intersection of nullspace of the operator $\mathfrak{X}_n$ with the matrix cone defined by the inequality $|||\boldsymbol{\Theta}|||_F^2 \leq \rho |||\boldsymbol{\Theta}|||_{\text{nuc}}^2$. Consequently, Theorem 10.12 has an immediate corollary for the exactness of the semidefinite programming relaxation (10.29):

Corollary 10.13 *Given $n > 2 c_0 d$ samples, the SDP* (10.29) *has the unique optimal solution $\widehat{\boldsymbol{\Theta}} = \boldsymbol{\Theta}^*$ with probability at least $1 - c_1 e^{-c_2 n}$.*

Proof Since $\widehat{\boldsymbol{\Theta}}$ and $\boldsymbol{\Theta}^*$ are optimal and feasible (respectively) for the convex program (10.29), we are guaranteed that $\text{trace}(\widehat{\boldsymbol{\Theta}}) \leq \text{trace}(\boldsymbol{\Theta}^*)$. Since both matrices must be positive semidefinite, this trace constraint is equivalent to $|||\widehat{\boldsymbol{\Theta}}|||_{\text{nuc}} \leq |||\boldsymbol{\Theta}^*|||_{\text{nuc}}$. This inequality, in conjunction with the rank-one nature of $\boldsymbol{\Theta}^*$ and the decomposability of the nuclear norm, implies that the error matrix $\widehat{\boldsymbol{\Delta}} = \widehat{\boldsymbol{\Theta}} - \boldsymbol{\Theta}^*$ satisfies the cone constraint $|||\widehat{\boldsymbol{\Delta}}|||_{\text{nuc}} \leq \sqrt{2}\, |||\widehat{\boldsymbol{\Delta}}|||_F$. Consequently, we can apply Theorem 10.12 with $\rho = 2$ to conclude that

$$0 = \frac{1}{n} \sum_{i=1}^n \langle\langle \mathbf{X}_i, \, \widehat{\boldsymbol{\Delta}} \rangle\rangle \geq \frac{1}{2} \|\widehat{\boldsymbol{\Delta}}\|_2^2,$$

from which we conclude that $\widehat{\boldsymbol{\Delta}} = 0$ with the claimed probability. □

Let us now return to prove Theorem 10.12.

Proof For each matrix $\boldsymbol{\Delta} \in \mathcal{S}^{d \times d}$, consider the (random) function $f_{\boldsymbol{\Delta}}(\mathbf{X}, v) = v \langle\langle \mathbf{X}, \, \boldsymbol{\Delta} \rangle\rangle$, where $v \in \{-1, 1\}$ is a Rademacher variable independent of $\mathbf{X}$. By construction, we then have $\mathbb{E}[f_{\Delta}(\mathbf{X}, v)] = 0$. Moreover, as shown in Exercise 10.9, we have

$$\|f_{\Delta}\|_2^2 = \mathbb{E}[\langle\!\langle \mathbf{X}, \Delta \rangle\!\rangle]^2 = |\!|\!|\Delta|\!|\!|_F^2 + 2(\operatorname{trace}(\Delta))^2. \tag{10.31a}$$

As a consequence, if we define the set $\mathbb{A}_1(\sqrt{\rho}) = \{\Delta \in \mathcal{S}^{d\times d} \mid |\!|\!|\Delta|\!|\!|_{\text{nuc}} \leq \sqrt{\rho}\,|\!|\!|\Delta|\!|\!|_F\}$, it suffices to show that

$$\underbrace{\frac{1}{n}\sum_{i=1}^{n} \langle\!\langle \mathbf{X}, \Delta \rangle\!\rangle^2}_{\|f_{\Delta}\|_n^2} \geq \frac{1}{2}\underbrace{\mathbb{E}[\langle\!\langle \mathbf{X}, \Delta \rangle\!\rangle^2]}_{\|f_{\Delta}\|_2^2} \qquad \text{for all } \Delta \in \mathbb{A}_1(\sqrt{\rho}) \tag{10.31b}$$

with probability at least $1 - c_1 e^{-c_2 n}$.

We prove claim (10.31b) as a corollary of a more general one-sided uniform law, stated as Theorem 14.12 in Chapter 14. First, observe that the function class $\mathscr{F} := \{f_{\Delta} \mid \Delta \in \mathbb{A}_1(\sqrt{\rho})\}$ is a cone, and so star-shaped around zero. Next we claim that the fourth-moment condition (14.22b) holds. From the result of Exercise 10.9, we can restrict attention to diagonal matrices without loss of generality. It suffices to show that $\mathbb{E}[f_{\mathbf{D}}^4(\mathbf{X}, v)] \leq C$ for all matrices such that $|\!|\!|\mathbf{D}|\!|\!|_F^2 = \sum_{j=1}^d D_{jj}^2 \leq 1$. Since the Gaussian variables have moments of all orders, by Rosenthal's inequality (see Exercise 2.20), there is a universal constant c such that

$$\mathbb{E}[f_{\mathbf{D}}^4(\mathbf{X}, v)] = \mathbb{E}\Big[\big(\sum_{j=1}^{d} D_{jj} x_j^2\big)^4\Big] \leq c\Big\{\sum_{j=1}^{d} D_{jj}^4 \mathbb{E}[x_j^8] + \big(\sum_{j=1}^{d} D_{jj}^2 \mathbb{E}[x_j^4]\big)^2\Big\}.$$

For standard normal variates, we have $\mathbb{E}[x_j^4] = 4$ and $\mathbb{E}[x_j^8] = 105$, whence

$$\mathbb{E}[f_{\mathbf{D}}^4(\mathbf{X}, v)] \leq c\Big\{105 \sum_{j=1}^{d} D_{jj}^4 + 16|\!|\!|\mathbf{D}|\!|\!|_F^4\Big\}.$$

Under the condition $\sum_{j=1}^d D_{jj}^2 \leq 1$, this quantity is bounded by a universal constant C, thereby verifying the moment condition (14.22b).

Next, we need to compute the local Rademacher complexity, and hence the critical radius δ_n. As shown by our previous calculation (10.31a), the condition $\|f_{\Delta}\|_2 \leq \delta$ implies that $|\!|\!|\Delta|\!|\!|_F \leq \delta$. Consequently, we have

$$\bar{\mathcal{R}}_n(\delta) \leq \mathbb{E}\Big[\sup_{\substack{\Delta \in \mathbb{A}_1(\sqrt{\rho}) \\ |\!|\!|\Delta|\!|\!|_F \leq \delta}} \Big|\frac{1}{n}\sum_{i=1}^{n} \varepsilon_i f_{\Delta}(\mathbf{X}_i; v_i)\Big|\Big],$$

where $\{\varepsilon_i\}_{i=1}^n$ is another i.i.d. Rademacher sequence. Using the definition of f_{Δ} and the duality between the operator and nuclear norms (see Exercise 9.4), we have

$$\bar{\mathcal{R}}_n(\delta) \leq \mathbb{E}\left[\sup_{\Delta \in \mathbb{A}_1(\sqrt{\rho})} |\!|\!|\big(\frac{1}{n}\sum_{i=1}^{n} \varepsilon_i (x_i \otimes x_i)\big)|\!|\!|_2 \, |\!|\!|\Delta|\!|\!|_{\text{nuc}}\right] \leq \sqrt{\rho}\,\delta\, \mathbb{E}\left[|\!|\!|\frac{1}{n}\sum_{i=1}^{n} \varepsilon_i (x_i \otimes x_i)|\!|\!|_2\right].$$

Finally, by our previous results on operator norms of random sub-Gaussian matrices (see Theorem 6.5), there is a constant c such that, in the regime $n > d$, we have

$$\mathbb{E}\Big[|\!|\!|\frac{1}{n}\sum_{i=1}^{n} v_i (x_i \otimes x_i)|\!|\!|_2\Big] \leq c\sqrt{\frac{d}{n}}.$$

Putting together the pieces, we conclude that inequality (14.24) is satisfied for any $\delta_n \succsim \sqrt{\rho}\sqrt{\frac{d}{n}}$. Consequently, as long as $n > c_0 \rho d$ for a sufficiently large constant c_0, we can set $\delta_n = 1/2$ in Theorem 14.12, which establishes the claim (10.31b). □

10.5 Multivariate regression with low-rank constraints

The problem of multivariate regression, as previously introduced in Example 10.1, involves estimating a prediction function, mapping covariate vectors $z \in \mathbb{R}^p$ to output vectors $y \in \mathbb{R}^T$. In the case of linear prediction, any such mapping can be parameterized by a matrix $\mathbf{\Theta}^* \in \mathbb{R}^{p\times T}$. A collection of n observations can be specified by the model

$$\mathbf{Y} = \mathbf{Z}\mathbf{\Theta}^* + \mathbf{W}, \tag{10.32}$$

where $(\mathbf{Y}, \mathbf{Z}) \in \mathbb{R}^{n\times T} \times \mathbb{R}^{p\times T}$ are observed, and $\mathbf{W} \in \mathbb{R}^{n\times T}$ is a matrix of noise variables. For this observation model, the least-squares cost takes the form $\mathcal{L}_n(\mathbf{\Theta}) = \frac{1}{2n}|||\mathbf{Y} - \mathbf{Z}\mathbf{\Theta}|||_{\mathrm{F}}^2$.

The following result is a corollary of Proposition 10.7 in application to this model. It is applicable to the case of fixed design and so involves the minimum and maximum eigenvalues of the sample covariance matrix $\widehat{\mathbf{\Sigma}} := \frac{\mathbf{Z}^{\mathrm{T}}\mathbf{Z}}{n}$.

Corollary 10.14 *Consider the observation model* (10.32) *in which* $\mathbf{\Theta}^* \in \mathbb{R}^{p\times T}$ *has rank at most r, and the noise matrix* $\mathbf{W}$ *has i.i.d. entries that are zero-mean and* σ*-sub-Gaussian. Then any solution to the program* (10.16) *with* $\lambda_n = 10\sigma\sqrt{\gamma_{\max}(\widehat{\mathbf{\Sigma}})}\,\big(\sqrt{\frac{p+T}{n}} + \delta\big)$ *satisfies the bound*

$$|||\widehat{\mathbf{\Theta}} - \mathbf{\Theta}^*|||_2 \leq 30\sqrt{2}\,\frac{\sigma\sqrt{\gamma_{\max}(\widehat{\mathbf{\Sigma}})}}{\gamma_{\min}(\widehat{\mathbf{\Sigma}})}\left(\sqrt{\frac{p+T}{n}} + \delta\right) \tag{10.33}$$

with probability at least $1 - 2e^{-2n\delta^2}$. *Moreover, we have*

$$|||\widehat{\mathbf{\Theta}} - \mathbf{\Theta}^*|||_{\mathrm{F}} \leq 4\sqrt{2r}|||\widehat{\mathbf{\Theta}} - \mathbf{\Theta}^*|||_2 \quad \textit{and} \quad |||\widehat{\mathbf{\Theta}} - \mathbf{\Theta}^*|||_{\mathrm{nuc}} \leq 32r|||\widehat{\mathbf{\Theta}} - \mathbf{\Theta}^*|||_2. \tag{10.34}$$

Note that the guarantee (10.33) is meaningful only when $n > p$, since the lower bound $\gamma_{\min}(\widehat{\mathbf{\Sigma}}) > 0$ cannot hold otherwise. However, even if the matrix $\mathbf{\Theta}^*$ were rank-one, it would have at least $p + T$ degrees of freedom, so this lower bound is unavoidable.

Proof We first claim that condition (10.20) holds with $\kappa = \gamma_{\min}(\widehat{\mathbf{\Sigma}})$ and $\tau_n = 0$. We have $\nabla\mathcal{L}_n(\mathbf{\Theta}) = \frac{1}{n}\mathbf{Z}^{\mathrm{T}}(y - \mathbf{Z}\mathbf{\Theta})$, and hence $\nabla\mathcal{L}_n(\mathbf{\Theta}^* + \mathbf{\Delta}) - \nabla\mathcal{L}_n(\mathbf{\Theta}^*) = \widehat{\mathbf{\Sigma}}\mathbf{\Delta}$ where $\widehat{\mathbf{\Sigma}} = \frac{\mathbf{Z}^{\mathrm{T}}\mathbf{Z}}{n}$ is the sample covariance. Thus, it suffices to show that

$$|||\widehat{\mathbf{\Sigma}}\mathbf{\Delta}|||_2 \geq \gamma_{\min}(\widehat{\mathbf{\Sigma}})|||\mathbf{\Delta}|||_2 \qquad \text{for all } \mathbf{\Delta} \in \mathbb{R}^{d\times T}.$$

For any vector $u \in \mathbb{R}^T$, we have $\|\widehat{\mathbf{\Sigma}}\mathbf{\Delta}u\|_2 \geq \gamma_{\min}(\widehat{\mathbf{\Sigma}})\|\mathbf{\Delta}u\|_2$, and thus

$$|||\widehat{\mathbf{\Sigma}}\mathbf{\Delta}|||_2 \sup_{\|u\|_2=1} \|\widehat{\mathbf{\Sigma}}\mathbf{\Delta}u\|_2 \geq \gamma_{\min}(\widehat{\mathbf{\Sigma}}) \sup_{\|u\|_2=1} \|\mathbf{\Delta}u\|_2 = \gamma_{\min}(\widehat{\mathbf{\Sigma}})\,|||\mathbf{\Delta}|||_2,$$

which establishes the claim.

It remains to verify that the inequality $|||\nabla \mathcal{L}_n(\Theta^*)|||_2 \leq \frac{\lambda_n}{2}$ holds with high probability under the stated choice of λ_n. For this model, we have $\nabla \mathcal{L}_n(\Theta^*) = \frac{1}{n}\mathbf{Z}^T\mathbf{W}$, where $\mathbf{W} \in \mathbb{R}^{n\times T}$ is a zero-mean matrix of i.i.d. σ-sub-Gaussian variates. As shown in Exercise 10.8, we have

$$\mathbb{P}\left[|||\frac{1}{n}\mathbf{Z}^T\mathbf{W}|||_2 \geq 5\sigma\sqrt{\gamma_{\max}(\widehat{\Sigma})}\,\Big(\sqrt{\frac{d+T}{n}} + \delta\Big) \geq\right] \leq 2\,e^{-2n\delta^2}, \tag{10.35}$$

from which the validity of λ_n follows. Thus, the bound (10.33) follows from Proposition 10.7.

Turning to the remaining bounds (10.34), with the given choice of λ_n, the cone inequality (10.15) guarantees that $|||\widehat{\Delta}_{\bar{\mathbb{M}}^\perp}|||_{\text{nuc}} \leq 3|||\widehat{\Delta}_{\bar{\mathbb{M}}}|||_{\text{nuc}}$. Since any matrix in $\bar{\mathbb{M}}$ has rank at most $2r$, we conclude that $|||\widehat{\Delta}|||_{\text{nuc}} \leq 4\sqrt{2r}|||\widehat{\Delta}|||_F$. Consequently, the nuclear norm bound in equation (10.34) follows from the Frobenius norm bound. We have

$$|||\widehat{\Delta}|||_F^2 = \langle\langle \widehat{\Delta},\, \widehat{\Delta}\rangle\rangle \overset{\text{(i)}}{\leq} |||\widehat{\Delta}|||_{\text{nuc}}|||\widehat{\Delta}|||_2 \overset{\text{(ii)}}{\leq} 4\sqrt{2r}|||\widehat{\Delta}|||_F\,|||\widehat{\Delta}|||_2,$$

where step (i) follows from Hölder's inequality, and step (ii) follows from our previous bound. Canceling out a factor of $|||\widehat{\Delta}|||_F$ from both sides yields the Frobenius norm bound in equation (10.34), thereby completing the proof. □

10.6 Matrix completion

Let us now return to analyze the matrix completion problem previously introduced in Example 10.2. Recall that it corresponds to a particular case of matrix regression: observations are of the form $y_i = \langle\langle \mathbf{X}_i,\, \Theta^*\rangle\rangle + w_i$, where $\mathbf{X}_i \in \mathbb{R}^{d_1\times d_2}$ is a sparse mask matrix, zero everywhere except for a single randomly chosen entry $(a(i), b(i))$, where it is equal to $\sqrt{d_1 d_2}$. The sparsity of these regression matrices introduces some subtlety into the analysis of the matrix completion problem, as will become clear in the analysis to follow.

Let us now clarify why we chose to use rescaled mask matrices $\mathbf{X}_i$—that is, equal to $\sqrt{d_1 d_2}$ instead of 1 in their unique non-zero entry. With this choice, we have the convenient relation

$$\mathbb{E}\left[\frac{\|\mathfrak{X}_n(\Theta^*)\|_2^2}{n}\right] = \frac{1}{n}\sum_{i=1}^{n}\mathbb{E}[\langle\langle \mathbf{X}_i,\, \Theta^*\rangle\rangle^2] = |||\Theta^*|||_F^2, \tag{10.36}$$

using the fact that each entry of Θ^* is picked out with probability $(d_1 d_2)^{-1}$.

The calculation (10.36) shows that, for any unit-norm matrix Θ^*, the squared Euclidean norm of $\|\mathfrak{X}_n(\Theta^*)\|_2/\sqrt{n}$ has mean one. Nonetheless, in the high-dimensional setting of interest, namely, when $n \ll d_1 d_2$, there are many non-zero matrices Θ^* of low rank such that $\mathfrak{X}_n(\Theta^*) = 0$ with high probability. This phenomenon is illustrated by the following example.

Example 10.15 (Troublesome cases for matrix completion) Consider the matrix

$$\Theta^{\text{bad}} := e_1 \otimes e_1 = \begin{bmatrix} 1 & 0 & 0 & \cdots & 0 \\ 0 & 0 & 0 & \cdots & 0 \\ 0 & 0 & 0 & \cdots & 0 \\ \vdots & \vdots & \vdots & \ddots & \vdots \\ 0 & 0 & 0 & \cdots & 0 \end{bmatrix}, \tag{10.37}$$

which is of rank one. Let $\mathfrak{X}_n : \mathbb{R}^{d \times d} \to \mathbb{R}^n$ be the random observation operation based on n i.i.d. draws (with replacement) of rescaled mask matrices $\mathbf{X}_i$. As we show in Exercise 10.3, we have $\mathfrak{X}_n(\Theta^{\text{bad}}) = 0$ with probability converging to one whenever $n = o(d^2)$. ♣

Consequently, if we wish to prove non-trivial results about matrix completion in the regime $n \ll d_1 d_2$, we need to exclude matrices of the form (10.37). One avenue for doing so is by imposing so-called matrix incoherence conditions directly on the singular vectors of the unknown matrix $\Theta^* \in \mathbb{R}^{d_1 \times d_2}$. These conditions were first introduced in the context of numerical linear algebra, in which context they are known as leverage scores (see the bibliographic section for further discussion). Roughly speaking, conditions on the leverage scores ensure that the singular vectors of Θ^* are relatively "spread out".

More specifically, consider the singular value decomposition $\Theta^* = \mathbf{U}\mathbf{D}\mathbf{V}^{\mathrm{T}}$, where $\mathbf{D}$ is a diagonal matrix of singular values, and the columns of $\mathbf{U}$ and $\mathbf{V}$ contain the left and right singular vectors, respectively. What does it mean for the singular values to be spread out? Consider the matrix $\mathbf{U} \in \mathbb{R}^{d_1 \times r}$ of left singular vectors. By construction, each of its d_1-dimensional columns is normalized to Euclidean norm one; thus, if each singular vector were perfectly spread out, then each entry would have magnitude of the order $1/\sqrt{d_1}$. As a consequence, in this ideal case, each r-dimensional row of $\mathbf{U}$ would have Euclidean norm exactly $\sqrt{r/d_1}$. Similarly, the rows of $\mathbf{V}$ would have Euclidean norm $\sqrt{r/d_2}$ in the ideal case.

In general, the Euclidean norms of the rows of $\mathbf{U}$ and $\mathbf{V}$ are known as the left and right *leverage scores* of the matrix Θ^*, and matrix incoherence conditions enforce that they are relatively close to the ideal case. More specifically, note that the matrix $\mathbf{U}\mathbf{U}^{\mathrm{T}} \in \mathbb{R}^{d_1 \times d_1}$ has diagonal entries corresponding to the squared left leverage scores, with a similar observation for the matrix $\mathbf{V}\mathbf{V}^{\mathrm{T}} \in \mathbb{R}^{d_2 \times d_2}$. Thus, one way in which to control the leverage scores is via bounds of the form

$$\|\mathbf{U}\mathbf{U}^{\mathrm{T}} - \frac{r}{d_1}\mathbf{I}_{d_1 \times d_1}\|_{\max} \le \mu \frac{\sqrt{r}}{d_1} \quad \text{and} \quad \|\mathbf{V}\mathbf{V}^{\mathrm{T}} - \frac{r}{d_2}\mathbf{I}_{d_2 d_2}\|_{\max} \le \mu \frac{\sqrt{r}}{d_2}, \tag{10.38}$$

where $\mu > 0$ is the *incoherence parameter*. When the unknown matrix Θ^* satisfies conditions of this type, it is possible to establish exact recovery results for the noiseless version of the matrix completion problem. See the bibliographic section for further discussion.

In the more realistic setting of noisy observations, the incoherence conditions (10.38) have an unusual property, in that they have no dependence on the singular values. In the presence of noise, one cannot expect to recover the matrix exactly, but rather only an estimate that captures all "significant" components. Here significance is defined relative to the noise level. Unfortunately, the incoherence conditions (10.38) are non-robust, and so less suitable in application to noisy problems. An example is helpful in understanding this issue.

Example 10.16 (Non-robustness of singular vector incoherence) Define the d-dimensional vector $z = \begin{bmatrix} 0 & 1 & 1 & \cdots & 1 \end{bmatrix}$, and the associated matrix $\mathbf{Z}^* := (z \otimes z)/d$. By construction, the matrix $\mathbf{Z}^*$ is rank-one, and satisfies the incoherence conditions (10.38) with constant μ. But now suppose that we "poison" this incoherent matrix with a small multiple of the "bad" matrix from Example 10.15, in particular forming the matrix

$$\mathbf{\Gamma}^* = (1 - \delta)\mathbf{Z}^* + \delta \mathbf{\Theta}^{\text{bad}} \qquad \text{for some } \delta \in (0, 1]. \tag{10.39}$$

As long as $\delta > 0$, then the matrix $\mathbf{\Gamma}^*$ has $e_1 \in \mathbb{R}^d$ as one of its eigenvectors, and so violates the incoherence conditions (10.38). But for the non-exact recovery results of interest in a statistical setting, very small values of δ need not be a concern, since the component $\delta\mathbf{\Theta}^{\text{bad}}$ has Frobenius norm δ, and so can be ignored. ♣

There are various ways of addressing this deficiency of the incoherence conditions (10.38). Possibly the simplest is by bounding the maximum absolute value of the matrix, or rather in order to preserve the scale of the problem, by bounding the ratio of the maximum value to its Frobenius norm. More precisely, for any non-zero matrix $\mathbf{\Theta} \in \mathbb{R}^{d_1 \times d_2}$, we define the *spikiness ratio*

$$\alpha_{\text{sp}}(\mathbf{\Theta}) = \frac{\sqrt{d_1 d_2}\, \|\mathbf{\Theta}\|_{\max}}{|||\mathbf{\Theta}|||_{\text{F}}}, \tag{10.40}$$

where $\|\cdot\|_{\max}$ denotes the elementwise maximum absolute value. By definition of the Frobenius norm, we have

$$|||\mathbf{\Theta}|||_{\text{F}}^2 = \sum_{j=1}^{d_1} \sum_{k=1}^{d_2} \Theta_{jk}^2 \leq d_1 d_2 \, \|\mathbf{\Theta}\|_{\max}^2,$$

so that the spikiness ratio is lower bounded by 1. On the other hand, it can also be seen that $\alpha_{\text{sp}}(\mathbf{\Theta}) \leq \sqrt{d_1 d_2}$, where this upper bound is achieved (for instance) by the previously constructed matrix (10.37). Recalling the "poisoned" matrix (10.39), note that unlike the incoherence condition, its spikiness ratio degrades as δ increases, but not in an abrupt manner. In particular, for any $\delta \in [0, 1]$, we have $\alpha_{\text{sp}}(\mathbf{\Gamma}^*) \leq \frac{(1-\delta)+\delta d}{1-2\delta}$.

The following theorem establishes a form of restricted strong convexity for the random operator that underlies matrix completion. To simplify the theorem statement, we adopt the shorthand $d = d_1 + d_2$.

Theorem 10.17 *Let $\mathfrak{X}_n \colon \mathbb{R}^{d_1 \times d_2} \to \mathbb{R}^n$ be the random matrix completion operator formed by n i.i.d. samples of rescaled mask matrices $\mathbf{X}_i$. Then there are universal positive constants (c_1, c_2) such that*

$$\left| \frac{1}{n} \frac{\|\mathfrak{X}_n(\mathbf{\Theta})\|_2^2}{|||\mathbf{\Theta}|||_{\text{F}}^2} - 1 \right| \leq c_1\, \alpha_{\text{sp}}(\mathbf{\Theta}) \frac{|||\mathbf{\Theta}|||_{\text{nuc}}}{|||\mathbf{\Theta}|||_{\text{F}}} \sqrt{\frac{d \log d}{n}} + c_2\, \alpha_{\text{sp}}^2(\mathbf{\Theta}) \left(\sqrt{\frac{d \log d}{n}} + \delta \right)^2 \tag{10.41}$$

for all non-zero $\mathbf{\Theta} \in \mathbb{R}^{d_1 \times d_2}$, uniformly with probability at least $1 - 2e^{-\frac{1}{2} d \log d - n\delta}$.

In order to interpret this claim, note that the ratio $\beta(\mathbf{\Theta}) := \frac{|||\mathbf{\Theta}|||_{\text{nuc}}}{|||\mathbf{\Theta}|||_{\text{F}}}$ serves as a "weak" measure of the rank. For any rank-r matrix, we have $\beta(\mathbf{\Theta}) \leq \sqrt{r}$, but in addition, there are many other higher-rank matrices that also satisfy this type of bound. On the other hand, recall the "bad" matrix $\mathbf{\Theta}^{\text{bad}}$ from Example 10.15. Although it has rank one, its spikiness ratio is maximal—that is, $\alpha_{\text{sp}}(\mathbf{\Theta}^{\text{bad}}) = d$. Consequently, the bound (10.41) does not provide any interesting guarantee until $n \gg d^2$. This prediction is consistent with the result of Exercise 10.3.

Before proving Theorem 10.17, let us state and prove one of its consequences for noisy matrix completion. Given n i.i.d. samples $\widetilde{y}_i$ from the noisy linear model (10.6), consider the nuclear norm regularized estimator

$$\widehat{\mathbf{\Theta}} \in \arg \min_{\|\mathbf{\Theta}\|_{\max} \leq \frac{\alpha}{\sqrt{d_1 d_2}}} \left\{ \frac{1}{2n} \sum_{i=1}^{n} d_1 d_2 \left\{\widetilde{y}_i - \Theta_{a(i),b(i)}\right\}^2 + \lambda_n |||\mathbf{\Theta}|||_{\text{nuc}} \right\}, \tag{10.42}$$

where Theorem 10.17 motivates the addition of the extra side constraint on the infinity norm of $\mathbf{\Theta}$. As before, we use the shorthand notation $d = d_1 + d_2$.

Corollary 10.18 *Consider the observation model* (10.6) *for a matrix* $\mathbf{\Theta}^*$ *with rank at most* r, *elementwise bounded as* $\|\mathbf{\Theta}^*\|_{\max} \leq \alpha/\sqrt{d_1 d_2}$, *and i.i.d. additive noise variables* $\{w_i\}_{i=1}^n$ *that satisfy the Bernstein condition with parameters* (σ, b). *Given a sample size* $n > \frac{100b^2}{\sigma^2} d \log d$, *if we solve the program* (10.42) *with* $\lambda_n^2 = 25 \frac{\sigma^2 d \log d}{n} + \delta^2$ *for some* $\delta \in (0, \frac{\sigma^2}{2b})$, *then any optimal solution* $\widehat{\mathbf{\Theta}}$ *satisfies the bound*

$$|||\widehat{\mathbf{\Theta}} - \mathbf{\Theta}^*|||_{\text{F}}^2 \leq c_1 \max\{\sigma^2, \alpha^2\}\, r \left\{ \frac{d \log d}{n} + \delta^2 \right\} \tag{10.43}$$

with probability at least $1 - e^{-\frac{n\delta^2}{16d}} - 2e^{-\frac{1}{2} d \log d - n\delta}$.

Remark: Note that the bound (10.43) implies that the squared Frobenius norm is small as long as (apart from a logarithmic factor) the sample size n is larger than the degrees of freedom in a rank-r matrix—namely, $r(d_1 + d_2)$.

Proof We first verify that the good event $\mathbb{G}(\lambda_n) = \{|||\nabla \mathcal{L}_n(\mathbf{\Theta}^*)|||_2 \leq \frac{\lambda_n}{2}\}$ holds with high probability. Under the observation model (10.6), the gradient of the least-squares objective (10.42) is given by

$$\nabla \mathcal{L}_n(\mathbf{\Theta}^*) = \frac{1}{n} \sum_{i=1}^{n} (d_1 d_2) \frac{w_i}{\sqrt{d_1 d_2}} \mathbf{E}_i = \frac{1}{n} \sum_{i=1}^{n} w_i \mathbf{X}_i,$$

where we recall the rescaled mask matrices $\mathbf{X}_i := \sqrt{d_1 d_2}\, \mathbf{E}_i$. From our calculations in Example 6.18, we have[3]

$$\mathbb{P}\Big[|||\frac{1}{n}\sum_{i=1}^{n} w_i \mathbf{X}_i|||_2 \geq \epsilon\Big] \leq 4d\, e^{-\frac{n\epsilon^2}{8d(\sigma^2+b\epsilon)}} \leq 4d\, e^{-\frac{n\epsilon^2}{16d\sigma^2}},$$

where the second inequality holds for any $\epsilon > 0$ such that $b\epsilon \leq \sigma^2$. Under the stated lower bound on the sample size, we are guaranteed that $b\lambda_n \leq \sigma^2$, from which it follows that the event $\mathbb{G}(\lambda_n)$ holds with the claimed probability.

Next we use Theorem 10.17 to verify a variant of the restricted strong convexity condition. Under the event $\mathbb{G}(\lambda_n)$, Proposition 9.13 implies that the error matrix $\widehat{\mathbf{\Delta}} = \widehat{\mathbf{\Theta}} - \mathbf{\Theta}^*$ satisfies the constraint $|||\widehat{\mathbf{\Delta}}|||_{\text{nuc}} \leq 4|||\widehat{\mathbf{\Delta}}_{\bar{\mathbb{M}}}|||_{\text{nuc}}$. As noted earlier, any matrix in $\bar{\mathbb{M}}$ has rank at most $2r$, whence $|||\widehat{\mathbf{\Delta}}|||_{\text{nuc}} \leq 4\sqrt{2r}\,|||\widehat{\mathbf{\Delta}}|||_{\text{F}}$. By construction, we also have $\|\widehat{\mathbf{\Delta}}\|_{\max} \leq \frac{2\alpha}{\sqrt{d_1 d_2}}$. Putting together the pieces, Theorem 10.17 implies that, with probability at least $1 - 2e^{-\frac{1}{2}d\log d - n\delta}$, the observation operator $\mathfrak{X}_n$ satisfies the lower bound

$$\begin{aligned}
\frac{\|\mathfrak{X}_n(\widehat{\mathbf{\Delta}})\|_2^2}{n} &\geq |||\widehat{\mathbf{\Delta}}|||_{\text{F}}^2 - 8\sqrt{2}\,c_1\alpha\sqrt{\frac{rd\log d}{n}}\,|||\widehat{\mathbf{\Delta}}|||_{\text{F}} - 4c_2\alpha^2\left(\sqrt{\frac{d\log d}{n}} + \delta\right)^2 \\
&\geq |||\widehat{\mathbf{\Delta}}|||_{\text{F}}\left\{|||\widehat{\mathbf{\Delta}}|||_{\text{F}} - 8\sqrt{2}\,c_1\alpha\sqrt{\frac{rd\log d}{n}}\right\} - 8c_2\alpha^2\left(\frac{d\log d}{n} + \delta^2\right). \qquad (10.44)
\end{aligned}$$

In order to complete the proof using this bound, we only need to consider two possible cases.

Case 1: On one hand, if either

$$|||\widehat{\mathbf{\Delta}}|||_{\text{F}} \leq 16\sqrt{2}c_1\alpha\sqrt{\frac{rd\log d}{n}} \quad \text{or} \quad |||\widehat{\mathbf{\Delta}}|||_{\text{F}}^2 \leq 64c_2\alpha^2\left(\frac{d\log d}{n} + \delta^2\right),$$

then the claim (10.43) follows.

Case 2: Otherwise, we must have

$$|||\widehat{\mathbf{\Delta}}|||_{\text{F}} - 8\sqrt{2}\,c_1\alpha\sqrt{\frac{rd\log d}{n}} > \frac{|||\widehat{\mathbf{\Delta}}|||_{\text{F}}}{2} \quad \text{and} \quad 8c_2\alpha^2\left(\frac{d\log d}{n} + \delta^2\right) < \frac{|||\widehat{\mathbf{\Delta}}|||_{\text{F}}^2}{4},$$

and hence the lower bound (10.44) implies that

$$\frac{\|\mathfrak{X}_n(\widehat{\mathbf{\Delta}})\|_2^2}{n} \geq \frac{1}{2}|||\widehat{\mathbf{\Delta}}|||_{\text{F}}^2 - \frac{1}{4}|||\widehat{\mathbf{\Delta}}|||_{\text{F}}^2 = \frac{1}{4}|||\widehat{\mathbf{\Delta}}|||_{\text{F}}^2.$$

This is the required restricted strong convexity condition, and so the proof is then complete. □

Finally, let us return to prove Theorem 10.17.

[3] Here we have included a factor of 8 (as opposed to 2) in the denominator of the exponent, to account for the possible need of symmetrizing the random variables w_i.

Proof Given the invariance of the inequality to rescaling, we may assume without loss of generality that $|||\mathbf{\Theta}|||_F = 1$. For given positive constants (α, ρ), define the set

$$\mathbb{S}(\alpha,\rho) = \left\{ \mathbf{\Theta} \in \mathbb{R}^{d_1 \times d_2} \mid |||\mathbf{\Theta}|||_F = 1, \quad \|\mathbf{\Theta}\|_{\max} \le \frac{\alpha}{\sqrt{d_1 d_2}} \quad \text{and} \quad |||\mathbf{\Theta}|||_{\text{nuc}} \le \rho \right\}, \tag{10.45}$$

as well as the associated random variable $Z(\alpha,\rho) := \sup_{\mathbf{\Theta}\in\mathbb{S}(\alpha,\rho)} \left|\frac{1}{n}\|\mathfrak{X}_n(\mathbf{\Theta})\|_2^2 - 1\right|$. We begin by showing that there are universal constants (c_1, c_2) such that

$$\mathbb{P}\left[Z(\alpha,\rho) \ge \frac{c_1}{4}\alpha\rho\sqrt{\frac{d\log d}{n}} + \frac{c_2}{4}(\alpha\sqrt{\frac{d\log d}{n}})^2\right] \le e^{-d\log d}. \tag{10.46}$$

Here our choice of the rescaling by $1/4$ is for later theoretical convenience. Our proof of this bound is divided into two steps.

Concentration around mean: Introducing the convenient shorthand notation $F_{\mathbf{\Theta}}(\mathbf{X}) := \langle\langle \mathbf{\Theta}, \mathbf{X}\rangle\rangle^2$, we can write

$$Z(\alpha, r) = \sup_{\mathbf{\Theta}\in\mathbb{S}(\alpha,\rho)} \left|\frac{1}{n}\sum_{i=1}^n F_{\mathbf{\Theta}}(\mathbf{X}_i) - \mathbb{E}[F_{\mathbf{\Theta}}(\mathbf{X}_i)]\right|,$$

so that concentration results for empirical processes from Chapter 3 can be applied. In particular, we will apply the Bernstein-type bound (3.86): in order to do, we need to bound $\|F_{\mathbf{\Theta}}\|_{\max}$ and $\text{var}(F_{\mathbf{\Theta}}(\mathbf{X}))$ uniformly over the class. On one hand, for any rescaled mask matrix $\mathbf{X}$ and parameter matrix $\mathbf{\Theta} \in \mathbb{S}(\alpha, r)$, we have

$$|F_{\mathbf{\Theta}}(\mathbf{X})| \le \|\mathbf{\Theta}\|_{\max}^2 \|\mathbf{X}\|_1^2 \le \frac{\alpha^2}{d_1 d_2} d_1 d_2 = \alpha^2,$$

where we have used the fact that $\|\mathbf{X}\|_1^2 = d_1 d_2$ for any rescaled mask matrix. Turning to the variance, we have

$$\text{var}(F_{\mathbf{\Theta}}(\mathbf{X})) \le \mathbb{E}[F_{\mathbf{\Theta}}^2(\mathbf{X})] \le \alpha^2\mathbb{E}[F_{\mathbf{\Theta}}(\mathbf{X})] = \alpha^2,$$

a bound which holds for any $\mathbf{\Theta} \in \mathbb{S}(\alpha,\rho)$. Consequently, applying the bound (3.86) with $\epsilon = 1$ and $t = d\log d$, we conclude that there are universal constants (c_1, c_2) such that

$$\mathbb{P}\left[Z(\alpha,\rho) \ge 2\mathbb{E}[Z(\alpha,r)] + \frac{c_1}{8}\alpha\sqrt{\frac{d\log d}{n}} + \frac{c_2}{4}\alpha^2\frac{d\log d}{n}\right] \le e^{-d\log d}. \tag{10.47}$$

Bounding the expectation: It remains to bound the expectation. By Rademacher symmetrization (see Proposition 4.11), we have

$$\mathbb{E}[Z(\alpha,\rho)] \le 2\mathbb{E}\left[\sup_{\mathbf{\Theta}\in\mathbb{S}(\alpha,\rho)}\left|\frac{1}{n}\sum_{i=1}^n \varepsilon_i\langle\langle \mathbf{X}_i, \mathbf{\Theta}\rangle\rangle^2\right|\right] \stackrel{\text{(ii)}}{\le} 4\alpha\,\mathbb{E}\left[\sup_{\mathbf{\Theta}\in\mathbb{S}(\alpha,\rho)}\left|\frac{1}{n}\sum_{i=1}^n \varepsilon_i\langle\langle \mathbf{X}_i, \mathbf{\Theta}\rangle\rangle\right|\right],$$

where inequality (ii) follows from the Ledoux–Talagrand contraction inequality (5.61) for Rademacher processes, using the fact that $|\langle\langle \mathbf{\Theta}, \mathbf{X}_i\rangle\rangle| \le \alpha$ for all pairs $(\mathbf{\Theta}, \mathbf{X}_i)$. Next we apply

Hölder's inequality to bound the remaining term: more precisely, since $|||\mathbf{\Theta}|||_{\text{nuc}} \leq \rho$ for any $\mathbf{\Theta} \in \mathbb{S}(\alpha, \rho)$, we have

$$\mathbb{E}\left[\sup_{\mathbf{\Theta}\in\mathbb{S}(\alpha,\rho)} \left|\langle\langle \frac{1}{n}\sum_{i=1}^{n} \varepsilon_i \mathbf{X}_i,\ \mathbf{\Theta}\rangle\rangle\right|\right] \leq \rho\ \mathbb{E}\left[|||\frac{1}{n}\sum_{i=1}^{n} \varepsilon_i \mathbf{X}_i|||_2\right].$$

Finally, note that each matrix $\varepsilon_i \mathbf{X}_i$ is zero-mean, has its operator norm upper bounded as $|||\varepsilon_i \mathbf{X}_i|||_2 \leq \sqrt{d_1 d_2} \leq d$, and its variance bounded as

$$|||\operatorname{var}(\varepsilon_i \mathbf{X}_i)|||_2 = \frac{1}{d_1 d_2} |||d_1 d_2 \,(1 \otimes 1)|||_2 = \sqrt{d_1 d_2}.$$

Consequently, the result of Exercise 6.10 implies that

$$\mathbb{P}\left[|||\frac{1}{n}\sum_{i=1}^{n} \varepsilon_i \mathbf{X}_i|||_2 \geq \delta\right] \leq 2d \exp\left\{\frac{n\delta^2}{2d(1+\delta)}\right\}.$$

Next, applying the result of Exercise 2.8(a) with $C = 2d$, $v^2 = \frac{d}{n}$ and $B = \frac{d}{n}$, we find that

$$\mathbb{E}\left[|||\frac{1}{n}\sum_{i=1}^{n} \varepsilon_i \mathbf{X}_i|||_2\right] \leq 2\sqrt{\frac{d}{n}}\left(\sqrt{\log(2d)} + \sqrt{\pi}\right) + \frac{4d\log(2d)}{n} \overset{\text{(i)}}{\leq} 16\sqrt{\frac{d\log d}{n}}.$$

Here the inequality (i) uses the fact that $n > d\log d$. Putting together the pieces, we conclude that

$$\mathbb{E}[Z(\alpha,\rho)] \leq \frac{c_1}{16}\alpha\rho\sqrt{\frac{d\log d}{n}},$$

for an appropriate definition of the universal constant c_1. Since $\rho \geq 1$, the claimed bound (10.46) follows.

Note that the bound (10.46) involves the fixed quantities (α, ρ), as opposed to the arbitrary quantities $(\sqrt{d_1 d_2}\|\mathbf{\Theta}\|_{\max}, |||\mathbf{\Theta}|||_{\text{nuc}})$ that would arise in applying the result to an arbitrary matrix. Extending the bound (10.46) to the more general bound (10.41) requires a technique known as peeling.

Extension via peeling: Let $\mathbb{B}_F(1)$ denote the Frobenius ball of norm one in $\mathbb{R}^{d_1\times d_2}$, and let $\mathcal{E}$ be the event that the bound (10.41) is violated for some $\mathbf{\Theta} \in \mathbb{B}_F(1)$. For $k, \ell = 1, 2, \ldots$, let us define the sets

$$\mathbb{S}_{k,\ell} := \left\{\mathbf{\Theta} \in \mathbb{B}_F(1) \mid 2^{k-1} \leq d\|\mathbf{\Theta}\|_{\max} \leq 2^k \text{ and } 2^{\ell-1} \leq |||\mathbf{\Theta}|||_{\text{nuc}} \leq 2^{\ell}\right\},$$

and let $\mathcal{E}_{k,\ell}$ be the event that the bound (10.41) is violated for some $\mathbf{\Theta} \in \mathbb{S}_{k,\ell}$. We first claim that

$$\mathcal{E} \subseteq \bigcup_{k,\ell=1}^{M} \mathcal{E}_{k,\ell}, \qquad \text{where } M = \lceil \log d \rceil. \tag{10.48}$$

Indeed, for any matrix $\Theta \in \mathbb{S}(\alpha, \rho)$, we have

$$|||\mathbf{\Theta}|||_{\text{nuc}} \geq |||\mathbf{\Theta}|||_{\text{F}} = 1 \quad \text{and} \quad |||\mathbf{\Theta}|||_{\text{nuc}} \leq \sqrt{d_1 d_2}|||\mathbf{\Theta}|||_{\text{F}} \leq d.$$

Thus, we may assume that $|||\mathbf{\Theta}|||_{\text{nuc}} \in [1, d]$ without loss of generality. Similarly, for any matrix of Frobenius norm one, we must have $d\|\mathbf{\Theta}\|_{\max} \geq \sqrt{d_1 d_2}\|\mathbf{\Theta}\|_{\max} \geq 1$ and $d\|\mathbf{\Theta}\|_{\max} \leq d$, showing that we may also assume that $d\|\mathbf{\Theta}\|_{\max} \in [1, d]$. Thus, if there exists a matrix $\mathbf{\Theta}$ of Frobenius norm one that violates the bound (10.41), then it must belong to some set $\mathbb{S}_{k,\ell}$ for $k, \ell = 1, 2 \ldots, M$, with $M = \lceil \log d \rceil$.

Next, for $\alpha = 2^k$ and $\rho = 2^\ell$, define the event

$$\widetilde{\mathcal{E}}_{k,\ell} := \left\{ Z(\alpha, \rho) \geq \frac{c_1}{4} \alpha\rho \sqrt{\frac{d \log d}{n}} + \frac{c_2}{4} \left(\alpha \sqrt{\frac{d \log d}{n}} \right)^2 \right\}.$$

We claim that $\mathcal{E}_{k,\ell} \subseteq \widetilde{\mathcal{E}}_{k,\ell}$. Indeed, if event $\mathcal{E}_{k,\ell}$ occurs, then there must exist some $\mathbf{\Theta} \in \mathbb{S}_{k,\ell}$ such that

$$\begin{aligned}
\left| \frac{1}{n} \|\mathfrak{X}_n(\mathbf{\Theta})\|_2^2 - 1 \right| &\geq c_1 d \|\mathbf{\Theta}\|_{\max} |||\mathbf{\Theta}|||_{\text{nuc}} \sqrt{\frac{d \log d}{n}} + c_2 \left(d \, \|\mathbf{\Theta}\|_{\max} \sqrt{\frac{d \log d}{n}} \right)^2 \\
&\geq c_1 2^{k-1} 2^{\ell-1} \sqrt{\frac{d \log d}{n}} + c_2 \left(2^{k-1} \sqrt{\frac{d \log d}{n}} \right)^2 \\
&\geq \frac{c_1}{4} 2^k 2^\ell \sqrt{\frac{d \log d}{n}} + \frac{c_2}{4} \left(2^k \sqrt{\frac{d \log d}{n}} \right)^2,
\end{aligned}$$

showing that $\widetilde{\mathcal{E}}_{k,\ell}$ occurs.

Putting together the pieces, we have

$$\mathbb{P}[\mathcal{E}] \overset{\text{(i)}}{\leq} \sum_{k,\ell=1}^{M} \mathbb{P}[\widetilde{\mathcal{E}}_{k,\ell}] \overset{\text{(ii)}}{\leq} M^2 e^{-d \log d} \leq e^{-\frac{1}{2} d \log d},$$

where inequality (i) follows from the union bound applied to the inclusion $\mathcal{E} \subseteq \bigcup_{k,\ell=1}^{M} \widetilde{\mathcal{E}}_{k,\ell}$; inequality (ii) is a consequence of the earlier tail bound (10.46); and inequality (iii) follows since $\log M^2 = 2 \log \log d \leq \frac{1}{2} d \log d$. □

10.7 Additive matrix decompositions

In this section, we turn to the problem of additive matrix decomposition. Consider a pair of matrices $\mathbf{\Lambda}^*$ and $\mathbf{\Gamma}^*$, and suppose that we observe a vector $y \in \mathbb{R}^n$ of the form

$$y = \mathfrak{X}_n(\mathbf{\Lambda}^* + \mathbf{\Gamma}^*) + w, \tag{10.49}$$

where $\mathfrak{X}_n$ is a known linear observation operator, mapping matrices in $\mathbb{R}^{d_1 \times d_2}$ to a vector in $\mathbb{R}^n$. In the simplest case, the observation operator performs a simple vectorization—that is, it maps a matrix $\mathbf{M}$ to the vectorized version $\text{vec}(\mathbf{M})$. In this case, the sample size n is equal to the product $d_1 d_2$ of the dimensions, and we observe noisy versions of the sum $\mathbf{\Lambda}^* + \mathbf{\Gamma}^*$.

How to recover the two components based on observations of this form? Of course, this problem is ill-posed without imposing any structure on the components. One type of structure that arises in various applications is the combination of a low-rank matrix $\mathbf{\Lambda}^*$ with a sparse matrix $\mathbf{\Gamma}^*$. We have already encountered one instance of this type of decomposition in our discussion of multivariate regression in Example 9.6. The problem of Gaussian graphical selection with hidden variables, to be discussed at more length in Section 11.4.2, provides another example of a low-rank and sparse decomposition. Here we consider some additional examples of such matrix decompositions.

Example 10.19 (Factor analysis with sparse noise) Factor analysis is a natural generalization of principal component analysis (see Chapter 8 for details on the latter). In factor analysis, we have i.i.d. random vectors $z \in \mathbb{R}^d$ assumed to be generated from the model

$$z_i = \mathbf{L}u_i + \varepsilon_i, \qquad \text{for } i = 1, 2, \ldots, N, \tag{10.50}$$

where $\mathbf{L} \in \mathbb{R}^{d \times r}$ is a loading matrix, and the vectors $u_i \sim \mathcal{N}(0, \mathbf{I}_r)$ and $\varepsilon_i \sim \mathcal{N}(0, \mathbf{\Gamma}^*)$ are independent. Given n i.i.d. samples from the model (10.50), the goal is to estimate the loading matrix $\mathbf{L}$, or the matrix $\mathbf{L}\mathbf{L}^{\mathrm{T}}$ that projects onto the column span of $\mathbf{L}$. A simple calculation shows that the covariance matrix of Z_i has the form $\mathbf{\Sigma} = \mathbf{L}\mathbf{L}^{\mathrm{T}} + \mathbf{\Gamma}^*$. Consequently, in the special case when $\mathbf{\Gamma}^* = \sigma^2 \mathbf{I}_d$, then the range of $\mathbf{L}$ is spanned by the top r eigenvectors of Σ, and so we can recover it via standard principal components analysis.

In other applications, we might no longer be guaranteed that $\mathbf{\Gamma}^*$ is the identity, in which case the top r eigenvectors of $\mathbf{\Sigma}$ need not be close to the column span of $\mathbf{L}$. Nonetheless, when $\mathbf{\Gamma}^*$ is a sparse matrix, the problem of estimating $\mathbf{L}\mathbf{L}^{\mathrm{T}}$ can be understood as an instance of our general observation model (10.3) with $n = d^2$. In particular, letting the observation vector $y \in \mathbb{R}^n$ be the vectorized version of the sample covariance matrix $\frac{1}{N}\sum_{i=1}^N z_i z_i^{\mathrm{T}}$, then some algebra shows that $y = \mathrm{vec}(\mathbf{\Lambda}^* + \mathbf{\Gamma}^*) + \mathrm{vec}(\mathbf{W})$, where $\mathbf{\Lambda}^* = \mathbf{L}\mathbf{L}^{\mathrm{T}}$ is of rank r, and the random matrix $\mathbf{W}$ is a Wishart-type noise—viz.

$$\mathbf{W} := \frac{1}{N}\sum_{i=1}^{N}(z_i \otimes z_i) - \{\mathbf{L}\mathbf{L}^{\mathrm{T}} + \mathbf{\Gamma}^*\}. \tag{10.51}$$

When $\mathbf{\Gamma}^*$ is assumed to be sparse, then this constraint can be enforced via the elementwise ℓ_1-norm. ♣

Other examples of matrix decomposition involve the combination of a low-rank matrix with a column or row-sparse matrix.

Example 10.20 (Matrix completion with corruptions) Recommender systems, as previously discussed in Example 10.2, are subject to various forms of corruption. For instance, in 2002, the Amazon recommendation system for books was compromised by a simple attack. Adversaries created a large number of false user accounts, amounting to additional rows in the matrix of user–book recommendations. These false user accounts were populated with strong positive ratings for a spiritual guide and a sex manual. Naturally enough, the end effect was that those users who liked the spiritual guide would also be recommended to read the sex manual.

If we again model the unknown true matrix of ratings as being low-rank, then such adversarial corruptions can be modeled in terms of the addition of a relatively sparse component. In the case of the false user attack described above, the adversarial component $\mathbf{\Gamma}^*$ would be relatively row-sparse, with the active rows corresponding to the false users. We are then led to the problem of recovering a low-rank matrix $\mathbf{\Lambda}^*$ based on partial observations of the sum $\mathbf{\Lambda}^* + \mathbf{\Gamma}^*$. ♣

As discussed in Chapter 6, the problem of covariance estimation is fundamental. A robust variant of the problem leads to another form of matrix decomposition, as discussed in the following example:

Example 10.21 (Robust covariance estimation) For $i = 1, 2, \ldots, N$, let $u_i \in \mathbb{R}^d$ be samples from a zero-mean distribution with unknown covariance matrix $\mathbf{\Lambda}^*$. When the vectors u_i are observed without any form of corruption, then it is straightforward to estimate $\mathbf{\Lambda}^*$ by performing PCA on the sample covariance matrix. Imagining that $j \in \{1, 2, \ldots, d\}$ indexes different individuals in the population, now suppose that the data associated with some subset S of individuals is arbitrarily corrupted. This adversarial corruption can be modeled by assuming that we observe the vectors $z_i = u_i + \gamma_i$ for $i = 1, \ldots, N$, where each $\gamma_i \in \mathbb{R}^d$ is a vector supported on the subset S. Letting $\widehat{\mathbf{\Sigma}} = \frac{1}{N}\sum_{i=1}^N (z_i \otimes z_i)$ be the sample covariance matrix of the corrupted samples, some algebra shows that it can be decomposed as $\widehat{\mathbf{\Sigma}} = \mathbf{\Lambda}^* + \mathbf{\Delta} + \mathbf{W}$, where $\mathbf{W} := \frac{1}{N}\sum_{i=1}^N (u_i \otimes u_i) - \mathbf{\Lambda}^*$ is again a type of recentered Wishart noise, and the remaining term can be written as

$$\mathbf{\Delta} := \frac{1}{N}\sum_{i=1}^N (\gamma_i \otimes \gamma_i) + \frac{1}{N}\sum_{i=1}^N (u_i \otimes \gamma_i + \gamma_i \otimes u_i). \tag{10.52}$$

Thus, defining $y = \mathrm{vec}(\widehat{\mathbf{\Sigma}})$, we have another instance of the general observation model with $n = d^2$—namely, $y = \mathrm{vec}(\mathbf{\Lambda}^* + \mathbf{\Delta}) + \mathrm{vec}(\mathbf{W})$.

Note that $\mathbf{\Delta}$ itself is not a column-sparse or row-sparse matrix; however, since each vector $v_i \in \mathbb{R}^d$ is supported only on some subset $S \subset \{1, 2, \ldots, d\}$, we can write $\mathbf{\Delta} = \mathbf{\Gamma}^* + (\mathbf{\Gamma}^*)^T$, where $\mathbf{\Gamma}^*$ is a column-sparse matrix with entries only in columns indexed by S. This structure can be enforced by the use of the column-sparse regularizer, as discussed in the sequel. ♣

Finally, as we discuss in Chapter 11 to follow, the problem of Gaussian graphical model selection with hidden variables also leads to a problem of additive matrix decomposition (see Section 11.4.2).

Having motivated additive matrix decompositions, let us now consider efficient methods for recovering them. For concreteness, we focus throughout on the case of low-rank plus elementwise-sparse matrices. First, it is important to note that—like the problem of matrix completion—we need somehow to exclude matrices that are simultaneously low-rank and sparse. Recall the matrix $\mathbf{\Theta}^{\mathrm{bad}}$ from Example 10.16: since it is both low-rank and sparse, it could be decomposed either as a low-rank matrix plus the all-zeros matrix as the sparse component, or as a sparse matrix plus the all-zeros matrix as the low-rank component.

Thus, it is necessary to impose further assumptions on the form of the decomposition. One possibility is to impose incoherence conditions (10.38) directly on the singular vectors of the low-rank matrix. As noted in Example 10.16, these bounds are not robust to small perturbations of this problem. Thus, in the presence of noise, it is more natural to consider a bound on the "spikiness" of the low-rank component, which can be enforced by bounding the maximum absolute value over its elements. Accordingly, we consider the following estimator:

$$(\widehat{\Gamma}, \widehat{\Lambda}) = \arg \min_{\substack{\Gamma \in \mathbb{R}^{d_1 \times d_2} \\ \|\Lambda\|_{\max} \leq \frac{\alpha}{\sqrt{d_1 d_2}}}} \left\{ \frac{1}{2} |||\mathbf{Y} - (\Gamma + \Lambda)|||_F^2 + \lambda_n \Big(\|\Gamma\|_1 + \omega_n |||\Lambda|||_2 \Big) \right\}. \tag{10.53}$$

It is parameterized by two regularization parameters, namely λ_n and ω_n. The following corollary provides suitable choices of these parameters that ensure the estimator is well behaved; the guarantee is stated in terms of the squared Frobenius norm error

$$e^2(\widehat{\Lambda} - \Lambda^*, \widehat{\Gamma} - \Gamma^*) := |||\widehat{\Lambda} - \Lambda^*|||_F^2 + |||\widehat{\Gamma} - \Gamma^*|||_F^2. \tag{10.54}$$

Corollary 10.22 *Suppose that we solve the convex program* (10.53) *with parameters*

$$\lambda_n \geq 2\,\|\mathbf{W}\|_{\max} + 4\frac{\alpha}{\sqrt{d_1 d_2}} \quad \text{and} \quad \omega_n \geq \frac{2|||\mathbf{W}|||_2}{\lambda_n}. \tag{10.55}$$

Then there are universal constants c_j such that for any matrix pair (Λ^*, Γ^*) with $\|\Lambda^*\|_{\max} \leq \frac{\alpha}{\sqrt{d_1 d_2}}$ and for all integers $r = 1, 2, \ldots, \min\{d_1, d_2\}$ and $s = 1, 2, \ldots, (d_1 d_2)$, the squared Frobenius error (10.54) *is upper bounded as*

$$c_1\, \omega_n^2 \lambda_n^2 \Big\{ r + \frac{1}{\omega_n \lambda_n} \sum_{j=r+1}^{\min\{d_1, d_2\}} \sigma_j(\Lambda^*) \Big\} + c_2\, \lambda_n^2 \Big\{ s + \frac{1}{\lambda_n} \sum_{(j,k) \notin S} |\Gamma^*_{jk}| \Big\}, \tag{10.56}$$

where S is an arbitrary subset of matrix indices of cardinality at most s.

As with many of our previous results, the bound (10.56) is a form of oracle inequality, meaning that the choices of target rank r and subset S can be optimized so as to achieve the tightest possible bound. For instance, when the matrix Λ^* is exactly low-rank and Γ^* is sparse, then setting $r = \text{rank}(\Lambda^*)$ and $S = \text{supp}(\Gamma^*)$ yields

$$e^2(\widehat{\Lambda} - \Lambda^*, \widehat{\Gamma} - \Gamma^*) \leq \lambda_n^2 \Big\{ c_1\, \omega_n^2\, \text{rank}(\Lambda^*) + c_2\, |\text{supp}(\Gamma^*)| \Big\}.$$

In many cases, this inequality yields optimal results for the Frobenius error of the low-rank plus sparse problem. We consider a number of examples in the exercises.

Proof We prove this claim as a corollary of Theorem 9.19. Doing so requires three steps: (i) verifying a form of restricted strong convexity; (ii) verifying the validity of the regularization parameters; and (iii) computing the subspace Lipschitz constant from Definition 9.18.

We begin with restricted strong convexity. Define the two matrices $\Delta_{\widehat{\Gamma}} = \widehat{\Gamma} - \Gamma^*$ and $\Delta_{\widehat{\Lambda}} := \widehat{\Lambda} - \Lambda^*$, corresponding to the estimation error in the sparse and low-rank components, respectively. By expanding out the quadratic form, we find that the first-order error in the Taylor series is given by

$$\mathcal{E}_n(\Delta_{\widehat{\Gamma}}, \Delta_{\widehat{\Lambda}}) = \tfrac{1}{2} |||\Delta_{\widehat{\Gamma}} + \Delta_{\widehat{\Lambda}}|||_F^2 = \underbrace{\tfrac{1}{2} \{ |||\Delta_{\widehat{\Gamma}}|||_F^2 + |||\Delta_{\widehat{\Lambda}}|||_F^2 \}}_{e^2(\Delta_{\widehat{\Lambda}}, \Delta_{\widehat{\Gamma}})} + \langle\!\langle \Delta_{\widehat{\Gamma}},\, \Delta_{\widehat{\Lambda}} \rangle\!\rangle.$$

By the triangle inequality and the construction of our estimator, we have

$$\|\Delta_{\widehat{\Lambda}}\|_{\max} \leq \|\widehat{\Lambda}\|_{\max} + \|\Lambda^*\|_{\max} \leq \frac{2\alpha}{\sqrt{d_1 d_2}}.$$

Combined with Hölder's inequality, we see that

$$\mathcal{E}_n(\Delta_{\widehat{\Gamma}}, \Delta_{\widehat{\Lambda}}) \geq \frac{1}{2} e^2(\Delta_{\widehat{\Gamma}}, \Delta_{\widehat{\Lambda}}) - \frac{2\alpha}{\sqrt{d_1 d_2}} \|\Delta_{\widehat{\Gamma}}\|_1,$$

so that restricted strong convexity holds with $\kappa = 1$, but along with an extra error term. Since it is proportional to $\|\Delta_{\widehat{\Gamma}}\|_1$, the proof of Theorem 9.19 shows that it can be absorbed without any consequence as long as $\lambda_n \geq \frac{4\alpha}{\sqrt{d_1 d_2}}$.

Verifying event $\mathbb{G}(\lambda_n)$: A straightforward calculation gives $\nabla \mathcal{L}_n(\Gamma^*, \Lambda^*) = (\mathbf{W}, \mathbf{W})$. From the dual norm pairs given in Table 9.1, we have

$$\Phi^*_{\omega_n}(\nabla \mathcal{L}_n(\Gamma^*, \Lambda^*)) = \max\left\{ \|\mathbf{W}\|_{\max}, \frac{|||\mathbf{W}|||_2}{\omega_n} \right\}, \tag{10.57}$$

so that the choices (10.55) guarantee that $\lambda_n \geq 2\Phi^*_{\omega_n}(\nabla \mathcal{L}_n(\Gamma^*, \Lambda^*))$.

Choice of model subspaces: For any subset S of matrix indices of cardinality at most s, define the subset $\mathbb{M}(S) := \{\Gamma \in \mathbb{R}^{d_1 \times d_2} \mid \Gamma_{ij} = 0 \text{ for all } (i, j) \notin S\}$. Similarly, for any $r = 1, \ldots, \min\{d_1, d_2\}$, let $\mathbb{U}_r$ and $\mathbb{V}_r$ be (respectively) the subspaces spanned by the top r left and right singular vectors of Λ^*, and recall the subspaces $\overline{\mathbb{M}}(\mathbb{U}_r, \mathbb{V}_r)$ and $\mathbb{M}^\perp(\mathbb{U}_r, \mathbb{V}_r)$ previously defined in equation (10.12). We are then guaranteed that the regularizer $\Phi_{\omega_n}(\Gamma, \Lambda) = \|\Gamma\|_1 + \omega_n |||\Lambda|||_{\text{nuc}}$ is decomposable with respect to the model subspace $\mathbb{M} := \mathbb{M}(S) \times \overline{\mathbb{M}}(\mathbb{U}_r, \mathbb{V}_r)$ and deviation space $\mathbb{M}^\perp(S) \times \mathbb{M}^\perp(\mathbb{U}_r, \mathbb{V}_r)$. It then remains to bound the subspace Lipschitz constant. We have

$$\begin{aligned}
\Psi(\mathbb{M}) = \sup_{(\Gamma,\Lambda) \in \mathbb{M}(S) \times \overline{\mathbb{M}}(\mathbb{U}_r, \mathbb{V}_r)} \frac{\|\Gamma\|_1 + \omega_n |||\Lambda|||_{\text{nuc}}}{\sqrt{|||\Gamma|||_F^2 + |||\Lambda|||_F^2}} &\leq \sup_{(\Gamma,\Lambda)} \frac{\sqrt{s}|||\Gamma|||_F + \omega_n \sqrt{2r}|||\Lambda|||_F}{\sqrt{|||\Gamma|||_F^2 + |||\Lambda|||_F^2}} \\
&\leq \sqrt{s} + \omega_n \sqrt{2r}.
\end{aligned}$$

Putting together the pieces, the overall claim (10.56) now follows as a corollary of Theorem 9.19. □

10.8 Bibliographic details and background

In her Ph.D. thesis, Fazel (2002) studied various applications of the nuclear norm as a surrogate for a rank constraint. Recht et al. (2010) studied the use of nuclear norm regularization for the compressed sensing variant of matrix regression, with noiseless observations and matrices $\mathbf{X}_i \in \mathbb{R}^{d_1 \times d_2}$ drawn independently, each with i.i.d. $\mathcal{N}(0, 1)$ entries. They established sufficient conditions for exact recovery in the noiseless setting (observation model (10.2) with $w_i = 0$) when the covariates $\mathbf{X}_i$ are drawn from the standard Gaussian ensemble (each entry of $\mathbf{X}_i$ distributed as $\mathcal{N}(0, 1)$, drawn independently). In the noisy setting, this particular ensemble was also studied by Candès and Plan (2010) and Negahban and Wainwright (2011a), who both gave sharp conditions on the required sample size. The former paper applies to sub-Gaussian but isotropic ensembles (identity covariance), whereas the latter paper established Theorem 10.8 that applies to Gaussian ensembles with arbitrary covariance matrices. Recht et al. (2009) provide precise results on the threshold behavior for the identity version of this ensemble.

Nuclear norm regularization has also been studied for more general problem classes. Rohde and Tsybakov (2011) impose a form of the restricted isometry condition (see Chapter 7), adapted to the matrix setting, whereas Negabahn and Wainwright (2011a) work with a milder lower curvature condition, corresponding to the matrix analog of a restricted eigenvalue condition in the special case of quadratic losses. Rohde and Tsybakov (2011) also provide bounds on the nuclear norm estimate in various other Schatten matrix norms. Bounds for multivariate (or multitask) regression, as in Corollary 10.14, have been proved by various authors (Lounici et al., 2011; Negahban and Wainwright, 2011a; Rohde and Tsybakov, 2011). The use of reduced rank estimators for multivariate regression has a lengthy history; see Exercise 10.1 for its explicit form as well as the references (Izenman, 1975, 2008; Reinsel and Velu, 1998) for some history and more details. See also Bunea et al. (2011) for non-asymptotic analysis of a class of reduced rank estimators in multivariate regression.

There are wide number of variants of the matrix completion problem; see the survey chapter by Laurent (2001) and references therein for more details. Srebro and his co-authors (2004; 2005a; 2005b) proposed low-rank matrix completion as a model for recommender systems, among them the Netflix problem described here. Srebro et al. (2005b) provide error bounds on the prediction error using nuclear norm regularization. Candès and Recht (2009) proved exact recovery guarantees for the nuclear norm estimator, assuming noiseless observations and certain incoherence conditions on the matrix involving the leverage scores. Leverage scores also play an important role in approximating low-rank matrices based on random subsamples of its rows or columns; see the survey by Mahoney (2011) and references therein. Gross (2011) provided a general scheme for exact recovery based on a dual witness construction, and making use of Ahlswede–Winter matrix bound from Section 6.4.4; see also Recht (2011) for a relatively simple argument for exact recovery. Keshavan et al. (2010a; 2010b) studied both methods based on the nuclear norm (SVD thresholding) as well as heuristic iterative methods for the matrix completion problem, providing guarantees in both the noiseless and noisy settings. Negahban and Wainwright (2012) study the more general setting of weighted sampling for both exactly low-rank and near-low-rank matrices, and provided minimax-optimal bounds for the ℓ_q-"balls" of matrices with control on the "spikiness" ratio (10.40). They proved a weighted form of Theorem 10.17; the proof given here for the uniformly sampled setting is more direct. Koltchinski et al. (2011) assume that the sampling design is known, and propose a variant of the matrix Lasso. In the case of uniform sampling, it corresponds to a form of SVD thresholding, an estimator that was also analyzed by Keshavan et al. (2010a; 2010b). See Exercise 10.11 for some analysis of this type of estimator.

The problem of phase retrieval from Section 10.4 has a lengthy history and various applications (e.g., Grechberg and Saxton, 1972; Fienup, 1982; Griffin and Lim, 1984; Fienup and Wackerman, 1986; Harrison, 1993). The idea of relaxing a non-convex quadratic program to a semidefinite program is a classical one (Shor, 1987; Lovász and Schrijver, 1991; Nesterov, 1998; Laurent, 2003). The semidefinite relaxation (10.29) for phase retrieval was proposed by Chai et al. (2011). Candès et al. (2013) provided the first theoretical guarantees on exact recovery, in particular for Gaussian measurement vectors. See also Waldspurger et al. (2015) for discussion and analysis of a closely related but different SDP relaxation.

The problem of additive matrix decompositions with sparse and low-rank matrices was first formalized by Chandrasekaran et al. (2011), who analyzed conditions for exact recovery based on deterministic incoherence conditions between the sparse and low-rank components. Candès et al. (2011) provided related guarantees for random ensembles with milder incoherence conditions. Chandrasekaran et al. (2012b) showed that the problem of Gaussian graphical model selection with hidden variables can be tackled within this framework; see Section 11.4.2 of Chapter 11 for more details on this problem. Agarwal et al. (2012) provide a general analysis of regularization-based methods for estimating matrix decompositions for noisy observations; their work uses the milder bounds on the maximum entry of the low-rank matrix, as opposed to incoherence conditions, but guarantees only approximate recovery. See Ren and Zhou (2012) for some two-stage approaches for estimating matrix decompositions. Fan et al. (2013) study a related class of models for covariance matrices involving both sparse and low-rank components.

10.9 Exercises

Exercise 10.1 (Reduced rank regression) Recall the model of multivariate regression from Example 10.1, and, for a target rank $r \leq T \leq p$, consider the reduced rank regression estimate

$$\widehat{\mathbf{\Theta}}_{\mathrm{RR}} := \arg \min_{\substack{\mathbf{\Theta} \in \mathbb{R}^{p \times T} \\ \mathrm{rank}(\mathbf{\Theta}) \leq r}} \left\{ \frac{1}{2n} |||\mathbf{Y} - \mathbf{Z}\mathbf{\Theta}|||_{\mathrm{F}}^2 \right\}.$$

Define the sample covariance matrix $\widehat{\mathbf{\Sigma}}_{ZZ} = \frac{1}{n}\mathbf{Z}^{\mathrm{T}}\mathbf{Z}$, and the sample cross-covariance matrix $\widehat{\mathbf{\Sigma}}_{ZY} = \frac{1}{n}\mathbf{Z}^{\mathrm{T}}\mathbf{Y}$. Assuming that $\widehat{\mathbf{\Sigma}}_{ZZ}$ is invertible, show that the reduced rank estimate has the explicit form

$$\widehat{\mathbf{\Theta}}_{\mathrm{RR}} = \widehat{\mathbf{\Sigma}}_{ZZ}^{-1}\widehat{\mathbf{\Sigma}}_{XY}\mathbf{V}\mathbf{V}^{\mathrm{T}},$$

where the matrix $\mathbf{V} \in \mathbb{R}^{T \times r}$ has columns consisting of the top r eigenvectors of the matrix $\widehat{\mathbf{\Sigma}}_{YZ}\widehat{\mathbf{\Sigma}}_{ZZ}^{-1}\widehat{\mathbf{\Sigma}}_{ZY}$.

Exercise 10.2 (Vector autogressive processes) Recall the vector autoregressive (VAR) model described in Example 10.5.

(a) Suppose that we initialize by choosing $z^1 \sim \mathcal{N}(0, \mathbf{\Sigma})$, where the symmetric matrix $\mathbf{\Sigma}$ satisfies the equation

$$\mathbf{\Sigma} - \mathbf{\Theta}^*\mathbf{\Sigma}(\mathbf{\Theta}^*)^{\mathrm{T}} - \mathbf{\Gamma} = 0. \tag{10.58}$$

Here $\mathbf{\Gamma} \succ 0$ is the covariance matrix of the driving noise. Show that the resulting stochastic process $\{z^t\}_{t=1}^{\infty}$ is stationary.

(b) Suppose that there exists a strictly positive definite solution $\mathbf{\Sigma}$ to equation (10.58). Show that $|||\mathbf{\Theta}^*|||_2 < 1$.

(c) Conversely, supposing that $|||\boldsymbol{\Theta}^*|||_2 < 1$, show that there exists a strictly positive definite solution $\boldsymbol{\Sigma}$ to equation (10.58).

Exercise 10.3 (Nullspace in matrix completion) Consider the random observation operator $\mathfrak{X}_n : \mathbb{R}^{d \times d} \to \mathbb{R}$ formed by n i.i.d. draws of rescaled mask matrices (zero everywhere except for d in an entry chosen uniformly at random). For the "bad" matrix $\boldsymbol{\Theta}^{\text{bad}}$ from equation (10.37), show that $\mathbb{P}[\mathfrak{X}_n(\boldsymbol{\Theta}^{\text{bad}}) = 0] = 1 - o(1)$ whenever $n = o(d^2)$.

Exercise 10.4 (Cone inequalities for nuclear norm) Suppose that $|||\widehat{\boldsymbol{\Theta}}|||_{\text{nuc}} \leq |||\boldsymbol{\Theta}^*|||_{\text{nuc}}$, where $\boldsymbol{\Theta}^*$ is a rank-r matrix. Show that $\widehat{\boldsymbol{\Delta}} = \widehat{\boldsymbol{\Theta}} - \boldsymbol{\Theta}^*$ satisfies the cone constraint $|||\widehat{\boldsymbol{\Delta}}_{\bar{\mathbb{M}}^\perp}|||_{\text{nuc}} \leq |||\widehat{\boldsymbol{\Delta}}_{\bar{\mathbb{M}}}|||_{\text{nuc}}$, where the subspace $\bar{\mathbb{M}}^\perp$ was defined in equation (10.14).

Exercise 10.5 (Operator norm bounds)

(a) Verify the specific form (10.20) of the Φ^*-curvature condition.
(b) Assume that $\boldsymbol{\Theta}^*$ has rank r, and that $\widehat{\boldsymbol{\Theta}} - \boldsymbol{\Theta}^*$ satisfies the cone constraint (10.15), where $\mathbb{M}(\mathbb{U}, \mathbb{V})$ is specified by subspace $\mathbb{U}$ and $\mathbb{V}$ of dimension r. Show that

$$|||\widehat{\boldsymbol{\Theta}} - \boldsymbol{\Theta}^*|||_{\text{F}} \leq 4\sqrt{2r}\, |||\widehat{\boldsymbol{\Theta}} - \boldsymbol{\Theta}^*|||_2.$$

Exercise 10.6 (Analysis of matrix compressed sensing) In this exercise, we work through part of the proof of Theorem 10.8 for the special case $\boldsymbol{\Sigma} = \mathbf{I}_D$, where $D = d_1 d_2$. In particular, defining the set

$$\mathbb{B}(t) := \left\{\boldsymbol{\Delta} \in \mathbb{R}^{d_1 \times d_1} \mid |||\boldsymbol{\Delta}|||_{\text{F}} = 1,\ |||\boldsymbol{\Delta}|||_{\text{nuc}} \leq t\right\},$$

for some $t > 0$, we show that

$$\inf_{\boldsymbol{\Delta} \in \mathbb{B}(t)} \sqrt{\frac{1}{n}\sum_{i=1}^{n} \langle\langle \mathbf{X}_i,\ \boldsymbol{\Delta}\rangle\rangle^2} \geq \frac{1}{2} - \delta - 2\left(\sqrt{\frac{d_1}{n}} + \sqrt{\frac{d_2}{n}}\right) t$$

with probability greater than $1 - e^{-n\delta^2/2}$. (This is a weaker result than Theorem 10.8, but the argument sketched here illustrates the essential ideas.)

(a) Reduce the problem to lower bounding the random variable

$$Z_n(t) := \inf_{\boldsymbol{\Delta} \in \mathbb{B}(t)} \sup_{\|u\|_2 = 1} \frac{1}{\sqrt{n}} \sum_{i=1}^{n} u_i \langle\langle \mathbf{X}_i,\ \boldsymbol{\Delta}\rangle\rangle.$$

(b) Show that the expectation can be lower bounded as

$$\mathbb{E}[Z_n(t)] \geq \frac{1}{\sqrt{n}}\left\{\mathbb{E}[\|w\|_2] - \mathbb{E}[|||\mathbf{W}|||_2]\, t\right\},$$

where $w \in \mathbb{R}^n$ and $\mathbf{W} \in \mathbb{R}^{d_1 \times d_2}$ are populated with i.i.d. $\mathcal{N}(0, 1)$ variables. (*Hint:* The Gordon–Slepian comparison principle from Chapter 5 could be useful here.)
(c) Complete the proof using concentration of measure and part (b).

Exercise 10.7 (Bounds for approximately low-rank matrices) Consider the observation model $y = \mathfrak{X}_n(\Theta^*) + w$ with $w \sim \mathcal{N}(0, \sigma^2 \mathbf{I}_n)$, and consider the nuclear norm constrained estimator

$$\widehat{\Theta} = \arg\min_{\Theta \in \mathbb{R}^{d \times d}} \left\{ \frac{1}{2n} \|y - \mathfrak{X}_n(\Theta)\|_2^2 \right\} \qquad \text{subject to } |||\Theta|||_{\text{nuc}} \leq |||\Theta^*|||_{\text{nuc}}.$$

Suppose that Θ^* belongs to the ℓ_q-"ball" of near-low-rank matrices (10.26).

In this exercise, we show that the estimate $\widehat{\Theta}$ satisfies an error bound of the form (10.27) when the random operator $\mathfrak{X}_n$ satisfies the lower bound of Theorem 10.8.

(a) For an arbitrary $r \in \{1, 2, \ldots, d\}$, let $\mathbb{U}$ and $\mathbb{V}$ be subspaces defined by the top r left and right singular vectors of Θ^*, and consider the subspace $\overline{\mathbb{M}}(\mathbb{U}, \mathbb{V})$. Prove that the error matrix $\widehat{\Delta}$ satisfies the inequality

$$|||\widehat{\Delta}_{\bar{\mathbb{M}}^\perp}|||_{\text{nuc}} \leq 2\sqrt{2r}|||\widehat{\Delta}|||_{\text{F}} + 2\sum_{j=r+1}^{d} \sigma_j(\Theta^*).$$

(b) Consider an integer $r \in \{1, \ldots, d\}$ such that $n > C\,rd$ for some sufficiently large but universal constant C. Using Theorem 10.8 and part (a), show that

$$|||\widehat{\Delta}|||_{\text{F}}^2 \precsim \underbrace{\max\{T_1(r), T_1^2(r)\}}_{\text{approximation error}} + \underbrace{\sigma\sqrt{\frac{rd}{n}}\,|||\widehat{\Delta}|||_{\text{F}}}_{\text{estimationerror}},$$

where $T_1(r) := \sigma\sqrt{\frac{d}{n}}\sum_{j=r+1}^{d} \sigma_j(\Theta^*)$. (*Hint:* You may assume that an inequality of the form $|||\frac{1}{n}\sum_{i=1}^{n} w_i \mathbf{X}_i|||_2 \precsim \sigma\sqrt{\frac{d}{n}}$ holds.)

(c) Specify a choice of r that trades off the estimation and approximation error optimally.

Exercise 10.8 Under the assumptions of Corollary 10.14, prove that the bound (10.35) holds.

Exercise 10.9 (Phase retrieval with Gaussian masks) Recall the real-valued phase retrieval problem, based on the functions $f_\Theta(\mathbf{X}) = \langle\langle \mathbf{X}, \Theta \rangle\rangle$, for a random matrix $\mathbf{X} = x \otimes x$ with $x \sim \mathcal{N}(0, \mathbf{I}_n)$.

(a) Letting $\Theta = \mathbf{U}^{\text{T}}\mathbf{D}\mathbf{U}$ denote the singular value decomposition of Θ, explain why the random variables $f_\Theta(\mathbf{X})$ and $f_{\mathbf{D}}(\mathbf{X})$ have the same distributions.

(b) Prove that

$$\mathbb{E}[f_\Theta^2(\mathbf{X})] = |||\Theta|||_{\text{F}}^2 + 2(\text{trace}(\Theta))^2.$$

Exercise 10.10 (Analysis of noisy matrix completion) In this exercise, we work through the proof of Corollary 10.18.

(a) Argue that with the setting $\lambda_n \geq |||\frac{1}{n}\sum_{i=1}^{n} w_i \mathbf{E}_i|||_2$, we are guaranteed that the error matrix $\widehat{\Delta} = \widehat{\Theta} - \Theta^*$ satisfies the bounds

$$\frac{|||\widehat{\Delta}|||_{\text{nuc}}}{|||\widehat{\Delta}|||_{\text{F}}} \leq 2\sqrt{2r} \quad \text{and} \quad \|\widehat{\Delta}\|_{\max} \leq 2\alpha.$$

(b) Use part (a) and results from the chapter to show that, with high probability, at least one of the following inequalities must hold:

$$|||\widehat{\boldsymbol{\Delta}}|||_{\mathrm{F}}^2 \le \frac{c_2}{2}\alpha^2 \frac{d \log d}{n} + 128\, c_1^2\, \alpha^2 \frac{rd \log d}{n} \quad \text{or} \quad \frac{\|\mathfrak{X}_n(\widehat{\boldsymbol{\Delta}})\|_2^2}{n} \ge \frac{|||\widehat{\boldsymbol{\Delta}}|||_{\mathrm{F}}^2}{4}.$$

(c) Use part (c) to establish the bound.

Exercise 10.11 (Alternative estimator for matrix completion) Consider the problem of noisy matrix completion, based on observations $y_i = \langle\langle \mathbf{X}_i, \boldsymbol{\Theta}^* \rangle\rangle + w_i$, where $\mathbf{X}_i \in \mathbb{R}^{d \times d}$ is a d-rescaled mask matrix (i.e., with a single entry of d in one location chosen uniformly at random, and zeros elsewhere). Consider the estimator

$$\widehat{\boldsymbol{\Theta}} = \arg \min_{\boldsymbol{\Theta} \in \mathbb{R}^{d \times d}} \left\{ \frac{1}{2} |||\boldsymbol{\Theta}|||_{\mathrm{F}}^2 - \langle\langle \boldsymbol{\Theta}, \frac{1}{n} \sum_{i=1}^{n} y_i \mathbf{X}_i \rangle\rangle + \lambda_n |||\boldsymbol{\Theta}|||_{\mathrm{nuc}} \right\}.$$

(a) Show that the optimal solution $\widehat{\boldsymbol{\Theta}}$ is unique, and can be obtained by soft thresholding the singular values of the matrix $\mathbf{M} := \frac{1}{n} \sum_{i=1}^{n} y_i \mathbf{X}_i$. In particular, if $\mathbf{U}\mathbf{D}\mathbf{V}^{\mathrm{T}}$ denotes the SVD of $\mathbf{M}$, then $\widehat{\boldsymbol{\Theta}} = \mathbf{U}\,[T_{\lambda_n}(\mathbf{D})]\,\mathbf{V}^{\mathrm{T}}$, where $T_{\lambda_n}(\mathbf{D})$ is the matrix formed by soft thresholding the diagonal matrix of singular values $\mathbf{D}$.

(b) Suppose that the unknown matrix $\boldsymbol{\Theta}^*$ has rank r. Show that, with the choice

$$\lambda_n \ge 2 \max_{|||\mathbf{U}|||_{\mathrm{nuc}} \le 1} \left| \frac{1}{n} \sum_{i=1}^{n} \langle\langle \mathbf{U}, \mathbf{X}_i \rangle\rangle \langle\langle \mathbf{X}_i, \boldsymbol{\Theta}^* \rangle\rangle - \langle\langle \mathbf{U}, \boldsymbol{\Theta}^* \rangle\rangle \right| + 2 ||| \frac{1}{n} \sum_{i=1}^{n} w_i \mathbf{X}_i |||_2,$$

the optimal solution $\widehat{\boldsymbol{\Theta}}$ satisfies the bound

$$|||\widehat{\boldsymbol{\Theta}} - \boldsymbol{\Theta}^*|||_{\mathrm{F}} \le \frac{3}{\sqrt{2}} \sqrt{r}\, \lambda_n.$$

(c) Suppose that the noise vector $w \in \mathbb{R}^n$ has i.i.d. σ-sub-Gaussian entries. Specify an appropriate choice of λ_n that yields a useful bound on $|||\widehat{\boldsymbol{\Theta}} - \boldsymbol{\Theta}^*|||_{\mathrm{F}}$.

11

Graphical models for high-dimensional data

Graphical models are based on a combination of ideas from both probability theory and graph theory, and are useful in modeling high-dimensional probability distributions. They have been developed and studied in a variety of fields, including statistical physics, spatial statistics, information and coding theory, speech processing, statistical image processing, computer vision, natural language processing, computational biology and social network analysis among others. In this chapter, we discuss various problems in high-dimensional statistics that arise in the context of graphical models.

11.1 Some basics

We begin with a brief introduction to some basic properties of graphical models, referring the reader to the bibliographic section for additional references. There are various types of graphical models, distinguished by the type of underlying graph used—directed, undirected, or a hybrid of the two. Here we focus exclusively on the case of *undirected graphical models*, also known as Markov random fields. These models are based on an undirected graph $G = (V, E)$, which consists of a set of vertices $V = \{1, 2, \ldots, d\}$ joined together by a collection of edges E. In the undirected case, an edge (j, k) is an unordered pair of distinct vertices $j, k \in V$.

In order to introduce a probabilistic aspect to our models, we associate to each vertex $j \in V$ a random variable X_j, taking values in some space $\mathcal{X}_j$. We then consider the distribution $\mathbb{P}$ of the d-dimensional random vector $X = (X_1, \ldots, X_d)$. Of primary interest to us are connections between the structure of $\mathbb{P}$, and the structure of the underlying graph G. There are two ways in which to connect the probabilistic and graphical structures: one based on factorization, and the second based on conditional independence properties. A classical result in the field, known as the Hammersley–Clifford theorem, asserts that these two characterizations are essentially equivalent.

11.1.1 Factorization

One way to connect the undirected graph G to the random variables is by enforcing a certain factorization of the probability distribution. A *clique* C is a subset of vertices that are all joined by edges, meaning that $(j, k) \in E$ for all distinct vertices $j, k \in C$. A maximal clique is a clique that is not a subset of any other clique. See Figure 11.1(b) for an illustration of these concepts. We use $\mathfrak{C}$ to denote the set of all cliques in G, and for each clique $C \in \mathfrak{C}$, we use ψ_C to denote a function of the subvector $x_C := (x_j, j \in C)$. This *clique compatibility function* takes inputs from the Cartesian product space $\mathcal{X}^C := \bigotimes_{j\in C} \mathcal{X}_j$, and returns non-negative real numbers. With this notation, we have the following:

Definition 11.1 The random vector $(X_1, \ldots, X_d)$ *factorizes according to the graph G* if its density function p can be represented as

$$p(x_1, \ldots, x_d) \propto \prod_{C\in\mathfrak{C}} \psi_C(x_C) \tag{11.1}$$

for some collection of clique compatibility functions $\psi_C : \mathcal{X}^C \to [0, \infty)$.

Here the density function is taken with respect either to the counting measure for discrete-valued random variables, or to some (possibly weighted) version of the Lebesgue measure for continuous random variables. As an illustration of Definition 11.1, any density that factorizes according to the graph shown in Figure 11.1(a) must have the form

$$p(x_1, \ldots, x_7) \propto \psi_{123}(x_1, x_2, x_3)\ \psi_{345}(x_3, x_4, x_5)\ \psi_{46}(x_4, x_6)\ \psi_{57}(x_5, x_7).$$

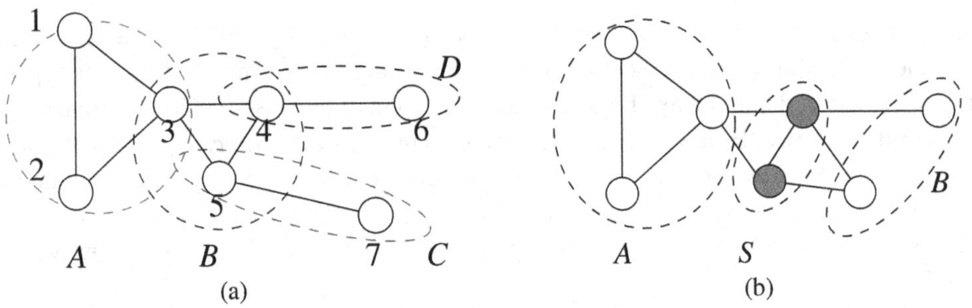

Figure 11.1 Illustration of basic graph-theoretic properties. (a) Subsets A and B are 3-cliques, whereas subsets C and D are 2-cliques. All of these cliques are maximal. Each vertex is a clique as well, but none of these singleton cliques are maximal for this graph. (b) Subset S is a vertex cutset, breaking the graph into two disconnected subgraphs with vertex sets A and B, respectively.

Without loss of generality—redefining the clique compatibility functions as necessary—the product over cliques can always be restricted to the set of all maximal cliques. However, in practice, it can be convenient to allow for terms associated with non-maximal cliques as well, as illustrated by the following.

Example 11.2 (Markov chain factorization) The standard way of factoring the distribution of a Markov chain on variables $(X_1, \ldots, X_d)$ is as

$$p(x_1, \ldots, x_d) = p_1(x_1)\ p_{2|1}(x_2 \mid x_1) \cdots p_{d|(d-1)}(x_d \mid x_{d-1}),$$

where p_1 denotes the marginal distribution of X_1, and for $j \in \{1, 2, \ldots, d-1\}$, the term $p_{j+1|j}$

denotes the conditional distribution of X_{j+1} given X_j. This representation can be understood as a special case of the factorization (11.1), using the vertex-based functions

$$\psi_1(x_1) = p_1(x_1) \quad \text{at vertex 1} \quad \text{and} \quad \psi_j(x_j) = 1 \quad \text{for all } j = 2, \ldots, d,$$

combined with the edge-based functions

$$\psi_{j,j+1}(x_j, x_{j+1}) = p_{j+1|j}(x_{j+1} \mid x_j) \qquad \text{for } j = 1, \ldots, d-1.$$

But this factorization is by no means unique. We could just as easily adopt the symmetrized factorization $\widetilde{\psi}_j(x_j) = p_j(x_j)$ for all $j = 1, \ldots, d$, and

$$\widetilde{\psi}_{jk}(x_j, x_k) = \frac{p_{jk}(x_j, x_k)}{p_j(x_j)p_k(x_k)} \qquad \text{for all } (j,k) \in E,$$

where p_{jk} denotes the joint distribution over the pair (X_j, X_k). ♣

Example 11.3 (Multivariate Gaussian factorization) Any non-degenerate Gaussian distribution with zero mean can be parameterized in terms of its inverse covariance matrix $\mathbf{\Theta}^* = \mathbf{\Sigma}^{-1}$, also known as the *precision matrix*. In particular, its density can be written as

$$p(x_1, \ldots, x_d; \mathbf{\Theta}^*) = \frac{\sqrt{\det(\mathbf{\Theta}^*)}}{(2\pi)^{d/2}} \, e^{-\frac{1}{2}x^{\mathrm{T}}\mathbf{\Theta}^* x}. \tag{11.2}$$

By expanding the quadratic form, we see that

$$e^{-\frac{1}{2}x^{\mathrm{T}}\mathbf{\Theta}^* x} = \exp\Big(-\frac{1}{2}\sum_{(j,k)\in E} \Theta^*_{jk}\, x_j x_k\Big) = \prod_{(j,k)\in E} \underbrace{e^{-\frac{1}{2}\Theta^*_{jk} x_j x_k}}_{\psi_{jk}(x_j,x_k)},$$

showing that any zero-mean Gaussian distribution can be factorized in terms of functions on edges, or cliques of size two. The Gaussian case is thus special: the factorization can always be restricted to cliques of size two, even if the underlying graph has higher-order cliques. ♣

We now turn to a non-Gaussian graphical model that shares a similar factorization:

Example 11.4 (Ising model) Consider a vector $X = (X_1, \ldots, X_d)$ of binary random variables, with each $X_j \in \{0, 1\}$. The *Ising model* is one of the earliest graphical models, first introduced in the context of statistical physics for modeling interactions in a magnetic field. Given an undirected graph $G = (V, E)$, it posits a factorization of the form

$$p(x_1, \ldots, x_d; \theta^*) = \frac{1}{Z(\theta^*)} \exp\Big\{\sum_{j\in V} \theta^*_j x_j + \sum_{(j,k)\in E} \theta^*_{jk} x_j x_k\Big\}, \tag{11.3}$$

where the parameter θ^*_j is associated with vertex $j \in V$, and the parameter θ^*_{jk} is associated with edge $(j,k) \in E$. The quantity $Z(\theta^*)$ is a constant that serves to enforce that the probability mass function p normalizes properly to one; more precisely, we have

$$Z(\theta^*) = \sum_{x\in\{0,1\}^d} \exp\Big\{\sum_{j\in V} \theta^*_j x_j + \sum_{(j,k)\in E} \theta^*_{jk} x_j x_k\Big\}.$$

See the bibliographic section for further discussion of the history and uses of this model. ♣

11.1.2 Conditional independence

We now turn to an alternative way in which to connect the probabilistic and graphical structures, involving certain conditional independence statements defined by the graph. These statements are based on the notion of a *vertex cutset* S, which (loosely stated) is a subset of vertices whose removal from the graph breaks it into two or more disjoint pieces. More formally, removing S from the vertex set V leads to the vertex-induced subgraph $G(V \setminus S)$, consisting of the vertex set $V \setminus S$, and the residual edge set

$$E(V \setminus S) := \{(j,k) \in E \mid j,k \in V \setminus S\}. \tag{11.4}$$

The set S is a vertex cutset if the residual graph $G(V\setminus S)$ consists of two or more disconnected non-empty components. See Figure 11.1(b) for an illustration.

We now define a conditional independence relationship associated with each vertex cutset of the graph. For any subset $A \subseteq V$, let $X_A := (X_j, j \in A)$ represent the subvector of random variables indexed by vertices in A. For any three disjoint subsets, say A, B and S, of the vertex set V, we use $X_A \perp\!\!\!\perp X_B \mid X_S$ to mean that the subvector X_A is conditionally independent of X_B given X_S.

Definition 11.5 A random vector $X = (X_1, \ldots, X_d)$ is *Markov with respect to a graph* G if, for all vertex cutsets S breaking the graph into disjoint pieces A and B, the conditional independence statement $X_A \perp\!\!\!\perp X_B \mid X_S$ holds.

Let us consider some examples to illustrate.

Example 11.6 (Markov chain conditional independence) The Markov chain provides the simplest (and most classical) illustration of this definition. A chain graph on vertex set $V = \{1, 2, \ldots, d\}$ contains the edges $(j, j+1)$ for $j = 1, 2, \ldots, d-1$; the case $d = 5$ is illustrated in Figure 11.2(a). For such a chain graph, each vertex $j \in \{2, 3, \ldots, d-1\}$ is a non-trivial cutset, breaking the graph into the "past" $P = \{1, 2, \ldots, j-1\}$ and "future" $F = \{j+1, \ldots, d\}$. These singleton cutsets define the essential Markov property of a Markov time-series model—namely, that the past X_P and future X_F are conditionally independent given the present X_j. ♣

Example 11.7 (Neighborhood-based cutsets) Another canonical type of vertex cutset is provided by the neighborhood structure of the graph. For any vertex $j \in V$, its *neighborhood set* is the subset of vertices

$$\mathcal{N}(j) := \{k \in V \mid (j,k) \in E\} \tag{11.5}$$

that are joined to j by an edge. It is easy to see that $\mathcal{N}(j)$ is always a vertex cutset, a non-trivial one as long as j is not connected to every other vertex; it separates the graph into the two disjoint components $A = \{j\}$ and $B = V \setminus (\mathcal{N}(j) \cup \{j\})$. This particular choice of vertex cutset plays an important role in our discussion of neighborhood-based methods for graphical model selection later in the chapter. ♣

11.1.3 Hammersley–Clifford equivalence

Thus far, we have introduced two (ostensibly distinct) ways of relating the random vector X to the underlying graph structure, namely the Markov property and the factorization property. We now turn to a fundamental theorem that establishes that these two properties are equivalent for any strictly positive distribution:

Theorem 11.8 (Hammersley–Clifford) *For a given undirected graph and any random vector $X = (X_1, \ldots, X_d)$ with strictly positive density p, the following two properties are equivalent:*

(a) *The random vector X factorizes according to the structure of the graph G, as in Definition 11.1.*
(b) *The random vector X is Markov with respect to the graph G, as in Definition 11.5.*

Proof Here we show that the factorization property (Definition 11.1) implies the Markov property (Definition 11.5). See the bibliographic section for references to proofs of the converse. Suppose that the factorization (11.1) holds, and let S be an arbitrary vertex cutset of the graph such that subsets A and B are separated by S. We may assume without loss of generality that both A and B are non-empty, and we need to show that $X_A \perp\!\!\!\perp X_B \mid X_S$. Let us define subsets of cliques by $\mathfrak{C}_A := \{C \in \mathfrak{C} \mid C \cap A \neq \emptyset\}$, $\mathfrak{C}_B := \{C \in \mathfrak{C} \mid C \cap B \neq \emptyset\}$ and $\mathfrak{C}_S := \{C \in \mathfrak{C} \mid C \subseteq S\}$. We claim that these three subsets form a disjoint partition of the full clique set—namely, $\mathfrak{C} = \mathfrak{C}_A \cup \mathfrak{C}_S \cup \mathfrak{C}_B$. Given any clique C, it is either contained entirely within S, or must have non-trivial intersection with either A or B, which proves the union property. To establish disjointedness, it is immediate that $\mathfrak{C}_S$ is disjoint from $\mathfrak{C}_A$ and $\mathfrak{C}_B$. On the other hand, if there were some clique $C \in \mathfrak{C}_A \cap \mathfrak{C}_B$, then there would exist nodes $a \in A$ and $b \in B$ with $\{a, b\} \in C$, which contradicts the fact that A and B are separated by the cutset S.

Given this disjoint partition, we may write

$$p(x_A, x_S, x_B) = \frac{1}{Z} \underbrace{\left[\prod_{C \in \mathfrak{C}_A} \psi_C(x_C)\right]}_{\Psi_A(x_A, x_S)} \underbrace{\left[\prod_{C \in \mathfrak{C}_S} \psi_C(x_C)\right]}_{\Psi_S(x_S)} \underbrace{\left[\prod_{C \in \mathfrak{C}_B} \psi_C(x_C)\right]}_{\Psi_B(x_B, x_S)}.$$

Defining the quantities

$$Z_A(x_S) := \sum_{x_A} \Psi_A(x_A, x_S) \quad \text{and} \quad Z_B(x_S) := \sum_{x_B} \Psi_B(x_B, x_S),$$

we then obtain the following expressions for the marginal distributions of interest:

$$p(x_S) = \frac{Z_A(x_S)\, Z_B(x_S)}{Z} \Psi_S(x_S) \quad \text{and} \quad p(x_A, x_S) = \frac{Z_B(x_S)}{Z} \Psi_A(x_A, x_S)\, \Psi_S(x_S),$$

with a similar expression for $p(x_B, x_S)$. Consequently, for any x_S for which $p(x_S) > 0$, we may write

$$\frac{p(x_A, x_S, x_B)}{p(x_S)} = \frac{\frac{1}{Z}\Psi_A(x_A, x_S)\Psi_S(x_S)\Psi_B(x_B, x_S)}{\frac{Z_A(x_S)\,Z_B(x_S)}{Z}\Psi_S(x_S)} = \frac{\Psi_A(x_A, x_S)\Psi_B(x_B, x_S)}{Z_A(x_S)\,Z_B(x_S)}. \tag{11.6}$$

Similar calculations yield the relations

$$\frac{p(x_A, x_S)}{p(x_S)} = \frac{\frac{Z_B(x_S)}{Z}\Psi_A(x_A, x_S)\,\Psi_S(x_S)}{\frac{Z_A(x_S)Z_B(x_S)}{Z}\Psi_S(x_S)} = \frac{\Psi_A(x_A, x_S)}{Z_A(x_S)} \tag{11.7a}$$

and

$$\frac{p(x_B, x_S)}{p(x_S)} = \frac{\frac{Z_A(x_S)}{Z}\Psi_B(x_B, x_S)\,\Psi_S(x_S)}{\frac{Z_A(x_S)Z_B(x_S)}{Z}\Psi_S(x_S)} = \frac{\Psi_B(x_B, x_S)}{Z_B(x_S)}. \tag{11.7b}$$

Combining equation (11.6) with equations (11.7a) and (11.7b) yields

$$p(x_A, x_B \mid x_S) = \frac{p(x_A, x_B, x_S)}{p(x_S)} = \frac{p(x_A, x_S)}{p(x_S)}\,\frac{p(x_B, x_S)}{p(x_S)} = p(x_A \mid x_S)\,p(x_B \mid x_S),$$

thereby showing that $X_A \perp\!\!\!\perp X_B \mid X_S$, as claimed. □

11.1.4 Estimation of graphical models

Typical applications of graphical models require solving some sort of inverse problem of the following type. Consider a collection of samples $\{x_i\}_{i=1}^n$, where each $x_i = (x_{i1}, \ldots, x_{id})$ is a d-dimensional vector, hypothesized to have been drawn from some graph-structured probability distribution. The goal is to estimate certain aspects of the underlying graphical model. In the problem of *graphical parameter estimation*, the graph structure itself is assumed to be known, and we want to estimate the compatibility functions $\{\psi_C, C \in \mathfrak{C}\}$ on the graph cliques. In the more challenging problem of *graphical model selection*, the graph structure itself is unknown, so that we need to estimate both it *and* the clique compatibility functions. In the following sections, we consider various methods for solving these problems for both Gaussian and non-Gaussian models.

11.2 Estimation of Gaussian graphical models

We begin our exploration of graph estimation for the case of Gaussian Markov random fields. As previously discussed in Example 11.3, for a Gaussian model, the factorization property is specified by the inverse covariance or precision matrix $\mathbf{\Theta}^*$. Consequently, the Hammersley–Clifford theorem is especially easy to interpret in this case: it ensures that $\Theta^*_{jk} = 0$ for any $(j, k) \notin E$. See Figure 11.2 for some illustrations of this correspondence between graph structure and the sparsity of the inverse covariance matrix.

Now let us consider some estimation problems that arise for Gaussian Markov random fields. Since the mean is easily estimated, we take it to be zero for the remainder of our development. Thus, the only remaining parameter is the precision matrix $\mathbf{\Theta}^*$. Given an estimate $\widehat{\mathbf{\Theta}}$ of $\mathbf{\Theta}^*$, its quality can be assessed in different ways. In the problem of graphical model selection, also known as *(inverse) covariance selection*, the goal is to recover the

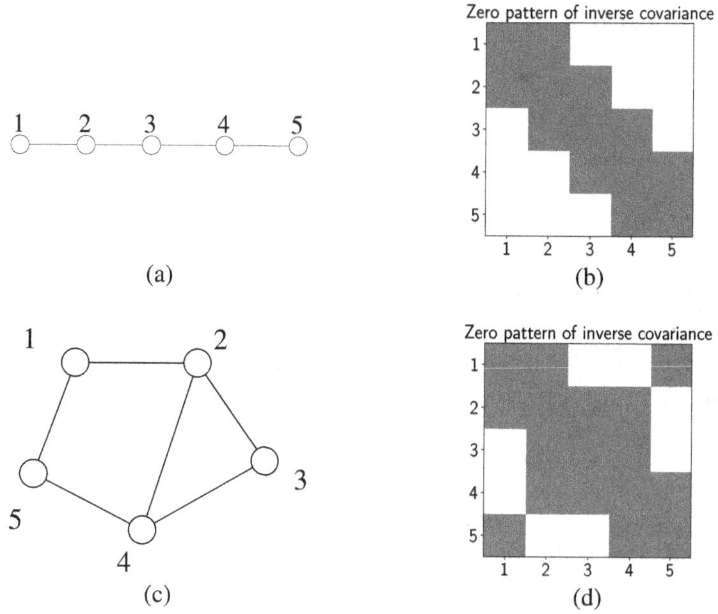

Figure 11.2 For Gaussian graphical models, the Hammersley–Clifford theorem guarantees a correspondence between the graph structure and the sparsity pattern of the inverse covariance matrix or precision matrix $\mathbf{\Theta}^*$. (a) Chain graph on five vertices. (b) Inverse covariance for a Gauss–Markov chain must have a tri-diagonal structure. (c), (d) More general Gauss–Markov random field and the associated inverse covariance matrix.

edge set E of the underlying graph G. More concretely, letting $\widehat{E}$ denote an estimate of the edge set based on $\widehat{\mathbf{\Theta}}$, one figure of merit is the error probability $\mathbb{P}[\widehat{E} \neq E]$, which assesses whether or not we have recovered the true underlying edge set. A related but more relaxed criterion would focus on the probability of recovering a fraction $1 - \delta$ of the edge set, where $\delta \in (0, 1)$ is a user-specified tolerance parameter. In other settings, we might be interested in estimating the inverse covariance matrix itself, and so consider various types of matrix norms, such as the operator norm $|||\widehat{\mathbf{\Theta}} - \mathbf{\Theta}^*|||_2$ or the Frobenius norm $|||\widehat{\mathbf{\Theta}} - \mathbf{\Theta}^*|||_{\mathrm{F}}$. In the following sections, we consider these different choices of metrics in more detail.

11.2.1 Graphical Lasso: ℓ_1-regularized maximum likelihood

We begin with a natural and direct method for estimating a Gaussian graphical model, namely one based on the global likelihood. In order to do so, let us first derive a convenient form of the rescaled negative log-likelihood, one that involves the log-determinant function. For any two symmetric matrices $\mathbf{A}$ and $\mathbf{B}$, recall that we use $\langle\langle \mathbf{A}, \mathbf{B} \rangle\rangle := \mathrm{trace}(\mathbf{AB})$ to denote the trace inner product. The negative log-determinant function is defined on the space $\mathcal{S}^{d\times d}$

of symmetric matrices as

$$-\log\det(\boldsymbol{\Theta}) := \begin{cases} -\sum_{j=1}^{d} \log \gamma_j(\boldsymbol{\Theta}) & \text{if } \boldsymbol{\Theta} \succ 0, \\ +\infty & \text{otherwise,} \end{cases} \tag{11.8}$$

where $\gamma_1(\boldsymbol{\Theta}) \geq \gamma_2(\boldsymbol{\Theta}) \geq \cdots \geq \gamma_d(\boldsymbol{\Theta})$ denote the ordered eigenvalues of the symmetric matrix Θ. In Exercise 11.1, we explore some basic properties of the log-determinant function, including its strict convexity and differentiability.

Using the parameterization (11.2) of the Gaussian distribution in terms of the precision matrix, the rescaled negative log-likelihood of the multivariate Gaussian, based on samples $\{x_i\}_{i=1}^n$, takes the form

$$\mathcal{L}_n(\boldsymbol{\Theta}) = \langle\langle \boldsymbol{\Theta}, \widehat{\boldsymbol{\Sigma}} \rangle\rangle - \log\det(\boldsymbol{\Theta}), \tag{11.9}$$

where $\widehat{\boldsymbol{\Sigma}} := \frac{1}{n}\sum_{i=1}^n x_i x_i^{\mathrm{T}}$ is the sample covariance matrix. Here we have dropped some constant factors in the log-likelihood that have no effect on the maximum likelihood solution, and also rescaled the log-likelihood by $-\frac{2}{n}$ for later theoretical convenience.

The unrestricted maximum likelihood solution $\widehat{\boldsymbol{\Theta}}_{\mathrm{MLE}}$ takes a very simple form for the Gaussian model. If the sample covariance matrix $\widehat{\boldsymbol{\Sigma}}$ is invertible, we have $\widehat{\boldsymbol{\Theta}}_{\mathrm{MLE}} = \widehat{\boldsymbol{\Sigma}}^{-1}$; otherwise, the maximum likelihood solution is undefined (see Exercise 11.2 for more details). Whenever $n < d$, the sample covariance matrix is always rank-deficient, so that the maximum likelihood estimate does not exist. In this setting, some form of regularization is essential. When the graph G is expected to have relatively few edges, a natural form of regularization is to impose an ℓ_1-constraint on the entries of $\boldsymbol{\Theta}$. (If computational considerations were not a concern, it would be natural to impose ℓ_0-constraint, but as in Chapter 7, we use the ℓ_1-norm as a convex surrogate.)

Combining ℓ_1-regularization with the negative log-likelihood yields the *graphical Lasso estimator*

$$\widehat{\boldsymbol{\Theta}} \in \arg\min_{\boldsymbol{\Theta} \in \mathcal{S}^{d\times d}} \Big\{ \underbrace{\langle\langle \boldsymbol{\Theta}, \widehat{\boldsymbol{\Sigma}} \rangle\rangle - \log\det\boldsymbol{\Theta}}_{\mathcal{L}_n(\boldsymbol{\Theta})} + \lambda_n |||\boldsymbol{\Theta}|||_{1,\mathrm{off}} \Big\}, \tag{11.10}$$

where $|||\boldsymbol{\Theta}|||_{1,\mathrm{off}} := \sum_{j \neq k} |\Theta_{jk}|$ corresponds to the ℓ_1-norm applied to the off-diagonal entries of $\boldsymbol{\Theta}$. One could also imagine penalizing the diagonal entries of $\boldsymbol{\Theta}$, but since they must be positive for any non-degenerate inverse covariance, doing so only introduces additional bias. The convex program (11.10) is a particular instance of a log-determinant program, and can be solved in polynomial time with various generic algorithms. Moreover, there is also a line of research on efficient methods specifically tailored to the graphical Lasso problem; see the bibliographic section for further discussion.

Frobenius norm bounds

We begin our investigation of the graphical Lasso (11.10) by deriving bounds on the Frobenius norm error $|||\widehat{\boldsymbol{\Theta}} - \boldsymbol{\Theta}^*|||_{\mathrm{F}}$. The following result is based on a sample covariance matrix $\widehat{\boldsymbol{\Sigma}}$ formed from n i.i.d. samples $\{x_i\}_{i=1}^n$ of a zero-mean random vector in which each coordinate has σ-sub-Gaussian tails (recall Definition 2.2 from Chapter 2).

Proposition 11.9 (Frobenius norm bounds for graphical Lasso) *Suppose that the inverse covariance matrix Θ^* has at most m non-zero entries per row, and we solve the graphical Lasso* (11.10) *with regularization parameter $\lambda_n = 8\sigma^2(\sqrt{\frac{\log d}{n}} + \delta)$ for some $\delta \in (0, 1]$. Then as long as $6(|||\Theta^*|||_2 + 1)^2 \lambda_n \sqrt{md} < 1$, the graphical Lasso estimate $\widehat{\Theta}$ satisfies*

$$|||\widehat{\Theta} - \Theta^*|||_F^2 \leq \frac{9}{(|||\Theta^*|||_2 + 1)^4} m\, d\, \lambda_n^2 \tag{11.11}$$

with probability at least $1 - 8e^{-\frac{1}{16}n\delta^2}$.

Proof We prove this result by applying Corollary 9.20 from Chapter 9. In order to do so, we need to verify the restricted strong convexity of the loss function (see Definition 9.15), as well as other technical conditions given in the corollary.

Let $\mathbb{B}_F(1) = \{\Delta \in \mathcal{S}^{d\times d} \mid |||\Delta|||_F \leq 1\}$ denote the set of symmetric matrices with Frobenius norm at most one. Using standard properties of the log-determinant function (see Exercise 11.1), the loss function underlying the graphical Lasso is twice differentiable, with

$$\nabla \mathcal{L}_n(\Theta) = \widehat{\Sigma} - \Theta^{-1} \quad \text{and} \quad \nabla^2 \mathcal{L}_n(\Theta) = \Theta^{-1} \otimes \Theta^{-1},$$

where $\otimes$ denotes the Kronecker product between matrices.

Verifying restricted strong convexity: Our first step is to establish that restricted strong convexity holds over the Frobenius norm ball $\mathbb{B}_F(1)$. Let $\text{vec}(\cdot)$ denote the vectorized form of a matrix. For any $\Delta \in \mathbb{B}_F(1)$, a Taylor-series expansion yields

$$\underbrace{\mathcal{L}_n(\Theta^* + \Delta) - \mathcal{L}_n(\Theta^*) - \langle\langle \nabla \mathcal{L}_n(\Theta^*),\ \Delta \rangle\rangle}_{\mathcal{E}_n(\Delta)} = \frac{1}{2} \text{vec}(\Delta)^{\mathrm{T}} \nabla^2 \mathcal{L}_n(\Theta^* + t\Delta)\, \text{vec}(\Delta)$$

for some $t \in [0, 1]$. Thus, we have

$$\mathcal{E}_n(\Delta) \geq \frac{1}{2} \gamma_{\min}(\nabla^2 \mathcal{L}_n(\Theta^* + t\Delta))\, \|\text{vec}(\Delta)\|_2^2 = \frac{1}{2} \frac{|||\Delta|||_F^2}{|||\Theta^* + t\Delta|||_2^2},$$

using the fact that $|||A^{-1} \otimes A^{-1}|||_2 = \frac{1}{|||A|||_2^2}$ for any symmetric invertible matrix. The triangle inequality, in conjunction with the bound $t|||\Delta|||_2 \leq t|||\Delta|||_F \leq 1$, implies that $|||\Theta^* + t\Delta|||_2^2 \leq (|||\Theta^*|||_2 + 1)^2$. Combining the pieces yields the lower bound

$$\mathcal{E}_n(\Delta) \geq \frac{\kappa}{2} |||\Delta|||_F^2 \qquad \text{where } \kappa := (|||\Theta^*|||_2 + 1)^{-2}, \tag{11.12}$$

showing that the RSC condition from Definition 9.15 holds over $\mathbb{B}_F(1)$ with tolerance $\tau_n^2 = 0$.

Computing the subspace Lipschitz constant: Next we introduce a subspace suitable for application of Corollary 9.20 to the graphical Lasso. Letting S denote the support set of $\mathbf{\Theta}^*$, we define the subspace

$$\mathbb{M}(S) = \{\mathbf{\Theta} \in \mathbb{R}^{d\times d} \mid \Theta_{jk} = 0 \text{ for all } (j,k) \notin S\}.$$

With this choice, we have

$$\Psi^2(\mathbb{M}(S)) = \sup_{\mathbf{\Theta}\in\mathbb{M}(S)} \frac{\left(\sum_{j\neq k} |\Theta_{jk}|\right)^2}{|||\mathbf{\Theta}|||_F^2} \leq |S| \stackrel{\text{(i)}}{\leq} md,$$

where inequality (i) follows since $\mathbf{\Theta}^*$ has at most m non-zero entries per row.

Verifying event $\mathbb{G}(\lambda_n)$*:* Next we verify that the stated choice of regularization parameter λ_n satisfies the conditions of Corollary 9.20 with high probability: in order to do so, we need to compute the score function and obtain a bound on its dual norm. Since $(\mathbf{\Theta}^*)^{-1} = \mathbf{\Sigma}$, the score function is given by $\nabla\mathcal{L}_n(\mathbf{\Theta}^*) = \widehat{\mathbf{\Sigma}} - \mathbf{\Sigma}$, corresponding to the deviations between the sample covariance and population covariance matrices. The dual norm defined by $|||\cdot|||_{1,\text{off}}$ is given by the ℓ_∞-norm applied to the off-diagonal matrix entries, which we denote by $|||\cdot|||_{\max,\text{off}}$. Using Lemma 6.26, we have

$$\mathbb{P}\left[|||\widehat{\mathbf{\Sigma}} - \mathbf{\Sigma}|||_{\max,\text{off}} \geq \sigma^2 t\right] \leq 8e^{-\frac{n}{16}\min\{t,t^2\}+2\log d} \qquad \text{for all } t > 0.$$

Setting $t = \lambda_n/\sigma^2$ shows that the event $\mathbb{G}(\lambda_n)$ from Corollary 9.20 holds with the claimed probability. Consequently, Proposition 9.13 implies that the error matrix $\widehat{\mathbf{\Delta}}$ satisfies the bound $\|\widehat{\mathbf{\Delta}}_{S^c}\|_1 \leq 3\|\widehat{\mathbf{\Delta}}_S\|_1$, and hence

$$\|\widehat{\mathbf{\Delta}}\|_1 \leq 4\|\widehat{\mathbf{\Delta}}_S\|_1 \leq 4\sqrt{md}|||\widehat{\mathbf{\Delta}}|||_F, \tag{11.13}$$

where the final inequality again uses the fact that $|S| \leq md$. In order to apply Corollary 9.20, the only remaining detail to verify is that $\widehat{\mathbf{\Delta}}$ belongs to the Frobenius ball $\mathbb{B}_F(1)$.

Localizing the error matrix: By an argument parallel to the earlier proof of RSC, we have

$$\mathcal{L}_n(\mathbf{\Theta}^*) - \mathcal{L}_n(\mathbf{\Theta}^* + \mathbf{\Delta}) + \langle\langle \nabla\mathcal{L}_n(\mathbf{\Theta}^* + \mathbf{\Delta}),\ -\mathbf{\Delta}\rangle\rangle \geq \frac{\kappa}{2}|||\mathbf{\Delta}|||_F^2.$$

Adding this lower bound to the inequality (11.12), we find that

$$\langle\langle \nabla\mathcal{L}_n(\mathbf{\Theta}^* + \mathbf{\Delta}) - \nabla\mathcal{L}_n(\mathbf{\Theta}^*),\ \mathbf{\Delta}\rangle\rangle \geq \kappa\, |||\mathbf{\Delta}|||_F^2.$$

The result of Exercise 9.10 then implies that

$$\langle\langle \nabla\mathcal{L}_n(\mathbf{\Theta}^* + \mathbf{\Delta}) - \nabla\mathcal{L}_n(\mathbf{\Theta}^*),\ \mathbf{\Delta}\rangle\rangle \geq \kappa\, |||\mathbf{\Delta}|||_F \qquad \text{for all } \mathbf{\Delta} \in \mathcal{S}^{d\times d} \setminus \mathbb{B}_F(1). \tag{11.14}$$

By the optimality of $\widehat{\mathbf{\Theta}}$, we have $0 = \langle\langle \nabla\mathcal{L}_n(\mathbf{\Theta}^* + \widehat{\mathbf{\Delta}}) + \lambda_n\widehat{\mathbf{Z}},\ \widehat{\mathbf{\Delta}}\rangle\rangle$, where $\widehat{\mathbf{Z}} \in \partial|||\widehat{\mathbf{\Theta}}|||_{1,\text{off}}$ is a subgradient matrix for the elementwise ℓ_1-norm. By adding and subtracting terms, we find that

$$\begin{aligned}
\langle\langle \nabla\mathcal{L}_n(\mathbf{\Theta}^* + \widehat{\mathbf{\Delta}}) - \nabla\mathcal{L}_n(\mathbf{\Theta}^*),\ \widehat{\mathbf{\Delta}}\rangle\rangle &\leq \lambda_n \left|\langle\langle \widehat{\mathbf{Z}},\ \widehat{\mathbf{\Delta}}\rangle\rangle\right| + \left|\langle\langle \nabla\mathcal{L}_n(\mathbf{\Theta}^*),\ \widehat{\mathbf{\Delta}}\rangle\rangle\right| \\
&\leq \left\{\lambda_n + \|\nabla\mathcal{L}_n(\mathbf{\Theta}^*)\|_{\max}\right\}\|\widehat{\mathbf{\Delta}}\|_1.
\end{aligned}$$

Since $\|\nabla \mathcal{L}_n(\Theta^*)\|_{\max} \leq \frac{\lambda_n}{2}$ under the previously established event $\mathbb{G}(\lambda_n)$, the right-hand side is at most

$$\frac{3\lambda_n}{2}\|\widehat{\Delta}\|_1 \leq 6\lambda_n \sqrt{md}\,|||\widehat{\Delta}|||_{\mathrm{F}},$$

where we have applied our earlier inequality (11.13). If $|||\widehat{\Delta}|||_{\mathrm{F}} > 1$, then our earlier lower bound (11.14) may be applied, from which we obtain

$$\kappa|||\widehat{\Delta}|||_{\mathrm{F}} \leq \frac{3\lambda_n}{2}\|\widehat{\Delta}\|_1 \leq 6\lambda_n \sqrt{md}|||\widehat{\Delta}|||_{\mathrm{F}}.$$

This inequality leads to a contradiction whenever $\frac{6\lambda_n \sqrt{md}}{\kappa} < 1$, which completes the proof. □

Edge selection and operator norm bounds

Proposition 11.9 is a relatively crude result, in that it only guarantees that the graphical Lasso estimate $\widehat{\Theta}$ is close in Frobenius norm, but not that the edge structure of the underlying graph is preserved. Moreover, the result actually precludes the setting $n < d$: indeed, the conditions of Proposition 11.9 imply that the sample size n must be lower bounded by a constant multiple of $md \log d$, which is larger than d.

Accordingly, we now turn to a more refined type of result, namely one that allows for high-dimensional scaling ($d \gg n$), and moreover guarantees that the graphical Lasso estimate $\widehat{\Theta}$ correctly selects all the edges of the graph. Such an edge selection result can be guaranteed by first proving that $\widehat{\Theta}$ is close to the true precision matrix Θ^* in the elementwise ℓ_∞-norm on the matrix elements (denoted by $\|\cdot\|_{\max}$). In turn, such max-norm control can also be converted to bounds on the ℓ_2-matrix operator norm, also known as the spectral norm.

The problem of edge selection in a Gaussian graphical model is closely related to the problem of variable selection in a sparse linear model. As previously discussed in Chapter 7, variable selection with an ℓ_1-norm penalty requires a certain type of incoherence condition, which limits the influence of irrelevant variables on relevant ones. In the case of least-squares regression, these incoherence conditions were imposed on the design matrix, or equivalently on the Hessian of the least-squares objective function. Accordingly, in a parallel manner, here we impose incoherence conditions on the Hessian of the objective function $\mathcal{L}_n$ in the graphical Lasso (11.10). As previously noted, this Hessian takes the form $\nabla^2 \mathcal{L}_n(\Theta) = \Theta^{-1} \otimes \Theta^{-1}$, a $d^2 \times d^2$ matrix that is indexed by ordered pairs of vertices (j, k).

More specifically, the incoherence condition must be satisfied by the d^2-dimensional matrix $\Gamma^* := \nabla^2 \mathcal{L}_n(\Theta^*)$, corresponding to the Hessian evaluated at the true precision matrix. We use $S := E \cup \{(j, j) \mid j \in V\}$ to denote the set of row/column indices associated with edges in the graph (including both (j, k) and (k, j)), along with all the self-edges (j, j). Letting $S^c = (V \times V) \setminus S$, we say that the matrix Γ^* is *α-incoherent* if

$$\max_{e \in S^c} \|\Gamma^*_{eS}(\Gamma^*_{SS})^{-1}\|_1 \leq 1 - \alpha \qquad \text{for some } \alpha \in (0, 1]. \tag{11.15}$$

With this definition, we have the following result:

Proposition 11.10 *Consider a zero-mean d-dimensional Gaussian distribution based on an α-incoherent inverse covariance matrix $\mathbf{\Theta}^*$. Given a sample size lower bounded as $n > c_0(1 + 8\alpha^{-1})^2 m^2 \log d$, suppose that we solve the graphical Lasso* (11.10) *with a regularization parameter $\lambda_n = \frac{c_1}{\alpha}\sqrt{\frac{\log d}{n}} + \delta$ for some $\delta \in (0, 1]$. Then with probability at least $1 - c_2 e^{-c_3 n \delta^2}$, we have the following:*

(a) *The graphical Lasso solution leads to no false inclusions—that is, $\widehat{\Theta}_{jk} = 0$ for all $(j, k) \notin E$.*
(b) *It satisfies the sup-norm bound*

$$\|\widehat{\mathbf{\Theta}} - \mathbf{\Theta}^*\|_{\max} \leq c_4 \Bigg\{ \underbrace{(1 + 8\alpha^{-1}) \sqrt{\frac{\log d}{n}}}_{\tau(n,d,\alpha)} + \lambda_n \Bigg\}. \tag{11.16}$$

Note that part (a) guarantees that the edge set estimate

$$\widehat{E} := \{(j, k) \in [d] \times [d] \mid j < k \text{ and } \widehat{\mathbf{\Theta}}_{jk} \neq 0\}$$

is always a subset of the true edge set E. Part (b) guarantees that $\widehat{\mathbf{\Theta}}$ is uniformly close to $\mathbf{\Theta}^*$ in an elementwise sense. Consequently, if we have a lower bound on the minimum nonzero entry of $|\mathbf{\Theta}^*|$—namely the quantity $\tau^*(\mathbf{\Theta}^*) = \min_{(j,k) \in E} |\Theta^*_{jk}|$—then we can guarantee that the graphical Lasso recovers the full edge set correctly. In particular, using the notation of part (b), as long as this minimum is lower bounded as $\tau^*(\mathbf{\Theta}^*) > c_4(\tau(n, d, \alpha) + \lambda_n)$, then the graphical Lasso recovers the correct edge set with high probability.

The proof of Proposition 11.10 is based on an extension of the primal–dual witness technique used to prove Theorem 7.21 in Chapter 7. In particular, it involves constructing a pair of matrices $(\widetilde{\mathbf{\Theta}}, \widetilde{\mathbf{Z}})$, where $\widetilde{\mathbf{\Theta}} \succ 0$ is a primal optimal solution and $\widetilde{\mathbf{Z}}$ a corresponding dual optimum. This pair of matrices is required to satisfy the zero subgradient conditions that define the optimum of the graphical Lasso (11.10)—namely

$$\widehat{\mathbf{\Sigma}} - \widetilde{\mathbf{\Theta}}^{-1} + \lambda_n \widetilde{\mathbf{Z}} = 0 \quad \text{or equivalently} \quad \widetilde{\mathbf{\Theta}}^{-1} = \widehat{\mathbf{\Sigma}} + \lambda_n \widetilde{\mathbf{Z}}.$$

The matrix $\widetilde{\mathbf{Z}}$ must belong to the subgradient of the $|||\cdot|||_{1,\text{off}}$ function, evaluated at $\widetilde{\mathbf{\Theta}}$, meaning that $|||\widetilde{\mathbf{Z}}|||_{\max,\text{off}} \leq 1$, and that $\widetilde{Z}_{jk} = \text{sign}(\widetilde{\Theta}_{jk})$ whenever $\widetilde{\Theta}_{jk} \neq 0$. We refer the reader to the bibliographic section for further details and references for the proof.

Proposition 11.10 also implies bounds on the operator norm error in the estimate $\widehat{\mathbf{\Theta}}$.

Corollary 11.11 (Operator norm bounds) *Under the conditions of Proposition 11.10, consider the graphical Lasso estimate $\widehat{\mathbf{\Theta}}$ with regularization parameter $\lambda_n = \frac{c_1}{\alpha}\sqrt{\frac{\log d}{n}} + \delta$*

for some $\delta \in (0,1]$. Then with probability at least $1 - c_2 e^{-c_3 n \delta^2}$, we have

$$|||\widehat{\Theta} - \Theta^*|||_2 \le c_4\, |||\mathbf{A}|||_2 \left\{(1 + 8\alpha^{-1}) \sqrt{\frac{\log d}{n}} + \lambda_n\right\}, \tag{11.17a}$$

where $\mathbf{A}$ denotes the adjacency matrix of the graph G (including ones on the diagonal). In particular, if the graph has maximum degree m, then

$$|||\widehat{\Theta} - \Theta^*|||_2 \le c_4 (m+1) \left\{(1 + 8\alpha^{-1}) \sqrt{\frac{\log d}{n}} + \lambda_n\right\}. \tag{11.17b}$$

Proof These claims follow in a straightforward way from Proposition 11.10 and certain properties of the operator norm exploited previously in Chapter 6. In particular, Proposition 11.10 guarantees that for any pair $(j,k) \notin E$, we have $|\widehat{\Theta}_{jk} - \Theta^*_{jk}| = 0$, whereas the bound (11.16) ensures that for any pair $(j,k) \in E$, we have $|\widehat{\Theta}_{jk} - \Theta^*_{jk}| \le c_4\{\tau(n,d,\alpha) + \lambda_n\}$. Note that the same bound holds whenever $j = k$. Putting together the pieces, we conclude that

$$|\widehat{\Theta}_{jk} - \Theta^*_{jk}| \le c_4\{\tau(n,d,\alpha) + \lambda_n\}\, A_{jk}, \tag{11.18}$$

where $\mathbf{A}$ is the adjacency matrix, including ones on the diagonal. Using the matrix-theoretic properties from Exercise 6.3(c), we conclude that

$$|||\widehat{\Theta} - \Theta^*|||_2 \le |||\,|\widehat{\Theta} - \Theta^*|\,|||_2 \le c_4\{\tau(n,d,\alpha) + \lambda_n\}\, |||\mathbf{A}|||_2,$$

thus establishing the bound (11.17a). The second inequality (11.17b) follows by noting that $|||\mathbf{A}|||_2 \le m+1$ for any graph of degree at most m. (See the discussion following Corollary 6.24 for further details.) □

As we noted in Chapter 6, the bound (11.17b) is not tight for a general graph with maximum degree m. In particular, a star graph with one hub connected to m other nodes (see Figure 6.1(b)) has maximum degree m, but satisfies $|||\mathbf{A}|||_2 = 1 + \sqrt{m-1}$, so that the bound (11.17a) implies the operator norm bound $|||\widehat{\Theta} - \Theta^*|||_2 \precsim \sqrt{\frac{m \log d}{n}}$. This guarantee is tighter by a factor of $\sqrt{m}$ than the conservative bound (11.17b).

It should also be noted that Proposition 11.10 also implies bounds on the Frobenius norm error. In particular, the elementwise bound (11.18) implies that

$$|||\widehat{\Theta} - \Theta^*|||_F \le c_3 \sqrt{2s + d} \left\{(1 + 8\alpha^{-1}) \sqrt{\frac{\log d}{n}} + \lambda_n\right\}, \tag{11.19}$$

where s is the total number of edges in the graph. We leave the verification of this claim as an exercise for the reader.

11.2.2 Neighborhood-based methods

The Gaussian graphical Lasso is a global method, one that estimates the full graph simultaneously. An alternative class of procedures, known as neighborhood-based methods, are instead local. They are based on the observation that recovering the full graph is equivalent to recovering the neighborhood set (11.5) of each vertex $j \in V$, and that these neighborhoods are revealed via the Markov properties of the graph.

Neighborhood-based regression

Recall our earlier Definition 11.5 of the Markov properties associated with a graph. In our discussion following this definition, we also noted that for any given vertex $j \in V$, the neighborhood $\mathcal{N}(j)$ is a vertex cutset that breaks the graph into the disjoint pieces $\{j\}$ and $V\backslash\mathcal{N}^+(j)$, where we have introduced the convenient shorthand $\mathcal{N}^+(j) := \{j\} \cup \mathcal{N}(j)$. Consequently, by applying the definition (11.5), we conclude that

$$X_j \perp\!\!\!\perp X_{V\backslash\mathcal{N}^+(j)} \mid X_{\mathcal{N}(j)}. \tag{11.20}$$

Thus, the neighborhood structure of each node is encoded in the structure of the conditional distribution. What is a good way to detect these conditional independence relationships and hence the neighborhood? A particularly simple method is based on the idea of neighborhood regression: for a given vertex $j \in V$, we use the random variables $X_{\backslash\{j\}} := \{X_k \mid k \in V \setminus \{j\}\}$ to predict X_j, and keep only those variables that turn out to be useful.

Let us now formalize this idea in the Gaussian case. In this case, by standard properties of multivariate Gaussian distributions, the conditional distribution of X_j given $X_{\backslash\{j\}}$ is also Gaussian. Therefore, the random variable X_j has a decomposition as the sum of the best linear prediction based on $X_{\backslash\{j\}}$ plus an error term—namely

$$X_j = \left\langle X_{\backslash\{j\}},\, \theta_j^* \right\rangle + W_j, \tag{11.21}$$

where $\theta_j^* \in \mathbb{R}^{d-1}$ is a vector of regression coefficients, and W_j is a zero-mean Gaussian variable, independent of $X_{\backslash\{j\}}$. (See Exercise 11.3 for the derivation of these and related properties.) Moreover, the conditional independence relation (11.20) ensures that $\theta_{jk}^* = 0$ for all $k \notin \mathcal{N}(j)$. In this way, we have reduced the problem of Gaussian graph selection to that of detecting the support in a sparse linear regression problem. As discussed in Chapter 7, the Lasso provides a computationally efficient approach to such support recovery tasks.

In summary, the neighborhood-based approach to Gaussian graphical selection proceeds as follows. Given n samples $\{x_1, \ldots, x_n\}$, we use $\mathbf{X} \in \mathbb{R}^{n\times d}$ to denote the design matrix with $x_i \in \mathbb{R}^d$ as its ith row, and then perform the following steps.

Lasso-based neighborhood regression:

1 For each node $j \in V$:

(a) Extract the column vector $X_j \in \mathbb{R}^n$ and the submatrix $\mathbf{X}_{\backslash\{j\}} \in \mathbb{R}^{n\times(d-1)}$.

(b) Solve the Lasso problem:

$$\widehat{\theta} = \arg\min_{\theta\in\mathbb{R}^{d-1}} \left\{ \frac{1}{2n}\|X_j - \mathbf{X}_{\backslash\{j\}}\theta\|_2^2 + \lambda_n\|\theta\|_1 \right\}. \tag{11.22}$$

(c) Return the neighborhood estimate $\widehat{\mathcal{N}}(j) = \{k \in V \setminus \{j\} \mid \widehat{\theta}_k \neq 0\}$.

2 Combine the neighborhood estimates to form an edge estimate $\widehat{E}$, using either the OR rule or the AND rule.

Note that the first step returns a neighborhood estimate $\widehat{\mathcal{N}}(j)$ for each vertex $j \in V$. These neighborhood estimates may be inconsistent, meaning that for a given pair of distinct vertices (j, k), it may be the case that $k \in \widehat{\mathcal{N}}(j)$ whereas $j \notin \widehat{\mathcal{N}}(k)$. Some rules to resolve this issue include:

- the OR rule that declares that $(j, k) \in \widehat{E}_{\text{OR}}$ if either $k \in \widehat{\mathcal{N}}(j)$ or $j \in \widehat{\mathcal{N}}(k)$;
- the AND rule that declares that $(j, k) \in \widehat{E}_{\text{AND}}$ if $k \in \widehat{\mathcal{N}}(j)$ and $j \in \widehat{\mathcal{N}}(k)$.

By construction, the AND rule is more conservative than the OR rule, meaning that $\widehat{E}_{\text{AND}} \subseteq \widehat{E}_{\text{OR}}$. The theoretical guarantees that we provide end up holding for either rule, since we control the behavior of each neighborhood regression problem.

Graph selection consistency

We now state a result that guarantees selection consistency of neighborhood regression. As with our previous analysis of the Lasso in Chapter 7 and the graphical Lasso in Section 11.2.1, we require an incoherence condition. Given a positive definite matrix $\mathbf{\Gamma}$ and a subset S of its columns, we say $\mathbf{\Gamma}$ is α-incoherent with respect to S if

$$\max_{k \notin S} \|\Gamma_{kS}(\mathbf{\Gamma}_{SS})^{-1}\|_1 \leq 1 - \alpha. \tag{11.23}$$

Here the scalar $\alpha \in (0, 1]$ is the incoherence parameter. As discussed in Chapter 7, if we view $\mathbf{\Gamma}$ as the covariance matrix of a random vector $Z \in \mathbb{R}^d$, then the row vector $\Gamma_{kS}(\mathbf{\Gamma}_{SS})^{-1}$ specifies the coefficients of the optimal linear predictor of Z_k given the variables $Z_S := \{Z_j, j \in S\}$. Thus, the incoherence condition (11.23) imposes a limit on the degree of dependence between the variables in the correct subset S and any variable outside of S.

The following result guarantees graph selection consistency of the Lasso-based neighborhood procedure, using either the AND or the OR rules, for a Gauss–Markov random field in which the covariance matrix $\mathbf{\Sigma}^* = (\mathbf{\Theta}^*)^{-1}$ has maximum degree m, and diagonals scaled such that $\text{diag}(\mathbf{\Sigma}^*) \leq 1$. This latter inequality entails no loss of generality, since it can always be guaranteed by rescaling the variables. Our statement involves the ℓ_∞-matrix-operator norm $|||A|||_2 := \max_{i=1,\ldots,d} \sum_{j=1}^d |A_{ij}|$.

Finally, in stating the result, we assume that the sample size is lower bounded as $n \gtrsim m \log d$. This assumption entails no loss of generality, because a sample size of this order is actually necessary for any method. See the bibliographic section for further details on such information-theoretic lower bounds for graphical model selection.

Theorem 11.12 (Graph selection consistency) *Consider a zero-mean Gaussian random vector with covariance $\mathbf{\Sigma}^*$ such that for each $j \in V$, the submatrix $\mathbf{\Sigma}^*_{\backslash\{j\}} := \text{cov}(\mathbf{X}_{\backslash\{j\}})$ is α-incoherent with respect to $\mathcal{N}(j)$, and $|||(\mathbf{\Sigma}^*_{\mathcal{N}(j),\mathcal{N}(j)})^{-1}|||_\infty \leq b$ for some $b \geq 1$. Suppose that the neighborhood Lasso selection method is implemented with $\lambda_n = c_0\{\frac{1}{\alpha}\sqrt{\frac{\log d}{n}} + \delta\}$*

for some $\delta \in (0, 1]$. Then with probability greater than $1 - c_2 e^{-c_3 n \min\{\delta^2, \frac{1}{m}\}}$, the estimated edge set $\widehat{E}$, based on either the AND or OR rules, has the following properties:

(a) *No false inclusions: it includes no false edges, so that $\widehat{E} \subseteq E$.*
(b) *All significant edges are captured: it includes all edges (j, k) for which $|\Theta^*_{jk}| \geq 7\, b\, \lambda_n$.*

Of course, if the non-zero entries of the precision matrix are bounded below in absolute value as $\min_{(j,k)\in E} |\Theta^*_{jk}| > 7\, b\, \lambda_n$, then in fact Theorem 11.12 guarantees that $\widehat{E} = E$ with high probability.

Proof It suffices to show that for each $j \in V$, the neighborhood $\mathcal{N}(j)$ is recovered with high probability; we can then apply the union bound over all the vertices. The proof requires an extension of the primal–dual witness technique used to prove Theorem 7.21. The main difference is that Theorem 11.12 applies to random covariates, as opposed to the case of deterministic design covered by Theorem 7.21. In order to reduce notational overhead, we adopt the shorthand $\mathbf{\Gamma}^* = \operatorname{cov}(X_{\setminus\{j\}})$ along with the two subsets $S = \mathcal{N}(j)$ and $S^c = V \setminus \mathcal{N}^+(j)$. In this notation, we can write our observation model as $X_j = \mathbf{X}_{\setminus\{j\}}\theta^* + W_j$, where $\mathbf{X}_{\setminus\{j\}} \in \mathbb{R}^{n\times(d-1)}$ while X_j and W_j are both n-vectors. In addition, we let $\widehat{\mathbf{\Gamma}} = \frac{1}{n}\mathbf{X}^{\mathrm{T}}_{\setminus\{j\}}\mathbf{X}_{\setminus\{j\}}$ denote the sample covariance defined by the design matrix, and we use $\widehat{\mathbf{\Gamma}}_{SS}$ to denote the submatrix indexed by the subset S, with the submatrix $\widehat{\mathbf{\Gamma}}_{S^cS}$ defined similarly.

Proof of part (a): We follow the proof of Theorem 7.21 until equation (7.53), namely

$$\widehat{z}_{S^c} = \underbrace{\widehat{\mathbf{\Gamma}}_{S^cS}(\widehat{\mathbf{\Gamma}}_{SS})^{-1}\widehat{z}_S}_{\mu\in\mathbb{R}^{d-s}} + \underbrace{\mathbf{X}^{\mathrm{T}}_{S^c}\left[\mathbf{I}_n - \mathbf{X}_S(\mathbf{X}^{\mathrm{T}}_S\mathbf{X}_S)^{-1}\mathbf{X}^{\mathrm{T}}_S\right]\left(\frac{W_j}{\lambda_n n}\right)}_{V_{S^c}\in\mathbb{R}^{d-s}}. \tag{11.24}$$

As argued in Chapter 7, in order to establish that the Lasso support is included within S, it suffices to establish the strict dual feasibility condition $\|\widehat{z}_{S^c}\|_\infty < 1$. We do so by establishing that

$$\mathbb{P}\left[\|\mu\|_\infty \geq 1 - \frac{3}{4}\alpha\right] \leq c_1 e^{-c_2 n\alpha^2 - \log d} \tag{11.25a}$$

and

$$\mathbb{P}\left[\|V_{S^c}\|_\infty \geq \frac{\alpha}{4}\right] \leq c_1 e^{-c_2 n\delta^2\alpha^2 - \log d}. \tag{11.25b}$$

Taken together, these bounds ensure that $\|\widehat{z}_{S^c}\|_\infty \leq 1 - \frac{\alpha}{2} < 1$, and hence that the Lasso support is contained within $S = \mathcal{N}(j)$, with probability at least $1 - c_1 e^{-c_2 n\delta^2\alpha^2 - \log d}$, where the values of the universal constants may change from line to line. Taking the union bound over all d vertices, we conclude that $\widehat{E} \subseteq E$ with probability at least $1 - c_1 e^{-c_2 n\delta^2\alpha^2}$.

Let us begin by establishing the bound (11.25a). By standard properties of multivariate Gaussian vectors, we can write

$$\mathbf{X}^{\mathrm{T}}_{S^c} = \mathbf{\Gamma}^*_{S^cS}(\mathbf{\Gamma}^*_{SS})^{-1}\mathbf{X}_S + \widetilde{\mathbf{W}}^{\mathrm{T}}_{S^c}, \tag{11.26}$$

where $\widetilde{\mathbf{W}}_{S^c} \in \mathbb{R}^{n\times|S^c|}$ is a zero-mean Gaussian random matrix that is independent of $\mathbf{X}_S$.

Observe moreover that

$$\operatorname{cov}(\widetilde{\mathbf{W}}_{S^c}) = \mathbf{\Gamma}^*_{S^cS^c} - \mathbf{\Gamma}^*_{S^cS}(\mathbf{\Gamma}^*_{SS})^{-1}\mathbf{\Gamma}^*_{SS^c} \preceq \mathbf{\Gamma}^*.$$

Recalling our assumption that $\operatorname{diag}(\mathbf{\Gamma}^*) \leq 1$, we see that the elements of $\widetilde{\mathbf{W}}_{S^c}$ have variance at most 1.

Using the decomposition (11.26) and the triangle inequality, we have

$$\begin{aligned}
\|\mu\|_\infty &= \left\|\mathbf{\Gamma}^*_{S^cS}(\mathbf{\Gamma}^*_{SS})^{-1}\widehat{z}_S + \frac{\widetilde{\mathbf{W}}^{\mathrm{T}}_{S^c}}{\sqrt{n}}\frac{\mathbf{X}_S}{\sqrt{n}}(\widehat{\mathbf{\Gamma}}_{SS})^{-1}\widehat{z}_S\right\|_\infty \\
&\overset{\text{(i)}}{\leq} (1-\alpha) + \underbrace{\left\|\frac{\widetilde{\mathbf{W}}^{\mathrm{T}}_{S^c}}{\sqrt{n}}\frac{\mathbf{X}_S}{\sqrt{n}}(\widehat{\mathbf{\Gamma}}_{SS})^{-1}\widehat{z}_S\right\|_\infty}_{\widetilde{V}\in\mathbb{R}^{|S^c|}}, \qquad (11.27)
\end{aligned}$$

where step (i) uses the population-level α-incoherence condition. Turning to the remaining stochastic term, conditioned on the design matrix, the vector $\widetilde{V}$ is a zero-mean Gaussian random vector, each entry of which has standard deviation at most

$$\begin{aligned}
\frac{1}{\sqrt{n}}\left\|\frac{\mathbf{X}_S}{\sqrt{n}}(\widehat{\mathbf{\Gamma}}_{SS})^{-1}\widehat{z}_S\right\|_2 &\leq \frac{1}{\sqrt{n}}\left|\!\left|\!\left|\frac{\mathbf{X}_S}{\sqrt{n}}(\widehat{\mathbf{\Gamma}}_{SS})^{-1}\right|\!\right|\!\right|_2 \|\widehat{z}_S\|_2 \\
&\leq \frac{1}{\sqrt{n}}\sqrt{|\!|\!|(\widehat{\mathbf{\Gamma}}_{SS})^{-1}|\!|\!|_2}\,\sqrt{m} \\
&\overset{\text{(i)}}{\leq} 2\sqrt{\frac{bm}{n}},
\end{aligned}$$

where inequality (i) follows with probability at least $1 - 4e^{-c_1 n}$, using standard bounds on Gaussian random matrices (see Theorem 6.1). Using this upper bound to control the conditional variance of $\widetilde{V}$, standard Gaussian tail bounds and the union bound then ensure that

$$\mathbb{P}[\|\widetilde{V}\|_\infty \geq t] \leq 2\,|S^c|\,e^{-\frac{nt^2}{8bm}} \leq 2\,e^{-\frac{nt^2}{8bm}+\log d}.$$

We now set $t = \left[\frac{64bm\log d}{n} + \frac{1}{64}\alpha^2\right]^{1/2}$, a quantity which is less than $\frac{\alpha}{4}$ as long as $n \geq c\,\frac{bm\log d}{\alpha}$ for a sufficiently large universal constant. Thus, we have established that $\|\widetilde{V}\|_\infty \leq \frac{\alpha}{4}$ with probability at least $1 - c_1 e^{-c_2 n\alpha^2 - \log d}$. Combined with the earlier bound (11.27), the claim (11.25a) follows.

Turning to the bound (11.25b), note that the matrix $\mathbf{\Pi} := \mathbf{I}_n - \mathbf{X}_S(\mathbf{X}_S^{\mathrm{T}}\mathbf{X}_S)^{-1}\mathbf{X}_S^{\mathrm{T}}$ has the range of $\mathbf{X}_S$ as its nullspace. Thus, using the decomposition (11.26), we have

$$V_{S^c} = \widetilde{\mathbf{W}}^{\mathrm{T}}_{S^c}\,\mathbf{\Pi}\left(\frac{W_j}{\lambda_n n}\right),$$

where $\widetilde{\mathbf{W}}_{S^c} \in \mathbb{R}^{|S^c|}$ is independent of $\mathbf{\Pi}$ and W_j. Since $\mathbf{\Pi}$ is a projection matrix, we have $\|\mathbf{\Pi}W_j\|_2 \leq \|W_j\|_2$. The vector $W_j \in \mathbb{R}^n$ has i.i.d. Gaussian entries with variance at most 1, and hence the event $\mathcal{E} = \{\frac{\|W_j\|_2}{\sqrt{n}} \leq 2\}$ holds with probability at least $1 - 2e^{-n}$. Conditioning on this event and its complement, we find that

$$\mathbb{P}[\|V_{S^c}\|_\infty \geq t] \leq \mathbb{P}\Big[\|V_{S^c}\|_\infty \geq t \mid \mathcal{E}\Big] + 2e^{-c_3 n}.$$

Conditioned on $\mathcal{E}$, each element of V_{S^c} has variance at most $\frac{4}{\lambda_n^2 n}$, and hence

$$\mathbb{P}[\|V_{S^c}\|_\infty \geq \frac{\alpha}{4}] \leq 2\, e^{-\frac{\lambda_n^2 n\alpha^2}{256} + \log|S^c|} + 2e^{-n},$$

where we have combined the union bound with standard Gaussian tail bounds. Since $\lambda_n = c_0\{\frac{1}{\alpha}\sqrt{\frac{\log d}{n}} + \delta\}$ for a universal constant c_0 that may be chosen, we can ensure that $\frac{\lambda_n^2 n\alpha^2}{256} \geq c_2 n\alpha^2\delta^2 + 2\log d$ for some constant c_2, for which it follows that

$$\mathbb{P}[\|V_{S^c}\|_\infty \geq \frac{\alpha}{4}] \leq c_1 e^{-c_2 n\delta^2\alpha^2 - \log d} + 2e^{-n}.$$

Proof of part (b): In order to prove part (b) of the theorem, it suffices to establish ℓ_∞-bounds on the error in the Lasso solution. Here we provide a proof in the case $m \leq \log d$, referring the reader to the bibliographic section for discussion of the general case. Again returning to the proof of Theorem 7.21, equation (7.54) guarantees that

$$\begin{aligned} \|\widehat{\theta}_S - \theta^*_S\|_\infty &\leq \left\|(\widehat{\mathbf{\Gamma}}_{SS})^{-1}\mathbf{X}_S^{\mathrm{T}}\frac{W_j}{n}\right\|_\infty + \lambda_n |||(\widehat{\mathbf{\Gamma}}_{SS})^{-1}|||_\infty \\ &\leq \left\|(\widehat{\mathbf{\Gamma}}_{SS})^{-1}\mathbf{X}_S^{\mathrm{T}}\frac{W_j}{n}\right\|_\infty + \lambda_n \left\{|||(\widehat{\mathbf{\Gamma}}_{SS})^{-1} - (\mathbf{\Gamma}^*_{SS})^{-1}|||_\infty + |||(\mathbf{\Gamma}^*_{SS})^{-1}|||_\infty\right\}. \end{aligned} \tag{11.28}$$

Now for any symmetric $m \times m$ matrix, we have

$$|||\mathbf{A}|||_\infty = \max_{i=1,\ldots,m} \sum_{\ell=1}^m |A_{i\ell}| \leq \sqrt{m} \max_{i=1,\ldots,m} \sqrt{\sum_{\ell=1}^m |A_{i\ell}|^2} \leq \sqrt{m}|||\mathbf{A}|||_2.$$

Applying this bound to the matrix $\mathbf{A} = (\widehat{\mathbf{\Gamma}}_{SS})^{-1} - (\mathbf{\Gamma}^*_{SS})^{-1}$, we find that

$$|||(\widehat{\mathbf{\Gamma}}_{SS})^{-1} - (\mathbf{\Gamma}^*_{SS})^{-1}|||_\infty \leq \sqrt{m}|||(\widehat{\mathbf{\Gamma}}_{SS})^{-1} - (\mathbf{\Gamma}^*_{SS})^{-1}|||_2. \tag{11.29}$$

Since $|||\mathbf{\Gamma}^*_{SS}|||_2 \leq |||\mathbf{\Gamma}^*_{SS}|||_\infty \leq b$, applying the random matrix bound from Theorem 6.1 allows us to conclude that

$$|||(\widehat{\mathbf{\Gamma}}_{SS})^{-1} - (\mathbf{\Gamma}^*_{SS})^{-1}|||_2 \leq 2b\left(\sqrt{\frac{m}{n}} + \frac{1}{\sqrt{m}} + 10\sqrt{\frac{\log d}{n}}\right),$$

with probability at least $1 - c_1 e^{-c_2\frac{n}{m} - \log d}$. Combined with the earlier bound (11.29), we find that

$$|||(\widehat{\mathbf{\Gamma}}_{SS})^{-1} - (\mathbf{\Gamma}^*_{SS})^{-1}|||_\infty \leq 2b\left(\sqrt{\frac{m^2}{n}} + 1 + 10\sqrt{\frac{m\log d}{n}}\right) \stackrel{\text{(i)}}{\leq} 6b, \tag{11.30}$$

where inequality (i) uses the assumed lower bound $n \gtrsim m\log d \geq m^2$. Putting together the pieces in the bound (11.28) leads to

$$\|\widehat{\theta}_S - \theta^*_S\|_\infty \leq \underbrace{\left\|(\widehat{\mathbf{\Gamma}}_{SS})^{-1}\mathbf{X}_S^{\mathrm{T}}\frac{W_j}{n}\right\|_\infty}_{U_S} + 7b\lambda_n. \tag{11.31}$$

Now the vector $W_j \in \mathbb{R}^n$ has i.i.d. Gaussian entries, each zero-mean with variance at most

$\mathrm{var}(X_j) \le 1$, and is independent of $\mathbf{X}_S$. Consequently, conditioned on $\mathbf{X}_S$, the quantity U_S is a zero-mean Gaussian m-vector, with maximal variance

$$\frac{1}{n}\|\,\mathrm{diag}(\widehat{\boldsymbol{\Gamma}}_{SS})^{-1}\|_\infty \le \frac{1}{n}\Big\{|||(\widehat{\boldsymbol{\Gamma}}_{SS})^{-1} - (\boldsymbol{\Gamma}^*_{SS})^{-1}|||_\infty + |||(\boldsymbol{\Gamma}^*_{SS})^{-1}|||_\infty\Big\} \le \frac{7b}{n},$$

where we have combined the assumed bound $|||(\boldsymbol{\Gamma}^*_{SS})^{-1}|||_\infty \le b$ with the inequality (11.30). Therefore, the union bound combined with Gaussian tail bounds implies that

$$\mathbb{P}[\|U_S\|_\infty \ge b\lambda_n] \le 2|S|e^{-\frac{n\lambda_n^2}{14}} \overset{\text{(i)}}{\le} c_1 e^{-c_2 n b \delta^2 - \log d},$$

where, as in our earlier argument, inequality (i) can be guaranteed by a sufficiently large choice of the pre-factor c_0 in the definition of λ_n. Substituting back into the earlier bound (11.31), we find that $\|\widehat{\theta}_S - \theta^*_S\|_\infty \le 7b\lambda_n$ with probability at least $1 - c_1 e^{-c_2 n\{\delta^2 \wedge \frac{1}{m}\} - \log d}$. Finally, taking the union bound over all vertices $j \in V$ causes a loss of at most a factor $\log d$ in the exponent. □

11.3 Graphical models in exponential form

Let us now move beyond the Gaussian case, and consider the graph estimation problem for a more general class of graphical models that can be written in an exponential form. In particular, for a given graph $G = (V, E)$, consider probability densities that have a pairwise factorization of the form

$$p_{\Theta^*}(x_1, \ldots, x_d) \propto \exp\Bigg\{\sum_{j\in V} \phi_j(x_j; \Theta^*_j) + \sum_{(j,k)\in E} \phi_{jk}(x_j, x_k; \boldsymbol{\Theta}^*_{jk})\Bigg\}, \tag{11.32}$$

where Θ^*_j is a vector of parameters for node $j \in V$, and $\boldsymbol{\Theta}^*_{jk}$ is a matrix of parameters for edge (j, k). For instance, the Gaussian graphical model is a special case in which $\Theta^*_j = \theta^*_j$ and $\boldsymbol{\Theta}^*_{jk} = \theta^*_{jk}$ are both scalars, the potential functions take the form

$$\phi_j(x_j; \theta^*_j) = \theta^*_j x_j, \qquad \phi_{jk}(x_j, x_k; \theta^*_{jk}) = \theta^*_{jk} x_j x_k, \tag{11.33}$$

and the density (11.32) is taken with respect to Lebesgue measure over $\mathbb{R}^d$. The Ising model (11.3) is another special case, using the same choice of potential functions (11.33), but taking the density with respect to the counting measure on the binary hypercube $\{0, 1\}^d$.

Let us consider a few more examples of this factorization:

Example 11.13 (Potts model) The *Potts model*, in which each variable X_s takes values in the discrete set $\{0, \ldots, M-1\}$ is another special case of the factorization (11.32). In this case, the parameter $\Theta^*_j = \{\Theta_{j;a}, a = 1, \ldots, M-1\}$ is an $(M-1)$-vector, whereas the parameter $\boldsymbol{\Theta}^*_{jk} = \{\Theta^*_{jk;ab}, a, b = 1, \ldots, M-1\}$ is an $(M-1) \times (M-1)$ matrix. The potential functions take the form

$$\phi_j(x_j; \Theta^*_j) = \sum_{a=1}^{M-1} \Theta^*_{j;a}\, \mathbb{I}[x_j = a] \tag{11.34a}$$

and

$$\phi_{jk}(x_j, x_k; \mathbf{\Theta}^*_{jk}) = \sum_{a=1}^{M-1} \sum_{b=1}^{M-1} \Theta^*_{jk;ab} \, \mathbb{I}[x_j = a, x_k = b]. \tag{11.34b}$$

Here $\mathbb{I}[x_j = a]$ is a zero–one indicator function for the event that $\{x_j = a\}$, with the indicator function $\mathbb{I}[x_j = a, x_k = b]$ defined analogously. Note that the Potts model is a generalization of the Ising model (11.3), to which it reduces for variables taking $M = 2$ states. ♣

Example 11.14 (Poisson graphical model) Suppose that we are interested in modeling a collection of random variables $(X_1, \ldots, X_d)$, each of which represents some type of count data taking values in the set of positive integers $\mathbb{Z}_+ = \{0, 1, 2, \ldots\}$. One way of building a graphical model for such variables is by specifying the conditional distribution of each variable given its neighbors. In particular, suppose that variable X_j, when conditioned on its neighbors, is a Poisson random variable with mean

$$\mu_j = \exp\left(\theta^*_j + \sum_{k \in N(j)} \theta^*_{jk} x_k\right).$$

This form of conditional distribution leads to a Markov random field of the form (11.32) with

$$\phi_j(x_j; \theta^*_j) = \theta^*_j x_j - \log(x!) \qquad \text{for all } j \in V, \tag{11.35a}$$

$$\phi_{jk}(x_j, x_k; \theta^*_{jk}) = \theta^*_{jk} x_j x_k \qquad \text{for all } (j,k) \in E. \tag{11.35b}$$

Here the density is taken with respect to the counting measure on $\mathbb{Z}_+$ for all variables. A potential deficiency of this model is that, in order for the density to be normalizable, we must necessarily have $\theta^*_{jk} \leq 0$ for all $(j,k) \in E$. Consequently, this model can only capture competitive interactions between variables. ♣

One can also consider various types of mixed graphical models, for instance in which some of the nodes take discrete values, whereas others are continuous-valued. Gaussian mixture models are one important class of such models.

11.3.1 A general form of neighborhood regression

We now consider a general form of neighborhood regression, applicable to any graphical model of the form (11.32). Let $\{x_i\}_{i=1}^n$ be a collection of n samples drawn i.i.d. from such a graphical model; here each x_i is a d-vector. Based on these samples, we can form a matrix $\mathbf{X} \in \mathbb{R}^{n \times d}$ with x_i^{T} as the ith row. For $j = 1, \ldots, d$, we let $X_j \in \mathbb{R}^n$ denote the jth column of $\mathbf{X}$. Neighborhood regression is based on predicting the column $X_j \in \mathbb{R}^n$ using the columns of the submatrix $\mathbf{X}_{\setminus\{j\}} \in \mathbb{R}^{n \times (d-1)}$.

Consider the conditional likelihood of $X_j \in \mathbb{R}^n$ given $\mathbf{X}_{\setminus\{j\}} \in \mathbb{R}^{n \times (d-1)}$. As we show in Exercise 11.6, for any distribution of the form (11.32), this conditional likelihood depends only on the vector of parameters

$$\mathbf{\Theta}_{j+} := \{\Theta_j, \mathbf{\Theta}_{jk}, \, k \in V \setminus \{j\}\} \tag{11.36}$$

that involve node j. Moreover, in the true model Θ^*, we are guaranteed that $\Theta^*_{jk} = 0$ whenever $(j,k) \notin E$, so that it is natural to impose some type of block-based sparsity penalty on Θ_{j+}. Letting $|||\cdot|||$ denote some matrix norm, we arrive at a general form of neighborhood regression:

$$\widehat{\Theta}_{j+} = \arg\min_{\Theta_{j+}} \Big\{ \underbrace{-\frac{1}{n}\sum_{i=1}^{n} \log p_{\Theta_{j+}}(x_{ij} \mid x_{i\,\backslash\{j\}})}_{\mathcal{L}_n(\Theta_{j+};\, x_j, x_{\backslash\{j\}})} + \lambda_n \sum_{k\in V\backslash\{j\}} |||\Theta_{jk}||| \Big\}. \tag{11.37}$$

This formulation actually describes a family of estimators, depending on which norm $|||\cdot|||$ that we impose on each matrix component Θ_{jk}. Perhaps the simplest is the Frobenius norm, in which case the estimator (11.37) is a general form of the group Lasso; for details, see equation (9.66) and the associated discussion in Chapter 9. Also, as we verify in Exercise 11.5, this formula reduces to ℓ_1-regularized linear regression (11.22) in the Gaussian case.

11.3.2 Graph selection for Ising models

In this section, we consider the graph selection problem for a particular type of non-Gaussian distribution, namely the Ising model. Recall that the Ising distribution is over binary variables, and takes the form

$$p_{\theta^*}(x_1, \ldots, x_d) \propto \exp\Big\{ \sum_{j\in V} \theta^*_j x_j + \sum_{(j,k)\in E} \theta^*_{jk} x_j x_k \Big\}. \tag{11.38}$$

Since there is only a single parameter per edge, imposing an ℓ_1-penalty suffices to encourage sparsity in the neighborhood regression. For any given node $j \in V$, we define the subset of coefficients associated with it—namely, the set

$$\theta_{j+} := \big\{\theta_j, \theta_{jk}, k \in V \setminus \{j\}\big\}.$$

For the Ising model, the neighborhood regression estimate reduces to a form of logistic regression—specifically

$$\widehat{\theta}_{j+} = \arg\min_{\theta_{j+}\in\mathbb{R}^d} \Big\{ \underbrace{\frac{1}{n}\sum_{i=1}^{n} f\Big(\theta_j x_{ij} + \sum_{k\in V\backslash\{j\}} \theta_{jk} x_{ij} x_{ik}\Big)}_{\mathcal{L}_n(\theta_{j+};\, ,x_j,\, x_{\backslash\{j\}})} + \lambda_n \sum_{k\in V\backslash\{j\}} |\theta_{jk}| \Big\}, \tag{11.39}$$

where $f(t) = \log(1 + e^t)$ is the logistic function. See Exercise 11.7 for details.

Under what conditions does the estimate (11.39) recover the correct neighborhood set $\mathcal{N}(j)$? As in our earlier analysis of neighborhood linear regression and the graphical Lasso, such a guarantee requires some form of incoherence condition, limiting the influence of irrelevant variables—those outside $\mathcal{N}(j)$—on variables inside the set. Recalling the cost function $\mathcal{L}_n$ in the optimization problem (11.39), let θ^*_{j+} denote the minimizer of the population objective function $\overline{\mathcal{L}}(\theta_{j+}) = \mathbb{E}[\mathcal{L}_n(\theta_{j+};\ X_j, \mathbf{X}_{\backslash\{j\}})]$. We then consider the Hessian of the cost function $\overline{\mathcal{L}}$ evaluated at the "true parameter" θ^*_{j+}—namely, the d-dimensional matrix $\mathbf{J} := \nabla^2\overline{\mathcal{L}}(\theta^*_{j+})$. For a given $\alpha \in (0, 1]$, we say that $\mathbf{J}$ satisfies an α-incoherence condition at node $j \in V$ if

$$\max_{k \notin S} \|J_{kS}(\mathbf{J}_{SS})^{-1}\|_1 \leq 1 - \alpha, \tag{11.40}$$

where we have introduced the shorthand $S = \mathcal{N}(j)$ for the neighborhood set of node j. In addition, we assume the submatrix $\mathbf{J}_{SS}$ has its smallest eigenvalue lower bounded by some $c_{\min} > 0$. With this set-up, the following result applies to an Ising model (11.38) defined on a graph G with d vertices and maximum degree at most m, with Fisher information $\mathbf{J}$ at node j satisfying the $c_{\min}$-eigenvalue bound, and the α-incoherence condition (11.40).

Theorem 11.15 *Given n i.i.d. samples with* $n > c_0 m^2 \log d$, *consider the estimator* (11.39) *with* $\lambda_n = \frac{32}{\alpha}\sqrt{\frac{\log d}{n}} + \delta$ *for some* $\delta \in [0, 1]$. *Then with probability at least* $1 - c_1 e^{-c_2(n\delta^2 + \log d)}$, *the estimate* $\widehat{\theta}_{j+}$ *has the following properties:*

(a) *It has a support* $\widehat{S} = \operatorname{supp}(\widehat{\theta})$ *that is contained within the neighborhood set* $\mathcal{N}(j)$.
(b) *It satisfies the* ℓ_∞*-bound* $\|\widehat{\theta}_{j+} - \theta^*_{j+}\|_\infty \leq \frac{c_3}{c_{\min}}\sqrt{m}\lambda_n$.

As with our earlier results on the neighborhood and graphical Lasso, part (a) guarantees that the method leads to *no false inclusions*. On the other hand, the ℓ_∞-bound in part (b) ensures that the method picks up all significant variables. The proof of Theorem 11.15 is based on the same type of primal–dual witness construction used in the proof of Theorem 11.12. See the bibliographic section for further details.

11.4 Graphs with corrupted or hidden variables

Thus far, we have assumed that the samples $\{x_i\}_{i=1}^n$ are observed perfectly. This idealized setting can be violated in a number of ways. The samples may be corrupted by some type of measurement noise, or certain entries may be missing. In the most extreme case, some subset of the variables are never observed, and so are known as hidden or latent variables. In this section, we discuss some methods for addressing these types of problems, focusing primarily on the Gaussian case for simplicity.

11.4.1 Gaussian graph estimation with corrupted data

Let us begin our exploration with the case of corrupted data. Letting $\mathbf{X} \in \mathbb{R}^{n \times d}$ denote the data matrix corresponding to the original samples, suppose that we instead observe a corrupted version $\mathbf{Z}$. In the simplest case, we might observe $\mathbf{Z} = \mathbf{X} + \mathbf{V}$, where the matrix $\mathbf{V}$ represents some type of measurement error. A naive approach would be simply to apply a standard Gaussian graph estimator to the observed data, but, as we will see, doing so typically leads to inconsistent estimates.

Correcting the Gaussian graphical Lasso

Consider the graphical Lasso (11.10), which is usually based on the sample covariance matrix $\widehat{\mathbf{\Sigma}}_x = \frac{1}{n}\mathbf{X}^{\mathrm{T}}\mathbf{X} = \frac{1}{n}\sum_{i=1} x_i x_i^{\mathrm{T}}$ of the raw samples. The naive approach would be instead to solve the convex program

$$\widehat{\mathbf{\Theta}}_{\mathrm{NAI}} = \arg\min_{\mathbf{\Theta}\in\mathcal{S}^{d\times d}} \left\{ \langle\langle \mathbf{\Theta},\, \widehat{\mathbf{\Sigma}}_z \rangle\rangle - \log\det\mathbf{\Theta} + \lambda_n |||\mathbf{\Theta}|||_{1,\mathrm{off}} \right\}, \tag{11.41}$$

where $\widehat{\mathbf{\Sigma}}_z = \frac{1}{n}\mathbf{Z}^{\mathrm{T}}\mathbf{Z} = \frac{1}{n}\sum_{i=1}^n z_i z_i^{\mathrm{T}}$ is now the sample covariance based on the observed data matrix $\mathbf{Z}$. However, as we explore in Exercise 11.8, the addition of noise does not preserve Markov properties, so that—at least in general—the estimate $\widehat{\mathbf{\Theta}}_{\mathrm{NAI}}$ will not lead to consistent estimates of either the edge set, or the underlying precision matrix $\mathbf{\Theta}^*$. In order to obtain a consistent estimator, we need to replace $\widehat{\mathbf{\Sigma}}_z$ with an unbiased estimator of $\mathrm{cov}(x)$ based on the observed data matrix $\mathbf{Z}$. In order to develop intuition, let us explore a few examples.

Example 11.16 (Unbiased covariance estimate for additive corruptions) In the additive noise setting ($\mathbf{Z} = \mathbf{X} + \mathbf{V}$), suppose that each row v_i of the noise matrix $\mathbf{V}$ is drawn i.i.d. from a zero-mean distribution, say with covariance $\mathbf{\Sigma}_v$. In this case, a natural estimate of $\mathbf{\Sigma}_x := \mathrm{cov}(x)$ is given by

$$\widehat{\mathbf{\Gamma}} := \frac{1}{n}\mathbf{Z}^{\mathrm{T}}\mathbf{Z} - \mathbf{\Sigma}_v. \tag{11.42}$$

As long as the noise matrix $\mathbf{V}$ is independent of $\mathbf{X}$, then $\widehat{\mathbf{\Gamma}}$ is an unbiased estimate of $\mathbf{\Sigma}_x$. Moreover, as we explore in Exercise 11.12, when both $\mathbf{X}$ and $\mathbf{V}$ have sub-Gaussian rows, then a deviation condition of the form $\|\widehat{\mathbf{\Gamma}} - \mathbf{\Sigma}_x\|_{\max} \precsim \sqrt{\frac{\log d}{n}}$ holds with high probability. ♣

Example 11.17 (Missing data) In other settings, some entries of the data matrix $\mathbf{X}$ might be missing, with the remaining entries observed. In the simplest model of missing data—known as missing completely at random—entry (i, j) of the data matrix is missing with some probability $\nu \in [0, 1)$. Based on the observed matrix $\mathbf{Z} \in \mathbb{R}^{n\times d}$, we can construct a new matrix $\widetilde{\mathbf{Z}} \in \mathbb{R}^{n\times d}$ with entries

$$\widetilde{Z}_{ij} = \begin{cases} \frac{Z_{ij}}{1-\nu} & \text{if entry } (i, j) \text{ is observed,} \\ 0 & \text{otherwise.} \end{cases}$$

With this choice, it can be verified that

$$\widehat{\mathbf{\Gamma}} = \frac{1}{n}\widetilde{\mathbf{Z}}^{\mathrm{T}}\widetilde{\mathbf{Z}} - \nu\,\mathrm{diag}\left(\frac{\widetilde{\mathbf{Z}}^{\mathrm{T}}\widetilde{\mathbf{Z}}}{n}\right) \tag{11.43}$$

is an unbiased estimate of the covariance matrix $\mathbf{\Sigma}_x = \mathrm{cov}(x)$, and moreover, under suitable tail conditions, it also satisfies the deviation condition $\|\widehat{\mathbf{\Gamma}} - \mathbf{\Sigma}_x\|_{\max} \precsim \sqrt{\frac{\log d}{n}}$ with high probability. See Exercise 11.13 for more details. ♣

More generally, any unbiased estimate $\widehat{\mathbf{\Gamma}}$ of $\mathbf{\Sigma}_x$ defines a form of the *corrected graphical Lasso* estimator

$$\widetilde{\mathbf{\Theta}} = \arg\min_{\mathbf{\Theta}\in\mathcal{S}_+^{d\times d}} \left\{ \langle\langle \mathbf{\Theta},\, \widehat{\mathbf{\Gamma}} \rangle\rangle - \log\det\mathbf{\Theta} + \lambda_n |||\mathbf{\Theta}|||_{1,\mathrm{off}} \right\}. \tag{11.44}$$

As with the usual graphical Lasso, this is a strictly convex program, so that the solution (when it exists) must be unique. However, depending on the nature of the covariance estimate $\widehat{\mathbf{\Gamma}}$, it need not be the case that the program (11.44) has any solution at all! In this case, equation (11.44) is nonsensical, since it presumes the existence of an optimal solution. However, in Exercise 11.9, we show that as long as $\lambda_n > \|\widehat{\mathbf{\Gamma}} - \mathbf{\Sigma}_x\|_{\max}$, then this optimization problem has a unique optimum that is achieved, so that the estimator is meaningfully defined. Moreover, by inspecting the proofs of the claims in Section 11.2.1, it can be seen that the estimator $\widetilde{\mathbf{\Theta}}$ obeys similar Frobenius norm and edge selection bounds as the usual graphical Lasso. Essentially, the only differences lie in the techniques used to bound the deviation $\|\widehat{\mathbf{\Gamma}} - \mathbf{\Sigma}_x\|_{\max}$.

Correcting neighborhood regression

We now describe how the method of neighborhood regression can be corrected to deal with corrupted or missing data. Here the underlying optimization problem is typically nonconvex, so that the analysis of the estimator becomes more interesting than the corrected graphical Lasso.

As previously described in Section 11.2.2, the neighborhood regression approach involves solving a linear regression problem, in which the observation vector $X_j \in \mathbb{R}^n$ at a given node j plays the role of the response variable, and the remaining $(d-1)$ variables play the role of the predictors. Throughout this section, we use $\mathbf{X}$ to denote the $n \times (d-1)$ matrix with $\{X_k, k \in V \setminus \{j\}\}$ as its columns, and we use $y = X_j$ to denote the response vector. With this notation, we have an instance of a corrupted linear regression model, namely

$$y = \mathbf{X}\theta^* + w \quad \text{and} \quad \mathbf{Z} \sim \mathbb{Q}(\cdot \mid \mathbf{X}), \tag{11.45}$$

where the conditional probability distribution $\mathbb{Q}$ varies according to the nature of the corruption. In application to graphical models, the response vector y might also be further corrupted, but this case can often be reduced to an instance of the previous model. For instance, if some entries of $y = X_j$ are missing, then we can simply discard those data points in performing the neighborhood regression at node j, or if y is subject to further noise, it can be incorporated into the model.

As before, the naive approach would be simply to solve a least-squares problem involving the cost function $\frac{1}{2n}\|y - \mathbf{Z}\theta\|_2^2$. As we explore in Exercise 11.10, doing so will lead to an inconsistent estimate of the neighborhood regression vector θ^*. However, as with the graphical Lasso, the least-squares estimator can also be corrected. What types of quantities need to be "corrected" in order to obtain a consistent form of linear regression? Consider the following population-level objective function

$$\overline{\mathcal{L}}(\theta) = \tfrac{1}{2}\theta^{\mathrm{T}}\mathbf{\Gamma}\theta - \langle \theta,\, \gamma \rangle, \tag{11.46}$$

where $\mathbf{\Gamma} := \operatorname{cov}(x)$ and $\gamma := \operatorname{cov}(x, y)$. By construction, the true regression vector is the unique global minimizer of $\overline{\mathcal{L}}$. Thus, a natural strategy is to solve a penalized regression problem in which the pair $(\gamma, \mathbf{\Gamma})$ are replaced by data-dependent estimates $(\widehat{\gamma}, \widehat{\mathbf{\Gamma}})$. Doing so leads to the empirical objective function

$$\mathcal{L}_n(\theta) = \tfrac{1}{2}\theta^{\mathrm{T}}\widehat{\mathbf{\Gamma}}\theta - \langle \theta,\, \widehat{\gamma} \rangle. \tag{11.47}$$

To be clear, the estimates $(\widehat{\gamma}, \widehat{\mathbf{\Gamma}})$ must be based on the observed data $(y, \mathbf{Z})$. In Examples 11.16

and 11.17, we described suitable unbiased estimators $\widehat{\boldsymbol{\Gamma}}$ for the cases of additive corruptions and missing entries, respectively. Exercises 11.12 and 11.13 discuss some unbiased estimators $\widehat{\gamma}$ of the cross-covariance vector γ.

Combining the ingredients, we are led to study the following *corrected Lasso* estimator

$$\min_{\|\theta\|_1 \leq \sqrt{\frac{n}{\log d}}} \left\{ \tfrac{1}{2}\theta^{\mathrm{T}}\widehat{\boldsymbol{\Gamma}}\theta - \langle \widehat{\gamma}, \theta \rangle + \lambda_n \|\theta\|_1 \right\}. \tag{11.48}$$

Note that it combines the objective function (11.47) with an ℓ_1-penalty, as well as an ℓ_1-constraint. At first sight, including both the penalty and constraint might seem redundant, but as shown in Exercise 11.11, this combination is actually needed when the objective function (11.47) is non-convex. Many of the standard choices of $\widehat{\boldsymbol{\Gamma}}$ lead to non-convex programs: for instance, in the high-dimensional regime ($n < d$), the previously described choices of $\widehat{\boldsymbol{\Gamma}}$ given in equations (11.42) and (11.43) both have negative eigenvalues, so that the associated optimization problem is non-convex.

When the optimization problem (11.48) is non-convex, it may have local optima in addition to global optima. Since standard algorithms such as gradient descent are only guaranteed to converge to local optima, it is desirable to have theory that applies them. More precisely, a *local optimum* for the program (11.48) is any vector $\widetilde{\theta} \in \mathbb{R}^d$ such that

$$\left\langle \nabla \mathcal{L}_n(\widetilde{\theta}),\, \theta - \widetilde{\theta} \right\rangle \geq 0 \qquad \text{for all } \theta \text{ such that } \|\theta\|_1 \leq \sqrt{\tfrac{n}{\log d}}. \tag{11.49}$$

When $\widetilde{\theta}$ belongs to the interior of the constraint set—that is, when it satisfies the inequality $\|\widetilde{\theta}\|_1 < \sqrt{\frac{n}{\log d}}$ strictly—then this condition reduces to the usual zero-gradient condition $\nabla \mathcal{L}_n(\widetilde{\theta}) = 0$. Thus, our specification includes both local minima, local maxima and saddle points.

We now establish an interesting property of the corrected Lasso (11.48): under suitable conditions—ones that still permit non-convexity—*any* local optimum is relatively close to the true regression vector. As in our analysis of the ordinary Lasso from Chapter 7, we impose a restricted eigenvalue (RE) condition on the covariance estimate $\widehat{\boldsymbol{\Gamma}}$: more precisely, we assume that there exists a constant $\kappa > 0$ such that

$$\left\langle \Delta,\, \widehat{\boldsymbol{\Gamma}}\Delta \right\rangle \geq \kappa \|\Delta\|_2^2 - c_0 \frac{\log d}{n} \|\Delta\|_1^2 \qquad \text{for all } \Delta \in \mathbb{R}^d. \tag{11.50}$$

Interestingly, such an RE condition can hold for matrices $\widehat{\boldsymbol{\Gamma}}$ that are indefinite (with both positive and negative eigenvalues), including our estimators for additive corruptions and missing data from Examples 11.16 and 11.17. See Exercises 11.12 and 11.13, respectively, for further details on these two cases.

Moreover, we assume that the minimizer θ^* of the population objective (11.46) has sparsity s and ℓ_2-norm at most one, and that the sample size n is lower bounded as $n \geq s \log d$. These assumptions ensure that $\|\theta^*\|_1 \leq \sqrt{s} \leq \sqrt{\frac{n}{\log d}}$, so that θ^* is feasible for the non-convex Lasso (11.48).

Proposition 11.18 *Under the RE condition* (11.50), *suppose that the pair* $(\widehat{\gamma}, \widehat{\mathbf{\Gamma}})$ *satisfy the deviation condition*

$$\|\widehat{\mathbf{\Gamma}}\theta^* - \widehat{\gamma}\|_{\max} \leq \varphi(\mathbb{Q}, \sigma_w)\sqrt{\frac{\log d}{n}}, \tag{11.51}$$

for a pre-factor $\varphi(\mathbb{Q}, \sigma_w)$ *depending on the conditional distribution* $\mathbb{Q}$ *and noise standard deviation* σ_w. *Then for any regularization parameter* $\lambda_n \geq 2(2c_0 + \varphi(\mathbb{Q}, \sigma_w))\sqrt{\frac{\log d}{n}}$, *any local optimum* $\widetilde{\theta}$ *to the program* (11.48) *satisfies the bound*

$$\|\widetilde{\theta} - \theta^*\|_2 \leq \frac{2}{\kappa}\sqrt{s}\,\lambda_n. \tag{11.52}$$

In order to gain intuition for the constraint (11.51), observe that the optimality of θ^* for the population-level objective (11.46) implies that $\nabla\overline{\mathcal{L}}(\theta^*) = \mathbf{\Gamma}\theta^* - \gamma = 0$. Consequently, condition (11.51) is the sample-based and approximate equivalent of this optimality condition. Moreover, under suitable tail conditions, it is satisfied with high probability by our previous choices of $(\widehat{\gamma}, \widehat{\mathbf{\Gamma}})$ for additively corrupted or missing data. Again, see Exercises 11.12 and 11.13 for further details.

Proof We prove this result in the special case when the optimum occurs in the interior of the set $\|\theta\|_1 \leq \sqrt{\frac{n}{\log d}}$. (See the bibliographic section for references to the general result.) In this case, any local optimum $\widetilde{\theta}$ must satisfy the condition $\nabla\mathcal{L}_n(\widetilde{\theta}) + \lambda_n\widetilde{z} = 0$, where $\widetilde{z}$ belongs to the subdifferential of the ℓ_1-norm at $\widetilde{\theta}$. Define the error vector $\widehat{\Delta} := \widetilde{\theta} - \theta^*$. Adding and subtracting terms and then taking inner products with $\widehat{\Delta}$ yields the inequality

$$\begin{aligned}\left\langle\widehat{\Delta},\, \nabla\mathcal{L}_n(\theta^* + \widehat{\Delta}) - \nabla\mathcal{L}_n(\theta^*)\right\rangle &\leq \left|\left\langle\widehat{\Delta},\, \nabla\mathcal{L}_n(\theta^*)\right\rangle\right| - \lambda_n\left\langle\widetilde{z},\, \widehat{\Delta}\right\rangle\\ &\leq \|\widehat{\Delta}\|_1\|\nabla\mathcal{L}_n(\theta^*)\|_\infty + \lambda_n\{\|\theta^*\|_1 - \|\widetilde{\theta}\|_1\},\end{aligned}$$

where we have used the facts that $\left\langle\widetilde{z},\, \widetilde{\theta}\right\rangle = \|\widetilde{\theta}\|_1$ and $\langle\widetilde{z},\, \theta^*\rangle \leq \|\theta^*\|_1$. From the proof of Theorem 7.8, since the vector θ^* is S-sparse, we have

$$\|\theta^*\|_1 - \|\widetilde{\theta}\|_1 \leq \|\widehat{\Delta}_S\|_1 - \|\widehat{\Delta}_{S^c}\|_1. \tag{11.53}$$

Since $\nabla\mathcal{L}_n(\theta) = \widehat{\mathbf{\Gamma}}\theta - \widehat{\gamma}$, the deviation condition (11.51) is equivalent to the bound

$$\|\nabla\mathcal{L}_n(\theta^*)\|_\infty \leq \varphi(\mathbb{Q}, \sigma_w)\sqrt{\frac{\log d}{n}},$$

which is less than $\lambda_n/2$ by our choice of regularization parameter. Consequently, we have

$$\left\langle\widehat{\Delta},\, \widehat{\mathbf{\Gamma}}\widehat{\Delta}\right\rangle \leq \frac{\lambda_n}{2}\|\widehat{\Delta}\|_1 + \lambda_n\{\|\widehat{\Delta}_S\|_1 - \|\widehat{\Delta}_{S^c}\|_1\} = \frac{3}{2}\lambda_n\|\widehat{\Delta}_S\|_1 - \frac{1}{2}\lambda_n\|\widehat{\Delta}_{S^c}\|_1. \tag{11.54}$$

Since θ^* is s-sparse, we have $\|\theta^*\|_1 \leq \sqrt{s}\|\theta^*\|_2 \leq \sqrt{\frac{n}{\log d}}$, where the final inequality follows from the assumption that $n \geq s\log d$. Consequently, we have

$$\|\widehat{\Delta}\|_1 \leq \|\widetilde{\theta}\|_1 + \|\theta^*\|_1 \leq 2\sqrt{\frac{n}{\log d}}.$$

Combined with the RE condition (11.50), we have

$$\left\langle \widehat{\Delta}, \widehat{\mathbf{\Gamma}\Delta} \right\rangle \geq \kappa \|\widehat{\Delta}\|_2^2 - c_0 \frac{\log d}{n} \|\widehat{\Delta}\|_1^2 \geq \kappa \|\widehat{\Delta}\|_2^2 - 2c_0 \sqrt{\frac{\log d}{n}} \|\widehat{\Delta}\|_1.$$

Recombining with our earlier bound (11.54), we have

$$\begin{aligned}
\kappa \|\widehat{\Delta}\|_2^2 &\leq 2c_0 \sqrt{\frac{\log d}{n}} \|\widehat{\Delta}\|_1 + \frac{3}{2} \lambda_n \|\widehat{\Delta}_S\|_1 - \frac{1}{2} \lambda_n \|\widehat{\Delta}_{S^c}\|_1 \\
&\leq \frac{1}{2} \lambda_n \|\widehat{\Delta}\|_1 + \frac{3}{2} \lambda_n \|\widehat{\Delta}_S\|_1 - \frac{1}{2} \lambda_n \|\widehat{\Delta}_{S^c}\|_1 \\
&= 2\lambda_n \|\widehat{\Delta}_S\|_1.
\end{aligned}$$

Since $\|\widehat{\Delta}_S\|_1 \leq \sqrt{s}\|\widehat{\Delta}\|_2$, the claim follows. □

11.4.2 Gaussian graph selection with hidden variables

In certain settings, a given set of random variables might not be accurately described using a sparse graphical model on their own, but can be when augmented with an additional set of hidden variables. The extreme case of this phenomenon is the distinction between independence and conditional independence: for instance, the random variables $X_1 = \texttt{Shoe size}$ and $X_2 = \texttt{Gray hair}$ are likely to be dependent, since few children have gray hair. However, it might be reasonable to model them as being conditionally independent given a third variable—namely $X_3 = \texttt{Age}$.

How to estimate a sparse graphical model when only a subset of the variables are observed? More precisely, consider a family of $d + r$ random variables—say written as $X := (X_1, \ldots, X_d, X_{d+1}, \ldots, X_{d+r})$—and suppose that this full vector can be modeled by a sparse graphical model with $d + r$ vertices. Now suppose that we observe only the subvector $X_{\mathrm{O}} := (X_1, \ldots, X_d)$, with the other components $X_{\mathrm{H}} := (X_{d+1}, \ldots, X_{d+r})$ staying hidden. Given this partial information, our goal is to recover useful information about the underlying graph.

In the Gaussian case, this problem has an attractive matrix-theoretic formulation. In particular, the observed samples of X_{O} give us information about the covariance matrix $\mathbf{\Sigma}^*_{\mathrm{OO}}$. On the other hand, since we have assumed that the full vector is Markov with respect to a sparse graph, the Hammersley–Clifford theorem implies that the inverse covariance matrix $\mathbf{\Theta}^\diamond$ of the full vector $X = (X_{\mathrm{O}}, X_{\mathrm{H}})$ is sparse. This $(d + r)$-dimensional matrix can be written in the block-partitioned form

$$\mathbf{\Theta}^\diamond = \begin{bmatrix} \mathbf{\Theta}^\diamond_{\mathrm{OO}} & \mathbf{\Theta}^\diamond_{\mathrm{OH}} \\ \mathbf{\Theta}^\diamond_{\mathrm{HO}} & \mathbf{\Theta}^\diamond_{\mathrm{HH}} \end{bmatrix}. \tag{11.55}$$

The block-matrix inversion formula (see Exercise 11.3) ensures that the inverse of the d-dimensional covariance matrix $\mathbf{\Sigma}^*_{\mathrm{OO}}$ has the decomposition

$$(\mathbf{\Sigma}^*_{\mathrm{OO}})^{-1} = \underbrace{\mathbf{\Theta}^\diamond_{\mathrm{OO}}}_{\mathbf{\Gamma}^*} - \underbrace{\mathbf{\Theta}^\diamond_{\mathrm{OH}} (\mathbf{\Theta}^\diamond_{\mathrm{HH}})^{-1} \mathbf{\Theta}^\diamond_{\mathrm{HO}}}_{\mathbf{\Lambda}^*}. \tag{11.56}$$

By our modeling assumptions, the matrix $\mathbf{\Gamma}^* := \mathbf{\Theta}^\diamond_{\mathrm{OO}}$ is sparse, whereas the second component $\mathbf{\Lambda}^* := \mathbf{\Theta}^\diamond_{\mathrm{OH}} (\mathbf{\Theta}^\diamond_{\mathrm{HH}})^{-1} \mathbf{\Theta}^\diamond_{\mathrm{HO}}$ has rank at most $\min\{r, d\}$. Consequently, it has low rank whenever the number of hidden variables r is substantially less than the number of observed variables d. In this way, the addition of hidden variables leads to an inverse covariance matrix that can be decomposed as the sum of a sparse and a low-rank matrix.

Now suppose that we are given n i.i.d. samples $x_i \in \mathbb{R}^d$ from a zero-mean Gaussian with covariance Σ^*_{OO}. In the absence of any sparsity in the low-rank component, we require $n > d$ samples to obtain any sort of reasonable estimate (recall our results on covariance estimation from Chapter 6). When $n > d$, then the sample covariance matrix $\widehat{\Sigma} = \frac{1}{n} \sum_{i=1}^n x_i x_i^{\mathrm{T}}$ will be invertible with high probability, and hence setting $\mathbf{Y} := (\widehat{\Sigma})^{-1}$, we can consider an observation model of the form

$$\mathbf{Y} = \mathbf{\Gamma}^* - \mathbf{\Lambda}^* + \mathbf{W}. \tag{11.57}$$

Here $\mathbf{W} \in \mathbb{R}^{d \times d}$ is a stochastic noise matrix, corresponding to the difference between the inverses of the population and sample covariances. This observation model (11.57) is a particular form of additive matrix decomposition, as previously discussed in Section 10.7.

How to estimate the components of this decomposition? In this section, we analyze a very simple two-step estimator, based on first computing a soft-thresholded version of the inverse sample covariance $\mathbf{Y}$ as an estimate of $\mathbf{\Gamma}^*$, and secondly, taking the residual matrix as an estimate of $\mathbf{\Lambda}^*$. In particular, for a threshold $\nu_n > 0$ to be chosen, we define the estimates

$$\widehat{\mathbf{\Gamma}} := T_{\nu_n}((\widehat{\Sigma})^{-1}) \quad \text{and} \quad \widehat{\mathbf{\Lambda}} := \widehat{\mathbf{\Gamma}} - (\widehat{\Sigma})^{-1}. \tag{11.58}$$

Here the hard-thresholding operator is given by $T_{\nu_n}(v) = v\, \mathbb{I}[|v| > \nu_n]$.

As discussed in Chapter 10, sparse-plus-low-rank decompositions are unidentifiable unless constraints are imposed on the pair $(\mathbf{\Gamma}^*, \mathbf{\Lambda}^*)$. As with our earlier study of matrix decompositions in Section 10.7, we assume here that the low-rank component satisfies a "spikiness" constraint, meaning that its elementwise max-norm is bounded as $\|\mathbf{\Lambda}^*\|_{\max} \leq \frac{\alpha}{d}$. In addition, we assume that the matrix square root of the true precision matrix $\mathbf{\Theta}^* = \mathbf{\Gamma}^* - \mathbf{\Lambda}^*$ has a bounded ℓ_∞-operator norm, meaning that

$$|||\sqrt{\mathbf{\Theta}^*}|||_\infty = \max_{j=1,\ldots,d} \sum_{k=1}^d |\sqrt{\mathbf{\Theta}^*}|_{jk} \leq \sqrt{M}. \tag{11.59}$$

In terms of the parameters (α, M), we then choose the threshold parameter ν_n in our estimates (11.58) as

$$\nu_n := M\left(4\sqrt{\frac{\log d}{n}} + \delta\right) + \frac{\alpha}{d} \qquad \text{for some } \delta \in [0, 1]. \tag{11.60}$$

Proposition 11.19 *Consider a precision matrix $\mathbf{\Theta}^*$ that can be decomposed as the difference $\mathbf{\Gamma}^* - \mathbf{\Lambda}^*$, where $\mathbf{\Gamma}^*$ has most s non-zero entries per row, and $\mathbf{\Lambda}^*$ is α-spiky. Given $n > d$ i.i.d. samples from the $\mathcal{N}(0, (\mathbf{\Theta}^*)^{-1})$ distribution and any $\delta \in (0, 1]$, the estimates $(\widehat{\mathbf{\Gamma}}, \widehat{\mathbf{\Lambda}})$ satisfy the bounds*

$$\|\widehat{\mathbf{\Gamma}} - \mathbf{\Gamma}^*\|_{\max} \leq 2M\left(4\sqrt{\frac{\log d}{n}} + \delta\right) + \frac{2\alpha}{d} \tag{11.61a}$$

and

$$|||\widehat{\Lambda} - \Lambda^*|||_2 \leq M\left(2\sqrt{\frac{d}{n}} + \delta\right) + s\,\|\widehat{\Gamma} - \Gamma^*\|_{\max} \tag{11.61b}$$

with probability at least $1 - c_1 e^{-c_2 n\delta^2}$.

Proof We first prove that the inverse sample covariance matrix $\mathbf{Y} := (\widehat{\Sigma})^{-1}$ is itself a good estimate of Θ^*, in the sense that, for all $\delta \in (0, 1]$,

$$|||\mathbf{Y} - \Theta^*|||_2 \leq M\Big(2\sqrt{\frac{d}{n}} + \delta\Big) \tag{11.62a}$$

and

$$\|\mathbf{Y} - \Theta^*\|_{\max} \leq M\Big(4\sqrt{\frac{\log d}{n}} + \delta\Big) \tag{11.62b}$$

with probability at least $1 - c_1 e^{-c_2 n\delta^2}$.

To prove the first bound (11.62a), we note that

$$(\widehat{\Sigma})^{-1} - \Theta^* = \sqrt{\Theta^*}\Big\{n^{-1}\mathbf{V}^{\mathrm{T}}\mathbf{V} - \mathbf{I}_d\Big\}\sqrt{\Theta^*}, \tag{11.63}$$

where $\mathbf{V} \in \mathbb{R}^{n\times d}$ is a standard Gaussian random matrix. Consequently, by sub-multiplicativity of the operator norm, we have

$$\begin{aligned}|||(\widehat{\Sigma})^{-1} - \Theta^*|||_2 &\leq |||\sqrt{\Theta^*}|||_2 |||n^{-1}\mathbf{V}^{\mathrm{T}}\mathbf{V} - I_d|||_2\, |||\sqrt{\Theta^*}|||_2 = |||\Theta^*|||_2\, |||n^{-1}\mathbf{V}^{\mathrm{T}}\mathbf{V} - I_d|||_2 \\ &\leq |||\Theta^*|||_2\left(2\sqrt{\frac{d}{n}} + \delta\right),\end{aligned}$$

where the final inequality holds with probability $1 - c_1 e^{-n\delta^2}$, via an application of Theorem 6.1. To complete the proof, we note that

$$|||\Theta^*|||_2 \leq |||\Theta^*|||_\infty \leq (|||\sqrt{\Theta^*}|||_\infty)^2 \leq M,$$

from which the bound (11.62a) follows.

Turning to the bound (11.62b), using the decomposition (11.63) and introducing the shorthand $\widetilde{\Sigma} = \frac{\mathbf{V}^{\mathrm{T}}\mathbf{V}}{n} - \mathbf{I}_d$, we have

$$\begin{aligned}\|(\widehat{\Sigma})^{-1} - \Theta^*\|_{\max} &= \max_{j,k=1,\ldots,d} \left|e_j^{\mathrm{T}} \sqrt{\Theta^*}\widetilde{\Sigma}\sqrt{\Theta^*}e_k\right| \\ &\leq \max_{j,k=1,\ldots,d} \|\sqrt{\Theta^*}e_j\|_1\, \|\widetilde{\Sigma}\sqrt{\Theta^*}e_k\|_\infty \\ &\leq \|\widetilde{\Sigma}\|_{\max} \max_{j=1,\ldots,d} \|\sqrt{\Theta^*}e_j\|_1^2.\end{aligned}$$

Now observe that

$$\max_{j=1,\ldots,d} \|\sqrt{\Theta^*}e_j\|_1 \leq \max_{\|u\|_1=1} \|\sqrt{\Theta^*}u\|_1 = \max_{\ell=1,\ldots,d} \sum_{k=1}^{d} |[\sqrt{\Theta^*}]_{k\ell}| = |||\sqrt{\Theta^*}|||_\infty,$$

where the final inequality uses the symmetry of $\sqrt{\Theta^*}$. Putting together the pieces yields that

$\|(\widehat{\boldsymbol{\Sigma}})^{-1} - \boldsymbol{\Theta}^*\|_{\max} \leq M\|\widetilde{\boldsymbol{\Sigma}}\|_{\max}$. Since $\widetilde{\boldsymbol{\Sigma}} = \mathbf{V}^{\mathrm{T}}\mathbf{V}/n - I$, where $\mathbf{V} \in \mathbb{R}^{n\times d}$ is a matrix of i.i.d. standard normal variates, we have $\|\widetilde{\boldsymbol{\Sigma}}\|_{\max} \leq 4\sqrt{\frac{\log d}{n}} + \delta$ with probability at least $1 - c_1 e^{-c_2 n\delta^2}$ for all $\delta \in [0, 1]$. This completes the proof of the bound (11.62b).

Next we establish bounds on the estimates $(\widehat{\boldsymbol{\Gamma}}, \widehat{\boldsymbol{\Lambda}})$ previously defined in equation (11.58). Recalling our shorthand $\mathbf{Y} = (\widehat{\boldsymbol{\Sigma}})^{-1}$, by the definition of $\widehat{\boldsymbol{\Gamma}}$ and the triangle inequality, we have

$$\begin{aligned}\|\widehat{\boldsymbol{\Gamma}} - \boldsymbol{\Gamma}^*\|_{\max} &\leq \|\mathbf{Y} - \boldsymbol{\Theta}^*\|_{\max} + \|\mathbf{Y} - T_{\nu_n}(\mathbf{Y})\|_{\max} + \|\boldsymbol{\Lambda}^*\|_{\max} \\ &\leq M\left(4\sqrt{\frac{\log d}{n}} + \delta\right) + \nu_n + \frac{\alpha}{d} \\ &\leq 2M\left(4\sqrt{\frac{\log d}{n}} + \delta\right) + \frac{2\alpha}{d},\end{aligned}$$

thereby establishing inequality (11.61a).

Turning to the operator norm bound, the triangle inequality implies that

$$|||\widehat{\boldsymbol{\Lambda}} - \boldsymbol{\Lambda}^*|||_2 \leq |||\mathbf{Y} - \boldsymbol{\Theta}^*|||_2 + |||\widehat{\boldsymbol{\Gamma}} - \boldsymbol{\Gamma}^*|||_2 \;\leq\; M\left(2\sqrt{\frac{d}{n}} + \delta\right) + |||\widehat{\boldsymbol{\Gamma}} - \boldsymbol{\Gamma}^*|||_2.$$

Recall that $\boldsymbol{\Gamma}^*$ has at most s-non-zero entries per row. For any index (j, k) such that $\Gamma^*_{jk} = 0$, we have $\Theta^*_{jk} = \Lambda^*_{jk}$, and hence

$$|Y_{jk}| \leq |Y_{jk} - \Theta^*_{jk}| + |\Lambda^*_{jk}| \leq M\left(4\sqrt{\frac{\log d}{n}} + \delta\right) + \frac{\alpha}{d} \leq \nu_n.$$

Consequently $\widehat{\Gamma}_{jk} = T_{\nu_n}(Y_{jk}) = 0$ by construction. Therefore, the error matrix $\widehat{\boldsymbol{\Gamma}} - \boldsymbol{\Gamma}^*$ has at most s non-zero entries per row, whence

$$|||\widehat{\boldsymbol{\Gamma}} - \boldsymbol{\Gamma}^*|||_2 \leq |||\widehat{\boldsymbol{\Gamma}} - \boldsymbol{\Gamma}^*|||_\infty = \max_{j=1,\ldots,d} \sum_{k=1}^{d} |\widehat{\Gamma}_{jk} - \Gamma^*_{jk}| \leq s\|\widehat{\boldsymbol{\Gamma}} - \boldsymbol{\Gamma}^*\|_{\max}.$$

Putting together the pieces yields the claimed bound (11.61b).

□

11.5 Bibliographic details and background

Graphical models have a rich history, with parallel developments taking place in statistical physics (Ising, 1925; Bethe, 1935; Baxter, 1982), information and coding theory (Gallager, 1968; Richardson and Urbanke, 2008), artificial intelligence (Pearl, 1988) and image processing (Geman and Geman, 1984), among other areas. See the books (Lauritzen, 1996; Mézard and Montanari, 2008; Wainwright and Jordan, 2008; Koller and Friedman, 2010) for further background. The Ising model from Example 11.4 was first proposed as a model for ferromagnetism in statistical physics (Ising, 1925), and has been extensively studied. The

Hammersley–Clifford theorem derives its name from the unpublished manuscript (Hammersley and Clifford, 1971). Grimmett (1973) and Besag (1974) were the first to publish proofs of the result; see Clifford (1990) for further discussion of its history. Lauritzen (1996) provides discussion of how the Markov factorization equivalence can break down when the strict positivity condition is not satisfied. There are a number of connections between the classical theory of exponential families (Barndorff-Nielson, 1978; Brown, 1986) and graphical models; see the monograph (Wainwright and Jordan, 2008) for further details.

The Gaussian graphical Lasso (11.10) has been studied by a large number of researchers (e.g., Friedman et al., 2007; Yuan and Lin, 2007; Banerjee et al., 2008; d'Aspremont et al., 2008; Rothman et al., 2008; Ravikumar et al., 2011), in terms of both its statistical and optimization-related properties. The Frobenius norm bounds in Proposition 11.9 were first proved by Rothman et al. (2008). Ravikumar et al. (2011) proved the model selection results given in Proposition 11.10; they also analyzed the estimator for more general non-Gaussian distributions, and under a variety of tail conditions. There are also related analyses of Gaussian maximum likelihood using various forms of non-convex penalties (e.g., Lam and Fan, 2009; Loh and Wainwright, 2017). Among others, Friedman et al. (2007) and d'Asprémont et al. (2008) have developed efficient algorithms for solving the Gaussian graphical Lasso.

Neighborhood-based methods for graph estimation have their roots in the notion of pseudo-likelihood, as studied in the classical work of Besag (1974; 1975; 1977). Besag (1974) discusses various neighbor-based specifications of graphical models, including the Gaussian graphical model from Example 11.3, the Ising (binary) graphical model from Example 11.4, and the Poisson graphical model from Example 11.14. Meinshausen and Bühlmann (2006) provided the first high-dimensional analysis of the Lasso as a method for neighborhood selection in Gaussian graphical models. Their analysis, and that of related work by Zhao and Yu (2006), was based on assuming that the design matrix itself satisfies the α-incoherence condition, whereas the result given in Theorem 11.12, adapted from Wainwright (2009b), imposes these conditions on the population, and then proves that the sample versions satisfy them with high probability. Whereas we only proved Theorem 11.12 when the maximum degree m is at most $\log d$, the paper (Wainwright, 2009b) provides a proof for the general case.

Meinshausen (2008) discussed the need for stronger incoherence conditions with the Gaussian graphical Lasso (11.10) as opposed to the neighborhood selection method; see also Ravikumar et al. (2011) for further comparison of these types of incoherence conditions. Other neighborhood-based methods have also been studied in the literature, including methods based on the Dantzig selector (Yuan, 2010) and the CLIME-based method (Cai et al., 2011). Exercise 11.4 works through some analysis for the CLIME estimator.

Ravikumar et al. (2010) analyzed the ℓ_1-regularized logistic regression method for Ising model selection using the primal–dual witness method; Theorem 11.15 is adapted from their work. Other authors have studied different methods for graphical model selection in discrete models, including various types of entropy tests, thresholding methods and greedy methods (e.g., Netrapalli et al., 2010; Anandkumar et al., 2012; Bresler et al., 2013; Bresler, 2014). Santhanam and Wainwright (2012) prove lower bounds on the number of samples required for Ising model selection; combined with the improved achievability results of Bento and Montanari (2009), these lower bounds show that ℓ_1-regularized logistic regression is an order-optimal method. It is more natural—as opposed to estimating each neighborhood separately—to perform a joint estimation of all neighborhoods simultaneously. One way in which to do so is to sum all of the conditional likelihoods associated with each node, and then optimize the sum jointly, ensuring that all edges use the same parameter value in each neighborhood. The resulting procedure is equivalent to the pseudo-likelihood method (Besag, 1975, 1977). Hoefling and Tibshirani (2009) compare the relative efficiency of various pseudo-likelihood-type methods for graph estimation.

The corrected least-squares cost (11.47) is a special case of a more general class of corrected likelihood methods (e.g., Carroll et al., 1995; Iturria et al., 1999; Xu and You, 2007). The corrected non-convex Lasso (11.48) was proposed and analyzed by Loh and Wainwright (2012; 2017). A related corrected form of the Dantzig selector was analyzed by Rosenbaum and Tsybakov (2010). Proposition 11.18 is a special case of more general results on non-convex M-estimators proved in the papers (Loh and Wainwright, 2015, 2017).

The matrix decomposition approach to Gaussian graph selection with hidden variables was pioneered by Chandrasekaran et al. (2012b), who proposed regularizing the global likelihood (log-determinant function) with nuclear and ℓ_1-norms. They provided sufficient conditions for exact recovery of sparsity and rank using the primal–dual witness method, previously used to analyze the standard graphical Lasso (Ravikumar et al., 2011). Ren and Zhou (2012) proposed more direct approaches for estimating such matrix decompositions, such as the simple estimator analyzed in Proposition 11.19. Agarwal et al. (2012) analyzed both a direct approach based on thresholding and truncated SVD, as well as regularization-based methods for more general problems of matrix decomposition. As with other work on matrix decomposition problems (Candès et al., 2011; Chandrasekaran et al., 2011), Chandrasekaran et al. (2012b) performed their analysis under strong incoherence conditions, essentially algebraic conditions that ensure perfect identifiability for the sparse-plus-low-rank problem. The milder constraint, namely of bounding the maximum entry of the low-rank component as in Proposition 11.19, was introduced by Agarwal et al. (2012).

In addition to the undirected graphical models discussed here, there is also a substantial literature on methods for directed graphical models; we refer the reader to the sources (Spirtes et al., 2000; Kalisch and Bühlmann, 2007; Bühlmann and van de Geer, 2011) and references therein for more details. Liu et al. (2009; 2012) propose and study the nonparanormal family, a nonparametric generalization of the Gaussian graphical model. Such models are obtained from Gaussian models by applying a univariate transformation to the random variable at each node. The authors discuss methods for estimating such models; see also Xue and Zou (2012) for related results.

11.6 Exercises

Exercise 11.1 (Properties of log-determinant function) Let $\mathcal{S}^{d\times d}$ denote the set of symmetric matrices, and $\mathcal{S}_+^{d\times d}$ denote the cone of symmetric and strictly positive definite matrices. In this exercise, we study properties of the (negative) log-determinant function $F : \mathcal{S}^{d\times d} \to \mathbb{R}$

given by

$$F(\mathbf{\Theta}) = \begin{cases} -\sum_{j=1}^{d} \log \gamma_j(\mathbf{\Theta}) & \text{if } \mathbf{\Theta} \in \mathcal{S}_+^{d\times d}, \\ +\infty & \text{otherwise,} \end{cases}$$

where $\gamma_j(\mathbf{\Theta}) > 0$ are the eigenvalues of $\mathbf{\Theta}$.

(a) Show that F is a strictly convex function on its domain $\mathcal{S}_+^{d\times d}$.
(b) For $\mathbf{\Theta} \in \mathcal{S}_+^{d\times d}$, show that $\nabla F(\mathbf{\Theta}) = -\mathbf{\Theta}^{-1}$.
(c) For $\mathbf{\Theta} \in \mathcal{S}_+^{d\times d}$, show that $\nabla F^2(\mathbf{\Theta}) = \mathbf{\Theta}^{-1} \otimes \mathbf{\Theta}^{-1}$.

Exercise 11.2 (Gaussian MLE) Consider the maximum likelihood estimate of the inverse covariance matrix $\mathbf{\Theta}^*$ for a zero-mean Gaussian. Show that it takes the form

$$\widehat{\mathbf{\Theta}}_{\text{MLE}} = \begin{cases} \widehat{\mathbf{\Sigma}}^{-1} & \text{if } \widehat{\mathbf{\Sigma}} \succ 0, \\ \text{not defined} & \text{otherwise,} \end{cases}$$

where $\widehat{\mathbf{\Sigma}} = \frac{1}{n}\sum_{i=1}^{n} x_i x_i^{\mathrm{T}}$ is the empirical covariance matrix for a zero-mean vector. (When $\widehat{\mathbf{\Sigma}}$ is rank-deficient, you need to show explicitly that there exists a sequence of matrices for which the likelihood diverges to infinity.)

Exercise 11.3 (Gaussian neighborhood regression) Let $X \in \mathbb{R}^d$ be a zero-mean jointly Gaussian random vector with strictly positive definite covariance matrix $\mathbf{\Sigma}^*$. Consider the conditioned random variable $Z := (X_j \mid X_{\backslash\{j\}})$, where we use the shorthand $\backslash\{j\} = V \setminus \{j\}$.

(a) Establish the validity of the decomposition (11.21).
(b) Show that $\theta_j^* = (\mathbf{\Sigma}^*_{\backslash\{j\},\backslash\{j\}})^{-1} \Sigma_{\backslash\{j\},j}$.
(c) Show that $\theta_{jk}^* = 0$ whenever $k \notin \mathcal{N}(j)$.
Hint: The following elementary fact could be useful: let $\mathbf{A}$ be an invertible matrix, given in the block-partitioned form

$$\mathbf{A} = \begin{bmatrix} \mathbf{A}_{11} & \mathbf{A}_{12} \\ \mathbf{A}_{21} & \mathbf{A}_{22} \end{bmatrix}.$$

Then letting $\mathbf{B} = \mathbf{A}^{-1}$, we have (see Horn and Johnson (1985))

$$\mathbf{B}_{22} = \left(\mathbf{A}_{22} - \mathbf{A}_{21}(\mathbf{A}_{11})^{-1}\mathbf{A}_{12}\right)^{-1} \quad \text{and} \quad \mathbf{B}_{12} = (\mathbf{A}_{11})^{-1}\mathbf{A}_{12}\left[\mathbf{A}_{21}(\mathbf{A}_{11})^{-1}\mathbf{A}_{12} - \mathbf{A}_{22}\right]^{-1}.$$

Exercise 11.4 (Alternative estimator of sparse precision matrix) Consider a d-variate Gaussian random vector with zero mean, and a sparse precision matrix $\mathbf{\Theta}^*$. In this exercise, we analyze the estimator

$$\widehat{\mathbf{\Theta}} = \arg\min_{\mathbf{\Theta}\in\mathbb{R}^{d\times d}} \{\|\mathbf{\Theta}\|_1\} \qquad \text{such that } \|\widehat{\mathbf{\Sigma}}\mathbf{\Theta} - \mathbf{I}_d\|_{\max} \le \lambda_n, \tag{11.64}$$

where $\widehat{\mathbf{\Sigma}}$ is the sample covariance based on n i.i.d. samples.

(a) For $j = 1, \ldots, d$, consider the linear program

$$\widehat{\Gamma}_j \in \arg\min_{\Gamma_j\in\mathbb{R}^d} \|\Gamma_j\|_1 \qquad \text{such that } \|\widehat{\mathbf{\Sigma}}\Gamma_j - e_j\|_{\max} \le \lambda_n, \tag{11.65}$$

where $e_j \in \mathbb{R}^d$ is the jth canonical basis vector. Show that $\widehat{\mathbf{\Theta}}$ is optimal for the original program (11.64) if and only if its jth column $\widehat{\Theta}_j$ is optimal for the program (11.65).

(b) Show that $\|\widehat{\Gamma}_j\|_1 \leq \|\Theta^*_j\|_1$ for each $j = 1, \ldots, d$ whenever the regularization parameter is lower bounded as $\lambda_n \geq |||\mathbf{\Theta}^*|||_1 \|\widehat{\mathbf{\Sigma}} - \mathbf{\Sigma}^*\|_{\max}$.

(c) State and prove a high-probability bound on $\|\widehat{\mathbf{\Sigma}} - \mathbf{\Sigma}^*\|_{\max}$. (For simplicity, you may assume that $\max_{j=1,\ldots,d} \Sigma^*_{jj} \leq 1$.)

(d) Use the preceding parts to show that, for an appropriate choice of λ_n, there is a universal constant c such that

$$\|\widehat{\mathbf{\Theta}} - \mathbf{\Theta}^*\|_{\max} \leq c\, |||\mathbf{\Theta}^*|||_1^2 \sqrt{\frac{\log d}{n}} \tag{11.66}$$

with high probability.

Exercise 11.5 (Special case of general neighborhood regression) Show that the general form of neighborhood regression (11.37) reduces to linear regression (11.22) in the Gaussian case. (*Note*: You may ignore constants, either pre-factors or additive ones, that do not depend on the data.)

Exercise 11.6 (Structure of conditional distribution) Given a density of the form (11.32), show that the conditional likelihood of X_j given $X_{\setminus\{j\}}$ depends only on

$$\mathbf{\Theta}_{j+} := \{\Theta_j, \mathbf{\Theta}_{jk}, k \in V \setminus \{j\}\}.$$

Prove that $\mathbf{\Theta}_{jk} = 0$ whenever $(j, k) \notin E$.

Exercise 11.7 (Conditional distribution for Ising model) For a binary random vector $X \in \{-1, 1\}^d$, consider the family of distributions

$$p_\theta(x_1, \ldots, x_d) = \exp\Big\{ \sum_{(j,k)\in E} \theta_{jk} x_j x_k - \Phi(\theta)\Big\}, \tag{11.67}$$

where E is the edge set of some undirected graph G on the vertices $V = \{1, 2, \ldots, d\}$.

(a) For each edge $(j, k) \in E$, show that $\frac{\partial \Phi(\theta)}{\partial \theta_{jk}} = \mathbb{E}_\theta[X_j X_k]$.

(b) Compute the conditional distribution of X_j given the subvector of random variables $\mathbf{X}_{\setminus\{j\}} := \{X_k, k \in V \setminus \{j\}\}$. Give an expression in terms of the logistic function $f(t) = \log(1 + e^t)$.

Exercise 11.8 (Additive noise and Markov properties) Let $X = (X_1, \ldots, X_d)$ be a zero-mean Gaussian random vector that is Markov with respect to some graph G, and let $Z = X + V$, where $V \sim \mathcal{N}(0, \sigma^2 \mathbf{I}_d)$ is an independent Gaussian noise vector. Supposing that $\sigma^2 |||\mathbf{\Theta}^*|||_2 < 1$, derive an expression for the inverse covariance of Z in terms of powers of $\sigma^2 \mathbf{\Theta}^*$. Interpret this expression in terms of weighted path lengths in the graph.

Exercise 11.9 (Solutions for corrected graphical Lasso) In this exercise, we explore properties of the corrected graphical Lasso from equation (11.44).

(a) Defining $\mathbf{\Sigma}_x := \text{cov}(x)$, show that as long as $\lambda_n > \|\widehat{\mathbf{\Gamma}} - \mathbf{\Sigma}_x\|_{\max}$, then the corrected graphical Lasso (11.44) has a unique optimal solution.
(b) Show what can go wrong when this condition is violated. (*Hint:* It suffices to consider a one-dimensional example.)

Exercise 11.10 (Inconsistency of uncorrected Lasso) Consider the linear regression model $y = \mathbf{X}\theta^* + w$, where we observe the response vector $y \in \mathbb{R}^n$ and the corrupted matrix $\mathbf{Z} = \mathbf{X} + \mathbf{V}$. A naive estimator of θ^* is

$$\widetilde{\theta} = \arg\min_{\theta \in \mathbb{R}^d} \left\{ \frac{1}{2n} \|y - \mathbf{Z}\theta\|_2^2 \right\},$$

where we regress y on the corrupted matrix $\mathbf{Z}$. Suppose that each row of $\mathbf{X}$ is drawn i.i.d. from a zero-mean distribution with covariance $\mathbf{\Sigma}$, and that each row of $\mathbf{V}$ is drawn i.i.d. (and independently from $\mathbf{X}$) from a zero-mean distribution with covariance $\sigma^2 I$. Show that $\widetilde{\theta}$ is inconsistent even if the sample size $n \to +\infty$ with the dimension fixed.

Exercise 11.11 (Solutions for corrected Lasso) Show by an example in two dimensions that the corrected Lasso (11.48) may not achieve its global minimum if an ℓ_1-bound of the form $\|\theta\|_1 \leq R$ for some radius R is not imposed.

Exercise 11.12 (Corrected Lasso for additive corruptions) In this exercise, we explore properties of corrected linear regression in the case of additive corruptions (Example 11.16), under the standard model $y = \mathbf{X}\theta^* + w$.

(a) Assuming that $\mathbf{X}$ and $\mathbf{V}$ are independent, show that $\widehat{\mathbf{\Gamma}}$ from equation (11.42) is an unbiased estimate of $\mathbf{\Sigma}_x = \text{cov}(x)$, and that $\widehat{\gamma} = \mathbf{Z}^{\mathrm{T}} y/n$ is an unbiased estimate of $\text{cov}(x, y)$.
(b) Now suppose that in addition both $\mathbf{X}$ and $\mathbf{V}$ are generated with i.i.d. rows from a zero-mean distribution, and that each element X_{ij} and V_{ij} is sub-Gaussian with parameter 1, and that the noise vector w is independent with i.i.d. $\mathcal{N}(0, \sigma^2)$ entries. Show that there is a universal constant c such that

$$\|\widehat{\mathbf{\Gamma}}\theta^* - \widehat{\gamma}\|_\infty \leq c(\sigma + \|\theta^*\|_2)\sqrt{\frac{\log d}{n}}$$

with high probability.
(c) In addition to the previous assumptions, suppose that $\mathbf{\Sigma}_v = \nu \mathbf{I}_d$ for some $\nu > 0$. Show that $\widehat{\mathbf{\Gamma}}$ satisfies the RE condition (11.50) with high probability. (*Hint:* The result of Exercise 7.10 may be helpful to you.)

Exercise 11.13 (Corrected Lasso for missing data) In this exercise, we explore properties of corrected linear regression in the case of missing data (Example 11.17). Throughout, we assume that the missing entries are removed completely independently at random, and that $\mathbf{X}$ has zero-mean rows, generated in an i.i.d. fashion from a 1-sub-Gaussian distribution.

(a) Show that the matrix $\widehat{\mathbf{\Gamma}}$ from equation (11.43) is an unbiased estimate of $\mathbf{\Sigma}_x := \text{cov}(x)$, and that the vector $\widehat{\gamma} = \frac{\mathbf{Z}^{\mathrm{T}} y}{n}$ is an unbiased estimate of $\text{cov}(x, y)$.

(b) Assuming that the noise vector $w \in \mathbb{R}^n$ has i.i.d. $\mathcal{N}(0, \sigma^2)$ entries, show there is a universal constant c such that

$$\|\widehat{\mathbf{\Gamma}}\theta^* - \widehat{\gamma}\|_\infty \leq c(\sigma + \|\theta^*\|_2)\sqrt{\frac{\log d}{n}}$$

with high probability.

(c) Show that $\widehat{\mathbf{\Gamma}}$ satisfies the RE condition (11.50) with high probability. (*Hint:* The result of Exercise 7.10 may be helpful to you.)

12

Reproducing kernel Hilbert spaces

Many problems in statistics—among them interpolation, regression and density estimation, as well as nonparametric forms of dimension reduction and testing—involve optimizing over function spaces. Hilbert spaces include a reasonably broad class of functions, and enjoy a geometric structure similar to ordinary Euclidean space. A particular class of function-based Hilbert spaces are those defined by reproducing kernels, and these spaces—known as reproducing kernel Hilbert spaces (RKHSs)—have attractive properties from both the computational and statistical points of view. In this chapter, we develop the basic framework of RKHSs, which are then applied to different problems in later chapters, including nonparametric least-squares (Chapter 13) and density estimation (Chapter 14).

12.1 Basics of Hilbert spaces

Hilbert spaces are particular types of vector spaces, meaning that they are endowed with the operations of addition and scalar multiplication. In addition, they have an inner product defined in the usual way:

Definition 12.1 An inner product on a vector space $\mathbb{V}$ is a mapping $\langle \cdot, \cdot \rangle_{\mathbb{V}} : \mathbb{V} \times \mathbb{V} \to \mathbb{R}$ such that

$$\langle f, g \rangle_{\mathbb{V}} = \langle g, f \rangle_{\mathbb{V}} \qquad \text{for all } f, g \in \mathbb{V}, \tag{12.1a}$$

$$\langle f, f \rangle_{\mathbb{V}} \geq 0 \qquad \text{for all } f \in \mathbb{V}, \text{ with equality iff } f = 0, \tag{12.1b}$$

$$\langle f + \alpha g, h \rangle_{\mathbb{V}} = \langle f, h \rangle_{\mathbb{V}} + \alpha \, \langle g, h \rangle_{\mathbb{V}} \qquad \text{for all } f, g, h \in \mathbb{V} \text{ and } \alpha \in \mathbb{R}. \tag{12.1c}$$

A vector space equipped with an inner product is known as an *inner product space*. Note that any inner product induces a norm via $\|f\|_{\mathbb{V}} := \sqrt{\langle f, f \rangle_{\mathbb{V}}}$. Given this norm, we can then define the usual notion of *Cauchy sequence*—that is, a sequence $(f_n)_{n=1}^{\infty}$ with elements in $\mathbb{V}$ is Cauchy if, for all $\epsilon > 0$, there exists some integer $N(\epsilon)$ such that

$$\|f_n - f_m\|_{\mathbb{V}} < \epsilon \qquad \text{for all } n, m \geq N(\epsilon).$$

Definition 12.2 A *Hilbert space* $\mathbb{H}$ is an inner product space $(\langle\cdot,\cdot\rangle_{\mathbb{H}}, \mathbb{H})$ in which every Cauchy sequence $(f_n)_{n=1}^{\infty}$ in $\mathbb{H}$ converges to some element $f^* \in \mathbb{H}$.

A metric space in which every Cauchy sequence $(f_n)_{n=1}^{\infty}$ converges to an element f^* of the space is known as *complete*. Thus, we can summarize by saying that a Hilbert space is a complete inner product space.

Example 12.3 (Sequence space $\ell^2(\mathbb{N})$) Consider the space of square-summable real-valued sequences, namely

$$\ell^2(\mathbb{N}) := \Big\{(\theta_j)_{j=1}^{\infty} \mid \sum_{j=1}^{\infty} \theta_j^2 < \infty\Big\}.$$

This set, when endowed with the usual inner product $\langle\theta, \gamma\rangle_{\ell^2(\mathbb{N})} = \sum_{j=1}^{\infty} \theta_j \gamma_j$, defines a classical Hilbert space. It plays an especially important role in our discussion of eigenfunctions for reproducing kernel Hilbert spaces. Note that the Hilbert space $\mathbb{R}^m$, equipped with the usual Euclidean inner product, can be obtained as a finite-dimensional subspace of $\ell^2(\mathbb{N})$: in particular, the space $\mathbb{R}^m$ is isomorphic to the "slice"

$$\{\theta \in \ell^2(\mathbb{N}) \mid \theta_j = 0 \qquad \text{for all } j \geq m+1\}.$$ ♣

Example 12.4 (The space $L^2[0,1]$) Any element of the space $L^2[0,1]$ is a function $f\colon [0,1] \to \mathbb{R}$ that is Lebesgue-integrable, and whose square satisfies the bound $\|f\|_{L^2[0,1]}^2 = \int_0^1 f^2(x)\,dx < \infty$. Since this norm does not distinguish between functions that differ only on a set of zero Lebesgue measure, we are implicitly identifying all such functions. The space $L^2[0,1]$ is a Hilbert space when equipped with the inner product $\langle f, g\rangle_{L^2[0,1]} = \int_0^1 f(x)g(x)\,dx$. When the space $L^2[0,1]$ is clear from the context, we omit the subscript in the inner product notation. In a certain sense, the space $L^2[0,1]$ is equivalent to the sequence space $\ell^2(\mathbb{N})$. In particular, let $(\phi_j)_{j=1}^{\infty}$ be any complete orthonormal basis of $L^2[0,1]$. By definition, the basis functions satisfy $\|\phi_j\|_{L^2[0,1]} = 1$ for all $j \in \mathbb{N}$, and $\langle\phi_i, \phi_j\rangle = 0$ for all $i \neq j$, and, moreover, any function $f \in L^2[0,1]$ has the representation $f = \sum_{j=1}^{\infty} a_j \phi_j$, where $a_j := \langle f, \phi_j\rangle$ is the jth basis coefficient. By Parseval's theorem, we have

$$\|f\|_{L^2[0,1]}^2 = \sum_{j=1}^{\infty} a_j^2,$$

so that $f \in L^2[0,1]$ if and only if the sequence $a = (a_j)_{j=1}^{\infty} \in \ell^2(\mathbb{N})$. The correspondence $f \leftrightarrow (a_j)_{j=1}^{\infty}$ thus defines an isomorphism between $L^2[0,1]$ and $\ell^2(\mathbb{N})$. ♣

All of the preceding examples are instances of *separable Hilbert spaces*, for which there is a countable dense subset. For such Hilbert spaces, we can always find a collection of functions $(\phi_j)_{j=1}^{\infty}$, orthonormal in the Hilbert space—meaning that $\langle\phi_i, \phi_j\rangle_{\mathbb{H}} = \delta_{ij}$ for all positive integers i, j—such that any $f \in \mathbb{H}$ can be written in the form $f = \sum_{j=1}^{\infty} a_j \phi_j$ for some sequence of coefficients $(a_j)_{j=1}^{\infty} \in \ell^2(\mathbb{N})$. Although there do exist non-separable Hilbert spaces, here we focus primarily on the separable case.

The notion of a linear functional plays an important role in characterizing reproducing kernel Hilbert spaces. A *linear functional* on a Hilbert space $\mathbb{H}$ is a mapping $L\colon \mathbb{H} \to \mathbb{R}$ that is linear, meaning that $L(f + \alpha g) = L(f) + \alpha L(g)$ for all $f, g \in \mathbb{H}$ and $\alpha \in \mathbb{R}$. A linear functional is said to be *bounded* if there exists some $M < \infty$ such that $|L(f)| \leq M\|f\|_{\mathbb{H}}$ for all $f \in \mathbb{H}$. Given any $g \in \mathbb{H}$, the mapping $f \mapsto \langle f, g\rangle_{\mathbb{H}}$ defines a linear functional. It is bounded, since by the Cauchy–Schwarz inequality we have $|\langle f, g\rangle_{\mathbb{H}}| \leq M\|f\|_{\mathbb{H}}$ for all $f \in \mathbb{H}$, where $M := \|g\|_{\mathbb{H}}$. The Riesz representation theorem guarantees that every bounded linear functional arises in exactly this way.

Theorem 12.5 (Riesz representation theorem) *Let L be a bounded linear functional on a Hilbert space. Then there exists a unique $g \in \mathbb{H}$ such that $L(f) = \langle f, g\rangle_{\mathbb{H}}$ for all $f \in \mathbb{H}$. (We refer to g as the representer of the functional L.)*

Proof Consider the nullspace $\mathbb{N}(L) = \{h \in \mathbb{H} \mid L(h) = 0\}$. Since L is a bounded linear operator, the nullspace is closed (see Exercise 12.1). Moreover, as we show in Exercise 12.3, for any such closed subspace, we have the direct sum decomposition $\mathbb{H} = \mathbb{N}(L) + [\mathbb{N}(L)]^\perp$, where $[\mathbb{N}(L)]^\perp$ consists of all $g \in \mathbb{H}$ such that $\langle h, g\rangle_{\mathbb{H}} = 0$ for all $h \in \mathbb{N}(L)$. If $\mathbb{N}(L) = \mathbb{H}$, then we take $g = 0$. Otherwise, there must exist a non-zero element $g_0 \in [\mathbb{N}(L)]^\perp$, and by rescaling appropriately, we may find some $g \in [\mathbb{N}(L)]^\perp$ such that $\|g\|_{\mathbb{H}} = L(g) > 0$. We then define $h := L(f)g - L(g)f$, and note that $L(h) = 0$ so that $h \in \mathbb{N}(L)$. Consequently, we must have $\langle g, h\rangle_{\mathbb{H}} = 0$, which implies that $L(f) = \langle g, f\rangle_{\mathbb{H}}$ as desired. As for uniqueness, suppose that there exist $g, g' \in \mathbb{H}$ such that $\langle g, f\rangle_{\mathbb{H}} = L(f) = \langle g', f\rangle_{\mathbb{H}}$ for all $f \in \mathbb{H}$. Rearranging yields $\langle g - g', f\rangle_{\mathbb{H}} = 0$ for all $f \in \mathbb{H}$, and setting $f = g - g'$ shows that $\|g - g'\|_{\mathbb{H}}^2 = 0$, and hence $g = g'$ as claimed. □

12.2 Reproducing kernel Hilbert spaces

We now turn to the notion of a reproducing kernel Hilbert space, or RKHS for short. These Hilbert spaces are particular types of function spaces—more specifically, functions f with domain $\mathcal{X}$ mapping to the real line $\mathbb{R}$. There are many different but equivalent ways in which to define an RKHS. One way is to begin with the notion of a positive semidefinite kernel function, and use it to construct a Hilbert space in an explicit way. A by-product of this construction is the reproducing property of the kernel. An alternative, and somewhat more abstract, way is by restricting attention to Hilbert spaces in which the evaluation functionals—that is, the mappings from the Hilbert space to the real line obtained by evaluating each function at a given point—are bounded. These functionals are particularly relevant in statistical settings, since many applications involve sampling a function at a subset of points on its domain. As our development will clarify, these two approaches are equivalent in that the kernel acts as the representer for the evaluation functional, in the sense of the Riesz representation theorem (Theorem 12.5).

12.2.1 Positive semidefinite kernel functions

Let us begin with the notion of a positive semidefinite kernel function. It is a natural generalization of the idea of a positive semidefinite matrix to the setting of general functions.

Definition 12.6 (Positive semidefinite kernel function) A symmetric bivariate function $\mathcal{K}: \mathcal{X} \times \mathcal{X} \to \mathbb{R}$ is positive semidefinite (PSD) if for all integers $n \geq 1$ and elements $\{x_i\}_{i=1}^n \subset \mathcal{X}$, the $n \times n$ matrix with elements $\mathbf{K}_{ij} := \mathcal{K}(x_i, x_j)$ is positive semidefinite.

This notion is best understood via some examples.

Example 12.7 (Linear kernels) When $\mathcal{X} = \mathbb{R}^d$, we can define the linear kernel function $\mathcal{K}(x, x') := \langle x, x' \rangle$. It is clearly a symmetric function of its arguments. In order to verify the positive semidefiniteness, let $\{x_i\}_{i=1}^n$ be an arbitrary collection of points in $\mathbb{R}^d$, and consider the matrix $\mathbf{K} \in \mathbb{R}^{n \times n}$ with entries $K_{ij} = \langle x_i, x_j \rangle$. For any vector $\alpha \in \mathbb{R}^n$, we have

$$\alpha^{\mathrm{T}} \mathbf{K} \alpha = \sum_{i,j=1}^{n} \alpha_i \alpha_j \langle x_i, x_j \rangle = \Big\| \sum_{i=1}^{n} a_i x_i \Big\|_2^2 \geq 0.$$

Since $n \in \mathbb{N}$, $\{x_i\}_{i=1}^n$ and $\alpha \in \mathbb{R}^n$ were all arbitrary, we conclude that $\mathcal{K}$ is positive semidefinite. ♣

Example 12.8 (Polynomial kernels) A natural generalization of the linear kernel on $\mathbb{R}^d$ is the *homogeneous polynomial kernel* $\mathcal{K}(x, z) = (\langle x, z \rangle)^m$ of degree $m \geq 2$, also defined on $\mathbb{R}^d$. Let us demonstrate the positive semidefiniteness of this function in the special case $m = 2$. Note that we have

$$\mathcal{K}(x, z) = \Big(\sum_{j=1}^{d} x_j z_j \Big)^2 = \sum_{j=1}^{d} x_j^2 z_j^2 + 2 \sum_{i<j} x_i x_j (z_i z_j).$$

Setting $D = d + \binom{d}{2}$, let us define a mapping $\Phi: \mathbb{R}^d \to \mathbb{R}^D$ with entries

$$\Phi(x) = \begin{bmatrix} x_j^2, & \text{for } j = 1, 2, \ldots, d \\ \sqrt{2} x_i x_j, & \text{for } i < j \end{bmatrix}, \tag{12.2}$$

corresponding to all polynomials of degree two in $(x_1, \ldots, x_d)$. With this definition, we see that $\mathcal{K}$ can be expressed as a Gram matrix—namely, in the form $\mathcal{K}(x, z) = \langle \Phi(x), \Phi(z) \rangle_{\mathbb{R}^D}$. Following the same argument as Example 12.7, it is straightforward to verify that this Gram representation ensures that $\mathcal{K}$ must be positive semidefinite.

An extension of the homogeneous polynomial kernel is the *inhomogeneous polynomial kernel* $\mathcal{K}(x, z) = (1 + \langle x, z \rangle)^m$, which is based on all polynomials of degree m or less. We leave it as an exercise for the reader to check that it is also a positive semidefinite kernel function. ♣

Example 12.9 (Gaussian kernels) As a more exotic example, given some compact subset $\mathcal{X} \subseteq \mathbb{R}^d$, consider the Gaussian kernel $\mathcal{K}(x, z) = \exp\big(-\frac{1}{2\sigma^2}\|x - z\|_2^2\big)$. Here, unlike the linear kernel and polynomial kernels, it is not immediately obvious that $\mathcal{K}$ is positive semidefinite, but it can be verified by building upon the PSD nature of the linear and polynomial kernels (see Exercise 12.19). The Gaussian kernel is a very popular choice in practice, and we return to study it further in the sequel. ♣

12.2.2 Feature maps in $\ell^2(\mathbb{N})$

The mapping $x \mapsto \Phi(x)$ defined for the polynomial kernel in equation (12.2) is often referred to as a *feature map*, since it captures the sense in which the polynomial kernel function embeds the original data into a higher-dimensional space. The notion of a feature mapping can be used to define a PSD kernel in far more generality. Indeed, any function $\Phi \colon \mathcal{X} \to \ell^2(\mathbb{N})$ can be viewed as mapping the original space $\mathcal{X}$ to some subset of the space $\ell^2(\mathbb{N})$ of all square-summable sequences. Our previously discussed mapping (12.2) for the polynomial kernel is a special case, since $\mathbb{R}^D$ is a finite-dimensional subspace of $\ell^2(\mathbb{N})$.

Given any such feature map, we can then define a symmetric kernel via the inner product $\mathcal{K}(x, z) = \langle \Phi(x), \Phi(z) \rangle_{\ell_2(\mathbb{N})}$. It is often the case, for suitably chosen feature maps, that this kernel has a closed-form expression in terms of the pair (x, z). Consequently, we can compute inner products between the embedded data pairs $(\Phi(x), \Phi(z))$ without actually having to work in $\ell^2(\mathbb{N})$, or some other high-dimensional space. This fact underlies the power of RKHS methods, and goes under the colloquial name of the "kernel trick". For example, in the context of the mth-degree polynomial kernel on $\mathbb{R}^d$ from Example 12.8, evaluating the kernel requires on the order of d basic operations, whereas the embedded data lies in a space of roughly d^m (see Exercise 12.11). Of course, there are other kernels that implicitly embed the data in some infinite-dimensional space, with the Gaussian kernel from Example 12.9 being one such case.

Let us consider a particular form of feature map that plays an important role in subsequent analysis:

Example 12.10 (PSD kernels from basis expansions) Consider the sinusoidal Fourier basis functions $\phi_j(x) := \sin\left(\frac{(2j-1)\pi x}{2}\right)$ for all $j \in \mathbb{N} = \{1, 2, \ldots\}$. By construction, we have

$$\langle \phi_j, \phi_k \rangle_{L^2[0,1]} = \int_0^1 \phi_j(x)\phi_k(x)\, dx = \begin{cases} 1 & \text{if } j = k, \\ 0 & \text{otherwise,} \end{cases}$$

so that these functions are orthonormal in $L^2[0, 1]$. Now given some sequence $(\mu_j)_{j=1}^{\infty}$ of non-negative weights for which $\sum_{j=1}^{\infty} \mu_j < \infty$, let us define the feature map

$$\Phi(x) := \left(\sqrt{\mu_1}\phi_1(x), \;\; \sqrt{\mu_2}\phi_2(x), \;\; \sqrt{\mu_3}\phi_3(x), \;\; \ldots \right).$$

By construction, the element $\Phi(x)$ belongs to $\ell^2(\mathbb{N})$, since

$$\|\Phi(x)\|^2_{\ell^2(\mathbb{N})} = \sum_{j=1}^{\infty} \mu_j \phi_j^2(x) \leq \sum_{j=1}^{\infty} \mu_j \; < \infty.$$

Consequently, this particular choice of feature map defines a PSD kernel of the form

$$\mathcal{K}(x, z) := \langle \Phi(x), \Phi(z) \rangle_{\ell^2(\mathbb{N})} = \sum_{j=1}^{\infty} \mu_j \phi_j(x) \phi_j(z).$$

As our development in the sequel will clarify, a very broad class of PSD kernel functions can be generated in this way. ♣

12.2.3 Constructing an RKHS from a kernel

In this section, we show how any positive semidefinite kernel function $\mathcal{K}$ defined on the Cartesian product space $\mathcal{X} \times \mathcal{X}$ can be used to construct a particular Hilbert space of functions on $\mathcal{X}$. This Hilbert space is unique, and has the following special property: for any $x \in \mathcal{X}$, the function $\mathcal{K}(\cdot, x)$ belongs to $\mathbb{H}$, and satisfies the relation

$$\langle f, \mathcal{K}(\cdot, x) \rangle_{\mathbb{H}} = f(x) \qquad \text{for all } f \in \mathbb{H}. \tag{12.3}$$

This property is known as the *kernel reproducing property* for the Hilbert space, and it underlies the power of RKHS methods in practice. More precisely, it allows us to think of the kernel itself as defining a feature map[1] $x \mapsto \mathcal{K}(\cdot, x) \in \mathbb{H}$. Inner products in the embedded space reduce to kernel evaluations, since the reproducing property ensures that $\langle \mathcal{K}(\cdot, x), \mathcal{K}(\cdot, z) \rangle_{\mathbb{H}} = \mathcal{K}(x, z)$ for all $x, z \in \mathcal{X}$. As mentioned earlier, this computational benefit of the RKHS embedding is often referred to as the kernel trick.

How does one use a kernel to define a Hilbert space with the reproducing property (12.3)? Recalling the definition of a Hilbert space, we first need to form a vector space of functions, and then we need to endow it with an appropriate inner product. Accordingly, let us begin by considering the set $\widetilde{\mathbb{H}}$ of functions of the form $f(\cdot) = \sum_{j=1}^{n} \alpha_j \mathcal{K}(\cdot, x_j)$ for some integer $n \geq 1$, set of points $\{x_j\}_{j=1}^{n} \subset \mathcal{X}$ and weight vector $\alpha \in \mathbb{R}^n$. It is easy to see that the set $\widetilde{\mathbb{H}}$ forms a vector space under the usual definitions of function addition and scalar multiplication.

Given any pair of functions $f, \bar{f}$ in our vector space—let us suppose that they take the form $f(\cdot) = \sum_{j=1}^{n} \alpha_j \mathcal{K}(\cdot, x_j)$ and $\bar{f}(\cdot) = \sum_{k=1}^{\bar{n}} \bar{\alpha}_k \mathcal{K}(\cdot, \bar{x}_k)$—we propose to define their inner product as

$$\left\langle f, \bar{f} \right\rangle_{\widetilde{\mathbb{H}}} := \sum_{j=1}^{n} \sum_{k=1}^{\bar{n}} \alpha_j \bar{\alpha}_k \mathcal{K}(x_j, \bar{x}_k). \tag{12.4}$$

It can be verified that this definition is independent of the particular representation of the functions f and $\bar{f}$. Moreover, this proposed inner product does satisfy the kernel reproducing property (12.3), since by construction, we have

$$\langle f, \mathcal{K}(\cdot, x) \rangle_{\widetilde{\mathbb{H}}} = \sum_{j=1}^{n} \alpha_j \mathcal{K}(x_j, x) = f(x).$$

Of course, we still need to verify that the definition (12.4) defines a valid inner product. Clearly, it satisfies the symmetry (12.1a) and linearity requirements (12.1c) of an inner

[1] This view—with the kernel itself defining an embedding from $\mathcal{X}$ to $\mathbb{H}$—is related to but slightly different than our earlier perspective, in which the feature map Φ was a mapping from $\mathcal{X}$ to $\ell^2(\mathbb{N})$. Mercer's theorem allows us to connect these two points of view; see equation (12.14) and the surrounding discussion.

product. However, we need to verify the condition (12.1b)—namely, that $\langle f, f\rangle_{\mathbb{H}} \geq 0$ with equality if and only if $f = 0$. After this step, we will have a valid inner product space, and the final step is to take closures of it (in a suitable sense) in order to obtain a Hilbert space. With this intuition in place, we now provide a formal statement, and then prove it:

Theorem 12.11 *Given any positive semidefinite kernel function $\mathcal{K}$, there is a unique Hilbert space $\mathbb{H}$ in which the kernel satisfies the reproducing property* (12.3). *It is known as the reproducing kernel Hilbert space associated with $\mathcal{K}$.*

Proof As outlined above, there are three remaining steps in the proof, and we divide our argument accordingly.

Verifying condition (12.1b)*:* The positive semidefiniteness of the kernel function $\mathcal{K}$ implies that $\|f\|_{\widetilde{\mathbb{H}}^2} = \langle f, f\rangle_{\widetilde{\mathbb{H}}} \geq 0$ for all f, so we need only show that $\|f\|^2_{\widetilde{\mathbb{H}}} = 0$ if and only if $f = 0$. Consider a function of the form $f(\cdot) = \sum_{i=1}^n \alpha_i \mathcal{K}(\cdot, x_i)$, and suppose that

$$\langle f, f\rangle_{\widetilde{\mathbb{H}}} = \sum_{i,j=1}^n \alpha_i \alpha_j \mathcal{K}(x_j, x_i) = 0.$$

We must then show that $f = 0$, or equivalently that $f(x) = \sum_{i=1}^n \alpha_i \mathcal{K}(x, x_i) = 0$ for all $x \in \mathcal{X}$. Let $(a, x) \in \mathbb{R} \times \mathcal{X}$ be arbitrary, and note that by the positive semidefiniteness of $\mathcal{K}$, we have

$$0 \leq \|a\mathcal{K}(\cdot, x) + \sum_{i=1}^n \alpha_i \mathcal{K}(\cdot, x_i)\|^2_{\widetilde{\mathbb{H}}} = a^2 \mathcal{K}(x, x) + 2a \sum_{i=1}^n \alpha_i \mathcal{K}(x, x_i).$$

Since $\mathcal{K}(x, x) \geq 0$ and the scalar $a \in \mathbb{R}$ is arbitrary, this inequality can hold only if $\sum_{i=1}^n \alpha_i \mathcal{K}(x, x_i) = 0$. Thus, we have shown that the pair $(\widetilde{\mathbb{H}}, \langle\cdot, \cdot\rangle_{\widetilde{\mathbb{H}}})$ is an inner product space.

Completing the space: It remains to extend $\widetilde{\mathbb{H}}$ to a complete inner product space—that is, a Hilbert space—with the given reproducing kernel. If $(f_n)_{n=1}^\infty$ is a Cauchy sequence in $\widetilde{\mathbb{H}}$, then for each $x \in \mathcal{X}$, the sequence $(f_n(x))_{n=1}^\infty$ is Cauchy in $\mathbb{R}$, and so must converge to some real number. We can thus define the pointwise limit function $f(x) := \lim_{n\to\infty} f_n(x)$, and we let $\mathbb{H}$ be the completion of $\widetilde{\mathbb{H}}$ by these objects. We define the norm of the limit function f as $\|f\|_{\mathbb{H}} := \lim_{n\to\infty} \|f_n\|_{\widetilde{\mathbb{H}}}$.

In order to verify that this definition is sensible, we need to show that for any Cauchy sequence $(g_n)_{n=1}^\infty$ in $\widetilde{\mathbb{H}}$ such that $\lim_{n\to\infty} g_n(x) = 0$ for all $x \in \mathcal{X}$, we also have $\lim_{n\to\infty} \|g_n\|_{\widetilde{\mathbb{H}}} = 0$. Taking subsequences as necessary, suppose that $\lim_{n\to\infty} \|g_n\|^2_{\widetilde{\mathbb{H}}} = 2\epsilon > 0$, so that for n, m sufficiently large, we have $\|g_n\|^2_{\widetilde{\mathbb{H}}} \geq \epsilon$ and $\|g_m\|^2_{\widetilde{\mathbb{H}}} > \epsilon$. Since the sequence $(g_n)_{n=1}^\infty$ is Cauchy, we also have $\|g_n - g_m\|_{\widetilde{\mathbb{H}}} < \epsilon/2$ for n, m sufficiently large. Now since $g_m \in \widetilde{\mathbb{H}}$, we can write $g_m(\cdot) = \sum_{i=1}^{N_m} \alpha_i \mathcal{K}(\cdot, x_i)$, for some finite positive integer N_m and vector $\alpha \in \mathbb{R}^{N_m}$. By the reproducing property, we have

$$\langle g_m, g_n \rangle_{\tilde{\mathbb{H}}} = \sum_{i=1}^{N_m} \alpha_i g_n(x_i) \to 0 \qquad \text{as } n \to +\infty,$$

since $g_n(x) \to 0$ for each fixed x. Hence, for n sufficiently large, we can ensure that $|\langle g_m, g_n \rangle_{\tilde{\mathbb{H}}}| \leq \epsilon/2$. Putting together the pieces, we have

$$\|g_n - g_m\|_{\tilde{\mathbb{H}}} = \|g_n\|_{\tilde{\mathbb{H}}}^2 + \|g_m\|_{\tilde{\mathbb{H}}}^2 - 2\,\langle g_n, g_m \rangle_{\tilde{\mathbb{H}}} \geq \epsilon + \epsilon - \epsilon = \epsilon.$$

But this lower bound contradicts the fact that $\|g_n - g_m\|_{\tilde{\mathbb{H}}} \leq \epsilon/2$.

Thus, the norm that we have defined is sensible, and it can be used to define an inner product on $\mathbb{H}$ via the polarization identity

$$\langle f, g \rangle_{\mathbb{H}} := \tfrac{1}{2}\left\{\|f + g\|_{\mathbb{H}}^2 - \|f\|_{\mathbb{H}}^2 + \|g\|_{\mathbb{H}}^2\right\}.$$

With this definition, it can be shown that $\langle \mathcal{K}(\cdot, x), f \rangle_{\mathbb{H}} = f(x)$ for all $f \in \mathbb{H}$, so that $\mathcal{K}(\cdot, x)$ is again reproducing over $\mathbb{H}$.

Uniqueness: Finally, let us establish uniqueness. Suppose that $\mathbb{G}$ is some other Hilbert space with $\mathcal{K}$ as its reproducing kernel, so that $\mathcal{K}(\cdot, x) \in \mathbb{G}$ for all $x \in \mathcal{X}$. Since $\mathbb{G}$ is complete and closed under linear operations, we must have $\mathbb{H} \subseteq \mathbb{G}$. Consequently, $\mathbb{H}$ is a closed linear subspace of $\mathbb{G}$, so that we can write $\mathbb{G} = \mathbb{H} \oplus \mathbb{H}^\perp$. Let $g \in \mathbb{H}^\perp$ be arbitrary, and note that $\mathcal{K}(\cdot, x) \in \mathbb{H}$. By orthogonality, we must have $0 = \langle \mathcal{K}(\cdot, x), g \rangle_{\mathbb{G}} = g(x)$, from which we conclude that $\mathbb{H}^\perp = \{0\}$, and hence that $\mathbb{H} = \mathbb{G}$ as claimed. □

12.2.4 A more abstract viewpoint and further examples

Thus far, we have seen how any positive semidefinite kernel function can be used to build a Hilbert space in which the kernel satisfies the reproducing property (12.3). In the context of the Riesz representation theorem (Theorem 12.5), the reproducing property is equivalent to asserting that the function $\mathcal{K}(\cdot, x)$ acts as the representer for the *evaluation functional* at x—namely, the linear functional $L_x\colon \mathbb{H} \to \mathbb{R}$ that performs the operation $f \mapsto f(x)$. Thus, it shows that in any reproducing kernel Hilbert space, the evaluation functionals are all bounded. This perspective leads to the natural question: How large is the class of Hilbert spaces for which the evaluation functional is bounded? It turns out that this class is exactly equivalent to the class of reproducing kernel Hilbert spaces defined in the proof of Theorem 12.11. Indeed, an alternative way in which to define an RKHS is as follows:

Definition 12.12 A *reproducing kernel Hilbert space* $\mathbb{H}$ is a Hilbert space of real-valued functions on $\mathcal{X}$ such that for each $x \in \mathcal{X}$, the evaluation functional $L_x\colon \mathbb{H} \to \mathbb{R}$ is bounded (i.e., there exists some $M < \infty$ such that $|L_x(f)| \leq M\|f\|_{\mathbb{H}}$ for all $f \in \mathbb{H}$).

Theorem 12.11 shows that any PSD kernel can be used to define a reproducing kernel Hilbert space in the sense of Definition 12.12. In order to complete the equivalence, we need to show that all Hilbert spaces specified by Definition 12.12 can be equipped with a reproducing kernel function. Let us state this claim formally, and then prove it:

Theorem 12.13 *Given any Hilbert space $\mathbb{H}$ in which the evaluation functionals are all bounded, there is a unique PSD kernel $\mathcal{K}$ that satisfies the reproducing property* (12.3).

Proof When L_x is a bounded linear functional, the Riesz representation (Theorem 12.5) implies that there must exist some element R_x of the Hilbert space $\mathbb{H}$ such that

$$f(x) = L_x(f) = \langle f, R_x \rangle_{\mathbb{H}} \qquad \text{for all } f \in \mathbb{H}. \tag{12.5}$$

Using these representers of evaluation, let us define a real-valued function $\mathcal{K}$ on the Cartesian product space $\mathcal{X} \times \mathcal{X}$ via $\mathcal{K}(x, z) := \langle R_x, R_z \rangle_{\mathbb{H}}$. Symmetry of the inner product ensures that $\mathcal{K}$ is a symmetric function, so that it remains to show that $\mathcal{K}$ is positive semidefinite. For any $n \geq 1$, let $\{x_i\}_{i=1}^n \subseteq \mathcal{X}$ be an arbitrary collection of points, and consider the $n \times n$ matrix $\mathbf{K}$ with elements $K_{ij} = \mathcal{K}(x_i, x_j)$. For an arbitrary vector $\alpha \in \mathbb{R}^n$, we have

$$\alpha^{\mathrm{T}} \mathbf{K} \alpha = \sum_{j,k=1}^{n} \alpha_j \alpha_k \mathcal{K}(x_j, x_k) = \Big\langle \sum_{j=1}^{n} \alpha_j R_{x_j}, \sum_{j=1}^{n} \alpha_j R_{x_j} \Big\rangle_{\mathbb{H}} = \| \sum_{j=1}^{n} \alpha_j R_{x_j} \|_{\mathbb{H}}^2 \geq 0,$$

which proves the positive semidefiniteness.

It remains to verify the reproducing property (12.3). It actually follows easily, since for any $x \in \mathcal{X}$, the function $\mathcal{K}(\cdot, x)$ is equivalent to $R_x(\cdot)$. In order to see this equivalence, note that for any $y \in \mathcal{X}$, we have

$$\mathcal{K}(y, x) \stackrel{\text{(i)}}{=} \langle R_y, R_x \rangle_{\mathbb{H}} \stackrel{\text{(ii)}}{=} R_x(y),$$

where step (i) follows from our original definition of the kernel function, and step (ii) follows since R_y is the representer of evaluation at y. It thus follows that our kernel satisfies the required reproducing property (12.3). Finally, in Exercise 12.4, we argue that the reproducing kernel of an RKHS must be unique. □

Let us consider some more examples to illustrate our different viewpoints on RKHSs.

Example 12.14 (Linear functions on $\mathbb{R}^d$) In Example 12.7, we showed that the linear kernel $\mathcal{K}(x, z) = \langle x, z \rangle$ is positive semidefinite on $\mathbb{R}^d$. The constructive proof of Theorem 12.11 dictates that the associated RKHS is generated by functions of the form

$$z \mapsto \sum_{i=1}^{n} \alpha_i \langle z, x_i \rangle = \Big\langle z, \sum_{i=1}^{n} \alpha_i x_i \Big\rangle.$$

Each such function is linear, and therefore the associated RKHS is the class of all linear functions—that is, functions of the form $f_\beta(\cdot) = \langle \cdot, \beta \rangle$ for some vector $\beta \in \mathbb{R}^m$. The induced inner product is given by $\big\langle f_\beta, f_{\widetilde{\beta}} \big\rangle_{\mathbb{H}} := \big\langle \beta, \widetilde{\beta} \big\rangle$. Note that for each $z \in \mathbb{R}^d$, the function

$\mathcal{K}(\cdot, z) = \langle \cdot, z \rangle \equiv f_z$ is linear. Moreover, for any linear function f_β, we have

$$\left\langle f_\beta, \mathcal{K}(\cdot, z) \right\rangle_{\mathbb{H}} = \langle \beta, z \rangle = f_\beta(z),$$

which provides an explicit verification of the reproducing property (12.3). ♣

Definition 12.12 and the associated Theorem 12.13 provide us with one avenue of verifying that a given Hilbert space is *not* an RKHS, and so cannot be equipped with a PSD kernel. In particular, the boundedness of the evaluation functionals R_x in an RKHS has a very important consequence: in particular, it ensures that convergence of a sequence of functions in an RKHS implies pointwise convergence. Indeed, if $f_n \to f^*$ in the Hilbert space norm, then for any $x \in \mathcal{X}$, we have

$$\left| f_n(x) - f^*(x) \right| = \left| \langle R_x, f_n - f^* \rangle_{\mathbb{H}} \right| \leq \|R_x\|_{\mathbb{H}} \, \|f_n - f^*\|_{\mathbb{H}} \; \to 0, \tag{12.6}$$

where we have applied the Cauchy–Schwarz inequality. This property is not shared by an arbitrary Hilbert space, with the Hilbert space $L^2[0, 1]$ from Example 12.4 being one case where this property fails.

Example 12.15 (The space $L^2[0, 1]$ is not an RKHS) From the argument above, it suffices to provide a sequence of functions $(f_n)_{n=1}^{\infty}$ that converge to the all-zero function in $L^2[0, 1]$, but do not converge to zero in a pointwise sense. Consider the sequence of functions $f_n(x) = x^n$ for $n = 1, 2, \ldots$. Since $\int_0^1 f_n^2(x)\, dx = \frac{1}{2n+1}$, this sequence is contained in $L^2[0, 1]$, and moreover $\|f_n\|_{L^2[0,1]} \to 0$. However, $f_n(1) = 1$ for all $n = 1, 2, \ldots$, so that this norm convergence does not imply pointwise convergence. Thus, if $L^2[0, 1]$ were an RKHS, then this would contradict inequality (12.6).

An alternative way to see that $L^2[0, 1]$ is not an RKHS is to ask whether it is possible to find a family of functions $\{R_x \in L^2[0, 1],\; x \in [0, 1]\}$ such that

$$\int_0^1 f(y) R_x(y)\, dy = f(x) \qquad \text{for all } f \in L^2[0, 1].$$

This identity will hold if we define R_x to be a "delta-function"—that is, infinite at x and zero elsewhere. However, such objects certainly do not belong to $L^2[0, 1]$, and exist only in the sense of generalized functions. ♣

Although $L^2[0, 1]$ itself is too large to be a reproducing kernel Hilbert space, we can obtain an RKHS by imposing further restrictions on our functions. One way to do so is by imposing constraints on functions and their derivatives. The *Sobolev spaces* form an important class that arise in this way: the following example describes a first-order Sobolev space that is an RKHS.

Example 12.16 (A simple Sobolev space) A function f over $[0, 1]$ is said to be *absolutely continuous* (or abs. cts. for short) if its derivative f' exists almost everywhere and is Lebesgue-integrable, and we have $f(x) = f(0) + \int_0^x f'(z)\, dz$ for all $x \in [0, 1]$. Now consider the set of functions

$$\mathbb{H}^1[0, 1] := \{f \colon [0, 1] \to \mathbb{R} \mid f(0) = 0, \text{ and } f \text{ is abs. cts. with } f' \in L^2[0, 1]\}. \tag{12.7}$$

Let us define an inner product on this space via $\langle f, g \rangle_{\mathbb{H}^1} := \int_0^1 f'(z) g'(z)\, dz$; we claim that the resulting Hilbert space is an RKHS.

One way to verify this claim is by exhibiting a representer of evaluation: for any $x \in [0, 1]$, consider the function $R_x(z) = \min\{x, z\}$. It is differentiable at every point $z \in [0, 1] \setminus \{x\}$, and we have $R'_x(z) = \mathbb{I}_{[0,x]}(z)$, corresponding to the binary-valued indicator function for membership in the interval $[0, x]$. Moreover, for any $z \in [0, 1]$, it is easy to verify that

$$\min\{x, z\} = \int_0^z \mathbb{I}_{[0,x]}(u)\, du, \tag{12.8}$$

so that R_x is absolutely continuous by definition. Since $R_x(0) = 0$, we conclude that R_x is an element of $\mathbb{H}^1[0, 1]$. Finally, to verify that R_x is the representer of evaluation, we calculate

$$\langle f, R_x \rangle_{\mathbb{H}^1} = \int_0^1 f'(z) R'_x(z)\, dz = \int_0^x f'(z)\, dz = f(x),$$

where the final equality uses the fundamental theorem of calculus.

As shown in the proof of Theorem 12.13, the function $\mathcal{K}(\cdot, x)$ is equivalent to the representer $R_x(\cdot)$. Thus, the kernel associated with the first-order Sobolev space on $[0, 1]$ is given by $\mathcal{K}(x, z) = R_x(z) = \min\{x, z\}$. To confirm that is positive semidefinite, note that equation (12.8) implies that

$$\mathcal{K}(x, z) = \int_0^1 \mathbb{I}_{[0,x]}(u) \mathbb{I}_{[0,z]}(u)\, du = \big\langle \mathbb{I}_{[0,x]}, \mathbb{I}_{[0,z]} \big\rangle_{L^2[0,1]},$$

thereby providing a Gram representation of the kernel that certifies its PSD nature. We conclude that $\mathcal{K}(x, z) = \min\{x, z\}$ is the unique positive semidefinite kernel function associated with this first-order Sobolev space. ♣

Let us now turn to some higher-order generalizations of the first-order Sobolev space from Example 12.16.

Example 12.17 (Higher-order Sobolev spaces and smoothing splines) For some fixed integer $\alpha \geq 1$, consider the class $\mathbb{H}^\alpha[0, 1]$ of real-valued functions on $[0, 1]$ that are α-times differentiable (almost everywhere), with the α-derivative $f^{(\alpha)}$ being Lebesgue-integrable, and such that $f(0) = f^{(1)}(0) = \cdots = f^{(\alpha-1)}(0) = 0$. (Here $f^{(k)}$ denotes the kth-order derivative of f.) We may define an inner product on this space via

$$\langle f, g \rangle_{\mathbb{H}} := \int_0^1 f^{(\alpha)}(z) g^{(\alpha)}(z)\, dz. \tag{12.9}$$

Note that this set-up generalizes Example 12.16, which corresponds to the case $\alpha = 1$.

We now claim that this inner product defines an RKHS, and more specifically, that the kernel is given by

$$\mathcal{K}(x, y) = \int_0^1 \frac{(x - z)_+^{\alpha-1}}{(\alpha - 1)!} \frac{(y - z)_+^{\alpha-1}}{(\alpha - 1)!}\, dz,$$

where $(t)_+ := \max\{0, t\}$. Note that the function $R_x(\cdot) := \mathcal{K}(\cdot, x)$ is α-times differentiable almost everywhere on $[0, 1]$ with $R_x^{(\alpha)}(y) = (x - y)_+^{\alpha-1}/(\alpha - 1)!$. To verify that R_x acts as the representer of evaluation, recall that any function $f \colon [0, 1] \to \mathbb{R}$ that is α-times differentiable almost everywhere has the Taylor-series expansion

$$f(x) = \sum_{\ell=0}^{\alpha-1} f^{(\ell)}(0)\frac{x^\ell}{\ell!} + \int_0^1 f^{(\alpha)}(z)\frac{(x-z)_+^{\alpha-1}}{(\alpha-1)!}\,dz. \tag{12.10}$$

Using the previously mentioned properties of R_x and the definition (12.9) of the inner product, we obtain

$$\langle R_x, f\rangle_{\mathbb{H}} = \int_0^1 f^{(\alpha)}(z)\frac{(x-z)_+^{\alpha-1}}{(\alpha-1)!}\,dz = f(x),$$

where the final equality uses the Taylor-series expansion (12.10), and the fact that the first $(\alpha - 1)$ derivatives of f vanish at 0.

In Example 12.29 to follow, we show how to augment the Hilbert space so as to remove the constraint on the first $(\alpha - 1)$ derivatives of the functions f. ♣

12.3 Mercer's theorem and its consequences

We now turn to a useful representation of a broad class of positive semidefinite kernel functions, namely in terms of their eigenfunctions. Recall from classical linear algebra that any positive semidefinite matrix has an orthonormal basis of eigenvectors, and the associated eigenvalues are non-negative. The abstract version of Mercer's theorem generalizes this decomposition to positive semidefinite kernel functions.

Let $\mathbb{P}$ be a non-negative measure over a compact metric space $\mathcal{X}$, and consider the function class $L^2(\mathcal{X}; \mathbb{P})$ with the usual squared norm

$$\|f\|^2_{L^2(\mathcal{X};\mathbb{P})} = \int_{\mathcal{X}} f^2(x)\,d\mathbb{P}(x).$$

Since the measure $\mathbb{P}$ remains fixed throughout, we frequently adopt the shorthand notation $L^2(\mathcal{X})$ or even just L^2 for this norm. Given a symmetric PSD kernel function $\mathcal{K}\colon \mathcal{X}\times\mathcal{X} \to \mathbb{R}$ that is continuous, we can define a linear operator $T_{\mathcal{K}}$ on $L^2(\mathcal{X})$ via

$$T_{\mathcal{K}}(f)(x) := \int_{\mathcal{X}} \mathcal{K}(x, z)f(z)\,d\mathbb{P}(z). \tag{12.11a}$$

We assume that the kernel function satisfies the inequality

$$\int_{\mathcal{X}\times\mathcal{X}} \mathcal{K}^2(x, z)\,d\mathbb{P}(x)\,d\mathbb{P}(z) < \infty, \tag{12.11b}$$

which ensures that $T_{\mathcal{K}}$ is a bounded linear operator on $L^2(\mathcal{X})$. Indeed, we have

$$\begin{aligned}\|T_{\mathcal{K}}(f)\|^2_{L^2(\mathcal{X})} &= \int_{\mathcal{X}}\Big(\int_{\mathcal{X}} \mathcal{K}(x, y)f(x)\,d\mathbb{P}(x)\Big)^2\,d\mathbb{P}(y)\\ &\le \|f\|^2_{L^2(\mathcal{X})}\int_{\mathcal{X}\times\mathcal{X}} \mathcal{K}^2(x,y)\,d\mathbb{P}(x)\,d\mathbb{P}(y),\end{aligned}$$

where we have applied the Cauchy–Schwarz inequality. Operators of this type are known as *Hilbert–Schmidt operators*.

Let us illustrate these definitions with some examples.

Example 12.18 (PSD matrices) Let $\mathcal{X} = [d] := \{1, 2, \ldots, d\}$ be equipped with the Hamming metric, and let $\mathbb{P}(\{j\}) = 1$ for all $j \in \{1, 2, \ldots, d\}$ be the counting measure on this discrete space. In this case, any function $f\colon \mathcal{X} \to \mathbb{R}$ can be identified with the d-dimensional vector $(f(1), \ldots, f(d))$, and a symmetric kernel function $\mathcal{K}\colon \mathcal{X} \times \mathcal{X} \to \mathbb{R}$ can be identified with the symmetric $d \times d$ matrix $\mathbf{K}$ with entries $K_{ij} = \mathcal{K}(i, j)$. Consequently, the integral operator (12.11a) reduces to ordinary matrix–vector multiplication

$$T_{\mathcal{K}}(f)(x) = \int_{\mathcal{X}} \mathcal{K}(x, z) f(z)\, d\mathbb{P}(z) = \sum_{z=1}^{d} \mathcal{K}(x, z) f(z).$$

By standard linear algebra, we know that the matrix $\mathbf{K}$ has an orthonormal collection of eigenvectors in $\mathbb{R}^d$, say $\{v_1, \ldots, v_d\}$, along with a set of non-negative eigenvalues $\mu_1 \geq \mu_2 \geq \cdots \geq \mu_d$, such that

$$\mathbf{K} = \sum_{j=1}^{d} \mu_j v_j v_j^{\mathrm{T}}. \tag{12.12}$$

Mercer's theorem, to be stated shortly, provides a substantial generalization of this decomposition to a general positive semidefinite kernel function. ♣

Example 12.19 (First-order Sobolev kernel) Now suppose that $\mathcal{X} = [0, 1]$, and that $\mathbb{P}$ is the Lebesgue measure. Recalling the kernel function $\mathcal{K}(x, z) = \min\{x, z\}$, we have

$$T_{\mathcal{K}}(f)(x) = \int_0^1 \min\{x, z\} f(z)\, dz = \int_0^x z f(z)\, dz + \int_x^1 x f(z)\, dz.$$

We return to analyze this particular integral operator in Example 12.23. ♣

Having gained some intuition for the general notion of a kernel integral operator, we are now ready for the statement of the abstract Mercer's theorem.

Theorem 12.20 (Mercer's theorem) *Suppose that $\mathcal{X}$ is compact, the kernel function $\mathcal{K}$ is continuous and positive semidefinite, and satisfies the Hilbert–Schmidt condition (12.11b). Then there exist a sequence of eigenfunctions $(\phi_j)_{j=1}^{\infty}$ that form an orthonormal basis of $L^2(\mathcal{X}; \mathbb{P})$, and non-negative eigenvalues $(\mu_j)_{j=1}^{\infty}$ such that*

$$T_{\mathcal{K}}(\phi_j) = \mu_j \phi_j \qquad \text{for } j = 1, 2, \ldots. \tag{12.13a}$$

Moreover, the kernel function has the expansion

$$\mathcal{K}(x, z) = \sum_{j=1}^{\infty} \mu_j \phi_j(x) \phi_j(z), \tag{12.13b}$$

where the convergence of the infinite series holds absolutely and uniformly.

Remarks: The original theorem proved by Mercer applied only to operators defined on $L^2([a, b])$ for some finite $a < b$. The more abstract version stated here follows as a consequence of more general results on the eigenvalues of compact operators on Hilbert spaces; we refer the reader to the bibliography section for references.

Among other consequences, Mercer's theorem provides intuition on how reproducing kernel Hilbert spaces can be viewed as providing a particular embedding of the function domain $\mathcal{X}$ into a subset of the sequence space $\ell^2(\mathbb{N})$. In particular, given the eigenfunctions and eigenvalues guaranteed by Mercer's theorem, we may define a mapping $\Phi \colon \mathcal{X} \to \ell^2(\mathbb{N})$ via

$$x \mapsto \Phi(x) := \Big(\sqrt{\mu_1}\,\phi_1(x), \quad \sqrt{\mu_2}\,\phi_2(x), \quad \sqrt{\mu_3}\,\phi_3(x), \quad \ldots \Big). \tag{12.14}$$

By construction, we have

$$\|\Phi(x)\|^2_{\ell^2(\mathbb{N})} = \sum_{j=1}^{\infty} \mu_j \phi_j^2(x) = \mathcal{K}(x, x) < \infty,$$

showing that the map $x \mapsto \Phi(x)$ is a type of (weighted) feature map that embeds the original vector into a subset of $\ell^2(\mathbb{N})$. Moreover, this feature map also provides an explicit inner product representation of the kernel over $\ell^2(\mathbb{N})$—namely

$$\langle \Phi(x), \Phi(z) \rangle_{\ell^2(\mathbb{N})} = \sum_{j=1}^{\infty} \mu_j\, \phi_j(x)\, \phi_j(z) \;=\; \mathcal{K}(x, z).$$

Let us illustrate Mercer's theorem by considering some examples:

Example 12.21 (Eigenfunctions for a symmetric PSD matrix) As discussed in Example 12.18, a symmetric PSD d-dimensional matrix can be viewed as a kernel function on the space $[d] \times [d]$, where we adopt the shorthand $[d] := \{1, 2, \ldots, d\}$. In this case, the eigenfunction $\phi_j \colon [d] \to \mathbb{R}$ can be identified with the vector $v_j := (\phi_j(1), \ldots, \phi_j(d)) \in \mathbb{R}^d$. Thus, in this special case, the eigenvalue equation $T_{\mathcal{K}}(\phi_j) = \mu_j \phi_j$ is equivalent to asserting that $v_j \in \mathbb{R}^d$ is an eigenvector of the kernel matrix. Consequently, the decomposition (12.13b) then reduces to the familiar statement that any symmetric PSD matrix has an orthonormal basis of eigenfunctions, with associated non-negative eigenvalues, as previously stated in equation (12.12). ♣

Example 12.22 (Eigenfunctions of a polynomial kernel) Let us compute the eigenfunctions of the second-order polynomial kernel $\mathcal{K}(x, z) = (1 + xz)^2$ defined over the Cartesian product $[-1, 1] \times [-1, 1]$, where the unit interval is equipped with the Lebesgue measure. Given a function $f \colon [-1, 1] \to \mathbb{R}$, we have

$$\begin{aligned}
\int_{-1}^{1} \mathcal{K}(x, z) f(z)\, dz &= \int_{-1}^{1} \big(1 + 2xz + x^2 z^2\big) f(z)\, dz \\
&= \left\{ \int_{-1}^{1} f(z)\, dz \right\} + \left\{ 2 \int_{-1}^{1} z f(z)\, dz \right\} x + \left\{ \int_{-1}^{1} z^2 f(z)\, dz \right\} x^2,
\end{aligned}$$

showing that any eigenfunction of the kernel integral operator must be a polynomial of degree at most two. Consequently, the eigenfunction problem can be reduced to an ordinary eigenvalue problem in terms of the coefficients in the expansion $f(x) = a_0 + a_1 x + a_2 x^2$. Following some simple algebra, we find that, if f is an eigenfunction with eigenvalue μ, then these coefficients must satisfy the linear system

$$\begin{bmatrix} 2 & 0 & 2/3 \\ 0 & 4/3 & 0 \\ 2/3 & 0 & 2/5 \end{bmatrix} \begin{bmatrix} a_0 \\ a_1 \\ a_2 \end{bmatrix} = \mu \begin{bmatrix} a_0 \\ a_1 \\ a_2 \end{bmatrix}.$$

Solving this ordinary eigensystem, we find the following eigenfunction–eigenvalue pairs

$$\phi_1(x) = -0.9403 - 0.3404x^2, \qquad \text{with } \mu_1 = 2.2414,$$
$$\phi_2(x) = x, \qquad \text{with } \mu_2 = 1.3333,$$
$$\phi_3(x) = -0.3404 + 0.9403x^2, \qquad \text{with } \mu_3 = 0.1586.$$

♣

Example 12.23 (Eigenfunctions for a first-order Sobolev space) In Example 12.16, we introduced the first-order Sobolev space $\mathbb{H}^1[0,1]$. In Example 12.19, we found that its kernel function takes the form $\mathcal{K}(x, z) = \min\{x, z\}$, and determined the form of the associated integral operator. Using this previous development, if $\phi\colon [0,1] \to \mathbb{R}$ is an eigenfunction of $T_{\mathcal{K}}$ with eigenvalue $\mu \neq 0$, then it must satisfy the relation $T_{\mathcal{K}}(\phi) = \mu\phi$, or equivalently

$$\int_0^x z\phi(z)\,dz + \int_x^1 x\phi(z)\,dz = \mu\phi(x) \qquad \text{for all } x \in [0,1].$$

Since this relation must hold for all $x \in [0,1]$, we may take derivatives with respect to x. Doing so twice yields the second-order differential equation $\mu\phi''(x) + \phi(x) = 0$. Combined with the boundary condition $\phi(0) = 0$, we obtain $\phi(x) = \sin(x/\sqrt{\mu})$ as potential eigenfunctions. Now using the boundary condition $\int_0^1 z\phi(z)\,dz = \mu\phi(1)$, we deduce that the eigenfunction–eigenvalue pairs are given by

$$\phi_j(t) = \sin\frac{(2j-1)\pi t}{2} \quad \text{and} \quad \mu_j = \left(\frac{2}{(2j-1)\pi}\right)^2 \qquad \text{for } j = 1, 2, \ldots.$$

♣

Example 12.24 (Translation-invariant kernels) An important class of kernels have a translation-invariant form. In particular, given a function $\psi\colon [-1,1] \to \mathbb{R}$ that is even (meaning that $\psi(u) = \psi(-u)$ for all $u \in [-1,1]$), let us extend its domain to the real line by the periodic extension $\psi(u + 2k) = \psi(u)$ for all $u \in [-1,1]$ and integers $k \in \mathbb{Z}$.

Using this function, we may define a *translation-invariant* kernel on the Cartesian product space $[-1,1] \times [-1,1]$ via $\mathcal{K}(x, z) = \psi(x - z)$. Note that the evenness of ψ ensures that this kernel is symmetric. Moreover, the kernel integral operator takes the form

$$T_{\mathcal{K}}(f)(x) = \underbrace{\int_{-1}^1 \psi(x - z) f(z)\,dz}_{(\psi * f)(x)},$$

and thus is a convolution operator.

A classical result from analysis is that the eigenfunctions of convolution operators are given by the Fourier basis; let us prove this fact here. We first show that the cosine functions

$\phi_j(x) = \cos(\pi jx)$ for $j = 0, 1, 2, \ldots$ are eigenfunctions of the operator $T_{\mathcal{K}}$. Indeed, we have

$$T_{\mathcal{K}}(\phi_j)(x) = \int_{-1}^{1} \psi(x-z)\cos(\pi jz)\,dz = \int_{-1-x}^{1-x} \psi(-u)\cos(2\pi j(x+u))\,du,$$

where we have made the change of variable $u = z - x$. Note that the interval of integration $[-1-x,\ 1-x]$ is of length 2, and since both $\psi(-u)$ and $\cos(2\pi(x+u))$ have period 2, we can shift the interval of integration to $[-1, 1]$. Combined with the evenness of ψ, we conclude that $T_{\mathcal{K}}(\phi_j)(x) = \int_{-1}^{1} \psi(u)\cos(2\pi j(x+u))\,du$. Using the elementary trigonometric identity

$$\cos(\pi j(x+u)) = \cos(\pi jx)\cos(\pi ju) - \sin(\pi jx)\sin(\pi ju),$$

we find that

$$\begin{aligned} T_{\mathcal{K}}(\phi_j)(x) &= \left\{\int_{-1}^{1} \psi(u)\cos(\pi ju)\,du\right\}\cos(\pi jx) - \left\{\int_{-1}^{1} \psi(u)\sin(\pi ju)\,du\right\}\sin(\pi jx) \\ &= c_j\cos(\pi jx), \end{aligned}$$

where $c_j = \int_{-1}^{1} \psi(u)\cos(\pi ju)\,du$ is the jth cosine coefficient of ψ. In this calculation, we have used the evenness of ψ to argue that the integral with the sine function vanishes.

A similar argument shows that each of the sinusoids

$$\widetilde{\phi}_j(x) = \sin(j\pi x) \qquad \text{for } j = 1, 2, \ldots$$

are also eigenfunctions with eigenvalue c_j. Since the functions $\{\phi_j,\ j = 0, 1, 2, \ldots\} \cup \{\widetilde{\phi}_j,\ j = 1, 2, \ldots\}$ form a complete orthogonal basis of $L^2[-1, 1]$, there are no other eigenfunctions that are not linear combinations of these functions. Consequently, by Mercer's theorem, the kernel function has the eigenexpansion

$$\mathcal{K}(x, z) = \sum_{j=0}^{\infty} c_j\{\cos(\pi jx)\cos(\pi jz) + \sin(\pi jx)\sin(\pi jz)\} = \sum_{j=0}^{\infty} c_j\cos(\pi j(x-z)),$$

where c_j are the (cosine) Fourier coefficients of ψ. Thus, we see that $\mathcal{K}$ is positive semidefinite if and only if $c_j \geq 0$ for $j = 0, 1, 2, \ldots$. ♣

Example 12.25 (Gaussian kernel) As previously introduced in Example 12.9, a popular choice of kernel on some subset $\mathcal{X} \subseteq \mathbb{R}^d$ is the Gaussian kernel given by $\mathcal{K}(x, z) = \exp(-\frac{\|x-z\|_2^2}{2\sigma^2})$, where $\sigma > 0$ is a bandwidth parameter. To keep our calculations relatively simple, let us focus here on the univariate case $d = 1$, and let $\mathcal{X}$ be some compact interval of the real line. By a rescaling argument, we can restrict ourselves to the case $\mathcal{X} = [-1, 1]$, so that we are considering solutions to the integral equation

$$\int_{-1}^{1} e^{-\frac{(x-z)^2}{2\sigma^2}}\phi_j(z)\,dz = \mu_j\phi_j(x). \tag{12.15}$$

Note that this problem cannot be tackled by the methods of the previous example, since we are *not* performing the periodic extension of our function.[2] Nonetheless, the eigenvalues of the Gaussian integral operator are very closely related to the Fourier transform.

[2] If we were to consider the periodically extended version, then the eigenvalues would be given by the cosine coefficients $c_j = \int_{-1}^{1} \exp\left(-\frac{u^2}{2\sigma^2}\right)\cos(\pi ju)\,du$, with the cosine functions as eigenfunctions.

In the remainder of our development, let us consider a slightly more general integral equation. Given a bounded, continuous and even function $\Psi\colon \mathbb{R} \to [0, \infty)$, we may define its (real-valued) Fourier transform $\psi(u) = \int_{-\infty}^{\infty} \Psi(\omega) e^{-i\omega u}\, d\omega$, and use it to define a translation-invariant kernel via $\mathcal{K}(x, z) := \psi(x - z)$. We are then led to the integral equation

$$\int_{-1}^{1} \psi(x - z)\phi_j(z)\, dz = \mu_j \phi_j(x). \tag{12.16}$$

Classical theory on integral operators can be used to characterize the spectrum of this integral operator. More precisely, for any operator such that $\log \Psi(\omega) \asymp -\omega^{\alpha}$ for some $\alpha > 1$, there is a constant c such that the eigenvalues $(\mu_j)_{j=1}^{\infty}$ associated with the integral equation (12.16) scale as $\mu_j \asymp e^{-cj \log j}$ as $j \to +\infty$. See the bibliographic section for further discussion of results of this type.

The Gaussian kernel is a special case of this set-up with the pair $\Psi(\omega) = \exp(-\frac{\sigma^2\omega^2}{2})$ and $\psi(u) = \exp(-\frac{u^2}{2\sigma^2})$. Applying the previous reasoning guarantees that the eigenvalues of the Gaussian kernel over a compact interval scale as $\mu_j \asymp \exp(-cj \log j)$ as $j \to +\infty$. We thus see that the Gaussian kernel class is relatively small, since its eigenvalues decay at exponential rate. (The reader should contrast this fast decay with the significantly slower $\mu_j \asymp j^{-2}$ decay rate of the first-order Sobolev class from Example 12.23.) ♣

An interesting consequence of Mercer's theorem is in giving a relatively explicit characterization of the RKHS associated with a given kernel.

Corollary 12.26 *Consider a kernel satisfying the conditions of Mercer's theorem with associated eigenfunctions $(\phi_j)_{j=1}^{\infty}$ and non-negative eigenvalues $(\mu_j)_{j=1}^{\infty}$. It induces the reproducing kernel Hilbert space*

$$\mathbb{H} := \left\{ f = \sum_{j=1}^{\infty} \beta_j \phi_j \mid \textit{for some } (\beta_j)_{j=1}^{\infty} \in \ell^2(\mathbb{N}) \textit{ with } \sum_{j=1}^{\infty} \frac{\beta_j^2}{\mu_j} < \infty \right\}, \tag{12.17a}$$

along with inner product

$$\langle f, g \rangle_{\mathbb{H}} := \sum_{j=1}^{\infty} \frac{\langle f, \phi_j \rangle \langle g, \phi_j \rangle}{\mu_j}, \tag{12.17b}$$

where $\langle \cdot, \cdot \rangle$ denotes the inner product in $L^2(\mathcal{X}; \mathbb{P})$.

Let us make a few comments on this claim. First, in order to assuage any concerns regarding division by zero, we can restrict all sums to only indices j for which $\mu_j > 0$. Second, note that Corollary 12.26 shows that the RKHS associated with a Mercer kernel is isomorphic to an infinite-dimensional ellipsoid contained with $\ell^2(\mathbb{N})$—namely, the set

$$\mathcal{E} := \left\{ (\beta_j)_{j=1}^{\infty} \in \ell^2(\mathbb{N}) \mid \sum_{j=1}^{\infty} \frac{\beta_j^2}{\mu_j} \leq 1 \right\}. \tag{12.18}$$

We study the properties of such ellipsoids at more length in Chapters 13 and 14.

Proof For the proof, we take $\mu_j > 0$ for all $j \in \mathbb{N}$. This assumption entails no loss of generality, since otherwise the same argument can be applied with relevant summations truncated to the positive eigenvalues of the kernel function. Recall that $\langle \cdot, \cdot \rangle$ denotes the inner product on $L^2(\mathcal{X}; \mathbb{P})$.

It is straightforward to verify that $\mathbb{H}$ along with the specified inner product $\langle \cdot, \cdot \rangle_{\mathbb{H}}$ is a Hilbert space. Our next step is to show that $\mathbb{H}$ is in fact a reproducing kernel Hilbert space, and satisfies the reproducing property with respect to the given kernel. We begin by showing that for each fixed $x \in \mathcal{X}$, the function $\mathcal{K}(\cdot, x)$ belongs to $\mathbb{H}$. By the Mercer expansion, we have $\mathcal{K}(\cdot, x) = \sum_{j=1}^{\infty} \mu_j \phi_j(x) \phi_j(\cdot)$, so that by definition (12.17a) of our Hilbert space, it suffices to show that $\sum_{j=1}^{\infty} \mu_j \phi_j^2(x) < \infty$. By the Mercer expansion, we have

$$\sum_{j=1}^{\infty} \mu_j \phi_j^2(x) = \mathcal{K}(x, x) < \infty,$$

so that $\mathcal{K}(\cdot, x) \in \mathbb{H}$.

Let us now verify the reproducing property. By the orthonormality of $(\phi_j)_{j=1}^{\infty}$ in $L^2(\mathcal{X}; \mathbb{P})$ and Mercer's theorem, we have $\langle \mathcal{K}(\cdot, x), \phi_j \rangle = \mu_j \phi_j(x)$ for each $j \in \mathbb{N}$. Thus, by definition (12.17b) of our Hilbert inner product, for any $f \in \mathbb{H}$, we have

$$\langle f, \mathcal{K}(\cdot, x) \rangle_{\mathbb{H}} = \sum_{j=1}^{\infty} \frac{\langle f, \phi_j \rangle \langle \mathcal{K}(\cdot, x), \phi_j \rangle}{\mu_j} = \sum_{j=1}^{\infty} \langle f, \phi_j \rangle \phi_j(x) = f(x),$$

where the final step again uses the orthonormality of $(\phi_j)_{j=1}^{\infty}$. Thus, we have shown that $\mathbb{H}$ is the RKHS with kernel $\mathcal{K}$. (As discussed in Theorem 12.11, the RKHS associated with any given kernel is unique.) □

12.4 Operations on reproducing kernel Hilbert spaces

In this section, we describe a number of operations on reproducing kernel Hilbert spaces that allow us to build new spaces.

12.4.1 Sums of reproducing kernels

Given two Hilbert spaces $\mathbb{H}_1$ and $\mathbb{H}_2$ of functions defined on domains $\mathcal{X}_1$ and $\mathcal{X}_2$, respectively, consider the space

$$\mathbb{H}_1 + \mathbb{H}_2 := \big\{ f_1 + f_2 \mid f_j \in \mathbb{H}_j,\ j = 1, 2 \big\},$$

corresponding to the set of all functions obtained as sums of pairs of functions from the two spaces.

Proposition 12.27 *Suppose that $\mathbb{H}_1$ and $\mathbb{H}_2$ are both RKHSs with kernels $\mathcal{K}_1$ and $\mathcal{K}_2$, respectively. Then the space $\mathbb{H} = \mathbb{H}_1 + \mathbb{H}_2$ with norm*

$$\|f\|_{\mathbb{H}}^2 := \min_{\substack{f=f_1+f_2 \\ f_1\in\mathbb{H}_1, f_2\in\mathbb{H}_2}} \{\|f_1\|_{\mathbb{H}_1}^2 + \|f_2\|_{\mathbb{H}_2}^2\} \tag{12.19}$$

is an RKHS with kernel $\mathcal{K} = \mathcal{K}_1 + \mathcal{K}_2$.

Remark: This construction is particularly simple when $\mathbb{H}_1$ and $\mathbb{H}_2$ share only the constant zero function, since any function $f \in \mathbb{H}$ can then be written as $f = f_1 + f_2$ for a unique pair (f_1, f_2), and hence $\|f\|_{\mathbb{H}}^2 = \|f_1\|_{\mathbb{H}_1}^2 + \|f_2\|_{\mathbb{H}_2}^2$. Let us illustrate the use of summation with some examples:

Example 12.28 (First-order Sobolev space and constant functions) Consider the kernel functions on $[0,1]\times[0,1]$ given by $\mathcal{K}_1(x,z) = 1$ and $\mathcal{K}_2(x,z) = \min\{x,z\}$. They generate the reproducing kernel Hilbert spaces

$$\mathbb{H}_1 = \text{span}\{1\} \quad \text{and} \quad \mathbb{H}_2 = \mathbb{H}^1[0,1],$$

where span{1} is the set of all constant functions, and $\mathbb{H}^1[0,1]$ is the first-order Sobolev space from Example 12.16. Note that $\mathbb{H}_1 \cap \mathbb{H}_2 = \{0\}$, since $f(0) = 0$ for any element of $\mathbb{H}_2$. Consequently, the RKHS with kernel $\mathcal{K}(x, z) = 1 + \min\{x,z\}$ consists of all functions

$$\bar{\mathbb{H}}^1[0,1] := \{f\colon [0,1] \to \mathbb{R} \mid f \text{ is absolutely continuous with } f' \in L^2[0,1]\},$$

equipped with the squared norm $\|f\|_{\bar{\mathbb{H}}^1[0,1]}^2 = f^2(0) + \int_0^1 (f'(z))^2\,dz$. ♣

As a continuation of the previous example, let us describe an extension of the higher-order Sobolev spaces from Example 12.17:

Example 12.29 (Extending higher-order Sobolev spaces) For an integer $\alpha \geq 1$, consider the kernel functions on $[0,1]\times[0,1]$ given by

$$\mathcal{K}_1(x,z) = \sum_{\ell=0}^{\alpha-1} \frac{x^\ell}{\ell!}\frac{z^\ell}{\ell!} \quad \text{and} \quad \mathcal{K}_2(x,z) = \int_0^1 \frac{(x-y)_+^{\alpha-1}}{(\alpha-1)!}\frac{(z-y)_+^{\alpha-1}}{(\alpha-1)!}\,dy.$$

The first kernel generates an RKHS $\mathbb{H}_1$ of polynomials of degree $\alpha - 1$, whereas the second kernel generates the α-order Sobolev space $\mathbb{H}_2 = \mathbb{H}^\alpha[0,1]$ previously defined in Example 12.17.

Letting $f^{(\ell)}$ denote the ℓth-order derivative, recall that any function $f \in \mathbb{H}^\alpha[0,1]$ satisfies the boundary conditions $f^{(\ell)}(0) = 0$ for $\ell = 0, 1, \ldots, \alpha - 1$. Consequently, we have $\mathbb{H}_1 \cap \mathbb{H}_2 = \{0\}$ so that Proposition 12.27 guarantees that the kernel

$$\mathcal{K}(x, z) = \sum_{\ell=0}^{\alpha-1} \frac{x^\ell}{\ell!}\frac{z^\ell}{\ell!} + \int_0^1 \frac{(x-y)_+^{\alpha-1}}{(\alpha-1)!}\frac{(z-y)_+^{\alpha-1}}{(\alpha-1)!}\,dy \tag{12.20}$$

generates the Hilbert space $\bar{\mathbb{H}}^\alpha[0,1]$ of all functions that are α-times differentiable almost everywhere, with $f^{(\alpha)}$ Lebesgue-integrable. As we verify in Exercise 12.15, the associated RKHS norm takes the form

$$\|f\|_{\mathbb{H}}^2 = \sum_{\ell=0}^{\alpha-1} (f^{(\ell)}(0))^2 + \int_0^1 (f^{(\alpha)}(z))^2\, dz. \tag{12.21}$$

♣

Example 12.30 (Additive models) It is often convenient to build up a multivariate function from simpler pieces, and additive models provide one way in which to do so. For $j = 1, 2, \ldots, M$, let $\mathbb{H}_j$ be a reproducing kernel Hilbert space, and let us consider functions that have an additive decomposition of the form $f = \sum_{j=1}^M f_j$, where $f_j \in \mathbb{H}_j$. By Proposition 12.27, the space $\mathbb{H}$ of all such functions is itself an RKHS equipped with the kernel function $\mathcal{K} = \sum_{j=1}^M \mathcal{K}_j$. A commonly used instance of such an additive model is when the individual Hilbert space $\mathbb{H}_j$ corresponds to functions of the jth coordinate of a d-dimensional vector, so that the space $\mathbb{H}$ consists of functions $f : \mathbb{R}^d \to \mathbb{R}$ that have the additive decomposition

$$f(x_1, \ldots, x_d) = \sum_{j=1}^d f_j(x_j),$$

where $f_j : \mathbb{R} \to \mathbb{R}$ is a univariate function for the jth coordinate. Since $\mathbb{H}_j \cap \mathbb{H}_k = \{0\}$ for all $j \neq k$, the associated Hilbert norm takes the form $\|f\|_{\mathbb{H}}^2 = \sum_{j=1}^d \|f_j\|_{\mathbb{H}_j}^2$. We provide some additional discussion of these additive decompositions in Exercise 13.9 and Example 14.11 to follow in later chapters.

More generally, it is natural to consider expansions of the form

$$f(x_1, \ldots, x_d) = \sum_{j=1}^d f_j(x_j) + \sum_{j \neq k} f_{jk}(x_j, x_k) + \cdots .$$

When the expansion functions are chosen to be mutually orthogonal, such expansions are known as *functional ANOVA* decompositions. ♣

We now turn to the proof of Proposition 12.27.

Proof Consider the direct sum $\mathbb{F} := \mathbb{H}_1 \oplus \mathbb{H}_2$ of the two Hilbert spaces; by definition, it is the Hilbert space $\{(f_1, f_2) \mid f_j \in \mathbb{H}_j, j = 1, 2\}$ of all ordered pairs, along with the norm

$$\|(f_1, f_2)\|_{\mathbb{F}}^2 := \|f_1\|_{\mathbb{H}_1}^2 + \|f_2\|_{\mathbb{H}_2}^2. \tag{12.22}$$

Now consider the linear operator $L : \mathbb{F} \to \mathbb{H}$ defined by $(f_1, f_2) \mapsto f_1 + f_2$, and note that it maps $\mathbb{F}$ onto $\mathbb{H}$. The nullspace $\mathbb{N}(L)$ of this operator is a subspace of $\mathbb{F}$, and we claim that it is closed. Consider some sequence $((f_n, -f_n))_{n=1}^\infty$ contained within the nullspace $\mathbb{N}(L)$ that converges to a point $(f, g) \in \mathbb{F}$. By the definition of the norm (12.22), this convergence implies that $f_n \to f$ in $\mathbb{H}_1$ (and hence pointwise) and $-f_n \to g$ in $\mathbb{H}_2$ (and hence pointwise). Overall, we conclude that $f = -g$, meaning $(f, g) \in \mathbb{N}(L)$.

Let $\mathbb{N}^\perp$ be the orthogonal complement of $\mathbb{N}(L)$ in $\mathbb{F}$, and let $L_\perp$ be the restriction of L to

$\mathbb{N}^\perp$. Since this map is a bijection between $\mathbb{N}^\perp$ and $\mathbb{H}$, we may define an inner product on $\mathbb{H}$ via

$$\langle f,\, g\rangle_{\mathbb{H}} := \langle L_\perp^{-1}(f),\, L_\perp^{-1}(g)\rangle_{\mathbb{F}}.$$

It can be verified that the space $\mathbb{H}$ with this inner product is a Hilbert space.

It remains to check that $\mathbb{H}$ is an RKHS with kernel $\mathcal{K} = \mathcal{K}_1 + \mathcal{K}_2$, and that the norm $\|\cdot\|_{\mathbb{H}}^2$ takes the given form (12.19). Since the functions $\mathcal{K}_1(\cdot, x)$ and $\mathcal{K}_2(\cdot, x)$ belong to $\mathbb{H}_1$ and $\mathbb{H}_2$, respectively, the function $\mathcal{K}(\cdot, x) = \mathcal{K}_1(\cdot, x) + \mathcal{K}_2(\cdot, x)$ belongs to $\mathbb{H}$. For a fixed $f \in \mathbb{F}$, let $(f_1, f_2) = L_\perp^{-1}(f) \in \mathbb{F}$, and for a fixed $x \in \mathcal{X}$, let $(g_1, g_2) = L_\perp^{-1}(\mathcal{K}(\cdot, x)) \in \mathbb{F}$. Since $(g_1 - \mathcal{K}_1(\cdot, x),\, g_2 - \mathcal{K}_2(\cdot, x))$ must belong to $\mathbb{N}(L)$, it must be orthogonal (in $\mathbb{F}$) to the element $(f_1, f_2) \in \mathbb{N}^\perp$. Consequently, we have $\langle (g_1 - \mathcal{K}_1(\cdot, x),\, g_2 - \mathcal{K}_2(\cdot, x)),\, (f_1, f_2)\rangle_{\mathbb{F}} = 0$, and hence

$$\begin{aligned}\langle f_1,\, \mathcal{K}_1(\cdot, x)\rangle_{\mathbb{H}_1} + \langle f_2,\, \mathcal{K}_2(\cdot, x)\rangle_{\mathbb{H}_2} &= \langle f_1,\, g_1\rangle_{\mathbb{H}_1} + \langle f_2,\, g_2\rangle_{\mathbb{H}_2} \\ &= \langle f,\, \mathcal{K}(\cdot, x)\rangle_{\mathbb{H}}.\end{aligned}$$

Since $\langle f_1,\, \mathcal{K}_1(\cdot, x)\rangle_{\mathbb{H}_1} \langle f_2,\, \mathcal{K}_2(\cdot, x)\rangle_{\mathbb{H}_2} = f_1(x) + f_2(x) = f(x)$, we have established that $\mathcal{K}$ has the reproducing property.

Finally, let us verify that the norm $\|f\|_{\mathbb{H}} := \|L_\perp^{-1}(f)\|_{\mathbb{F}}$ that we have defined is equivalent to the definition (12.19). For a given $f \in \mathbb{H}$, consider some pair $(f_1, f_2) \in \mathbb{F}$ such that $f = f_1 + f_2$, and define $(v_1, v_2) = (f_1, f_2) - L_\perp^{-1}(f)$. We have

$$\|f_1\|_{\mathbb{H}_1}^2 + \|f_2\|_{\mathbb{H}_2}^2 \overset{\text{(i)}}{=} \|(f_1, f_2)\|_{\mathbb{F}}^2 \overset{\text{(ii)}}{=} \|(v_1, v_2)\|_{\mathbb{F}}^2 + \|L_\perp^{-1}(f)\|_{\mathbb{F}}^2 \overset{\text{(iii)}}{=} \|(v_1, v_2)\|_{\mathbb{F}}^2 + \|f\|_{\mathbb{H}}^2,$$

where step (i) uses the definition (12.22) of the norm in $\mathbb{F}$, step (ii) follows from the Pythagorean property, as applied to the pair $(v_1, v_2) \in \mathbb{N}(L)$ and $L_\perp^{-1}(f) \in \mathbb{N}^\perp$, and step (iii) uses our definition of the norm $\|f\|_{\mathbb{H}}$. Consequently, we have shown that for any pair f_1, f_2 such that $f = f_1 + f_2$, we have

$$\|f\|_{\mathbb{H}}^2 \leq \|f_1\|_{\mathbb{H}_1}^2 + \|f_2\|_{\mathbb{H}_2}^2,$$

with equality holding if and only if $(v_1, v_2) = (0, 0)$, or equivalently $(f_1, f_2) = L_\perp^{-1}(f)$. This establishes the equivalence of the definitions. □

12.4.2 Tensor products

Consider two separable Hilbert spaces $\mathbb{H}_1$ and $\mathbb{H}_2$ of functions, say with domains $\mathcal{X}_1$ and $\mathcal{X}_2$, respectively. They can be used to define a new Hilbert space, denoted by $\mathbb{H}_1 \otimes \mathbb{H}_2$, known as the tensor product of $\mathbb{H}_1$ and $\mathbb{H}_2$. Consider the set of functions $h\colon \mathcal{X}_1 \times \mathcal{X}_2 \to \mathbb{R}$ that have the form

$$\{h = \sum_{j=1}^{n} f_j g_j \mid \text{for some } n \in \mathbb{N} \text{ and such that } f_j \in \mathbb{H}_1,\ g_j \in \mathbb{H}_2 \text{ for all } j \in [n]\}.$$

If $h = \sum_{j=1}^n f_j g_j$ and $\widetilde{h} = \sum_{k=1}^m \widetilde{f}_k \widetilde{g}_k$ are two members of this set, we define their inner product

$$\langle h,\, \widetilde{h}\rangle_{\mathbb{H}} := \sum_{j=1}^{n} \sum_{k=1}^{m} \langle f_j,\, \widetilde{f}_k\rangle_{\mathbb{H}_1} \langle g_j,\, \widetilde{g}_k\rangle_{\mathbb{H}_2}. \tag{12.23}$$

Note that the value of the inner product depends neither on the chosen representation of h nor on that of $\widetilde{h}$; indeed, using linearity of the inner product, we have

$$\langle h, \widetilde{h}\rangle_{\mathbb{H}} = \sum_{k=1}^{m} \langle (h \odot \widetilde{f}_k), \widetilde{g}_k\rangle_{\mathbb{H}_2},$$

where $(h \odot \widetilde{f}_k) \in \mathbb{H}_2$ is the function given by $x_2 \mapsto \langle h(\cdot, x_2), \widetilde{f}_k\rangle_{\mathbb{H}_1}$. A similar argument shows that the inner product does not depend on the representation of $\widetilde{h}$, so that the inner product (12.23) is well defined.

It is straightforward to check that the inner product (12.23) is bilinear and symmetric, and that $\langle h, h\rangle^2_{\mathbb{H}} = \|h\|^2_{\mathbb{H}} \geq 0$ for all $h \in \mathbb{H}$. It remains to check that $\|h\|_{\mathbb{H}} = 0$ if and only if $h = 0$. Consider some $h \in \mathbb{H}$ with the representation $h = \sum_{j=1}^{n} f_j g_j$. Let $(\phi_j)_{j=1}^{\infty}$ and $(\psi_k)_{k=1}^{\infty}$ be complete orthonormal bases of $\mathbb{H}_1$ and $\mathbb{H}_2$, respectively, ordered such that

$$\text{span}\{f_1, \ldots, f_n\} \subseteq \text{span}\{\phi_1, \ldots, \phi_n\} \quad \text{and} \quad \text{span}\{g_1, \ldots, g_n\} \subseteq \text{span}\{\psi_1, \ldots, \psi_n\}.$$

Consequently, we can write f equivalently as the double summation $f = \sum_{j,j=1}^{n} \alpha_{j,k}\phi_j\psi_k$ for some set of real numbers $\{\alpha_{j,k}\}_{j,k=1}^{n}$. Using this representation, we are guaranteed the equality $\|f\|^2_{\mathbb{H}} = \sum_{j=1}^{n}\sum_{k=1}^{n} \alpha^2_{j,k}$, which shows that $\|f\|_{\mathbb{H}} = 0$ if and only if $\alpha_{j,k} = 0$ for all (j, k), or equivalently $f = 0$.

In this way, we have defined the tensor product $\mathbb{H} = \mathbb{H}_1 \otimes \mathbb{H}_2$ of two Hilbert spaces. The next result asserts that when the two component spaces have reproducing kernels, then the tensor product space is also a reproducing kernel Hilbert space:

Proposition 12.31 *Suppose that $\mathbb{H}_1$ and $\mathbb{H}_2$ are reproducing kernel Hilbert spaces of real-valued functions with domains $\mathcal{X}_1$ and $\mathcal{X}_2$, and equipped with kernels $\mathcal{K}_1$ and $\mathcal{K}_2$, respectively. Then the tensor product space $\mathbb{H} = \mathbb{H}_1 \otimes \mathbb{H}_2$ is an RKHS of real-valued functions with domain $\mathcal{X}_1 \times \mathcal{X}_2$, and with kernel function*

$$\mathcal{K}((x_1, x_2), (x_1', x_2')) = \mathcal{K}_1(x_1, x_1')\,\mathcal{K}_2(x_2, x_2'). \tag{12.24}$$

Proof In Exercise 12.16, it is shown that $\mathcal{K}$ defined in equation (12.24) is a positive semidefinite function. By definition of the tensor product space $\mathbb{H} = \mathbb{H}_1 \otimes \mathbb{H}_2$, for each pair $(x_1, x_2) \in \mathcal{X}_1 \times \mathcal{X}_2$, the function $\mathcal{K}((\cdot, \cdot), (x_1, x_2)) = \mathcal{K}_1(\cdot, x_1)\mathcal{K}_2(\cdot, x_2)$ is an element of the tensor product space $\mathbb{H}$. Let $f = \sum_{j,k=1}^{n} \alpha_{j,k}\phi_j\psi_k$ be an arbitrary element of $\mathbb{H}$. By definition of the inner product (12.23), we have

$$\begin{aligned}\langle f, \mathcal{K}((\cdot, \cdot), (x_1, x_2))\rangle_{\mathbb{H}} &= \sum_{j,k=1}^{n} \alpha_{j,k}\langle \phi_j, \mathcal{K}_1(\cdot, x_1)\rangle_{\mathbb{H}_1}\, \langle \psi_k, \mathcal{K}_2(\cdot, x_2)\rangle_{\mathbb{H}_2} \\ &= \sum_{j,k=1}^{n} \alpha_{j,k}\phi_j(x_1)\psi_k(x_2) = f(x_1, x_2),\end{aligned}$$

thereby verifying the reproducing property. □

12.5 Interpolation and fitting

Reproducing kernel Hilbert spaces are useful for the classical problems of interpolating and fitting functions. An especially attractive property is the ease of computation: in particular, the representer theorem allows many optimization problems over the RKHS to be reduced to relatively simple calculations involving the kernel matrix.

12.5.1 Function interpolation

Let us begin with the problem of function interpolation. Suppose that we observe n samples of an unknown function $f^*: \mathcal{X} \to \mathbb{R}$, say of the form $y_i = f^*(x_i)$ for $i = 1, 2, \ldots, n$, where the design sequence $\{x_i\}_{i=1}^n$ is known to us. Note that we are assuming for the moment that the function values are observed without any noise or corruption. In this context, some questions of interest include:

- For a given function class $\mathscr{F}$, does there exist a function $f \in \mathscr{F}$ that exactly fits the data, meaning that $f(x_i) = y_i$ for all $i = 1, 2, \ldots, n$?
- Of all functions in $\mathscr{F}$ that exactly fit the data, which does the "best" job of interpolating the data?

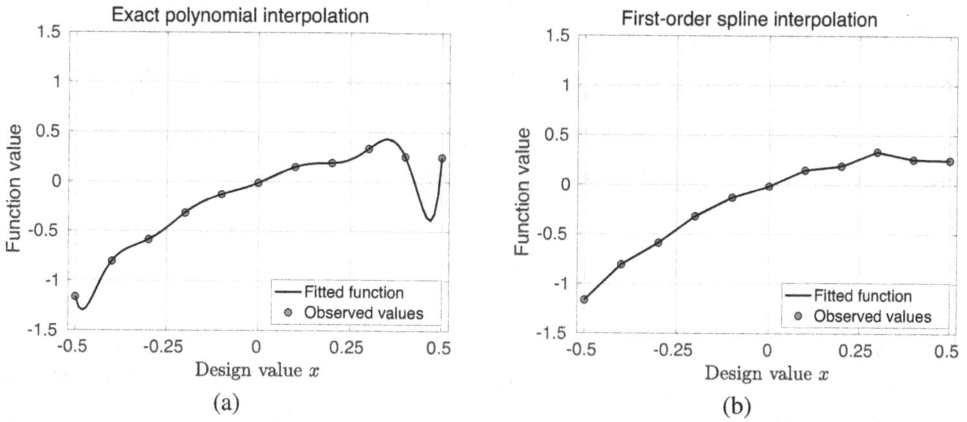

Figure 12.1 Exact interpolation of $n = 11$ equally sampled function values using RKHS methods. (a) Polynomial kernel $\mathcal{K}(x, z) = (1 + x\,z)^{12}$. (b) First-order Sobolev kernel $\mathcal{K}(x, z) = 1 + \min\{x, z\}$.

The first question can often be answered in a definitive way—in particular, by producing a function that exactly fits the data. The second question is vaguely posed and can be answered in multiple ways, depending on our notion of "best". In the context of a reproducing kernel Hilbert space, the underlying norm provides a way of ordering functions, and so we are led to the following formalization: of all the functions that exactly fit the data, choose the one with minimal RKHS norm. This approach can be formulated as an optimization problem in Hilbert space—namely,

$$\text{choose} \quad \widehat{f} \in \arg\min_{f \in \mathbb{H}} \|f\|_{\mathbb{H}} \quad \text{such that} \quad f(x_i) = y_i \text{ for } i = 1, 2, \ldots, n. \tag{12.25}$$

This method is known as *minimal norm interpolation*, and it is feasible whenever there exists at least one function $f \in \mathbb{H}$ that fits the data exactly. We provide necessary and sufficient conditions for such feasibility in the result to follow. Figure 12.1 illustrates this minimal Hilbert norm interpolation method, using the polynomial kernel from Example 12.8 in Figure 12.1(a), and the first-order Sobolev kernel from Example 12.23 in Figure 12.1(b).

For a general Hilbert space, the optimization problem (12.25) may not be well defined, or may be computationally challenging to solve. Hilbert spaces with reproducing kernels are attractive in this regard, as the computation can be reduced to simple linear algebra involving the kernel matrix $\mathbf{K} \in \mathbb{R}^{n \times n}$ with entries $K_{ij} = \mathcal{K}(x_i, x_j)/n$. The following result provides one instance of this general phenomenon:

Proposition 12.32 *Let $\mathbf{K} \in \mathbb{R}^{n \times n}$ be the kernel matrix defined by the design points $\{x_i\}_{i=1}^n$. The convex program (12.25) is feasible if and only if $y \in$ range($\mathbf{K}$), in which case any optimal solution can be written as*

$$\widehat{f}(\cdot) = \frac{1}{\sqrt{n}} \sum_{i=1}^{n} \widehat{\alpha}_i \mathcal{K}(\cdot, x_i), \qquad \textit{where } \mathbf{K}\widehat{\alpha} = y/\sqrt{n}.$$

Remark: Our choice of normalization by $1/\sqrt{n}$ is for later theoretical convenience.

Proof For a given vector $\alpha \in \mathbb{R}^n$, define the function $f_\alpha(\cdot) := \frac{1}{\sqrt{n}} \sum_{i=1}^n \alpha_i \mathcal{K}(\cdot, x_i)$, and consider the set $\mathbb{L} := \{f_\alpha \mid \alpha \in \mathbb{R}^n\}$. Note that for any $f_\alpha \in \mathbb{L}$, we have

$$f_\alpha(x_j) = \frac{1}{\sqrt{n}} \sum_{i=1}^{n} \alpha_i \mathcal{K}(x_j, x_i) = \sqrt{n}(\mathbf{K}\alpha)_j,$$

where $(\mathbf{K}\alpha)_j$ is the jth component of the vector $\mathbf{K}\alpha \in \mathbb{R}^n$. Thus, the function $f_\alpha \in \mathbb{L}$ satisfies the interpolation condition if and only if $\mathbf{K}\alpha = y/\sqrt{n}$. Consequently, the condition $y \in$ range($\mathbf{K}$) is sufficient. It remains to show that this range condition is necessary, and that the optimal interpolating function must lie in $\mathbb{L}$.

Note that $\mathbb{L}$ is a finite-dimensional (hence closed) linear subspace of $\mathbb{H}$. Consequently, any function $f \in \mathbb{H}$ can be decomposed uniquely as $f = f_\alpha + f_\perp$, where $f_\alpha \in \mathbb{L}$ and $f_\perp$ is orthogonal to $\mathbb{L}$. (See Exercise 12.3 for details of this direct sum decomposition.) Using this decomposition and the reproducing property, we have

$$f(x_j) = \langle f, \mathcal{K}(\cdot, x_j)\rangle_{\mathbb{H}} = \langle f_\alpha + f_\perp, \mathcal{K}(\cdot, x_j)\rangle_{\mathbb{H}} = f_\alpha(x_j),$$

where the final equality follows because $\mathcal{K}(\cdot, x_j)$ belongs to $\mathbb{L}$, and we have $\langle f_\perp, \mathcal{K}(\cdot, x_j)\rangle_{\mathbb{H}} = 0$ due to the orthogonality of $f_\perp$ and $\mathbb{L}$. Thus, the component $f_\perp$ has no effect on the interpolation property, showing that the condition $y \in$ range($\mathbf{K}$) is also a necessary condition. Moreover, since f_α and $f_\perp$ are orthogonal, we are guaranteed to have $\|f_\alpha + f_\perp\|_{\mathbb{H}}^2 = \|f_\alpha\|_{\mathbb{H}}^2 + \|f_\perp\|_{\mathbb{H}}^2$. Consequently, for any Hilbert norm interpolant, we must have $f_\perp = 0$. □

12.5.2 Fitting via kernel ridge regression

In a statistical setting, it is usually unrealistic to assume that we observe noiseless observations of function values. Rather, it is more natural to consider a noisy observation model, say of the form

$$y_i = f^*(x_i) + w_i, \qquad \text{for } i = 1, 2, \ldots, n,$$

where the coefficients $\{w_i\}_{i=1}^n$ model noisiness or disturbance in the measurement model. In the presence of noise, the exact constraints in our earlier interpolation method (12.25) are no longer appropriate; instead, it is more sensible to minimize some trade-off between the fit to the data and the Hilbert norm. For instance, we might only require that the mean-squared differences between the observed data and fitted values be small, which then leads to the optimization problem

$$\min_{f \in \mathbb{H}} \|f\|_{\mathbb{H}} \qquad \text{such that } \frac{1}{2n} \sum_{i=1}^n (y_i - f(x_i))^2 \leq \delta^2, \tag{12.26}$$

where $\delta > 0$ is some type of tolerance parameter. Alternatively, we might minimize the mean-squared error subject to a bound on the Hilbert radius of the solution, say

$$\min_{f \in \mathbb{H}} \frac{1}{2n} \sum_{i=1}^n (y_i - f(x_i))^2 \qquad \text{such that } \|f\|_{\mathbb{H}} \leq R \tag{12.27}$$

for an appropriately chosen radius $R > 0$. Both of these problems are convex, and so by Lagrangian duality, they can be reformulated in the penalized form

$$\widehat{f} = \arg\min_{f \in \mathbb{H}} \Big\{ \frac{1}{2n} \sum_{i=1}^n (y_i - f(x_i))^2 + \lambda_n \|f\|_{\mathbb{H}}^2 \Big\}. \tag{12.28}$$

Here, for a fixed set of observations $\{(x_i, y_i)\}_{i=1}^n$, the regularization parameter $\lambda_n \geq 0$ is a function of the tolerance δ or radius R. This form of function estimate is most convenient to implement, and in the case of a reproducing kernel Hilbert space considered here, it is known as the *kernel ridge regression* estimate, or KRR estimate for short. The following result shows how the KRR estimate is easily computed in terms of the kernel matrix $\mathbf{K} \in \mathbb{R}^{n \times n}$ with entries $K_{ij} = \mathcal{K}(x_i, x_j)/n$.

Proposition 12.33 *For all $\lambda_n > 0$, the kernel ridge regression estimate* (12.28) *can be written as*

$$\widehat{f}(\cdot) = \frac{1}{\sqrt{n}} \sum_{i=1}^n \widehat{\alpha}_i \mathcal{K}(\cdot, x_i), \tag{12.29}$$

where the optimal weight vector $\widehat{\alpha} \in \mathbb{R}^n$ is given by

$$\widehat{\alpha} = (\mathbf{K} + \lambda_n \mathbf{I}_n)^{-1} \frac{y}{\sqrt{n}}. \tag{12.30}$$

Remarks: Note that Proposition 12.33 is a natural generalization of Proposition 12.32, to which it reduces when $\lambda_n = 0$ (and the kernel matrix is invertible). Given the kernel matrix $\mathbf{K}$, computing $\widehat{\alpha}$ via equation (12.30) requires at most $O(n^3)$ operations, using standard routines in numerical linear algebra (see the bibliography for more details). Assuming that the kernel function can be evaluated in constant time, computing the $n \times n$ matrix requires an additional $O(n^2)$ operations. See Figure 12.2 for some illustrative examples.

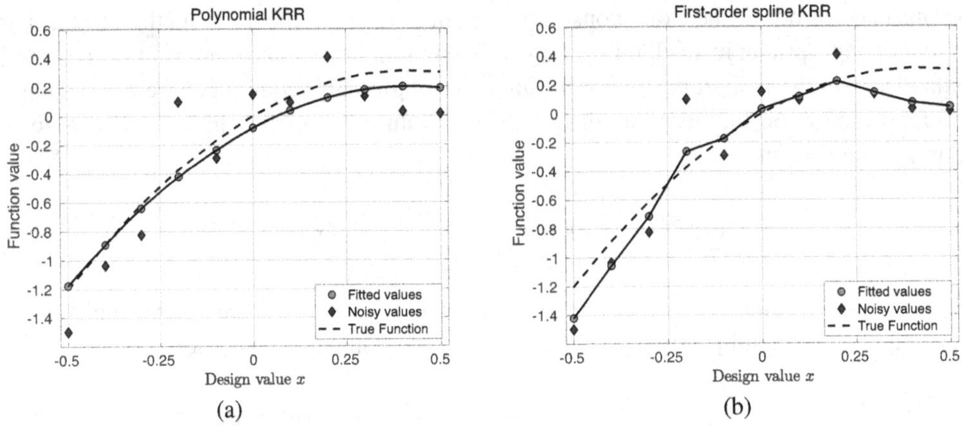

Figure 12.2 Illustration of kernel ridge regression estimates of function $f^*(x) = \frac{3x}{2} - \frac{9}{5}x^2$ based on $n = 11$ samples, located at design points $x_i = -0.5 + 0.10\,(i-1)$ over the interval $[-0.5, 0.5]$. (a) Kernel ridge regression estimate using the second-order polynomial kernel $\mathcal{K}(x, z) = (1 + xz)^2$ and regularization parameter $\lambda_n = 0.10$. (b) Kernel ridge regression estimate using the first-order Sobolev kernel $\mathcal{K}(x, z) = 1 + \min\{x, z\}$ and regularization parameter $\lambda_n = 0.10$.

We now turn to the proof of Proposition 12.33.

Proof Recall the argument of Proposition 12.32, and the decomposition $f = f_\alpha + f_\perp$. Since $f_\perp(x_i) = 0$ for all $i = 1, 2, \ldots, n$, it can have no effect on the least-squares data component of the objective function (12.28). Consequently, following a similar line of reasoning to the proof of Proposition 12.32, we again see that any optimal solution must be of the specified form (12.29).

It remains to prove the specific form (12.30) of the optimal $\widehat{\alpha}$. Given a function f of the form (12.29), for each $j = 1, 2, \ldots, n$, we have

$$f(x_j) = \frac{1}{\sqrt{n}} \sum_{i=1}^{n} \alpha_i \mathcal{K}(x_j, x_i) = \sqrt{n}\, e_j^{\mathrm{T}} \mathbf{K} \alpha,$$

where $e_j \in \mathbb{R}^n$ is the canonical basis vector with 1 in position j, and we have recalled that $K_{ji} = \mathcal{K}(x_j, x_i)/n$. Similarly, we have the representation

$$\|f\|_{\mathbb{H}}^2 = \frac{1}{n} \left\langle \sum_{i=1}^{n} \alpha_i \mathcal{K}(\cdot, x_i), \sum_{j=1}^{n} \alpha_j \mathcal{K}(\cdot, x_j) \right\rangle_{\mathbb{H}} = \alpha^{\mathrm{T}} \mathbf{K} \alpha.$$

Substituting these relations into the cost function, we find that it is a quadratic in the vector α, given by

$$\frac{1}{n}\|y - \sqrt{n}\mathbf{K}\alpha\|_2^2 + \lambda\,\alpha^{\mathrm{T}}\mathbf{K}\alpha = \frac{1}{n}\|y\|_2^2 + \alpha^{\mathrm{T}}(\mathbf{K}^2 + \lambda\mathbf{K})\alpha - \frac{2}{\sqrt{n}}y^{\mathrm{T}}\mathbf{K}\alpha.$$

In order to find the minimum of this quadratic function, we compute the gradient and set it equal to zero, thereby obtaining the stationary condition

$$\mathbf{K}\,(\mathbf{K} + \lambda\mathbf{I}_n)\alpha = \mathbf{K}\frac{y}{\sqrt{n}}.$$

Thus, we see that the vector $\widehat{\alpha}$ previously defined in equation (12.30) is optimal. Note that any vector $\beta \in \mathbb{R}^n$ such that $\mathbf{K}\beta = 0$ has no effect on the optimal solution. □

We return in Chapter 13 to study the statistical properties of the kernel ridge regression estimate.

12.6 Distances between probability measures

There are various settings in which it is important to construct distances between probability measures, and one way in which to do so is via measuring mean discrepancies over a given function class. More precisely, let $\mathbb{P}$ and $\mathbb{Q}$ be a pair of probability measures on a space $\mathcal{X}$, and let $\mathscr{F}$ be a class of functions $f\colon \mathcal{X} \to \mathbb{R}$ that are integrable with respect to $\mathbb{P}$ and $\mathbb{Q}$. We can then define the quantity

$$\rho_{\mathscr{F}}(\mathbb{P}, \mathbb{Q}) := \sup_{f\in\mathscr{F}} \left|\int f(d\mathbb{P} - d\mathbb{Q})\right| = \sup_{f\in\mathscr{F}} \left|\mathbb{E}_{\mathbb{P}}[f(X)] - \mathbb{E}_{\mathbb{Q}}[f(Z)]\right|. \tag{12.31}$$

It can be verified that, for any choice of function class $\mathscr{F}$, this always defines a pseudometric, meaning that $\rho_{\mathscr{F}}$ satisfies all the metric properties, except that there may exist pairs $\mathbb{P} \neq \mathbb{Q}$ such that $\rho_{\mathscr{F}}(\mathbb{P}, \mathbb{Q}) = 0$. When $\mathscr{F}$ is sufficiently rich, then $\rho_{\mathscr{F}}$ becomes a metric, known as an *integral probability metric*. Let us provide some classical examples to illustrate:

Example 12.34 (Kolmogorov metric) Suppose that $\mathbb{P}$ and $\mathbb{Q}$ are measures on the real line. For each $t \in \mathbb{R}$, let $\mathbb{I}_{(-\infty,t]}$ denote the $\{0, 1\}$-valued indicator function for the event $\{x \leq t\}$, and consider the function class $\mathscr{F} = \{\mathbb{I}_{(-\infty,t]} \mid t \in \mathbb{R}\}$. We then have

$$\rho_{\mathscr{F}}(\mathbb{P}, \mathbb{Q}) = \sup_{t\in\mathbb{R}} \left|\mathbb{P}(X \leq t) - \mathbb{Q}(X \leq t)\right| = \|F_{\mathbb{P}} - F_{\mathbb{Q}}\|_\infty,$$

where $F_{\mathbb{P}}$ and $F_{\mathbb{Q}}$ are the cumulative distribution functions of $\mathbb{P}$ and $\mathbb{Q}$, respectively. Thus, this choice leads to the *Kolmogorov distance* between $\mathbb{P}$ and $\mathbb{Q}$. ♣

Example 12.35 (Total variation distance) Consider the class $\mathscr{F} = \{f\colon \mathcal{X} \to \mathbb{R} \mid \|f\|_\infty \leq 1\}$ of real-valued functions bounded by one in the supremum norm. With this choice, we have

$$\rho_{\mathscr{F}}(\mathbb{P}, \mathbb{Q}) = \sup_{\|f\|_\infty \leq 1} \left|\int f(d\mathbb{P} - d\mathbb{Q})\right|.$$

As we show in Exercise 12.17, this metric corresponds to (two times) the total variation distance

$$\|\mathbb{P} - \mathbb{Q}\|_1 = \sup_{A \subset \mathcal{X}} |\mathbb{P}(A) - \mathbb{Q}(A)|,$$

where the supremum ranges over all measurable subsets of $\mathcal{X}$. ♣

When we choose $\mathscr{F}$ to be the unit ball of an RKHS, we obtain a mean discrepancy pseudometric that is easy to compute. In particular, given an RKHS with kernel function $\mathcal{K}$, consider the associated pseudometric

$$\rho_{\mathbb{H}}(\mathbb{P}, \mathbb{Q}) := \sup_{\|f\|_{\mathbb{H}} \leq 1} \left|\mathbb{E}_{\mathbb{P}}[f(X)] - \mathbb{E}_{\mathbb{Q}}[f(Z)]\right|.$$

As verified in Exercise 12.18, the reproducing property allows us to obtain a simple closed-form expression for this pseudometric—namely,

$$\rho^2_{\mathbb{H}}(\mathbb{P}, \mathbb{Q}) = \mathbb{E}[\mathcal{K}(X, X') + \mathcal{K}(Z, Z') - 2\mathcal{K}(X, Z)], \tag{12.32}$$

where $X, X' \sim \mathbb{P}$ and $Z, Z' \sim \mathbb{Q}$ are all mutually independent random vectors. We refer to this pseudometric as a *kernel means discrepancy*, or KMD for short.

Example 12.36 (KMD for linear and polynomial kernels) Let us compute the KMD for the linear kernel $\mathcal{K}(x, z) = \langle x, z\rangle$ on $\mathbb{R}^d$. Letting $\mathbb{P}$ and $\mathbb{Q}$ be two distributions on $\mathbb{R}^d$ with mean vectors $\mu_p = \mathbb{E}_{\mathbb{P}}[X]$ and $\mu_q = \mathbb{E}_{\mathbb{Q}}[Z]$, respectively, we have

$$\begin{aligned}
\rho^2_{\mathbb{H}}(\mathbb{P}, \mathbb{Q}) &= \mathbb{E}\Big[\langle X, X'\rangle + \langle Z, Z'\rangle - 2\langle X, Z\rangle\Big] \\
&= \|\mu_p\|_2^2 + \|\mu_q\|_2^2 - 2\left\langle \mu_p, \mu_q\right\rangle \\
&= \|\mu_p - \mu_q\|_2^2.
\end{aligned}$$

Thus, we see that the KMD pseudometric for the linear kernel simply computes the Euclidean distance of the associated mean vectors. This fact demonstrates that KMD in this very special case is not actually a metric (but rather just a pseudometric), since $\rho_{\mathbb{H}}(\mathbb{P}, \mathbb{Q}) = 0$ for any pair of distributions with the same means (i.e., $\mu_p = \mu_q$).

Moving onto polynomial kernels, let us consider the homogeneous polynomial kernel of degree two, namely $\mathcal{K}(x, z) = \langle x, z\rangle^2$. For this choice of kernel, we have

$$\mathbb{E}[\mathcal{K}(X, X')] = \mathbb{E}\left[\Big(\sum_{j=1}^d X_j X'_j\Big)^2\right] = \sum_{i,j=1}^d \mathbb{E}[X_i X_j]\mathbb{E}[X'_i X'_j] = |||\mathbf{\Gamma}_p|||_F^2,$$

where $\mathbf{\Gamma}_p \in \mathbb{R}^{d\times d}$ is the second-order moment matrix with entries $[\mathbf{\Gamma}_p]_{ij} = \mathbb{E}[X_i X_j]$, and the squared Frobenius norm corresponds to the sum of the squared matrix entries. Similarly, we have $\mathbb{E}[\mathcal{K}(Z, Z')] = |||\mathbf{\Gamma}_q|||_F^2$, where $\mathbf{\Gamma}_q$ is the second-order moment matrix for $\mathbb{Q}$. Finally, similar calculations yield that

$$\mathbb{E}[\mathcal{K}(X, Z)] = \sum_{i,j=1}^d [\mathbf{\Gamma}_p]_{ij}[\mathbf{\Gamma}_q]_{ij} = \langle\!\langle \mathbf{\Gamma}_p, \mathbf{\Gamma}_q\rangle\!\rangle,$$

where $\langle\langle \cdot, \cdot \rangle\rangle$ denotes the trace inner product between symmetric matrices. Putting together the pieces, we conclude that, for the homogeneous second-order polynomial kernel, we have

$$\rho^2_{\mathbb{H}}(\mathbb{P}, \mathbb{Q}) = |||\boldsymbol{\Gamma}_p - \boldsymbol{\Gamma}_q|||^2_F. \qquad \clubsuit$$

Example 12.37 (KMD for a first-order Sobolev kernel) Let us now consider the KMD induced by the kernel function $\mathcal{K}(x, z) = \min\{x, z\}$, defined on the Cartesian product $[0, 1] \times [0, 1]$. As seen previously in Example 12.16, this kernel function generates the first-order Sobolev space

$$\mathbb{H}^1[0, 1] = \left\{ f \colon \mathbb{R}[0, 1] \to \mathbb{R} \mid f(0) = 0 \text{ and } \int_0^1 (f'(x))^2 \, dx < \infty \right\},$$

with Hilbert norm $\|f\|^2_{\mathbb{H}^1[0,1]} = \int_0^1 (f'(x))^2 \, dx$. With this choice, we have

$$\rho^2_{\mathbb{H}}(\mathbb{P}, \mathbb{Q}) = \mathbb{E}\Big[\min\{X, X'\} + \min\{Z, Z'\} - 2 \min\{X, Z\} \Big]. \qquad \clubsuit$$

12.7 Bibliographic details and background

The notion of a reproducing kernel Hilbert space emerged from the study of positive semidefinite kernels and their links to Hilbert space structure. The seminal paper by Aronszajn (1950) develops a number of the basic properties from first principles, including Propositions 12.27 and 12.31 as well as Theorem 12.11 from this chapter. The use of the kernel trick for computing inner products via kernel evaluations dates back to Aizerman et al. (1964), and underlies the success of the support vector machine developed by Boser et al. (1992), and discussed in Exercise 12.20. The book by Wahba (1990) contains a wealth of information on RKHSs, as well as the connections between splines and penalized methods for regression. See also the books by Berlinet and Thomas-Agnan (2004) as well as Gu (2002). The book by Schölkopf and Smola (2002) provides a number of applications of kernels in the setting of machine learning, including the support vector machine (Exercise 12.20) and related methods for classification, as well as kernel principal components analysis. The book by Steinwart and Christmann (2008) also contains a variety of theoretical results on kernels and reproducing kernel Hilbert spaces.

The argument underlying the proofs of Propositions 12.32 and 12.33 is known as the *representer theorem*, and is due to Kimeldorf and Wahba (1971). From the computational point of view, it is extremely important, since it allows the infinite-dimensional problem of optimizing over an RKHS to be reduced to an n-dimensional convex program. Bochner's theorem relates the positive semidefiniteness of kernel functions to the non-negativity of Fourier coefficients. In its classical formulation, it applies to the Fourier transform over $\mathbb{R}^d$, but it can be generalized to all locally compact Abelian groups (Rudin, 1990). The results used to compute the asymptotic scaling of the eigenvalues of the Gaussian kernel in Example 12.25 are due to Widom (1963; 1964).

There are a number of papers that study the approximation-theoretic properties of various types of reproducing kernel Hilbert spaces. For a given Hilbert space $\mathbb{H}$ and norm $\|\cdot\|$, such results are often phrased in terms of the function

$$A(f^*; R) := \inf_{\|f\|_{\mathbb{H}} \leq R} \|f - f^*\|_p, \tag{12.33}$$

where $\|g\|_p := (\int_{\mathcal{X}} g^p(x)\,dx)^{1/p}$ is the usual L^p-norm on a compact space $\mathcal{X}$. This function measures how quickly the $L^p(\mathcal{X})$-error in approximating some function f^* decays as the Hilbert radius R is increased. See the papers (Smale and Zhou, 2003; Zhou, 2013) for results on this form of the approximation error. A reproducing kernel Hilbert space is said to be $L^p(\mathcal{X})$-*universal* if $\lim_{R\to\infty} A(f^*; R) = 0$ for any $f^* \in L^p(\mathcal{X})$. There are also various other forms of universality; see the book by Steinwart and Christmann (2008) for further details.

Integral probability metrics of the form (12.31) have been studied extensively (Müller, 1997; Rachev et al., 2013). The particular case of RKHS-based distances are computationally convenient, and have been studied in the context of proper scoring rules (Dawid, 2007; Gneiting and Raftery, 2007) and two-sample testing (Borgwardt et al., 2006; Gretton et al., 2012).

12.8 Exercises

Exercise 12.1 (Closedness of nullspace) Let L be a bounded linear functional on a Hilbert space. Show that the subspace $\text{null}(L) = \{f \in \mathbb{H} \mid L(f) = 0\}$ is closed.

Exercise 12.2 (Projections in a Hilbert space) Let $\mathbb{G}$ be a closed convex subset of a Hilbert space $\mathbb{H}$. In this exercise, we show that for any $f \in \mathbb{H}$, there exists a unique $\widehat{g} \in \mathbb{G}$ such that

$$\|\widehat{g} - f\|_{\mathbb{H}} = \underbrace{\inf_{g\in\mathbb{G}} \|\widehat{g} - f\|_{\mathbb{H}}}_{p^*}.$$

This element $\widehat{g}$ is known as the projection of f onto $\mathbb{G}$.

(a) By the definition of infimum, there exists a sequence $(g_n)_{n=1}^\infty$ contained in $\mathbb{G}$ such that $\|g_n - f\|_{\mathbb{H}} \to p^*$. Show that this sequence is a Cauchy sequence. (*Hint:* First show that $\|f - \frac{g_n+g_m}{2}\|_{\mathbb{H}}$ converges to p^*.)
(b) Use this Cauchy sequence to establish the existence of $\widehat{g}$.
(c) Show that the projection must be unique.
(d) Does the same claim hold for an arbitrary convex set $\mathbb{G}$?

Exercise 12.3 (Direct sum decomposition in Hilbert space) Let $\mathbb{H}$ be a Hilbert space, and let $\mathbb{G}$ be a closed linear subspace of $\mathbb{H}$. Show that any $f \in \mathbb{H}$ can be decomposed uniquely as $g + g^\perp$, where $g \in \mathbb{G}$ and $g^\perp \in \mathbb{G}^\perp$. In brief, we say that $\mathbb{H}$ has the direct sum decomposition $\mathbb{G} \oplus \mathbb{G}^\perp$. (*Hint:* The notion of a projection onto a closed convex set from Exercise 12.2 could be helpful to you.)

Exercise 12.4 (Uniqueness of kernel) Show that the kernel function associated with any reproducing kernel Hilbert space must be unique.

Exercise 12.5 (Kernels and Cauchy–Schwarz)

(a) For any positive semidefinite kernel $\mathcal{K}: \mathcal{X} \times \mathcal{X} \to \mathbb{R}$, prove that

$$\mathcal{K}(x, z) \leq \sqrt{\mathcal{K}(x, x)\mathcal{K}(z, z)} \qquad \text{for all } x, z \in \mathcal{X}.$$

(b) Show how the classical Cauchy–Schwarz inequality is a special case.

Exercise 12.6 (Eigenfunctions for linear kernels) Consider the ordinary linear kernel $\mathcal{K}(x, z) = \langle x, z \rangle$ on $\mathbb{R}^d$ equipped with a probability measure $\mathbb{P}$. Assuming that a random vector $X \sim \mathbb{P}$ has all its second moments finite, show how to compute the eigenfunctions of the associated kernel operator acting on $L^2(\mathcal{X}; \mathbb{P})$ in terms of linear algebraic operations.

Exercise 12.7 (Different kernels for polynomial functions) For an integer $m \geq 1$, consider the kernel functions $\mathcal{K}_1(x, z) = (1 + xz)^m$ and $\mathcal{K}_2(x, z) = \sum_{\ell=0}^{m} \frac{x^\ell}{\ell!} \frac{z^\ell}{\ell!}$.

(a) Show that they are both PSD, and generate RKHSs of polynomial functions of degree at most m.
(b) Why does this not contradict the result of Exercise 12.4?

Exercise 12.8 True or false? If true, provide a short proof; if false, give an explicit counterexample.

(a) Given two PSD kernels $\mathcal{K}_1$ and $\mathcal{K}_2$, the bivariate function $\mathcal{K}(x, z) = \min_{j=1,2} \mathcal{K}_j(x, z)$ is also a PSD kernel.
(b) Let $f: \mathcal{X} \to \mathbb{H}$ be a function from an arbitrary space $\mathcal{X}$ to a Hilbert space $\mathbb{H}$. The bivariate function

$$\mathcal{K}(x, z) = \frac{\langle f(x), f(z) \rangle_{\mathbb{H}}}{\|f(x)\|_{\mathbb{H}} \|f(z)\|_{\mathbb{H}}}$$

defines a PSD kernel on $\mathcal{X} \times \mathcal{X}$.

Exercise 12.9 (Left–right multiplication and kernels) Let $\mathcal{K}: \mathcal{X} \times \mathcal{X} \to \mathbb{R}$ be a positive semidefinite kernel, and let $f: \mathcal{X} \to \mathbb{R}$ be an arbitrary function. Show that $\widetilde{\mathcal{K}}(x, z) = f(x)\mathcal{K}(x, z)f(z)$ is also a positive semidefinite kernel.

Exercise 12.10 (Kernels and power sets) Given a finite set S, its power set $\mathcal{P}(S)$ is the set of all the subsets of S. Show that the function $\mathcal{K}: \mathcal{P}(S) \times \mathcal{P}(S) \to \mathbb{R}$ given by $\mathcal{K}(A, B) = 2^{|A \cap B|}$ is a positive semidefinite kernel function.

Exercise 12.11 (Feature map for polynomial kernel) Recall from equation (12.14) the notion of a feature map. Show that the polynomial kernel $\mathcal{K}(x, z) = (1 + \langle x, z \rangle)^m$ defined on the Cartesian product space $\mathbb{R}^d \times \mathbb{R}^d$ can be realized by a feature map $x \mapsto \Phi(x) \in \mathbb{R}^D$, where $D = \binom{d+m}{m}$.

Exercise 12.12 (Probability spaces and kernels) Consider a probability space with events $\mathcal{E}$ and probability law $\mathbb{P}$. Show that the real-valued function

$$\mathcal{K}(A, B) := \mathbb{P}[A \cap B] - \mathbb{P}[A]\mathbb{P}[B]$$

is a positive semidefinite kernel function on $\mathcal{E} \times \mathcal{E}$.

Exercise 12.13 (From sets to power sets) Suppose that $\mathcal{K}\colon S \times S \to \mathbb{R}$ is a symmetric PSD kernel function on a finite set S. Show that

$$\mathcal{K}'(A, B) = \sum_{x \in A,\, z \in B} \mathcal{K}(x, z)$$

is a symmetric PSD kernel on the power set $\mathcal{P}(S)$.

Exercise 12.14 (Kernel and function boundedness) Consider a PSD kernel $\mathcal{K}\colon \mathcal{X} \times \mathcal{X} \to \mathbb{R}$ such that $\mathcal{K}(x, z) \leq b^2$ for all $x, z \in \mathcal{X}$. Show that $\|f\|_\infty \leq b$ for any function f in the unit ball of the associated RKHS.

Exercise 12.15 (Sobolev kernels and norms) Show that the Sobolev kernel defined in equation (12.20) generates the norm given in equation (12.21).

Exercise 12.16 (Hadamard products and kernel products) In this exercise, we explore properties of product kernels and the Hadamard product of matrices.

(a) Given two $n \times n$ matrices $\mathbf{\Gamma}$ and $\mathbf{\Sigma}$ that are symmetric and positive semidefinite, show that the Hadamard product matrix $\mathbf{\Sigma} \odot \mathbf{\Gamma} \in \mathbb{R}^{n \times n}$ is also positive semidefinite. (The Hadamard product is simply the elementwise product—that is, $(\mathbf{\Sigma} \odot \mathbf{\Gamma})_{ij} = \Sigma_{ij}\Gamma_{ij}$ for all $i, j = 1, 2, \ldots, n$.)

(b) Suppose that $\mathcal{K}_1$ and $\mathcal{K}_2$ are positive semidefinite kernel functions on $\mathcal{X} \times \mathcal{X}$. Show that the function $\mathcal{K}(x, z) := \mathcal{K}_1(x, z)\mathcal{K}_2(x, z)$ is a positive semidefinite kernel function. (*Hint:* The result of part (a) could be helpful.)

Exercise 12.17 (Total variation norm) Given two probability measures $\mathbb{P}$ and $\mathbb{Q}$ on $\mathcal{X}$, show that

$$\sup_{\|f\|_\infty \leq 1} \left| \int f(d\mathbb{P} - d\mathbb{Q}) \right| = 2 \sup_{A \subset \mathcal{X}} |\mathbb{P}(A) - \mathbb{Q}(A)|,$$

where the left supremum ranges over all measurable functions $f\colon \mathcal{X} \to \mathbb{R}$, and the right supremum ranges over all measurable subsets A of $\mathcal{X}$.

Exercise 12.18 (RKHS-induced semi-metrics) Let $\mathbb{H}$ be a reproducing kernel Hilbert space of functions with domain $\mathcal{X}$, and let $\mathbb{P}$ and $\mathbb{Q}$ be two probability distributions on $\mathcal{X}$. Show that

$$\sup_{\|f\|_{\mathbb{H}} \leq 1} \left|\mathbb{E}_{\mathbb{P}}[f(X)] - \mathbb{E}_{\mathbb{Q}}[f(Z)]\right|^2 = \mathbb{E}[\mathcal{K}(X, X') + \mathcal{K}(Z, Z') - 2\mathcal{K}(X, Z)],$$

where $X, X' \sim \mathbb{P}$ and $Z, Z' \sim \mathbb{Q}$ are jointly independent.

Exercise 12.19 (Positive semidefiniteness of Gaussian kernel) Let $\mathcal{X}$ be a compact subset of $\mathbb{R}^d$. In this exercise, we work through a proof of the fact that the Gaussian kernel $\mathcal{K}(x, z) = e^{-\frac{\|x-z\|_2^2}{2\sigma^2}}$ on $\mathcal{X} \times \mathcal{X}$ is positive semidefinite.

(a) Let $\widetilde{\mathcal{K}}$ be a PSD kernel, and let p be a polynomial with non-negative coefficients. Show that $\mathcal{K}(x, z) = p(\widetilde{\mathcal{K}}(x, z))$ is a PSD kernel.

(b) Show that the kernel $\mathcal{K}_1(x, z) = e^{\langle x, z \rangle/\sigma^2}$ is positive semidefinite. (*Hint:* Part (a) and the fact that a pointwise limit of PSD kernels is also PSD could be useful.)

(c) Show that the Gaussian kernel is PSD. (*Hint:* The result of Exercise 12.9 could be useful.)

Exercise 12.20 (Support vector machines and kernel methods) In the problem of binary classification, one observes a collection of pairs $\{(x_i, y_i)\}_{i=1}^n$, where each feature vector $x_i \in \mathbb{R}^d$ is associated with a label $y_i \in \{-1, +1\}$, and the goal is derive a classification function that can be applied to unlabelled feature vectors. In the context of reproducing kernel Hilbert spaces, one way of doing so is by minimizing a criterion of the form

$$\widehat{f} = \arg\min_{f \in \mathbb{H}} \left\{ \frac{1}{n} \sum_{i=1}^{n} \max\{0, 1 - y_i f(x_i)\} + \frac{1}{2} \lambda_n \|f\|_{\mathbb{H}}^2 \right\}, \tag{12.34}$$

where $\mathbb{H}$ is a reproducing kernel Hilbert space, and $\lambda_n > 0$ is a user-defined regularization parameter. The classification rule is then given by $x \mapsto \text{sign}(\widehat{f}(x))$.

(a) Prove that $\widehat{f}$ can be written in the form $\widehat{f}(\cdot) = \frac{1}{\sqrt{n}} \sum_{i=1}^n \widehat{\alpha}_i \mathcal{K}(\cdot, x_i)$, for some vector $\widehat{\alpha} \in \mathbb{R}^n$.
(b) Use part (a) and duality theory to show that an optimal coefficient vector $\widehat{\alpha}$ can be obtained by solving the problem

$$\widehat{\alpha} \in \arg\max_{\alpha \in \mathbb{R}^n} \left\{ \frac{1}{n} \sum_{i=1}^{n} \alpha_i - \frac{1}{2} \alpha^{\mathrm{T}} \widetilde{\mathbf{K}} \alpha \right\} \qquad \text{s.t. } \alpha_i \in [0, \tfrac{1}{\lambda_n \sqrt{n}}] \text{ for all } i = 1, \ldots, n,$$

and where $\widetilde{\mathbf{K}} \in \mathbb{R}^{n \times n}$ has entries $\widetilde{K}_{ij} := y_i y_j \mathcal{K}(x_i, x_j)/n$.

13

Nonparametric least squares

In this chapter, we consider the problem of nonparametric regression, in which the goal is to estimate a (possibly nonlinear) function on the basis of noisy observations. Using results developed in previous chapters, we analyze the convergence rates of procedures based on solving nonparametric versions of least-squares problems.

13.1 Problem set-up

A regression problem is defined by a set of predictors or covariates $x \in \mathcal{X}$, along with a response variable $y \in \mathcal{Y}$. Throughout this chapter, we focus on the case of real-valued response variables, in which the space $\mathcal{Y}$ is the real line or some subset thereof. Our goal is to estimate a function $f \colon \mathcal{X} \to \mathcal{Y}$ such that the error $y - f(x)$ is as small as possible over some range of pairs (x, y). In the *random design* version of regression, we model both the response and covariate as random quantities, in which case it is reasonable to measure the quality of f in terms of its *mean-squared error* (MSE)

$$\bar{\mathcal{L}}_f := \mathbb{E}_{X,Y}[(Y - f(X))^2]. \tag{13.1}$$

The function f^* minimizing this criterion is known as the *Bayes' least-squares estimate* or the *regression function*, and it is given by the conditional expectation

$$f^*(x) = \mathbb{E}[Y \mid X = x], \tag{13.2}$$

assuming that all relevant expectations exist. See Exercise 13.1 for further details.

In practice, the expectation defining the MSE (13.1) cannot be computed, since the joint distribution over (X, Y) is not known. Instead, we are given a collection of samples $\{(x_i, y_i)\}_{i=1}^n$, which can be used to compute an empirical analog of the mean-squared error, namely

$$\widehat{\mathcal{L}}_f := \frac{1}{n} \sum_{i=1}^n (y_i - f(x_i))^2. \tag{13.3}$$

The method of *nonparametric least squares*, to be discussed in detail in this chapter, is based on minimizing this least-squares criterion over some suitably controlled function class.

13.1.1 Different measures of quality

Given an estimate f of the regression function, it is natural to measure its quality in terms of the *excess risk*—namely, the difference between the optimal MSE $\bar{\mathcal{L}}_{f^*}$ achieved by the regression function f^*, and that achieved by the estimate f. In the special case of the least-squares cost function, it can be shown (see Exercise 13.1) that this excess risk takes the form

$$\bar{\mathcal{L}}_f - \bar{\mathcal{L}}_{f^*} = \underbrace{\mathbb{E}_X[(f(X) - f^*(X))^2]}_{\|f^*-f\|^2_{L^2(\mathbb{P})}}, \tag{13.4}$$

where $\mathbb{P}$ denotes the distribution over the covariates. When this underlying distribution is clear from the context, we frequently adopt the shorthand notation $\|f - f^*\|_2$ for the $L^2(\mathbb{P})$-norm.

In this chapter, we measure the error using a closely related but slightly different measure, one that is defined by the samples $\{x_i\}_{i=1}^n$ of the covariates. In particular, they define the empirical distribution $\mathbb{P}_n := \frac{1}{n}\sum_{i=1}^n \delta_{x_i}$ that places a weight $1/n$ on each sample, and the associated $L^2(\mathbb{P}_n)$-norm is given by

$$\|f - f^*\|_{L^2(\mathbb{P}_n)} := \Big[\frac{1}{n}\sum_{i=1}^n (f(x_i) - f^*(x_i))^2\Big]^{1/2}. \tag{13.5}$$

In order to lighten notation, we frequently use $\|\widehat{f} - f^*\|_n$ as a shorthand for the more cumbersome $\|\widehat{f} - f^*\|_{L^2(\mathbb{P}_n)}$. Throughout the remainder of this chapter, we will view the samples $\{x_i\}_{i=1}^n$ as being fixed, a set-up known as regression with a *fixed design*. The theory in this chapter focuses on error bounds in terms of the empirical $L^2(\mathbb{P}_n)$-norm. Results from Chapter 14 to follow can be used to translate these bounds into equivalent results in the population $L^2(\mathbb{P})$-norm.

13.1.2 Estimation via constrained least squares

Given a fixed collection $\{x_i\}_{i=1}^n$ of fixed design points, the associated response variables $\{y_i\}_{i=1}^n$ can always be written in the generative form

$$y_i = f^*(x_i) + v_i, \qquad \text{for } i = 1, 2, \ldots, n, \tag{13.6}$$

where v_i is a random variable representing the "noise" in the ith response variable. Note that these noise variables must have zero mean, given the form (13.2) of the regression function f^*. Apart from this zero-mean property, their structure in general depends on the distribution of the conditioned random variable $(Y \mid X = x)$. In the *standard nonparametric regression* model, we assume the noise variables are drawn in an i.i.d. manner from the $\mathcal{N}(0, \sigma^2)$ distribution, where $\sigma > 0$ is a standard deviation parameter. In this case, we can write $v_i = \sigma w_i$, where $w_i \sim \mathcal{N}(0, 1)$ is a Gaussian random variable.

Given this set-up, one way in which to estimate the regression function f^* is by constrained least squares—that is, by solving the problem[1]

$$\widehat{f} \in \arg\min_{f\in\mathscr{F}} \left\{\frac{1}{n}\sum_{i=1}^n (y_i - f(x_i))^2\right\}, \tag{13.7}$$

[1] Although the renormalization by n^{-1} in the definition (13.7) has no consequence on $\widehat{f}$, we do so in order to emphasize the connection between this method and the $L^2(\mathbb{P}_n)$-norm.

where $\mathscr{F}$ is a suitably chosen subset of functions. When $v_i \sim \mathcal{N}(0, \sigma^2)$, note that the estimate defined by the criterion (13.7) is equivalent to the constrained maximum likelihood estimate. However, as with least-squares regression in the parametric setting, the estimator is far more generally applicable.

Typically, we restrict the optimization problem (13.7) to some appropriately chosen subset of $\mathscr{F}$—for instance, a ball of radius R in an underlying norm $\|\cdot\|_{\mathscr{F}}$. Choosing $\mathscr{F}$ to be a reproducing kernel Hilbert space, as discussed in Chapter 12, can be useful for computational reasons. It can also be convenient to use regularized estimators of the form

$$\widehat{f} \in \arg\min_{f \in \mathscr{F}} \left\{ \frac{1}{n} \sum_{i=1}^{n} (y_i - f(x_i))^2 + \lambda_n \|f\|_{\mathscr{F}}^2 \right\}, \tag{13.8}$$

where $\lambda_n > 0$ is a suitably chosen regularization weight. We return to analyze such estimators in Section 13.4.

13.1.3 Some examples

Let us illustrate the estimators (13.7) and (13.8) with some examples.

Example 13.1 (Linear regression) For a given vector $\theta \in \mathbb{R}^d$, define the linear function $f_\theta(x) = \langle \theta, x \rangle$. Given a compact subset $C \subseteq \mathbb{R}^d$, consider the function class

$$\mathscr{F}_C := \{f_\theta \colon \mathbb{R}^d \to \mathbb{R} \mid \theta \in C\}.$$

With this choice, the estimator (13.7) reduces to a constrained form of least-squares estimation, more specifically

$$\widehat{\theta} \in \arg\min_{\theta \in C} \left\{ \frac{1}{n} \|y - \mathbf{X}\theta\|_2^2 \right\},$$

where $\mathbf{X} \in \mathbb{R}^{n \times d}$ is the design matrix with the vector $x_i \in \mathbb{R}^d$ in its ith row. Particular instances of this estimator include *ridge regression*, obtained by setting

$$C = \left\{ \theta \in \mathbb{R}^d \mid \|\theta\|_2^2 \leq R_2 \right\}$$

for some (squared) radius $R_2 > 0$. More generally, this class of estimators contains all the *constrained ℓ_q-ball* estimators, obtained by setting

$$C = \left\{ \theta \in \mathbb{R}^d \mid \sum_{j=1}^{d} |\theta_j|^q \leq R_q \right\}$$

for some $q \in [0, 2]$ and radius $R_q > 0$. See Figure 7.1 for an illustration of these sets for $q \in (0, 1]$. The constrained form of the Lasso (7.19), as analyzed in depth in Chapter 7, is a special but important case, obtained by setting $q = 1$.

Whereas the previous example was a parametric problem, we now turn to some nonparametric examples:

Example 13.2 (Cubic smoothing spline) Consider the class of twice continuously differentiable functions $f\colon [0,1] \to \mathbb{R}$, and for a given squared radius $R > 0$, define the function class

$$\mathscr{F}(R) := \left\{ f\colon [0,1] \to \mathbb{R} \mid \int_0^1 (f''(x))^2\, dx \le R \right\}, \tag{13.9}$$

where f'' denotes the second derivative of f. The integral constraint on f'' can be understood as a Hilbert norm bound in the second-order Sobolev space $\mathbb{H}^\alpha[0,1]$ introduced in Example 12.17. In this case, the penalized form of the nonparametric least-squares estimate is given by

$$\widehat{f} \in \arg\min_f \left\{ \frac{1}{n} \sum_{i=1}^n (y_i - f(x_i))^2 + \lambda_n \int_0^1 (f''(x))^2\, dx \right\}, \tag{13.10}$$

where $\lambda_n > 0$ is a user-defined regularization parameter. It can be shown that any minimizer $\widehat{f}$ is a cubic spline, meaning that it is a piecewise cubic function, with the third derivative changing at each of the distinct design points x_i. In the limit as $R \to 0$ (or equivalently, as $\lambda_n \to +\infty$), the cubic spline fit $\widehat{f}$ becomes a linear function, since we have $f'' = 0$ only for a linear function. ♣

The spline estimator in the previous example turns out to be a special case of a more general class of estimators, based on regularization in a reproducing kernel Hilbert space (see Chapter 12 for background). Let us consider this family more generally:

Example 13.3 (Kernel ridge regression) Let $\mathbb{H}$ be a reproducing kernel Hilbert space, equipped with the norm $\|\cdot\|_{\mathbb{H}}$. Given some regularization parameter $\lambda_n > 0$, consider the estimator

$$\widehat{f} \in \arg\min_{f \in \mathbb{H}} \left\{ \frac{1}{n} \sum_{i=1}^n (y_i - f(x_i))^2 + \lambda_n \|f\|_{\mathbb{H}}^2 \right\}.$$

As discussed in Chapter 12, the computation of this estimate can be reduced to solving a quadratic program involving the empirical kernel matrix defined by the design points $\{x_i\}_{i=1}^n$. In particular, if we define the kernel matrix with entries $K_{ij} = \mathcal{K}(x_i, x_j)/n$, then the solution takes the form $\widehat{f}(\cdot) = \frac{1}{\sqrt{n}} \sum_{i=1}^n \widehat{\alpha}_i \mathcal{K}(\cdot, x_i)$, where $\widehat{\alpha} := (\mathbf{K} + \lambda_n \mathbf{I}_n)^{-1} \frac{y}{\sqrt{n}}$. In Exercise 13.3, we show how the spline estimator from Example 13.2 can be understood in the context of kernel ridge regression. ♣

Let us now consider an example of what is known as *shape-constrained* regression.

Example 13.4 (Convex regression) Suppose that $f^*\colon C \to \mathbb{R}$ is known to be a convex function over its domain C, some convex and open subset of $\mathbb{R}^d$. In this case, it is natural to consider the least-squares estimator with a convexity constraint—namely

$$\widehat{f} \in \arg \min_{\substack{f\colon C \to \mathbb{R} \\ f \text{ is convex}}} \left\{ \frac{1}{n} \sum_{i=1}^n (y_i - f(x_i))^2 \right\}.$$

As stated, this optimization problem is infinite-dimensional in nature. Fortunately, by exploiting the structure of convex functions, it can be converted to an equivalent finite-dimensional problem. In particular, any convex function f is subdifferentiable at each point in the (relative) interior of its domain C. More precisely, at any interior point $x \in C$, there exists at least one vector $z \in \mathbb{R}^d$ such that

$$f(y) \geq f(x) + \langle z,\ y - x \rangle \qquad \text{for all } y \in C. \tag{13.11}$$

Any such vector is known as a *subgradient*, and each point $x \in C$ can be associated with the set $\partial f(x)$ of its subgradients, which is known as the *subdifferential* of f at x. When f is actually differentiable at x, then the lower bound (13.11) holds if and only if $z = \nabla f(x)$, so that we have $\partial f(x) = \{\nabla f(x)\}$. See the bibliographic section for some standard references in convex analysis.

Applying this fact to each of the sampled points $\{x_i\}_{i=1}^n$, we find that there must exist subgradient vectors $\widetilde{z}_i \in \mathbb{R}^d$ such that

$$f(x) \geq f(x_i) + \langle \widetilde{z}_i,\ x - x_i \rangle \qquad \text{for all } x \in C. \tag{13.12}$$

Since the cost function depends only on the values $\widetilde{y}_i := f(x_i)$, the optimum does not depend on the function behavior elsewhere. Consequently, it suffices to consider the collection $\{(\widetilde{y}_i, \widetilde{z}_i)\}_{i=1}^n$ of function value and subgradient pairs, and solve the optimization problem

$$\min_{\{(\widetilde{y}_i, z_i)\}_{i=1}^n} \frac{1}{n} \sum_{i=1}^n (y_i - \widetilde{y}_i)^2 \tag{13.13}$$

$$\text{such that} \quad \widetilde{y}_j \geq \widetilde{y}_i + \langle \widetilde{z}_i,\ x_j - x_i \rangle \qquad \text{for all } i, j = 1, 2, \ldots, n.$$

Note that this is a convex program in $N = n(d + 1)$ variables, with a quadratic cost function and a total of $2\binom{n}{2}$ linear constraints.

An optimal solution $\{(\widehat{y}_i, \widehat{z}_i)\}_{i=1}^n$ can be used to define the estimate $\widehat{f} \colon C \to \mathbb{R}$ via

$$\widehat{f}(x) := \max_{i=1,\ldots,n} \{\widehat{y}_i + \langle \widehat{z}_i,\ x - x_i \rangle\}. \tag{13.14}$$

As the maximum of a collection of linear functions, the function $\widehat{f}$ is convex. Moreover, a short calculation—using the fact that $\{(\widehat{y}_i, \widehat{z}_i)\}_{i=1}^n$ are feasible for the program (13.13)—shows that $\widehat{f}(x_i) = \widehat{y}_i$ for all $i = 1, 2, \ldots, n$. Figure 13.1(a) provides an illustration of the convex regression estimate (13.14), showing its piecewise linear nature.

There are various extensions to the basic convex regression estimate. For instance, in the one-dimensional setting ($d = 1$), it might be known *a priori* that f is a non-decreasing function, so that its derivative (or, more generally, subgradients) are non-negative. In this case, it is natural to impose additional non-negativity constraints ($\widetilde{z}_j \geq 0$) on the subgradients in the estimator (13.13). Figure 13.1(b) compares the standard convex regression estimate with the estimator that imposes these additional monotonicity constraints. ♣

13.2 Bounding the prediction error

From a statistical perspective, an essential question associated with the nonparametric least-squares estimate (13.7) is how well it approximates the true regression function f^*. In this

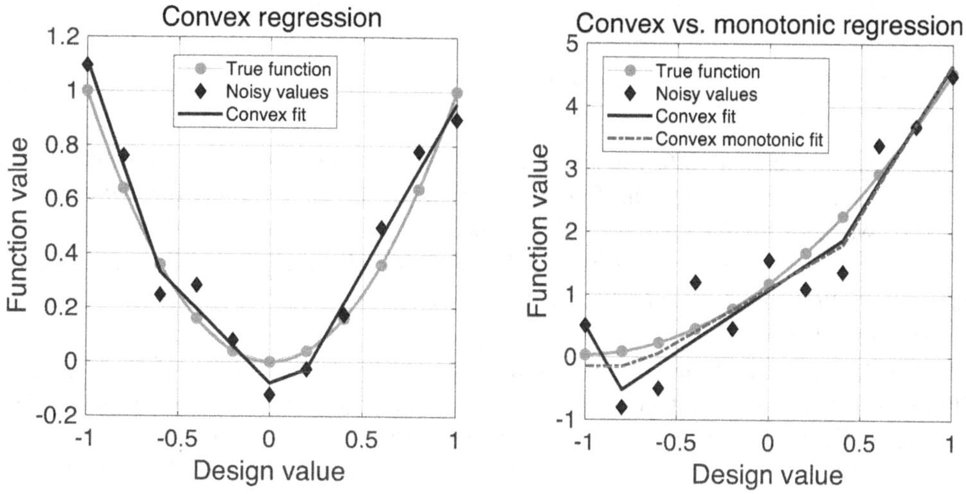

(a) Convex regression (b) Convex vs convex/monotonic regression

Figure 13.1 (a) Illustration of the convex regression estimate (13.14) based on a fixed design with $n = 11$ equidistant samples over the interval $C = [-1, 1]$. (b) Ordinary convex regression compared with convex and monotonic regression estimate.

section, we develop some techniques to bound the error $\|\widehat{f} - f^*\|_n$, as measured in the $L^2(\mathbb{P}_n)$-norm. In Chapter 14, we develop results that allow such bounds to be translated into bounds in the $L^2(\mathbb{P})$-norm.

Intuitively, the difficulty of estimating the function f^* should depend on the complexity of the function class $\mathscr{F}$ in which it lies. As discussed in Chapter 5, there are a variety of ways of measuring the complexity of a function class, notably by its metric entropy or its Gaussian complexity. We make use of both of these complexity measures in the results to follow.

Our first main result is defined in terms of a *localized form* of Gaussian complexity: it measures the complexity of the function class $\mathscr{F}$, locally in a neighborhood around the true regression function f^*. More precisely, we define the set

$$\mathscr{F}^* := \mathscr{F} - \{f^*\} = \{f - f^* \mid f \in \mathscr{F}\}, \tag{13.15}$$

corresponding to an f^*-shifted version of the original function class $\mathscr{F}$. For a given radius $\delta > 0$, the *local Gaussian complexity* around f^* at scale δ is given by

$$\mathcal{G}_n(\delta; \mathscr{F}^*) := \mathbb{E}_w\Big[\sup_{\substack{g \in \mathscr{F}^* \\ \|g\|_n \leq \delta}} \Big|\frac{1}{n}\sum_{i=1}^n w_i g(x_i)\Big|\Big], \tag{13.16}$$

where the variables $\{w_i\}_{i=1}^n$ are i.i.d. $\mathcal{N}(0, 1)$ variates. Throughout this chapter, this complexity measure should be understood as a deterministic quantity, since we are considering the case of fixed covariates $\{x_i\}_{i=1}^n$.

A central object in our analysis is the set of positive scalars δ that satisfy the *critical inequality*

$$\frac{\mathcal{G}_n(\delta; \mathcal{F}^*)}{\delta} \leq \frac{\delta}{2\sigma}. \tag{13.17}$$

As we verify in Lemma 13.6, whenever the shifted function class $\mathcal{F}^*$ is star-shaped,[2] the left-hand side is a non-increasing function of δ, which ensures that the inequality can be satisfied. We refer to any $\delta_n > 0$ satisfying inequality (13.17) as being *valid*, and we use $\delta_n^* > 0$ to denote the smallest positive radius for which inequality (13.17) holds. See the discussion following Theorem 13.5 for more details on the star-shaped property and the existence of valid radii δ_n.

Figure 13.2 illustrates the non-increasing property of the function $\delta \mapsto \mathcal{G}_n(\delta)/\delta$ for two different function classes: a first-order Sobolev space in Figure 13.2(a), and a Gaussian kernel space in Figure 13.2(b). Both of these function classes are convex, so that the star-shaped property holds for any f^*. Setting $\sigma = 1/2$ for concreteness, the critical radius δ_n^* can be determined by finding where this non-increasing function crosses the line with slope one, as illustrated. As will be clarified later, the Gaussian kernel class is much smaller than the first-order Sobolev space, so that its critical radius is correspondingly smaller. This ordering reflects the natural intuition that it should be easier to perform regression over a smaller function class.

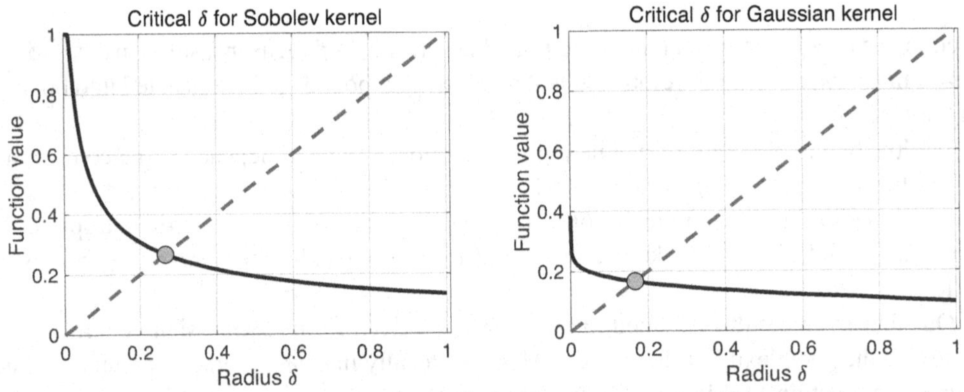

(a) (b)

Figure 13.2 Illustration of the critical radius for sample size $n = 100$ and two different function classes. (a) A first-order Sobolev space. (b) A Gaussian kernel class. In both cases, the function $\delta \mapsto \frac{\mathcal{G}_n(\delta;\mathcal{F})}{\delta}$, plotted as a solid line, is non-increasing, as guaranteed by Lemma 13.6. The critical radius δ_n^*, marked by a gray dot, is determined by finding its intersection with the line of slope $1/(2\sigma)$ with $\sigma = 1$, plotted as the dashed line. The set of all valid δ_n consists of the interval $[\delta_n^*, \infty)$.

Some intuition: Why should the inequality (13.17) be relevant to the analysis of the nonparametric least-squares estimator? A little calculation is helpful in gaining intuition. Since $\widehat{f}$ and f^* are optimal and feasible, respectively, for the constrained least-squares prob-

[2] A function class $\mathcal{H}$ is star-shaped if for any $h \in \mathcal{H}$ and $\alpha \in [0, 1]$, the rescaled function αh also belongs to $\mathcal{H}$.

lem (13.7), we are guaranteed that

$$\frac{1}{2n}\sum_{i=1}^{n}(y_i - \widehat{f}(x_i))^2 \leq \frac{1}{2n}\sum_{i=1}^{n}(y_i - f^*(x_i))^2.$$

Recalling that $y_i = f^*(x_i) + \sigma w_i$, some simple algebra leads to the equivalent expression

$$\frac{1}{2}\|\widehat{f} - f^*\|_n^2 \leq \frac{\sigma}{n}\sum_{i=1}^{n} w_i(\widehat{f}(x_i) - f^*(x_i)), \tag{13.18}$$

which we call the *basic inequality for nonparametric least squares.*

Now, by definition, the difference function $\widehat{f} - f^*$ belongs to $\mathscr{F}^*$, so that we can bound the right-hand side by taking the supremum over all functions $g \in \mathscr{F}^*$ with $\|g\|_n \leq \|\widehat{f} - f^*\|_n$. Reasoning heuristically, this observation suggests that the squared error $\delta^2 := \mathbb{E}[\|\widehat{f} - f^*\|_n^2]$ should satisfy a bound of the form

$$\frac{\delta^2}{2} \leq \sigma\, \mathcal{G}_n(\delta; \mathscr{F}^*) \quad \text{or equivalently} \quad \frac{\delta}{2\sigma} \leq \frac{\mathcal{G}_n(\delta; \mathscr{F}^*)}{\delta}. \tag{13.19}$$

By definition (13.17) of the critical radius δ_n^*, this inequality can only hold for values of $\delta \leq \delta_n^*$. In summary, this heuristic argument suggests a bound of the form $\mathbb{E}[\|\widehat{f} - f^*\|_n^2] \leq (\delta_n^*)^2$.

To be clear, the step from the basic inequality (13.18) to the bound (13.19) is *not* rigorously justified for various reasons, but the underlying intuition is correct. Let us now state a rigorous result, one that applies to the least-squares estimator (13.7) based on observations from the standard Gaussian noise model $y_i = f^*(x_i) + \sigma w_i$.

Theorem 13.5 *Suppose that the shifted function class $\mathscr{F}^*$ is star-shaped, and let δ_n be any positive solution to the critical inequality* (13.17). *Then for any $t \geq \delta_n$, the nonparametric least-squares estimate $\widehat{f}_n$ satisfies the bound*

$$\mathbb{P}\Big[\|\widehat{f}_n - f^*\|_n^2 \geq 16\, t\, \delta_n\Big] \leq e^{-\frac{nt\delta_n}{2\sigma^2}}. \tag{13.20}$$

Remarks: The bound (13.20) provides non-asymptotic control on the regression error $\|\widehat{f} - f^*\|_2^2$. By integrating this tail bound, it follows that the mean-squared error in the $L^2(\mathbb{P}_n)$-semi-norm is upper bounded as

$$\mathbb{E}[\|\widehat{f}_n - f^*\|_n^2] \leq c\left\{\delta_n^2 + \frac{\sigma^2}{n}\right\} \qquad \text{for some universal constant } c.$$

As shown in Exercise 13.5, for any function class $\mathscr{F}$ that contains the constant function $f \equiv 1$, we necessarily have $\delta_n^2 \geq \frac{2}{\pi}\frac{\sigma^2}{n}$, so that (disregarding constants) the δ_n^2 term is always the dominant one.

For concreteness, we have stated the result for the case of additive Gaussian noise ($v_i = \sigma w_i$). However, as the proof will clarify, all that is required is an upper tail bound on the random variable

$$Z_n(\delta) := \sup_{\substack{g\in\mathscr{F}^* \\ \|g\|_n\le\delta}} \Big|\frac{1}{n}\sum_{i=1}^{n} \frac{v_i}{\sigma} g(x_i)\Big|$$

in terms of its expectation. The expectation $\mathbb{E}[Z_n(\delta)]$ defines a more general form of (potentially non-Gaussian) noise complexity that then determines the critical radius.

The star-shaped condition on the shifted function class $\mathscr{F}^* = \mathscr{F} - f^*$ is needed at various parts of the proof, including in ensuring the existence of valid radii δ_n (see Lemma 13.6 to follow). In explicit terms, the function class $\mathscr{F}^*$ is star-shaped if for any $g \in \mathscr{F}$ and $\alpha \in [0, 1]$, the function αg also belongs to $\mathscr{F}^*$. Equivalently, we say that $\mathscr{F}$ is star-shaped around f^*. For instance, if $\mathscr{F}$ is convex, then as illustrated in Figure 13.3 it is necessarily star-shaped around any $f^* \in \mathscr{F}$. Conversely, if $\mathscr{F}$ is not convex, then there must exist choices $f^* \in \mathscr{F}$ such that $\mathscr{F}^*$ is not star-shaped. However, for a general non-convex set $\mathscr{F}$, it is still possible that $\mathscr{F}^*$ is star-shaped for *some* choices of f^*. See Figure 13.3 for an illustration of these possibilities, and Exercise 13.4 for further details.

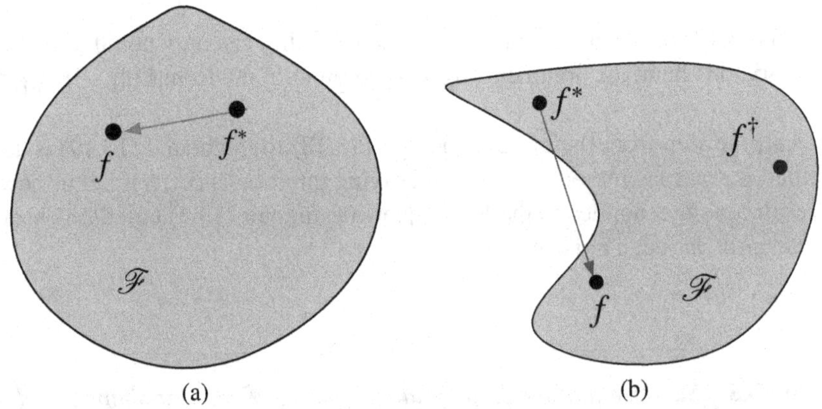

Figure 13.3 Illustration of star-shaped properties of sets. (a) The set $\mathscr{F}$ is convex, and hence is star-shaped around any of its points. The line between f^* and f is contained within $\mathscr{F}$, and the same is true for any line joining any pair of points in $\mathscr{F}$. (b) A set $\mathscr{F}$ that is not star-shaped around all its points. It fails to be star-shaped around the point f^*, since the line drawn to $f \in \mathscr{F}$ does not lie within the set. However, this set is star-shaped around the point $f^\dagger$.

If the star-shaped condition fails to hold, then Theorem 13.5 can instead by applied with δ_n defined in terms of the *star hull*

$$\mathrm{star}(\mathscr{F}^*; 0) := \{\alpha g \mid g \in \mathscr{F}^*, \alpha \in [0, 1]\} = \{\alpha(f - f^*) \mid f \in \mathscr{F}, \alpha \in [0, 1]\}. \tag{13.21}$$

Moreover, since the function f^* is not known to us, we often replace $\mathscr{F}^*$ with the larger class

$$\partial\mathscr{F} := \mathscr{F} - \mathscr{F} = \{f_1 - f_2 \mid f_1, f_2 \in \mathscr{F}\}, \tag{13.22}$$

or its star hull when necessary. We illustrate these considerations in the concrete examples to follow.

Let us now verify that the star-shaped condition ensures existence of the critical radius:

Lemma 13.6 *For any star-shaped function class $\mathscr{H}$, the function $\delta \mapsto \frac{\mathcal{G}_n(\delta;\mathscr{H})}{\delta}$ is non-increasing on the interval $(0, \infty)$. Consequently, for any constant $c > 0$, the inequality*

$$\frac{\mathcal{G}_n(\delta; \mathscr{H})}{\delta} \leq c\,\delta \tag{13.23}$$

has a smallest positive solution.

Proof So as to ease notation, we drop the dependence of $\mathcal{G}_n$ on the function class $\mathscr{H}$ throughout this proof. Given a pair $0 < \delta \leq t$, it suffices to show that $\frac{\delta}{t}\mathcal{G}_n(t) \leq \mathcal{G}_n(\delta)$. Given any function $h \in \mathscr{H}$ with $\|h\|_n \leq t$, we may define the rescaled function $\widetilde{h} = \frac{\delta}{t}h$, and write

$$\frac{1}{n}\Big\{\frac{\delta}{t}\sum_{i=1}^{n} w_i h(x_i)\Big\} = \frac{1}{n}\Big\{\sum_{i=1}^{n} w_i \widetilde{h}(x_i)\Big\}.$$

By construction, we have $\|\widetilde{h}\|_n \leq \delta$; moreover, since $\delta \leq t$, the star-shaped assumption guarantees that $\widetilde{h} \in \mathscr{H}$. Consequently, for any $\widetilde{h}$ formed in this way, the right-hand side is at most $\mathcal{G}_n(\delta)$ in expectation. Taking the supremum over the set $\mathscr{H} \cap \{\|h\|_n \leq t\}$ followed by expectations yields $\mathcal{G}_n(t)$ on the left-hand side. Combining the pieces yields the claim. □

In practice, determining the exact value of the critical radius δ_n^* may be difficult, so that we seek reasonable upper bounds on it. As shown in Exercise 13.5, we always have $\delta_n^* \leq \sigma$, but this is a very crude result. By bounding the local Gaussian complexity, we will obtain much finer results, as illustrated in the examples to follow.

13.2.1 Bounds via metric entropy

Note that the localized Gaussian complexity corresponds to the expected absolute maximum of a Gaussian process. As discussed in Chapter 5, Dudley's entropy integral can be used to upper bound such quantities.

In order to do so, let us begin by introducing some convenient notation. For any function class $\mathscr{H}$, we define $\mathbb{B}_n(\delta; \mathscr{H}) := \{h \in \text{star}(\mathscr{H}) \mid \|h\|_n \leq \delta\}$, and we let $N_n(t; \mathbb{B}_n(\delta; \mathscr{H}))$ denote the t-covering number of $\mathbb{B}_n(\delta; \mathscr{H})$ in the norm $\|\cdot\|_n$. With this notation, we have the following corollary:

Corollary 13.7 *Under the conditions of Theorem 13.5, any $\delta \in (0, \sigma]$ such that*

$$\frac{16}{\sqrt{n}} \int_{\frac{\delta^2}{4\sigma}}^{\delta} \sqrt{\log N_n(t;\ \mathbb{B}_n(\delta; \mathscr{F}^*))}\, dt \leq \frac{\delta^2}{4\sigma} \tag{13.24}$$

satisfies the critical inequality (13.17), *and hence can be used in the conclusion of Theorem 13.5.*

Proof For any $\delta \in (0, \sigma]$, we have $\frac{\delta^2}{4\sigma} < \delta$, so that we can construct a minimal $\frac{\delta^2}{4\sigma}$-covering of the set $\mathbb{B}_n(\delta; \mathscr{F}^*)$ in the $L^2(\mathbb{P}_n)$-norm, say $\{g^1, \ldots, g^M\}$. For any function $g \in \mathbb{B}_n(\delta; \mathscr{F}^*)$, there is an index $j \in [M]$ such that $\|g^j - g\|_n \leq \frac{\delta^2}{4\sigma}$. Consequently, we have

$$\begin{aligned}
\Big|\frac{1}{n}\sum_{i=1}^n w_i g(x_i)\Big| &\overset{\text{(i)}}{\leq} \Big|\frac{1}{n}\sum_{i=1}^n w_i g^j(x_i)\Big| + \Big|\frac{1}{n}\sum_{i=1}^n w_i (g(x_i) - g^j(x_i))\Big| \\
&\overset{\text{(ii)}}{\leq} \max_{j=1,\ldots,M} \Big|\frac{1}{n}\sum_{i=1}^n w_i g^j(x_i)\Big| + \sqrt{\frac{\sum_{i=1}^n w_i^2}{n}}\sqrt{\frac{\sum_{i=1}^n (g(x_i) - g^j(x_i))^2}{n}} \\
&\overset{\text{(iii)}}{\leq} \max_{j=1,\ldots,M} \Big|\frac{1}{n}\sum_{i=1}^n w_i g^j(x_i)\Big| + \sqrt{\frac{\sum_{i=1}^n w_i^2}{n}}\,\frac{\delta^2}{4\sigma},
\end{aligned}$$

where step (i) follows from the triangle inequality, step (ii) follows from the Cauchy–Schwarz inequality and step (iii) uses the covering property. Taking the supremum over $g \in \mathbb{B}_n(\delta; \mathscr{F}^*)$ on the left-hand side and then expectation over the noise, we obtain

$$\mathcal{G}_n(\delta) \leq \mathbb{E}_w\Big[\max_{j=1,\ldots,M} \Big|\frac{1}{n}\sum_{i=1}^n w_i g^j(x_i)\Big|\Big] + \frac{\delta^2}{4\sigma}, \tag{13.25}$$

where we have used the fact that $\mathbb{E}_w \sqrt{\frac{\sum_{i=1}^n w_i^2}{n}} \leq 1$.

It remains to upper bound the expected maximum over the M functions in the cover, and we do this by using the chaining method from Chapter 5. Define the family of Gaussian random variables $Z(g^j) := \frac{1}{\sqrt{n}} \sum_{i=1}^n w_i g^j(x_i)$ for $j = 1, \ldots, M$. Some calculation shows that they are zero-mean, and their associated semi-metric is given by

$$\rho_Z^2(g^j, g^k) := \text{var}(Z(g^j) - Z(g^k)) = \|g^j - g^k\|_n^2.$$

Since $\|g\|_n \leq \delta$ for all $g \in \mathbb{B}_n(\delta; \mathscr{F}^*)$, the coarsest resolution of the chaining can be set to δ, and we can terminate it at $\frac{\delta^2}{4\sigma}$, since any member of our finite set can be reconstructed exactly at this resolution. Working through the chaining argument, we find that

$$\begin{aligned}
\mathbb{E}_w\Big[\max_{j=1,\ldots,M} \Big|\frac{1}{n}\sum_{i=1}^n w_i g^j(x_i)\Big|\Big] &= \mathbb{E}_w\Big[\max_{j=1,\ldots,M} \frac{|Z(g^j)|}{\sqrt{n}}\Big] \\
&\leq \frac{16}{\sqrt{n}} \int_{\frac{\delta^2}{4\sigma}}^{\delta} \sqrt{\log N_n(t; \mathbb{B}_n(\delta; \mathscr{F}^*))}\, dt.
\end{aligned}$$

Combined with our earlier bound (13.25), this establishes the claim. □

Some examples are helpful in understanding the uses of Theorem 13.5 and Corollary 13.7, and we devote the following subsections to such illustrations.

13.2.2 Bounds for high-dimensional parametric problems

We begin with some bounds for parametric problems, allowing for a general dimension.

Example 13.8 (Bound for linear regression) As a warm-up, consider the standard linear regression model $y_i = \langle \theta^*, x_i \rangle + w_i$, where $\theta^* \in \mathbb{R}^d$. Although it is a parametric model, some insight can be gained by analyzing it using our general theory. The usual least-squares estimate corresponds to optimizing over the function class

$$\mathscr{F}_{\text{lin}} = \{f_\theta(\cdot) = \langle \theta, \cdot \rangle \mid \theta \in \mathbb{R}^d\}.$$

Let $\mathbf{X} \in \mathbb{R}^{n \times d}$ denote the design matrix, with $x_i \in \mathbb{R}^d$ as its ith row. In this example, we use our general theory to show that the least-squares estimate satisfies a bound of the form

$$\|f_{\widehat{\theta}} - f_{\theta^*}\|_n^2 = \frac{\|\mathbf{X}(\widehat{\theta} - \theta^*)\|_2^2}{n} \precsim \sigma^2 \frac{\operatorname{rank}(\mathbf{X})}{n} \tag{13.26}$$

with high probability. To be clear, in this special case, this bound (13.26) can be obtained by a direct linear algebraic argument, as we explore in Exercise 13.2. However, it is instructive to see how our general theory leads to concrete predictions in a special case.

We begin by observing that the shifted function class $\mathscr{F}^*_{\text{lin}}$ is equal to $\mathscr{F}_{\text{lin}}$ for any choice of f^*. Moreover, the set $\mathscr{F}_{\text{lin}}$ is convex and hence star-shaped around any point (see Exercise 13.4), so that Corollary 13.7 can be applied. The mapping $\theta \mapsto \|f_\theta\|_n = \frac{\|\mathbf{X}\theta\|_2}{\sqrt{n}}$ defines a norm on the subspace range($\mathbf{X}$), and the set $\mathbb{B}_n(\delta; \mathscr{F}_{\text{lin}})$ is isomorphic to a δ-ball within the space range($\mathbf{X}$). Since this range space has dimension given by rank($\mathbf{X}$), by a volume ratio argument (see Example 5.8), we have

$$\log N_n(t; \mathbb{B}_n(\delta; \mathscr{F}_{\text{lin}})) \leq r \log\Big(1 + \frac{2\delta}{t}\Big), \qquad \text{where } r := \operatorname{rank}(\mathbf{X}).$$

Using this upper bound in Corollary 13.7, we find that

$$\begin{aligned}
\frac{1}{\sqrt{n}} \int_0^\delta \sqrt{\log N_n(t; \mathbb{B}_n(\delta; \mathscr{F}_{\text{lin}})}\, dt &\leq \sqrt{\frac{r}{n}} \int_0^\delta \sqrt{\log(1 + \frac{2\delta}{t})}\, dt \\
&\overset{\text{(i)}}{=} \delta \sqrt{\frac{r}{n}} \int_0^1 \sqrt{\log(1 + \frac{2}{u})}\, du \\
&\overset{\text{(ii)}}{=} c\, \delta \sqrt{\frac{r}{n}},
\end{aligned}$$

where we have made the change of variables $u = t/\delta$ in step (i), and the final step (ii) follows since the integral is a constant. Putting together the pieces, an application of Corollary 13.7 yields the claim (13.26). In fact, the bound (13.26) is minimax-optimal up to constant factors, as we will show in Chapter 15. ♣

Let us now consider another high-dimensional parametric problem, namely that of sparse linear regression.

Example 13.9 (Bounds for linear regression over ℓ_q-"balls") Consider the case of sparse linear regression, where the d-variate regression vector θ is assumed to lie within the ℓ_q-ball of radius R_q—namely, the set

$$\mathbb{B}_q(R_q) := \Big\{\theta \in \mathbb{R}^d \mid \sum_{j=1}^{d} |\theta_j|^q \leq R_q\Big\}. \tag{13.27}$$

See Figure 7.1 for an illustration of these sets for different choices of $q \in (0, 1]$. Consider class of linear functions $f_\theta(x) = \langle \theta, x \rangle$ given by

$$\mathscr{F}_q(R_q) := \big\{f_\theta \mid \theta \in \mathbb{B}_q(R_q)\big\}. \tag{13.28}$$

We adopt the shorthand $\mathscr{F}_q$ when the radius R_q is clear from context.

In this example, we focus on the range $q \in (0, 1)$. Suppose that we solve the least-squares problem with ℓ_q regularization—that is, we compute the estimate

$$\widehat{\theta} \in \arg \min_{\theta \in \mathbb{B}_q(R_q)} \left\{\frac{1}{n} \sum_{i=1}^{n} \big(y_i - \langle x_i, \theta \rangle\big)^2\right\}. \tag{13.29}$$

Unlike the ℓ_1-constrained Lasso analyzed in Chapter 7, note that this is *not* a convex program. Indeed, for $q \in (0, 1)$, the function class $\mathscr{F}_q(R_q)$ is not convex, so that there exists $\theta^* \in \mathbb{B}_q(R_q)$ such that the shifted class $\mathscr{F}_q^* = \mathscr{F}_q - f_{\theta^*}$ is not star-shaped. Accordingly, we instead focus on bounding the metric entropy of the function class $\mathscr{F}_q(R_q) - \mathscr{F}_q(R_q) = 2\mathscr{F}_q(R_q)$. Note that for all $q \in (0, 1)$ and numbers $a, b \in \mathbb{R}$, we have $|a + b|^q \leq |a|^q + |b|^q$, which implies that $2\mathscr{F}_q(R_q)$ is contained with $\mathscr{F}_q(2R_q)$.

It is known that for $q \in (0, 1)$, and under mild conditions on the choice of t relative to the triple (n, d, R_q), the metric entropy of the ℓ_q-ball with respect to ℓ_2-norm is upper bounded by

$$\log N_{2,q}(t) \leq C_q \Big[R_q^{\frac{2}{2-q}} \big(\frac{1}{t}\big)^{\frac{2q}{2-q}} \log d\Big], \tag{13.30}$$

where C_q is a constant depending only on q.

Given our design vectors $\{x_i\}_{i=1}^n$, consider the $n \times d$ design matrix $\mathbf{X}$ with x_i^{T} as its ith row, and let $X_j \in \mathbb{R}^n$ denote its jth column. Our objective is to bound the metric entropy of the set of all vectors of the form

$$\frac{\mathbf{X}\theta}{\sqrt{n}} = \frac{1}{\sqrt{n}} \sum_{j=1}^{d} X_j \theta_j \tag{13.31}$$

as θ ranges over $\mathbb{B}_q(R_q)$, an object known as the *q-convex hull* of the renormalized column vectors $\{X_1, \ldots, X_d\}/\sqrt{n}$. Letting C denote a numerical constant such that $\max_{j=1,\ldots,d} \|X_j\|_2/\sqrt{n} \leq C$, it is known that the metric entropy of this q-convex hull has the same scaling as the original ℓ_q-ball. See the bibliographic section for further discussion of these facts about metric entropy.

Exploiting this fact and our earlier bound (13.30) on the metric entropy of the ℓ_q-ball, we find that

$$\frac{1}{\sqrt{n}} \int_{\frac{\delta^2}{4\sigma}}^{\delta} \sqrt{\log N_n\big(t;\ \mathbb{B}_n(\delta; \mathscr{F}_q(2R_q))\big)}\, dt \precsim R_q^{\frac{1}{2-q}} \sqrt{\frac{\log d}{n}} \int_0^{\delta} \Big(\frac{1}{t}\Big)^{\frac{q}{2-q}} dt$$
$$\precsim R_q^{\frac{1}{2-q}} \sqrt{\frac{\log d}{n}}\, \delta^{1-\frac{q}{2-q}},$$

a calculation valid for all $q \in (0, 1)$. Corollary 13.7 now implies that the critical condition (13.17) is satisfied as long as

$$R_q^{\frac{1}{2-q}} \sqrt{\frac{\sigma^2 \log d}{n}} \precsim \delta^{1+\frac{q}{2-q}} \quad \text{or equivalently} \quad R_q\Big(\frac{\sigma^2 \log d}{n}\Big)^{1-\frac{q}{2}} \precsim \delta^2.$$

Theorem 13.5 then implies that

$$\|f_{\widehat{\theta}} - f_{\theta^*}\|_n^2 = \frac{\|\mathbf{X}(\widehat{\theta} - \theta^*)\|_2^2}{n} \precsim R_q\Big(\frac{\sigma^2 \log d}{n}\Big)^{1-\frac{q}{2}},$$

with high probability. Although this result is a corollary of our general theorem, this rate is minimax-optimal up to constant factors, meaning that no estimator can achieve a faster rate. See the bibliographic section for further discussion and references of these connections. ♣

13.2.3 Bounds for nonparametric problems

Let us now illustrate the use of our techniques for some nonparametric problems.

Example 13.10 (Bounds for Lipschitz functions) Consider the class of functions

$$\mathscr{F}_{\mathrm{Lip}}(L) := \{f\colon [0, 1] \to \mathbb{R} \mid f(0) = 0,\ f \text{ is } L\text{-Lipschitz}\}. \tag{13.32}$$

Recall that f is L-Lipschitz means that $|f(x) - f(x')| \leq L|x - x'|$ for all $x, x' \in [0, 1]$. Let us analyze the prediction error associated with nonparametric least squares over this function class.

Noting the inclusion

$$\mathscr{F}_{\mathrm{Lip}}(L) - \mathscr{F}_{\mathrm{Lip}}(L) = 2\mathscr{F}_{\mathrm{Lip}}(L) \subseteq \mathscr{F}_{\mathrm{Lip}}(2L),$$

it suffices to upper bound the metric entropy of $\mathscr{F}_{\mathrm{Lip}}(2L)$. Based on our discussion from Example 5.10, the metric entropy of this class in the supremum norm scales as $\log N_\infty(\epsilon; \mathscr{F}_{\mathrm{Lip}}(2L)) \asymp (L/\epsilon)$. Consequently, we have

$$\frac{1}{\sqrt{n}} \int_0^{\delta} \sqrt{\log N_n(t;\ \mathbb{B}_n(\delta; \mathscr{F}_{\mathrm{Lip}}(2L)))}\, dt \precsim \int_0^{\delta} \sqrt{\log N_\infty(t;\ \mathscr{F}_{\mathrm{Lip}}(2L))}\, dt$$
$$\precsim \frac{1}{\sqrt{n}} \int_0^{\delta} (L/t)^{\frac{1}{2}}\, dt$$
$$\precsim \frac{1}{\sqrt{n}} \sqrt{L\delta},$$

where $\precsim$ denotes an inequality holding apart from constants not dependent on the triplet (δ, L, n). Thus, it suffices to choose $\delta_n > 0$ such that $\frac{\sqrt{L\delta_n}}{\sqrt{n}} \precsim \frac{\delta_n^2}{\sigma}$, or equivalently $\delta_n^2 \asymp \big(\frac{L\sigma^2}{n}\big)^{\frac{2}{3}}$.

Putting together the pieces, Corollary 13.7 implies that the error in the nonparametric least-squares estimate satisfies the bound

$$\|\widehat{f} - f^*\|_n^2 \precsim \Big(\frac{L\sigma^2}{n}\Big)^{2/3} \tag{13.33}$$

with probability at least $1 - c_1 e^{-c_2\left(\frac{n}{L\sigma^2}\right)^{1/3}}$. ♣

Example 13.11 (Bounds for convex regression) As a continuation of the previous example, let us consider the class of *convex* 1-Lipschitz functions, namely

$$\mathscr{F}_{\text{conv}}([0,1];1) := \{f\colon [0,1] \to \mathbb{R} \mid f(0) = 0 \text{ and } f \text{ is convex and 1-Lipschitz}\}.$$

As discussed in Example 13.4, computation of the nonparametric least-squares estimate over such convex classes can be reduced to a type of quadratic program. Here we consider the statistical rates that are achievable by such an estimator.

It is known that the metric entropy of $\mathscr{F}_{\text{conv}}$, when measured in the infinity norm, satisfies the upper bound

$$\log N(\epsilon; \mathscr{F}_{\text{conv}}, \|\cdot\|_\infty) \precsim \Big(\frac{1}{\epsilon}\Big)^{1/2} \tag{13.34}$$

for all $\epsilon > 0$ sufficiently small. (See the bibliographic section for details.) Thus, we can again use an entropy integral approach to derive upper bounds on the prediction error. In particular, calculations similar to those in the previous example show that the conditions of Corollary 13.7 hold for $\delta_n^2 \simeq (\frac{\sigma^2}{n})^{\frac{4}{5}}$, and so we are guaranteed that

$$\|\widehat{f} - f^*\|_n^2 \precsim \Big(\frac{\sigma^2}{n}\Big)^{4/5} \tag{13.35}$$

with probability at least $1 - c_1 e^{-c_2\left(\frac{n}{\sigma^2}\right)^{1/5}}$.

Note that our error bound (13.35) for convex Lipschitz functions is substantially faster than our earlier bound (13.33) for Lipschitz functions *without a convexity constraint*—in particular, the respective rates are $n^{-4/5}$ versus $n^{-2/3}$. In Chapter 15, we show that both of these rates are minimax-optimal, meaning that, apart from constant factors, they cannot be improved substantially. Thus, we see that the additional constraint of convexity is significant from a statistical point of view. In fact, as we explore in Exercise 13.8, in terms of their estimation error, convex Lipschitz functions behave exactly like the class of all twice-differentiable functions with bounded second derivative, so that the convexity constraint amounts to imposing an extra degree of smoothness. ♣

13.2.4 Proof of Theorem 13.5

We now turn to the proof of our previously stated theorem.

Establishing a basic inequality

Recall the basic inequality (13.18) established in our earlier discussion. In terms of the shorthand notation $\hat{\Delta} = \widehat{f} - f^*$, it can be written as

$$\frac{1}{2}\|\hat{\Delta}\|_n^2 \leq \frac{\sigma}{n}\sum_{i=1}^{n} w_i \hat{\Delta}(x_i). \tag{13.36}$$

By definition, the error function $\hat{\Delta} = \widehat{f} - f^*$ belongs to the shifted function class $\mathscr{F}^*$.

Controlling the right-hand side

In order to control the stochastic component on the right-hand side, we begin by stating an auxiliary lemma in a somewhat more general form, since it is useful for subsequent arguments. Let $\mathscr{H}$ be an arbitrary star-shaped function class, and let $\delta_n > 0$ satisfy the inequality $\frac{\mathcal{G}_n(\delta;\mathscr{H})}{\delta} \leq \frac{\delta}{2\sigma}$. For a given scalar $u \geq \delta_n$, define the event

$$\mathcal{A}(u) := \left\{\exists\, g \in \mathscr{H} \cap \{\|g\|_n \geq u\} \mid \left|\frac{\sigma}{n}\sum_{i=1}^{n} w_i g(x_i)\right| \geq 2\|g\|_n u\right\}. \tag{13.37}$$

The following lemma provides control on the probability of this event:

Lemma 13.12 *For all $u \geq \delta_n$, we have*

$$\mathbb{P}[\mathcal{A}(u)] \leq e^{-\frac{nu^2}{2\sigma^2}}. \tag{13.38}$$

Let us prove the main result by exploiting this lemma, in particular with the settings $\mathscr{H} = \mathscr{F}^*$ and $u = \sqrt{t\delta_n}$ for some $t \geq \delta_n$, so that we have

$$\mathbb{P}[\mathcal{A}^c(\sqrt{t\delta_n})] \geq 1 - e^{-\frac{nt\delta_n}{2\sigma^2}}.$$

If $\|\hat{\Delta}\|_n < \sqrt{t\,\delta_n}$, then the claim is immediate. Otherwise, we have $\hat{\Delta} \in \mathscr{F}^*$ and $\|\hat{\Delta}\|_n \geq \sqrt{t\,\delta_n}$, so that we may condition on $\mathcal{A}^c(\sqrt{t\delta_n})$ so as to obtain the bound

$$\left|\frac{\sigma}{n}\sum_{i=1}^{n} w_i \hat{\Delta}(x_i)\right| \leq 2\,\|\hat{\Delta}\|_n\,\sqrt{t\delta_n}.$$

Consequently, the basic inequality (13.36) implies that $\|\hat{\Delta}\|_n^2 \leq 4\|\hat{\Delta}\|_n\,\sqrt{t\delta_n}$, or equivalently that $\|\hat{\Delta}\|_n^2 \leq 16t\delta_n$, a bound that holds with probability at least $1 - e^{-\frac{nt\delta_n}{2\sigma^2}}$.

In order to complete the proof of Theorem 13.5, it remains to prove Lemma 13.12.

Proof of Lemma 13.12

Our first step is to reduce the problem to controlling a supremum over a subset of functions satisfying the upper bound $\|\widetilde{g}\|_n \leq u$. Suppose that there exists some $g \in \mathscr{H}$ with $\|g\|_n \geq u$

such that

$$\Big|\frac{\sigma}{n}\sum_{i=1}^{n} w_i g(x_i)\Big| \geq 2\|g\|_n u. \tag{13.39}$$

Defining the function $\widetilde{g} := \frac{u}{\|g\|_n} g$, we observe that $\|\widetilde{g}\|_n = u$. Since $g \in \mathscr{H}$ and $\frac{u}{\|g\|_n} \in (0, 1]$, the star-shaped assumption implies that $\widetilde{g} \in \mathscr{H}$. Consequently, we have shown that if there exists a function g satisfying the inequality (13.39), which occurs whenever the event $\mathcal{A}(u)$ is true, then there exists a function $\widetilde{g} \in \mathscr{H}$ with $\|\widetilde{g}\|_n = u$ such that

$$\Big|\frac{1}{n}\sum_{i=1}^{n} w_i \widetilde{g}(x_i)\Big| = \frac{u}{\|g\|_n}\Big|\frac{\sigma}{n}\sum_{i=1}^{n} w_i g(x_i)\Big| \geq 2u^2.$$

We thus conclude that

$$\mathbb{P}[\mathcal{A}(u)] \leq \mathbb{P}[Z_n(u) \geq 2u^2], \quad \text{where} \quad Z_n(u) := \sup_{\substack{\widetilde{g} \in \mathscr{H} \\ \|\widetilde{g}\|_n \leq u}} \Big|\frac{\sigma}{n}\sum_{i=1}^{n} w_i \widetilde{g}(x_i)\Big|. \tag{13.40}$$

Since the noise variables $w_i \sim \mathcal{N}(0, 1)$ are i.i.d., the variable $\frac{\sigma}{n}\sum_{i=1}^{n} w_i \widetilde{g}(x_i)$ is zero-mean and Gaussian for each fixed $\widetilde{g}$. Therefore, the variable $Z_n(u)$ corresponds to the supremum of a Gaussian process. If we view this supremum as a function of the standard Gaussian vector $(w_1, \ldots, w_n)$, then it can be verified that the associated Lipschitz constant is at most $\frac{\sigma u}{\sqrt{n}}$. Consequently, Theorem 2.26 guarantees the tail bound $\mathbb{P}[Z_n(u) \geq \mathbb{E}[Z_n(u)] + s] \leq e^{-\frac{ns^2}{2u^2\sigma^2}}$, valid for any $s > 0$. Setting $s = u^2$ yields

$$\mathbb{P}[Z_n(u) \geq \mathbb{E}[Z_n(u)] + u^2] \leq e^{-\frac{nu^2}{2\sigma^2}}. \tag{13.41}$$

Finally, by definition of $Z_n(u)$ and $\mathcal{G}_n(u)$, we have $\mathbb{E}[Z_n(u)] = \sigma\mathcal{G}_n(u)$. By Lemma 13.6, the function $v \mapsto \frac{\mathcal{G}_n(v)}{v}$ is non-decreasing, and since $u \geq \delta_n$ by assumption, we have

$$\sigma\frac{\mathcal{G}_n(u)}{u} \leq \sigma\frac{\mathcal{G}_n(\delta_n)}{\delta_n} \overset{\text{(i)}}{\leq} \delta_n/2 \leq \delta_n,$$

where step (i) uses the critical condition (13.17). Putting together the pieces, we have shown that $\mathbb{E}[Z_n(u)] \leq u\delta_n$. Combined with the tail bound (13.41), we obtain

$$\mathbb{P}[Z_n(u) \geq 2u^2] \overset{\text{(ii)}}{\leq} \mathbb{P}[Z_n(u) \geq u\delta_n + u^2] \leq e^{-\frac{nu^2}{2\sigma^2}},$$

where step (ii) uses the inequality $u^2 \geq u\delta_n$.

13.3 Oracle inequalities

In our analysis thus far, we have assumed that the regression function f^* belongs to the function class $\mathscr{F}$ over which the constrained least-squares estimator (13.7) is defined. In practice, this assumption might be violated, but it is nonetheless of interest to obtain bounds on the performance of the nonparametric least-squares estimator. In such settings, we expect its performance to involve both the *estimation error* that arises in Theorem 13.5, and some additional form of *approximation error*, arising from the fact that $f^* \notin \mathscr{F}$.

A natural way in which to measure approximation error is in terms of the best approximation to f^* using functions from $\mathscr{F}$. In the setting of interest in this chapter, the error in this best approximation is given by $\inf_{f \in \mathscr{F}} \|f - f^*\|_n^2$. Note that this error can only be achieved by an "oracle" that has direct access to the samples $\{f^*(x_i)\}_{i=1}^n$. For this reason, results that involve this form of approximation error are referred to as *oracle inequalities*. With this setup, we have the following generalization of Theorem 13.5. As before, we assume that we observe samples $\{(y_i, x_i)\}_{i=1}^n$ from the model $y_i = f^*(x_i) + \sigma w_i$, where $w_i \sim \mathcal{N}(0, 1)$. The reader should also recall the shorthand notation $\partial \mathscr{F} = \{f_1 - f_2 \mid f_1, f_2 \in \mathscr{F}\}$. We assume that this set is star-shaped; if not, it should be replaced by its star hull in the results to follow.

Theorem 13.13 *Let δ_n be any positive solution to the inequality*

$$\frac{\mathcal{G}_n(\delta; \partial\mathscr{F})}{\delta} \leq \frac{\delta}{2\sigma}. \tag{13.42a}$$

There are universal positive constants (c_0, c_1, c_2) such that for any $t \geq \delta_n$, the nonparametric least-squares estimate $\widehat{f}_n$ satisfies the bound

$$\|\widehat{f} - f^*\|_n^2 \leq \inf_{\gamma \in (0,1)} \left\{ \frac{1+\gamma}{1-\gamma} \|f - f^*\|_n^2 + \frac{c_0}{\gamma(1-\gamma)} \, t\delta_n \right\} \quad \text{for all } f \in \mathscr{F} \tag{13.42b}$$

with probability greater than $1 - c_1 e^{-c_2 \frac{nt\delta_n}{\sigma^2}}$.

Remarks: Note that the guarantee (13.42b) is actually a family of bounds, one for each $f \in \mathscr{F}$. When $f^* \in \mathscr{F}$, then we can set $f = f^*$, so that the bound (13.42b) reduces to asserting that $\|\widehat{f} - f^*\|_n^2 \precsim t\delta_n$ with high probability, where δ_n satisfies our previous critical inequality (13.17). Thus, up to constant factors, we recover Theorem 13.5 as a special case of Theorem 13.13. In the more general setting when $f^* \notin \mathscr{F}$, setting $t = \delta_n$ and taking the infimum over $f \in \mathscr{F}$ yields an upper bound of the form

$$\|\widehat{f} - f^*\|_n^2 \precsim \inf_{f \in \mathscr{F}} \|f - f^*\|_n^2 + \delta_n^2. \tag{13.43a}$$

Similarly, by integrating the tail bound, we are guaranteed that

$$\mathbb{E}\Big[\|\widehat{f} - f^*\|_n^2\Big] \precsim \inf_{f \in \mathscr{F}} \|f - f^*\|_n^2 + \delta_n^2 + \frac{\sigma^2}{n}. \tag{13.43b}$$

These forms of the bound clarify the terminology *oracle inequality*: more precisely, the quantity $\inf_{f \in \mathscr{F}} \|f - f^*\|_n^2$ is the error achievable only by an oracle that has access to uncorrupted samples of the function f^*. The bound (13.43a) guarantees that the least-squares estimate $\widehat{f}$ has prediction error that is at most a constant multiple of the oracle error, plus a term proportional to δ_n^2. The term $\inf_{f \in \mathscr{F}} \|f - f^*\|_n^2$ can be viewed a form of *approximation error* that decreases as the function class $\mathscr{F}$ grows, whereas the term δ_n^2 is the *estimation error* that increases as $\mathscr{F}$ becomes more complex. This upper bound can thus be used to choose $\mathscr{F}$ as a function of the sample size so as to obtain a desirable trade-off between the two types of error. We will see specific instantiations of this procedure in the examples to follow.

13.3.1 Some examples of oracle inequalities

Theorem 13.13 as well as oracle inequality (13.43a) are best understood by applying them to derive explicit rates for some particular examples.

Example 13.14 (Orthogonal series expansion) Let $(\phi_m)_{m=1}^{\infty}$ be an orthonormal basis of $L^2(\mathbb{P})$, and for each integer $T = 1, 2, \ldots$, consider the function class

$$\mathscr{F}_{\text{ortho}}(1;T) := \Big\{ f = \sum_{m=1}^{T} \beta_m \phi_m \mid \sum_{m=1}^{T} \beta_m^2 \leq 1 \Big\}, \tag{13.44}$$

and let $\widehat{f}$ be the constrained least-squares estimate over this class. Its computation is straightforward: it reduces to a version of linear ridge regression (see Exercise 13.10).

Let us consider the guarantees of Theorem 13.13 for $\widehat{f}$ as an estimate of some function f^* in the unit ball of $L^2(\mathbb{P})$. Since $(\phi_m)_{m=1}^{\infty}$ is an orthonormal basis of $L^2(\mathbb{P})$, we have $f^* = \sum_{m=1}^{\infty} \theta_m^* \phi_m$ for some coefficient sequence $(\theta_m^*)_{m=1}^{\infty}$. Moreover, by Parseval's theorem, we have the equivalence $\|f^*\|_2^2 = \sum_{m=1}^{\infty} (\theta_m^*)^2 \leq 1$, and a straightforward calculation yields that

$$\inf_{f \in \mathscr{F}_{\text{ortho}}(1;T)} \|f - f^*\|_2^2 = \sum_{m=T+1}^{\infty} (\theta_m^*)^2, \qquad \text{for each } T = 1, 2, \ldots.$$

Moreover, this infimum is achieved by the truncated function $\widetilde{f}_T = \sum_{m=1}^{T} \theta_m^* \phi_m$; see Exercise 13.10 for more details.

On the other hand, since the estimator over $\mathscr{F}_{\text{ortho}}(1;T)$ corresponds to a form of ridge regression in dimension T, the calculations from Example 13.8 imply that the critical equation (13.42a) is satisfied by $\delta_n^2 \simeq \sigma^2 \frac{T}{n}$. Setting $f = \widetilde{f}_T$ in the oracle inequality (13.43b) and then taking expectations over the covariates $\mathbf{X} = \{x_i\}_{i=1}^n$ yields that the least-squares estimate $\widehat{f}$ over $\mathscr{F}_{\text{ortho}}(1;T)$ satisfies the bound

$$\mathbb{E}_{\mathbf{X},w}\big[\|\widehat{f} - f^*\|_n^2\big] \precsim \sum_{m=T+1}^{\infty} (\theta_m^*)^2 + \sigma^2 \frac{T}{n}. \tag{13.45}$$

This oracle inequality allows us to choose the parameter T, which indexes the number of coefficients used in our basis expansion, so as to balance the approximation and estimation errors.

The optimal choice of T will depend on the rate at which the basis coefficients $(\theta_m^*)_{m=1}^{\infty}$ decay to zero. For example, suppose that they exhibit a polynomial decay, say $|\theta_m^*| \leq C m^{-\alpha}$ for some $\alpha > 1/2$. In Example 13.15 to follow, we provide a concrete instance of such polynomial decay using Fourier coefficients and α-times-differentiable functions. Figure 13.4(a) shows a plot of the upper bound (13.45) as a function of T, with one curve for each of the sample sizes $n \in \{100, 250, 500, 1000\}$. The solid markers within each curve show the point $T^* = T^*(n)$ at which the upper bound is minimized, thereby achieving the optimal trade-off between approximation and estimation errors. Note how this optimum grows with the sample size, since more samples allow us to reliably estimate a larger number of coefficients. ♣

As a more concrete instantiation of the previous example, let us consider the approximation of differentiable functions over the space $L^2[0, 1]$.

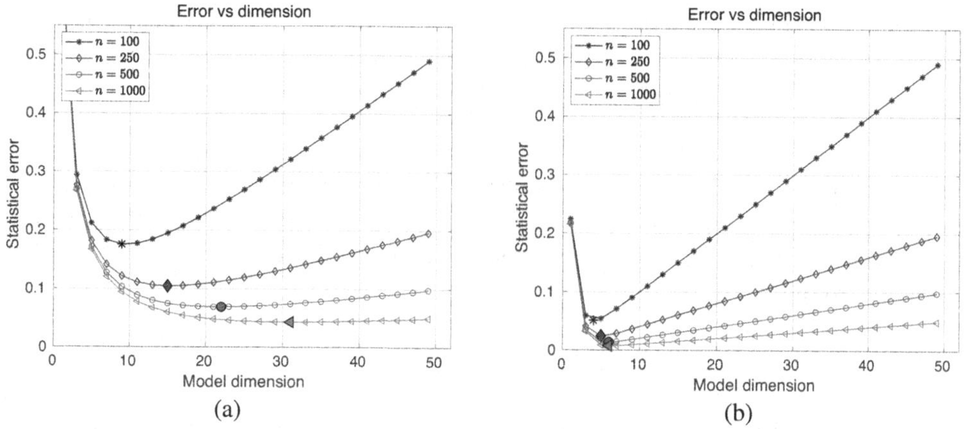

Figure 13.4 Plot of upper bound (13.45) versus the model dimension T, in all cases with noise variance $\sigma^2 = 1$. Each of the four curves corresponds to a different sample size $n \in \{100, 250, 500, 1000\}$. (a) Polynomial decaying coefficients $|\theta^*_m| \leq m^{-1}$. (b) Exponential decaying coefficients $|\theta^*_m| \leq e^{-m/2}$.

Example 13.15 (Fourier bases and differentiable functions) Define the constant function $\phi_0(x) = 1$ for all $x \in [0, 1]$, and the sinusoidal functions

$$\phi_m(x) := \sqrt{2}\cos(2m\pi x) \quad \text{and} \quad \widetilde{\phi}_m(x) := \sqrt{2}\sin(2m\pi x) \qquad \text{for } m = 1, 2, \ldots.$$

It can be verified that the collection $\{\phi_0\} \cup \{\phi_m\}_{m=1}^\infty \cup \{\widetilde{\phi}_m\}_{m=1}^\infty$ forms an orthonormal basis of $L^2[0, 1]$. Consequently, any function $f^* \in L^2[0, 1]$ has the series expansion

$$f^* = \theta^*_0 + \sum_{m=1}^{\infty} \{\theta^*_m \phi_m + \widetilde{\theta}^*_m \widetilde{\phi}_m\}.$$

For each $M = 1, 2, \ldots$, define the function class

$$\mathscr{G}(1; M) = \Big\{\beta_0 + \sum_{m=1}^{M}(\beta_m\phi_m + \widetilde{\beta}_m\widetilde{\phi}_m) \mid \beta_0^2 + \sum_{m=1}^{M}(\beta_m^2 + \widetilde{\beta}_m^2) \leq 1\Big\}. \tag{13.46}$$

Note that this is simply a re-indexing of a function class $\mathscr{F}_{\text{ortho}}(1; T)$ of the form (13.44) with $T = 2M + 1$.

Now suppose that for some integer $\alpha \geq 1$, the target function f^* is α-times differentiable, and suppose that $\int_0^1 [(f^*)^{(\alpha)}(x)]^2\, dx \leq R$ for some radius R. It can be verified that there is a constant c such that $(\beta^*_m)^2 + (\widetilde{\beta}^*_m)^2 \leq \frac{c}{m^{2\alpha}}$ for all $m \geq 1$, and, moreover, we can find a function $f \in \mathscr{G}(1; M)$ such that

$$\|f - f^*\|_2^2 \leq \frac{c' R}{M^{2\alpha}}. \tag{13.47}$$

See Exercise 13.11 for details on these properties.

Putting together the pieces, the bound (13.45) combined with the approximation-theoretic guarantee (13.47) implies that the least-squares estimate $\widehat{f}_M$ over $\mathscr{G}(1; M)$ satisfies the bound

$$\mathbb{E}_{X,w}\big[\|\widehat{f}_M - f^*\|_n^2\big] \precsim \frac{1}{M^{2\alpha}} + \sigma^2\,\frac{(2M+1)}{n}.$$

Thus, for a given sample size n and assuming knowledge of the smoothness α and noise variance σ^2, we can choose $M = M(n, \alpha, \sigma^2)$ so as to balance the approximation and estimation error terms. A little algebra shows that the optimal choice is $M \simeq (n/\sigma^2)^{\frac{1}{2\alpha+1}}$, which leads to the overall rate

$$\mathbb{E}_{X,w}\big[\|\widehat{f}_M - f^*\|_n^2\big] \precsim \Big(\frac{\sigma^2}{n}\Big)^{\frac{2\alpha}{2\alpha+1}}.$$

As will be clarified in Chapter 15, this $n^{-\frac{2\alpha}{2\alpha+1}}$ decay in mean-squared error is the best that can be expected for general univariate α-smooth functions. ♣

We now turn to the use of oracle inequalities in high-dimensional sparse linear regression.

Example 13.16 (Best sparse approximation) Consider the standard linear model $y_i = f_{\theta^*}(x_i) + \sigma w_i$, where $f_{\theta^*}(x) := \langle \theta^*, x\rangle$ is an unknown linear regression function, and $w_i \sim \mathcal{N}(0, 1)$ is an i.i.d. noise sequence. For some sparsity index $s \in \{1, 2, \ldots, d\}$, consider the class of all linear regression functions based on s-sparse vectors—namely, the class

$$\mathscr{F}_{\text{spar}}(s) := \{f_\theta \mid \theta \in \mathbb{R}^d,\ \|\theta\|_0 \leq s\},$$

where $\|\theta\|_0 = \sum_{j=1}^d \mathbb{I}[\theta_j \neq 0]$ counts the number of non-zero coefficients in the vector $\theta \in \mathbb{R}^d$.

Disregarding computational considerations, a natural estimator is given by

$$\widehat{\theta} \in \arg\min_{\theta \in \mathscr{F}_{\text{spar}}(s)} \|y - \mathbf{X}\theta\|_n^2, \tag{13.48}$$

corresponding to performing least squares over the set of all regression vectors with at most s non-zero coefficients. As a corollary of Theorem 13.13, we claim that the $L^2(\mathbb{P}_n)$-error of this estimator is upper bounded as

$$\|f_{\widehat{\theta}} - f_{\theta^*}\|_n^2 \precsim \inf_{\theta \in \mathscr{F}_{\text{spar}}(s)} \|f_\theta - f_{\theta^*}\|_n^2 + \underbrace{\sigma^2\frac{s\log(\frac{ed}{s})}{n}}_{\delta_n^2} \tag{13.49}$$

with high probability. Consequently, up to constant factors, its error is as good as the best s-sparse predictor plus the penalty term δ_n^2, arising from the estimation error. Note that the penalty term grows linearly with the sparsity s, but only logarithmically in the dimension d, so that it can be very small even when the dimension is exponentially larger than the sample size n. In essence, this result guarantees that we pay a relatively small price for not knowing in advance the best s-sized subset of coefficients to use.

In order to derive this result as a corollary of Theorem 13.13, we need to compute the local Gaussian complexity (13.42a) for our function class. Making note of the inclusion $\partial\mathscr{F}_{\text{spar}}(s) \subset \mathscr{F}_{\text{spar}}(2s)$, we have $\mathcal{G}_n(\delta; \partial\mathscr{F}_{\text{spar}}(s)) \leq \mathcal{G}_n(\delta; \mathscr{F}_{\text{spar}}(2s))$. Now let $S \subset \{1, 2, \ldots, d\}$ be an arbitrary $2s$-sized subset of indices, and let $\mathbf{X}_S \in \mathbb{R}^{n\times 2s}$ denote the submatrix with columns indexed by S. We can then write

$$\mathcal{G}_n(\delta; \mathscr{F}_{\text{spar}}(2s)) = \mathbb{E}_w[\max_{|S|=2s} Z_n(S)], \quad \text{where} \quad Z_n(S) := \sup_{\substack{\theta_S \in \mathbb{R}^{2s} \\ \|\mathbf{X}_S \theta_S\|_2/\sqrt{n} \leq \delta}} \left|\frac{w^{\mathrm{T}} \mathbf{X}_S \theta_S}{n}\right|.$$

Viewed as a function of the standard Gaussian vector w, the variable $Z_n(S)$ is Lipschitz with constant at most $\frac{\delta}{\sqrt{n}}$, from which Theorem 2.26 implies the tail bound

$$\mathbb{P}[Z_n(S) \geq \mathbb{E}[Z_n(S)] + t\delta] \leq e^{-\frac{nt^2}{2}} \qquad \text{for all } t > 0. \tag{13.50}$$

We now upper bound the expectation. Consider the singular value decomposition $\mathbf{X}_S = \mathbf{U}\mathbf{D}\mathbf{V}^{\mathrm{T}}$, where $\mathbf{U} \in \mathbb{R}^{n \times 2s}$ and $\mathbf{V} \in \mathbb{R}^{d \times 2s}$ are matrices of left and right singular vectors, respectively, and $\mathbf{D} \in \mathbb{R}^{2s \times 2s}$ is a diagonal matrix of the singular values. Noting that $\|\mathbf{X}_S \theta_S\|_2 = \|\mathbf{D}\mathbf{V}^{\mathrm{T}} \theta_S\|_2$, we arrive at the upper bound

$$\mathbb{E}[Z_n(S)] \leq \mathbb{E}\Big[\sup_{\substack{\beta \in \mathbb{R}^{2s} \\ \|\beta\|_2 \leq \delta}} \Big|\frac{1}{\sqrt{n}} \langle \mathbf{U}^{\mathrm{T}} w, \beta \rangle\Big|\Big] \leq \frac{\delta}{\sqrt{n}} \mathbb{E}\big[\|\mathbf{U}^{\mathrm{T}} w\|_2\big].$$

Since $w \sim \mathcal{N}(0, \mathbf{I}_n)$ and the matrix $\mathbf{U}$ has orthonormal columns, we have $\mathbf{U}^{\mathrm{T}} w \sim \mathcal{N}(0, \mathbf{I}_{2s})$, and therefore $\mathbb{E}\|\mathbf{U}^{\mathrm{T}} w\|_2 \leq \sqrt{2s}$. Combining this upper bound with the earlier tail bound (13.50), an application of the union bound yields

$$\mathbb{P}\left[\max_{|S|=2s} Z_n(S) \geq \delta\Big(\sqrt{\frac{2s}{n}} + t\Big)\right] \leq \binom{d}{2s} e^{-\frac{nt^2}{2}}, \qquad \text{valid for all } t \geq 0.$$

By integrating this tail bound, we find that

$$\frac{\mathbb{E}[\max_{|S|=2s} Z_n(S)]}{\delta} = \frac{\mathcal{G}_n(\delta)}{\delta} \precsim \sqrt{\frac{s}{n}} + \sqrt{\frac{\log \binom{d}{2s}}{n}} \precsim \sqrt{\frac{s \log(\frac{ed}{s})}{n}},$$

so that the critical inequality (13.17) is satisfied for $\delta_n^2 \simeq \sigma^2 \frac{s \log(ed/s)}{n}$, as claimed. ♣

13.3.2 Proof of Theorem 13.13

We now turn to the proof of our oracle inequality; it is a relatively straightforward extension of the proof of Theorem 13.5. Given an arbitrary $\widetilde{f} \in \mathscr{F}$, since it is feasible and $\widehat{f}$ is optimal, we have

$$\frac{1}{2n} \sum_{i=1}^{n} (y_i - \widehat{f}(x_i))^2 \leq \frac{1}{2n} \sum_{i=1}^{n} (y_i - \widetilde{f}(x_i))^2.$$

Using the relation $y_i = f^*(x_i) + \sigma w_i$, some algebra then yields

$$\frac{1}{2}\|\widehat{\Delta}\|_n^2 \leq \frac{1}{2}\|\widetilde{f} - f^*\|_n^2 + \Big|\frac{\sigma}{n} \sum_{i=1}^{n} w_i \widetilde{\Delta}(x_i)\Big|, \tag{13.51}$$

where we have defined $\widehat{\Delta} := \widehat{f} - f^*$ and $\widetilde{\Delta} = \widehat{f} - \widetilde{f}$.

It remains to analyze the term on the right-hand side involving $\widetilde{\Delta}$. We break our analysis into two cases.

Case 1: First suppose that $\|\widetilde{\Delta}\|_n \leq \sqrt{t\delta_n}$. We then have

$$\begin{aligned}\|\widehat{\Delta}\|_n^2 = \|\widehat{f} - f^*\|_n^2 = \|(\widetilde{f} - f^*) + \widetilde{\Delta}\|_n^2 \\ &\overset{\text{(i)}}{\leq} \left\{\|\widetilde{f} - f^*\|_n + \sqrt{t\delta_n}\right\}^2 \\ &\overset{\text{(ii)}}{\leq} (1 + 2\beta)\|\widetilde{f} - f^*\|_n^2 + (1 + \frac{2}{\beta})t\delta_n,\end{aligned}$$

where step (i) follows from the triangle inequality, and step (ii) is valid for any $\beta > 0$, using the Fenchel–Young inequality. Now setting $\beta = \frac{\gamma}{1-\gamma}$ for some $\gamma \in (0, 1)$, observe that $1 + 2\beta = \frac{1+\gamma}{1-\gamma}$, and $1 + \frac{2}{\beta} = \frac{2-\gamma}{\gamma} \leq \frac{2}{\gamma(1-\gamma)}$, so that the stated claim (13.42b) follows.

Case 2: Otherwise, we may assume that $\|\widetilde{\Delta}\|_n > \sqrt{t\delta_n}$. Noting that the function $\widetilde{\Delta}$ belongs to the difference class $\partial\mathscr{F} := \mathscr{F} - \mathscr{F}$, we then apply Lemma 13.12 with $u = \sqrt{t\delta_n}$ and $\mathscr{H} = \partial\mathscr{F}$. Doing so yields that

$$\mathbb{P}\Big[2\Big|\frac{\sigma}{n}\sum_{i=1}^{n} w_i\widetilde{\Delta}(x_i)\Big| \geq 4\sqrt{t\delta_n}\,\|\widetilde{\Delta}\|_n\Big] \leq e^{-\frac{nt\delta_n}{2\sigma^2}}.$$

Combining with the basic inequality (13.51), we find that, with probability at least $1-2e^{-\frac{nt\delta_n}{2\sigma^2}}$, the squared error is bounded as

$$\begin{aligned}\|\widehat{\Delta}\|_n^2 &\leq \|\widetilde{f} - f^*\|_n^2 + 4\sqrt{t\delta_n}\,\|\widetilde{\Delta}\|_n \\ &\leq \|\widetilde{f} - f^*\|_n^2 + 4\sqrt{t\delta_n}\left\{\|\widehat{\Delta}\|_n + \|\widetilde{f} - f^*\|_n\right\},\end{aligned}$$

where the second step follows from the triangle inequality. Applying the Fenchel–Young inequality with parameter $\beta > 0$, we find that

$$4\sqrt{t\delta_n}\,\|\widehat{\Delta}\|_n \leq 4\beta\|\widehat{\Delta}\|_n^2 + \frac{4}{\beta}t\delta_n$$

and

$$4\sqrt{t\delta_n}\|\widetilde{f} - f^*\|_n \leq 4\beta\|\widetilde{f} - f^*\|_n^2 + \frac{4}{\beta}t\delta_n.$$

Combining the pieces yields

$$\|\widehat{\Delta}\|_n^2 \leq (1 + 4\beta)\|\widetilde{f} - f^*\|_n^2 + 4\beta\|\widehat{\Delta}\|_n^2 + \frac{8}{\beta}t\delta_n.$$

For all $\beta \in (0, 1/4)$, rearranging yields the bound

$$\|\widehat{\Delta}\|_n^2 \leq \frac{1 + 4\beta}{1 - 4\beta}\|\widetilde{f} - f^*\|_n^2 + \frac{8}{\beta(1 - 4\beta)}t\delta_n.$$

Setting $\gamma = 4\beta$ yields the claim.

13.4 Regularized estimators

Up to this point, we have analyzed least-squares estimators based on imposing explicit constraints on the function class. From the computational point of view, it is often more convenient to implement estimators based on explicit penalization or regularization terms. As we will see, these estimators enjoy statistical behavior similar to their constrained analogs.

More formally, given a space $\mathscr{F}$ of real-valued functions with an associated semi-norm $\|\cdot\|_{\mathscr{F}}$, consider the family of regularized least-squares problems

$$\widehat{f} \in \arg\min_{f\in\mathscr{F}} \Big\{\frac{1}{2n}\sum_{i=1}^{n}(y_i - f(x_i))^2 + \lambda_n \|f\|_{\mathscr{F}}^2\Big\}, \tag{13.52}$$

where $\lambda_n \geq 0$ is a regularization weight to be chosen by the statistician. We state a general oracle-type result that does not require f^* to be a member of $\mathscr{F}$.

13.4.1 Oracle inequalities for regularized estimators

Recall the compact notation $\partial\mathscr{F} = \mathscr{F} - \mathscr{F}$. As in our previous theory, the statistical error involves a local Gaussian complexity over this class, which in this case takes the form

$$\mathcal{G}_n(\delta; \mathbb{B}_{\partial\mathscr{F}}(3)) := \mathbb{E}_w\Big[\sup_{\substack{g\in\partial\mathscr{F}\\ \|g\|_{\mathscr{F}}\leq 3,\ \|g\|_n\leq\delta}} \Big|\frac{1}{n}\sum_{i=1}^{n} w_i f(x_i)\Big|\Big], \tag{13.53}$$

where $w_i \sim \mathcal{N}(0,1)$ are i.i.d. variates. When the function class $\mathscr{F}$ and rescaled ball $\mathbb{B}_{\partial\mathscr{F}}(3) = \{g \in \partial\mathscr{F} \mid \|g\|_{\mathscr{F}} \leq 3\}$ are clear from the context, we adopt $\mathcal{G}_n(\delta)$ as a convenient shorthand. For a user-defined radius $R > 0$, we let $\delta_n > 0$ be any number satisfying the inequality

$$\frac{\mathcal{G}_n(\delta)}{\delta} \leq \frac{R}{2\sigma}\delta. \tag{13.54}$$

Theorem 13.17 *Given the previously described observation model and a convex function class $\mathscr{F}$, suppose that we solve the convex program (13.52) with some regularization parameter $\lambda_n \geq 2\delta_n^2$. Then there are universal positive constants (c_j, c_j') such that*

$$\|\widehat{f} - f^*\|_n^2 \leq c_0 \inf_{\|f\|_{\mathscr{F}}\leq R} \|f - f^*\|_n^2 + c_1 R^2\{\delta_n^2 + \lambda_n\} \tag{13.55a}$$

with probability greater than $1 - c_2 e^{-c_3\frac{nR^2\delta_n^2}{\sigma^2}}$. Similarly, we have

$$\mathbb{E}\|\widehat{f} - f^*\|_n^2 \leq c_0' \inf_{\|f\|_{\mathscr{F}}\leq R} \|f - f^*\|_n^2 + c_1'\, R^2\, \{\delta_n^2 + \lambda_n\}. \tag{13.55b}$$

We return to prove this claim in Section 13.4.4.

13.4.2 Consequences for kernel ridge regression

Recall from Chapter 12 our discussion of the kernel ridge regression estimate (12.28). There we showed that this KRR estimate has attractive computational properties, in that it only requires computing the empirical kernel matrix, and then solving a linear system (see Proposition 12.33). Here we turn to the complementary question of understanding its statistical behavior. Since it is a special case of the general estimator (13.52), Theorem 13.17 can be used to derive upper bounds on the prediction error. Interestingly, these bounds have a very intuitive interpretation, one involving the eigenvalues of the empirical kernel matrix.

From our earlier definition, the (rescaled) empirical kernel matrix $\mathbf{K} \in \mathbb{R}^{n \times n}$ is symmetric and positive semidefinite, with entries of the form $K_{ij} = \mathcal{K}(x_i, x_j)/n$. It is thus diagonalizable with non-negative eigenvalues, which we take to be ordered as $\hat{\mu}_1 \geq \hat{\mu}_2 \geq \cdots \geq \hat{\mu}_n \geq 0$. The following corollary of Theorem 13.17 provides bounds on the performance of the kernel ridge regression estimate in terms of these eigenvalues:

Corollary 13.18 *For the KRR estimate* (12.28), *the bounds of Theorem 13.17 hold for any $\delta_n > 0$ satisfying the inequality*

$$\sqrt{\frac{2}{n}} \sqrt{\sum_{j=1}^{n} \min\{\delta^2, \hat{\mu}_j\}} \leq \frac{R}{4\sigma} \delta^2. \tag{13.56}$$

We provide the proof in Section 13.4.3. Before doing so, let us examine the implications of Corollary 13.18 for some specific choices of kernels.

Example 13.19 (Rates for polynomial regression) Given some integer $m \geq 2$, consider the kernel function $\mathcal{K}(x, z) = (1 + x z)^{m-1}$. The associated RKHS corresponds to the space of all polynomials of degree at most $m - 1$, which is a vector space with dimension m. Consequently, the empirical kernel matrix $\mathbf{K} \in \mathbb{R}^{n \times n}$ can have rank at most $\min\{n, m\}$. Therefore, for any sample size n larger than m, we have

$$\frac{1}{\sqrt{n}} \sqrt{\sum_{j=1}^{n} \min\{\delta^2, \hat{\mu}_j\}} \leq \frac{1}{\sqrt{n}} \sqrt{\sum_{j=1}^{m} \min\{\delta^2, \hat{\mu}_j\}} \leq \delta \sqrt{\frac{m}{n}}.$$

Consequently, the critical inequality (13.56) is satisfied for all $\delta \gtrsim \frac{\sigma}{R} \sqrt{\frac{m}{n}}$, so that the KRR estimate satisfies the bound

$$\|\widehat{f} - f^*\|_n^2 \precsim \inf_{\|f\|_{\mathbb{H}} \leq R} \|f - f^*\|_n^2 + \sigma^2 \frac{m}{n},$$

both in high probability and in expectation. This bound is intuitively reasonable: since the space of $m - 1$ polynomials has a total of m free parameters, we expect that the ratio m/n should converge to zero in order for consistent estimation to be possible. More generally, this same bound with $m = r$ holds for any kernel function that has some finite rank $r \geq 1$. ♣

We now turn to a kernel function with an infinite number of eigenvalues:

Example 13.20 (First-order Sobolev space) Previously, we introduced the kernel function $\mathcal{K}(x, z) = \min\{x, z\}$ defined on the unit square $[0, 1] \times [0, 1]$. As discussed in Example 12.16, the associated RKHS corresponds to a first-order Sobolev space

$$\mathbb{H}^1[0,1] := \Big\{f\colon [0,1] \to \mathbb{R} \mid f(0) = 0, \text{ and } f \text{ is abs. cts. with } f' \in L^2[0,1]\Big\}.$$

As shown in Example 12.23, the kernel integral operator associated with this space has the eigendecomposition

$$\phi_j(x) = \sin(x/\sqrt{\mu_j}), \quad \mu_j = \Big(\frac{2}{(2j-1)\pi}\Big)^2 \qquad \text{for } j = 1, 2, \ldots,$$

so that the eigenvalues drop off at the rate j^{-2}. As the sample size increases, the eigenvalues of the empirical kernel matrix $\mathbf{K}$ approach those of the population kernel operator. For the purposes of calculation, Figure 13.5(a) suggests the heuristic of assuming that $\hat{\mu}_j \leq \frac{c}{j^2}$ for some universal constant c. Our later analysis in Chapter 14 will provide a rigorous way of making such an argument.[3]

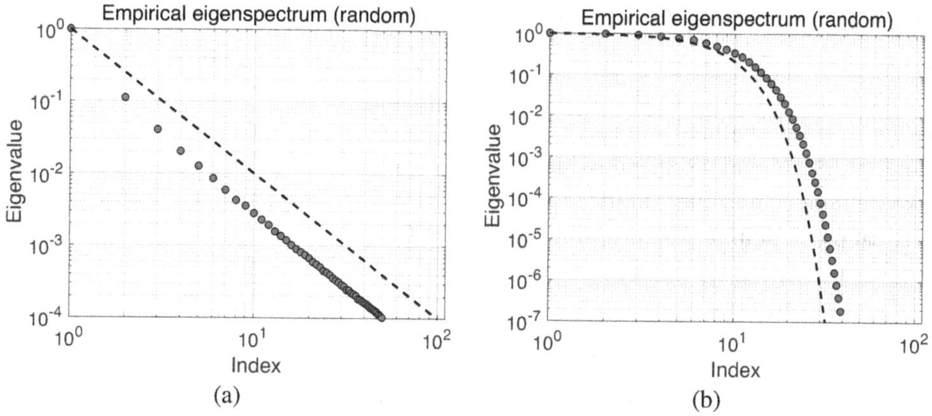

Figure 13.5 Log–log behavior of the eigenspectrum of the empirical kernel matrix based on $n = 2000$ samples drawn i.i.d. from the uniform distribution over the interval $\mathcal{X}$ for two different kernel functions. The plotted circles correspond to empirical eigenvalues, whereas the dashed line shows the theoretically predicted drop-off of the population operator. (a) The first-order Sobolev kernel $\mathcal{K}(x, z) = \min\{x, z\}$ on the interval $\mathcal{X} = [0, 1]$. (b) The Gaussian kernel $\mathcal{K}(x, z) = \exp(-\frac{(x-z)^2}{2\sigma^2})$ with $\sigma = 0.5$ on the interval $\mathcal{X} = [-1, 1]$.

Under our heuristic assumption, we have

$$\frac{1}{\sqrt{n}}\sqrt{\sum_{j=1}^{n} \min\{\delta^2, \hat{\mu}_j\}} \leq \frac{1}{\sqrt{n}}\sqrt{\sum_{j=1}^{n} \min\{\delta^2, c\, j^{-2}\}} \leq \frac{1}{\sqrt{n}}\sqrt{k\delta^2 + c\sum_{j=k+1}^{n} j^{-2}},$$

where k is the smallest positive integer such that $ck^{-2} \leq \delta^2$. Upper bounding the final sum

[3] In particular, Proposition 14.25 shows that the critical radii computed using the population and empirical kernel eigenvalues are equivalent up to constant factors.

by an integral, we have $c \sum_{j=k+1}^{n} j^{-2} \leq c \int_{k+1}^{\infty} t^{-2}\, dt \leq ck^{-1} \leq k\delta^2$, and hence

$$\frac{1}{\sqrt{n}} \sqrt{\sum_{j=1}^{n} \min\{\delta^2, \hat{\mu}_j\}} \leq c' \sqrt{\frac{k}{n}}\, \delta \leq c'' \sqrt{\frac{\delta}{n}}.$$

Consequently, the critical inequality (13.56) is satisfied by $\delta_n^{3/2} \simeq \frac{\sigma}{R\sqrt{n}}$, or equivalently $\delta_n^2 \simeq (\frac{\sigma^2}{R^2} \frac{1}{n})^{2/3}$. Putting together the pieces, Corollary 13.18 implies that the KRR estimate will satisfy the upper bound

$$\|\widehat{f} - f^*\|_n^2 \precsim \inf_{\|f\|_{\mathbb{H}} \leq R} \|f - f^*\|_n^2 + R^2\delta_n^2 \simeq \inf_{\|f\|_{\mathbb{H}} \leq R} \|f - f^*\|_n^2 + R^{2/3} \left(\frac{\sigma^2}{n}\right)^{2/3},$$

both with high probability and in expectation. As will be seen later in Chapter 15, this rate is minimax-optimal for the first-order Sobolev space. ♣

Example 13.21 (Gaussian kernel) Now let us consider the same issues for the Gaussian kernel $\mathcal{K}(x, z) = e^{-\frac{(x-z)^2}{2\sigma^2}}$ on the square $[-1, 1] \times [-1, 1]$. As discussed in Example 12.25, the eigenvalues of the associated kernel operator scale as $\mu_j \simeq e^{-cj \log j}$ as $j \to +\infty$. Accordingly, let us adopt the heuristic that the empirical eigenvalues satisfy a bound of the form $\hat{\mu}_j \leq c_0 e^{-c_1 j \log j}$. Figure 13.5(b) provides empirical justification of this scaling for the Gaussian kernel: notice how the empirical plots on the log–log scale agree qualitatively with the theoretical prediction. Again, Proposition 14.25 in Chapter 14 allows us to make a rigorous argument that reaches the conclusion sketched here.

Under our heuristic assumption, for a given $\delta > 0$, we have

$$\begin{aligned} \frac{1}{\sqrt{n}} \sqrt{\sum_{j=1}^{n} \min\{\delta^2, \hat{\mu}_j\}} &\leq \frac{1}{\sqrt{n}} \sqrt{\sum_{j=1}^{n} \min\{\delta^2, c_0\, e^{-c_1 j \log j}\}} \\ &\leq \frac{1}{\sqrt{n}} \sqrt{k\delta^2 + c_0 \sum_{j=k+1}^{n} e^{-c_1 j \log j}}, \end{aligned}$$

where k is the smallest positive integer such that $c_0 e^{-c_1 k \log k} \leq \delta^2$.

Some algebra shows that the critical inequality will be satisfied by $\delta_n^2 \simeq \frac{\sigma^2}{R^2} \frac{\log(\frac{Rn}{\sigma})}{n}$, so that nonparametric regression over the Gaussian kernel class satisfies the bound

$$\|\widehat{f} - f^*\|_n^2 \precsim \inf_{\|f\|_{\mathbb{H}} \leq R} \|f - f^*\|_n^2 + R^2\delta_n^2 = \inf_{\|f\|_{\mathbb{H}} \leq R} \|f - f^*\|_n^2 + c\,\sigma^2 \frac{\log(\frac{Rn}{\sigma})}{n},$$

for some universal constant c. The estimation error component of this upper bound is very fast—within a logarithmic factor of the n^{-1} parametric rate—thereby revealing that the Gaussian kernel class is much smaller than the first-order Sobolev space from Example 13.20. However, the trade-off is that the approximation error decays very slowly as a function of the radius R. See the bibliographic section for further discussion of this important trade-off. ♣

13.4.3 Proof of Corollary 13.18

The proof of this corollary is based on a bound on the local Gaussian complexity (13.53) of the unit ball of an RKHS. Since it is of independent interest, let us state it as a separate result:

Lemma 13.22 *Consider an RKHS with kernel function $\mathcal{K}$. For a given set of design points $\{x_i\}_{i=1}^n$, let $\hat{\mu}_1 \geq \hat{\mu}_2 \geq \cdots \geq \hat{\mu}_n \geq 0$ be the eigenvalues of the normalized kernel matrix $\mathbf{K}$ with entries $K_{ij} = \mathcal{K}(x_i, x_j)/n$. Then for all $\delta > 0$, we have*

$$\mathbb{E}\left[\sup_{\substack{\|f\|_{\mathbb{H}} \leq 1 \\ \|f\|_n \leq \delta}} \left|\frac{1}{n}\sum_{i=1}^n w_i f(x_i)\right|\right] \leq \sqrt{\frac{2}{n}}\sqrt{\sum_{j=1}^n \min\{\delta^2, \hat{\mu}_j\}}, \tag{13.57}$$

where $w_i \sim \mathcal{N}(0, 1)$ are i.i.d. Gaussian variates.

Proof It suffices to restrict our attention to functions of the form

$$g(\cdot) = \frac{1}{\sqrt{n}}\sum_{i=1}^n \alpha_i \mathcal{K}(\cdot, x_i), \tag{13.58}$$

some vector of coefficients $\alpha \in \mathbb{R}^n$. Indeed, as argued in our proof of Proposition 12.33, any function f in the Hilbert space can be written in the form $f = g + g_\perp$, where $g_\perp$ is a function orthogonal to all functions of the form (13.58). Thus, we must have $g_\perp(x_i) = \langle g_\perp, \mathcal{K}(\cdot, x_i)\rangle_{\mathbb{H}} = 0$, so that neither the objective nor the constraint $\|f\|_n \leq \delta$ have any dependence on $g_\perp$. Lastly, by the Pythagorean theorem, we have $\|f\|_{\mathbb{H}}^2 = \|g\|_{\mathbb{H}}^2 + \|g_\perp\|_{\mathbb{H}}^2$, so that we may assume without loss of generality that $g_\perp = 0$.

In terms of the coefficient vector $\alpha \in \mathbb{R}^n$ and kernel matrix $\mathbf{K}$, the constraint $\|g\|_n \leq \delta$ is equivalent to $\|\mathbf{K}\alpha\|_2 \leq \delta$, whereas the inequality $\|g\|_{\mathbb{H}}^2 \leq 1$ corresponds to $\|g\|_{\mathbb{H}}^2 = \alpha^{\mathrm{T}}\mathbf{K}\alpha \leq 1$. Thus, we can write the local Gaussian complexity as an optimization problem in the vector $\alpha \in \mathbb{R}^n$ with a linear cost function and quadratic constraints—namely,

$$\mathcal{G}_n(\delta) = \frac{1}{\sqrt{n}}\mathbb{E}_w\Big[\sup_{\substack{\alpha^{\mathrm{T}}\mathbf{K}\alpha \leq 1 \\ \alpha\mathbf{K}^2\alpha \leq \delta^2}} \big|w^{\mathrm{T}}\mathbf{K}\alpha\big|\Big].$$

Since the kernel matrix $\mathbf{K}$ is symmetric and positive semidefinite, it has an eigendecomposition[4] of the form $\mathbf{K} = \mathbf{U}^{\mathrm{T}}\mathbf{\Lambda}\mathbf{U}$, where $\mathbf{U}$ is orthogonal and $\mathbf{\Lambda}$ is diagonal with entries $\hat{\mu}_1 \geq \hat{\mu}_2 \geq \cdots \geq \hat{\mu}_n > 0$. If we then define the transformed vector $\beta = \mathbf{K}\alpha$, we find (following some algebra) that the complexity can be written as

$$\mathcal{G}_n(\delta) = \frac{1}{\sqrt{n}}\mathbb{E}_w[\sup_{\beta \in \mathcal{D}} |w^{\mathrm{T}}\beta|], \quad \text{where} \quad \mathcal{D} := \{\beta \in \mathbb{R}^n \mid \|\beta\|_2^2 \leq \delta^2, \sum_{j=1}^n \frac{\beta_j^2}{\hat{\mu}_j} \leq 1\}$$

[4] In this argument, so as to avoid potential division by zero, we assume that $\mathbf{K}$ has strictly positive eigenvalues; otherwise, we can simply repeat the argument given here while restricting the relevant summations to positive eigenvalues.

is the intersection of two ellipses. Now define the ellipse

$$\mathcal{E} := \Big\{\beta \in \mathbb{R}^n \mid \sum_{j=1}^n \eta_j \beta_j^2 \leq 2\Big\}, \qquad \text{where } \eta_j = \max\{\delta^{-2}, \hat{\mu}_j^{-1}\}.$$

We claim that $\mathcal{D} \subset \mathcal{E}$; indeed, for any $\beta \in \mathcal{D}$, we have

$$\sum_{j=1}^n \max\{\delta^{-2}, \hat{\mu}_j^{-1}\}\beta_j^2 \leq \sum_{j=1}^n \frac{\beta_j^2}{\delta^2} + \sum_{j=1}^n \frac{\beta_j^2}{\hat{\mu}_j} \leq 2.$$

Applying Hölder's inequality with the norm induced by $\mathcal{E}$ and its dual, we find that

$$\mathcal{G}_n(\delta) \leq \frac{1}{\sqrt{n}} \mathbb{E}\big[\sup_{\beta \in \mathcal{E}} |\langle w, \beta\rangle|\big] \leq \sqrt{\frac{2}{n}}\, \mathbb{E}\sqrt{\sum_{j=1}^n \frac{w_j^2}{\eta_j}}.$$

Jensen's inequality allows us to move the expectation inside the square root, so that

$$\mathcal{G}_n(\delta) \leq \sqrt{\frac{2}{n}} \sqrt{\sum_{j=1}^n \frac{\mathbb{E}[w_j^2]}{\eta_j}} = \sqrt{\frac{2}{n}} \sqrt{\sum_{j=1}^n \frac{1}{\eta_j}},$$

and substituting $(\eta_j)^{-1} = (\max\{\delta^{-2}, \hat{\mu}_j^{-1}\})^{-1} = \min\{\delta^2, \hat{\mu}_j\}$ yields the claim. □

13.4.4 Proof of Theorem 13.17

Finally, we turn to the proof of our general theorem on regularized M-estimators. By rescaling the observation model by R, we can analyze an equivalent model with noise variance $(\frac{\sigma}{R})^2$, and with the rescaled approximation error $\inf_{\|f\|_{\mathcal{F}} \leq 1} \|f - f^*\|_n^2$. Our final mean-squared error then should be multiplied by R^2 so as to obtain a result for the original problem.

In order to keep the notation streamlined, we introduce the shorthand $\tilde{\sigma} = \sigma/R$. Let $\widetilde{f}$ be any element of $\mathcal{F}$ such that $\|\widetilde{f}\|_{\mathcal{F}} \leq 1$. At the end of the proof, we optimize this choice. Since $\widehat{f}$ and $\widetilde{f}$ are optimal and feasible (respectively) for the program (13.52), we have

$$\frac{1}{2}\sum_{i=1}^n (y_i - \widehat{f}(x_i))^2 + \lambda_n \|\widehat{f}\|_{\mathcal{F}}^2 \leq \frac{1}{2}\sum_{i=1}^n (y_i - \widetilde{f}(x_i))^2 + \lambda_n \|\widetilde{f}\|_{\mathcal{F}}^2.$$

Defining the errors $\widehat{\Delta} = \widehat{f} - f^*$ and $\widetilde{\Delta} = \widehat{f} - \widetilde{f}$ and recalling that $y_i = f^*(x_i) + \tilde{\sigma} w_i$, performing some algebra yields the *modified basic inequality*

$$\frac{1}{2}\|\widehat{\Delta}\|_n^2 \leq \frac{1}{2}\|\widetilde{f} - f^*\|_n^2 + \frac{\tilde{\sigma}}{n}\Big|\sum_{i=1}^n w_i \widetilde{\Delta}(x_i)\Big| + \lambda_n\{\|\widetilde{f}\|_{\mathcal{F}}^2 - \|\widehat{f}\|_{\mathcal{F}}^2\}, \tag{13.59}$$

where $w_i \sim \mathcal{N}(0, 1)$ are i.i.d. Gaussian variables.

Since $\|\widetilde{f}\|_{\mathcal{F}} \leq 1$ by assumption, we certainly have the possibly weaker bound

$$\frac{1}{2}\|\widehat{\Delta}\|_n^2 \leq \frac{1}{2}\|\widetilde{f} - f^*\|_n^2 + \frac{\tilde{\sigma}}{n}\Big|\sum_{i=1}^n w_i \widetilde{\Delta}(x_i)\Big| + \lambda_n. \tag{13.60}$$

Consequently, if $\|\widetilde{\Delta}\|_n \leq \sqrt{t\delta_n}$, we can then follow the same argument as in the proof of Theorem 13.13, thereby establishing the bound (along with the extra term λ_n from our modified basic inequality).

Otherwise, we may assume that $\|\widetilde{\Delta}\|_n > \sqrt{t\delta_n}$, and we do so throughout the remainder of the proof. We now split the argument into two cases.

Case 1: First, suppose that $\|\widehat{f}\|_{\mathscr{F}} \leq 2$. The bound $\|\widetilde{f}\|_{\mathscr{F}} \leq 1$ together with the inequality $\|\widehat{f}\|_{\mathscr{F}} \leq 2$ implies that $\|\widetilde{\Delta}\|_{\mathscr{F}} \leq 3$. Consequently, by applying Lemma 13.12 over the set of functions $\{g \in \partial\mathscr{F} \mid \|g\|_{\mathscr{F}} \leq 3\}$, we conclude that

$$\frac{\widetilde{\sigma}}{n}\Big|\sum_{i=1}^{n} w_i \widetilde{\Delta}(x_i)\Big| \leq c_0 \sqrt{t\delta_n}\|\widetilde{\Delta}\|_n \qquad \text{with probability at least } 1 - e^{-\frac{t^2}{2\sigma^2}}.$$

By the triangle inequality, we have

$$\begin{aligned} 2\sqrt{t\delta_n}\|\widetilde{\Delta}\|_n &\leq 2\sqrt{t\delta_n}\|\widehat{\Delta}\|_n + 2\sqrt{t\delta_n}\|\widetilde{f} - f^*\|_n \\ &\leq 2\sqrt{t\delta_n}\|\widehat{\Delta}\|_n + 2t\delta_n + \frac{\|\widetilde{f} - f^*\|_n^2}{2}, \end{aligned} \tag{13.61}$$

where the second step uses the Fenchel–Young inequality. Substituting these upper bounds into the basic inequality (13.60), we find that

$$\tfrac{1}{2}\|\widehat{\Delta}\|_n^2 \leq \tfrac{1}{2}(1 + c_0)\|\widetilde{f} - f^*\|_n^2 + 2c_0 t\delta_n + 2c_0\sqrt{t\delta_n}\|\widehat{\Delta}\|_n + \lambda_n,$$

so that the claim follows by the quadratic formula, modulo different values of the numerical constants.

Case 2: Otherwise, we may assume that $\|\widehat{f}\|_{\mathscr{F}} > 2 > 1 \geq \|\widetilde{f}\|_{\mathscr{F}}$. In this case, we have

$$\|\widetilde{f}\|_{\mathscr{F}}^2 - \|\widehat{f}\|_{\mathscr{F}}^2 = \underbrace{\{\|\widetilde{f}\|_{\mathscr{F}} + \|\widehat{f}\|_{\mathscr{F}}\}}_{>1}\underbrace{\{\|\widetilde{f}\|_{\mathscr{F}} - \|\widehat{f}\|_{\mathscr{F}}\}}_{<0} \leq \underbrace{\{\|\widetilde{f}\|_{\mathscr{F}} - \|\widehat{f}\|_{\mathscr{F}}\}}_{<0}.$$

Writing $\widehat{f} = \widetilde{f} + \widetilde{\Delta}$ and noting that $\|\widehat{f}\|_{\mathscr{F}} \geq \|\widetilde{\Delta}\|_{\mathscr{F}} - \|\widetilde{f}\|_{\mathscr{F}}$ by the triangle inequality, we obtain

$$\begin{aligned} \lambda_n\{\|\widetilde{f}\|_{\mathscr{F}}^2 - \|\widehat{f}\|_{\mathscr{F}}^2\} &\leq \lambda_n\{\|\widetilde{f}\|_{\mathscr{F}} - \|\widehat{f}\|_{\mathscr{F}}\} \\ &\leq \lambda_n\{2\|\widetilde{f}\|_{\mathscr{F}} - \|\widetilde{\Delta}\|_{\mathscr{F}}\} \\ &\leq \lambda_n\{2 - \|\widetilde{\Delta}\|_{\mathscr{F}}\}, \end{aligned}$$

where we again use the bound $\|\widetilde{f}\|_{\mathscr{F}} \leq 1$ in the final step.

Substituting this upper bound into our modified basic inequality (13.59) yields the upper bound

$$\frac{1}{2}\|\widehat{\Delta}\|_n^2 \leq \frac{1}{2}\|\widetilde{f} - f^*\|_n^2 + \Big|\frac{\widetilde{\sigma}}{n}\sum_{i=1}^{n} w_i \widetilde{\Delta}(x_i)\Big| + 2\lambda_n - \lambda_n\|\widetilde{\Delta}\|_{\mathscr{F}}. \tag{13.62}$$

Our next step is to upper bound the stochastic component in the inequality (13.62).

Lemma 13.23 *There are universal positive constants (c_1, c_2) such that, with probability greater than $1 - c_1 e^{-\frac{n\delta_n^2}{c_2\sigma^2}}$, we have*

$$\left|\frac{\widetilde{\sigma}}{n}\sum_{i=1}^{n} w_i \Delta(x_i)\right| \leq 2\delta_n \|\Delta\|_n + 2\delta_n^2 \|\Delta\|_{\mathscr{F}} + \frac{1}{16}\|\Delta\|_n^2, \tag{13.63}$$

a bound that holds uniformly for all $\Delta \in \partial\mathscr{F}$ with $\|\Delta\|_{\mathscr{F}} \geq 1$.

We now complete the proof of the theorem using this lemma. We begin by observing that, since $\|\widetilde{f}\|_{\mathscr{F}} \leq 1$ and $\|\widehat{f}\|_{\mathscr{F}} > 2$, the triangle inequality implies that $\|\widetilde{\Delta}\|_{\mathscr{F}} \geq \|\widehat{f}\|_{\mathscr{F}} - \|\widetilde{f}\|_{\mathscr{F}} > 1$, so that Lemma 13.23 may be applied. Substituting the upper bound (13.63) into the inequality (13.62) yields

$$\begin{aligned}
\frac{1}{2}\|\widehat{\Delta}\|_n^2 &\leq \frac{1}{2}\|\widetilde{f} - f^*\|_n^2 + 2\delta_n\|\widetilde{\Delta}\|_n + \{2\delta_n^2 - \lambda_n\}\|\widetilde{\Delta}\|_{\mathscr{F}} + 2\lambda_n + \frac{\|\widetilde{\Delta}\|_n^2}{16} \\
&\leq \frac{1}{2}\|\widetilde{f} - f^*\|_n^2 + 2\delta_n\|\widetilde{\Delta}\|_n + 2\lambda_n + \frac{\|\widetilde{\Delta}\|_n^2}{16},
\end{aligned} \tag{13.64}$$

where the second step uses the fact that $2\delta_n^2 - \lambda_n \leq 0$ by assumption.

Our next step is to convert the terms involving $\widetilde{\Delta}$ into quantities involving $\widehat{\Delta}$: in particular, by the triangle inequality, we have $\|\widetilde{\Delta}\|_n \leq \|\widetilde{f} - f^*\|_n + \|\widehat{\Delta}\|_n$. Thus, we have

$$2\delta_n\|\widetilde{\Delta}\|_n \leq 2\delta_n\|\widetilde{f} - f^*\|_n + 2\delta_n\|\widehat{\Delta}\|_n, \tag{13.65a}$$

and in addition, combined with the inequality $(a+b)^2 \leq 2a^2 + 2b^2$, we find that

$$\frac{\|\widetilde{\Delta}\|_n^2}{16} \leq \frac{1}{8}\left\{\|\widetilde{f} - f^*\|_n^2 + \|\widehat{\Delta}\|_n^2\right\}. \tag{13.65b}$$

Substituting inequalities (13.65a) and (13.65b) into the earlier bound (13.64) and performing some algebra yields

$$\{\tfrac{1}{2} - \tfrac{1}{8}\}\|\widehat{\Delta}\|_n^2 \leq \{\tfrac{1}{2} + \tfrac{1}{8}\}\|\widetilde{f} - f^*\|_n^2 + 2\delta_n\|\widetilde{f} - f^*\|_n + 2\delta_n\|\widehat{\Delta}\|_n + 2\lambda_n.$$

The claim (13.55a) follows by applying the quadratic formula to this inequality.

It remains to prove Lemma 13.23. We claim that it suffices to prove the bound (13.63) for functions $g \in \partial\mathscr{F}$ such that $\|g\|_{\mathscr{F}} = 1$. Indeed, suppose that it holds for all such functions, and that we are given a function Δ with $\|\Delta\|_{\mathscr{F}} > 1$. By assumption, we can apply the inequality (13.63) to the new function $g := \Delta/\|\Delta\|_{\mathscr{F}}$, which belongs to $\partial\mathscr{F}$ by the star-shaped assumption. Applying the bound (13.63) to g and then multiplying both sides by $\|\Delta\|_{\mathscr{F}}$, we obtain

$$\begin{aligned}
\left|\frac{\widetilde{\sigma}}{n}\sum_{i=1}^{n} w_i\Delta(x_i)\right| &\leq c_1\,\delta_n\|\Delta\|_n + c_2\,\delta_n^2\|\Delta\|_{\mathscr{F}} + \frac{1}{16}\frac{\|\Delta\|_n^2}{\|\Delta\|_{\mathscr{F}}} \\
&\leq c_1\,\delta_n\|\Delta\|_n + c_2\,\delta_n^2\|\Delta\|_{\mathscr{F}} + \frac{1}{16}\|\Delta\|_n^2,
\end{aligned}$$

where the second inequality uses the fact that $\|\Delta\|_{\mathscr{F}} > 1$ by assumption.

In order to establish the bound (13.63) for functions with $\|g\|_{\mathscr{F}} = 1$, we first consider it over the ball $\{\|g\|_n \leq t\}$, for some fixed radius $t > 0$. Define the random variable

$$Z_n(t) := \sup_{\substack{\|g\|_{\mathscr{F}} \leq 1 \\ \|g\|_n \leq t}} \left| \frac{\tilde{\sigma}}{n} \sum_{i=1}^n w_i g(x_i) \right|.$$

Viewed as a function of the standard Gaussian vector w, it is Lipschitz with parameter at most $\tilde{\sigma} t/\sqrt{n}$. Consequently, Theorem 2.26 implies that

$$\mathbb{P}[Z_n(t) \geq \mathbb{E}[Z_n(t)] + u] \leq e^{-\frac{nu^2}{2\tilde{\sigma}^2 t^2}}. \tag{13.66}$$

We first derive a bound for $t = \delta_n$. By the definitions of $\mathcal{G}_n$ and the critical radius, we have $\mathbb{E}[Z_n(\delta_n)] \leq \tilde{\sigma}\mathcal{G}_n(\delta_n) \leq \delta_n^2$. Setting $u = \delta_n$ in the tail bound (13.66), we find that

$$\mathbb{P}[Z_n(\delta_n) \geq 2\delta_n^2] \leq e^{-\frac{n\delta_n^2}{2\tilde{\sigma}^2}}. \tag{13.67a}$$

On the other hand, for any $t > \delta_n$, we have

$$\mathbb{E}[Z_n(t)] = \tilde{\sigma}\mathcal{G}_n(t) = t\frac{\tilde{\sigma}\mathcal{G}_n(t)}{t} \overset{\text{(i)}}{\leq} t\frac{\tilde{\sigma}\mathcal{G}_n(\delta_n)}{\delta_n} \overset{\text{(ii)}}{\leq} t\delta_n,$$

where inequality (i) follows from Lemma 13.6, and inequality (ii) follows by our choice of δ_n. Using this upper bound on the mean and setting $u = t^2/32$ in the tail bound (13.66) yields

$$\mathbb{P}\left[Z_n(t) \geq t\delta_n + \frac{t^2}{32}\right] \leq e^{-c_2\frac{nt^2}{\tilde{\sigma}^2}} \qquad \text{for each } t > \delta_n. \tag{13.67b}$$

We are now equipped to complete the proof by a "peeling" argument. Let $\mathcal{E}$ denote the event that the bound (13.63) is violated for some function $g \in \partial\mathscr{F}$ with $\|g\|_{\mathscr{F}} = 1$. For real numbers $0 \leq a < b$, let $\mathcal{E}(a, b)$ denote the event that it is violated for some function such that $\|g\|_n \in [a, b]$ and $\|g\|_{\mathscr{F}} = 1$. For $m = 0, 1, 2, \ldots$, define $t_m = 2^m\delta_n$. We then have the decomposition $\mathcal{E} = \mathcal{E}(0, t_0) \cup \left(\bigcup_{m=0}^{\infty} \mathcal{E}(t_m, t_{m+1})\right)$ and hence, by the union bound,

$$\mathbb{P}[\mathcal{E}] \leq \mathbb{P}[\mathcal{E}(0, t_0)] + \sum_{m=0}^{\infty} \mathbb{P}[\mathcal{E}(t_m, t_{m+1})]. \tag{13.68}$$

The final step is to bound each of the terms in this summation. Since $t_0 = \delta_n$, we have

$$\mathbb{P}[\mathcal{E}(0, t_0)] \leq \mathbb{P}[Z_n(\delta_n) \geq 2\delta_n^2] \leq e^{-\frac{n\delta_n^2}{2\tilde{\sigma}^2}}, \tag{13.69}$$

using our earlier tail bound (13.67a). On the other hand, suppose that $\mathcal{E}(t_m, t_{m+1})$ holds, meaning that there exists some function g with $\|g\|_{\mathscr{F}} = 1$ and $\|g\|_n \in [t_m, t_{m+1}]$ such that

$$\begin{aligned}
\left|\frac{\tilde{\sigma}}{n}\sum_{i=1}^n w_i g(x_i)\right| &\overset{\text{(i)}}{\geq} 2\delta_n\|g\|_n + 2\delta_n^2 + \frac{1}{16}\|g\|_n^2 \\
&\overset{\text{(i)}}{\geq} 2\delta_n t_m + 2\delta_n^2 + \frac{1}{8}t_m^2 \\
&\overset{\text{(ii)}}{=} \delta_n t_{m+1} + 2\delta_n^2 + \frac{1}{32}t_{m+1}^2,
\end{aligned}$$

where step (i) follows since $\|g\|_n \geq t_m$, and step (ii) follows since $t_{m+1} = 2t_m$. This lower bound implies that $Z_n(t_{m+1}) \geq \delta_n t_{m+1} + \frac{t_{m+1}^2}{32}$, and applying the tail bound (13.67b) yields

$$\mathbb{P}\Big[\mathcal{E}(t_m, t_{m+1})\Big] \leq e^{-c_2 \frac{n t_{m+1}^2}{\sigma^2}} = e^{-c_2 \frac{n\, 2^{2m+2} \delta_n^2}{\sigma^2}}.$$

Substituting this inequality and our earlier bound (13.69) into equation (13.68) yields

$$\mathbb{P}[\mathcal{E}] \leq e^{-\frac{n\delta_n^2}{2\sigma^2}} + \sum_{m=0}^{\infty} e^{-c_2 \frac{n\, 2^{2m+2}\delta_n^2}{\sigma^2}} \leq c_1 e^{-c_2 \frac{n\delta_n^2}{\sigma^2}},$$

where the reader should recall that the precise values of universal constants may change from line to line.

13.5 Bibliographic details and background

Nonparametric regression is a classical problem in statistics with a lengthy and rich history. Although this chapter is limited to the method of nonparametric least squares, there are a variety of other cost functions that can be used for regression, which might be preferable for reasons of robustness. The techniques described this chapter are relevant for analyzing any such M-estimator—that is, any method based on minimizing or maximizing some criterion of fit. In addition, nonparametric regression can be tackled via methods that are not most naturally viewed as M-estimators, including orthogonal function expansions, local polynomial representations, kernel density estimators, nearest-neighbor methods and scatterplot smoothing methods, among others. We refer the reader to the books (Gyorfi et al., 2002; Härdle et al., 2004; Wasserman, 2006; Eggermont and LaRiccia, 2007; Tsybakov, 2009) and references therein for further background on these and other methods.

An extremely important idea in this chapter was the use of localized forms of Gaussian or Rademacher complexity, as opposed to the global forms studied in Chapter 4. These localized complexity measures are needed in order to obtain optimal rates for nonparametric estimation problems. The idea of localization plays an important role in empirical process theory, and we embark on a more in-depth study of it in Chapter 14 to follow. Local function complexities of the form given in Corollary 13.7 are used extensively by van de Geer (2000), whereas other authors have studied localized forms of the Rademacher and Gaussian complexities (Koltchinskii, 2001, 2006; Bartlett et al., 2005). The bound on the localized Rademacher complexity of reproducing kernel Hilbert spaces, as stated in Lemma 13.22, is due to Mendelson (2002); see also the paper by Bartlett and Mendelson (2002) for related results. The peeling technique used in the proof of Lemma 13.23 is widely used in empirical process theory (Alexander, 1987; van de Geer, 2000).

The ridge regression estimator from Examples 13.1 and 13.8 was introduced by Hoerl and Kennard (1970). The Lasso estimator from Example 13.1 is treated in detail in Chapter 7. The cubic spline estimator from Example 13.2, as well as the kernel ridge regression estimator from Example 13.3, are standard methods; see Chapter 12 as well as the books (Wahba, 1990; Gu, 2002) for more details. The ℓ_q-ball constrained estimators from Examples 13.1 and 13.9 were analyzed by Raskutti et al. (2011), who also used information-theoretic methods, to be discussed in Chapter 15, in order to derive matching lower bounds. The results on metric entropies of q-convex hulls in this example are based on results from Carl and

Pajor (1988), as well as Guédon and Litvak (2000); see also the arguments given by Raskutti et al. (2011) for details on the specific claims given here.

The problems of convex and/or monotonic regression from Example 13.4 are particular examples of what is known as shape-constrained estimation. It has been the focus of classical work (Hildreth, 1954; Brunk, 1955, 1970; Hanson and Pledger, 1976), as well as much recent and on-going work (e.g., Balabdaoui et al., 2009; Cule et al., 2010; Dümbgen et al., 2011; Seijo and Sen, 2011; Chatterjee et al., 2015), especially in the multivariate setting. The books (Rockafellar, 1970; Hiriart-Urruty and Lemaréchal, 1993; Borwein and Lewis, 1999; Bertsekas, 2003; Boyd and Vandenberghe, 2004) contain further information on subgradients and other aspects of convex analysis. The bound (13.34) on the sup-norm (L_∞) metric entropy for bounded convex Lipschitz functions is due to Bronshtein (1976); see also Section 8.4 of Dudley (1999) for more details. On the other hand, the class of all convex functions $f\colon [0,1] \to [0,1]$ without any Lipschitz constraint is *not* totally bounded in the sup-norm metric; see Exercise 5.1 for details. Guntuboyina and Sen (2013) provide bounds on the entropy in the L_p-metrics over the range $p \in [1, \infty)$ for convex functions without the Lipschitz condition.

Stone (1985) introduced the class of additive nonparametric regression models discussed in Exercise 13.9, and subsequent work has explored many extensions and variants of these models (e.g., Hastie and Tibshirani, 1986; Buja et al., 1989; Meier et al., 2009; Ravikumar et al., 2009; Koltchinskii and Yuan, 2010; Raskutti et al., 2012). Exercise 13.9 in this chapter and Exercise 14.8 in Chapter 14 explore some properties of the standard additive model.

13.6 Exercises

Exercise 13.1 (Characterization of the Bayes least-squares estimate)

(a) Given a random variable Z with finite second moment, show that the function $G(t) = \mathbb{E}[(Z-t)^2]$ is minimized at $t = \mathbb{E}[Z]$.

(b) Assuming that all relevant expectations exist, show that the minimizer of the population mean-squared error (13.1) is given by the conditional expectation $f^*(x) = \mathbb{E}[Y \mid X = x]$. (*Hint*: The tower property and part (a) may be useful to you.)

(c) Let f be any other function for which the mean-squared error $\mathbb{E}_{X,Y}[(Y - f(X))^2]$ is finite. Show that the excess risk of f is given by $\|f - f^*\|_2^2$, as in equation (13.4).

Exercise 13.2 (Prediction error in linear regression) Recall the linear regression model from Example 13.8 with fixed design. Show via a direct argument that

$$\mathbb{E}[\|f_{\widehat{\theta}} - f_{\theta^*}\|_n^2] \le \sigma^2 \frac{\operatorname{rank}(\mathbf{X})}{n},$$

valid for any observation noise that is zero-mean with variance σ^2.

Exercise 13.3 (Cubic smoothing splines) Recall the cubic spline estimate (13.10) from Example 13.2, as well as the kernel function $\mathcal{K}(x, z) = \int_0^1 (x-y)_+ (z-y)_+ \, dy$ from Example 12.29.

(a) Show that the optimal solution must take the form

$$\widehat{f}(x) = \widehat{\theta}_0 + \widehat{\theta}_1 x + \frac{1}{\sqrt{n}} \sum_{i=1} \widehat{\alpha}_i \mathcal{K}(x, x_i)$$

for some vectors $\widehat{\theta} \in \mathbb{R}^2$ and $\widehat{\alpha} \in \mathbb{R}^n$.

(b) Show that these vectors can be obtained by solving the quadratic program

$$(\widehat{\theta}, \widehat{\alpha}) = \arg \min_{(\theta,\alpha)\in\mathbb{R}^2\times\mathbb{R}^n} \left\{ \frac{1}{2n} \|y - \mathbf{X}\theta - \sqrt{n}\mathbf{K}\alpha\|_2^2 + \lambda_n \alpha^{\mathrm{T}} \mathbf{K}\alpha \right\},$$

where $\mathbf{K} \in \mathbb{R}^{n\times n}$ is the kernel matrix defined by the kernel function in part (a), and $\mathbf{X} \in \mathbb{R}^{n\times 2}$ is a design matrix with ith row given by $[1 \quad x_i]$.

Exercise 13.4 (Star-shaped sets and convexity) In this exercise, we explore some properties of star-shaped sets.

(a) Show that a set C is star-shaped around one of its points x^* if and only if the point $\alpha x + (1 - \alpha)x^*$ belongs to C for any $x \in C$ and any $\alpha \in [0, 1]$.

(b) Show that a set C is convex if and only if it is star-shaped around each one of its points.

Exercise 13.5 (Lower bounds on the critical inequality) Consider the critical inequality (13.17) in the case $f^* = 0$, so that $\mathscr{F}^* = \mathscr{F}$.

(a) Show that the critical inequality (13.17) is always satisfied for $\delta^2 = 4\sigma^2$.

(b) Suppose that a convex function class $\mathscr{F}$ contains the constant function $f \equiv 1$. Show that any $\delta \in (0, 1]$ satisfying the critical inequality (13.17) must be lower bounded as $\delta^2 \geq \min\{1, \frac{8}{\pi}\frac{\sigma^2}{n}\}$.

Exercise 13.6 (Local Gaussian complexity and adaptivity) This exercise illustrates how, even for a fixed base function class, the local Gaussian complexity $\mathcal{G}_n(\delta; \mathscr{F}^*)$ of the shifted function class can vary dramatically as the target function f^* is changed. For each $\theta \in \mathbb{R}^n$, let $f_\theta(x) = \langle \theta, x \rangle$ be a linear function, and consider the class $\mathscr{F}_{\ell_1}(1) = \{f_\theta \mid \|\theta\|_1 \leq 1\}$. Suppose that we observe samples of the form

$$y_i = f_{\theta^*}(e_i) + \frac{\sigma}{\sqrt{n}} w_i \;=\; \theta_i^* + \frac{\sigma}{\sqrt{n}} w_i,$$

where $w_i \sim \mathcal{N}(0, 1)$ is an i.i.d. noise sequence. Let us analyze the performance of the ℓ_1-constrained least-squares estimator

$$\widehat{\theta} = \arg \min_{f_\theta\in\mathscr{F}_{\ell_1}(1)} \left\{\frac{1}{n} \sum_{i=1}^{n} (y_i - f_\theta(e_i))^2\right\} \;=\; \arg \min_{\substack{\theta\in\mathbb{R}^d \\ \|\theta\|_1\leq 1}} \left\{\frac{1}{n} \sum_{i=1}^{n} (y_i - \theta_i)^2\right\}.$$

(a) For any $f_{\theta^*} \in \mathscr{F}_{\ell_1}(1)$, show that $\mathcal{G}_n(\delta; \mathscr{F}^*_{\ell_1}(1)) \leq c_1 \sqrt{\frac{\log n}{n}}$ for some universal constant c_1, and hence that $\|\widehat{\theta} - \theta^*\|_2^2 \leq c_1' \sigma \sqrt{\frac{\log n}{n}}$ with high probability.

(b) Now consider some f_{θ^*} with $\theta^* \in \{e_1, \ldots, e_n\}$—that is, one of the canonical basis vectors. Show that there is a universal constant c_2 such that the local Gaussian complexity is bounded as $\mathcal{G}_n(\delta; \mathscr{F}^*_{\ell_1}(1)) \leq c_2\, \delta \frac{\sqrt{\log n}}{n}$, and hence that $\|\widehat{\theta} - \theta^*\|_2^2 \leq c_2' \frac{\sigma^2 \log n}{n}$ with high probability.

Exercise 13.7 (Rates for polynomial regression) Consider the class of all $(m-1)$-degree polynomials

$$\mathcal{P}_m = \{f_\theta : \mathbb{R} \to \mathbb{R} \mid \theta \in \mathbb{R}^m\}, \qquad \text{where } f_\theta(x) = \sum_{j=0}^{m-1} \theta_j x^j,$$

and suppose that $f^* \in \mathcal{P}_m$. Show that there are universal positive constants (c_0, c_1, c_2) such that the least-squares estimator satisfies

$$\mathbb{P}\Big[\|\widehat{f} - f^*\|_n^2 \geq c_0 \frac{\sigma^2 m \log n}{n}\Big] \leq c_1 e^{-c_2 m \log n}.$$

Exercise 13.8 (Rates for twice-differentiable functions) Consider the function class $\mathscr{F}$ of functions $f : [0,1] \to \mathbb{R}$ that are twice differentiable with $\|f\|_\infty + \|f'\|_\infty + \|f''\|_\infty \leq C$ for some constant $C < \infty$. Show that there are positive constants (c_0, c_1, c_2), which may depend on C but not on (n, σ^2), such that the non-parametric least-squares estimate satisfies

$$\mathbb{P}\Big[\|\widehat{f} - f^*\|_n^2 \geq c_0 \big(\frac{\sigma^2}{n}\big)^{\frac{4}{5}}\Big] \leq c_1 e^{-c_2 (n/\sigma^2)^{1/5}}.$$

(*Hint:* Results from Chapter 5 may be useful to you.)

Exercise 13.9 (Rates for additive nonparametric models) Given a convex and symmetric class $\mathscr{G}$ of univariate functions $g : \mathbb{R} \to \mathbb{R}$ equipped with a norm $\|\cdot\|_{\mathscr{G}}$, consider the class of additive functions over $\mathbb{R}^d$, namely

$$\mathscr{F}_{\text{add}} = \{f : \mathbb{R}^d \to \mathbb{R} \mid f = \sum_{j=1}^{d} g_j \quad \text{for some } g_j \in \mathscr{G} \text{ with } \|g_j\|_{\mathscr{G}} \leq 1\}. \tag{13.70}$$

Suppose that we have n i.i.d. samples of the form $y_i = f^*(x_i) + \sigma w_i$, where each $x_i = (x_{i1}, \ldots, x_{id}) \in \mathbb{R}^d$, $w_i \sim \mathcal{N}(0,1)$, and $f^* := \sum_{j=1}^d g_j^*$ is some function in $\mathscr{F}_{\text{add}}$, and that we estimate f^* by the constrained least-squares estimate

$$\widehat{f} := \arg\min_{f \in \mathscr{F}_{\text{add}}} \Big\{\frac{1}{n} \sum_{i=1}^{n} (y_i - f(x_i))^2\Big\}.$$

For each $j = 1, \ldots, d$, define the jth-coordinate Gaussian complexity

$$\mathcal{G}_{n,j}(\delta; 2\mathscr{G}) = \mathbb{E}\Big[\sup_{\substack{\|g_j\|_{\mathscr{G}} \leq 2 \\ \|g_j\|_n \leq \delta}} \Big|\frac{1}{n} \sum_{i=1}^{n} w_i g_j(x_{ij})\Big|\Big],$$

and let $\delta_{n,j} > 0$ be the smallest positive solution to the inequality $\frac{\mathcal{G}_{n,j}(\delta; 2\mathscr{G})}{\delta} \leq \frac{\delta}{2\sigma}$.

(a) Defining $\delta_{n,\max} = \max_{j=1,\ldots,d} \delta_{n,j}$, show that, for each $t \geq \delta_{n,\max}$, we have

$$\frac{\sigma}{n}\Big|\sum_{i=1}^{n} w_i \widehat{\Delta}(x_i)\Big| \leq dt\delta_{n,\max} + 2\sqrt{t\delta_{n,\max}}\Big(\sum_{j=1}^{d} \|\widehat{\Delta}_j\|_n\Big)$$

with probability at least $1 - c_1 d e^{-c_2 n t \delta_{n,\max}}$. (Note that $\widehat{f} = \sum_{j=1}^d \widehat{g}_j$ for some $\widehat{g}_j \in \mathscr{G}$, so that the function $\widehat{\Delta}_j = \widehat{g}_j - g_j^*$ corresponds to the error in coordinate j, and $\widehat{\Delta} := \sum_{j=1}^d \widehat{\Delta}_j$ is the full error function.)

(b) Suppose that there is a universal constant $K \geq 1$ such that

$$\sqrt{\sum_{j=1}^{n} \|g_j\|_n^2} \leq \sqrt{K} \|\sum_{j=1}^{d} g_j\|_n \qquad \text{for all } g_j \in \mathscr{G}.$$

Use this bound and part (a) to show that $\|\widehat{f} - f^*\|_n^2] \leq c_3\, K\, d\, \delta_{n,\max}^2$ with high probability.

Exercise 13.10 (Orthogonal series expansions) Recall the function class $\mathscr{F}_{\text{ortho}}(1; T)$ from Example 13.14 defined by orthogonal series expansion with T coefficients.

(a) Given a set of design points $\{x_1, \ldots, x_n\}$, define the $n \times T$ matrix $\mathbf{\Phi} \equiv \mathbf{\Phi}(x_1^n)$ with (i, j)th entry $\Phi_{ij} = \phi_j(x_i)$. Show that the nonparametric least-squares estimate $\widehat{f}$ over $\mathscr{F}_{\text{ortho}}(1; T)$ can be obtained by solving the ridge regression problem

$$\min_{\theta \in \mathbb{R}^T} \left\{ \frac{1}{n} \|y - \mathbf{\Phi}\, \theta\|_2^2 + \lambda_n \|\theta\|_2^2 \right\}$$

for a suitable choice of regularization parameter $\lambda_n \geq 0$.

(b) Show that $\inf_{f \in \mathscr{F}_{\text{ortho}}(1;T)} \|f - f^*\|_2^2 = \sum_{j=T+1}^{\infty} \theta_j^2$.

Exercise 13.11 (Differentiable functions and Fourier coefficients) For a given integer $\alpha \geq 1$ and radius $R > 0$, consider the class of functions $\mathscr{F}_\alpha(R) \subset L^2[0, 1]$ such that:

- The function f is α-times differentiable, with $\int_0^1 (f^{(\alpha)}(x))^2\, dx \leq R$.
- It and its derivatives satisfy the boundary conditions $f^{(j)}(0) = f^{(j)}(1) = 0$ for all $j = 0, 1, \ldots, \alpha$.

(a) For a function $f \in \mathscr{F}_\alpha(R) \cap \{\|f\|_2 \leq 1\}$, let $\{\beta_0, (\beta_m, \widetilde{\beta}_m)_{m=1}^{\infty}\}$ be its Fourier coefficients as previously defined in Example 13.15. Show that there is a constant c such that $\beta_m^2 + \widetilde{\beta}_m^2 \leq \frac{cR}{m^{2\alpha}}$ for all $m \geq 1$.

(b) Verify the approximation-theoretic guarantee (13.47).

14

Localization and uniform laws

As discussed previously in Chapter 4, uniform laws of large numbers concern the deviations between sample and population averages, when measured in a uniform sense over a given function class. The classical forms of uniform laws are asymptotic in nature, guaranteeing that the deviations converge to zero in probability or almost surely. The more modern approach is to provide non-asymptotic guarantees that hold for all sample sizes, and provide sharp rates of convergence. In order to achieve the latter goal, an important step is to localize the deviations to a small neighborhood of the origin. We have already encountered a form of localization in our discussion of nonparametric regression from Chapter 13. In this chapter, we turn to a more in-depth study of this technique and its use in establishing sharp uniform laws for various types of processes.

14.1 Population and empirical L^2-norms

We begin our exploration with a detailed study of the relation between the population and empirical L^2-norms. Given a function $f\colon \mathcal{X} \to \mathbb{R}$ and a probability distribution $\mathbb{P}$ over $\mathcal{X}$, the usual $L^2(\mathbb{P})$-norm is given by

$$\|f\|_{L^2(\mathbb{P})}^2 := \int_{\mathcal{X}} f^2(x)\mathbb{P}(dx) = \mathbb{E}[f^2(X)], \tag{14.1}$$

and we say that $f \in L^2(\mathbb{P})$ whenever this norm is finite. When the probability distribution $\mathbb{P}$ is clear from the context, we adopt $\|f\|_2$ as a convenient shorthand for $\|f\|_{L^2(\mathbb{P})}$.

Given a set of n samples $\{x_i\}_{i=1}^n := \{x_1, x_2, \ldots, x_n\}$, each drawn i.i.d. according to $\mathbb{P}$, consider the empirical distribution

$$\mathbb{P}_n(x) := \frac{1}{n}\sum_{i=1}^n \delta_{x_i}(x)$$

that places mass $1/n$ at each sample. It induces the *empirical L^2-norm*

$$\|f\|_{L^2(\mathbb{P}_n)}^2 := \frac{1}{n}\sum_{i=1}^n f^2(x_i) = \int_{\mathcal{X}} f^2(x)\mathbb{P}_n(dx). \tag{14.2}$$

Again, to lighten notation, when the underlying empirical distribution $\mathbb{P}_n$ is clear from context, we adopt the convenient shorthand $\|f\|_n$ for $\|f\|_{L^2(\mathbb{P}_n)}$.

In our analysis of nonparametric least squares from Chapter 13, we provided bounds on the $L^2(\mathbb{P}_n)$-error in which the samples $\{x_i\}_{i=1}^n$ were viewed as fixed. By contrast, throughout this chapter, we view the samples as being random variables, so that the empirical norm is itself a random variable. Since each $x_i \sim \mathbb{P}$, the linearity of expectation guarantees that

$$\mathbb{E}[\|f\|_n^2] = \mathbb{E}\left[\frac{1}{n}\sum_{i=1}^n f^2(x_i)\right] = \|f\|_2^2 \qquad \text{for any function } f \in L^2(\mathbb{P}).$$

Consequently, under relatively mild conditions on the random variable $f(x)$, the law of large numbers implies that $\|f\|_n^2$ converges to $\|f\|_2^2$. Such a limit theorem has its usual non-asymptotic analogs: for instance, if the function f is uniformly bounded, that is, if

$$\|f\|_\infty := \sup_{x \in \mathcal{X}} |f(x)| \leq b \qquad \text{for some } b < \infty,$$

then Hoeffding's inequality (cf. Proposition 2.5 and equation (2.11)) implies that

$$\mathbb{P}\Big[\big|\|f\|_n^2 - \|f\|_2^2\big| \geq t\Big] \leq 2e^{-\frac{nt^2}{2b^4}}.$$

As in Chapter 4, our interest is in extending this type of tail bound—valid for a single function f—to a result that applies uniformly to all functions in a certain function class $\mathscr{F}$. Our analysis in this chapter, however, will be more refined: by using localized forms of complexity, we obtain optimal bounds.

14.1.1 A uniform law with localization

We begin by stating a theorem that controls the deviations in the random variable $\big|\|f\|_n - \|f\|_2\big|$, when measured in a uniform sense over a function class $\mathscr{F}$. We then illustrate some consequences of this result in application to nonparametric regression.

As with our earlier results on nonparametric least squares from Chapter 13, our result is stated in terms of a localized form of Rademacher complexity. For the current purposes, it is convenient to define the complexity at the population level. For a given radius $\delta > 0$ and function class $\mathscr{F}$, consider the *localized population Rademacher complexity*

$$\bar{\mathcal{R}}_n(\delta; \mathscr{F}) = \mathbb{E}_{\varepsilon, x}\left[\sup_{\substack{f \in \mathscr{F} \\ \|f\|_2 \leq \delta}} \left|\frac{1}{n}\sum_{i=1}^n \varepsilon_i f(x_i)\right|\right], \tag{14.3}$$

where $\{x_i\}_{i=1}^n$ are i.i.d. samples from some underlying distribution $\mathbb{P}$, and $\{\varepsilon_i\}_{i=1}^n$ are i.i.d. Rademacher variables taking values in $\{-1, +1\}$ equiprobably, independent of the sequence $\{x_i\}_{i=1}^n$.

In the following result, we assume that $\mathscr{F}$ is star-shaped around the origin, meaning that, for any $f \in \mathscr{F}$ and scalar $\alpha \in [0, 1]$, the function αf also belongs to $\mathscr{F}$. In addition, we require the function class to be b-uniformly bounded, meaning that there is a constant $b < \infty$ such that $\|f\|_\infty \leq b$ for all $f \in \mathscr{F}$.

Theorem 14.1 *Given a star-shaped and b-uniformly bounded function class $\mathscr{F}$, let δ_n be any positive solution of the inequality*

$$\bar{\mathcal{R}}_n(\delta; \mathscr{F}) \leq \frac{\delta^2}{b}. \tag{14.4}$$

Then for any $t \geq \delta_n$, we have

$$\Big| \|f\|_n^2 - \|f\|_2^2 \Big| \leq \frac{1}{2}\|f\|_2^2 + \frac{t^2}{2} \qquad \textit{for all } f \in \mathscr{F} \tag{14.5a}$$

with probability at least $1 - c_1 e^{-c_2 \frac{nt^2}{b^2}}$. If in addition $n\delta_n^2 \geq \frac{2}{c_2}\log(4\log(1/\delta_n))$, then

$$\Big| \|f\|_n - \|f\|_2 \Big| \leq c_0\, \delta_n \qquad \textit{for all } f \in \mathscr{F} \tag{14.5b}$$

with probability at least $1 - c_1' e^{-c_2' \frac{n\delta_n^2}{b^2}}$.

It is worth noting that a similar result holds in terms of the localized *empirical Rademacher complexity*, namely the data-dependent quantity

$$\widehat{\mathcal{R}}_n(\delta) \equiv \widehat{\mathcal{R}}_n(\delta; \mathscr{F}) := \mathbb{E}_\varepsilon\Big[\sup_{\substack{f \in \mathscr{F} \\ \|f\|_n \leq \delta}} \Big|\frac{1}{n}\sum_{i=1}^n \varepsilon_i f(x_i)\Big|\Big], \tag{14.6}$$

and any positive solution $\hat{\delta}_n$ to the inequality

$$\widehat{\mathcal{R}}_n(\delta) \leq \frac{\delta^2}{b}. \tag{14.7}$$

Since the Rademacher complexity $\widehat{\mathcal{R}}_n$ depends on the data, this critical radius $\hat{\delta}_n$ is a random quantity, but it is closely related to the deterministic radius δ_n defined in terms of the population Rademacher complexity (14.3). More precisely, let δ_n and $\hat{\delta}_n$ denote the smallest positive solutions to inequalities (14.4) and (14.7), respectively. Then there are universal constants $c < 1 < C$ such that, with probability at least $1 - c_1 e^{-c_2 \frac{n\delta_n^2}{b}}$, we are guaranteed that $\hat{\delta}_n \in [c\delta_n, C\delta_n]$, and hence

$$\Big| \|f\|_n - \|f\|_2 \Big| \leq \frac{c_0}{c}\, \hat{\delta}_n \qquad \text{for all } f \in \mathscr{F}. \tag{14.8}$$

See Proposition 14.25 in the Appendix (Section 14.5) for the details and proof.

Theorem 14.1 is best understood by considering some concrete examples.

Example 14.2 (Bounds for quadratic functions) For a given coefficient vector $\theta \in \mathbb{R}^3$, define the quadratic function $f_\theta(x) := \theta_0 + \theta_1 x + \theta_2 x^2$, and let us consider the set of all bounded quadratic functions over the unit interval $[-1, 1]$, that is, the function class

$$\mathcal{P}_2 := \{f_\theta \text{ for some } \theta \in \mathbb{R}^3 \text{ such that } \max_{x \in [-1,1]} |f_\theta(x)| \leq 1\}. \tag{14.9}$$

Suppose that we are interested in relating the population and empirical L^2-norms uniformly over this family, when the samples are drawn from the uniform distribution over $[-1, 1]$.

We begin by exploring a naive approach, one that ignores localization and hence leads to a sub-optimal rate. From our results on VC dimension in Chapter 4—in particular, see Proposition 4.20—it is straightforward to see that $\mathcal{P}_2$ has VC dimension at most 3. In conjunction with the boundedness of the function class, Lemma 4.14 guarantees that for any $\delta > 0$, we have

$$\mathbb{E}_\varepsilon\Big[\sup_{\substack{f_\theta\in\mathcal{P}_2\\ \|f_\theta\|_2\le\delta}}\Big|\frac{1}{n}\sum_{i=1}^n \varepsilon_i f(x_i)\Big|\Big] \overset{\text{(i)}}{\le} 2\sqrt{\frac{3\log(n+1)}{n}} \le 4\sqrt{\frac{\log(n+1)}{n}} \tag{14.10}$$

for any set of samples $\{x_i\}_{i=1}^n$. As we will see, this upper bound is actually rather loose for small values of δ, since inequality (i) makes no use of the localization condition $\|f_\theta\|_2 \le \delta$.

Based on the naive upper bound (14.10), we can conclude that there is a constant c_0 such that inequality (14.4) is satisfied with $\delta_n = c_0(\frac{\log(n+1)}{n})^{1/4}$. Thus, for any $t \ge c_0(\frac{\log(n+1)}{n})^{1/4}$, Theorem 14.1 guarantees that

$$\Big|\|f\|_n^2 - \|f\|_2^2\Big| \le \tfrac{1}{2}\|f\|_2^2 + t^2 \qquad \text{for all } f \in \mathcal{P}_2 \tag{14.11}$$

with probability at least $1 - c_1 e^{-c_2 t^2}$. This bound establishes that $\|f\|_2^2$ and $\|f\|_n^2$ are of the same order for all functions with norm $\|f\|_2 \ge c_0(\frac{\log(n+1)}{n})^{1/4}$, but this order of fluctuation is sub-optimal. As we explore in Exercise 14.3, an entropy integral approach can be used to remove the superfluous logarithm from this result, but the slow $n^{-1/4}$ rate remains.

Let us now see how localization can be exploited to yield the optimal scaling $n^{-1/2}$. In order to do so, it is convenient to re-parameterize our quadratic functions in terms of an orthonormal basis of $L^2[-1,1]$. In particular, the first three functions in the Legendre basis take the form

$$\phi_0(x) = \frac{1}{\sqrt{2}}, \quad \phi_1(x) = \sqrt{\frac{3}{2}}\,x \quad \text{and} \quad \phi_2(x) = \sqrt{\frac{5}{8}}\,(3x^2-1).$$

By construction, these functions are orthonormal in $L^2[-1,1]$, meaning that the inner product $\langle\phi_j,\phi_k\rangle_{L^2[-1,1]} := \int_{-1}^1 \phi_j(x)\phi_k(x)\,dx$ is equal to one if $j = k$, and zero otherwise. Using these basis functions, any polynomial function in $\mathcal{P}_2$ then has an expansion of the form $f_\gamma(x) = \gamma_0\phi_0(x) + \gamma_1\phi_1(x) + \gamma_2\phi_2(x)$, where $\|f_\gamma\|_2 = \|\gamma\|_2$ by construction. Given a set of n samples, let us define an $n\times 3$ matrix $\mathbf{M}$ with entries $M_{ij} = \phi_j(x_i)$. In terms of this matrix, we then have

$$\begin{aligned}\mathbb{E}\Big[\sup_{\substack{f_\gamma\in\mathcal{P}_2\\ \|f_\gamma\|_2\le\delta}}\Big|\frac{1}{n}\sum_{i=1}^n \varepsilon_i f_\gamma(x_i)\Big|\Big] &\le \mathbb{E}\Big[\sup_{\|\gamma\|_2\le\delta}\Big|\frac{1}{n}\varepsilon^{\mathrm{T}}\mathbf{M}\gamma\Big|\Big]\\ &\overset{\text{(i)}}{\le} \frac{\delta}{n}\,\mathbb{E}\big[\|\varepsilon^{\mathrm{T}}\mathbf{M}\|_2\big]\\ &\overset{\text{(ii)}}{\le} \frac{\delta}{n}\,\sqrt{\mathbb{E}[\|\varepsilon^{\mathrm{T}}\mathbf{M}\|_2^2]},\end{aligned}$$

where step (i) follows from the Cauchy–Schwarz inequality, and step (ii) follows from Jensen's inequality and concavity of the square-root function. Now since the Rademacher variables are independent, we have

$$\mathbb{E}_\varepsilon[\|\varepsilon^{\mathrm{T}}\mathbf{M}\|_2^2] = \mathrm{trace}(\mathbf{M}\mathbf{M}^{\mathrm{T}}) = \mathrm{trace}(\mathbf{M}^{\mathrm{T}}\mathbf{M}).$$

By the orthonormality of the basis $\{\phi_0, \phi_1, \phi_2\}$, we have $\mathbb{E}_x[\text{trace}(\mathbf{M}^T\mathbf{M})] = 3n$. Putting together the pieces yields the upper bound

$$\mathbb{E}\Big[\sup_{\substack{f_\gamma \in \mathcal{P}_2 \\ \|f_\gamma\|_2 \le \delta}} \Big|\frac{1}{n}\sum_{i=1}^n \varepsilon_i f_\gamma(x_i)\Big|\Big] \le \frac{\sqrt{3}\,\delta}{\sqrt{n}}.$$

Based on this bound, we see that there is a universal constant c such that inequality (14.4) is satisfied with $\delta_n = \frac{c}{\sqrt{n}}$. Applying Theorem 4.10 then guarantees that for any $t \ge \frac{c}{\sqrt{n}}$, we have

$$\Big|\|f\|_n^2 - \|f\|_2^2\Big| \le \frac{\|f\|_2^2}{2} + \frac{1}{2}t^2 \qquad \text{for all } f \in \mathcal{P}_2, \tag{14.12}$$

a bound that holds with probability at least $1 - c_1 e^{-c_2 n t^2}$. Unlike the earlier bound (14.11), this result has exploited the localization and thereby increased the rate from the slow one of $(\frac{\log n}{n})^{1/4}$ to the optimal one of $(\frac{1}{n})^{1/2}$. ♣

Whereas the previous example concerned a parametric class of functions, Theorem 14.1 also applies to nonparametric function classes. Since metric entropy has been computed for many such classes, it provides one direct route for obtaining upper bounds on the solutions of inequalities (14.4) or (14.7). One such avenue is summarized in the following:

Corollary 14.3 *Let $N_n(t; \mathbb{B}_n(\delta; \mathscr{F}))$ denote the t-covering number of the set $\mathbb{B}_n(\delta; \mathscr{F}) = \{f \in \mathscr{F} \mid \|f\|_n \le \delta\}$ in the empirical $L^2(\mathbb{P}_n)$-norm. Then the empirical version of critical inequality* (14.7) *is satisfied for any $\delta > 0$ such that*

$$\frac{64}{\sqrt{n}} \int_{\frac{\delta^2}{2b}}^{\delta} \sqrt{\log N_n(t; \mathbb{B}_n(\delta; \mathscr{F}))}\, dt \le \frac{\delta^2}{b}. \tag{14.13}$$

The proof of this result is essentially identical to the proof of Corollary 13.7, so that we leave the details to the reader.

In order to make use of Corollary 14.3, we need to control the covering number N_n in the empirical $L^2(\mathbb{P}_n)$-norm. One approach is based on observing that the covering number N_n can always bounded by the covering number N_{sup} in the supremum norm $\|\cdot\|_\infty$. Let us illustrate this approach with an example.

Example 14.4 (Bounds for convex Lipschitz functions) Recall from Example 13.11 the class of convex 1-Lipschitz functions

$$\mathscr{F}_{\text{conv}}([0,1]; 1) := \{f\colon [0,1] \to \mathbb{R} \mid f(0) = 0, \text{ and } f \text{ is convex and 1-Lipschitz}\}.$$

From known results, the metric entropy of this function class in the sup-norm is upper bounded as $\log N_{\text{sup}}(t; \mathscr{F}_{\text{conv}}) \precsim t^{-1/2}$ for all $t > 0$ sufficiently small (see the bibliographic section for details). Thus, in order to apply Corollary 14.3, it suffices to find $\delta > 0$ such that

$$\frac{1}{\sqrt{n}} \int_0^\delta (1/t)^{1/4}\, dt = \frac{1}{\sqrt{n}} \frac{4}{3} \delta^{3/4} \precsim \delta^2.$$

Setting $\delta = c n^{-2/5}$ for a sufficiently large constant $c > 0$ is suitable, and applying Theorem 14.1 with this choice yields

$$\Big| \|f\|_2 - \|f\|_n \Big| \leq c'\, n^{-2/5} \qquad \text{for all } f \in \mathscr{F}_{\text{conv}}([0,1]; 1)$$

with probability greater than $1 - c_1 e^{-c_2 n^{1/5}}$. ♣

In the exercises at the end of this chapter, we explore various other results that can be derived using Corollary 14.3.

14.1.2 Specialization to kernel classes

As discussed in Chapter 12, reproducing kernel Hilbert spaces (RKHSs) have a number of attractive computational properties in application to nonparametric estimation. In this section, we discuss the specialization of Theorem 14.1 to the case of a function class $\mathscr{F}$ that corresponds to the unit ball of an RKHS.

Recall that any RKHS is specified by a symmetric, positive semidefinite kernel function $\mathcal{K}\colon \mathcal{X} \times \mathcal{X} \to \mathbb{R}$. Under mild conditions, Mercer's theorem (as stated previously in Theorem 12.20) ensures that $\mathcal{K}$ has a countable collection of non-negative eigenvalues $(\mu_j)_{j=1}^\infty$. The following corollary shows that the population form of the localized Rademacher complexity for an RKHS is determined by the decay rate of these eigenvalues, and similarly, the empirical version is determined by the eigenvalues of the empirical kernel matrix.

Corollary 14.5 *Let $\mathscr{F} = \{f \in \mathbb{H} \mid \|f\|_{\mathbb{H}} \leq 1\}$ be the unit ball of an RKHS with eigenvalues $(\mu_j)_{j=1}^\infty$. Then the localized population Rademacher complexity* (14.3) *is upper bounded as*

$$\bar{\mathcal{R}}_n(\delta; \mathscr{F}) \leq \sqrt{\frac{2}{n}} \sqrt{\sum_{j=1}^\infty \min\{\mu_j, \delta^2\}}. \tag{14.14a}$$

Similarly, letting $(\widehat{\mu}_j)_{j=1}^n$ denote the eigenvalues of the renormalized kernel matrix $\mathbf{K} \in \mathbb{R}^{n\times n}$ with entries $K_{ij} = \mathcal{K}(x_i, x_j)/n$, the localized empirical Rademacher complexity (14.6) *is upper bounded as*

$$\widehat{\mathcal{R}}_n(\delta; \mathscr{F}) \leq \sqrt{\frac{2}{n}} \sqrt{\sum_{j=1}^n \min\{\widehat{\mu}_j, \delta^2\}}. \tag{14.14b}$$

Given knowledge of the eigenvalues of the kernel (operator or matrix), these upper bounds on the localized Rademacher complexities allow us to specify values δ_n that satisfy the inequalities (14.4) and (14.7), in the population and empirical cases, respectively. Lemma 13.22

from Chapter 13 provides an upper bound on the empirical Gaussian complexity for a kernel class, which yields the claim (14.14b). The proof of inequality (14.14a) is based on techniques similar to the proof of Lemma 13.22; we work through the details in Exercise 14.4.

Let us illustrate the use of Corollary 14.5 with some examples.

Example 14.6 (Bounds for first-order Sobolev space) Consider the first-order Sobolev space

$$\mathbb{H}^1[0,1] := \{f\colon [0,1] \to \mathbb{R} \mid f(0) = 0, \text{ and } f \text{ is abs. cts. with } f' \in L^2[0,1]\}.$$

Recall from Example 12.16 that it is a reproducing kernel Hilbert space with kernel function $\mathcal{K}(x,z) = \min\{x,z\}$. From the result of Exercise 12.14, the unit ball $\{f \in \mathbb{H}^1[0,1] \mid \|f\|_{\mathbb{H}} \leq 1\}$ is uniformly bounded with $b = 1$, so that Corollary 14.5 may be applied. Moreover, from Example 12.23, the eigenvalues of this kernel function are given by $\mu_j = (\frac{2}{(2j-1)\pi})^2$ for $j = 1, 2, \ldots$. Using calculations analogous to those from Example 13.20, it can be shown that

$$\frac{1}{\sqrt{n}} \sqrt{\sum_{j=1}^{\infty} \min\{\delta^2, \mu_j\}} \leq c' \sqrt{\frac{\delta}{n}}$$

for some universal constant $c' > 0$. Consequently, Corollary 14.5 implies that the critical inequality (14.4) is satisfied for $\delta_n = cn^{-1/3}$. Applying Theorem 14.1, we conclude that

$$\sup_{\|f\|_{\mathbb{H}^1[0,1]} \leq 1} \big|\|f\|_2 - \|f\|_n\big| \leq c_0\, n^{-1/3}$$

with probability greater than $1 - c_1 e^{-c_2 n^{1/3}}$. ♣

Example 14.7 (Bounds for Gaussian kernels) Consider the RKHS generated by the Gaussian kernel $\mathcal{K}(x,z) = e^{-\frac{1}{2}(x-z)^2}$ defined on the unit square $[-1,1] \times [-1,1]$. As discussed in Example 13.21, there are universal constants (c_0, c_1) such that the eigenvalues of the associated kernel operator satisfy a bound of the form

$$\mu_j \leq c_0\, e^{-c_1 j \log j} \qquad \text{for } j = 1, 2, \ldots.$$

Following the same line of calculation as in Example 13.21, it is straightforward to show that inequality (14.14a) is satisfied by $\delta_n = c_0 \sqrt{\frac{\log(n+1)}{n}}$ for a sufficiently large but universal constant c_0. Consequently, Theorem 14.1 implies that, for the unit ball of the Gaussian kernel RKHS, we have

$$\sup_{\|f\|_{\mathbb{H}} \leq 1} \big|\|f\|_2 - \|f\|_n\big| \leq c_0 \sqrt{\frac{\log(n+1)}{n}}$$

with probability greater than $1 - 2e^{-c_1 \log(n+1)}$. By comparison to the parametric function class discussed in Example 14.2, we see that the unit ball of a Gaussian kernel RKHS obeys a uniform law with a similar rate. This fact illustrates that the unit ball of the Gaussian kernel RKHS—even though nonparametric in nature—is still relatively small. ♣

14.1.3 Proof of Theorem 14.1

Let us now return to prove Theorem 14.1. By a rescaling argument, it suffices to consider the case $b = 1$. Moreover, it is convenient to redefine δ_n as a positive solution to the inequality

$$\bar{\mathcal{R}}_n(\delta; \mathscr{F}) \leq \frac{\delta^2}{16}. \tag{14.15}$$

This new δ_n is simply a rescaled version of the original one, and we shall use it to prove a version of the theorem with $c_0 = 1$.

With these simplifications, our proof is based on the family of random variables

$$Z_n(r) := \sup_{f \in \mathbb{B}_2(r; \mathscr{F})} \left| \|f\|_2^2 - \|f\|_n^2 \right|, \qquad \text{where } \mathbb{B}_2(r; \mathscr{F}) = \{f \in \mathscr{F} \mid \|f\|_2 \leq r\}, \tag{14.16}$$

indexed by $r \in (0, 1]$. We let $\mathcal{E}_0$ and $\mathcal{E}_1$, respectively, denote the events that inequality (14.5a) or inequality (14.5b) are violated. We also define the auxiliary events $\mathcal{A}_0(r) := \{Z_n(r) \geq r^2/2\}$, and

$$\mathcal{A}_1 := \left\{ Z_n(\|f\|_2) \geq \delta_n \|f\|_2 \text{ for some } f \in \mathscr{F} \text{ with } \|f\|_2 \geq \delta_n \right\}.$$

The following lemma shows that it suffices to control these two auxiliary events:

Lemma 14.8 *For any star-shaped function class, we have*

$$\mathcal{E}_0 \overset{\text{(i)}}{\subseteq} \mathcal{A}_0(t) \quad \text{and} \quad \mathcal{E}_1 \overset{\text{(ii)}}{\subseteq} \mathcal{A}_0(\delta_n) \cup \mathcal{A}_1. \tag{14.17}$$

Proof Beginning with the inclusion (i), we divide the analysis into two cases. First, suppose that there exists some function with norm $\|f\|_2 \leq t$ that violates inequality (14.5a). For this function, we must have $\left|\|f\|_n^2 - \|f\|_2^2\right| > \frac{t^2}{2}$, showing that $Z_n(t) > \frac{t^2}{2}$ so that $\mathcal{A}_0(t)$ must hold. Otherwise, suppose that the inequality (14.5a) is violated by some function with $\|f\|_2 > t$. Any such function satisfies the inequality $\left|\|f\|_2^2 - \|f\|_n^2\right| > \|f\|_2^2/2$. We may then define the rescaled function $\widetilde{f} = \frac{t}{\|f\|_2} f$; by construction, it has $\|\widetilde{f}\|_2 = t$, and also belongs to $\mathscr{F}$ due to the star-shaped condition. Hence, reasoning as before, we find that $\mathcal{A}_0(t)$ must also hold in this case.

Turning to the inclusion (ii), it is equivalent to show that $\mathcal{A}_0^c(\delta_n) \cap \mathcal{A}_1^c \subseteq \mathcal{E}_1^c$. We split the analysis into two cases:

Case 1: Consider a function $f \in \mathscr{F}$ with $\|f\|_2 \leq \delta_n$. Then on the complement of $\mathcal{A}_0(\delta_n)$, either we have $\|f\|_n \leq \delta_n$, in which case $\left|\|f\|_n - \|f\|_2\right| \leq \delta_n$, or we have $\|f\|_n \geq \delta_n$, in which case

$$\left|\|f\|_n - \|f\|_2\right| = \frac{\left|\|f\|_2^2 - \|f\|_n^2\right|}{\|f\|_n + \|f\|_2} \leq \frac{\delta_n^2}{\delta_n} = \delta_n.$$

Case 2: Next consider a function $f \in \mathscr{F}$ with $\|f\|_2 > \delta_n$. In this case, on the complement of $\mathcal{A}_1$, we have

$$\big|\|f\|_n - \|f\|_2\big| = \frac{\big|\|f\|_n^2 - \|f\|_2^2\big|}{\|f\|_n + \|f\|_2} \leq \frac{\|f\|_2\, \delta_n}{\|f\|_n + \|f\|_2} \leq \delta_n,$$

which completes the proof. □

In order to control the events $\mathcal{A}_0(r)$ and $\mathcal{A}_1$, we need to control the tail behavior of the random variable $Z_n(r)$.

Lemma 14.9 *For all $r, s \geq \delta_n$, we have*

$$\mathbb{P}\Big[Z_n(r) \geq \frac{r\delta_n}{4} + \frac{s^2}{4}\Big] \leq 2e^{-c_2 n \min\{\frac{s^4}{r^2}, s^2\}}. \tag{14.18}$$

Setting both r and s equal to $t \geq \delta_n$ in Lemma 14.9 yields the bound $\mathbb{P}[\mathcal{A}_0(t)] \leq 2e^{-c_2 n t^2}$. Using inclusion (i) in Lemma 14.8, this completes the proof of inequality (14.5a).

Let us now prove Lemma 14.9.

Proof Beginning with the expectation, we have

$$\mathbb{E}[Z_n(r)] \overset{\text{(i)}}{\leq} 2\mathbb{E}\Big[\sup_{f \in \mathbb{B}_2(r;\mathscr{F})} \Big|\frac{1}{n}\sum_{i=1}^n \varepsilon_i f^2(x_i)\Big|\Big] \overset{\text{(ii)}}{\leq} 4\mathbb{E}\Big[\sup_{f \in \mathbb{B}_2(r;\mathscr{F})} \Big|\frac{1}{n}\sum_{i=1}^n \varepsilon_i f(x_i)\Big|\Big] = 4\mathcal{R}_n(r),$$

where step (i) uses a standard symmetrization argument (in particular, see the proof of Theorem 4.10 in Chapter 4); and step (ii) follows from the boundedness assumption ($\|f\|_\infty \leq 1$ uniformly for all $f \in \mathscr{F}$) and the Ledoux–Talagrand contraction inequality (5.61) from Chapter 5. Given our star-shaped condition on the function class, Lemma 13.6 guarantees that the function $r \mapsto \mathcal{R}_n(r)/r$ is non-increasing on the interval $(0, \infty)$. Consequently, for any $r \geq \delta_n$, we have

$$\frac{\mathcal{R}_n(r)}{r} \overset{\text{(iii)}}{\leq} \frac{\mathcal{R}_n(\delta_n)}{\delta_n} \overset{\text{(iv)}}{\leq} \frac{\delta_n}{16}, \tag{14.19}$$

where step (iii) follows from the non-increasing property, and step (iv) follows from our definition of δ_n. Putting together the pieces, we find that the expectation is upper bounded as $\mathbb{E}[Z_n(r)] \leq \frac{r\delta_n}{4}$.

Next we establish a tail bound above the expectation using Talagrand's inequality from Theorem 3.27. Let f be an arbitrary member of $\mathbb{B}_2(r;\mathscr{F})$. Since $\|f\|_\infty \leq 1$ for all $f \in \mathscr{F}$, the recentered functions $g = f^2 - \mathbb{E}[f^2(X)]$ are bounded as $\|g\|_\infty \leq 1$, and moreover

$$\text{var}(g) \leq \mathbb{E}[f^4] \leq \mathbb{E}[f^2] \leq r^2,$$

using the fact that $f \in \mathbb{B}_2(r;\mathscr{F})$. Consequently, by applying Talagrand's concentration inequality (3.83), we find that there is a universal constant c such that

$$\mathbb{P}\Big[Z_n(r) \geq \mathbb{E}[Z(r)] + \frac{s^2}{4}\Big] \leq 2\exp\Big(-\frac{ns^4}{c\,(r^2 + r\delta_n + s^2)}\Big) \leq e^{-c_2 n \min\{\frac{s^4}{r^2}, s^2\}},$$

where the final step uses the fact that $r \geq \delta_n$. □

It remains to use Lemmas 14.8 and 14.9 to establish inequality (14.5b). By combining inclusion (ii) in Lemma 14.8 with the union bound, it suffices to bound the sum $\mathbb{P}[\mathcal{A}_0(\delta_n)] + \mathbb{P}[\mathcal{A}_1]$. Setting $r = s = \delta_n$ in the bound (14.18) yields the bound $\mathbb{P}[\mathcal{A}_0(\delta_n)] \leq e^{-c_1 n\delta_n^2}$, whereas setting $s^2 = r\delta_n$ yields the bound

$$\mathbb{P}[Z_n(r) \geq \frac{r\delta_n}{2}] \leq 2e^{-c_2 n\delta_n^2}. \tag{14.20}$$

Given this bound, one is tempted to "complete" the proof by setting $r = \|f\|_2$, and applying the tail bound (14.20) to the variable $Z_n(\|f\|_2)$. The delicacy here is that the tail bound (14.20) applies only to a *deterministic* radius r, as opposed to the random[1] radius $\|f\|_2$. This difficulty can be addressed by using a so-called "peeling" argument. For $m = 1, 2, \ldots$, define the events

$$\mathcal{S}_m := \{f \in \mathscr{F} \mid 2^{m-1}\delta_n \leq \|f\|_2 \leq 2^m \delta_n\}.$$

Since $\|f\|_2 \leq \|f\|_\infty \leq 1$ by assumption, any function $\mathscr{F} \cap \{\|f\|_2 \geq \delta_n\}$ belongs to some $\mathcal{S}_m$ for $m \in \{1, 2, \ldots, M\}$, where $M \leq 4\log(1/\delta_n)$.

By the union bound, we have $\mathbb{P}(\mathcal{A}_1) \leq \sum_{m=1}^{M} \mathbb{P}(\mathcal{A}_1 \cap \mathcal{S}_m)$. Now if the event $\mathcal{A}_1 \cap \mathcal{S}_m$ occurs, then there is a function f with $\|f\|_2 \leq r_m := 2^m\delta_n$ such that

$$\left|\|f\|_n^2 - \|f\|_2^2\right| \geq \|f\|_2\, \delta_n \geq \tfrac{1}{2} r_m \delta_n.$$

Consequently, we have $\mathbb{P}[\mathcal{S}_m \cap \mathcal{E}_1] \leq \mathbb{P}[Z(r_m) \geq \frac{1}{2} r_m \delta_n] \leq e^{-c_2 n\delta_n^2}$, and putting together the pieces yields

$$\mathbb{P}[\mathcal{A}_1] \leq \sum_{m=1}^{M} e^{-n\delta_n^2/16} \leq e^{-c_2 n\delta_n^2 + \log M} \leq e^{-\frac{c_2 n\delta_n^2}{2}},$$

where the final step follows from the assumed inequality $\frac{c_2}{2} n\delta_n^2 \geq \log(4\log(1/\delta_n))$. □

14.2 A one-sided uniform law

A potentially limiting aspect of Theorem 14.1 is that it requires the underlying function class to be b-uniformly bounded. To a certain extent, this condition can be relaxed by instead imposing tail conditions of the sub-Gaussian or sub-exponential type. See the bibliographic discussion for references to results of this type.

However, in many applications—including the problem of nonparametric least squares from Chapter 13—it is the *lower bound* on $\|f\|_n^2$ that is of primary interest. As discussed in Chapter 2, for ordinary scalar random variables, such one-sided tail bounds can often be obtained under much milder conditions than their corresponding two-sided analogs. Concretely, in the current context, for any fixed function $f \in \mathscr{F}$, applying the lower tail bound (2.23) to the i.i.d. sequence $\{f(x_i)\}_{i=1}^n$ yields the guarantee

$$\mathbb{P}[\|f\|_n^2 \leq \|f\|_2^2 - t] \leq e^{-\frac{nt^2}{2\,\mathbb{E}[f^4(x)]}}. \tag{14.21}$$

[1] It is random because the norm of the function f that violates the bound is a random variable.

Consequently, whenever the fourth moment can be controlled by some multiple of the second moment, then we can obtain non-trivial lower tail bounds.

Our goal in this section is to derive lower tail bounds of this type that hold uniformly over a given function class. Let us state more precisely the type of fourth-moment control that is required. In particular, suppose that there exists a constant C such that

$$\mathbb{E}[f^4(x)] \leq C^2\, \mathbb{E}[f^2(x)] \qquad \text{for all } f \in \mathscr{F} \text{ with } \|f\|_2 \leq 1. \tag{14.22a}$$

When does a bound of this type hold? It is certainly implied by the global condition

$$\mathbb{E}[f^4(x)] \leq C^2(\mathbb{E}[f^2(x)])^2 \qquad \text{for all } f \in \mathscr{F}. \tag{14.22b}$$

However, as illustrated in Example 14.11 below, there are other function classes for which the milder condition (14.22a) can hold while the stronger condition (14.22b) fails.

Let us illustrate these fourth-moment conditions with some examples.

Example 14.10 (Linear functions and random matrices) For a given vector $\theta \in \mathbb{R}^d$, define the linear function $f_\theta(x) = \langle x, \theta \rangle$, and consider the class of all linear functions $\mathscr{F}_{\text{lin}} = \{f_\theta \mid \theta \in \mathbb{R}^d\}$. As discussed in more detail in Example 14.13 to follow shortly, uniform laws for $\|f\|_n^2$ over such a function class are closely related to random matrix theory. Note that the linear function class $\mathscr{F}_{\text{lin}}$ is never uniformly bounded in a meaningful way. Nonetheless, it is still possible for the strong moment condition (14.22b) to hold under certain conditions on the zero-mean random vector x.

For instance, suppose that for each $\theta \in \mathbb{R}^d$, the random variable $f_\theta(x) = \langle x, \theta \rangle$ is Gaussian. In this case, using the standard formula (2.54) for the moments of a Gaussian random vector, we have $\mathbb{E}[f_\theta^4(x)] = 3(\mathbb{E}[f_\theta^2(x)])^2$, showing that condition (14.22b) holds uniformly with $C^2 = 3$. Note that C does not depend on the variance of $f_\theta(x)$, which can be arbitrarily large. Exercise 14.6 provides some examples of non-Gaussian variables for which the fourth-moment condition (14.22b) holds in application to linear functions. ♣

Example 14.11 (Additive nonparametric models) Given a univariate function class $\mathscr{G}$, consider the class of functions on $\mathbb{R}^d$ given by

$$\mathscr{F}_{\text{add}} = \{f : \mathbb{R}^d \to \mathbb{R} \mid f = \sum_{j=1}^d g_j \quad \text{for some } g_j \in \mathscr{G}\}. \tag{14.23}$$

The problem of estimating a function of this type is known as *additive regression*, and it provides one avenue for escaping the curse of dimension; see the bibliographic section for further discussion.

Suppose that the univariate function class $\mathscr{G}$ is uniformly bounded, say $\|g_j\|_\infty \leq b$ for all $g_j \in \mathscr{G}$, and consider a distribution over $x \in \mathbb{R}^d$ under which each $g_j(x_j)$ is a zero-mean random variable. (This latter assumption can always be ensured by a recentering step.) Assume moreover that the design vector $x \in \mathbb{R}^d$ has four-way independent components—that is, for any distinct quadruple (j, k, ℓ, m), the random variables (x_j, x_k, x_ℓ, x_m) are jointly independent. For a given $\delta \in (0, 1]$, consider a function $f = \sum_{j=1}^d g_j \in \mathscr{F}$ such that $\mathbb{E}[f^2(x)] = \delta^2$, or equivalently, using our independence conditions, such that

$$\mathbb{E}[f^2(x)] = \sum_{j=1}^{d} \|g_j\|_2^2 = \delta^2.$$

For any such function, the fourth moment can be bounded as

$$\begin{aligned}\mathbb{E}[f^4(x)] = \mathbb{E}\Big[\Big(\sum_{j=1}^{d} g_j(x_j)\Big)^4\Big] &= \sum_{j=1}^{d} \mathbb{E}[g_j^4(x_j)] + 6 \sum_{j \neq k} \mathbb{E}[g_j^2(x_j)]\mathbb{E}[g_k^2(x_k)] \\ &\le \sum_{j=1}^{d} \mathbb{E}[g_j^4(x_j)] + 6\delta^4,\end{aligned}$$

where we have used the zero-mean property, and the four-way independence of the coordinates. Since $\|g_j\|_\infty \le b$ for each $g_j \in \mathscr{G}$, we have $\mathbb{E}[g_j^4(x_j)] \le b^2\mathbb{E}[g_j^2(x_j)]$, and putting together the pieces yields

$$\mathbb{E}[f^4(x)] \le b^2\delta^2 + 6\delta^4 \le (b^2 + 6)\,\delta^2,$$

where the final step uses the fact that $\delta \le 1$ by assumption. Consequently, for any $\delta \in (0, 1]$, the weaker condition (14.22a) holds with $C^2 = b^2 + 6$. ♣

Having seen some examples of function classes that satisfy the moment conditions (14.22a) and/or (14.22b), let us now state a one-sided uniform law. Recalling that $\bar{\mathcal{R}}_n$ denotes the population Rademacher complexity, consider the usual type of inequality

$$\frac{\bar{\mathcal{R}}_n(\delta; \mathscr{F})}{\delta} \le \frac{\delta}{128C}, \tag{14.24}$$

where the constant C appears in the fourth-moment condition (14.22a). Our statement also involves the convenient shorthand $\mathbb{B}_2(\delta) := \{f \in \mathscr{F} \mid \|f\|_2 \le \delta\}$.

Theorem 14.12 *Consider a star-shaped class $\mathscr{F}$ of functions, each zero-mean under $\mathbb{P}$, and such that the fourth-moment condition* (14.22a) *holds uniformly over $\mathscr{F}$, and suppose that the sample size n is large enough to ensure that there is a solution $\delta_n \le 1$ to the inequality* (14.24). *Then for any $\delta \in [\delta_n, 1]$, we have*

$$\|f\|_n^2 \ge \tfrac{1}{2}\|f\|_2^2 \qquad \text{for all } f \in \mathscr{F} \setminus \mathbb{B}_2(\delta) \tag{14.25}$$

with probability at least $1 - e^{-c_1 \frac{n\delta^2}{C^2}}$.

Remark: The set $\mathscr{F} \setminus \mathbb{B}_2(\delta)$ can be replaced with $\mathscr{F}$ whenever the set $\mathscr{F} \cap \mathbb{B}_2(\delta)$ is cone-like—that is, whenever any non-zero function $f \in \mathbb{B}_2(\delta) \cap \mathscr{F}$ can be rescaled by $\alpha := \delta/\|f\|_2 \ge 1$, thereby yielding a new function $g := \alpha f$ that *remains within* $\mathscr{F}$.

In order to illustrate Theorem 14.12, let us revisit our earlier examples.

Example 14.13 (Linear functions and random matrices, *continued*) Recall the linear function class $\mathscr{F}_{\text{lin}}$ introduced previously in Example 14.10. Uniform laws over this function class are closely related to earlier results on non-asymptotic random matrix theory from Chapter 6. In particular, supposing that the design vector x has a zero-mean distribution with covariance matrix $\mathbf{\Sigma}$, the function $f_\theta(x) = \langle x, \theta \rangle$ has $L^2(\mathbb{P})$-norm

$$\|f_\theta\|_2^2 = \theta^{\mathrm{T}} \mathbb{E}[xx^{\mathrm{T}}]\theta = \|\sqrt{\mathbf{\Sigma}}\theta\|_2^2 \qquad \text{for each } f_\theta \in \mathscr{F}. \tag{14.26}$$

On the other hand, given a set of n samples $\{x_i\}_{i=1}^n$, we have

$$\|f_\theta\|_n^2 = \frac{1}{n}\sum_{i=1}^n \langle x_i, \theta\rangle^2 = \frac{1}{n}\|\mathbf{X}\theta\|_2^2, \tag{14.27}$$

where the design matrix $\mathbf{X} \in \mathbb{R}^{n\times d}$ has the vector x_i^{T} as its ith row. Consequently, in application to this function class, Theorem 14.12 provides a uniform lower bound on the quadratic forms $\frac{1}{n}\|\mathbf{X}\theta\|_2^2$: in particular, as long as the sample size n is large enough to ensure that $\delta_n \le 1$, we have

$$\frac{1}{n}\|\mathbf{X}\theta\|_2^2 \ge \frac{1}{2}\|\sqrt{\mathbf{\Sigma}}\theta\|_2^2 \qquad \text{for all } \theta \in \mathbb{R}^d. \tag{14.28}$$

As one concrete example, suppose that the covariate vector x follows a $\mathcal{N}(0, \mathbf{\Sigma})$ distribution. For any $\theta \in \mathbb{S}^{d-1}$, the random variable $\langle x, \theta\rangle$ is sub-Gaussian with parameter at most $|||\sqrt{\mathbf{\Sigma}}|||_2$, but this quantity could be very large, and potentially growing with the dimension d. However, as discussed in Example 14.10, the strong moment condition (14.22b) always holds with $C^2 = 3$, regardless of the size of $|||\sqrt{\mathbf{\Sigma}}|||_2$. In order to apply Theorem 14.12, we need to determine a positive solution δ_n to the inequality (14.24). Writing each $x = \sqrt{\mathbf{\Sigma}}w$, where $w \sim \mathcal{N}(0, \mathbf{\Sigma})$, note that we have $\|f_\theta(x)\|_2 = \|\sqrt{\mathbf{\Sigma}}\theta\|_2$. Consequently, by definition of the local Rademacher complexity, we have

$$\bar{\mathcal{R}}_n(\delta; \mathscr{F}_{\text{lin}}) = \mathbb{E}\Big[\sup_{\substack{\theta\in\mathbb{R}^d \\ \|\sqrt{\mathbf{\Sigma}}\theta\|_2 \le \delta}} \Big|\Big\langle \frac{1}{n}\sum_{i=1}^n \varepsilon_i w_i,\ \sqrt{\mathbf{\Sigma}}\theta\Big\rangle\Big|\Big] = \delta\, \mathbb{E}\|\frac{1}{n}\sum_{i=1}^n \varepsilon_i w_i\|_2.$$

Note that the random variables $\{\varepsilon_i w_i\}_{i=1}^n$ are i.i.d. and standard Gaussian (since the symmetrization by independent Rademacher variables has no effect). Consequently, previous results from Chapter 2 guarantee that $\mathbb{E}\|\frac{1}{n}\sum_{i=1}^n \varepsilon_i w_i\|_2 \le \sqrt{\frac{d}{n}}$. Putting together the pieces, we conclude that $\delta_n^2 \precsim \frac{d}{n}$. Therefore, for this particular ensemble, Theorem 14.12 implies that, as long as $n \succsim d$, then

$$\frac{\|\mathbf{X}\theta\|_2^2}{n} \ge \frac{1}{2}\|\sqrt{\mathbf{\Sigma}}\theta\|_2^2 \qquad \text{for all } \theta \in \mathbb{R}^d \tag{14.29}$$

with high probability. The key part of this lower bound is that the maximum eigenvalue $|||\sqrt{\mathbf{\Sigma}}|||_2$ never enters the result.

As another concrete example, the four-way independent and B-bounded random variables described in Exercise 14.6 also satisfy the moment condition (14.22b) with $C^2 = B + 6$. A similar calculation then shows that, with high probability, this ensemble also satisfies a lower bound of the form (14.29) where $\mathbf{\Sigma} = \mathbf{I}_d$. Note that these random variables need not be sub-Gaussian—in fact, the condition does not even require the existence of moments larger than four. ♣

In Exercise 14.7, we illustrate the use of Theorem 14.12 for controlling the restricted eigenvalues (RE) of some random matrix ensembles.

Let us now return to a nonparametric example:

Example 14.14 (Additive nonparametric models, *continued*) In this example, we return to the class $\mathscr{F}_{\text{add}}$ of additive nonparametric models previously introduced in Example 14.11. We let ε_n be the critical radius for the univariate function class $\mathscr{G}$ in the definition (14.23); thus, the scalar ε_n satisfies an inequality of the form $\bar{\mathcal{R}}_n(\varepsilon; \mathscr{F}) \precsim \varepsilon^2$. In Exercise 14.8, we prove that the critical radius δ_n for the d-dimensional additive family $\mathscr{F}_{\text{add}}$ satisfies the upper bound $\delta_n \precsim \sqrt{d}\, \varepsilon_n$. Consequently, Theorem 14.12 guarantees that

$$\|f\|_n^2 \geq \tfrac{1}{2}\|f\|_2^2 \qquad \text{for all } f \in \mathscr{F}_{\text{add}} \text{ with } \|f\|_2 \geq c_0 \sqrt{d}\, \varepsilon_n \tag{14.30}$$

with probability at least $1 - e^{-c_1 n d \varepsilon_n^2}$.

As a concrete example, suppose that the univariate function class $\mathscr{G}$ is given by a first-order Sobolev space; for such a family, the univariate rate scales as $\varepsilon_n^2 \asymp n^{-2/3}$ (see Example 13.20 for details). For this particular class of additive models, with probability at least $1 - e^{-c_1 d n^{1/3}}$, we are guaranteed that

$$\underbrace{\|\sum_{j=1}^{d} g_j\|_n^2}_{\|f\|_n^2} \geq \frac{1}{2} \underbrace{\sum_{j=1}^{d} \|g_j\|_2^2}_{\|f\|_2^2} \tag{14.31}$$

uniformly over all functions of the form $f = \sum_{j=1}^{d} g_j$ with $\|f\|_2 \succsim \sqrt{d} n^{-1/3}$. ♣

14.2.1 Consequences for nonparametric least squares

Theorem 14.12, in conjunction with our earlier results from Chapter 13, has some immediate corollaries for nonparametric least squares. Recall the standard model for nonparametric regression, in which we observe noisy samples of the form $y_i = f^*(x_i) + \sigma w_i$, where $f^* \in \mathscr{F}$ is the unknown regression function. Our corollary involves the local complexity of the shifted function class $\mathscr{F}^* = \mathscr{F} - f^*$.

We let δ_n and ε_n (respectively) be any positive solutions to the inequalities

$$\frac{\bar{\mathcal{R}}_n(\delta; \mathscr{F})}{\delta} \overset{\text{(i)}}{\leq} \frac{\delta}{128C} \quad \text{and} \quad \frac{\mathcal{G}_n(\varepsilon; \mathscr{F}^*)}{\varepsilon} \overset{\text{(ii)}}{\leq} \frac{\varepsilon}{2\sigma}, \tag{14.32}$$

where the localized Gaussian complexity $\mathcal{G}_n(\varepsilon; \mathscr{F}^*)$ was defined in equation (13.16), prior to the statement of Theorem 13.5. To be clear, the quantity ε_n is a random variable, since it depends on the covariates $\{x_i\}_{i=1}^n$, which are modeled as random in this chapter.

Corollary 14.15 *Under the conditions of Theorems 13.5 and 14.12, there are universal positive constants* (c_0, c_1, c_2) *such that the nonparametric least-squares estimate* $\widehat{f}$ *satisfies*

$$\mathbb{P}_{w,x}\Big[\|\widehat{f} - f^*\|_2^2 \geq c_0\,(\varepsilon_n^2 + \delta_n^2)\Big] \leq c_1 e^{-c_2 \frac{n\delta_n^2}{\sigma^2 + C^2}}. \tag{14.33}$$

Proof We split the argument into two cases:

Case 1: Suppose that $\delta_n \geq \varepsilon_n$. We are then guaranteed that δ_n is a solution to inequality (ii) in equation (14.32). Consequently, we may apply Theorem 13.5 with $t = \delta_n$ to find that

$$\mathbb{P}_w[\|\widehat{f} - f^*\|_n \geq 16\delta_n^2] \leq e^{-\frac{n\delta_n^2}{2\sigma^2}}.$$

On the other hand, Theorem 14.12 implies that

$$\mathbb{P}_{x,w}\Big[\|\widehat{f} - f^*\|_2^2 \geq 2\delta_n^2 + 2\|\widehat{f} - f^*\|_n^2\Big] \leq e^{-c_2 \frac{n\delta_n^2}{C^2}}.$$

Putting together the pieces yields that

$$\mathbb{P}_{x,w}\Big[\|\widehat{f} - f^*\|_2^2 \geq c_0\delta_n^2\Big] \leq c_1 e^{-c_2 \frac{n\delta_n^2}{\sigma^2 + C^2}},$$

which implies the claim.

Case 2: Otherwise, we may assume that the event $\mathcal{A} := \{\delta_n < \varepsilon_n\}$ holds. Note that this event depends on the random covariates $\{x_i\}_{i=1}^n$ via the random quantity ε_n. It suffices to bound the probability of the event $\mathcal{E} \cap \mathcal{A}$, where

$$\mathcal{E} := \Big\{\|\widehat{f} - f^*\|_2^2 \geq 16\varepsilon_n^2 + 2\delta_n^2\Big\}.$$

In order to do so, we introduce a third event, namely $\mathcal{B} := \{\|\widehat{f} - f^*\|_n^2 \leq 8\varepsilon_n^2\}$, and make note of the upper bound

$$\mathbb{P}[\mathcal{E} \cap \mathcal{A}] \leq \mathbb{P}[\mathcal{E} \cap \mathcal{B}] + \mathbb{P}[\mathcal{A} \cap \mathcal{B}^c].$$

On one hand, we have

$$\mathbb{P}[\mathcal{E} \cap \mathcal{B}] \leq \mathbb{P}\Big[\|\widehat{f} - f^*\|_2^2 \geq 2\|\widehat{f} - f^*\|_n^2 + 2\delta_n^2\Big] \leq e^{-c_2 \frac{n\delta_n^2}{C^2}},$$

where the final inequality follows from Theorem 14.12.

On the other hand, let $\mathbb{I}[\mathcal{A}]$ be a zero–one indicator for the event $\mathcal{A} := \{\delta_n < \varepsilon_n\}$. Then applying Theorem 13.5 with $t = \varepsilon_n$ yields

$$\mathbb{P}[\mathcal{A} \cap \mathcal{B}^c] \leq \mathbb{E}_x\Big[e^{-\frac{n\varepsilon_n^2}{2\sigma^2}}\,\mathbb{I}[\mathcal{A}]\Big] \leq e^{-\frac{n\delta_n^2}{2\sigma^2}}.$$

Putting together the pieces yields the claim. □

14.2.2 Proof of Theorem 14.12

Let us now turn to the proof of Theorem 14.12. We first claim that it suffices to consider functions belonging to the boundary of the δ-ball—namely, the set $\partial\mathbb{B}_2(\delta) = \{f \in \mathscr{F} \mid \|f\|_2 = \delta\}$. Indeed, suppose that the inequality (14.25) is violated for some $g \in \mathscr{F}$ with $\|g\|_2 > \delta$. By the star-shaped condition, the function $f := \frac{\delta}{\|g\|_2} g$ belongs to $\mathscr{F}$ and has norm $\|f\|_2 = \delta$. Finally, by rescaling, the inequality $\|g\|_n^2 < \frac{1}{2}\|g\|_2^2$ is equivalent to $\|f\|_n^2 < \frac{1}{2}\|f\|_2^2$.

For any function $f \in \partial\mathbb{B}_2(\delta)$, it is equivalent to show that

$$\|f\|_n^2 \geq \frac{3}{4}\|f\|_2^2 - \frac{\delta^2}{4}. \tag{14.34}$$

In order to prove this bound, we make use of a truncation argument. For a level $\tau > 0$ to be chosen, consider the truncated quadratic

$$\varphi_\tau(u) := \begin{cases} u^2 & \text{if } |u| \leq \tau, \\ \tau^2 & \text{otherwise,} \end{cases} \tag{14.35}$$

and define $f_\tau(x) = \text{sign}(f(x))\sqrt{\varphi_\tau(f(x))}$. By construction, for any $f \in \partial\mathbb{B}_2(\delta)$, we have $\|f\|_n^2 \geq \|f_\tau\|_n^2$, and hence

$$\|f\|_n^2 \geq \|f_\tau\|_2^2 - \sup_{f \in \partial\mathbb{B}_2(\delta)} \left| \|f_\tau\|_n^2 - \|f_\tau\|_2^2 \right|. \tag{14.36}$$

The remainder of the proof consists of showing that a suitable choice of truncation level τ ensures that

$$\|f_\tau\|_2^2 \geq \tfrac{3}{4}\|f\|_2^2 \qquad \text{for all } f \in \partial\mathbb{B}_2(\delta) \tag{14.37a}$$

and

$$\mathbb{P}[Z_n \geq \tfrac{1}{4}\delta^2] \leq c_1 e^{-c_2 n\delta^2} \qquad \text{where } Z_n := \sup_{f \in \partial\mathbb{B}_2(\delta)} \left| \|f_\tau\|_n^2 - \|f_\tau\|_2^2 \right|. \tag{14.37b}$$

These two bounds in conjunction imply that the lower bound (14.34) holds with probability at least $1 - c_1 e^{-c_2 n\delta^2}$, uniformly all f with $\|f\|_2 = \delta$.

Proof of claim (14.37a): Letting $\mathbb{I}[|f(x)| \geq \tau]$ be a zero–one indicator for the event $|f(x)| \geq \tau$, we have

$$\|f\|_2^2 - \|f_\tau\|_2^2 \leq \mathbb{E}\left[f^2(x)\, \mathbb{I}[|f(x)| \geq \tau]\right] \leq \sqrt{\mathbb{E}[f^4(x)]}\, \sqrt{\mathbb{P}[|f(x)| \geq \tau]},$$

where the last step uses the Cauchy–Schwarz inequality. Combining the moment bound (14.22a) with Markov's inequality yields

$$\|f\|_2^2 - \|f_\tau\|_2^2 \leq C\,\|f\|_2 \sqrt{\frac{\mathbb{E}[f^4(x)]}{\tau^4}} \leq C^2 \frac{\|f\|_2^2}{\tau^2},$$

where the final inequality uses the moment bound (14.22a) again. Setting $\tau^2 = 4C^2$ yields the bound $\|f\|_2^2 - \|f_\tau\|_2^2 \leq \frac{1}{4}\|f\|_2^2$, which is equivalent to the claim (14.37a).

Proof of claim (14.37b): Beginning with the expectation, a standard symmetrization argument (see Proposition 4.11) guarantees that

$$\mathbb{E}_x[Z_n] \le 2\,\mathbb{E}_{x,\varepsilon}\Big[\sup_{f\in\mathbb{B}_2(\delta;\mathscr{F})}\Big|\frac{1}{n}\sum_{i=1}^n \varepsilon_i f_\tau^2(x_i)\Big|\Big].$$

Our truncation procedure ensures that $f_\tau^2(x) = \varphi_\tau(f(x))$, where φ_τ is a Lipschitz function with constant $L = 2\tau$. Consequently, the Ledoux–Talagrand contraction inequality (5.61) guarantees that

$$\mathbb{E}_x[Z_n] \le 8\tau\mathbb{E}_{x,\varepsilon}\Big[\sup_{f\in\mathbb{B}_2(\delta;\mathscr{F})}\Big|\frac{1}{n}\sum_{i=1}^n \varepsilon_i f(x_i)\Big|\Big] \le 8\tau\bar{\mathcal{R}}_n(\delta;\mathscr{F}) \le 8\tau\frac{\delta^2}{128C},$$

where the final step uses the assumed inequality $\bar{\mathcal{R}}_n(\delta;\mathscr{F}) \le \frac{\delta^2}{128C}$. Our previous choice $\tau = 2C$ ensures that $\mathbb{E}_x[Z_n] \le \frac{1}{8}\delta^2$.

Next we prove an upper tail bound on the random variable Z_n, in particular using Talagrand's theorem for empirical processes (Theorem 3.27). By construction, we have $\|f_\tau^2\|_\infty \le \tau^2 = 4C^2$, and

$$\mathrm{var}(f_\tau^2(x)) \le \mathbb{E}[f_\tau^4(x)] \le \tau^2\|f\|_2^2 = 4C^2\,\delta^2.$$

Consequently, Talagrand's inequality (3.83) implies that

$$\mathbb{P}[Z_n \ge \mathbb{E}[Z_n] + u] \le c_1\exp\Big(-\frac{c_2 n u^2}{C\delta^2 + C^2 u}\Big). \tag{14.38}$$

Since $\mathbb{E}[Z_n] \le \frac{\delta^2}{8}$, the claim (14.37b) follows by setting $u = \frac{\delta^2}{8}$.

14.3 A uniform law for Lipschitz cost functions

Up to this point, we have considered uniform laws for the difference between the empirical squared norm $\|f\|_n^2$ and its expectation $\|f\|_2^2$. As formalized in Corollary 14.15, such results are useful, for example, in deriving bounds on the $L^2(\mathbb{P})$-error of the nonparametric least-squares estimator. In this section, we turn to a more general class of prediction problems, and a type of uniform law that is useful for many of them.

14.3.1 General prediction problems

A general prediction problem can be specified in terms of a space $\mathcal{X}$ of covariates or predictors, and a space $\mathcal{Y}$ of response variables. A predictor is a function f that maps a covariate $x \in \mathcal{X}$ to a prediction $\widehat{y} = f(x) \in \widetilde{\mathcal{Y}}$. Here the space $\widetilde{\mathcal{Y}}$ may be either the same as the response space $\mathcal{Y}$, or a superset thereof. The goodness of a predictor f is measured in terms of a cost function $\mathcal{L}\colon \widetilde{\mathcal{Y}}\times\mathcal{Y} \to \mathbb{R}$, whose value $\mathcal{L}(\widehat{y}, y)$ corresponds to the cost of predicting $\widehat{y} \in \widetilde{\mathcal{Y}}$ when the underlying true response is some $y \in \mathcal{Y}$. Given a collection of n samples $\{(x_i, y_i)\}_{i=1}^n$, a natural way in which to determine a predictor is by minimizing the empirical cost

$$\mathbb{P}_n(\mathcal{L}(f(x), y)) := \frac{1}{n}\sum_{i=1}^n \mathcal{L}(f(x_i), y_i). \tag{14.39}$$

Although the estimator $\widehat{f}$ is obtained by minimizing the empirical cost (14.39), our ultimate goal is in assessing its quality when measured in terms of the population cost function

$$\mathbb{P}(\mathcal{L}(f(x), y)) := \mathbb{E}_{x,y}[\mathcal{L}(f(x), y)], \tag{14.40}$$

and our goal is thus to understand when a minimizer of the empirical cost (14.39) is a near-minimizer of the population cost.

As discussed previously in Chapter 4, this question can be addressed by deriving a suitable type of uniform law of large numbers. More precisely, for each $f \in \mathscr{F}$, let us define the function $\mathcal{L}_f : \mathcal{X} \times \mathcal{Y} \to \mathbb{R}_+$ via $\mathcal{L}_f(x, y) = \mathcal{L}(f(x), y)$, and let us write

$$\mathbb{P}_n(\mathcal{L}_f) = \mathbb{P}_n(\mathcal{L}(f(x), y)) \quad \text{and} \quad \overline{\mathcal{L}}_f := \mathbb{P}(\mathcal{L}_f) = \mathbb{P}(\mathcal{L}(f(x), y)).$$

In terms of this convenient shorthand, our question can be understand as deriving a Glivenko–Cantelli law for the so-called *cost class* $\{\mathcal{L}_f \mid f \in \mathscr{F}\}$.

Throughout this section, we study prediction problems for which $\mathcal{Y}$ is some subset of the real line $\mathbb{R}$. For a given constant $L > 0$, we say that the cost function $\mathcal{L}$ is *L-Lipschitz in its first argument* if

$$|\mathcal{L}(z, y) - \mathcal{L}(\widetilde{z}, y)| \leq L|z - \widetilde{z}| \tag{14.41}$$

for all pairs $z, \widetilde{z} \in \widetilde{\mathcal{Y}}$ and $y \in \mathcal{Y}$. We say that the population cost function $f \mapsto \mathbb{P}(\mathcal{L}_f)$ is *γ-strongly convex* with respect to the $L^2(\mathbb{P})$-norm at f^* if there is some $\gamma > 0$ such that

$$\mathbb{P}\Big[\underbrace{\mathcal{L}_f}_{\mathcal{L}(f(x),y)} - \underbrace{\mathcal{L}_f^*}_{\mathcal{L}(f^*(x),y)} - \underbrace{\frac{\partial \mathcal{L}}{\partial z}\Big|_{f^*}}_{\frac{\partial \mathcal{L}}{\partial z}(f^*(x),y)} \underbrace{(f - f^*)}_{f(x)-f^*(x)}\Big] \geq \frac{\gamma}{2} \|f - f^*\|_2^2 \tag{14.42}$$

for all $f \in \mathscr{F}$. Note that it is sufficient (but not necessary) for the function $z \mapsto \mathcal{L}(z, y)$ to be γ-strongly convex in a pointwise sense for each $y \in \mathcal{Y}$. Let us illustrate these conditions with some examples.

Example 14.16 (Least-squares regression) In a standard regression problem, the response space $\mathcal{Y}$ is the real line or some subset thereof, and our goal is to estimate a regression function $x \mapsto f(x) \in \mathbb{R}$. In Chapter 13, we studied methods for nonparametric regression based on the least-squares cost $\mathcal{L}(z, y) = \frac{1}{2}(y-z)^2$. This cost function is *not* globally Lipschitz in general; however, it does become Lipschitz in certain special cases. For instance, consider the standard observation model $y = f^*(x) + \varepsilon$ in the special case of bounded noise—say $|\varepsilon| \leq c$ for some constant c. If we perform nonparametric regression over a b-uniformly bounded function class $\mathscr{F}$, then for all $f, g \in \mathscr{F}$, we have

$$\begin{aligned}
\big|\mathcal{L}(f(x), y) - \mathcal{L}(g(x), y)\big| &= \tfrac{1}{2}\big|(y - f(x))^2 - (y - g(x))^2\big| \\
&\leq \tfrac{1}{2}\big|f^2(x) - g^2(x)\big| + |y|\,|f(x) - g(x)| \\
&\leq (b + (b + c))|f(x) - g(x)|,
\end{aligned}$$

so that the least squares satisfies the Lipschitz condition (14.41) with $L = 2b + c$. Of course, this example is rather artificial since it excludes any types of non-bounded noise variables ε, including the canonical case of Gaussian noise.

In terms of strong convexity, note that, for any $y \in \mathbb{R}$, the function $z \mapsto \frac{1}{2}(y-z)^2$ is strongly convex with parameter $\gamma = 1$, so that $f \mapsto \mathcal{L}_f$ satisfies the strong convexity condition (14.42) with $\gamma = 1$. ♣

Example 14.17 (Robust forms of regression) A concern with the use of the squared cost function in regression is its potential lack of robustness: if even a very small subset of observations are corrupted, then they can have an extremely large effect on the resulting solution. With this concern in mind, it is interesting to consider a more general family of cost functions, say of the form

$$\mathcal{L}(z, y) = \Psi(y - z), \tag{14.43}$$

where $\Psi\colon \mathbb{R} \to [0, \infty]$ is a function that is a symmetric around zero with $\Psi(0) = 0$, and almost everywhere differentiable with $\|\Psi'\|_\infty \leq L$. Note that the least-squares cost fails to satisfy the required derivative bound, so it does *not* fall within this class.

Examples of cost functions in the family (14.43) include the ℓ_1-norm $\Psi_{\ell_1}(u) = |u|$, as well as Huber's robust function

$$\Psi_{\text{Huber}}(u) = \begin{cases} \dfrac{u^2}{2} & \text{if } |u| \leq \tau, \\ \tau u - \dfrac{\tau^2}{2} & \text{otherwise}, \end{cases} \tag{14.44}$$

where $\tau > 0$ is a parameter to be specified. The Huber cost function offers some sort of compromise between the least-squares cost and the ℓ_1-norm cost function.

By construction, the function Ψ_{ℓ_1} is almost everywhere differentiable with $\|\Psi'_{\ell_1}\|_\infty \leq 1$, whereas the Huber cost function is everywhere differentiable with $\|\Psi'_{\text{Huber}}\|_\infty \leq \tau$. Consequently, the ℓ_1-norm and Huber cost functions satisfy the Lipschitz condition (14.41) with parameters $L = 1$ and $L = \tau$, respectively. Moreover, since the Huber cost function is locally equivalent to the least-squares cost, the induced cost function (14.43) is locally strongly convex under fairly mild tail conditions on the random variable $y - f(x)$. ♣

Example 14.18 (Logistic regression) The goal of binary classification is to predict a label $y \in \{-1, +1\}$ on the basis of a covariate vector $x \in \mathcal{X}$. Suppose that we model the conditional distribution of the label $y \in \{-1, +1\}$ as

$$\mathbb{P}_f(y \mid x) = \frac{1}{1 + e^{-2yf(x)}}, \tag{14.45}$$

where $f\colon \mathcal{X} \to \mathbb{R}$ is the discriminant function to be estimated. The method of maximum likelihood then corresponds to minimizing the cost function

$$\mathcal{L}_f(x, y) := \mathcal{L}(f(x), y) = \log\left(1 + e^{-2yf(x)}\right). \tag{14.46}$$

It is easy to see that the function $\mathcal{L}$ is 1-Lipschitz in its first argument. Moreover, at the population level, we have

$$\mathbb{P}(\mathcal{L}_f - \mathcal{L}_{f^*}) = \mathbb{E}_{x,y}\Big[\log \frac{1 + e^{-2f(x)y}}{1 + e^{-2f^*(x)y}}\Big] = \mathbb{E}_x[D(\mathbb{P}_{f^*}(\cdot \mid x) \,\|\, \mathbb{P}_f(\cdot \mid x))],$$

corresponding to the expected value of the Kullback–Leibler divergence between the two conditional distributions indexed by f^* and f. Under relatively mild conditions on the behavior of the random variable $f(x)$ as f ranges over $\mathscr{F}$, this cost function will be γ-strongly convex. ♣

Example 14.19 (Support vector machines and hinge cost) Support vector machines are another method for binary classification, again based on estimating discriminant functions $f: \mathcal{X} \to \mathbb{R}$. In their most popular instantiation, the discriminant functions are assumed to belong to some reproducing kernel Hilbert space $\mathbb{H}$, equipped with the norm $\|\cdot\|_{\mathbb{H}}$. The support vector machine is based on the *hinge cost function*

$$\mathcal{L}(f(x), y) = \max\{0, 1 - yf(x)\}, \tag{14.47}$$

which is 1-Lipschitz by inspection. Again, the strong convexity properties of the population cost $f \mapsto \mathbb{P}(\mathcal{L}_f)$ depend on the distribution of the covariates x, and the function class $\mathscr{F}$ over which we optimize.

Given a set of n samples $\{(x_i, y_i)\}_{i=1}^n$, a common choice is to minimize the empirical risk

$$\mathbb{P}_n(\mathcal{L}(f(x), y)) = \frac{1}{n}\sum_{i=1}^n \max\{0, 1 - y_i f(x_i)\}$$

over a ball $\|f\|_{\mathbb{H}} \leq R$ in some reproducing kernel Hilbert space. As explored in Exercise 12.20, this optimization problem can be reformulated as a quadratic program in n dimensions, and so can be solved easily. ♣

14.3.2 Uniform law for Lipschitz cost functions

With these examples as underlying motivation, let us now turn to stating a general uniform law for Lipschitz cost functions. Let $f^* \in \mathscr{F}$ minimize the population cost function $f \mapsto \mathbb{P}(\mathcal{L}_f)$, and consider the shifted function class.

$$\mathscr{F}^* := \{f - f^* \mid f \in \mathscr{F}\}. \tag{14.48}$$

Our uniform law involves the population version of the localized Rademacher complexity

$$\bar{\mathcal{R}}_n(\delta; \mathscr{F}^*) := \mathbb{E}_{x,\varepsilon}\Bigg[\sup_{\substack{g\in\mathscr{F}^* \\ \|g\|_2\leq\delta}} \Big|\frac{1}{n}\sum_{i=1}^n \varepsilon_i\, g(x_i)\Big|\Bigg]. \tag{14.49}$$

Theorem 14.20 (Uniform law for Lipschitz cost functions) *Given a uniformly 1-bounded function class $\mathscr{F}$ that is star-shaped around the population minimizer f^*, let $\delta_n^2 \geq \frac{c}{n}$ be any solution to the inequality*

$$\bar{\mathcal{R}}_n(\delta; \mathscr{F}^*) \leq \delta^2. \tag{14.50}$$

(a) *Suppose that the cost function is L-Lipschitz in its first argument. Then we have*

$$\sup_{f\in\mathscr{F}} \frac{\big|\mathbb{P}_n(\mathcal{L}_f - \mathcal{L}_{f^*}) - \mathbb{P}(\mathcal{L}_f - \mathcal{L}_{f^*})\big|}{\|f - f^*\|_2 + \delta_n} \leq 10L\delta_n \tag{14.51}$$

with probability greater than $1 - c_1 e^{-c_2 n\delta_n^2}$.

(b) *Suppose that the cost function is L-Lipschitz and γ-strongly convex. Then for any* *function $\widehat{f} \in \mathscr{F}$ such that $\mathbb{P}_n(\mathcal{L}_{\widehat{f}} - \mathcal{L}_{f^*}) \le 0$, we have*

$$\|\widehat{f} - f^*\|_2 \le \left(\frac{20L}{\gamma} + 1\right)\delta_n \tag{14.52a}$$

and

$$\mathbb{P}(\mathcal{L}_{\widehat{f}} - \mathcal{L}_{f^*}) \le 10L\left(\frac{20L}{\gamma} + 2\right)\delta_n^2, \tag{14.52b}$$

where both inequalities hold with the same probability as in part (a).

Under certain additional conditions on the function class, part (a) can be used to guarantee consistency of a procedure that chooses $\widehat{f} \in \mathscr{F}$ to minimize the empirical cost $f \mapsto \mathbb{P}_n(\mathcal{L}_f)$ over $\mathscr{F}$. In particular, since $f^* \in \mathscr{F}$ by definition, this procedure ensures that $\mathbb{P}_n(\mathcal{L}_{\widehat{f}} - \mathcal{L}_{f^*}) \le 0$. Consequently, for any function class $\mathscr{F}$ with[2] $\|\cdot\|_2$-diameter at most D, the inequality (14.51) implies that

$$\mathbb{P}(\mathcal{L}_{\widehat{f}}) \le \mathbb{P}(\mathcal{L}_{f^*}) + 10L\delta_n\{2D + \delta_n\} \tag{14.53}$$

with high probability. Thus, the bound (14.53) implies the consistency of the empirical cost minimization procedure in the following sense: up to a term of order δ_n, the value $\mathbb{P}(\mathcal{L}_{\widehat{f}})$ is as small as the optimum $\mathbb{P}(\mathcal{L}_{f^*}) = \min_{f \in \mathscr{F}} \mathbb{P}(\mathcal{L}_f)$.

Proof of Theorem 14.20

The proof is based on an analysis of the family of random variables

$$Z_n(r) = \sup_{\|f-f^*\|_2 \le r} \left|\mathbb{P}_n(\mathcal{L}_f - \mathcal{L}_{f^*}) - \mathbb{P}(\mathcal{L}_f - \mathcal{L}_{f^*})\right|,$$

where $r > 0$ is a radius to be varied. The following lemma provides suitable control on the upper tails of these random variables:

Lemma 14.21 *For each $r \ge \delta_n$, the variable $Z_n(r)$ satisfies the tail bound*

$$\mathbb{P}[Z_n(r) \ge 8L\,r\delta_n + u] \le c_1 \exp\left(-\frac{c_2 n u^2}{L^2 r^2 + Lu}\right). \tag{14.54}$$

Deferring the proof of this intermediate claim for the moment, let us use it to complete the proof of Theorem 14.20; the proof itself is similar to that of Theorem 14.1. Define the events $\mathcal{E}_0 := \{Z_n(\delta_n) \ge 9L\delta_n^2\}$, and

$$\mathcal{E}_1 := \{\exists\, f \in \mathscr{F} \mid |\mathbb{P}_n(\mathcal{L}_f - \mathcal{L}_{f^*}) - \mathbb{P}(\mathcal{L}_f - \mathcal{L}_{f^*})| \ge 10L\delta_n \|f - f^*\|_2 \text{ and } \|f - f^*\|_2 \ge \delta_n\}.$$

If there is some function $f \in \mathscr{F}$ that violates the bound (14.51), then at least one of the events $\mathcal{E}_0$ or $\mathcal{E}_1$ must occur. Applying Lemma 14.21 with $u = L\delta_n^2$ guarantees that $\mathbb{P}[\mathcal{E}_0] \le c_1 e^{-c_2 n \delta_n^2}$. Moreover, using the same peeling argument as in Theorem 14.1, we find

[2] A function class $\mathscr{F}$ has $\|\cdot\|_2$-diameter at most D if $\|f\|_2 \le D$ for all $f \in \mathscr{F}$. In this case, we have $\|\widehat{f} - f^*\|_2 \le 2D$.

that $\mathbb{P}[\mathcal{E}_1] \leq c_1 e^{-c_2' n\delta_n^2}$, valid for all $\delta_n^2 \geq \frac{c}{n}$. Putting together the pieces completes the proof of the claim (14.51) in part (a).

Let us now prove the claims in part (b). By examining the proof of part (a), we see that it actually implies that either $\|\widehat{f} - f^*\|_2 \leq \delta_n$, or

$$\left|\mathbb{P}_n(\mathcal{L}_{\widehat{f}} - \mathcal{L}_{f^*}) - \mathbb{P}(\mathcal{L}_{\widehat{f}} - \mathcal{L}_{f^*})\right| \leq 10L\delta_n \|\widehat{f} - f^*\|_2.$$

Since $\mathbb{P}_n(\mathcal{L}_{\widehat{f}} - \mathcal{L}_{f^*}) \leq 0$ by assumption, we see that any minimizer must satisfy either the bound $\|\widehat{f} - f^*\|_2 \leq \delta_n$, or the bound $\mathbb{P}(\mathcal{L}_{\widehat{f}} - \mathcal{L}_{f^*}) \leq 10L\delta_n\|\widehat{f} - f^*\|_2$. On one hand, if the former inequality holds, then so does inequality (14.52a). On the other hand, if the latter inequality holds, then, combined with the strong convexity condition (14.42), we obtain $\|\widehat{f} - f^*\|_2 \leq \frac{10L}{\gamma}$, which also implies inequality (14.52a).

In order to establish the bound (14.52b), we make use of inequality (14.52a) within the original inequality (14.51); we then perform some algebra, recalling that $\widehat{f}$ satisfies the inequality $\mathbb{P}_n(\mathcal{L}_{\widehat{f}} - \mathcal{L}_{f^*}) \leq 0$.

It remains to prove Lemma 14.21. By a rescaling argument, we may assume that $b = 1$. In order to bound the upper tail of $Z_n(r)$, we need to control the differences $\mathcal{L}_f - \mathcal{L}_{f^*}$ uniformly over all functions $f \in \mathscr{F}$ such that $\|f - f^*\|_2 \leq r$. By the Lipschitz condition on the cost function and the boundedness of the functions f, we have $|\mathcal{L}_f - \mathcal{L}_{f^*}|_\infty \leq L\|f - f^*\|_\infty \leq 2L$. Moreover, we have

$$\operatorname{var}(\mathcal{L}_f - \mathcal{L}_{f^*}) \leq \mathbb{P}[(\mathcal{L}_f - \mathcal{L}_{f^*})^2] \overset{(i)}{\leq} L^2\|f - f^*\|_2^2 \overset{(ii)}{\leq} L^2 r^2,$$

where inequality (i) follows from the Lipschitz condition on the cost function, and inequality (ii) follows since $\|f - f^*\|_2 \leq r$. Consequently, by Talagrand's concentration theorem for empirical processes (Theorem 3.27), we have

$$\mathbb{P}\Big[Z_n(r) \geq 2\mathbb{E}[Z_n(r)] + u\Big] \leq c_1 \exp\left\{-\frac{c_2 n u^2}{L^2 r^2 + Lu}\right\}. \tag{14.55}$$

It remains to upper bound the expectation: in particular, we have

$$\begin{aligned}
\mathbb{E}[Z_n(r)] &\overset{(i)}{\leq} 2\mathbb{E}\Big[\sup_{\|f-f^*\|_2 \leq r} \Big|\frac{1}{n}\sum_{i=1}^n \varepsilon_i\big\{\mathcal{L}(f(x_i), y_i) - \mathcal{L}(f^*(x_i), y_i)\big\}\Big|\Big] \\
&\overset{(ii)}{\leq} 4L\,\mathbb{E}\Big[\sup_{\|f-f^*\|_2 \leq r} \Big|\frac{1}{n}\sum_{i=1}^n \varepsilon_i(f(x_i) - f^*(x_i))\Big|\Big] \\
&= 4L\bar{\mathcal{R}}_n(r; \mathscr{F}^*) \\
&\overset{(iii)}{\leq} 4Lr\delta_n, \qquad \text{valid for all } r \geq \delta_n,
\end{aligned}$$

where step (i) follows from a symmetrization argument; step (ii) follows from the L-Lipschitz condition on the first argument of the cost function, and the Ledoux–Talagrand contraction inequality (5.61); and step (iii) uses the fact that the function $r \mapsto \frac{\bar{\mathcal{R}}_n(r;\mathscr{F}^*)}{r}$ is non-increasing, and our choice of δ_n. Combined with the tail bound (14.55), the proof of Lemma 14.21 is complete.

14.4 Some consequences for nonparametric density estimation

The results and techniques developed thus far have some useful applications to the problem of nonparametric density estimation. The problem is easy to state: given a collection of i.i.d. samples $\{x_i\}_{i=1}^n$, assumed to have been drawn from an unknown distribution with density f^*, how do we estimate the unknown density? The density estimation problem has been the subject of intensive study, and there are many methods for tackling it. In this section, we restrict our attention to two simple methods that are easily analyzed using the results from this and preceding chapters.

14.4.1 Density estimation via the nonparametric maximum likelihood estimate

Perhaps the most easily conceived method for density estimation is via a nonparametric analog of maximum likelihood. In particular, suppose that we fix some base class of densities $\mathscr{F}$, and then maximize the likelihood of the observed samples over this class. Doing so leads to a constrained form of the nonparametric maximum likelihood estimate (MLE)—namely

$$\widehat{f} \in \arg\min_{f \in \mathscr{F}} \mathbb{P}_n(-\log f(x)) = \arg\min_{f \in \mathscr{F}} \left\{ -\frac{1}{n} \sum_{i=1}^n \log f(x_i) \right\}. \tag{14.56}$$

To be clear, the class of densities $\mathscr{F}$ must be suitably restricted for this estimator to be well defined, which we assume to be the case for the present discussion. (See Exercise 14.9 for an example in which the nonparametric MLE $\widehat{f}$ fails to exist.) As an alternative to constraining the estimate, it also possible to define a regularized form of the nonparametric MLE.

In order to illustrate the use of some bounds from this chapter, let us analyze the estimator (14.56) in the simple case when the true density f^* is assumed to belong to $\mathscr{F}$. Given an understanding of this case, it is relatively straightforward to derive a more general result, in which the error is bounded by a combination of estimation error and approximation error terms, with the latter being non-zero when $f^* \notin \mathscr{F}$.

For reasons to be clarified, it is convenient to measure the error in terms of the squared *Hellinger distance*. For densities f and g with respect to a base measure μ, it is given by

$$H^2(f \,\|\, g) := \frac{1}{2} \int_{\mathcal{X}} (\sqrt{f} - \sqrt{g})^2 \, d\mu. \tag{14.57a}$$

As we explore in Exercise 14.10, a useful connection here is that the Kullback–Leibler (KL) divergence is lower bounded by (a multiple of) the squared Hellinger distance—viz.

$$D(f \,\|\, g) \geq 2H^2(f \,\|\, g). \tag{14.57b}$$

Up to a constant pre-factor, the squared Hellinger distance is equivalent to the $L^2(\mu)$-norm difference of the square-root densities. For this reason, the square-root function class $\mathscr{G} = \{g = \sqrt{f} \text{ for some } f \in \mathscr{F}\}$ plays an important role in our analysis, as does the shifted square-root function class $\mathscr{G}^* := \mathscr{G} - \sqrt{f^*}$.

In the relatively simple result to be given here, we assume that there are positive constants (b, ν) such that the square-root density class $\mathscr{G}$ is $\sqrt{b}$-uniformly bounded, and star-shaped around $\sqrt{f^*}$, and moreover that the unknown density $f^* \in \mathscr{F}$ is uniformly lower bounded as

$$f^*(x) \geq \nu > 0 \qquad \text{for all } x \in \mathcal{X}.$$

In terms of the population Rademacher complexity $\bar{\mathcal{R}}_n$, our result involves the critical inequality

$$\bar{\mathcal{R}}_n(\delta; \mathscr{G}^*) \leq \frac{\delta^2}{\sqrt{b+\nu}}. \tag{14.58}$$

With this set-up, we have the following guarantee:

Corollary 14.22 *Given a class of densities satisfying the previous conditions, let δ_n be any solution to the critical inequality (14.58) such that $\delta_n^2 \geq (1 + \frac{b}{\nu})\frac{1}{n}$. Then the nonparametric density estimate $\widehat{f}$ satisfies the Hellinger bound*

$$H^2(\widehat{f} \,\|\, f^*) \leq c_0\, \delta_n^2 \tag{14.59}$$

with probability greater than $1 - c_1 e^{-c_2 \frac{\nu}{b+\nu} n \delta_n^2}$.

Proof Our proof is based on applying Theorem 14.20(b) to the transformed function class

$$\mathcal{H} = \left\{ \sqrt{\frac{f + f^*}{2f^*}} \;\middle|\; f \in \mathscr{F} \right\}$$

equipped with the cost functions $\mathcal{L}_h(x) = -\log h(x)$. Since $\mathscr{F}$ is b-uniformly bounded and $f^*(x) \geq \nu$ for all $x \in \mathcal{X}$, for any $h \in \mathcal{H}$, we have

$$\|h\|_\infty = \left\| \sqrt{\frac{f + f^*}{2f^*}} \right\|_\infty \leq \sqrt{\frac{1}{2}\Big(\frac{b}{\nu} + 1\Big)} = \frac{1}{\sqrt{2\nu}} \sqrt{b + \nu}.$$

Moreover, for any $h \in \mathcal{H}$, we have $h(x) \geq 1/\sqrt{2}$ for all $x \in \mathcal{X}$ and whence the mean value theorem applied to the logarithm, combined with the triangle inequality, implies that

$$\left|\mathcal{L}_h(x) - \mathcal{L}_{\widetilde{h}}(x)\right| \leq \sqrt{2}\, |h(x) - \widetilde{h}(x)| \qquad \text{for all } x \in \mathcal{X}, \text{ and } h, \widetilde{h} \in \mathcal{H},$$

showing that the logarithmic cost function is L-Lipschitz with $L = \sqrt{2}$. Finally, by construction, for any $h \in \mathcal{H}$ and with $h^* := \frac{f^* + f^*}{2f^*} = 1$, we have

$$\|h - h^*\|_2^2 = \mathbb{E}_{f^*}\left[\Big\{\Big(\frac{f + f^*}{2f^*}\Big)^{\frac{1}{2}} - 1\Big\}^2\right] = 2H^2\left(\frac{f + f^*}{2} \,\|\, f^*\right).$$

Therefore, the lower bound (14.57b) on the squared Hellinger distance in terms of the KL divergence is equivalent to asserting that $\mathbb{P}(\mathcal{L}_h - \mathcal{L}_h^*) \geq \|h - h^*\|_2^2$, meaning that the cost function is 2-strongly convex around h^*. Consequently, the claim (14.59) follows via an application of Theorem 14.20(b). □

14.4.2 Density estimation via projections

Another very simple method for density estimation is via projection onto a function class $\mathscr{F}$. Concretely, again given n samples $\{x_i\}_{i=1}^n$, assumed to have been drawn from an unknown density f^* on a space $\mathcal{X}$, consider the projection-based estimator

$$\widehat{f} \in \arg\min_{f\in\mathscr{F}} \left\{\frac{1}{2}\|f\|_2^2 - \mathbb{P}_n(f)\right\} = \arg\min_{f\in\mathscr{F}} \left\{\frac{1}{2}\|f\|_2^2 - \frac{1}{n}\sum_{i=1}^n f(x_i)\right\}. \tag{14.60}$$

For many choices of the underlying function class $\mathscr{F}$, this estimator can be computed in closed form. Let us consider some examples to illustrate.

Example 14.23 (Density estimation via series expansion) This is a follow-up on Example 13.14, where we considered the use of series expansion for regression. Here we consider the use of such expansions for density estimation—say, for concreteness, of univariate densities supported on $[0, 1]$. For a given integer $T \geq 1$, consider a collection of functions $\{\phi_m\}_{m=1}^T$, taken to be orthogonal in $L^2[0, 1]$, and consider the linear function class

$$\mathscr{F}_{\text{ortho}}(T) := \Big\{f = \sum_{m=1}^T \beta_m\phi_m \mid \beta \in \mathbb{R}^T,\ \beta_1 = 1\Big\}. \tag{14.61}$$

As one concrete example, we might define the indicator functions

$$\phi_m(x) = \begin{cases} 1 & \text{if } x \in (m-1, m]/T, \\ 0 & \text{otherwise.} \end{cases} \tag{14.62}$$

With this choice, an expansion of the form $f = \sum_{m=1}^T \beta_m\phi_m(T)$ yields a piecewise constant function that is non-negative and integrates to 1. When used for density estimation, it is known as a *histogram estimate*, and is perhaps the simplest type of density estimate.

Another example is given by truncating the Fourier basis previously described in Example 13.15. In this case, since the first function $\phi_1(x) = 1$ for all $x \in [0, 1]$ and the remaining functions are orthogonal, we are guaranteed that the function expansion integrates to one. The resulting density estimate is known as a *projected Fourier-series estimate*. A minor point is that, since the sinusoidal functions are not non-negative, it is possible that the projected Fourier-series density estimate could take negative values; this concern could be alleviated by projecting the function values back onto the orthant.

For the function class $\mathscr{F}_{\text{ortho}}(T)$, the density estimate (14.60) is straightforward to compute: some calculation shows that

$$\widehat{f}_T = \sum_{m=1}^T \widehat{\beta}_m\phi_m, \qquad \text{where } \widehat{\beta}_m = \frac{1}{n}\sum_{i=1}^n \phi_m(x_i). \tag{14.63}$$

For example, when using the histogram basis (14.62), the coefficient $\widehat{\beta}_m$ corresponds to the fraction of samples that fall into the interval $(m-1, m]/T$. When using a Fourier basis expansion, the estimate $\widehat{\beta}_m$ corresponds to an empirical Fourier-series coefficient. In either case, the estimate $\widehat{f}_T$ is easy to compute.

Figure 14.1 shows plots of histogram estimates of a Gaussian density $N(1/2, (0.15)^2)$, with the plots in Figure 14.1(a) and (b) corresponding to sample sizes $n = 100$ and $n = 2000$, respectively. In addition to the true density in light gray, each plot shows the histogram estimate for $T \in \{5, 20\}$. By construction, each histogram estimate is piecewise constant, and the parameter T determines the length of the pieces, and hence how quickly the estimate varies. For sample size $n = 100$, the estimate with $T = 20$ illustrates the phenomenon of overfitting, whereas for $n = 2000$, the estimate with $T = 5$ leads to oversmoothing.

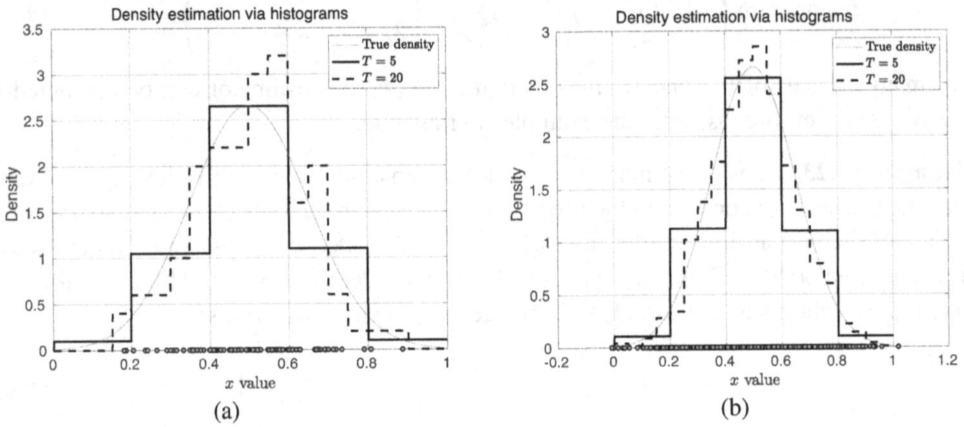

Figure 14.1 Plots of the behavior of the histogram density estimate. Each plot shows the true function (in this case, a Gaussian distribution $\mathcal{N}(1/2, (0.15)^2)$) in light gray and two density estimates using $T = 5$ bins (solid line) and $T = 20$ bins (dashed line). (a) Estimates based on $n = 100$ samples. (b) Estimates based on $n = 2000$ samples.

Figure 14.2 shows some plots of the Fourier-series estimator for estimating the density

$$f^*(x) = \begin{cases} 3/2 & \text{for } x \in [0, 1/2], \\ 1/2 & \text{for } x \in (1/2, 1]. \end{cases} \tag{14.64}$$

As in Figure 14.1, the plots in Figure 14.2(a) and (b) are for sample sizes $n = 100$ and $n = 2000$, respectively, with the true density f^* shown in a gray line. The solid and dashed lines show the truncated Fourier-series estimator with $T = 5$ and $T = 20$ coefficients, respectively. Again, we see overfitting by the estimator with $T = 20$ coefficients when the sample size is small ($n = 100$). For the larger sample size ($n = 2000$), the estimator with $T = 20$ is more accurate than the $T = 5$ estimator, which suffers from oversmoothing. ♣

Having considered some examples of the density estimate (14.60), let us now state a theoretical guarantee on its behavior. As with our earlier results, this guarantee applies to the estimate based on a star-shaped class of densities $\mathscr{F}$, which we assume to be uniformly bounded by some b. Recalling that $\bar{\mathcal{R}}_n$ denotes the (localized) Rademacher complexity, we let $\delta_n > 0$ be any positive solution to the inequality $\bar{\mathcal{R}}_n(\delta; \mathscr{F}) \leq \frac{\delta^2}{b}$.

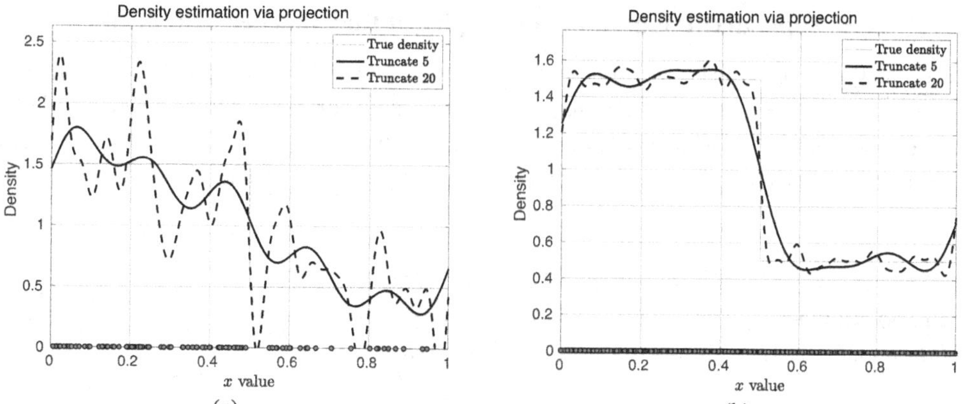

Figure 14.2 Plots of the behavior of the orthogonal series density estimate (14.63) using Fourier series as the orthonormal basis. Each plot shows the true function f^* from equation (14.64) in light gray, and two density estimates for $T = 5$ (solid line) and $T = 20$ (dashed line). (a) Estimates based on $n = 100$ samples. (b) Estimates based on $n = 2000$ samples.

Corollary 14.24 *There are universal constants c_j, $j = 0, 1, 2, 3$, such that for any density f^* uniformly bounded by b, the density estimate* (14.60) *satisfies the oracle inequality*

$$\|\widehat{f} - f^*\|_2^2 \leq c_0 \inf_{f \in \mathscr{F}} \|f - f^*\|_2^2 + c_1 \delta_n^2 \tag{14.65}$$

with probability at least $1 - c_2 e^{-c_3 n \delta_n^2}$.

The proof of this result is very similar to our oracle inequality for nonparametric regression (Theorem 13.13). Accordingly, we leave the details as an exercise for the reader.

14.5 Appendix: Population and empirical Rademacher complexities

Let $\delta_n > 0$ and $\hat{\delta}_n > 0$ be the smallest positive solutions to the inequalities $\bar{\mathcal{R}}_n(\delta_n) \leq \delta_n^2$ and $\widehat{\mathcal{R}}_n(\hat{\delta}_n) \leq \hat{\delta}_n^2$, respectively. Note that these inequalities correspond to our previous definitions (14.4) and (14.7), with $b = 1$. (The general case $b \neq 1$ can be recovered by a rescaling argument.) In this appendix, we show that these quantities satisfy a useful sandwich relation:

Proposition 14.25 *For any 1-bounded and star-shaped function class $\mathscr{F}$, the population and empirical radii satisfy the sandwich relation*

$$\frac{\delta_n}{4} \overset{\text{(i)}}{\leq} \hat{\delta}_n \overset{\text{(ii)}}{\leq} 3\delta_n, \tag{14.66}$$

with probability at least $1 - c_1 e^{-c_2 n \delta_n^2}$.

Proof For each $t > 0$, let us define the random variable

$$\bar{Z}_n(t) := \mathbb{E}_\epsilon\Big[\sup_{\substack{f \in \mathscr{F} \\ \|f\|_2 \leq t}} \Big|\frac{1}{n}\sum_{i=1}^n \varepsilon_i f(x_i)\Big|\Big],$$

so that $\bar{\mathcal{R}}_n(t) = \mathbb{E}_x[\bar{Z}_n(t)]$ by construction. Define the events

$$\mathcal{E}_0(t) := \Big\{\big|\bar{Z}_n(t) - \bar{\mathcal{R}}_n(t)\big| \leq \frac{\delta_n t}{8}\Big\} \quad \text{and} \quad \mathcal{E}_1 := \Big\{\sup_{f \in \mathscr{F}} \frac{\big|\|f\|_n^2 - \|f\|_2^2\big|}{\|f\|_2^2 + \delta_n^2} \leq \frac{1}{2}\Big\}.$$

Note that, conditioned on $\mathcal{E}_1$, we have

$$\|f\|_n \leq \sqrt{\tfrac{3}{2}\|f\|_2^2 + \tfrac{1}{2}\delta_n^2} \leq 2\|f\|_2 + \delta_n \tag{14.67a}$$

and

$$\|f\|_2 \leq \sqrt{2\|f\|_n^2 + \delta_n^2} \leq 2\|f\|_n + \delta_n, \tag{14.67b}$$

where both inequalities hold for all $f \in \mathscr{F}$. Consequently, conditioned on $\mathcal{E}_1$, we have

$$\bar{Z}_n(t) \leq \mathbb{E}_\epsilon\left[\sup_{\substack{f \in \mathscr{F} \\ \|f\|_n \leq 2t + \delta_n}} \Big|\frac{1}{n}\sum_{i=1}^n \varepsilon_i f(x_i)\Big|\right] = \widehat{\mathcal{R}}_n(2t + \delta_n) \tag{14.68a}$$

and

$$\widehat{\mathcal{R}}_n(t) \leq \bar{Z}_n(2t + \delta_n). \tag{14.68b}$$

Equipped with these inequalities, we now proceed to prove our claims.

Proof of upper bound (ii) in (14.66)*:* Conditioned on the events $\mathcal{E}_0(7\delta_n)$ and $\mathcal{E}_1$, we have

$$\widehat{\mathcal{R}}_n(3\delta_n) \overset{\text{(i)}}{\leq} \bar{Z}_n(7\delta_n) \overset{\text{(ii)}}{\leq} \bar{\mathcal{R}}_n(7\delta_n) + \tfrac{7}{8}\delta_n^2,$$

where step (i) follows from inequality (14.68b) with $t = 3\delta_n$, and step (ii) follows from

$\mathcal{E}_0(7\delta_n)$. Since $7\delta_n \geq \delta_n$, the argument used to establish the bound (14.19) guarantees that $\mathcal{R}_n(7\delta_n) \leq 7\delta_n^2$. Putting together the pieces, we have proved that

$$\widehat{\mathcal{R}}_n(3\delta_n) \leq 8\delta_n^2 < (3\delta_n)^2.$$

By definition, the quantity $\hat{\delta}_n$ is the smallest positive number satisfying this inequality, so that we conclude that $\hat{\delta}_n \leq 3\delta_n$, as claimed.

Proof of lower bound (i) in (14.66)*:* Conditioning on the events $\mathcal{E}_0(\delta_n)$ and $\mathcal{E}_1$, we have

$$\delta_n^2 = \bar{\mathcal{R}}_n(\delta_n) \overset{\text{(i)}}{\leq} \bar{Z}_n(\delta_n) + \tfrac{1}{8}\delta_n^2 \overset{\text{(ii)}}{\leq} \widehat{\mathcal{R}}_n(3\delta_n) + \tfrac{1}{8}\delta_n^2 \overset{\text{(iii)}}{\leq} 3\delta_n\hat{\delta}_n + \tfrac{1}{8}\delta_n^2,$$

where step (i) follows $\mathcal{E}_0(\delta_n)$, step (ii) follows from inequality (14.68a) with $t = \delta_n$, and step (iii) follows from the same argument leading to equation (14.19). Rearranging yields that $\frac{7}{8}\delta_n^2 \leq 3\delta_n\hat{\delta}_n$, which implies that $\hat{\delta}_n \geq \delta_n/4$.

Bounding the probabilities of $\mathcal{E}_0(t)$ and $\mathcal{E}_1$: On one hand, Theorem 14.1 implies that $\mathbb{P}[\mathcal{E}_1^c] \leq c_1 e^{-c_2 n\delta_n^2}$.

Otherwise, we need to bound the probability $\mathbb{P}[\mathcal{E}_0^c(\alpha\delta_n)]$ for an arbitrary constant $\alpha \geq 1$. In particular, our proof requires control for the choices $\alpha = 1$ and $\alpha = 7$. From theorem 16 of Bousquet et al. (2003), we have

$$\mathbb{P}[\mathcal{E}_0^c(\alpha\delta_n)] = \mathbb{P}\Big[\big|\bar{Z}_n(\alpha\delta_n) - \bar{\mathcal{R}}_n(\alpha\delta_n)\big| \geq \frac{\alpha\delta_n^2}{8}\Big] \leq 2\exp\Big(-\frac{1}{64}\,\frac{n\alpha\delta_n^4}{2\bar{\mathcal{R}}_n(\alpha\delta_n) + \frac{\alpha\delta_n^2}{12}}\Big).$$

For any $\alpha \geq 1$, we have $\bar{\mathcal{R}}_n(\alpha\delta_n) \geq \bar{\mathcal{R}}_n(\delta_n) = \delta_n^2$, whence $\mathbb{P}[\mathcal{E}_0^c(\alpha\delta_n)] \leq 2e^{-c_2 n\delta_n^2}$. □

14.6 Bibliographic details and background

The localized forms of the Rademacher and Gaussian complexities used in this chapter are standard objects in mathematical statistics (Koltchinskii, 2001, 2006; Bartlett et al., 2005). Localized entropy integrals, such as the one underlying Corollary 14.3, were introduced by van de Geer (2000). The two-sided results given in Section 14.1 are based on b-uniform boundedness conditions on the functions. This assumption, common in much of non-asymptotic empirical process theory, allows for the use of standard concentration inequalities for empirical processes (e.g., Theorem 3.27) and the Ledoux–Talagrand contraction inequality (5.61). For certain classes of unbounded functions, two-sided bounds can also be obtained based on sub-Gaussian and/or sub-exponential tail conditions; for instance, see the papers (Mendelson et al., 2007; Adamczak, 2008; Adamczak et al., 2010; Mendelson, 2010) for results of this type. One-sided uniform laws related to Theorem 14.12 have been proved by various authors (Raskutti et al., 2012; Oliveira, 2013; Mendelson, 2015). The proof given here is based on a truncation argument.

Results on the localized Rademacher complexities, as stated in Corollary 14.5, can be found in Mendelson (2002). The class of additive regression models from Example 14.11 were introduced by Stone (1985), and have been studied in great depth (e.g., Hastie and Tibshirani, 1986; Buja et al., 1989). An interesting extension is the class of sparse additive models, in which the function f is restricted to have a decomposition using at most $s \ll d$

univariate functions; such models have been the focus of more recent study (e.g., Meier et al., 2009; Ravikumar et al., 2009; Koltchinskii and Yuan, 2010; Raskutti et al., 2012).

The support vector machine from Example 14.19 is a popular method for classification introduced by Boser et al. (1992); see the book by Steinwart and Christmann (2008) for further details. The problem of density estimation treated briefly in Section 14.4 has been the subject of intensive study; we refer the reader to the books (Devroye and Györfi, 1986; Silverman, 1986; Scott, 1992; Eggermont and LaRiccia, 2001) and references therein for more details. Good and Gaskins (1971) proposed a roughness-penalized form of the nonparametric maximum likelihood estimate; see Geman and Hwang (1982) and Silverman (1982) for analysis of this and some related estimators. We analyzed the constrained form of the nonparametric MLE under the simplifying assumption that the true density f^* belongs to the density class $\mathscr{F}$. In practice, this assumption may not be satisfied, and there would be an additional form of approximation error in the analysis, as in the oracle inequalities discussed in Chapter 13.

14.7 Exercises

Exercise 14.1 (Bounding the Lipschitz constant) In the setting of Proposition 14.25, show that $\mathbb{E}\big[\sup_{\|f\|_2 \leq t} \|f\|_n\big] \leq \sqrt{5}t$ for all $t \geq \delta_n$.

Exercise 14.2 (Properties of local Rademacher complexity) Recall the localized Rademacher complexity

$$\bar{\mathcal{R}}_n(\delta) := \mathbb{E}_{x,\varepsilon}\Big[\sup_{\substack{f \in \mathscr{F} \\ \|f\|_2 \leq \delta}} \Big|\frac{1}{n}\sum_{i=1}^n \varepsilon_i f(x_i)\Big|\Big],$$

and let δ_n be the smallest positive solution to the inequality $\bar{\mathcal{R}}_n(\delta) \leq \delta^2$. Assume that function class $\mathscr{F}$ is star-shaped around the origin (so that $f \in \mathscr{F}$ implies $\alpha f \in \mathscr{F}$ for all $\alpha \in [0, 1]$).

(a) Show that $\bar{\mathcal{R}}_n(s) \leq \max\{\delta_n^2, s\delta_n\}$. (*Hint:* Lemma 13.6 could be useful.)
(b) For some constant $C \geq 1$, let $t_n > 0$ be the small positive solution to the inequality $\bar{\mathcal{R}}_n(t) \leq Ct^2$. Show that $t_n \leq \frac{\delta_n}{\sqrt{C}}$. (*Hint*: Part (a) could be useful.)

Exercise 14.3 (Sharper rates via entropy integrals) In the setting of Example 14.2, show that there is a universal constant c' such that

$$\mathbb{E}_\varepsilon\Big[\sup_{\substack{f_\theta \in \mathcal{P}_2 \\ \|f_\theta\|_2 \leq \delta}} \Big|\frac{1}{n}\sum_{i=1}^n \varepsilon_i f(x_i)\Big|\Big] \leq c'\sqrt{\frac{1}{n}}.$$

Exercise 14.4 (Uniform laws for kernel classes) In this exercise, we work through the proof of the bound (14.14a) from Corollary 14.5.

(a) Letting $(\phi_j)_{j=1}^\infty$ be the eigenfunctions of the kernel operator, show that

$$\sup_{\substack{\|f\|_{\mathbb{H}} \leq 1 \\ \|f\|_2 \leq \delta}} \Big|\sum_{i=1}^n \varepsilon_i f(x_i)\Big| = \sup_{\theta \in \mathbb{K}} \Big|\sum_{j=1}^\infty \theta_j z_j\Big|,$$

where $z_j := \sum_{i=1}^n \varepsilon_i \phi_j(x_i)$ and

$$\mathcal{D} := \Big\{(\theta)_{j=1}^{\infty} \mid \sum_{j=1}^{\infty} \theta_j^2 \le \delta,\ \sum_{j=1}^{\infty} \frac{\theta_j^2}{\mu_j} \le 1\Big\}.$$

(b) Defining the sequence $\eta_j = \min\{\delta^2, \mu_j\}$ for $j = 1, 2, \ldots$, show that $\mathcal{D}$ is contained within the ellipse $\mathcal{E} := \{(\theta)_{j=1}^{\infty} \mid \sum_{j=1}^{\infty} \theta_j^2/\eta_j \le 2\}$.

(c) Use parts (a) and (b) to show that

$$\mathbb{E}_{\varepsilon,x}\Big[\sup_{\substack{\|f\|_{\mathbb{H}}\le 1\\ \|f\|_2 \le \delta}} \Big|\frac{1}{n}\sum_{i=1}^n \varepsilon_i f(x_i)\Big|\Big] \le \sqrt{\frac{2}{n}}\sqrt{\sum_{j=1}^{\infty} \min\{\delta^2, \mu_j\}}.$$

Exercise 14.5 (Empirical approximations of kernel integral operators) Let $\mathcal{K}$ be a PSD kernel function satisfying the conditions of Mercer's theorem (Theorem 12.20), and define the associated representer $R_x(\cdot) = \mathcal{K}(\cdot, x)$. Letting $\mathbb{H}$ be the associated reproducing kernel Hilbert space, consider the integral operator $T_{\mathcal{K}}$ as defined in equation (12.11a).

(a) Letting $\{x_i\}_{i=1}^n$ denote i.i.d. samples from $\mathbb{P}$, define the random linear operator $\widehat{T}_{\mathcal{K}}\colon \mathbb{H} \to \mathbb{H}$ via

$$f \mapsto \widehat{T}_{\mathcal{K}}(f) := \frac{1}{n}\sum_{i=1}^n [R_{x_i} \otimes R_{x_i}](f) \;=\; \frac{1}{n}\sum_{i=1}^n f(x_i) R_{x_i}.$$

Show that $\mathbb{E}[\widehat{T}_{\mathcal{K}}] = T_{\mathcal{K}}$.

(b) Use techniques from this chapter to bound the operator norm

$$|||\widehat{T}_{\mathcal{K}} - T_{\mathcal{K}}|||_{\mathbb{H}} := \sup_{\|f\|_{\mathbb{H}} \le 1} \|(\widehat{T}_{\mathcal{K}} - T_{\mathcal{K}})(f)\|_{\mathbb{H}}.$$

(c) Letting ϕ_j denote the jth eigenfunction of $T_{\mathcal{K}}$, with associated eigenvalue $\mu_j > 0$, show that

$$\|\widehat{T}_{\mathcal{K}}(\phi_j) - \mu_j \phi_j\|_{\mathbb{H}} \le \frac{|||\widehat{T}_{\mathcal{K}} - T_{\mathcal{K}}|||_{\mathbb{H}}}{\mu_j}.$$

Exercise 14.6 (Linear functions and four-way independence) Recall the class $\mathscr{F}_{\text{lin}}$ of linear functions from Example 14.10. Consider a random vector $x \in \mathbb{R}^d$ with four-way independent components—i.e., the variables (x_j, x_k, x_ℓ, x_m) are independent for all distinct quadruples of indices. Assume, moreover, that each component has mean zero and variance one, and that $\mathbb{E}[x_j^4] \le B$. Show that the strong moment condition (14.22b) is satisfied with $C = B + 6$.

Exercise 14.7 (Uniform laws and sparse eigenvalues) In this exercise, we explore the use of Theorem 14.12 for bounding sparse restricted eigenvalues (see Chapter 7). Let $\mathbf{X} \in \mathbb{R}^{n\times d}$ be a random matrix with i.i.d. $\mathcal{N}(0, \mathbf{\Sigma})$ rows. For a given parameter $s > 0$, define the function class $\mathscr{F}_{\text{spcone}} = \{f_\theta \mid \|\theta\|_1 \le \sqrt{s}\|\theta\|_2\}$, where $f_\theta(x) = \langle x, \theta\rangle$. Letting $\rho^2(\mathbf{\Sigma})$ denote the maximal diagonal entry of $\mathbf{\Sigma}$, show that, as long as

$$n > c_0 \frac{\rho^2(\mathbf{\Sigma})}{\gamma_{\min}(\mathbf{\Sigma})}\, s \log\Big(\frac{ed}{s}\Big)$$

for a sufficiently large constant c, then we are guaranteed that

$$\underbrace{\|f_\theta\|_n^2}_{\|\mathbf{X}\theta\|_2^2/n} \geq \frac{1}{2} \underbrace{\|f_\theta\|_2^2}_{\|\sqrt{\Sigma}\theta\|_2^2} \qquad \text{for all } f_\theta \in \mathscr{F}_{\text{spcone}}$$

with probability at least $1 - e^{-c_1 n}$. Thus, we have proved a somewhat sharper version of Theorem 7.16. (*Hint:* Exercise 7.15 could be useful to you.)

Exercise 14.8 (Estimation of nonparametric additive models) Recall from Example 14.11 the class $\mathscr{F}_{\text{add}}$ of additive models formed by some base class $\mathscr{G}$ that is convex and 1-uniformly bounded ($\|g\|_\infty \leq 1$ for all $g \in \mathscr{G}$). Let δ_n be the smallest positive solution to the inequality $\bar{\mathcal{R}}_n(\delta; \mathscr{F}) \leq \delta^2$. Letting ϵ_n be the smallest positive solution to the inequality $\bar{\mathcal{R}}_n(\epsilon; \mathscr{G}) \leq \epsilon^2$, show that $\delta_n^2 \precsim d\, \epsilon_n^2$.

Exercise 14.9 (Nonparametric maximum likelihood) Consider the nonparametric density estimate (14.56) over the class of all differentiable densities. Show that the minimum is not achieved. (*Hint:* Consider a sequence of differentiable approximations to the density function placing mass $1/n$ at each of the data points.)

Exercise 14.10 (Hellinger distance and Kullback–Leibler divergence) Prove the lower bound (14.57b) on the Kullback–Leibler divergence in terms of the squared Hellinger distance.

Exercise 14.11 (Bounds on histogram density estimation) Recall the histogram estimator defined by the basis (14.62), and suppose that we apply it to estimate a density f^* on the unit interval $[0, 1]$ that is differentiable with $\|f'\|_\infty \leq 1$. Use the oracle inequality from Corollary 14.24 to show that there is a universal constant c such that

$$\|\widehat{f} - f^*\|_2^2 \leq c n^{-2/3} \tag{14.69}$$

with high probability.

15

Minimax lower bounds

In the preceding chapters, we have derived a number of results on the convergence rates of different estimation procedures. In this chapter, we turn to the complementary question: Can we obtain matching lower bounds on estimation rates? This question can be asked both in the context of a specific procedure or algorithm, and in an algorithm-independent sense. We focus on the latter question in this chapter. In particular, our goal is to derive lower bounds on the estimation error achievable by *any procedure*, regardless of its computational complexity and/or storage.

Lower bounds of this type can yield two different but complementary types of insight. A first possibility is that they can establish that known—and possibly polynomial-time—estimators are statistically "optimal", meaning that they have estimation error guarantees that match the lower bounds. In this case, there is little purpose in searching for estimators with lower statistical error, although it might still be interesting to study optimal estimators that enjoy lower computational and/or storage costs, or have other desirable properties such as robustness. A second possibility is that the lower bounds do not match the best known upper bounds. In this case, assuming that the lower bounds are tight, one has a strong motivation to study alternative estimators.

In this chapter, we develop various techniques for establishing such lower bounds. Of particular relevance to our development are the properties of packing sets and metric entropy, as discussed in Chapter 5. In addition, we require some basic aspects of information theory, including entropy and the Kullback–Leibler divergence, as well as other types of divergences between probability measures, which we provide in this chapter.

15.1 Basic framework

Given a class of distributions $\mathcal{P}$, we let θ denote a functional on the space $\mathcal{P}$—that is, a mapping from a distribution $\mathbb{P}$ to a parameter $\theta(\mathbb{P})$ taking values in some space Ω. Our goal is to estimate $\theta(\mathbb{P})$ based on samples drawn from the unknown distribution $\mathbb{P}$.

In certain cases, the quantity $\theta(\mathbb{P})$ uniquely determines the underlying distribution $\mathbb{P}$, meaning that $\theta(\mathbb{P}_0) = \theta(\mathbb{P}_1)$ if and only if $\mathbb{P}_0 = \mathbb{P}_1$. In such cases, we can think of θ as providing a parameterization of the family of distributions. Such classes include most of the usual finite-dimensional parametric classes, as well as certain nonparametric problems, among them nonparametric regression problems. For such classes, we can write $\mathcal{P} = \{\mathbb{P}_\theta \mid \theta \in \Omega\}$, as we have done in previous chapters.

In other settings, however, we might be interested in estimating a functional $\mathbb{P} \mapsto \theta(\mathbb{P})$ that does *not* uniquely specify the distribution. For instance, given a class of distributions $\mathcal{P}$

on the unit interval $[0, 1]$ with differentiable density functions f, we might be interested in estimating the quadratic functional $\mathbb{P} \mapsto \theta(\mathbb{P}) = \int_0^1 (f'(t))^2 \, dt \in \mathbb{R}$. Alternatively, for a class of unimodal density functions f on the unit interval $[0, 1]$, we might be interested in estimating the mode of the density $\theta(\mathbb{P}) = \arg\max_{x \in [0,1]} f(x)$. Thus, the viewpoint of estimating functionals adopted here is considerably more general than a parameterized family of distributions.

15.1.1 Minimax risks

Suppose that we are given a random variable X drawn according to a distribution $\mathbb{P}$ for which $\theta(\mathbb{P}) = \theta^*$. Our goal is to estimate the unknown quantity θ^* on the basis of the data X. An estimator $\widehat{\theta}$ for doing so can be viewed as a measurable function from the domain $\mathcal{X}$ of the random variable X to the parameter space Ω. In order to assess the quality of any estimator, we let $\rho \colon \Omega \times \Omega \to [0, \infty)$ be a semi-metric,[1] and we consider the quantity $\rho(\widehat{\theta}, \theta^*)$. Here the quantity θ^* is fixed but unknown, whereas the quantity $\widehat{\theta} \equiv \widehat{\theta}(X)$ is a random variable, so that $\rho(\widehat{\theta}, \theta^*)$ is random. By taking expectations over the observable X, we obtain the deterministic quantity $\mathbb{E}_{\mathbb{P}}[\rho(\widehat{\theta}, \theta^*)]$. As the parameter θ^* is varied, we obtain a function, typically referred to as the risk function, associated with the estimator.

The first property to note is that it makes no sense to consider the set of estimators that are good in a pointwise sense. For any *fixed* θ^*, there is always a very good way in which to estimate it: simply ignore the data, and return θ^*. The resulting deterministic estimator has zero risk when evaluated at the fixed θ^*, but of course is likely to behave very poorly for other choices of the parameter. There are various ways in which to circumvent this and related difficulties. The Bayesian approach is to view the unknown parameter θ^* as a random variable; when endowed with some prior distribution, we can then take expectations over the risk function with respect to this prior. A closely related approach is to model the choice of θ^* in an adversarial manner, and to compare estimators based on their worst-case performance. More precisely, for each estimator $\widehat{\theta}$, we compute the worst-case risk $\sup_{\mathbb{P} \in \mathcal{P}} \mathbb{E}_{\mathbb{P}}[\rho(\widehat{\theta}, \theta(\mathbb{P}))]$, and rank estimators according to this ordering. The estimator that is optimal in this sense defines a quantity known as the *minimax risk*—namely,

$$\mathfrak{M}(\theta(\mathcal{P}); \rho) := \inf_{\widehat{\theta}} \sup_{\mathbb{P} \in \mathcal{P}} \mathbb{E}_{\mathbb{P}}[\rho(\widehat{\theta}, \theta(\mathbb{P}))], \tag{15.1}$$

where the infimum ranges over all possible estimators, by which we mean measurable functions of the data. When the estimator is based on n i.i.d. samples from $\mathbb{P}$, we use $\mathfrak{M}_n$ to denote the associated minimax risk.

We are often interested in evaluating minimax risks defined not by a norm, but rather by a squared norm. This extension is easily accommodated by letting $\Phi \colon [0, \infty) \to [0, \infty)$ be an increasing function on the non-negative real line, and then defining a slight generalization of the ρ-minimax risk—namely

$$\mathfrak{M}(\theta(\mathcal{P}); \Phi \circ \rho) := \inf_{\widehat{\theta}} \sup_{\mathbb{P} \in \mathcal{P}} \mathbb{E}_{\mathbb{P}}[\Phi(\rho(\widehat{\theta}, \theta(\mathbb{P})))]. \tag{15.2}$$

[1] In our usage, a semi-metric satisfies all properties of a metric, except that there may exist pairs $\theta \neq \theta'$ for which $\rho(\theta, \theta') = 0$.

A particularly common choice is $\Phi(t) = t^2$, which can be used to obtain minimax risks for the mean-squared error associated with ρ.

15.1.2 From estimation to testing

With this set-up, we now turn to the primary goal of this chapter: developing methods for lower bounding the minimax risk. Our first step is to show how lower bounds can be obtained via "reduction" to the problem of obtaining lower bounds for the probability of error in a certain testing problem. We do so by constructing a suitable packing of the parameter space (see Chapter 5 for background on packing numbers and metric entropy).

More precisely, suppose that $\{\theta^1, \ldots, \theta^M\}$ is a 2δ-separated set[2] contained in the space $\theta(\mathcal{P})$, meaning a collection of elements $\rho(\theta^j, \theta^k) \geq 2\delta$ for all $j \neq k$. For each θ^j, let us choose some representative distribution $\mathbb{P}_{\theta^j}$—that is, a distribution such that $\theta(\mathbb{P}_{\theta^j}) = \theta^j$—and then consider the M-ary hypothesis testing problem defined by the family of distributions $\{\mathbb{P}_{\theta^j}, j = 1, \ldots, M\}$. In particular, we generate a random variable Z by the following procedure:

(1) Sample a random integer J from the uniform distribution over the index set $[M] := \{1, \ldots, M\}$.
(2) Given $J = j$, sample $Z \sim \mathbb{P}_{\theta^j}$.

We let $\mathbb{Q}$ denote the joint distribution of the pair (Z, J) generated by this procedure. Note that the marginal distribution over Z is given by the uniformly weighted mixture distribution $\bar{\mathbb{Q}} := \frac{1}{M} \sum_{j=1}^{M} \mathbb{P}_{\theta^j}$. Given a sample Z from this mixture distribution, we consider the M-ary hypothesis testing problem of determining the randomly chosen index J. A *testing function* for this problem is a mapping $\psi \colon \mathcal{Z} \to [M]$, and the associated probability of error is given by $\mathbb{Q}[\psi(Z) \neq J]$, where the probability is taken jointly over the pair (Z, J). This error probability may be used to obtain a lower bound on the minimax risk as follows:

Proposition 15.1 (From estimation to testing) *For any increasing function Φ and choice of 2δ-separated set, the minimax risk is lower bounded as*

$$\mathfrak{M}(\theta(\mathcal{P}), \Phi \circ \rho) \geq \Phi(\delta) \inf_{\psi} \mathbb{Q}[\psi(Z) \neq J], \tag{15.3}$$

where the infimum ranges over test functions.

Note that the right-hand side of the bound (15.3) involves two terms, both of which depend on the choice of δ. By assumption, the function Φ is increasing in δ, so that it is maximized by choosing δ as large as possible. On the other hand, the testing error $\mathbb{Q}[\psi(Z) \neq J]$ is defined in terms of a collection of 2δ-separated distributions. As $\delta \to 0^+$, the underlying testing problem becomes more difficult, and so that, at least in general, we should expect that $\mathbb{Q}[\psi(Z) \neq J]$ grows as δ decreases. If we choose a value δ^* sufficiently small to ensure

[2] Here we enforce only the milder requirement $\rho(\theta^j, \theta^k) \geq 2\delta$, as opposed to the strict inequality required for a packing set. This looser requirement turns out to be convenient in later calculations.

that this testing error is at least 1/2, then we may conclude that $\mathfrak{M}(\theta(\mathcal{P}), \Phi \circ \rho) \geq \frac{1}{2}\Phi(\delta^*)$. For a given choice of δ, the other additional degree of freedom is our choice of packing set, and we will see a number of different constructions in the sequel.

We now turn to the proof of the proposition.

Proof For any $\mathbb{P} \in \mathcal{P}$ with parameter $\theta = \theta(\mathbb{P})$, we have

$$\mathbb{E}_{\mathbb{P}}[\Phi(\rho(\widehat{\theta}, \theta))] \overset{\text{(i)}}{\geq} \Phi(\delta)\ \mathbb{P}[\Phi(\rho(\widehat{\theta}, \theta)) \geq \Phi(\delta)] \overset{\text{(ii)}}{\geq} \Phi(\delta)\ \mathbb{P}[\rho(\widehat{\theta}, \theta) \geq \delta],$$

where step (i) follows from Markov's inequality, and step (ii) follows from the increasing nature of Φ. Thus, it suffices to lower bound the quantity

$$\sup_{\mathbb{P} \in \mathcal{P}} \mathbb{P}[\rho(\widehat{\theta}, \theta(\mathbb{P})) \geq \delta].$$

Recall that $\mathbb{Q}$ denotes the joint distribution over the pair (Z, J) defined by our construction. Note that

$$\sup_{\mathbb{P} \in \mathcal{P}} \mathbb{P}[\rho(\widehat{\theta}, \theta(\mathbb{P})) \geq \delta] \geq \frac{1}{M} \sum_{j=1}^{M} \mathbb{P}_{\theta^j}[\rho(\widehat{\theta}, \theta^j) \geq \delta] = \mathbb{Q}[\rho(\widehat{\theta}, \theta^J) \geq \delta],$$

so we have reduced the problem to lower bounding the quantity $\mathbb{Q}[\rho(\widehat{\theta}, \theta^J) \geq \delta]$.

Now observe that any estimator $\widehat{\theta}$ can be used to define a test—namely, via

$$\psi(Z) := \arg\min_{\ell \in [M]} \rho(\theta^\ell, \widehat{\theta}). \tag{15.4}$$

(If there are multiple indices that achieve the minimizing argument, then we break such ties in an arbitrary but well-defined way.) Suppose that the true parameter is θ^j: we then claim that the event $\{\rho(\theta^j, \widehat{\theta}) < \delta\}$ ensures that the test (15.4) is correct. In order to see this implication, note that, for any other index $k \in [M]$, an application of the triangle inequality

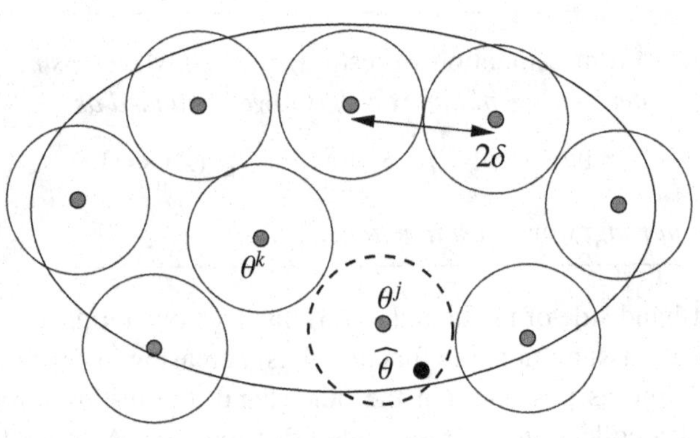

Figure 15.1 Reduction from estimation to testing using a 2δ-separated set in the space Ω in the semi-metric ρ. If an estimator $\widehat{\theta}$ satisfies the bound $\rho(\widehat{\theta}, \theta^j) < \delta$ whenever the true parameter is θ^j, then it can be used to determine the correct index j in the associated testing problem.

guarantees that

$$\rho(\theta^k, \widehat{\theta}) \geq \underbrace{\rho(\theta^k, \theta^j)}_{\geq 2\delta} - \underbrace{\rho(\theta^j, \widehat{\theta})}_{<\delta} > 2\delta - \delta = \delta,$$

where the lower bound $\rho(\theta^j, \theta^k) \geq 2\delta$ follows by the 2δ-separated nature of our set. Consequently, we have $\rho(\theta^k, \widehat{\theta}) > \rho(\theta^j, \widehat{\theta})$ for all $k \neq j$, so that, by the definition (15.4) of our test, we must have $\psi(Z) = j$. See Figure 15.1 for the geometry of this argument.

Therefore, conditioned on $J = j$, the event $\{\rho(\widehat{\theta}, \theta^j) < \delta\}$ is contained within the event $\{\psi(Z) = j\}$, which implies that $\mathbb{P}_{\theta^j}[\rho(\widehat{\theta}, \theta^j) \geq \delta] \geq \mathbb{P}_{\theta^j}[\psi(Z) \neq j]$. Taking averages over the index j, we find that

$$\mathbb{Q}[\rho(\widehat{\theta}, \theta^J) \geq \delta] = \frac{1}{M} \sum_{j=1}^{M} \mathbb{P}_{\theta^j}[\rho(\widehat{\theta}, \theta^j) \geq \delta] \geq \mathbb{Q}[\psi(Z) \neq J].$$

Combined with our earlier argument, we have shown that

$$\sup_{\mathbb{P} \in \mathcal{P}} \mathbb{E}_{\mathbb{P}}[\Phi(\rho(\widehat{\theta}, \theta))] \geq \Phi(\delta)\, \mathbb{Q}[\psi(Z) \neq J].$$

Finally, we may take the infimum over all estimators $\widehat{\theta}$ on the left-hand side, and the infimum over the induced set of tests on the right-hand side. The full infimum over all tests can only be smaller, from which the claim follows. □

15.1.3 Some divergence measures

Thus far, we have established a connection between minimax risks and error probabilities in testing problems. Our next step is to develop techniques for lower bounding the error probability, for which we require some background on different types of divergence measures between probability distributions. Three such measures of particular importance are the total variation (TV) distance, the Kullback–Leibler (KL) divergence and the Hellinger distance.

Let $\mathbb{P}$ and $\mathbb{Q}$ be two distributions on $\mathcal{X}$ with densities p and q with respect to some underlying base measure ν. Note that there is no loss of generality in assuming the existence of densities, since any pair of distributions have densities with respect to the base measure $\nu = \frac{1}{2}(\mathbb{P} + \mathbb{Q})$. The *total variation (TV) distance* between two distributions $\mathbb{P}$ and $\mathbb{Q}$ is defined as

$$\|\mathbb{P} - \mathbb{Q}\|_{\mathrm{TV}} := \sup_{A \subseteq \mathcal{X}} |\mathbb{P}(A) - \mathbb{Q}(A)|. \tag{15.5}$$

In terms of the underlying densities, we have the equivalent definition

$$\|\mathbb{P} - \mathbb{Q}\|_{\mathrm{TV}} = \frac{1}{2} \int_{\mathcal{X}} |p(x) - q(x)|\, \nu(dx), \tag{15.6}$$

corresponding to one-half the $L^1(\nu)$-norm between the densities. (See Exercise 3.13 from Chapter 3 for details on this equivalence.) In the sequel, we will see how the total variation distance is closely connected to the Bayes error in binary hypothesis testing.

A closely related measure of the "distance" between distributions is the *Kullback–Leibler divergence*. When expressed in terms of the densities q and p, it takes the form

$$D(\mathbb{Q} \,\|\, \mathbb{P}) = \int_{\mathcal{X}} q(x) \log \frac{q(x)}{p(x)} \nu(dx), \tag{15.7}$$

where ν is some underlying base measure defining the densities. Unlike the total variation distance, it is not actually a metric, since, for example, it fails to be symmetric in its arguments in general (i.e., there are pairs for which $D(\mathbb{Q} \,\|\, \mathbb{P}) \neq D(\mathbb{P} \,\|\, \mathbb{Q})$). However, it can be used to upper bound the TV distance, as stated in the following classical result:

Lemma 15.2 (Pinsker–Csiszár–Kullback inequality) *For all distributions* $\mathbb{P}$ *and* $\mathbb{Q}$,

$$\|\mathbb{P} - \mathbb{Q}\|_{\mathrm{TV}} \leq \sqrt{\tfrac{1}{2} D(\mathbb{Q} \,\|\, \mathbb{P})}. \tag{15.8}$$

Recall that this inequality arose in our study of the concentration of measure phenomenon (Chapter 3). This inequality is also useful here, but instead in the context of establishing minimax lower bounds. See Exercise 15.6 for an outline of the proof of this bound.

A third distance that plays an important role in statistical problems is the *squared Hellinger distance*, given by

$$H^2(\mathbb{P} \,\|\, \mathbb{Q}) := \int \left(\sqrt{p(x)} - \sqrt{q(x)} \right)^2 \nu(dx). \tag{15.9}$$

It is simply the $L^2(\nu)$-norm between the square-root density functions, and an easy calculation shows that it takes values in the interval $[0, 2]$. When the base measure is clear from the context, we use the notation $H^2(p \,\|\, q)$ and $H^2(\mathbb{P} \,\|\, \mathbb{Q})$ interchangeably.

Like the KL divergence, the Hellinger distance can also be used to upper bound the TV distance:

Lemma 15.3 (Le Cam's inequality) *For all distributions* $\mathbb{P}$ *and* $\mathbb{Q}$,

$$\|\mathbb{P} - \mathbb{Q}\|_{\mathrm{TV}} \leq H(\mathbb{P} \,\|\, \mathbb{Q}) \sqrt{1 - \frac{H^2(\mathbb{P} \,\|\, \mathbb{Q})}{4}}. \tag{15.10}$$

We work through the proof of this inequality in Exercise 15.5.

Let $(\mathbb{P}_1, \ldots, \mathbb{P}_n)$ be a collection of n probability measures, each defined on $\mathcal{X}$, and let $\mathbb{P}^{1:n} = \bigotimes_{i=1}^n \mathbb{P}_i$ be the product measure defined on $\mathcal{X}^n$. If we define another product measure $\mathbb{Q}^{1:n}$ in a similar manner, then it is natural to ask whether the divergence between $\mathbb{P}^{1:n}$ and $\mathbb{Q}^{1:n}$ has a "nice" expression in terms of divergences between the individual pairs.

In this context, the total variation distance behaves badly: in general, it is difficult to express the distance $\|\mathbb{P}^{1:n} - \mathbb{Q}^{1:n}\|_{\mathrm{TV}}$ in terms of the individual distances $\|\mathbb{P}_i - \mathbb{Q}_i\|_{\mathrm{TV}}$. On the other hand, the Kullback–Leibler divergence exhibits a very attractive decoupling property, in that we have

$$D(\mathbb{P}^{1:n} \,\|\, \mathbb{Q}^{1:n}) = \sum_{i=1}^{n} D(\mathbb{P}_i \,\|\, \mathbb{Q}_i). \tag{15.11a}$$

This property is straightforward to verify from the definition. In the special case of i.i.d. product distributions—meaning that $\mathbb{P}_i = \mathbb{P}_1$ and $\mathbb{Q}_i = \mathbb{Q}_1$ for all i—then we have

$$D(\mathbb{P}^{1:n} \,\|\, \mathbb{Q}^{1:n}) = nD(\mathbb{P}_1 \,\|\, \mathbb{Q}_1). \tag{15.11b}$$

Although the squared Hellinger distance does not decouple in quite such a simple way, it does have the following property:

$$\tfrac{1}{2}H^2(\mathbb{P}^{1:n} \,\|\, \mathbb{Q}^{1:n}) = 1 - \prod_{i=1}^{n}\Big(1 - \tfrac{1}{2}H^2(\mathbb{P}_i \,\|\, \mathbb{Q}_i)\Big). \tag{15.12a}$$

Thus, in the i.i.d. case, we have

$$\tfrac{1}{2}H^2(\mathbb{P}^{1:n} \,\|\, \mathbb{Q}^{1:n}) = 1 - \Big(1 - \tfrac{1}{2}H^2(\mathbb{P}_1 \,\|\, \mathbb{Q}_1)\Big)^n \leq \tfrac{1}{2}nH^2(\mathbb{P}_1 \,\|\, \mathbb{Q}_1). \tag{15.12b}$$

See Exercises 15.3 and 15.7 for verifications of these and related properties, which play an important role in the sequel.

15.2 Binary testing and Le Cam's method

The simplest type of testing problem, known as a binary hypothesis test, involves only two distributions. In this section, we describe the connection between binary testing and the total variation norm, and use it to develop various lower bounds, culminating in a general technique known as Le Cam's method.

15.2.1 Bayes error and total variation distance

In a binary testing problem with equally weighted hypotheses, we observe a random variable Z drawn according to the mixture distribution $\bar{\mathbb{Q}} := \frac{1}{2}\mathbb{P}_0 + \frac{1}{2}\mathbb{P}_1$. For a given decision rule $\psi\colon \mathcal{Z} \to \{0, 1\}$, the associated probability of error is given by

$$\mathbb{Q}[\psi(Z) \neq J] = \tfrac{1}{2}\mathbb{P}_0[\psi(Z) \neq 0] + \tfrac{1}{2}\mathbb{P}_1[\psi(Z) \neq 1].$$

If we take the infimum of this error probability over all decision rules, we obtain a quantity known as the *Bayes risk* for the problem. In the binary case, the Bayes risk can actually be expressed explicitly in terms of the total variation distance $\|\mathbb{P}_1 - \mathbb{P}_0\|_{\mathrm{TV}}$, as previously defined in equation (15.5)—more precisely, we have

$$\inf_{\psi} \mathbb{Q}[\psi(Z) \neq J] = \tfrac{1}{2}\big\{1 - \|\mathbb{P}_1 - \mathbb{P}_0\|_{\mathrm{TV}}\big\}. \tag{15.13}$$

Note that the worst-case value of the Bayes risk is one-half, achieved when $\mathbb{P}_1 = \mathbb{P}_0$, so that the hypotheses are completely indistinguishable. At the other extreme, the best-case Bayes risk is zero, achieved when $\|\mathbb{P}_1 - \mathbb{P}_0\|_{\mathrm{TV}} = 1$. This latter equality occurs, for instance, when $\mathbb{P}_0$ and $\mathbb{P}_1$ have disjoint supports.

In order to verify the equivalence (15.13), note that there is a one-to-one correspondence between decision rules ψ and measurable partitions (A, A^c) of the space $\mathcal{X}$; more precisely, any decision rule ψ is uniquely determined by the set $A = \{x \in \mathcal{X} \mid \psi(x) = 1\}$. Thus, we have

$$\sup_{\psi} \mathbb{Q}[\psi(Z) = J] = \sup_{A \subseteq \mathcal{X}} \left\{\tfrac{1}{2}\mathbb{P}_1(A) + \tfrac{1}{2}\mathbb{P}_0(A^c)\right\} = \tfrac{1}{2} \sup_{A \subseteq \mathcal{X}} \{\mathbb{P}_1(A) - \mathbb{P}_0(A)\} + \tfrac{1}{2}.$$

Since $\sup_{\psi} \mathbb{Q}[\psi(Z) = J] = 1 - \inf_{\psi} \mathbb{Q}[\psi(Z) \neq J]$, the claim (15.13) then follows from the definition (15.5) of the total variation distance.

The representation (15.13), in conjunction with Proposition 15.1, provides one avenue for deriving lower bounds. In particular, for any pair of distributions $\mathbb{P}_0, \mathbb{P}_1 \in \mathcal{P}$ such that $\rho(\theta(\mathbb{P}_0), \theta(\mathbb{P}_1)) \geq 2\delta$, we have

$$\mathfrak{M}(\theta(\mathcal{P}), \Phi \circ \rho) \geq \frac{\Phi(\delta)}{2} \{1 - \|\mathbb{P}_1 - \mathbb{P}_0\|_{\mathrm{TV}}\}. \tag{15.14}$$

Let us illustrate the use of this simple lower bound with some examples.

Example 15.4 (Gaussian location family) For a fixed variance σ^2, let $\mathbb{P}_\theta$ be the distribution of a $\mathcal{N}(\theta, \sigma^2)$ variable; letting the mean θ vary over the real line defines the Gaussian location family $\{\mathbb{P}_\theta, \theta \in \mathbb{R}\}$. Here we consider the problem of estimating θ under either the absolute error $|\widehat{\theta} - \theta|$ or the squared error $(\widehat{\theta} - \theta)^2$ using a collection $Z = (Y_1, \ldots, Y_n)$ of n i.i.d. samples drawn from a $\mathcal{N}(\theta, \sigma^2)$ distribution. We use $\mathbb{P}_\theta^n$ to denote this product distribution.

Let us apply the two-point Le Cam bound (15.14) with the distributions $\mathbb{P}_0^n$ and $\mathbb{P}_\theta^n$. We set $\theta = 2\delta$, for some δ to be chosen later in the proof, which ensures that the two means are 2δ-separated. In order to apply the two-point Le Cam bound, we need to bound the total variation distance $\|\mathbb{P}_\theta^n - \mathbb{P}_0^n\|_{\mathrm{TV}}$. From the second-moment bound in Exercise 15.10(b), we have

$$\|\mathbb{P}_\theta^n - \mathbb{P}_0^n\|_{\mathrm{TV}}^2 \leq \tfrac{1}{4}\left\{e^{n\theta^2/\sigma^2} - 1\right\} = \tfrac{1}{4}\left\{e^{4n\delta^2/\sigma^2} - 1\right\}. \tag{15.15}$$

Setting $\delta = \frac{1}{2}\frac{\sigma}{\sqrt{n}}$ thus yields

$$\inf_{\widehat{\theta}} \sup_{\theta \in \mathbb{R}} \mathbb{E}_\theta[|\widehat{\theta} - \theta|] \geq \frac{\delta}{2}\left\{1 - \frac{1}{2}\sqrt{e - 1}\right\} \geq \frac{\delta}{6} = \frac{1}{12}\frac{\sigma}{\sqrt{n}} \tag{15.16a}$$

and

$$\inf_{\widehat{\theta}} \sup_{\theta \in \mathbb{R}} \mathbb{E}_\theta[(\widehat{\theta} - \theta)^2] \geq \frac{\delta^2}{2}\left\{1 - \frac{1}{2}\sqrt{e - 1}\right\} \geq \frac{\delta^2}{6} = \frac{1}{24}\frac{\sigma^2}{n}. \tag{15.16b}$$

Although the pre-factors $1/12$ and $1/24$ are not optimal, the scalings $\sigma/\sqrt{n}$ and σ^2/n are sharp. For instance, the sample mean $\widetilde{\theta}_n := \frac{1}{n}\sum_{i=1}^n Y_i$ satisfies the bounds

$$\sup_{\theta \in \mathbb{R}} \mathbb{E}_\theta[|\widetilde{\theta}_n - \theta|] = \sqrt{\frac{2}{\pi}}\frac{\sigma}{\sqrt{n}} \quad \text{and} \quad \sup_{\theta \in \mathbb{R}} \mathbb{E}_\theta[(\widetilde{\theta}_n - \theta)^2] = \frac{\sigma^2}{n}.$$

In Exercise 15.8, we explore an alternative approach, based on using the Pinsker–Csiszár–Kullback inequality from Lemma 15.2 to upper bound the TV distance in terms of the KL divergence. This approach yields a result with sharper constants. ♣

Mean-squared error decaying as n^{-1} is typical for parametric problems with a certain type of regularity, of which the Gaussian location model is the archetypal example. For other "non-regular" problems, faster rates become possible, and the minimax lower bounds take a different form. The following example provides one illustration of this phenomenon:

Example 15.5 (Uniform location family) Let us consider the uniform location family, in which, for each $\theta \in \mathbb{R}$, the distribution $\mathbb{U}_\theta$ is uniform over the interval $[\theta, \theta + 1]$. We let $\mathbb{U}_\theta^n$ denote the product distribution of n i.i.d. samples from $\mathbb{U}_\theta$. In this case, it is not possible to use Lemma 15.2 to control the total variation norm, since the Kullback–Leibler divergence between $\mathbb{U}_\theta$ and $\mathbb{U}_{\theta'}$ is infinite whenever $\theta \neq \theta'$. Accordingly, we need to use an alternative distance measure: in this example, we illustrate the use of the Hellinger distance (see equation (15.9)).

Given a pair $\theta, \theta' \in \mathbb{R}$, let us compute the Hellinger distance between $\mathbb{U}_\theta$ and $\mathbb{U}_{\theta'}$. By symmetry, it suffices to consider the case $\theta' > \theta$. If $\theta' > \theta + 1$, then we have $H^2(\mathbb{U}_\theta \,\|\, \mathbb{U}_{\theta'}) = 2$. Otherwise, when $\theta' \in (\theta, \theta + 1]$, we have

$$H^2(\mathbb{U}_\theta \,\|\, \mathbb{U}_{\theta'}) = \int_\theta^{\theta'} dt + \int_{\theta+1}^{\theta'+1} dt = 2\,|\theta' - \theta|.$$

Consequently, if we take a pair θ, θ' such that $|\theta' - \theta| = 2\delta := \frac{1}{4n}$, then the relation (15.12b) guarantees that

$$\frac{1}{2} H^2(\mathbb{U}_\theta^n \,\|\, \mathbb{U}_{\theta'}^n) \leq \frac{n}{2}\, 2\,|\theta' - \theta| = \frac{1}{4}.$$

In conjunction with Lemma 15.3, we find that

$$\|\mathbb{U}_\theta^n - \mathbb{U}_{\theta'}^n\|_{\mathrm{TV}}^2 \leq H^2(\mathbb{U}_\theta^n \,\|\, \mathbb{U}_{\theta'}^n) \leq \tfrac{1}{2}.$$

From the lower bound (15.14) with $\Phi(t) = t^2$, we conclude that, for the uniform location family, the minimax risk is lower bounded as

$$\inf_{\widehat{\theta}} \sup_{\theta \in \mathbb{R}} \mathbb{E}_\theta[(\widehat{\theta} - \theta)^2] \geq \frac{(1 - \frac{1}{\sqrt{2}})}{128} \frac{1}{n^2}.$$

The significant aspect of this lower bound is the faster n^{-2} rate, which should be contrasted with the n^{-1} rate in the regular situation. In fact, this n^{-2} rate is optimal for the uniform location model, achieved for instance by the estimator $\widetilde{\theta} = \min\{Y_1, \ldots, Y_n\}$; see Exercise 15.9 for details. ♣

Le Cam's method is also useful for various nonparametric problems, for instance those in which our goal is to estimate some functional $\theta\colon \mathscr{F} \to \mathbb{R}$ defined on a class of densities $\mathscr{F}$. For instance, a standard example is the problem of estimating a density at a point, say $x = 0$, in which case $\theta(f) := f(0)$ is known as an evaluation functional.

An important quantity in the Le Cam approach to such problems is the Lipschitz constant of the functional θ with respect to the Hellinger norm, given by

$$\omega(\epsilon; \theta, \mathscr{F}) := \sup_{f, g \in \mathscr{F}} \left\{ |\theta(f) - \theta(g)| \mid H^2(f \,\|\, g) \leq \epsilon^2 \right\}. \tag{15.17}$$

Here we use $H^2(f \,\|\, g)$ to mean the squared Hellinger distance between the distributions associated with the densities f and g. Note that the quantity ω measures the size of the fluctuations of $\theta(f)$ when f is perturbed in a Hellinger neighborhood of radius ϵ. The following corollary reveals the importance of this Lipschitz constant (15.17):

Corollary 15.6 (Le Cam for functionals) *For any increasing function Φ on the non-negative real line and any functional $\theta\colon \mathscr{F} \to \mathbb{R}$, we have*

$$\inf_{\widehat{\theta}} \sup_{f \in \mathscr{F}} \mathbb{E}\Big[\Phi(\widehat{\theta} - \theta(f))\Big] \geq \frac{1}{4}\Phi\Big(\frac{1}{2}\,\omega\Big(\frac{1}{2\sqrt{n}}; \theta, \mathscr{F}\Big)\Big). \tag{15.18}$$

Proof We adopt the shorthand $\omega(t) \equiv \omega(t; \theta, \mathscr{F})$ throughout the proof. Setting $\epsilon^2 = \frac{1}{4n}$, choose a pair f, g that achieve[3] the supremum defining $\omega(1/(2\sqrt{n}))$. By a combination of Le Cam's inequality (Lemma 15.3) and the decoupling property (15.12b) for the Hellinger distance, we have

$$\|\mathbb{P}^n_f - \mathbb{P}^n_g\|^2_{\mathrm{TV}} \leq H^2(\mathbb{P}^n_f \,\|\, \mathbb{P}^n_g) \leq nH^2(\mathbb{P}_f \,\|\, \mathbb{P}_g) \leq \tfrac{1}{4}.$$

Consequently, Le Cam's bound (15.14) with $\delta = \frac{1}{2}\omega(\frac{1}{2\sqrt{n}})$ implies that

$$\inf_{\widehat{\theta}} \sup_{f \in \mathscr{F}} \mathbb{E}\Big[\Phi(\widehat{\theta} - \theta(f))\Big] \geq \frac{1}{4}\Phi\Big(\frac{1}{2}\,\omega\Big(\frac{1}{2\sqrt{n}}\Big)\Big),$$

as claimed. □

The elegance of Corollary 15.6 is in that it reduces the calculation of lower bounds to a geometric object—namely, the Lipschitz constant (15.17). Some concrete examples are helpful to illustrate the basic ideas.

Example 15.7 (Pointwise estimation of Lipschitz densities) Let us consider the family of densities on $[-\frac{1}{2}, \frac{1}{2}]$ that are bounded uniformly away from zero, and are Lipschitz with constant one—that is, $|f(x) - f(y)| \leq |x - y|$ for all $x, y \in [-\frac{1}{2}, \frac{1}{2}]$. Suppose that our goal is to estimate the linear functional $f \mapsto \theta(f) := f(0)$. In order to apply Corollary 15.6, it suffices to lower bound $\omega(\frac{1}{2\sqrt{n}}; \theta, \mathscr{F})$ and we can do so by choosing a pair $f_0, g \in \mathscr{F}$ with $H^2(f_0 \,\|\, g) = \frac{1}{4n}$, and then evaluating the difference $|\theta(f_0) - \theta(g)|$. Let $f_0 \equiv 1$ be the uniform density on $[-\frac{1}{2}, \frac{1}{2}]$. For a parameter $\delta \in (0, \frac{1}{6}]$ to be chosen, consider the function

$$\phi(x) = \begin{cases} \delta - |x| & \text{for } |x| \leq \delta, \\ |x - 2\delta| - \delta & \text{for } x \in [\delta, 3\delta], \\ 0 & \text{otherwise.} \end{cases} \tag{15.19}$$

See Figure 15.2 for an illustration. By construction, the function ϕ is 1-Lipschitz, uniformly

[3] If the supremum is not achieved, then we can choose a pair that approximate it to any desired accuracy, and repeat the argument.

bounded with $\|\phi\|_\infty = \delta \leq \frac{1}{6}$, and integrates to zero—that is, $\int_{-1/2}^{1/2} \phi(x)\,dx = 0$. Consequently, the perturbed function $g := f_0 + \phi$ is a density function belonging to our class, and by construction, we have the equality $|\theta(f_0) - \theta(g)| = \delta$.

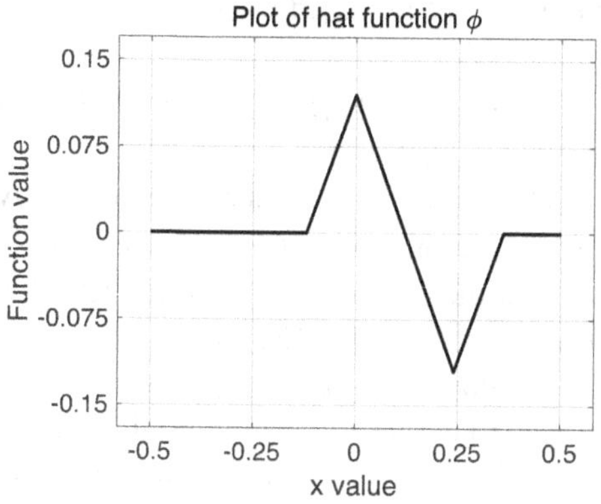

Figure 15.2 Illustration of the hat function ϕ from equation (15.19) for $\delta = 0.12$. It is 1-Lipschitz, uniformly bounded as $\|\phi\|_\infty \leq \delta$, and it integrates to zero.

It remains to control the squared Hellinger distance. By definition, we have

$$\tfrac{1}{2}H^2(f_0 \,\|\, g) = 1 - \int_{-1/2}^{1/2} \sqrt{1 + \phi(t)}\,dt.$$

Define the function $\Psi(u) = \sqrt{1+u}$, and note that $\sup_{u\in\mathbb{R}} |\Psi''(u)| \leq \frac{1}{4}$. Consequently, by a Taylor-series expansion, we have

$$\tfrac{1}{2}H^2(f_0 \,\|\, g) = \int_{-1/2}^{1/2} \{\Psi(0) - \Psi(\phi(t))\}\,dt \leq \int_{-1/2}^{1/2} \left\{-\Psi'(0)\phi(t) + \tfrac{1}{8}\phi^2(t)\right\} dt. \tag{15.20}$$

Observe that

$$\int_{-1/2}^{1/2} \phi(t)\,dt = 0 \quad \text{and} \quad \int_{-1/2}^{1/2} \phi^2(t)\,dt = 4\int_0^\delta (\delta - x)^2\,dx = \tfrac{4}{3}\delta^3.$$

Combined with our Taylor-series bound (15.20), we find that

$$H^2(f_0 \,\|\, g) \leq \tfrac{2}{8} \cdot \tfrac{4}{3}\delta^3 = \tfrac{1}{3}\delta^3.$$

Consequently, setting $\delta^3 = \frac{3}{4n}$ ensures that $H^2(f_0 \,\|\, g) \leq \frac{1}{4n}$. Putting together the pieces, Corollary 15.6 with $\Phi(t) = t^2$ implies that

$$\inf_{\widehat{\theta}} \sup_{f\in\mathscr{F}} \mathbb{E}\left[(\widehat{\theta} - f(0))^2\right] \geq \frac{1}{16}\omega^2\Big(\frac{1}{2\sqrt{n}}\Big) \gtrsim n^{-2/3}.$$

This $n^{-2/3}$ lower bound for the Lipschitz family can be achieved by various estimators, so that we have derived a sharp lower bound. ♣

We now turn to the use of the two-class lower bound for a nonlinear functional in a nonparametric problem. Although the resulting bound is non-trivial, it is *not* a sharp result—unlike in the previous examples. Later, we will develop Le Cam's refinement of the two-class approach so as to obtain sharp rates.

Example 15.8 (Lower bounds for quadratic functionals) Given positive constants $c_0 < 1 < c_1$ and $c_2 > 1$, consider the class of twice-differentiable density functions

$$\mathscr{F}_2([0,1]) := \left\{ f\colon [0,1] \to [c_0, c_1] \,\Big|\, \|f''\|_\infty \le c_2 \text{ and } \int_0^1 f(x)\,dx = 1 \right\} \tag{15.21}$$

that are uniformly bounded above and below, and have a uniformly bounded second derivative. Consider the quadratic functional $f \mapsto \theta(f) := \int_0^1 (f'(x))^2\,dx$. Note that $\theta(f)$ provides a measure of the "smoothness" of the density: it is zero for the uniform density, and becomes large for densities with more erratic behavior. Estimation of such quadratic functionals arises in a variety of applications; see the bibliographic section for further discussion.

We again use Corollary 15.6 to derive a lower bound. Let f_0 denote the uniform distribution on $[0,1]$, which clearly belongs to $\mathscr{F}_2$. As in Example 15.7, we construct a perturbation g of f_0 such that $H^2(f_0 \,\|\, g) = \frac{1}{4n}$; Corollary 15.6 then gives a minimax lower bound of the order $(\theta(f_0) - \theta(g))^2$.

In order to construct the perturbation, let $\phi\colon [0,1] \to \mathbb{R}$ be a fixed twice-differentiable function that is uniformly bounded as $\|\phi\|_\infty \le \frac{1}{2}$, and such that

$$\int_0^1 \phi(x)\,dx = 0 \quad \text{and} \quad b_\ell := \int_0^1 (\phi^{(\ell)}(x))^2\,dx > 0 \qquad \text{for } \ell = 0, 1. \tag{15.22}$$

Now divide the unit interval $[0,1]$ into m sub-intervals $[x_j, x_{j+1}]$, with $x_j = \frac{j}{m}$ for $j = 0, \ldots, m-1$. For a suitably small constant $C > 0$, define the shifted and rescaled functions

$$\phi_j(x) := \begin{cases} \dfrac{C}{m^2}\phi(m(x - x_j)) & \text{if } x \in [x_j, x_{j+1}], \\ 0 & \text{otherwise.} \end{cases} \tag{15.23}$$

We then consider the density $g(x) := 1 + \sum_{j=1}^m \phi_j(x)$. It can be seen that $g \in \mathscr{F}_2$ as long as the constant C is chosen sufficiently small. See Figure 15.3 for an illustration of this construction.

Let us now control the Hellinger distance. Following the same Taylor-series argument as in Example 15.7, we have

$$\begin{aligned} \frac{1}{2}H^2(f_0 \,\|\, g) = 1 - \int_0^1 \sqrt{1 + \sum_{j=1}^m \phi_j(x)}\,dx &\le \frac{1}{8}\int_0^1 \Big(\sum_{j=1}^m \phi_j(x)\Big)^2 dx \\ &= \frac{1}{8}\sum_{j=1}^m \int_0^1 \phi_j^2(x)\,dx \\ &= c\,b_0 \frac{1}{m^4}, \end{aligned}$$

where $c > 0$ is a universal constant. Consequently, the choice $m^4 := 2c\,b_0\,n$ ensures that $H^2(f_0 \,\|\, g) \le \frac{1}{n}$, as required for applying Corollary 15.6.

It remains to evaluate the difference $\theta(f_0)$ and $\theta(g)$. On one hand, we have $\theta(f_0) = 0$,

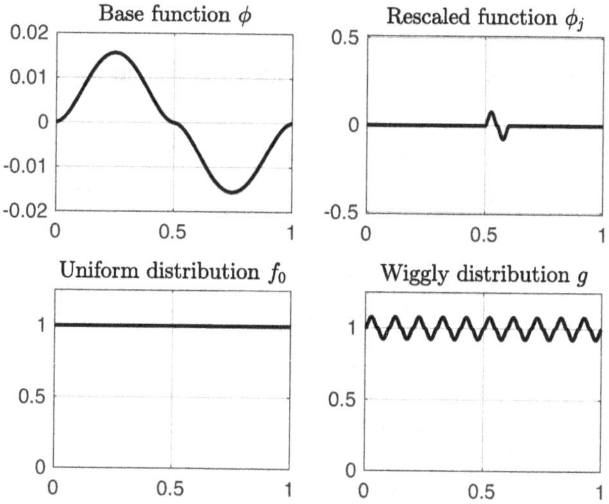

Figure 15.3 Illustration of the construction of the density g. Upper left: an example of a base function ϕ. Upper right: function ϕ_j is a rescaled and shifted version of ϕ. Lower left: original uniform distribution. Lower right: final density g is the superposition of the uniform density f_0 with the sum of the shifted functions $\{\phi_j\}_{j=1}^m$.

whereas on the other hand, we have

$$\theta(g) = \int_0^1 \left(\sum_{j=1}^m \phi_j'(x) \right)^2 dx = m \int_0^1 (\phi_j'(x))^2 \, dx = \frac{C^2 b_1}{m^2}.$$

Recalling the specified choice of m, we see that $|\theta(g) - \theta(f_0)| \geq \frac{K}{\sqrt{n}}$ for some universal constant K independent of n. Consequently, Corollary 15.6 with $\Phi(t) = t$ implies that

$$\sup_{f \in \mathscr{F}_2} \mathbb{E}[|\widehat{\theta}(f) - \theta(f)|] \gtrsim n^{-1/2}. \tag{15.24}$$

This lower bound, while valid, is *not optimal*—there is no estimator that can achieve error of the order of $n^{-1/2}$ uniformly over $\mathscr{F}_2$. Indeed, we will see that the minimax risk scales as $n^{-4/9}$, but proving this optimal lower bound requires an extension of the basic two-point technique, as we describe in the next section. ♣

15.2.2 Le Cam's convex hull method

Our discussion up until this point has focused on lower bounds obtained by single pairs of hypotheses. As we have seen, the difficulty of the testing problem is controlled by the total variation distance between the two distributions. Le Cam's method is an elegant generalization of this idea, one which allows us to take the convex hulls of two classes of distributions. In many cases, the separation in total variation norm as measured over the convex hulls is much smaller than the pointwise separation between two classes, and so leads to better lower bounds.

More concretely, consider two subsets $\mathcal{P}_0$ and $\mathcal{P}_1$ of $\mathcal{P}$ that are 2δ-separated, in the sense

that

$$\rho(\theta(\mathbb{P}_0), \theta(\mathbb{P}_1)) \geq 2\delta \qquad \text{for all } \mathbb{P}_0 \in \mathcal{P}_0 \text{ and } \mathbb{P}_1 \in \mathcal{P}_1. \tag{15.25}$$

Lemma 15.9 (Le Cam) *For any 2δ-separated classes of distributions $\mathcal{P}_0$ and $\mathcal{P}_1$ contained within $\mathcal{P}$, any estimator $\widehat{\theta}$ has worst-case risk at least*

$$\sup_{\mathbb{P}\in\mathcal{P}} \mathbb{E}_{\mathbb{P}}\left[\rho(\widehat{\theta}, \theta(\mathbb{P}))\right] \geq \frac{\delta}{2} \sup_{\substack{\mathbb{P}_0\in\text{conv}(\mathcal{P}_0)\\ \mathbb{P}_1\in\text{conv}(\mathcal{P}_1)}} \left\{1 - \|\mathbb{P}_0 - \mathbb{P}_1\|_{\text{TV}}\right\}. \tag{15.26}$$

Proof For any estimator $\widehat{\theta}$, let us define the random variables

$$V_j(\widehat{\theta}) = \frac{1}{2\delta} \inf_{\mathbb{P}_j\in\mathcal{P}_j} \rho(\widehat{\theta}, \theta(\mathbb{P}_j)), \qquad \text{for } j = 0, 1.$$

We then have

$$\begin{aligned}\sup_{\mathbb{P}\in\mathcal{P}} \mathbb{E}_{\mathbb{P}}[\rho(\widehat{\theta}, \theta(\mathbb{P}))] &\geq \tfrac{1}{2}\left\{\mathbb{E}_{\mathbb{P}_0}[\rho(\widehat{\theta}, \theta(\mathbb{P}_0))] + \mathbb{E}_{\mathbb{P}_1}[\rho(\widehat{\theta}, \theta(\mathbb{P}_1))]\right\}\\ &\geq \delta\left\{\mathbb{E}_{\mathbb{P}_0}[V_0(\widehat{\theta})] + \mathbb{E}_{\mathbb{P}_1}[V_1(\widehat{\theta})]\right\}.\end{aligned}$$

Since the right-hand side is linear in $\mathbb{P}_0$ and $\mathbb{P}_1$, we can take suprema over the convex hulls, and thus obtain the lower bound

$$\sup_{\mathbb{P}\in\mathcal{P}} \mathbb{E}_{\mathbb{P}}[\rho(\widehat{\theta}, \theta(\mathbb{P}))] \geq \delta \sup_{\substack{\mathbb{P}_0\in\text{conv}(\mathcal{P}_0)\\ \mathbb{P}_1\in\text{conv}(\mathcal{P}_1)}} \left\{\mathbb{E}_{\mathbb{P}_0}[V_0(\widehat{\theta})] + \mathbb{E}_{\mathbb{P}_1}[V_1(\widehat{\theta})]\right\}.$$

By the triangle inequality, we have

$$\rho(\widehat{\theta}, \theta(\mathbb{P}_0)) + \rho(\widehat{\theta}, \theta(\mathbb{P}_1)) \geq \rho(\theta(\mathbb{P}_0), \theta(\mathbb{P}_1)) \geq 2\delta.$$

Taking infima over $\mathbb{P}_j \in \mathcal{P}_j$ for each $j = 0, 1$, we obtain

$$\inf_{\mathbb{P}_0\in\mathcal{P}_0} \rho(\widehat{\theta}, \theta(\mathbb{P}_0)) + \inf_{\mathbb{P}_1\in\mathcal{P}_1} \rho(\widehat{\theta}, \theta(\mathbb{P}_1)) \geq 2\delta,$$

which is equivalent to $V_0(\widehat{\theta}) + V_1(\widehat{\theta}) \geq 1$. Since $V_j(\widehat{\theta}) \geq 0$ for $j = 0, 1$, the variational representation of the TV distance (see Exercise 15.1) implies that, for any $\mathbb{P}_j \in \text{conv}(\mathcal{P}_j)$, we have

$$\mathbb{E}_{\mathbb{P}_0}[V_0(\widehat{\theta})] + \mathbb{E}_{\mathbb{P}_1}[V_1(\widehat{\theta})] \geq 1 - \|\mathbb{P}_1 - \mathbb{P}_0\|_{\text{TV}},$$

which completes the proof. □

In order to see how taking the convex hulls can decrease the total variation norm, it is instructive to return to the Gaussian location model previously introduced in Example 15.4:

Example 15.10 (Sharpened bounds for Gaussian location family) In Example 15.4, we used a two-point form of Le Cam's method to prove a lower bound on mean estimation in the Gaussian location family. A key step was to upper bound the TV distance $\|\mathbb{P}_\theta^n - \mathbb{P}_0^n\|_{\text{TV}}$ between the n-fold product distributions based on the Gaussian models $\mathcal{N}(\theta, \sigma^2)$ and $\mathcal{N}(0, \sigma^2)$, respectively. Here let us show how the convex hull version of Le Cam's method can be used to sharpen this step, so as obtain a bound with tighter constants. In particular, setting $\theta = 2\delta$ as before, consider the two families $\mathcal{P}_0 = \{\mathbb{P}_0^n\}$ and $\mathcal{P}_1 = \{\mathbb{P}_\theta^n, \mathbb{P}_{-\theta}^n\}$. Note that the mixture distribution $\bar{\mathbb{P}} := \frac{1}{2}\mathbb{P}_\theta^n + \frac{1}{2}\mathbb{P}_{-\theta}^n$ belongs to $\text{conv}(\mathcal{P}_1)$. From the second-moment bound explored in Exercise 15.10(c), we have

$$\|\bar{\mathbb{P}} - \mathbb{P}_0^n\|_{\text{TV}}^2 \le \tfrac{1}{4}\Big\{e^{\frac{1}{2}(\frac{\sqrt{n}\theta}{\sigma})^4} - 1\Big\} = \tfrac{1}{4}\Big\{e^{\frac{1}{2}(\frac{2\sqrt{n}\delta}{\sigma})^4} - 1\Big\}. \tag{15.27}$$

Setting $\delta = \frac{\sigma t}{2\sqrt{n}}$ for some parameter $t > 0$ to be chosen, the convex hull Le Cam bound (15.26) yields

$$\min_{\widehat{\theta}} \sup_{\theta \in \mathbb{R}} \mathbb{E}_\theta[|\widehat{\theta} - \theta|] \ge \frac{\sigma}{4\sqrt{n}} \sup_{t>0} \Big\{t\big(1 - \frac{1}{2}\sqrt{e^{\frac{1}{2}t^4} - 1}\big)\Big\} \ge \frac{3}{20}\frac{\sigma}{\sqrt{n}}.$$

This bound is an improvement over our original bound (15.16a) from Example 15.4, which has the pre-factor of $\frac{1}{12} \approx 0.08$, as opposed to $\frac{3}{20} = 0.15$ obtained from this analysis. Thus, even though we used the same base separation δ, our use of mixture distributions reduced the TV distance—compare the bounds (15.27) and (15.15)—thereby leading to a sharper result. ♣

In the previous example, the gains from extending to the convex hull are only in terms of the constant pre-factors. Let us now turn to an example in which the gain is more substantial. Recall Example 15.8 in which we investigated the problem of estimating the quadratic functional $f \mapsto \theta(f) = \int_0^1 (f'(x))^2\,dx$ over the class $\mathscr{F}_2$ from equation (15.21). Let us now demonstrate how the use of Le Cam's method in its full convex hull form allows for the derivation of an optimal lower bound for the minimax risk.

Example 15.11 (Optimal bounds for quadratic functionals) For each binary vector $\alpha \in \{-1, +1\}^m$, define the distribution $\mathbb{P}_\alpha$ with density given by

$$f_\alpha(x) = 1 + \sum_{j=1}^m \alpha_j \phi_j(x).$$

Note that the perturbed density g constructed in Example 15.8 is a special member of this family, generated by the binary vector $\alpha = (1, 1, \ldots, 1)$. Let $\mathbb{P}_\alpha^n$ denote the product distribution on $\mathcal{X}^n$ formed by sampling n times independently from $\mathbb{P}_\alpha$, and define the two classes $\mathcal{P}_0 := \{\mathbb{U}^n\}$ and $\mathcal{P}_1 := \{\mathbb{P}_\alpha^n, \alpha \in \{-1, +1\}^m\}$. With these choices, we then have

$$\inf_{\substack{\mathbb{P}_j \in \text{conv}(\mathcal{P}_j) \\ j=0,1}} \|\mathbb{P}_0 - \mathbb{P}_1\|_{\text{TV}} \le \|\mathbb{U}^n - \mathbb{Q}\|_{\text{TV}} \le H(\mathbb{U}^n \,\|\, \mathbb{Q}),$$

where $\mathbb{Q} := 2^{-m} \sum_{\alpha \in \{-1,+1\}^m} \mathbb{P}_\alpha^n$ is the uniformly weighted mixture over all 2^m choices of $\mathbb{P}_\alpha^n$.

In this case, since $\mathbb{Q}$ is not a product distribution, we can no longer apply the decomposition (15.12a) so as to bound the Hellinger distance $H(\mathbb{U}^n \,\|\, \mathbb{Q})$ by a univariate version. Instead, some more technical calculations are required. One possible upper bound is given by

$$H^2(\mathbb{U}^n \,\|\, \mathbb{Q}) \le n^2 \sum_{j=1}^m \Big(\int_0^1 \phi_j^2(x)\,dx\Big)^2. \tag{15.28}$$

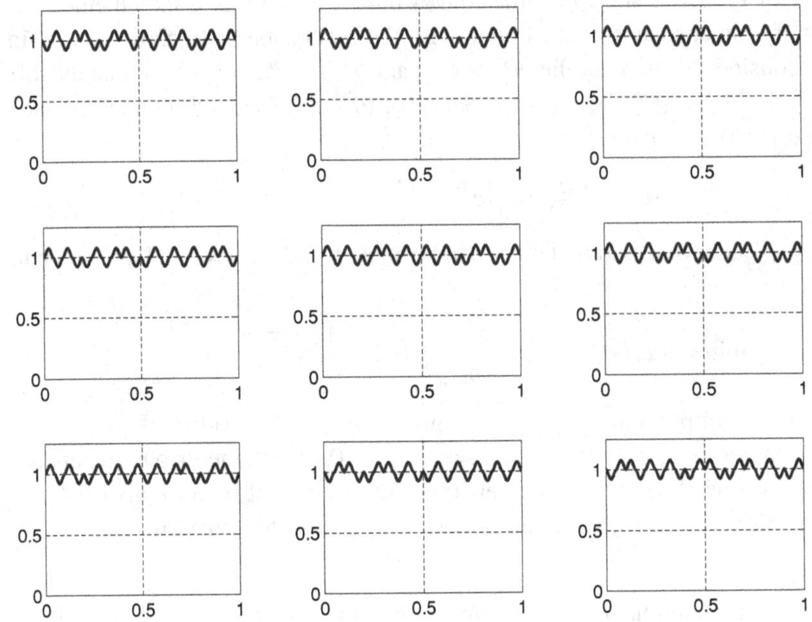

Figure 15.4 Illustration of some densities of the form $f_\alpha(x) = 1 + \sum_{j=1}^m \alpha_j \phi_j(x)$ for different choices of sign vectors $\alpha \in \{-1, 1\}^m$. Note that there are 2^m such densities in total.

See the bibliographic section for discussion of this upper bound as well as related results. If we take the upper bound (15.28) as given, then using the calculations from Example 15.8 —in particular, recall the definition of the constants b_ℓ from equation (15.22)—we find that

$$H^2(\mathbb{U}^n \,\|\, \mathbb{Q}) \leq mn^2 \frac{b_0^2}{m^{10}} = b_0^2 \frac{n^2}{m^9}.$$

Setting $m^9 = 4b_0^2 n^2$ yields that $\|\mathbb{U}^{1:n} - \mathbb{Q}\|_{\mathrm{TV}} \leq H(\mathbb{U}^{1:n} \,\|\, \mathbb{P}^{1:n}) \leq 1/2$, and hence Lemma 15.9 implies that

$$\sup_{f \in \mathscr{F}_2} \mathbb{E}|\widehat{\theta}(f) - \theta(f)| \geq \delta/4 = \frac{C^2 b_1}{8m^2} \gtrsim n^{-4/9}.$$

Thus, by using the full convex form of Le Cam's method, we have recovered a better lower bound on the minimax risk ($n^{-4/9} \gg n^{-1/2}$). This lower bound turns out to be unimprovable; see the bibliographic section for further discussion. ♣

15.3 Fano's method

In this section, we describe an alternative method for deriving lower bounds, one based on a classical result from information theory known as Fano's inequality.

15.3.1 Kullback–Leibler divergence and mutual information

Recall our basic set-up: we are interested in lower bounding the probability of error in an M-ary hypothesis testing problem, based on a family of distributions $\{\mathbb{P}_{\theta^1}, \ldots, \mathbb{P}_{\theta^M}\}$. A sample Z is generated by choosing an index J uniformly at random from the index set $[M] := \{1, \ldots, M\}$, and then generating data according to $\mathbb{P}_{\theta^J}$. In this way, the observation follows the *mixture distribution* $\mathbb{Q}_Z = \bar{\mathbb{Q}} := \frac{1}{M} \sum_{j=1}^M \mathbb{P}_{\theta^j}$. Our goal is to identify the index J of the probability distribution from which a given sample has been drawn.

Intuitively, the difficulty of this problem depends on the amount of dependence between the observation Z and the unknown random index J. In the extreme case, if Z were actually independent of J, then observing Z would have no value whatsoever. How to measure the amount of dependence between a pair of random variables? Note that the pair (Z, J) are independent if and only if their joint distribution $\mathbb{Q}_{Z,J}$ is equal to the product of its marginals—namely, $\mathbb{Q}_Z \mathbb{Q}_J$. Thus, a natural way in which to measure dependence is by computing some type of divergence measure between the joint distribution and the product of marginals. The *mutual information* between the random variables (Z, J) is defined in exactly this way, using the Kullback–Leibler divergence as the underlying measure of distance—that is

$$I(Z, J) := D(\mathbb{Q}_{Z,J} \,\|\, \mathbb{Q}_Z \, \mathbb{Q}_J). \tag{15.29}$$

By standard properties of the KL divergence, we always have $I(Z, J) \geq 0$, and moreover $I(Z, J) = 0$ if and only if Z and J are independent.

Given our set-up and the definition of the KL divergence, the mutual information can be written in terms of component distributions $\{\mathbb{P}_{\theta^j}, \, j \in [M]\}$ and the mixture distribution $\bar{\mathbb{Q}} \equiv \mathbb{Q}_Z$—in particular as

$$I(Z; J) = \frac{1}{M} \sum_{j=1}^{M} D(\mathbb{P}_{\theta^j} \,\|\, \bar{\mathbb{Q}}), \tag{15.30}$$

corresponding to the mean KL divergence between $\mathbb{P}_{\theta^j}$ and $\bar{\mathbb{Q}}$, averaged over the choice of index j. Consequently, the mutual information is small if the distributions $\mathbb{P}_{\theta^j}$ are hard to distinguish from the mixture distribution $\bar{\mathbb{Q}}$ on average.

15.3.2 Fano lower bound on minimax risk

Let us now return to the problem at hand: namely, obtaining lower bounds on the minimax error. The Fano method is based on the following lower bound on the error probability in an M-ary testing problem, applicable when J is uniformly distributed over the index set:

$$\mathbb{P}[\psi(Z) \neq J] \geq 1 - \frac{I(Z; J) + \log 2}{\log M}. \tag{15.31}$$

When combined with the reduction from estimation to testing given in Proposition 15.1, we obtain the following lower bound on the minimax error:

Proposition 15.12 *Let $\{\theta^1, \ldots, \theta^M\}$ be a 2δ-separated set in the ρ semi-metric on $\Theta(\mathcal{P})$, and suppose that J is uniformly distributed over the index set $\{1, \ldots, M\}$, and $(Z \mid J = j) \sim \mathbb{P}_{\theta^j}$. Then for any increasing function $\Phi\colon [0, \infty) \to [0, \infty)$, the minimax risk is lower bounded as*

$$\mathfrak{M}(\theta(\mathcal{P}); \Phi \circ \rho) \geq \Phi(\delta) \left\{1 - \frac{I(Z; J) + \log 2}{\log M}\right\}, \tag{15.32}$$

where $I(Z; J)$ is the mutual information between Z and J.

We provide a proof of the Fano bound (15.31), from which Proposition 15.12 follows, in the sequel (see Section 15.4). For the moment, in order to gain intuition for this result, it is helpful to consider the behavior of the different terms of $\delta \to 0^+$. As we shrink δ, then the 2δ-separation criterion becomes milder, so that the cardinality $M \equiv M(2\delta)$ in the denominator increases. At the same time, in a generic setting, the mutual information $I(Z; J)$ will decrease, since the random index $J \in [M(2\delta)]$ can take on a larger number of potential values. By decreasing δ sufficiently, we may thereby ensure that

$$\frac{I(Z; J) + \log 2}{\log M} \leq \frac{1}{2}, \tag{15.33}$$

so that the lower bound (15.32) implies that $\mathfrak{M}(\theta(\mathcal{P}); \Phi \circ \rho) \geq \frac{1}{2}\Phi(\delta)$. Thus, we have a generic scheme for deriving lower bounds on the minimax risk.

In order to derive lower bounds in this way, there remain two technical and possibly challenging steps. The first requirement is to specify 2δ-separated sets with large cardinality $M(2\delta)$. Here the theory of metric entropy developed in Chapter 5 plays an important role, since any 2δ-packing set is (by definition) 2δ-separated in the ρ semi-metric. The second requirement is to compute—or more realistically to upper bound—the mutual information $I(Z; J)$. In general, this second step is non-trivial, but various avenues are possible.

The simplest upper bound on the mutual information is based on the convexity of the Kullback–Leibler divergence (see Exercise 15.3). Using this convexity and the mixture representation (15.30), we find that

$$I(Z; J) \leq \frac{1}{M^2} \sum_{j,k=1}^{M} D(\mathbb{P}_{\theta^j} \,\|\, \mathbb{P}_{\theta^k}). \tag{15.34}$$

Consequently, if we can construct a 2δ-separated set such that all pairs of distributions $\mathbb{P}_{\theta^j}$ and $\mathbb{P}_{\theta^k}$ are close on average, the mutual information can be controlled. Let us illustrate the use of this upper bound for a simple parametric problem.

Example 15.13 (Normal location model via Fano method) Recall from Example 15.4 the normal location family, and the problem of estimating $\theta \in \mathbb{R}$ under the squared error. There we showed how to lower bound the minimax error using Le Cam's method; here let us derive a similar lower bound using Fano's method.

Consider the 2δ-separated set of real-valued parameters $\{\theta^1, \theta^2, \theta^3\} = \{0, 2\delta, -2\delta\}$. Since

$\mathbb{P}_{\theta^j} = \mathcal{N}(\theta^j, \sigma^2)$, we have

$$D(\mathbb{P}^{1:n}_{\theta^j} \,\|\, \mathbb{P}^{1:n}_{\theta^k}) = \frac{n}{2\sigma^2}(\theta^j - \theta^k)^2 \leq \frac{2n\delta^2}{\sigma^2} \qquad \text{for all } j, k = 1, 2, 3.$$

The bound (15.34) then ensures that $I(Z; J_\delta) \leq \frac{2n\delta^2}{\sigma^2}$, and choosing $\delta^2 = \frac{\sigma^2}{20n}$ ensures that $\frac{2n\delta^2/\sigma^2 + \log 2}{\log 3} < 0.75$. Putting together the pieces, the Fano bound (15.32) with $\Phi(t) = t^2$ implies that

$$\sup_{\theta \in \mathbb{R}} \mathbb{E}_\theta[(\widehat{\theta} - \theta)^2] \geq \frac{\delta^2}{4} = \frac{1}{80}\frac{\sigma^2}{n}.$$

In this way, we have re-derived a minimax lower bound of the order σ^2/n, which, as discussed in Example 15.4, is of the correct order. ♣

15.3.3 Bounds based on local packings

Let us now formalize the approach that was used in the previous example. It is based on a local packing of the parameter space Ω, which underlies what is called the "generalized Fano" method in the statistics literature. (As a sidenote, this nomenclature is very misleading, because the method is actually based on a substantial weakening of the Fano bound, obtained from the inequality (15.34).)

The local packing approach proceeds as follows. Suppose that we can construct a 2δ-separated set contained within Ω such that, for some quantity c, the Kullback–Leibler divergences satisfy the uniform upper bound

$$\sqrt{D(\mathbb{P}_{\theta^j} \,\|\, \mathbb{P}_{\theta^k})} \leq c\sqrt{n}\,\delta \qquad \text{for all } j \neq k. \tag{15.35a}$$

The bound (15.34) then implies that $I(Z; J) \leq c^2 n\delta^2$, and hence the bound (15.33) will hold as long as

$$\log M(2\delta) \geq 2\{c^2 n\delta^2 + \log 2\}. \tag{15.35b}$$

In summary, if we can find a 2δ-separated family of distributions such that conditions (15.35a) and (15.35b) both hold, then we may conclude that the minimax risk is lower bounded as $\mathfrak{M}(\theta(\mathcal{P}), \Phi \circ \rho) \geq \frac{1}{2}\Phi(\delta)$.

Let us illustrate the local packing approach with some examples.

Example 15.14 (Minimax risks for linear regression) Consider the standard linear regression model $y = \mathbf{X}\theta^* + w$, where $\mathbf{X} \in \mathbb{R}^{n \times d}$ is a fixed design matrix, and the vector $w \sim \mathcal{N}(0, \sigma^2 \mathbf{I}_n)$ is observation noise. Viewing the design matrix $\mathbf{X}$ as fixed, let us obtain lower bounds on the minimax risk in the prediction (semi-)norm $\rho_{\mathbf{X}}(\widehat{\theta}, \theta^*) := \frac{\|\mathbf{X}(\widehat{\theta} - \theta^*)\|_2}{\sqrt{n}}$, assuming that θ^* is allowed to vary over $\mathbb{R}^d$.

For a tolerance $\delta > 0$ to be chosen, consider the set

$$\{\gamma \in \text{range}(\mathbf{X}) \mid \|\gamma\|_2 \leq 4\delta\sqrt{n}\},$$

and let $\{\gamma^1, \ldots, \gamma^M\}$ be a $2\delta\sqrt{n}$-packing in the ℓ_2-norm. Since this set sits in a space of dimension $r = \text{rank}(\mathbf{X})$, Lemma 5.7 implies that we can find such a packing with $\log M \geq r \log 2$

elements. We thus have a collection of vectors of the form $\gamma^j = \mathbf{X}\theta^j$ for some $\theta^j \in \mathbb{R}^d$, and such that

$$\frac{\|\mathbf{X}\theta^j\|_2}{\sqrt{n}} \leq 4\delta, \qquad \text{for each } j \in [M], \tag{15.36a}$$

$$2\delta \leq \frac{\|\mathbf{X}(\theta^j - \theta^k)\|_2}{\sqrt{n}} \leq 8\delta \qquad \text{for each } j \neq k \in [M] \times [M]. \tag{15.36b}$$

Let $\mathbb{P}_{\theta^j}$ denote the distribution of y when the true regression vector is θ^j; by the definition of the model, under $\mathbb{P}_{\theta^j}$, the observed vector $y \in \mathbb{R}^n$ follows a $\mathcal{N}(\mathbf{X}\theta^j, \sigma^2\mathbf{I}_n)$ distribution. Consequently, the result of Exercise 15.13 ensures that

$$D(\mathbb{P}_{\theta^j} \,\|\, \mathbb{P}_{\theta^k}) = \frac{1}{2\sigma^2}\|\mathbf{X}(\theta^j - \theta^k)\|_2^2 \leq \frac{32n\delta^2}{\sigma^2}, \tag{15.37}$$

where the inequality follows from the upper bound (15.36b). Consequently, for r sufficiently large, the lower bound (15.35b) can be satisfied by setting $\delta^2 = \frac{\sigma^2}{64}\frac{r}{n}$, and we conclude that

$$\inf_{\widehat{\theta}} \sup_{\theta \in \mathbb{R}^d} \mathbb{E}\Big[\frac{1}{n}\|\mathbf{X}(\widehat{\theta} - \theta)\|_2^2\Big] \geq \frac{\sigma^2}{128}\,\frac{\text{rank}(\mathbf{X})}{n}.$$

This lower bound is sharp up to constant pre-factors: as shown by our analysis in Example 13.8 and Exercise 13.2, it can be achieved by the usual linear least-squares estimate. ♣

Let us now see how the upper bound (15.34) and Fano's method can be applied to a non-parametric problem.

Example 15.15 (Minimax risk for density estimation) Recall from equation (15.21) the family $\mathscr{F}_2$ of twice-smooth densities on $[0, 1]$, bounded uniformly above, bounded uniformly away from zero, and with uniformly bounded second derivative. Let us consider the problem of estimating the entire density function f, using the Hellinger distance as our underlying metric ρ.

In order to construct a local packing, we make use of the family of perturbed densities from Example 15.11, each of the form $f_\alpha(x) = 1 + \sum_{j=1}^m \alpha_j \phi_j(x)$, where $\alpha \in \{-1, +1\}^m$ and the function ϕ_j was defined in equation (15.23). Although there are 2^m such perturbed densities, it is convenient to use only a well-separated subset of them. Let $M_H(\frac{1}{4}; \mathbb{H}^m)$ denote the $\frac{1}{4}$-packing number of the binary hypercube $\{-1, +1\}^m$ in the rescaled Hamming metric. From our calculations in Example 5.3, we know that

$$\log M_H(\tfrac{1}{4}; \mathbb{H}^m) \geq m\, D(\tfrac{1}{4} \,\|\, \tfrac{1}{2}) \geq \frac{m}{10}.$$

(See in particular equation (5.3).) Consequently, we can find a subset $\mathbb{T} \subset \{-1, +1\}^m$ with cardinality at least $e^{m/10}$ such that

$$d_H(\alpha, \beta) = \frac{1}{m}\sum_{j=1}^m \mathbb{I}[\alpha_j \neq \beta_j] \geq 1/4 \qquad \text{for all } \alpha \neq \beta \in \mathbb{T}. \tag{15.38}$$

We then consider the family of $M = e^{m/10}$ distributions $\{\mathbb{P}_\alpha, \alpha \in \mathbb{T}\}$, where $\mathbb{P}_\alpha$ has density f_α.

We first lower bound the Hellinger distance between distinct pairs f_α and f_β. Since ϕ_j is non-zero only on the interval $I_j = [x_j, x_{j+1}]$, we can write

$$\int_0^1 \left(\sqrt{f_\alpha(x)} - \sqrt{f_\beta(x)}\right)^2 dx = \sum_{j=0}^{m-1} \int_{I_j} \left(\sqrt{f_\alpha(x)} - \sqrt{f_\beta(x)}\right)^2 dx.$$

But on the interval I_j, we have

$$\left(\sqrt{f_\alpha(x)} + \sqrt{f_\beta(x)}\right)^2 = 2\left(f_\alpha(x) + f_\beta(x)\right) \leq 4,$$

and therefore

$$\begin{aligned}\int_{I_j} \left(\sqrt{f_\alpha(x)} - \sqrt{f_\beta(x)}\right)^2 dx &\geq \frac{1}{4} \int_{I_j} \left(f_\alpha(x) - f_\beta(x)\right)^2 \\ &\geq \int_{I_j} \phi_j^2(x)\, dx \qquad \text{whenever } \alpha_j \neq \beta_j.\end{aligned}$$

Since $\int_{I_j} \phi_j^2(x)\, dx = \int_0^1 \phi^2(x)\, dx = \frac{b_0}{m^5}$ and any distinct $\alpha \neq \beta$ differ in at least $m/4$ positions, we find that $H^2(\mathbb{P}_\alpha \,\|\, \mathbb{P}_\beta) \geq \frac{m}{4} \frac{b_0}{m^5} = \frac{b_0}{m^4} \equiv 4\delta^2$. Consequently, we have constructed a 2δ-separated set with $\delta^2 = \frac{b_0}{4m^4}$.

Next we upper bound the pairwise KL divergence. By construction, we have $f_\alpha(x) \geq 1/2$ for all $x \in [0, 1]$, and thus

$$\begin{aligned}D(\mathbb{P}_\alpha \,\|\, \mathbb{P}_\beta) &\leq \int_0^1 \frac{(\sqrt{f_\alpha(x)} - \sqrt{f_\beta(x)})^2}{f_\alpha(x)} dx \\ &\leq 2 \int_0^1 (\sqrt{f_\alpha(x)} - \sqrt{f_\beta(x)})^2 dx \leq \frac{4b_0}{m^4},\end{aligned} \tag{15.39}$$

where the final inequality follows by a similar sequence of calculations. Overall, we have established the upper bound $D(\mathbb{P}_\alpha^n \,\|\, \mathbb{P}_\beta^n) = nD(\mathbb{P}_\alpha \,\|\, \mathbb{P}_\beta) \leq 4b_0 \frac{n}{m^4} = 4n\delta^2$. Finally, we must ensure that

$$\log M = \frac{m}{10} \geq 2\left\{4n\delta^2 + \log 2\right\} = 2\left\{4b_0 \frac{n}{m^4} + \log 2\right\}.$$

This equality holds if we choose $m = \frac{n^{1/5}}{C}$ for a sufficiently small constant C. With this choice, we have $\delta^2 \asymp m^{-4} \asymp n^{-4/5}$, and hence conclude that

$$\sup_{f \in \mathscr{F}_2} H^2(\widehat{f} \,\|\, f) \gtrsim n^{-4/5}.$$

This rate is minimax-optimal for densities with two orders of smoothness; recall that we encountered the same rate for the closely related problem of nonparametric regression in Chapter 13. ♣

As a third example, let us return to the high-dimensional parametric setting, and study minimax risks for the problem of sparse linear regression, which we studied in detail in Chapter 7.

Example 15.16 (Minimax risk for sparse linear regression) Consider the high-dimensional linear regression model $y = \mathbf{X}\theta^* + w$, where the regression vector θ^* is known *a priori* to be sparse, say with at most $s < d$ non-zero coefficients. It is then natural to consider the minimax risk over the set

$$\mathbb{S}^d(s) := \mathbb{B}_0^d(s) \cap \mathbb{B}_2(1) = \left\{\theta \in \mathbb{R}^d \mid \|\theta\|_0 \le s, \|\theta\|_2 \le 1\right\} \tag{15.40}$$

of s-sparse vectors within the Euclidean unit ball.

Let us first construct a $1/2$-packing of the set $\mathbb{S}^d(s)$. From our earlier results in Chapter 5 (in particular, see Exercise 5.8), there exists a $1/2$-packing of this set with log cardinality at least $\log M \ge \frac{s}{2} \log \frac{d-s}{s}$. We follow the same rescaling procedure as in Example 15.14 to form a 2δ-packing such that $\|\theta^j - \theta^k\|_2 \le 4\delta$ for all pairs of vectors in our packing set. Since the vector $\theta^j - \theta^k$ is at most $2s$-sparse, we have

$$\sqrt{D(\mathbb{P}_{\theta^j} \,\|\, \mathbb{P}_{\theta^k})} = \frac{1}{\sqrt{2}\sigma} \|\mathbf{X}(\theta^j - \theta^k)\|_2 \le \frac{\gamma_{2s}}{\sqrt{2}\sigma} 4\delta,$$

where $\gamma_{2s} := \max_{|T|=2s} \sigma_{\max}(\mathbf{X}_T)/\sqrt{n}$. Putting together the pieces, we see that the minimax risk is lower bounded by any $\delta > 0$ for which

$$\frac{s}{2} \log \frac{d-s}{s} \ge 128 \frac{\gamma_{2s}^2}{\sigma^2} n\delta^2 + 2\log 2.$$

As long as $s \le d/2$ and $s \ge 10$, the choice $\delta^2 = \frac{\sigma^2}{400\gamma_{2s}^2} s \log \frac{d-s}{s}$ suffices. Putting together the pieces, we conclude that in the range $10 \le s \le d/2$, the minimax risk is lower bounded as

$$\mathfrak{M}(\mathbb{S}^d(s); \|\cdot\|_2) \gtrsim \frac{\sigma^2}{\gamma_{2s}^2} \frac{s \log \frac{ed}{s}}{n}. \tag{15.41}$$

The constant obtained by this argument is not sharp, but this lower bound is otherwise unimprovable: see the bibliographic section for further details. ♣

15.3.4 Local packings with Gaussian entropy bounds

Our previous examples have also used the convexity-based upper bound (15.34) on the mutual information. We now turn to a different upper bound on the mutual information, applicable when the conditional distribution of Z given J is Gaussian.

Lemma 15.17 *Suppose J is uniformly distributed over $[M] = \{1, \ldots, M\}$ and that Z conditioned on $J = j$ has a Gaussian distribution with covariance $\mathbf{\Sigma}^j$. Then the mutual information is upper bounded as*

$$I(Z; J) \le \frac{1}{2} \left\{ \log \det \operatorname{cov}(Z) - \frac{1}{M} \sum_{j=1}^{M} \log \det(\mathbf{\Sigma}^j) \right\}. \tag{15.42}$$

This upper bound is a consequence of the maximum entropy property of the multivariate Gaussian distribution; see Exercise 15.14 for further details. In the special case when $\mathbf{\Sigma}^j = \mathbf{\Sigma}$ for all $j \in [M]$, it takes on the simpler form

$$I(Z; J) \leq \frac{1}{2} \log \left(\frac{\det \operatorname{cov}(Z)}{\det(\mathbf{\Sigma})} \right). \tag{15.43}$$

Let us illustrate the use of these bounds with some examples.

Example 15.18 (Variable selection in sparse linear regression) Let us return to the model of sparse linear regression from Example 15.16, based on the standard linear model $y = \mathbf{X}\theta^* + w$, where the unknown regression vector $\theta^* \in \mathbb{R}^d$ is s-sparse. Here we consider the problem of lower bounding the minimax risk for the problem of variable selection—namely, determining the support set $S = \{j \in \{1, 2, \ldots, d\} \mid \theta_j^* \neq 0\}$, which is assumed to have cardinality $s \ll d$.

In this case, the problem of interest is itself a multiway hypothesis test—namely, that of choosing from all $\binom{d}{s}$ possible subsets. Consequently, a direct application of Fano's inequality leads to lower bounds, and we can obtain different such bounds by constructing various ensembles of subproblems. These subproblems are parameterized by the pair (d, s), as well as the quantity $\theta_{\min} = \min_{j \in S} |\theta_j^*|$. In this example, we show that, in order to achieve a probability of error below $1/2$, any method requires a sample size of at least

$$n > \max \left\{ 8 \frac{\log(d + s - 1)}{\log(1 + \frac{\theta_{\min}^2}{\sigma^2})}, \; 8 \frac{\log \binom{d}{s}}{\log(1 + s \frac{\theta_{\min}^2}{\sigma^2})} \right\}, \tag{15.44}$$

as long as $\min \left\{ \log(d + s - 1), \log \binom{d}{s} \right\} \geq 4 \log 2$.

For this problem, our observations consist of the response vector $y \in \mathbb{R}^n$ and design matrix $\mathbf{X} \in \mathbb{R}^{n \times d}$. We derive lower bounds by first conditioning on a particular instantiation $\mathbf{X} = \{x_i\}_{i=1}^n$ of the design matrix, and using a form of Fano's inequality that involves the mutual information $I_{\mathbf{X}}(y; J)$ between the response vector y and the random index J with the design matrix $\mathbf{X}$ held fixed. In particular, we have

$$\mathbb{P}[\psi(y, \mathbf{X}) \neq J \mid \mathbf{X} = \{x_i\}_{i=1}^n] \geq 1 - \frac{I_{\mathbf{X}}(y; J) + \log 2}{\log M},$$

so that by taking averages over $\mathbf{X}$, we can obtain lower bounds on $\mathbb{P}[\psi(y, \mathbf{X}) \neq J]$ that involve the quantity $\mathbb{E}_{\mathbf{X}}[I_{\mathbf{X}}(y; J)]$.

Ensemble A: Consider the class $M = \binom{d}{s}$ of all possible subsets of cardinality s, enumerated in some fixed way. For the ℓth subset S^ℓ, let $\theta^\ell \in \mathbb{R}^d$ have values $\theta_{\min}$ for all indices $j \in S^\ell$, and zeros in all other positions. For a fixed covariate vector $x_i \in \mathbb{R}^d$, an observed response $y_i \in \mathbb{R}$ then follows the mixture distribution $\frac{1}{M} \sum_{\ell=1}^M \mathbb{P}_{\theta^\ell}$, where $\mathbb{P}_{\theta^\ell}$ is the distribution of a $\mathcal{N}(\langle x_i, \theta^\ell \rangle, \sigma^2)$ random variable.

By the definition of mutual information, we have

$$
\begin{aligned}
I_{\mathbf{X}}(y; J) &= H_{\mathbf{X}}(y) - H_{\mathbf{X}}(y \mid J) \\
&\stackrel{\text{(i)}}{\leq} \left[\sum_{i=1}^{n} H_{\mathbf{X}}(y_i)\right] - H_{\mathbf{X}}(y \mid J) \\
&\stackrel{\text{(ii)}}{=} \sum_{i=1}^{n} \{H_{\mathbf{X}}(y_1) - H_{\mathbf{X}}(y_1 \mid J)\} \\
&= \sum_{i=1}^{n} I_{\mathbf{X}}(y_i; J),
\end{aligned} \tag{15.45}
$$

where step (i) follows since independent random vectors have larger entropy than dependent ones (see Exercise 15.4), and step (ii) follows since $(y_1, \ldots, y_n)$ are independent conditioned on J. Next, applying Lemma 15.17 repeatedly for each $i \in [n]$ with $Z = y_i$, conditionally on the matrix $\mathbf{X}$ of covariates, yields

$$
I_{\mathbf{X}}(y; J) \leq \frac{1}{2} \sum_{i=1}^{n} \log \frac{\operatorname{var}(y_i \mid x_i)}{\sigma^2}.
$$

Now taking averages over $\mathbf{X}$ and using the fact that the pairs (y_i, x_i) are jointly i.i.d., we find that

$$
\mathbb{E}_{\mathbf{X}}\left[I_{\mathbf{X}}(y; J)\right] \leq \frac{n}{2} \mathbb{E}[\log \frac{\operatorname{var}(y_1 \mid x_1)}{\sigma^2}] \leq \frac{n}{2} \log \frac{\mathbb{E}_{x_1}[\operatorname{var}(y_1 \mid x_1)]}{\sigma^2},
$$

where the last inequality follows Jensen's inequality, and concavity of the logarithm.

It remains to upper bound the variance term. Since the random vector y_1 follows a mixture distribution with M components, we have

$$
\begin{aligned}
\mathbb{E}_{x_1}[\operatorname{var}(y_1 \mid x_1)] \; &\leq \mathbb{E}_{x_1}[\mathbb{E}[y_1^2 \mid x_1]] = \mathbb{E}_{x_1}\Big[x_1^{\mathrm{T}}\{\frac{1}{M}\sum_{j=1}^{M} \theta^j \otimes \theta^j\}x_1 + \sigma^2\Big] \\
&= \operatorname{trace}\Big(\frac{1}{M}\sum_{j=1}^{M}(\theta^j \otimes \theta^j)\Big) + \sigma^2.
\end{aligned}
$$

Now each index $j \in \{1, 2, \ldots, d\}$ appears in $\binom{d-1}{s-1}$ of the total number of subsets $M = \binom{d}{s}$, so that

$$
\operatorname{trace}\Big(\frac{1}{M}\sum_{j=1}^{M} \theta^j \otimes \theta^j\Big) = d\frac{\binom{d-1}{s-1}}{\binom{d}{s}}\theta_{\min}^2 = s\theta_{\min}^2.
$$

Putting together the pieces, we conclude that

$$
\mathbb{E}_{\mathbf{X}}[I_{\mathbf{X}}(y; J)] \leq \frac{n}{2} \log\left(1 + \frac{s\theta_{\min}^2}{\sigma^2}\right),
$$

and hence the Fano lower bound implies that

$$
\mathbb{P}[\psi(y, \mathbf{X}) \neq J] \geq 1 - \frac{\frac{n}{2}\log(1 + \frac{s\theta_{\min}^2}{\sigma^2}) + \log 2}{\log\binom{d}{s}},
$$

from which the first lower bound in equation (15.44) follows as long as $\log\binom{d}{s} \geq 4\log 2$, as assumed.

Ensemble B: Let $\bar{\theta} \in \mathbb{R}^d$ be a vector with $\theta_{\min}$ in its first $s-1$ coordinates, and zero in all remaining $d-s+1$ coordinates. For each $j = 1,\ldots,d$, let $e_j \in \mathbb{R}^d$ denote the jth standard basis vector with a single one in position j. Define the family of $M = d-s+1$ vectors $\theta^j := \bar{\theta} + \theta_{\min} e_j$ for $j = s,\ldots,d$. By a straightforward calculation, we have $\mathbb{E}[Y \mid x] = \langle x, \gamma\rangle$, where $\gamma := \bar{\theta} + \frac{1}{M}\theta_{\min} e_{s\to d}$, and the vector $e_{s\to d} \in \mathbb{R}^d$ has ones in positions s through d, and zeros elsewhere. By the same argument as for ensemble A, it suffices to upper bound the quantity $\mathbb{E}_{x_1}[\operatorname{var}(y_1 \mid x_1)]$. Using the definition of our ensemble, we have

$$\mathbb{E}_{x_1}[\operatorname{var}(y_1 \mid x_1)] = \sigma^2 + \operatorname{trace}\left\{\frac{1}{M}\sum_{j=1}^{M}(\theta^j \otimes \theta^j - \gamma \otimes \gamma)\right\} \leq \sigma^2 + \theta^2_{\min}. \tag{15.46}$$

Recall that we have assumed that $\log(d-s+1) > 4\log 2$. Using Fano's inequality and the upper bound (15.46), the second term in the lower bound (15.44) then follows. ♣

Let us now turn to a slightly different problem, namely that of lower bounds for principal component analysis. Recall from Chapter 8 the spiked covariance ensemble, in which a random vector $x \in \mathbb{R}^d$ is generated via

$$x \stackrel{\mathrm{d}}{=} \sqrt{\nu}\xi\theta^* + w. \tag{15.47}$$

Here $\nu > 0$ is a given signal-to-noise ratio, θ^* is a fixed vector with unit Euclidean norm, and the random quantities $\xi \sim \mathcal{N}(0,1)$ and $w \sim \mathcal{N}(0, \mathbf{I}_d)$ are independent. Observe that the d-dimensional random vector x is zero-mean Gaussian with a covariance matrix of the form $\mathbf{\Sigma} := \mathbf{I}_d + \nu(\theta^* \otimes \theta^*)$. Moreover, by construction, the vector θ^* is the unique maximal eigenvector of the covariance matrix $\mathbf{\Sigma}$.

Suppose that our goal is to estimate θ^* based on n i.i.d. samples of the random vector x. In the following example, we derive lower bounds on the minimax risk in the squared Euclidean norm $\|\widehat{\theta} - \theta^*\|_2^2$. (As discussed in Chapter 8, recall that there is always a sign ambiguity in estimating eigenvectors, so that in computing the Euclidean norm, we implicitly assume that the correct direction is chosen.)

Example 15.19 (Lower bounds for PCA) Let $\{\Delta^1,\ldots,\Delta^M\}$ be a $1/2$-packing of the unit sphere in $\mathbb{R}^{d-1}$; from Example 5.8, for all $d \geq 3$, there exists such a set with cardinality $\log M \geq (d-1)\log 2 \geq d/2$. For a given orthonormal matrix $\mathbf{U} \in \mathbb{R}^{(d-1)\times(d-1)}$ and tolerance $\delta \in (0,1)$ to be chosen, consider the family of vectors

$$\theta^j(\mathbf{U}) = \sqrt{1-\delta^2}\begin{bmatrix}1\\0_{d-1}\end{bmatrix} + \delta\begin{bmatrix}0\\\mathbf{U}\Delta^j\end{bmatrix} \qquad \text{for } j \in [M], \tag{15.48}$$

where 0_{d-1} denotes the $(d-1)$-dimensional vector of zeros. By construction, each vector $\theta^j(\mathbf{U})$ lies on the unit sphere in $\mathbb{R}^d$, and the collection of all M vectors forms a $\delta/2$-packing set. Consequently, we can lower bound the minimax risk by constructing a testing problem based on the family of vectors (15.48). In fact, so as to make the calculations clean, we construct one testing problem for each choice of orthonormal matrix $\mathbf{U}$, and then take averages over a randomly chosen matrix.

Let $\mathbb{P}_{\theta^j(\mathbf{U})}$ denote the distribution of a random vector from the spiked ensemble (15.47) with leading eigenvector $\theta^* := \theta^j(\mathbf{U})$. By construction, it is a zero-mean Gaussian random vector with covariance matrix

$$\mathbf{\Sigma}^j(\mathbf{U}) := \mathbf{I}_d + \nu\big(\theta^j(\mathbf{U}) \otimes \theta^j(\mathbf{U})\big).$$

Now for a fixed $\mathbf{U}$, suppose that we choose an index $J \in [M]$ uniformly at random, and then drawn n i.i.d. samples from the distribution $\mathbb{P}_{\theta^J(\mathbf{U})}$. Letting $Z_1^n(\mathbf{U})$ denote the samples thus obtained, Fano's inequality then implies that the testing error is lower bounded as

$$\mathbb{P}[\psi(Z_1^n(\mathbf{U})) \neq J \mid \mathbf{U}] \geq 1 - \frac{I(Z_1^n(\mathbf{U}); J) + \log 2}{d/2}, \tag{15.49}$$

where we have used the fact that $\log M \geq d/2$. For each fixed $\mathbf{U}$, the samples $Z_1^n(\mathbf{U})$ are conditionally independent given J. Consequently, following the same line of reasoning leading to equation (15.45), we can conclude that $I(Z_1^n(\mathbf{U}); J) \leq nI(Z(\mathbf{U}); J)$, where $Z(\mathbf{U})$ denotes a single sample.

Since the lower bound (15.49) holds for each fixed choice of orthonormal matrix $\mathbf{U}$, we can take averages when $\mathbf{U}$ is chosen uniformly at random. Doing so simplifies the task of bounding the mutual information, since we need only bound the averaged mutual information $\mathbb{E}_{\mathbf{U}}[I(Z(\mathbf{U}); J)]$. Since $\det(\mathbf{\Sigma}^j(\mathbf{U})) = 1 + \nu$ for each $j \in [M]$, Lemma 15.17 implies that

$$\begin{aligned}\mathbb{E}_{\mathbf{U}}[I(Z(\mathbf{U}); J) &\leq \tfrac{1}{2}\big\{\mathbb{E}_{\mathbf{U}} \log\det(\operatorname{cov}(Z(\mathbf{U}))) - \log(1+\nu)\big\} \\ &\leq \tfrac{1}{2}\Big\{\log\det \underbrace{\mathbb{E}_{\mathbf{U}}(\operatorname{cov}(Z(\mathbf{U})))}_{:=\mathbf{\Gamma}} - \log(1+\nu)\Big\},\end{aligned} \tag{15.50}$$

where the second step uses the concavity of the log-determinant function, and Jensen's inequality. Let us now compute the entries of the expected covariance matrix $\mathbf{\Gamma}$. It can be seen that $\Gamma_{11} = 1 + \nu - \nu\delta^2$; moreover, using the fact that $\mathbf{U}\Delta^j$ is uniformly distributed over the unit sphere in dimension $(d-1)$, the first column is equal to

$$\Gamma_{(2\to d),1} = \nu\delta\sqrt{1-\delta^2}\,\frac{1}{M}\sum_{j=1}^{M} \mathbb{E}_{\mathbf{U}}[\mathbf{U}\Delta^j] = 0.$$

Letting $\mathbf{\Gamma}_{\mathrm{low}}$ denote the lower square block of side length $(d-1)$, we have

$$\mathbf{\Gamma}_{\mathrm{low}} = \mathbf{I}_{d-1} + \frac{\delta^2\nu}{M}\sum_{j=1}^{M}\mathbb{E}\big[(\mathbf{U}\Delta^j)\otimes(\mathbf{U}\Delta^j)\big] = \Big(1 + \frac{\delta^2\nu}{d-1}\Big)\mathbf{I}_{d-1},$$

again using the fact that the random vector $\mathbf{U}\Delta^j$ is uniformly distributed over the sphere in dimension $d-1$. Putting together the pieces, we have shown that $\mathbf{\Gamma} = \operatorname{blkdiag}(\Gamma_{11}, \mathbf{\Gamma}_{\mathrm{low}})$, and hence

$$\log\det\mathbf{\Gamma} = (d-1)\log\Big(1 + \frac{\nu\delta^2}{d-1}\Big) + \log(1 + \nu - \nu\delta^2).$$

Combining our earlier bound (15.50) with the elementary inequality $\log(1+t) \leq t$, we find

that

$$2\mathbb{E}_{\mathbf{U}}\Big[I(Z(\mathbf{U}); J)\Big] \leq (d-1)\log\Big(1+\frac{\nu\delta^2}{d-1}\Big) + \log\Big(1-\frac{\nu}{1+\nu}\delta^2\Big)$$
$$\leq \Big(\nu - \frac{\nu}{1+\nu}\Big)\delta^2$$
$$= \frac{\nu^2}{1+\nu}\delta^2.$$

Taking averages over our earlier Fano bound (15.49) and using this upper bound on the averaged mutual information, we find that the minimax risk for estimating the spiked eigenvector in squared Euclidean norm is lower bounded as

$$\mathfrak{M}(\text{PCA}; \mathbb{S}^{d-1}, \|\cdot\|_2^2) \gtrsim \min\Big\{\frac{1+\nu}{\nu^2}\frac{d}{n}, 1\Big\}.$$

In Corollary 8.7, we proved that the maximum eigenvector of the sample covariance achieves this squared Euclidean error up to constant pre-factors, so that we have obtained a sharp characterization of the minimax risk. ♣

As a follow-up to the previous example, we now turn to the sparse variant of principal components analysis. As discussed in Chapter 8, there are a number of motivations for studying sparsity in PCA, including the fact that it allows eigenvectors to be estimated at substantially faster rates. Accordingly, let us now prove some lower bounds for variable selection in sparse PCA, again working under the spiked model (15.47).

Example 15.20 (Lower bounds for variable selection in sparse PCA) Suppose that our goal is to determine the scaling of the sample size required to ensure that the support set of an s-sparse eigenvector θ^* can be recovered. Of course, the difficulty of the problem depends on the minimum value $\theta_{\min} = \min_{j \in S} |\theta^*_j|$. Here we show that if $\theta_{\min} \gtrsim \frac{1}{\sqrt{s}}$, then any method requires $n \gtrsim \frac{1+\nu}{\nu^2} s \log(d-s+1)$ samples to correctly recover the support. In Exercise 15.15, we prove a more general lower bound for arbitrary scalings of $\theta_{\min}$.

Recall our analysis of variable selection in sparse linear regression from Example 15.18: here we use an approach similar to ensemble B from that example. In particular, fix a subset S of size $s-1$, and let $\varepsilon \in \{-1, 1\}^d$ be a vector of sign variables. For each $j \in S^c := [d] \setminus S$, we then define the vector

$$[\theta^j(\varepsilon)]_\ell = \begin{cases} \frac{1}{\sqrt{s}} & \text{if } \ell \in S, \\ \frac{\varepsilon_j}{\sqrt{s}} & \text{if } \ell = j, \\ 0 & \text{otherwise.} \end{cases}$$

In Example 15.18, we computed averages over a randomly chosen orthonormal matrix $\mathbf{U}$; here instead we average over the choice of random sign vectors ε.

Let $\mathbb{P}_{\theta^j(\varepsilon)}$ denote the distribution of the spiked vector (15.47) with $\theta^* = \theta^j(\varepsilon)$, and let $Z(\varepsilon)$ be a sample from the mixture distribution $\frac{1}{M}\sum_{j \in S^c} \mathbb{P}_{\theta^j(\varepsilon)}$. Following a similar line of calculation as Example 15.19, we have

$$\mathbb{E}_\varepsilon[I(Z(\varepsilon); J)] \leq \tfrac{1}{2}\Big\{\log\det(\mathbf{\Gamma}) - \log(1+\nu)\Big\},$$

where $\mathbf{\Gamma} := \mathbb{E}_\varepsilon[\text{cov}(Z(\varepsilon))]$ is the averaged covariance matrix, taken over the uniform distribution over all Rademacher vectors. Letting $\mathbf{E}_{s-1}$ denote a square matrix of all ones with side length $s-1$, a straightforward calculation yields that $\mathbf{\Gamma}$ is a block diagonal matrix with $\mathbf{\Gamma}_{SS} = \mathbf{I}_{s-1} + \frac{\nu}{s}\mathbf{E}_{s-1}$ and $\mathbf{\Gamma}_{S^cS^c} = (1 + \frac{\nu}{s(d-s+1)})\,\mathbf{I}_{d-s+1}$. Consequently, we have

$$\begin{aligned}
2\mathbb{E}_\varepsilon[I(Z(\varepsilon); J)] &\leq \log\Big(1 + \nu\frac{s-1}{s}\Big) + (d-s+1)\log\Big(1 + \frac{\nu}{s(d-s+1)}\Big) - \log(1+\nu) \\
&= \log\Big(1 - \frac{\nu}{1+\nu}\frac{1}{s}\Big) + (d-s+1)\log\Big(1 + \frac{\nu}{s(d-s+1)}\Big) \\
&\leq \frac{1}{s}\Big\{-\frac{\nu}{1+\nu} + \nu\Big\} \\
&= \frac{1}{s}\frac{\nu^2}{1+\nu}.
\end{aligned}$$

Recalling that we have n samples and that $\log M = \log(d-s-1)$, Fano's inequality implies that the probability of error is bounded away from zero as long as the ratio

$$\frac{n}{s\log(d-s+1)}\,\frac{\nu^2}{1+\nu}$$

is upper bounded by a sufficiently small but universal constant, as claimed. ♣

15.3.5 Yang–Barron version of Fano's method

Our analysis thus far has been based on relatively naive upper bounds on the mutual information. These upper bounds are useful whenever we are able to construct a local packing of the parameter space, as we have done in the preceding examples. In this section, we develop an alternative upper bound on the mutual information. It is particularly useful for nonparametric problems, since it obviates the need for constructing a local packing.

Lemma 15.21 (Yang–Barron method) *Let $N_{\mathrm{KL}}(\epsilon; \mathcal{P})$ denote the ϵ-covering number of $\mathcal{P}$ in the square-root KL divergence. Then the mutual information is upper bounded as*

$$I(Z; J) \leq \inf_{\epsilon > 0}\{\epsilon^2 + \log N_{\mathrm{KL}}(\epsilon; \mathcal{P})\}. \tag{15.51}$$

Proof Recalling the form (15.30) of the mutual information, we observe that for any distribution $\mathbb{Q}$, the mutual information is upper bounded by

$$I(Z; J) = \frac{1}{M}\sum_{j=1}^{M} D(\mathbb{P}_{\theta^j} \,\|\, \bar{\mathbb{Q}}) \overset{\text{(i)}}{\leq} \frac{1}{M}\sum_{j=1}^{M} D(\mathbb{P}_{\theta^j} \,\|\, \mathbb{Q}) \leq \max_{j=1,\ldots,M} D(\mathbb{P}_{\theta^j} \,\|\, \mathbb{Q}), \tag{15.52}$$

where inequality (i) uses the fact that the mixture distribution $\bar{\mathbb{Q}} := \frac{1}{M}\sum_{j=1}^{M} \mathbb{P}_{\theta^j}$ minimizes the average Kullback–Leibler divergence over the family $\{\mathbb{P}_{\theta^1}, \ldots, \mathbb{P}_{\theta^M}\}$—see Exercise 15.11 for details.

Since the upper bound (15.52) holds for any distribution $\mathbb{Q}$, we are free to choose it: in particular, we let $\{\gamma^1, \ldots, \gamma^N\}$ be an ϵ-covering of Ω in the square-root KL pseudo-distance, and then set $\mathbb{Q} = \frac{1}{N}\sum_{k=1}^{N} \mathbb{P}_{\gamma^k}$. By construction, for each θ^j with $j \in [M]$, we can find some γ^k such that $D(\mathbb{P}_{\theta^j} \,\|\, \mathbb{P}_{\gamma^k}) \leq \epsilon^2$. Therefore, we have

$$\begin{aligned} D(\mathbb{P}_{\theta^j} \,\|\, \mathbb{Q}) &= \mathbb{E}_{\theta^j}\left[\log \frac{d\mathbb{P}_{\theta_j}}{\frac{1}{N}\sum_{\ell=1}^{N} d\mathbb{P}_{\gamma^\ell}}\right] \\ &\leq \mathbb{E}_{\theta^j}\left[\log \frac{d\mathbb{P}_{\theta_j}}{\frac{1}{N} d\mathbb{P}_{\gamma^k}}\right] \\ &= D(\mathbb{P}_{\theta^j} \,\|\, \mathbb{P}_{\gamma^k}) + \log N \\ &\leq \epsilon^2 + \log N. \end{aligned}$$

Since this bound holds for any choice of $j \in [M]$ and any choice of $\epsilon > 0$, the claim (15.51) follows. □

In conjunction with Proposition 15.12, Lemma 15.21 allows us to prove a minimax lower bound of the order δ as long as the pair $(\delta, \epsilon) \in \mathbb{R}_+^2$ are chosen such that

$$\log M(\delta; \rho, \Omega) \geq 2\{\epsilon^2 + \log N_{\mathrm{KL}}(\epsilon; \mathcal{P}) + \log 2\}.$$

Finding such a pair can be accomplished via a two-step procedure:

(A) First, choose $\epsilon_n > 0$ such that

$$\epsilon_n^2 \geq \log N_{\mathrm{KL}}(\epsilon_n; \mathcal{P}). \tag{15.53a}$$

Since the KL divergence typically scales with n, it is usually the case that ϵ_n^2 also grows with n, hence the subscript in our notation.

(B) Second, choose the largest $\delta_n > 0$ that satisfies the lower bound

$$\log M(\delta_n; \rho, \Omega) \geq 4\epsilon_n^2 + 2\log 2. \tag{15.53b}$$

As before, this two-step procedure is best understood by working through some examples.

Example 15.22 (Density estimation revisited) In order to illustrate the use of the Yang–Barron method, let us return to the problem of density estimation in the Hellinger metric, as previously considered in Example 15.15. Our analysis involved the class $\mathscr{F}_2$, as defined in equation (15.21), of densities on $[0, 1]$, bounded uniformly above, bounded uniformly away from zero, and with uniformly bounded second derivative. Using the local form of Fano's method, we proved that the minimax risk in squared Hellinger distance is lower bounded as $n^{-4/5}$. In this example, we recover the same result more directly by using known results about the metric entropy.

For uniformly bounded densities on the interval $[0, 1]$, the squared Hellinger metric is sandwiched above and below by constant multiples of the $L^2([0, 1])$-norm:

$$\|p - q\|_2^2 := \int_0^1 (p(x) - q(x))^2 \, dx.$$

Moreover, again using the uniform lower bound, the Kullback–Leibler divergence between any pair of distributions in this family is upper bounded by a constant multiple of the squared Hellinger distance, and hence by a constant multiple of the squared Euclidean distance. (See equation (15.39) for a related calculation.) Consequently, in order to apply the Yang–Barron method, we need only understand the scaling of the metric entropy in the L^2-norm. From classical theory, it is known that the metric entropy of the class $\mathscr{F}_2$ in L^2-norm scales as $\log N(\delta; \mathscr{F}_2, \|\cdot\|_2) \asymp (1/\delta)^{1/2}$ for $\delta > 0$ sufficiently small.

Step A: Given n i.i.d. samples, the square-root Kullback–Leibler divergence is multiplied by a factor of $\sqrt{n}$, so that the inequality (15.53a) can be satisfied by choosing $\epsilon_n > 0$ such that

$$\epsilon_n^2 \gtrsim \left(\frac{\sqrt{n}}{\epsilon_n}\right)^{1/2}.$$

In particular, the choice $\epsilon_n^2 \asymp n^{1/5}$ is sufficient.

Step B: With this choice of ϵ_n, the second condition (15.53b) can be satisfied by choosing $\delta_n > 0$ such that

$$\left(\frac{1}{\delta_n}\right)^{1/2} \gtrsim n^{2/5},$$

or equivalently $\delta_n^2 \asymp n^{-4/5}$. In this way, we have a much more direct re-derivation of the $n^{-4/5}$ lower bound on the minimax risk. ♣

As a second illustration of the Yang–Barron approach, let us now derive some minimax risks for the problem of nonparametric regression, as discussed in Chapter 13. Recall that the standard regression model is based on i.i.d. observations of the form

$$y_i = f^*(x_i) + \sigma w_i, \qquad \text{for } i = 1, 2, \ldots, n,$$

where $w_i \sim \mathcal{N}(0, 1)$. Assuming that the design points $\{x_i\}_{i=1}^n$ are drawn in an i.i.d. fashion from some distribution $\mathbb{P}$, let us derive lower bounds in the $L^2(\mathbb{P})$-norm:

$$\|\widehat{f} - f^*\|_2^2 = \int_{\mathcal{X}} \left[\widehat{f}(x) - f^*(x)\right]^2 \mathbb{P}(dx).$$

Example 15.23 (Minimax risks for generalized Sobolev families) For a smoothness parameter $\alpha > 1/2$, consider the ellipsoid $\ell^2(\mathbb{N})$ given by

$$\mathcal{E}_\alpha = \Big\{(\theta_j)_{j=1}^\infty \mid \sum_{j=1}^\infty j^{2\alpha}\, \theta_j^2 \leq 1\Big\}. \tag{15.54a}$$

Given an orthonormal sequence $(\phi_j)_{j=1}^\infty$ in $L^2(\mathbb{P})$, we can then define the function class

$$\mathscr{F}_\alpha := \left\{ f = \sum_{j=1}^\infty \theta_j \phi_j \mid (\theta_j)_{j=1}^\infty \in \mathcal{E}_\alpha \right\}. \tag{15.54b}$$

As discussed in Chapter 12, these function classes can be viewed as particular types of reproducing kernel Hilbert spaces, where α corresponds to the degree of smoothness. For any such function class, we claim that the minimax risk in squared $L^2(\mathbb{P})$-norm is lower bounded as

$$\inf_{\widehat{f}} \sup_{f \in \mathscr{F}_\alpha} \mathbb{E}[\|\widehat{f} - f\|_2^2] \gtrsim \min\left\{1, \left(\frac{\sigma^2}{n}\right)^{\frac{2\alpha}{2\alpha+1}}\right\}, \tag{15.55}$$

and here we prove this claim via the Yang–Barron technique.

Consider a function of the form $f = \sum_{j=1}^{\infty} \theta_j \phi_j$ for some $\theta \in \ell^2(\mathbb{N})$, and observe that by the orthonormality of $(\phi_j)_{j=1}^{\infty}$, Parseval's theorem implies that $\|f\|_2^2 = \sum_{j=1}^{\infty} \theta_j^2$. Consequently, based on our calculations from Example 5.12, the metric entropy of $\mathscr{F}_\alpha$ scales as $\log N(\delta; \mathscr{F}_\alpha, \|\cdot\|_2) \asymp (1/\delta)^{1/\alpha}$. Accordingly, we can find a δ-packing $\{f^1, \ldots, f^M\}$ of $\mathscr{F}_\alpha$ in the $\|\cdot\|_2$-norm with $\log M \gtrsim (1/\delta)^{1/\alpha}$ elements.

Step A: For this part of the calculation, we first need to upper bound the metric entropy in the KL divergence. For each $j \in [M]$, let $\mathbb{P}_{f^j}$ denote the distribution of y given $\{x_i\}_{i=1}^n$ when the true regression function is f^j, and let $\mathbb{Q}$ denote the n-fold product distribution over the covariates $\{x_i\}_{i=1}^n$. When the true regression function is f^j, the joint distribution over $(y, \{x_i\}_{i=1}^n)$ is given by $\mathbb{P}_{f^j} \times \mathbb{Q}$, and hence for any distinct pair of indices $j \neq k$, we have

$$\begin{aligned} D(\mathbb{P}_{f^j} \times \mathbb{Q} \,\|\, \mathbb{P}_{f^k} \times \mathbb{Q}) = \mathbb{E}_x[D(\mathbb{P}_{f^j} \,\|\, \mathbb{P}_{f^k})] &= \mathbb{E}_x\Big[\frac{1}{2\sigma^2} \sum_{i=1}^n (f^j(x_i) - f^k(x_i))^2\Big] \\ &= \frac{n}{2\sigma^2}\|f^j - f^k\|_2^2. \end{aligned}$$

Consequently, we find that

$$\log N_{\mathrm{KL}}(\epsilon) = \log N\Big(\frac{\sigma\sqrt{2}}{\sqrt{n}}\epsilon; \mathscr{F}_\alpha, \|\cdot\|_2\Big) \lesssim \Big(\frac{\sqrt{n}}{\sigma\epsilon}\Big)^{1/\alpha},$$

where the final inequality again uses the result of Example 5.12. Consequently, inequality (15.53a) can be satisfied by setting $\epsilon_n^2 \asymp (\frac{n}{\sigma^2})^{\frac{1}{2\alpha+1}}$.

Step B: It remains to choose $\delta > 0$ to satisfy the inequality (15.53b). Given our choice of ε_n and the scaling of the packing entropy, we require

$$(1/\delta)^{1/\alpha} \geq c\left\{\left(\frac{n}{\sigma^2}\right)^{\frac{1}{2\alpha+1}} + 2\log 2\right\}. \tag{15.56}$$

As long as n/σ^2 is larger than some universal constant, the choice $\delta_n^2 \asymp (\frac{\sigma^2}{n})^{\frac{2\alpha}{2\alpha+1}}$ satisfies the condition (15.56). Putting together the pieces yields the claim (15.55). ♣

In the exercises, we explore a number of other applications of the Yang–Barron method.

15.4 Appendix: Basic background in information theory

This appendix is devoted to some basic information-theoretic background, including a proof of Fano's inequality. The most fundamental concept is that of the *Shannon entropy*: it is a functional on the space of probability distributions that provides a measure of their dispersion.

Definition 15.24 Let $\mathbb{Q}$ be a probability distribution with density $q = \frac{d\mathbb{Q}}{d\mu}$ with respect to some base measure μ. The Shannon entropy is given by

$$H(\mathbb{Q}) := -\mathbb{E}[\log q(X)] = -\int_{\mathcal{X}} q(x) \log q(x) \mu(dx), \tag{15.57}$$

when this integral is finite.

The simplest form of entropy arises when $\mathbb{Q}$ is supported on a discrete set $\mathcal{X}$, so that q can be taken as a probability mass function—hence a density with respect to the counting measure on $\mathcal{X}$. In this case, the definition (15.57) yields the discrete entropy

$$H(\mathbb{Q}) = -\sum_{x \in \mathcal{X}} q(x) \log q(x). \tag{15.58}$$

It is easy to check that the discrete entropy is always non-negative. Moreover, when $\mathcal{X}$ is a finite set, it satisfies the upper bound $H(\mathbb{Q}) \leq \log |\mathcal{X}|$, with equality achieved when $\mathbb{Q}$ is uniform over $\mathcal{X}$. See Exercise 15.2 for further discussion of these basic properties.

An important remark on notation is needed before proceeding: Given a random variable $X \sim \mathbb{Q}$, one often writes $H(X)$ in place of $H(\mathbb{Q})$. From a certain point of view, this is abusive use of notation, since the entropy is a functional of the distribution $\mathbb{Q}$ as opposed to the random variable X. However, as it is standard practice in information theory, we make use of this convenient notation in this appendix.

Definition 15.25 Given a pair of random variables (X, Y) with joint distribution $\mathbb{Q}_{X,Y}$, the conditional entropy of $X \mid Y$ is given by

$$H(X \mid Y) := \mathbb{E}_Y[H(\mathbb{Q}_{X|Y})] = \mathbb{E}_Y\left[\int_{\mathcal{X}} q(x \mid Y) \log q(x \mid Y) \mu(dx)\right]. \tag{15.59}$$

We leave the reader to verify the following elementary properties of entropy and mutual information. First, conditioning can only reduce entropy:

$$H(X \mid Y) \leq H(X). \tag{15.60a}$$

As will be clear below, this inequality is equivalent to the non-negativity of the mutual information $I(X; Y)$. Secondly, the joint entropy can be decomposed into a sum of singleton and conditional entropies as

$$H(X, Y) = H(Y) + H(X \mid Y). \tag{15.60b}$$

This decomposition is known as the chain rule for entropy. The conditional entropy also satisfies a form of chain rule:

$$H(X, Y \mid Z) = H(X \mid Z) + H(X \mid Y, Z). \tag{15.60c}$$

Finally, it is worth noting the connections between entropy and mutual information. By expanding the definition of mutual information, we see that

$$I(X; Y) = H(X) + H(Y) - H(X, Y). \tag{15.60d}$$

By replacing the joint entropy with its chain rule decomposition (15.60b), we obtain

$$I(X; Y) = H(Y) - H(Y \mid X). \tag{15.60e}$$

With these results in hand, we are now ready to prove the Fano bound (15.31). We do so by first establishing a slightly more general result. Introducing the shorthand notation $q_e = \mathbb{P}[\psi(Z) \neq J]$, we let $h(q_e) = -q_e \log q_e - (1 - q_e) \log(1 - q_e)$ denote the binary entropy. With this notation, the standard form of Fano's inequality is that the error probability in any M-ary testing problem is lower bounded as

$$h(q_e) + q_e \log(M - 1) \geq H(J \mid Z). \tag{15.61}$$

To see how this lower bound implies the stated claim (15.31), we note that

$$H(J \mid Z) \stackrel{\text{(i)}}{=} H(J) - I(Z; J) \stackrel{\text{(ii)}}{=} \log M - I(Z; J),$$

where equality (i) follows from the representation of mutual information in terms of entropy, and equality (ii) uses our assumption that J is uniformly distributed over the index set. Since $h(q_e) \leq \log 2$, we find that

$$\log 2 + q_e \log M \geq \log M - I(Z; J),$$

which is equivalent to the claim (15.31).

It remains to prove the lower bound (15.61). Define the $\{0, 1\}$-valued random variable $V := \mathbb{I}[\psi(Z) \neq J]$, and note that $H(V) = h(q_e)$ by construction. We now proceed to expand the conditional entropy $H(V, J \mid Z)$ in two different ways. On one hand, by the chain rule, we have

$$H(V, J \mid Z) = H(J \mid Z) + H(V \mid J, Z) = H(J \mid Z), \tag{15.62}$$

where the second equality follows since V is a function of Z and J. By an alternative application of the chain rule, we have

$$H(V, J \mid Z) = H(V \mid Z) + H(J \mid V, Z) \leq h(q_e) + H(J \mid V, Z),$$

where the inequality follows since conditioning can only reduce entropy. By the definition of conditional entropy, we have

$$H(J \mid V, Z) = \mathbb{P}[V = 1] H(J \mid Z, V = 1) + \mathbb{P}[V = 0] H(J \mid Z, V = 0).$$

If $V = 0$, then $J = \psi(Z)$, so that $H(J \mid Z, V = 0) = 0$. On the other hand, if $V = 1$, then we know that $J \neq \psi(Z)$, so that the conditioned random variable $(J \mid Z, V = 1)$ can take at most $M - 1$ values, which implies that

$$H(J \mid Z, V = 1) \leq \log(M - 1),$$

since entropy is maximized by the uniform distribution. We have thus shown that

$$H(V, J \mid Z) \leq h(q_e) + \log(M - 1),$$

and combined with the earlier equality (15.62), the claim (15.61) follows.

15.5 Bibliographic details and background

Information theory was introduced in the seminal work of Shannon (1948; 1949); see also Shannon and Weaver (1949). Kullback and Leibler (1951) introduced the Kullback–Leibler divergence, and established various connections to both large-deviation theory and testing problems. Early work by Lindley (1956) also established connections between information and statistical estimation. Kolmogorov was the first to connect information theory and metric entropy; in particular, see appendix II of the paper by Kolmogorov and Tikhomirov (1959). The book by Cover and Thomas (1991) is a standard introductory-level text on information theory. The proof of Fano's inequality given here follows their book.

The parametric problems discussed in Examples 15.4 and 15.5 were considered in Le Cam (1973), where he described the lower bounding approach now known as Le Cam's method. In this same paper, Le Cam also shows how a variety of nonparametric problems can also be treated by this method, using results on metric entropy. The paper by Hasminskii (1978) used the weakened form of the Fano method, based on the upper bound (15.34) on the mutual information, to derive lower bounds on density estimation in the uniform metric; see also the book by Hasminskii and Ibragimov (1981), as well as their survey paper (Hasminskii and Ibragimov, 1990). Assouad (1983) developed a method for deriving lower bounds based on placing functions at vertices of the binary hypercube. See also Birgé (1983; 1987; 2005) for further refinements on methods for deriving both lower and upper bounds. The chapter by Yu (1996) provides a comparison of both Le Cam's and Fano's method, as well Assouad's method (Assouad, 1983). Examples 15.8, 15.11 and 15.15 follow parts of her development. Birgé and Massart (1995) prove the upper bound (15.28) on the squared Hellinger distance; see theorem 1 in their paper for further details. In their paper, they study the more general problem of estimating functionals of the density and its first k derivatives under general smoothness conditions of order α. The quadratic functional problem considered in Examples 15.8 and 15.11 correspond to the special case with $k = 1$ and $\alpha = 2$. The refined upper bound on mutual information from Lemma 15.21 is due to Yang and Barron (1999). Their work showed how Fano's method can be applied directly with global metric entropies, as opposed to constructing specific local packings of the function class, as in the local packing version of Fano's method discussed in Section 15.3.3.

Guntuboyina (2011) proves a generalization of Fano's inequality to an arbitrary f-divergence. See Exercise 15.12 for further background on f-divergences and their properties. His result reduces to the classical Fano's inequality when the underlying f-divergence is the Kullback–Leibler divergence. He illustrates how such generalized Fano bounds can be used to derive minimax bounds for various classes of problems, including covariance estimation.

Lower bounds on variable selection in sparse linear regression using the Fano method, as considered in Example 15.18, were derived by Wainwright (2009a). See also the papers (Reeves and Gastpar, 2008; Fletcher et al., 2009; Akcakaya and Tarokh, 2010; Wang et al., 2010) for further results of this type. The lower bound on variable selection in sparse PCA from Example 15.20 was derived in Amini and Wainwright (2009); the proof given here is somewhat more streamlined due to the symmetrization with Rademacher variables.

The notion of minimax risk discussed in this chapter is the classical one, in which no additional constraints (apart from measurability) are imposed on the estimators. Consequently, the theory allows for estimators that may involve prohibitive computational, storage or communication costs to implement. A more recent line of work has been studying constrained forms of statistical minimax theory, in which the infimum over estimators is suitably restricted (Wainwright, 2014). In certain cases, there can be substantial gaps between the classical minimax risk and their computationally constrained analogs (e.g., Berthet and Rigollet, 2013; Ma and Wu, 2013; Wang et al., 2014; Zhang et al., 2014; Cai et al., 2015; Gao et al., 2015). Similarly, privacy constraints can lead to substantial differences in the classical and private minimax risks (Duchi et al., 2014, 2013).

15.6 Exercises

Exercise 15.1 (Alternative representation of TV norm) Show that the total variation norm has the equivalent variational representation

$$\|\mathbb{P}_1 - \mathbb{P}_0\|_{\text{TV}} = 1 - \inf_{f_0+f_1\geq 1} \big\{\mathbb{E}_0[f_0] + \mathbb{E}_1[f_1]\big\},$$

where the infimum runs over all non-negative measurable functions, and the inequality is taken pointwise.

Exercise 15.2 (Basics of discrete entropy) Let $\mathbb{Q}$ be the distribution of a discrete random variable on a finite set $\mathcal{X}$. Letting q denote the associated probability mass function, its Shannon entropy has the explicit formula

$$H(\mathbb{Q}) \equiv H(X) = -\sum_{x\in\mathcal{X}} q(x)\log q(x),$$

where we interpret $0 \log 0 = 0$.

(a) Show that $H(X) \geq 0$.
(b) Show that $H(X) \leq \log|\mathcal{X}|$, with equality achieved when X has the uniform distribution over $\mathcal{X}$.

Exercise 15.3 (Properties of Kullback–Leibler divergence) In this exercise, we study some properties of the Kullback–Leibler divergence. Let $\mathbb{P}$ and $\mathbb{Q}$ be two distributions having densities p and q with respect to a common base measure.

(a) Show that $D(\mathbb{P}\,\|\,\mathbb{Q}) \geq 0$ with equality if and only if the equality $p(x) = q(x)$ holds $\mathbb{P}$-almost everywhere.
(b) Given a collection of non-negative weights such that $\sum_{j=1}^m \lambda_j = 1$, show that

$$D\Big(\sum_{j=1}^m \lambda_j \mathbb{P}_j\,\|\,\mathbb{Q}\Big) \leq \sum_{j=1}^m \lambda_j D(\mathbb{P}_j\,\|\,\mathbb{Q}) \tag{15.63a}$$

and

$$D(\mathbb{Q} \,\|\, \sum_{j=1}^{m} \lambda_j \mathbb{P}_j) \leq \sum_{j=1}^{m} \lambda_j D(\mathbb{Q} \,\|\, \mathbb{P}_j). \tag{15.63b}$$

(c) Prove that the KL divergence satisfies the decoupling property (15.11a) for product measures.

Exercise 15.4 (More properties of Shannon entropy) Let (X, Y, Z) denote a triplet of random variables, and recall the definition (15.59) of the conditional entropy.

(a) Prove that conditioning reduces entropy—that is, $H(X \mid Y) \leq H(X)$.
(b) Prove the chain rule for entropy:

$$H(X, Y, Z) = H(X) + H(Y \mid X) + H(Z \mid Y, X).$$

(c) Conclude from the previous parts that

$$H(X, Y, Z) \leq H(X) + H(Y) + H(Z),$$

so that joint entropy is maximized by independent variables.

Exercise 15.5 (Le Cam's inequality) Prove the upper bound (15.10) on the total variation norm in terms of the Hellinger distance. (*Hint:* The Cauchy–Schwarz inequality could be useful.)

Exercise 15.6 (Pinsker–Csiszár–Kullback inequality) In this exercise, we work through a proof of the Pinsker–Csiszár–Kullback inequality (15.8) from Lemma 15.2.

(a) When $\mathbb{P}$ and $\mathbb{Q}$ are Bernoulli distributions with parameters $\delta_p \in [0, 1]$ and $\delta_q \in [0, 1]$, show that inequality (15.8) reduces to

$$2\,(\delta_p - \delta_q)^2 \leq \delta_p \log \frac{\delta_p}{\delta_q} + (1 - \delta_p) \log \frac{1 - \delta_p}{1 - \delta_q}. \tag{15.64}$$

Prove the inequality in this special case.
(b) Use part (a) and Jensen's inequality to prove the bound in the general case. (*Hint:* Letting p and q denote densities, consider the set $A := \{x \in \mathcal{X} \mid p(x) \geq q(x)\}$, and try to reduce the problem to a version of part (a) with $\delta_p = \mathbb{P}[A]$ and $\delta_q = \mathbb{Q}[A]$.)

Exercise 15.7 (Decoupling for Hellinger distance) Show that the Hellinger distance satisfies the decoupling relation (15.12a) for product measures.

Exercise 15.8 (Sharper bounds for Gaussian location family) Recall the normal location model from Example 15.4. Use the two-point form of Le Cam's method and the Pinsker–Csiszár–Kullback inequality from Lemma 15.2 to derive the sharper lower bounds

$$\inf_{\widehat{\theta}} \sup_{\theta \in \mathbb{R}} \mathbb{E}_\theta[|\widehat{\theta} - \theta|] \geq \frac{1}{8} \frac{\sigma}{\sqrt{n}} \quad \text{and} \quad \inf_{\widehat{\theta}} \sup_{\theta \in \mathbb{R}} \mathbb{E}_\theta[(\widehat{\theta} - \theta)^2] \geq \frac{1}{16} \frac{\sigma^2}{n}.$$

Exercise 15.9 (Achievable rates for uniform shift family) In the context of the uniform shift family (Example 15.5), show that the estimator $\widetilde{\theta} = \min\{Y_1, \ldots, Y_n\}$ satisfies the bound $\sup_{\theta \in \mathbb{R}} \mathbb{E}[(\widetilde{\theta} - \theta)^2] \leq \frac{2}{n^2}$.

Exercise 15.10 (Bounds on the TV distance)

(a) Prove that the squared total variation distance is upper bounded as

$$\|\mathbb{P} - \mathbb{Q}\|_{\mathrm{TV}}^2 \le \frac{1}{4}\left\{\int_{\mathcal{X}} \frac{p^2(x)}{q(x)}\nu(dx) - 1\right\},$$

where p and q are densities with respect to the base measure ν.

(b) Use part (a) to show that

$$\|\mathbb{P}^n_{\theta,\sigma} - \mathbb{P}^n_{0,\sigma}\|_{\mathrm{TV}}^2 \le \tfrac{1}{4}\left\{e^{\left(\frac{\sqrt{n}\theta}{\sigma}\right)^2} - 1\right\}, \tag{15.65}$$

where, for any $\gamma \in \mathbb{R}^n$, we use $\mathbb{P}^n_{\gamma,\sigma}$ to denote the n-fold product distribution of a $\mathcal{N}(\gamma, \sigma^2)$ variate.

(c) Use part (a) to show that

$$\|\bar{\mathbb{P}} - \mathbb{P}^n_{0,\sigma}\|_{\mathrm{TV}}^2 \le \tfrac{1}{4}\left\{e^{\frac{1}{2}\left(\frac{\sqrt{n}\theta}{\sigma}\right)^4} - 1\right\}, \tag{15.66}$$

where $\bar{\mathbb{P}} = \frac{1}{2}\mathbb{P}^n_{\theta,\sigma} + \frac{1}{2}\mathbb{P}^n_{-\theta,\sigma}$ is a mixture distribution.

Exercise 15.11 (Mixture distributions and KL divergence) Given a collection of distributions $\{\mathbb{P}_1, \ldots, \mathbb{P}_M\}$, consider the mixture distribution $\bar{\mathbb{Q}} = \frac{1}{M}\sum_{j=1}^M \mathbb{P}_j$. Show that

$$\frac{1}{M}\sum_{j=1}^M D(\mathbb{P}_j \,\|\, \bar{\mathbb{Q}}) \le \frac{1}{M}\sum_{j=1}^M D(\mathbb{P}_j \,\|\, \mathbb{Q})$$

for any other distribution $\mathbb{Q}$.

Exercise 15.12 (f-divergences) Let $f\colon \mathbb{R}_+ \to \mathbb{R}$ be a strictly convex function. Given two distributions $\mathbb{P}$ and $\mathbb{Q}$ (with densities p and q, respectively), their f-divergence is given by

$$D_f(\mathbb{P} \,\|\, \mathbb{Q}) := \int q(x) f\big(p(x)/q(x)\big)\nu(dx). \tag{15.67}$$

(a) Show that the Kullback–Leibler divergence corresponds to the f-divergence defined by $f(t) = t\log t$.
(b) Compute the f-divergence generated by $f(t) = -\log(t)$.
(c) Show that the squared Hellinger divergence $H^2(\mathbb{P} \,\|\, \mathbb{Q})$ is also an f-divergence for an appropriate choice of f.
(d) Compute the f-divergence generated by the function $f(t) = 1 - \sqrt{t}$.

Exercise 15.13 (KL divergence for multivariate Gaussian) For $j = 1, 2$, let $\mathbb{Q}_j$ be a d-variate normal distribution with mean vector $\mu_j \in \mathbb{R}^d$ and covariance matrix $\mathbf{\Sigma}_j \succ 0$.

(a) If $\mathbf{\Sigma}_1 = \mathbf{\Sigma}_2 = \mathbf{\Sigma}$, show that

$$D(\mathbb{Q}_1 \,\|\, \mathbb{Q}_2) = \tfrac{1}{2}\langle \mu_1 - \mu_2,\, \mathbf{\Sigma}^{-1}(\mu_1 - \mu_2)\rangle.$$

(b) In the general setting, show that

$$D(\mathbb{Q}_1 \,\|\, \mathbb{Q}_2) = \tfrac{1}{2}\left\{\langle \mu_1 - \mu_2,\, \mathbf{\Sigma}_2^{-1}(\mu_1 - \mu_2)\rangle + \log\frac{\det(\mathbf{\Sigma}_2)}{\det(\mathbf{\Sigma}_1)} + \operatorname{trace}(\mathbf{\Sigma}_2^{-1}\mathbf{\Sigma}_1) - d\right\}.$$

Exercise 15.14 (Gaussian distributions and maximum entropy) For a given $\sigma > 0$, let Q_σ be the class of all densities q with respect to Lebesgue measure on the real line such that $\int_{-\infty}^{\infty} xq(x) = 0$, and $\int_{-\infty}^{\infty} q(x)x^2\, dx \leq \sigma^2$. Show that the maximum entropy distribution over this family is the Gaussian $\mathcal{N}(0, \sigma^2)$.

Exercise 15.15 (Sharper bound for variable selection in sparse PCA) In the context of Example 15.20, show that for a given $\theta_{\min} = \min_{j \in S} |\theta^*_j| \in (0, 1)$, support recovery in sparse PCA is not possible whenever

$$n < c_0 \frac{1 + \nu}{\nu^2} \frac{\log(d - s + 1)}{\theta^2_{\min}}$$

for some constant $c_0 > 0$. (*Note:* This result sharpens the bound from Example 15.20, since we must have $\theta^2_{\min} \leq \frac{1}{s}$ due to the unit norm and s-sparsity of the eigenvector.)

Exercise 15.16 (Lower bounds for sparse PCA in ℓ_2-error) Consider the problem of estimating the maximal eigenvector θ^* based on n i.i.d. samples from the spiked covariance model (15.47). Assuming that θ^* is s-sparse, show that any estimator $\widehat{\theta}$ satisfies the lower bound

$$\sup_{\theta^* \in \mathbb{B}_0(s) \cap \mathbb{S}^{d-1}} \mathbb{E}[\|\widehat{\theta} - \theta^*\|_2^2] \geq c_0 \frac{\nu + 1}{\nu^2} \frac{s \log\left(\frac{ed}{s}\right)}{n}$$

for some universal constant $c_0 > 0$. (*Hint:* The packing set from Example 15.16 may be useful to you. Moreover, you might consider a construction similar to Example 15.19, but with the random orthonormal matrix $\mathbf{U}$ replaced by a random permutation matrix along with random sign flips.)

Exercise 15.17 (Lower bounds for generalized linear models) Consider the problem of estimating a vector $\theta^* \in \mathbb{R}^d$ with Euclidean norm at most one, based on regression with a fixed set of design vectors $\{x_i\}_{i=1}^n$, and responses $\{y_i\}_{i=1}^n$ drawn from the distribution

$$\mathbb{P}_\theta(y_1, \ldots, y_n) = \prod_{i=1}^n \left[h(y_i) \exp\left(\frac{y_i \langle x_i, \theta \rangle - \Phi(\langle x_i, \theta \rangle)}{s(\sigma)} \right) \right],$$

where $s(\sigma) > 0$ is a known scale factor, and $\Phi \colon \mathbb{R} \to \mathbb{R}$ is the cumulant function of the generalized linear model.

(a) Compute an expression for the Kullback–Leibler divergence between $\mathbb{P}_\theta$ and $\mathbb{P}_{\theta'}$ involving Φ and its derivatives.
(b) Assuming that $\|\Phi''\|_\infty \leq L < \infty$, give an upper bound on the Kullback–Leibler divergence that scales quadratically in the Euclidean norm $\|\theta - \theta'\|_2$.
(c) Use part (b) and previous arguments to show that there is a universal constant $c > 0$ such that

$$\inf_{\widehat{\theta}} \sup_{\theta \in \mathbb{B}_2^d(1)} \mathbb{E}\left[\|\widehat{\theta} - \theta\|_2^2\right] \geq \min\left\{1,\ c\frac{s(\sigma)}{L\eta^2_{\max}} \frac{d}{n}\right\},$$

where $\eta_{\max} = \sigma_{\max}(\mathbf{X}/\sqrt{n})$ is the maximum singular value. (Here as usual $\mathbf{X} \in \mathbb{R}^{n \times d}$ is the design matrix with x_i as its ith row.)
(d) Explain how part (c) yields our lower bound on linear regression as a special case.

Exercise 15.18 (Lower bounds for additive nonparametric regression) Recall the class of additive functions first introduced in Exercise 13.9, namely

$$\mathscr{F}_{\text{add}} = \left\{ f\colon \mathbb{R}^d \to \mathbb{R} \;\middle|\; f = \sum_{j=1}^{d} g_j \text{ for } g_j \in \mathscr{G} \right\},$$

where $\mathscr{G}$ is some fixed class of univariate functions. In this exercise, we assume that the base class has metric entropy scaling as $\log N(\delta; \mathscr{G}, \|\cdot\|_2) \asymp \left(\frac{1}{\delta}\right)^{1/\alpha}$ for some $\alpha > 1/2$, and that we compute $L^2(\mathbb{P})$-norms using a product measure over $\mathbb{R}^d$.

(a) Show that

$$\inf_{\widehat{f}} \sup_{f \in \mathscr{F}_{\text{add}}} \mathbb{E}[\|\widehat{f} - f\|_2^2] \gtrsim d \left(\frac{\sigma^2}{n}\right)^{\frac{2\alpha}{2\alpha+1}}.$$

By comparison with the result of Exercise 14.8, we see that the least-squares estimator is minimax-optimal up to constant factors.

(b) Now consider the sparse variant of this model, namely based on the sparse additive model (SPAM) class

$$\mathscr{F}_{\text{spam}} = \left\{ f\colon \mathbb{R}^d \to \mathbb{R} \;\middle|\; f = \sum_{j \in S} g_j \text{ for } g_j \in \mathscr{G}, \text{ and a subset } |S| \leq s \right\}.$$

Show that

$$\inf_{\widehat{f}} \sup_{f \in \mathscr{F}_{\text{spam}}} \mathbb{E}[\|\widehat{f} - f\|_2^2] \gtrsim s \left(\frac{\sigma^2}{n}\right)^{\frac{2\alpha}{2\alpha+1}} + \sigma^2 \frac{s \log\left(\frac{ed}{s}\right)}{n}.$$

References

Adamczak, R. 2008. A tail inequality for suprema of unbounded empirical processes with applications to Markov chains. *Electronic Journal of Probability*, **34**, 1000–1034.

Adamczak, R., Litvak, A. E., Pajor, A., and Tomczak-Jaegermann, N. 2010. Quantitative estimations of the convergence of the empirical covariance matrix in log-concave ensembles. *Journal of the American Mathematical Society*, **23**, 535–561.

Agarwal, A., Negahban, S., and Wainwright, M. J. 2012. Noisy matrix decomposition via convex relaxation: Optimal rates in high dimensions. *Annals of Statistics*, **40**(2), 1171–1197.

Ahlswede, R., and Winter, A. 2002. Strong converse for identification via quantum channels. *IEEE Transactions on Information Theory*, **48**(3), 569–579.

Aizerman, M. A., Braverman, E. M., and Rozonoer, L. I. 1964. Theoretical foundations of the potential function method in pattern recognition learning. *Automation and Remote Control*, **25**, 821–837.

Akcakaya, M., and Tarokh, V. 2010. Shannon theoretic limits on noisy compressive sampling. *IEEE Transactions on Information Theory*, **56**(1), 492–504.

Alexander, K. S. 1987. Rates of growth and sample moduli for weighted empirical processes indexed by sets. *Probability Theory and Related Fields*, **75**, 379–423.

Alliney, S., and Ruzinsky, S. A. 1994. An algorithm for the minimization of mixed ℓ_1 and ℓ_2 norms with application to Bayesian estimation. *IEEE Transactions on Signal Processing*, **42**(3), 618–627.

Amini, A. A., and Wainwright, M. J. 2009. High-dimensional analysis of semdefinite relaxations for sparse principal component analysis. *Annals of Statistics*, **5B**, 2877–2921.

Anandkumar, A., Tan, V. Y. F., Huang, F., and Willsky, A. S. 2012. High-dimensional structure learning of Ising models: Local separation criterion. *Annals of Statistics*, **40**(3), 1346–1375.

Anderson, T. W. 1984. *An Introduction to Multivariate Statistical Analysis*. Wiley Series in Probability and Mathematical Statistics. New York, NY: Wiley.

Ando, R. K., and Zhang, T. 2005. A framework for learning predictive structures from multiple tasks and unlabeled data. *Journal of Machine Learning Research*, **6**(December), 1817–1853.

Aronszajn, N. 1950. Theory of reproducing kernels. *Transactions of the American Mathematical Society*, **68**, 337–404.

Assouad, P. 1983. Deux remarques sur l'estimation. *Comptes Rendus de l'Académie des Sciences, Paris*, **296**, 1021–1024.

Azuma, K. 1967. Weighted sums of certain dependent random variables. *Tohoku Mathematical Journal*, **19**, 357–367.

Bach, F., Jenatton, R., Mairal, J., and Obozinski, G. 2012. Optimization with sparsity-inducing penalties. *Foundations and Trends in Machine Learning*, **4**(1), 1–106.

Bahadur, R. R., and Rao, R. R. 1960. On deviations of the sample mean. *Annals of Mathematical Statistics*, **31**, 1015–1027.

Bai, Z., and Silverstein, J. W. 2010. *Spectral Analysis of Large Dimensional Random Matrices*. New York, NY: Springer. Second edition.

Baik, J., and Silverstein, J. W. 2006. Eigenvalues of large sample covariance matrices of spiked populations models. *Journal of Multivariate Analysis*, **97**(6), 1382–1408.

Balabdaoui, F., Rufibach, K., and Wellner, J. A. 2009. Limit distribution theory for maximum likelihood estimation of a log-concave density. *Annals of Statistics*, **62**(3), 1299–1331.

Ball, K. 1997. An elementary introduction to modern convex geometry. Pages 1–55 of: *Flavors of Geometry*. MSRI Publications, vol. 31. Cambridge, UK: Cambridge University Press.

Banerjee, O., El Ghaoui, L., and d'Aspremont, A. 2008. Model selection through sparse maximum likelihood estimation for multivariate Gaussian or binary data. *Journal of Machine Learning Research*, **9**(March), 485–516.

Baraniuk, R. G., Cevher, V., Duarte, M. F., and Hegde, C. 2010. Model-based compressive sensing. *IEEE Transactions on Information Theory*, **56**(4), 1982–2001.

Barndorff-Nielson, O. E. 1978. *Information and Exponential Families*. Chichester, UK: Wiley.

Bartlett, P. L., and Mendelson, S. 2002. Gaussian and Rademacher complexities: Risk bounds and structural results. *Journal of Machine Learning Research*, **3**, 463–482.

Bartlett, P. L., Bousquet, O., and Mendelson, S. 2005. Local Rademacher complexities. *Annals of Statistics*, **33**(4), 1497–1537.

Baxter, R. J. 1982. *Exactly Solved Models in Statistical Mechanics*. New York, NY: Academic Press.

Bean, D., Bickel, P. J., El Karoui, N., and Yu, B. 2013. Optimal M-estimation in high-dimensional regression. *Proceedings of the National Academy of Sciences of the USA*, **110**(36), 14563–14568.

Belloni, A., Chernozhukov, V., and Wang, L. 2011. Square-root lasso: pivotal recovery of sparse signals via conic programming. *Biometrika*, **98**(4), 791–806.

Bennett, G. 1962. Probability inequalities for the sum of independent random variables. *Journal of the American Statistical Association*, **57**(297), 33–45.

Bento, J., and Montanari, A. 2009 (December). Which graphical models are difficult to learn? In: *Proceedings of the NIPS Conference*.

Berlinet, A., and Thomas-Agnan, C. 2004. *Reproducing Kernel Hilbert Spaces in Probability and Statistics*. Norwell, MA: Kluwer Academic.

Bernstein, S. N. 1937. On certain modifications of Chebyshev's inequality. *Doklady Akademii Nauk SSSR*, **16**(6), 275–277.

Berthet, Q., and Rigollet, P. 2013 (June). Computational lower bounds for sparse PCA. In: *Conference on Computational Learning Theory*.

Bertsekas, D. P. 2003. *Convex Analysis and Optimization*. Boston, MA: Athena Scientific.

Besag, J. 1974. Spatial interaction and the statistical analysis of lattice systems. *Journal of the Royal Statistical Society, Series B*, **36**, 192–236.

Besag, J. 1975. Statistical analysis of non-lattice data. *The Statistician*, **24**(3), 179–195.

Besag, J. 1977. Efficiency of pseudolikelihood estimation for simple Gaussian fields. *Biometrika*, **64**(3), 616–618.

Bethe, H. A. 1935. Statistics theory of superlattices. *Proceedings of the Royal Society of London, Series A*, **150**(871), 552–575.

Bhatia, R. 1997. *Matrix Analysis*. Graduate Texts in Mathematics. New York, NY: Springer.

Bickel, P. J., and Doksum, K. A. 2015. *Mathematical Statistics: Basic Ideas and Selected Topics*. Boca Raton, FL: CRC Press.

Bickel, P. J., and Levina, E. 2008a. Covariance regularization by thresholding. *Annals of Statistics*, **36**(6), 2577–2604.

Bickel, P. J., and Levina, E. 2008b. Regularized estimation of large covariance matrices. *Annals of Statistics*, **36**(1), 199–227.

Bickel, P. J., Ritov, Y., and Tsybakov, A. B. 2009. Simultaneous analysis of lasso and Dantzig selector. *Annals of Statistics*, **37**(4), 1705–1732.

Birgé, L. 1983. Approximation dans les espaces metriques et theorie de l'estimation. *Z. Wahrsch. verw. Gebiete*, **65**, 181–327.

Birgé, L. 1987. Estimating a density under order restrictions: Non-asymptotic minimax risk. *Annals of Statistics*, **15**(3), 995–1012.

Birgé, L. 2005. A new lower bound for multiple hypothesis testing. *IEEE Transactions on Information Theory*, **51**(4), 1611–1614.

Birgé, L., and Massart, P. 1995. Estimation of integral functionals of a density. *Annals of Statistics*, **23**(1), 11–29.

Birnbaum, A., Johnstone, I. M., Nadler, B., and Paul, D. 2012. Minimax bounds for sparse PCA with noisy high-dimensional data. *Annals of Statistics*, **41**(3), 1055–1084.

Bobkov, S. G. 1999. Isoperimetric and analytic inequalities for log-concave probability measures. *Annals of Probability*, **27**(4), 1903–1921.

Bobkov, S. G., and Götze, F. 1999. Exponential integrability and transportation cost related to logarithmic Sobolev inequalities. *Journal of Functional Analysis*, **163**, 1–28.

Bobkov, S. G., and Ledoux, M. 2000. From Brunn-Minkowski to Brascamp-Lieb and to logarithmic Sobolev inequalities. *Geometric and Functional Analysis*, **10**, 1028–1052.

Borgwardt, K., Gretton, A., Rasch, M., Kriegel, H. P., Schölkopf, B., and Smola, A. J. 2006. Integrating structured biological data by kernel maximum mean discrepancy. *Bioinformatics*, **22**(14), 49–57.

Borwein, J., and Lewis, A. 1999. *Convex Analysis*. New York, NY: Springer.

Boser, B. E., Guyon, I. M., and Vapnik, V. N. 1992. A training algorithm for optimal margin classifiers. Pages 144–152 of: *Proceedings of the Conference on Learning Theory (COLT)*. New York, NY: ACM.

Boucheron, S., Lugosi, G., and Massart, P. 2003. Concentration inequalities using the entropy method. *Annals of Probability*, **31**(3), 1583–1614.

Boucheron, S., Lugosi, G., and Massart, P. 2013. *Concentration inequalities: A nonasymptotic theory of independence*. Oxford, UK: Oxford University Press.

Bourgain, J., Dirksen, S., and Nelson, J. 2015. Toward a unified theory of sparse dimensionality reduction in Euclidean space. *Geometric and Functional Analysis*, **25**(4).

Bousquet, O. 2002. A Bennett concentration inequality and its application to suprema of empirical processes. *Comptes Rendus de l'Académie des Sciences, Paris, Série I*, **334**, 495–500.

Bousquet, O. 2003. Concentration inequalities for sub-additive functions using the entropy method. *Stochastic Inequalities and Applications*, **56**, 213–247.

Boyd, S., and Vandenberghe, L. 2004. *Convex optimization*. Cambridge, UK: Cambridge University Press.

Brascamp, H. J., and Lieb, E. H. 1976. On extensions of the Brunn–Minkowski and Prékopa–Leindler theorems, including inequalities for log concave functions, and with an application to the diffusion equation. *Journal of Functional Analysis*, **22**, 366–389.

Breiman, L. 1992. *Probability*. Classics in Applied Mathematics. Philadelphia, PA: S IAM.

Bresler, G. 2014. *Efficiently learning Ising models on arbitrary graphs*. Tech. rept. MIT.

Bresler, G., Mossel, E., and Sly, A. 2013. Reconstruction of Markov Random Fields from samples: Some observations and algorithms. *SIAM Journal on Computing*, **42**(2), 563–578.

Bronshtein, E. M. 1976. ϵ-entropy of convex sets and functions. *Siberian Mathematical Journal*, **17**, 393–398.

Brown, L. D. 1986. *Fundamentals of statistical exponential families*. Hayward, CA: Institute of Mathematical Statistics.

Brunk, H. D. 1955. Maximum likelihood estimates of monotone parameters. *Annals of Math. Statistics*, **26**, 607–616.

Brunk, H. D. 1970. Estimation of isotonic regression. Pages 177–197 of: *Nonparametric techniques in statistical inference*. New York, NY: Cambridge University Press.

Bühlmann, P., and van de Geer, S. 2011. *Statistics for high-dimensional data*. Springer Series in Statistics. Springer.

Buja, A., Hastie, T. J., and Tibshirani, R. 1989. Linear smoothers and additive models. *Annals of Statistics*, **17**(2), 453–510.

Buldygin, V. V., and Kozachenko, Y. V. 2000. *Metric characterization of random variables and random processes*. Providence, RI: American Mathematical Society.

Bunea, F., Tsybakov, A. B., and Wegkamp, M. 2007. Sparsity oracle inequalities for the Lasso. *Electronic Journal of Statistics*, 169–194.

Bunea, F., She, Y., and Wegkamp, M. 2011. Optimal selection of reduced rank estimators of high-dimensional matrices. *Annals of Statistics*, **39**(2), 1282–1309.

Cai, T. T., Zhang, C. H., and Zhou, H. H. 2010. Optimal rates of convergence for covariance matrix estimation. *Annals of Statistics*, **38**(4), 2118–2144.

Cai, T. T., Liu, W., and Luo, X. 2011. A constrained ℓ_1-minimization approach to sparse precision matrix estimation. *Journal of the American Statistical Association*, **106**, 594–607.

Cai, T. T., Liang, T., and Rakhlin, A. 2015. *Computational and statistical boundaries for submatrix localization in a large noisy matrix*. Tech. rept. Univ. Penn.

Candès, E. J., and Plan, Y. 2010. Matrix completion with noise. *Proceedings of the IEEE*, **98**(6), 925–936.

Candès, E. J., and Recht, B. 2009. Exact matrix completion via convex optimization. *Foundations of Computational Mathematics*, **9**(6), 717–772.

Candès, E. J., and Tao, T. 2005. Decoding by linear programming. *IEEE Transactions on Information Theory*, **51**(12), 4203–4215.

Candès, E. J., and Tao, T. 2007. The Dantzig selector: statistical estimation when p is much larger than n. *Annals of Statistics*, **35**(6), 2313–2351.

Candès, E. J., Li, X., Ma, Y., and Wright, J. 2011. Robust principal component analysis? *Journal of the ACM*, **58**(3), 11 (37pp).

Candès, E. J., Strohmer, T., and Voroninski, V. 2013. PhaseLift: exact and stable signal recovery from magnitude measurements via convex programming. *Communications on Pure and Applied Mathematics*, **66**(8), 1241–1274.

Cantelli, F. P. 1933. Sulla determinazione empirica della legge di probabilita. *Giornale dell'Istituto Italiano degli Attuari*, **4**, 421–424.

Carl, B., and Pajor, A. 1988. Gelfand numbers of operators with values in a Hilbert space. *Inventiones Mathematicae*, **94**, 479–504.

Carl, B., and Stephani, I. 1990. *Entropy, Compactness and the Approximation of Operators*. Cambridge Tracts in Mathematics. Cambridge, UK: Cambridge University Press.

Carlen, E. 2009. Trace inequalities and quantum entropy: an introductory course. In: *Entropy and the Quantum*. Providence, RI: American Mathematical Society.

Carroll, R. J., Ruppert, D., and Stefanski, L. A. 1995. *Measurement Error in Nonlinear Models*. Boca Raton, FL: Chapman & Hall/CRC.

Chai, A., Moscoso, M., and Papanicolaou, G. 2011. Array imaging using intensity-only measurements. *Inverse Problems*, **27**(1), 1—15.

Chandrasekaran, V., Sanghavi, S., Parrilo, P. A., and Willsky, A. S. 2011. Rank-Sparsity Incoherence for Matrix Decomposition. *SIAM Journal on Optimization*, **21**, 572–596.

Chandrasekaran, V., Recht, B., Parrilo, P. A., and Willsky, A. S. 2012a. The convex geometry of linear inverse problems. *Foundations of Computational Mathematics*, **12**(6), 805–849.

Chandrasekaran, V., Parrilo, P. A., and Willsky, A. S. 2012b. Latent variable graphical model selection via convex optimization. *Annals of Statistics*, **40**(4), 1935–1967.

Chatterjee, S. 2005 (October). *An error bound in the Sudakov-Fernique inequality*. Tech. rept. UC Berkeley. arXiv:math.PR/0510424.

Chatterjee, S. 2007. Stein's method for concentration inequalities. *Probability Theory and Related Fields*, **138**(1–2), 305–321.

Chatterjee, S., Guntuboyina, A., and Sen, B. 2015. On risk bounds in isotonic and other shape restricted regression problems. *Annals of Statistics*, **43**(4), 1774–1800.

Chen, S., Donoho, D. L., and Saunders, M. A. 1998. Atomic decomposition by basis pursuit. *SIAM J. Sci. Computing*, **20**(1), 33–61.

Chernoff, H. 1952. A measure of asymptotic efficiency for tests of a hypothesis based on a sum of observations. *Annals of Mathematical Statistics*, **23**, 493–507.

Chernozhukov, V., Chetverikov, D., and Kato, K. 2013. *Comparison and anti-concentration bounds for maxima of Gaussian random vectors*. Tech. rept. MIT.

Chung, F.R.K. 1991. *Spectral Graph Theory*. Providence, RI: American Mathematical Society.

Clifford, P. 1990. Markov random fields in statistics. In: Grimmett, G.R., and Welsh, D. J. A. (eds), *Disorder in physical systems*. Oxford Science Publications.

Cohen, A., Dahmen, W., and DeVore, R. A. 2008. Compressed sensing and best k-term approximation. *J. of. American Mathematical Society*, **22**(1), 211–231.

Cormode, G. 2012. Synopses for massive data: Samples, histograms, wavelets and sketches. *Foundations and Trends in Databases*, **4**(2), 1–294.

Cover, T.M., and Thomas, J.A. 1991. *Elements of Information Theory*. New York, NY: Wiley.

Cule, M., Samworth, R. J., and Stewart, M. 2010. Maximum likelihood estimation of a multi-dimensional log-concave density. *J. R. Stat. Soc. B*, **62**, 545–607.

Dalalyan, A. S., Hebiri, M., and Lederer, J. 2014. *On the prediction performance of the Lasso*. Tech. rept. ENSAE. arxiv:1402,1700, to appear in Bernoulli.

d'Aspremont, A., El Ghaoui, L., Jordan, M. I., and Lanckriet, G. R. 2007. A direct formulation for sparse PCA using semidefinite programming. *SIAM Review*, **49**(3), 434–448.

d'Aspremont, A., Banerjee, O., and El Ghaoui, L. 2008. First order methods for sparse covariance selection. *SIAM Journal on Matrix Analysis and Its Applications*, **30**(1), 55–66.

Davidson, K. R., and Szarek, S. J. 2001. Local operator theory, random matrices, and Banach spaces. Pages 317–336 of: *Handbook of Banach Spaces*, vol. 1. Amsterdam, NL: Elsevier.

Dawid, A. P. 2007. The geometry of proper scoring rules. *Annals of the Institute of Statistical Mathematics*, **59**, 77–93.

de La Pena, V., and Giné, E. 1999. *Decoupling: From dependence to independence*. New York, NY: Springer.

Dembo, A. 1997. Information inequalities and concentration of measure. *Annals of Probability*, **25**(2), 927–939.

Dembo, A., and Zeitouni, O. 1996. Transportation approach to some concentration inequalities in product spaces. *Electronic Communications in Probability*, **1**, 83–90.

DeVore, R. A., and Lorentz, G. G. 1993. *Constructive Approximation*. New York, NY: Springer.

Devroye, L., and Györfi, L. 1986. *Nonparametric density estimation: the L_1 view*. New York, NY: Wiley.

Donoho, D. L. 2006a. For most large underdetermined systems of linear equations, the minimal ℓ_1-norm near-solution approximates the sparsest near-solution. *Communications on Pure and Applied Mathematics*, **59**(7), 907–934.

Donoho, D. L. 2006b. For most large underdetermined systems of linear equations, the minimal ℓ_1-norm solution is also the sparsest solution. *Communications on Pure and Applied Mathematics*, **59**(6), 797–829.

Donoho, D. L., and Huo, X. 2001. Uncertainty principles and ideal atomic decomposition. *IEEE Transactions on Information Theory*, **47**(7), 2845–2862.

Donoho, D. L., and Johnstone, I. M. 1994. Minimax risk over ℓ_p-balls for ℓ_q-error. *Probability Theory and Related Fields*, **99**, 277–303.

Donoho, D. L., and Montanari, A. 2013. *High dimensional robust M-estimation: asymptotic variance via approximate message passing*. Tech. rept. Stanford University. Posted as arxiv:1310.7320.

Donoho, D. L., and Stark, P. B. 1989. Uncertainty principles and signal recovery. *SIAM Journal of Applied Mathematics*, **49**, 906–931.

Donoho, D. L., and Tanner, J. M. 2008. Counting faces of randomly-projected polytopes when the projection radically lowers dimension. *Journal of the American Mathematical Society*, July.

Duchi, J. C., Wainwright, M. J., and Jordan, M. I. 2013. *Local privacy and minimax bounds: Sharp rates for probability estimation*. Tech. rept. UC Berkeley.

Duchi, J. C., Wainwright, M. J., and Jordan, M. I. 2014. Privacy-aware learning. *Journal of the ACM*, **61**(6), Article 37.

Dudley, R. M. 1967. The sizes of compact subsets of Hilbert spaces and continuity of Gaussian processes. *Journal of Functional Analysis*, **1**, 290–330.

Dudley, R. M. 1978. Central limit theorems for empirical measures. *Annals of Probability*, **6**, 899–929.

Dudley, R. M. 1999. *Uniform central limit theorems*. Cambridge, UK: Cambridge University Press.

Dümbgen, L., Samworth, R. J., and Schuhmacher, D. 2011. Approximation by log-concave distributions with applications to regression. *Annals of Statistics*, **39**(2); 702–730.

Durrett, R. 2010. *Probability: Theory and examples*. Cambridge, UK: Cambridge University Press.

Dvoretsky, A., Kiefer, J., and Wolfowitz, J. 1956. Asymptotic minimax character of the sample distribution function and of the classical multinomial estimator. *Annals of Mathematical Statistics*, **27**, 642–669.

Eggermont, P. P. B., and LaRiccia, V. N. 2001. *Maximum penalized likelihood estimation: V. I Density estimation*. Springer Series in Statistics, vol. 1. New York, NY: Springer.

Eggermont, P. P. B., and LaRiccia, V. N. 2007. *Maximum penalized likelihood estimation: V. II Regression*. Springer Series in Statistics, vol. 2. New York, NY: Springer.

El Karoui, N. 2008. Operator norm consistent estimation of large-dimensional sparse covariance matrices. *Annals of Statistics*, **36**(6), 2717–2756.

El Karoui, N. 2013. *Asymptotic behavior of unregularized and ridge-regularized high-dimensional robust regression estimators : rigorous results*. Tech. rept. UC Berkeley. Posted as arxiv:1311.2445.

El Karoui, N., Bean, D., Bickel, P. J., and Yu, B. 2013. On robust regression with high-dimensional predictors. *Proceedings of the National Academy of Sciences of the USA*, **110**(36), 14557–14562.

Elad, M., and Bruckstein, A. M. 2002. A generalized uncertainty principle and sparse representation in pairs of bases. *IEEE Transactions on Information Theory*, **48**(9), 2558–2567.

Fan, J., and Li, R. 2001. Variable selection via non-concave penalized likelihood and its oracle properties. *Journal of the American Statistical Association*, **96**(456), 1348–1360.

Fan, J., and Lv, J. 2011. Nonconcave penalized likelihood with NP-dimensionality. *IEEE Transactions on Information Theory*, **57**(8), 5467–5484.

Fan, J., Liao, Y., and Mincheva, M. 2013. Large covariance estimation by thresholding principal orthogonal components. *Journal of the Royal Statistical Society B*, **75**, 603–680.

Fan, J., Xue, L., and Zou, H. 2014. Strong oracle optimality of folded concave penalized estimation. *Annals of Statistics*, **42**(3), 819–849.

Fazel, M. 2002. *Matrix Rank Minimization with Applications*. Ph.D. thesis, Stanford. Available online: http://faculty.washington.edu/mfazel/thesis-final.pdf.

Fernique, X. M. 1974. Des resultats nouveaux sur les processus Gaussiens. *Comptes Rendus de l'Académie des Sciences, Paris*, **278**, A363–A365.

Feuer, A., and Nemirovski, A. 2003. On sparse representation in pairs of bases. *IEEE Transactions on Information Theory*, **49**(6), 1579–1581.

Fienup, J. R. 1982. Phase retrieval algorithms: a comparison. *Applied Optics*, **21**(15), 2758–2769.

Fienup, J. R., and Wackerman, C. C. 1986. Phase-retrieval stagnation problems and solutions. *Journal of the Optical Society of America A*, **3**, 1897–1907.

Fletcher, A. K., Rangan, S., and Goyal, V. K. 2009. Necessary and Sufficient Conditions for Sparsity Pattern Recovery. *IEEE Transactions on Information Theory*, **55**(12), 5758–5772.

Foygel, R., and Srebro, N. 2011. *Fast rate and optimistic rate for ℓ_1-regularized regression*. Tech. rept. Toyoto Technological Institute. arXiv:1108.037v1.

Friedman, J. H., and Stuetzle, W. 1981. Projection pursuit regression. *Journal of the American Statistical Association*, **76**(376), 817–823.

Friedman, J. H., and Tukey, J. W. 1994. A projection pursuit algorithm for exploratory data analysis. *IEEE Transactions on Computers*, **C-23**, 881–889.

Friedman, J. H., Hastie, T. J., and Tibshirani, R. 2007. Sparse inverse covariance estimation with the graphical Lasso. *Biostatistics*.

Fuchs, J. J. 2004. Recovery of exact sparse representations in the presence of noise. Pages 533–536 of: *ICASSP*, vol. 2.

Gallager, R. G. 1968. *Information theory and reliable communication*. New York, NY: Wiley.

Gao, C., Ma, Z., and Zhou, H. H. 2015. *Sparse CCA: Adaptive estimation and computational barriers*. Tech. rept. Yale University.

Gardner, R. J. 2002. The Brunn-Minkowski inequality. *Bulletin of the American Mathematical Society*, **39**, 355–405.

Geman, S. 1980. A limit theorem for the norm of random matrices. *Annals of Probability*, **8**(2), 252–261.

Geman, S., and Geman, D. 1984. Stochastic Relaxation, Gibbs Distributions, and the Bayesian Restoration of Images. *IEEE Transactions on Pattern Analysis and Machine Intelligence*, **6**, 721–741.

Geman, S., and Hwang, C. R. 1982. Nonparametric maximum likelihood estimation by the method of sieves. *Annals of Statistics*, **10**(2), 401–414.

Glivenko, V. 1933. Sulla determinazione empirica della legge di probabilita. *Giornale dell'Istituto Italiano degli Attuari*, **4**, 92–99.

Gneiting, T., and Raftery, A. E. 2007. Strictly proper scoring rules, prediction, and estimation. *Journal of the American Statistical Association*, **102**(477), 359–378.

Goldberg, K., Roeder, T., Gupta, D., and Perkins, C. 2001. Eigentaste: A constant time collaborative filtering algorithm. *Information Retrieval*, **4**(2), 133–151.

Good, I. J., and Gaskins, R. A. 1971. Nonparametric roughness penalties for probability densities. *Biometrika*, **58**, 255–277.

Gordon, Y. 1985. Some inequalities for Gaussian processes and applications. *Israel Journal of Mathematics*, **50**, 265–289.

Gordon, Y. 1986. On Milman's inequality and random subspaces which escape through a mesh in $\mathbb{R}^n$. Pages 84–106 of: *Geometric aspects of functional analysis*. Lecture Notes in Mathematics, vol. 1317. Springer-Verlag.

Gordon, Y. 1987. Elliptically contoured distributions. *Probability Theory and Related Fields*, **76**, 429–438.

Götze, F., and Tikhomirov, A. 2004. Rate of convergence in probability to the Marčenko-Pastur law. *Bernoulli*, **10**(3), 503–548.

Grechberg, R. W., and Saxton, W. O. 1972. A practical algorithm for the determination of phase from image and diffraction plane intensities. *Optik*, **35**, 237–246.

Greenshtein, E., and Ritov, Y. 2004. Persistency in high dimensional linear predictor-selection and the virtue of over-parametrization. *Bernoulli*, **10**, 971–988.

Gretton, A., Borgwardt, K., Rasch, M., Schölkopf, B., and Smola, A. 2012. A kernel two-sample test. *Journal of Machine Learning Research*, **13**, 723–773.

Griffin, D., and Lim, J. 1984. Signal estimation from modified short-time Fourier transforms. *IEEE Transactions on Acoustics, Speech, and Signal Processing*, **32**(2), 236–243.

Grimmett, G. R. 1973. A theorem about random fields. *Bulletin of the London Mathematical Society*, **5**, 81–84.

Gross, D. 2011. Recovering low-rank matrices from few coefficients in any basis. *IEEE Transactions on Information Theory*, **57**(3), 1548–1566.

Gross, L. 1975. Logarithmic Sobolev inequalities. *American Journal Math.*, **97**, 1061–1083.

Gu, C. 2002. *Smoothing spline ANOVA models*. Springer Series in Statistics. New York, NY: Springer.

Guédon, O., and Litvak, A. E. 2000. Euclidean projections of a p-convex body. Pages 95–108 of: *Geometric aspects of functional analysis*. Springer.

Guntuboyina, A. 2011. Lower bounds for the minimax risk using f-divergences and applications. *IEEE Transactions on Information Theory*, **57**(4), 2386–2399.

Guntuboyina, A., and Sen, B. 2013. Covering numbers for convex functions. *IEEE Transactions on Information Theory*, **59**, 1957–1965.

Gyorfi, L., Kohler, M., Krzyzak, A., and Walk, H. 2002. *A Distribution-Free Theory of Nonparametric Regression*. Springer Series in Statistics. Springer.

Hammersley, J. M., and Clifford, P. 1971. *Markov fields on finite graphs and lattices*. Unpublished.

Hanson, D. L., and Pledger, G. 1976. Consistency in concave regression. *Annals of Statistics*, **4**, 1038–1050.

Hanson, D. L., and Wright, F. T. 1971. A bound on tail probabilities for quadratic forms in independent random variables. *Annals of Mathematical Statistics*, **42**(3), 1079–1083.

Härdle, W. K., and Stoker, T. M. 1989. Investigating smooth multiple regression by the method of average derivatives. *Journal of the American Statistical Association*, **84**, 986–995.

Härdle, W. K., Hall, P., and Ichimura, H. 1993. Optimal smoothing in single-index models. *Annals of Statistics*, **21**, 157–178.

Härdle, W. K., Müller, M., Sperlich, S., and Werwatz, A. 2004. *Nonparametric and semiparametric models*. Springer Series in Statistics. New York, NY: Springer.

Harper, L. H. 1966. Optimal numberings and isoperimetric problems on graphs. *Journal of Combinatorial Theory*, **1**, 385–393.

Harrison, R. W. 1993. Phase problem in crystallography. *Journal of the Optical Society of America A*, **10**(5), 1046–1055.

Hasminskii, R. Z. 1978. A lower bound on the risks of nonparametric estimates of densities in the uniform metric. *Theory of Probability and Its Applications*, **23**, 794–798.

Hasminskii, R. Z., and Ibragimov, I. 1981. *Statistical estimation: Asymptotic theory*. New York, NY: Springer.

Hasminskii, R. Z., and Ibragimov, I. 1990. On density estimation in the view of Kolmogorov's ideas in approximation theory. *Annals of Statistics*, **18**(3), 999–1010.

Hastie, T. J., and Tibshirani, R. 1986. Generalized additive models. *Statistical Science*, **1**(3), 297–310.

Hastie, T. J., and Tibshirani, R. 1990. *Generalized Additive Models*. Boca Raton, FL: Chapman & Hall/CRC.

Hildreth, C. 1954. Point estimates of ordinates of concave functions. *Journal of the American Statistical Association*, **49**, 598–619.

Hiriart-Urruty, J., and Lemaréchal, C. 1993. *Convex Analysis and Minimization Algorithms*. Vol. 1. New York, NY: Springer.

Hoeffding, W. 1963. Probability inequalities for sums of bounded random variables. *Journal of the American Statistical Association*, **58**, 13–30.

Hoerl, A. E., and Kennard, R. W. 1970. Ridge Regression: Biased Estimation for Nonorthogonal Problems. *Technometrics*, **12**, 55–67.

Hölfing, H., and Tibshirani, R. 2009. Estimation of sparse binary pairwise Markov networks using pseudo-likelihoods. *Journal of Machine Learning Research*, **19**, 883–906.

Holley, R., and Stroock, D. 1987. Log Sobolev inequalities and stochastic Ising models. *Journal of Statistical Physics*, **46**(5), 1159–1194.

Horn, R. A., and Johnson, C. R. 1985. *Matrix Analysis*. Cambridge, UK: Cambridge University Press.

Horn, R. A., and Johnson, C. R. 1991. *Topics in Matrix Analysis*. Cambridge, UK: Cambridge University Press.

Hristache, M., Juditsky, A., and Spokoiny, V. 2001. Direct estimation of the index coefficient in a single index model. *Annals of Statistics*, **29**, 595–623.

Hsu, D., Kakade, S. M., and Zhang, T. 2012a. Tail inequalities for sums of random matrices that depend on the intrinsic dimension. *Electronic Communications in Probability*, **17**(14), 1–13.

Hsu, D., Kakade, S. M., and Zhang, T. 2012b. A tail inequality for quadratic forms of sub-Gaussian random vectors. *Electronic Journal of Probability*, **52**, 1–6.

Huang, J., and Zhang, T. 2010. The benefit of group sparsity. *Annals of Statistics*, **38**(4), 1978–2004.

Huang, J., Ma, S., and Zhang, C. H. 2008. Adaptive Lasso for sparse high-dimensional regression models. *Statistica Sinica*, **18**, 1603–1618.

Huber, P. J. 1973. Robust regression: Asymptotics, conjectures and Monte Carlo. *Annals of Statistics*, **1**(5), 799–821.

Huber, P. J. 1985. Projection pursuit. *Annals of Statistics*, **13**(2), 435–475.

Ichimura, H. 1993. Semiparametric least squares (SLS) and weighted (SLS) estimation of single index models. *Journal of Econometrics*, **58**, 71–120.

Ising, E. 1925. Beitrag zur Theorie der Ferromagnetismus. *Zeitschrift für Physik*, **31**(1), 253–258.

Iturria, S. J., Carroll, R. J., and Firth, D. 1999. Polynomial Regression and Estimating Functions in the Presence of Multiplicative Measurement Error. *Journal of the Royal Statistical Society B*, **61**, 547–561.

Izenman, A. J. 1975. Reduced-rank regression for the multivariate linear model. *Journal of Multivariate Analysis*, **5**, 248–264.

Izenman, A. J. 2008. *Modern multivariate statistical techniques: Regression, classification and manifold learning*. New York, NY: Springer.

Jacob, L., Obozinski, G., and Vert, J. P. 2009. Group Lasso with overlap and graph Lasso. Pages 433–440 of: *International Conference on Machine Learning (ICML)*.

Jalali, A., Ravikumar, P., Sanghavi, S., and Ruan, C. 2010. A Dirty Model for Multi-task Learning. Pages 964–972 of: *Advances in Neural Information Processing Systems 23*.

Johnson, W. B., and Lindenstrauss, J. 1984. Extensions of Lipschitz mappings into a Hilbert space. *Contemporary Mathematics*, **26**, 189–206.

Johnstone, I. M. 2001. On the distribution of the largest eigenvalue in principal components analysis. *Annals of Statistics*, **29**(2), 295–327.

Johnstone, I. M. 2015. *Gaussian estimation: Sequence and wavelet models*. New York, NY: Springer.

Johnstone, I. M., and Lu, A. Y. 2009. On consistency and sparsity for principal components analysis in high dimensions. *Journal of the American Statistical Association*, **104**, 682–693.

Jolliffe, I. T. 2004. *Principal Component Analysis*. New York, NY: Springer.

Jolliffe, I. T., Trendafilov, N. T., and Uddin, M. 2003. A modified principal component technique based on the LASSO. *Journal of Computational and Graphical Statistics*, **12**, 531–547.

Juditsky, A., and Nemirovski, A. 2000. Functional aggregation for nonparametric regression. *Annals of Statistics*, **28**, 681–712.

Kahane, J. P. 1986. Une inequalité du type de Slepian et Gordon sur les processus Gaussiens. *Israel Journal of Mathematics*, **55**, 109–110.

Kalisch, M., and Bühlmann, P. 2007. Estimating high-dimensional directed acyclic graphs with the PC algorithm. *Journal of Machine Learning Research*, **8**, 613–636.

Kane, D. M., and Nelson, J. 2014. Sparser Johnson-Lindenstrauss transforms. *Journal of the ACM*, **61**(1).

Kantorovich, L. V., and Rubinstein, G. S. 1958. On the space of completely additive functions. *Vestnik Leningrad Univ. Ser. Math. Mekh. i. Astron*, **13**(7), 52–59. In Russian.

Keener, R. W. 2010. *Theoretical Statistics: Topics for a Core Class*. New York, NY: Springer.

Keshavan, R. H., Montanari, A., and Oh, S. 2010a. Matrix Completion from Few Entries. *IEEE Transactions on Information Theory*, **56**(6), 2980–2998.

Keshavan, R. H., Montanari, A., and Oh, S. 2010b. Matrix Completion from Noisy Entries. *Journal of Machine Learning Research*, **11**(July), 2057–2078.

Kim, Y., Kim, J., and Kim, Y. 2006. Blockwise sparse regression. *Statistica Sinica*, **16**(2).

Kimeldorf, G., and Wahba, G. 1971. Some results on Tchebycheffian spline functions. *Journal of Mathematical Analysis and Applications*, **33**, 82–95.

Klein, T., and Rio, E. 2005. Concentration around the mean for maxima of empirical processes. *Annals of Probability*, **33**(3), 1060–1077.

Koller, D., and Friedman, N. 2010. *Graphical Models*. New York, NY: MIT Press.

Kolmogorov, A. N. 1956. Asymptotic characterization of some completely bounded metric spaces. *Doklady Akademii Nauk SSSR*, **108**, 585–589.

Kolmogorov, A. N. 1958. Linear dimension of topological vector spaces. *Doklady Akademii Nauk SSSR*, **120**, 239–241–589.

Kolmogorov, A. N., and Tikhomirov, B. 1959. ϵ-entropy and ϵ-capacity of sets in functional spaces. *Uspekhi Mat. Nauk.*, **86**, 3–86. Appeared in English as *1961. American Mathematical Society Translations*, **17**, 277–364.

Koltchinskii, V. 2001. Rademacher penalities and structural risk minimization. *IEEE Transactions on Information Theory*, **47**(5), 1902–1914.

Koltchinskii, V. 2006. Local Rademacher complexities and oracle inequalities in risk minimization. *Annals of Statistics*, **34**(6), 2593–2656.

Koltchinskii, V., and Panchenko, D. 2000. Rademacher processes and bounding the risk of function learning. Pages 443–459 of: *High-dimensional probability II*. Springer.

Koltchinskii, V., and Yuan, M. 2010. Sparsity in multiple kernel learning. *Annals of Statistics*, **38**, 3660–3695.

Koltchinskii, V., Lounici, K., and Tsybakov, A. B. 2011. Nuclear-norm penalization and optimal rates for noisy low-rank matrix completion. *Annals of Statistics*, **39**, 2302–2329.

Kontorovich, L. A., and Ramanan, K. 2008. Concentration inequalities for dependent random variables via the martingale method. *Annals of Probability*, **36**(6), 2126–2158.

Kruskal, J. B. 1969. Towards a practical method which helps uncover the structure of a set of multivariate observation by finding the linear transformation which optimizes a new 'index of condensation'. In: *Statistical computation*. New York, NY: Academic Press.

Kühn, T. 2001. A lower estimate for entropy numbers. *Journal of Approximation Theory*, **110**, 120–124.

Kullback, S., and Leibler, R. A. 1951. On information and sufficiency. *Annals of Mathematical Statistics*, **22**(1), 79–86.

Lam, C., and Fan, J. 2009. Sparsistency and Rates of Convergence in Large Covariance Matrix Estimation. *Annals of Statistics*, **37**, 4254–4278.

Laurent, M. 2001. Matrix Completion Problems. Pages 221—229 of: *The Encyclopedia of Optimization*. Kluwer Academic.

Laurent, M. 2003. A comparison of the Sherali-Adams, Lovász-Schrijver and Lasserre relaxations for 0-1 programming. *Mathematics of Operations Research*, **28**, 470–496.

Lauritzen, S. L. 1996. *Graphical Models*. Oxford: Oxford University Press.

Le Cam, L. 1973. Convergence of estimates under dimensionality restrictions. *Annals of Statistics*, January.

Ledoux, M. 1996. On Talagrand's deviation inequalities for product measures. *ESAIM: Probability and Statistics*, **1**(July), 63–87.

Ledoux, M. 2001. *The Concentration of Measure Phenomenon*. Mathematical Surveys and Monographs. Providence, RI: American Mathematical Society.

Ledoux, M., and Talagrand, M. 1991. *Probability in Banach Spaces: Isoperimetry and Processes*. New York, NY: Springer.

Lee, J. D., Sun, Y., and Taylor, J. 2013. *On model selection consistency of M-estimators with geometrically decomposable penalties*. Tech. rept. Stanford University. arxiv1305.7477v4.

Leindler, L. 1972. On a certain converse of Hölder's inequality. *Acta Scientiarum Mathematicarum (Szeged)*, **33**, 217–223.

Levy, S., and Fullagar, P. K. 1981. Reconstruction of a sparse spike train from a portion of its spectrum and application to high-resolution deconvolution. *Geophysics*, **46**(9), 1235–1243.

Lieb, E. H. 1973. Convex trace functions and the Wigner-Yanase-Dyson conjecture. *Advances in Mathematics*, **11**, 267–288.

Lindley, D. V. 1956. On a measure of the information provided by an experiment. *Annals of Mathematical Statistics*, **27**(4), 986–1005.

Liu, H., Lafferty, J. D., and Wasserman, L. A. 2009. The nonparanormal: Semiparametric estimation of high-dimensional undirected graphs. *Journal of Machine Learning Research*, **10**, 1–37.

Liu, H., Han, F., Yuan, M., Lafferty, J. D., and Wasserman, L. A. 2012. High-dimensional semiparametric Gaussian copula graphical models. *Annals of Statistics*, **40**(4), 2293–2326.

Loh, P., and Wainwright, M. J. 2012. High-dimensional regression with noisy and missing data: Provable guarantees with non-convexity. *Annals of Statistics*, **40**(3), 1637–1664.

Loh, P., and Wainwright, M. J. 2013. Structure estimation for discrete graphical models: Generalized covariance matrices and their inverses. *Annals of Statistics*, **41**(6), 3022–3049.

Loh, P., and Wainwright, M. J. 2015. Regularized M-estimators with nonconvexity: Statistical and algorithmic theory for local optima. *Journal of Machine Learning Research*, **16**(April), 559–616.

Loh, P., and Wainwright, M. J. 2017. Support recovery without incoherence: A case for nonconvex regularization. *Annals of Statistics*, **45**(6), 2455–2482. Appeared as arXiv:1412.5632.

Lorentz, G. G. 1966. Metric entropy and approximation. *Bulletin of the AMS*, **72**(6), 903–937.

Lounici, K., Pontil, M., Tsybakov, A. B., and van de Geer, S. 2011. Oracle inequalities and optimal inference under group sparsity. *Annals of Statistics*, **39**(4), 2164–2204.

Lovász, L., and Schrijver, A. 1991. Cones of matrices and set-functions and $0-1$ optimization. *SIAM Journal of Optimization*, **1**, 166–190.

Ma, Z. 2010. *Contributions to high-dimensional principal component analysis*. Ph.D. thesis, Department of Statistics, Stanford University.

Ma, Z. 2013. Sparse principal component analysis and iterative thresholding. *Annals of Statistics*, **41**(2), 772–801.

Ma, Z., and Wu, Y. 2013. Computational barriers in minimax submatrix detection. *arXiv preprint arXiv:1309.5914*.

Mackey, L. W., Jordan, M. I., Chen, R. Y., Farrell, B., and Tropp, J. A. 2014. Matrix concentration inequalities via the method of exchangeable pairs. *Annals of Probability*, **42**(3), 906–945.

Mahoney, M. W. 2011. Randomized algorithms for matrices and data. *Foundations and Trends in Machine Learning*, **3**(2), 123–224.

Marton, K. 1996a. Bounding $\bar{d}$-distance by information divergence: a method to prove measure concentration. *Annals of Probability*, **24**, 857–866.

Marton, K. 1996b. A measure concentration inequality for contracting Markov chains. *Geometric and Functional Analysis*, **6**(3), 556–571.

Marton, K. 2004. Measure concentration for Euclidean distance in the case of dependent random variables. *Annals of Probability*, **32**(3), 2526–2544.

Marčenko, V. A., and Pastur, L. A. 1967. Distribution of eigenvalues for some sets of random matrices. *Annals of Probability*, **4**(1), 457–483.

Massart, P. 1990. The tight constant in the Dvoretzky-Kiefer-Wolfowitz inequality. *Annals of Probability*, **18**, 1269–1283.

Massart, P. 2000. Some applications of concentration inequalities to statistics. *Annales de la Faculté des Sciences de Toulouse*, **IX**, 245–303.

Maurey, B. 1991. Some deviation inequalities. *Geometric and Functional Analysis*, **1**, 188–197.

McDiarmid, C. 1989. On the method of bounded differences. Pages 148–188 of: *Surveys in Combinatorics*. London Mathematical Society Lecture Notes, no. 141. Cambridge, UK: Cambridge University Press.

Mehta, M. L. 1991. *Random Matrices*. New York, NY: Academic Press.

Meier, L., van de Geer, S., and Bühlmann, P. 2009. High-dimensional additive modeling. *Annals of Statistics*, **37**, 3779–3821.

Meinshausen, N. 2008. A note on the lasso for graphical Gaussian model selection. *Statistics and Probability Letters*, **78**(7), 880–884.

Meinshausen, N., and Bühlmann, P. 2006. High-dimensional graphs and variable selection with the Lasso. *Annals of Statistics*, **34**, 1436–1462.

Mendelson, S. 2002. Geometric parameters of kernel machines. Pages 29–43 of: *Proceedings of COLT*.

Mendelson, S. 2010. Empirical processes with a bounded ψ_1-diameter. *Geometric and Functional Analysis*, **20**(4), 988–1027.

Mendelson, S. 2015. Learning without concentration. *Journal of the ACM*, **62**(3), 1–25.

Mendelson, S., Pajor, A., and Tomczak-Jaegermann, N. 2007. Reconstruction of subgaussian operators. *Geometric and Functional Analysis*, **17**(4), 1248–1282.

Mézard, M., and Montanari, A. 2008. *Information, Physics and Computation*. New York, NY: Oxford University Press.

Milman, V., and Schechtman, G. 1986. *Asymptotic Theory of Finite Dimensional Normed Spaces*. Lecture Notes in Mathematics, vol. 1200. New York, NY: Springer.

Minsker, S. 2011. *On some extensions of Bernstein's inequality for self-adjoint operators*. Tech. rept. Duke University.

Mitjagin, B. S. 1961. The approximation dimension and bases in nuclear spaces. *Uspekhi. Mat. Naut.*, **61**(16), 63–132.

Muirhead, R. J. 2008. *Aspects of multivariate statistical theory*. Wiley Series in Probability and Mathematical Statistics. New York, NY: Wiley.

Müller, A. 1997. Integral probability metrics and their generating classes of functions. *Advances in Applied Probability*, **29**(2), 429–443.

Negahban, S., and Wainwright, M. J. 2011a. Estimation of (near) low-rank matrices with noise and high-dimensional scaling. *Annals of Statistics*, **39**(2), 1069–1097.

Negahban, S., and Wainwright, M. J. 2011b. Simultaneous support recovery in high-dimensional regression: Benefits and perils of $\ell_{1,\infty}$-regularization. *IEEE Transactions on Information Theory*, **57**(6), 3481–3863.

Negahban, S., and Wainwright, M. J. 2012. Restricted strong convexity and (weighted) matrix completion: Optimal bounds with noise. *Journal of Machine Learning Research*, **13**(May), 1665–1697.

Negahban, S., Ravikumar, P., Wainwright, M. J., and Yu, B. 2010 (October). *A unified framework for high-dimensional analysis of M-estimators with decomposable regularizers*. Tech. rept. UC Berkeley. Arxiv pre-print 1010.2731v1, Version 1.

Negahban, S., Ravikumar, P., Wainwright, M. J., and Yu, B. 2012. A unified framework for high-dimensional analysis of M-estimators with decomposable regularizers. *Statistical Science*, **27**(4), 538–557.

Nemirovski, A. 2000. Topics in non-parametric statistics. In: Bernard, P. (ed), *Ecole d'Été de Probabilities de Saint-Flour XXVIII*. Lecture Notes in Mathematics. Berlin, Germany: Springer.

Nesterov, Y. 1998. Semidefinite relaxation and nonconvex quadratic optimization. *Optimization methods and software*, **9**(1), 141–160.

Netrapalli, P., Banerjee, S., Sanghavi, S., and Shakkottai, S. 2010. Greedy learning of Markov network structure. Pages 1295–1302 of: *Communication, Control, and Computing (Allerton), 2010 48th Annual Allerton Conference on*. IEEE.

Obozinski, G., Wainwright, M. J., and Jordan, M. I. 2011. Union support recovery in high-dimensional multivariate regression. *Annals of Statistics*, **39**(1), 1–47.

Oldenburg, D. W., Scheuer, T., and Levy, S. 1983. Recovery of the acoustic impedance from reflection seismograms. *Geophysics*, **48**(10), 1318–1337.

Oliveira, R. I. 2010. Sums of random Hermitian matrices and an inequality by Rudelson. *Electronic Communications in Probability*, **15**, 203–212.

Oliveira, R. I. 2013. *The lower tail of random quadratic forms, with applicaitons to ordinary least squares and restricted eigenvalue properties*. Tech. rept. IMPA, Rio de Janeiro, Brazil.

Ortega, J. M., and Rheinboldt, W. C. 2000. *Iterative Solution of Nonlinear Equations in Several Variables*. Classics in Applied Mathematics. New York, NY: SIAM.

Pastur, L. A. 1972. On the spectrum of random matrices. *Theoretical and Mathematical Physics*, **10**, 67–74.

Paul, D. 2007. Asymptotics of sample eigenstructure for a large-dimensional spiked covariance model. *Statistica Sinica*, **17**, 1617–1642.

Pearl, J. 1988. *Probabilistic Reasoning in Intelligent Systems*. San Mateo, CA: Morgan Kaufmann.

Petrov, V. V. 1995. *Limit theorems of probability theory: Sequence of independent random variables*. Oxford, UK: Oxford University Press.

Pilanci, M., and Wainwright, M. J. 2015. Randomized sketches of convex programs with sharp guarantees. *IEEE Transactions on Information Theory*, **9**(61), 5096–5115.

Pinkus, A. 1985. *N-Widths in Approximation Theory*. New York: Springer.

Pisier, G. 1989. *The Volume of Convex Bodies and Banach Space Geometry*. Cambridge Tracts in Mathematics, vol. 94. Cambridge, UK: Cambridge University Press.

Pollard, D. 1984. *Convergence of Stochastic Processes*. New York, NY: Springer.

Portnoy, S. 1984. Asymptotic behavior of M-estimators of p regression parameters when p^2/n is large: I. Consistency. *Annals of Statistics*, **12**(4), 1296–1309.

Portnoy, S. 1985. Asymptotic behavior of M-estimators of p regression parameters when p^2/n is large: II. Normal approximation. *Annals of Statistics*, **13**(4), 1403–1417.

Portnoy, S. 1988. Asymptotic behavior of likelhoood methods for exponential families when the number of parameters tends to infinity. *Annals of Statistics*, **16**(1), 356–366.

Prékopa, A. 1971. Logarithmic concave measures with application to stochastic programming. *Acta Scientiarum Mathematicarum (Szeged)*, **32**, 301–315.

Prékopa, A. 1973. On logarithmic concave measures and functions. *Acta Scientiarum Mathematicarum (Szeged)*, **33**, 335–343.

Rachev, S. T., and Ruschendorf, L. 1998. *Mass Transportation Problems, Volume II, Applications*. New York, NY: Springer.

Rachev, S. T., Klebanov, L., Stoyanov, S. V., and Fabozzi, F. 2013. *The Method of Distances in the Theory of Probability and Statistics*. New York, NY: Springer.

Rao, C. R. 1949. On some problems arising out of discrimination with multiple characters. *Sankhya (Indian Journal of Statistics)*, **9**(4), 343–366.

Raskutti, G., Wainwright, M. J., and Yu, B. 2010. Restricted eigenvalue conditions for correlated Gaussian designs. *Journal of Machine Learning Research*, **11**(August), 2241–2259.

Raskutti, G., Wainwright, M. J., and Yu, B. 2011. Minimax rates of estimation for high-dimensional linear regression over ℓ_q-balls. *IEEE Transactions on Information Theory*, **57**(10), 6976—6994.

Raskutti, G., Wainwright, M. J., and Yu, B. 2012. Minimax-optimal rates for sparse additive models over kernel classes via convex programming. *Journal of Machine Learning Research*, **12**(March), 389–427.

Raudys, V., and Young, D. M. 2004. Results in Statistical Discriminant Analysis: A Review of the Former Soviet Union Literature. *Journal of Multivariate Analysis*, **89**(1), 1–35.

Ravikumar, P., Liu, H., Lafferty, J. D., and Wasserman, L. A. 2009. SpAM: sparse additive models. *Journal of the Royal Statistical Society, Series B*, **71**(5), 1009–1030.

Ravikumar, P., Wainwright, M. J., and Lafferty, J. D. 2010. High-dimensional Ising model selection using ℓ_1-regularized logistic regression. *Annals of Statistics*, **38**(3), 1287–1319.

Ravikumar, P., Wainwright, M. J., Raskutti, G., and Yu, B. 2011. High-dimensional covariance estimation by minimizing ℓ_1-penalized log-determinant divergence. *Electronic Journal of Statistics*, **5**, 935–980.

Recht, B. 2011. A Simpler Approach to Matrix Completion. *Journal of Machine Learning Research*, **12**, 3413–3430.

Recht, B., Xu, W., and Hassibi, B. 2009. *Null space conditions and thresholds for rank minimization*. Tech. rept. U. Madison. Available at http://pages.cs.wisc.edu/ brecht/papers/10.RecXuHas.Thresholds.pdf.

Recht, B., Fazel, M., and Parrilo, P. A. 2010. Guaranteed Minimum-Rank Solutions of Linear Matrix Equations via Nuclear Norm Minimization. *SIAM Review*, **52**(3), 471–501.

Reeves, G., and Gastpar, M. 2008 (July). Sampling Bounds for Sparse Support Recovery in the Presence of Noise. In: *International Symposium on Information Theory*.

Reinsel, G. C., and Velu, R. P. 1998. *Multivariate Reduced-Rank Regression*. Lecture Notes in Statistics, vol. 136. New York, NY: Springer.

Ren, Z., and Zhou, H. H. 2012. Discussion: Latent variable graphical model selection via convex optimization. *Annals of Statistics*, **40**(4), 1989–1996.

Richardson, T., and Urbanke, R. 2008. *Modern Coding Theory*. Cambridge University Press.

Rockafellar, R. T. 1970. *Convex Analysis*. Princeton: Princeton University Press.

Rohde, A., and Tsybakov, A. B. 2011. Estimation of high-dimensional low-rank matrices. *Annals of Statistics*, **39**(2), 887–930.

Rosenbaum, M., and Tsybakov, A. B. 2010. Sparse recovery under matrix uncertainty. *Annals of Statistics*, **38**, 2620–2651.

Rosenthal, H. P. 1970. On the subspaces of ℓ^p $(p > 2)$ spanned by sequences of independent random variables. *Israel Journal of Mathematics*, **8**, 1546–1570.

Rothman, A. J., Bickel, P. J., Levina, E., and Zhu, J. 2008. Sparse permutation invariant covariance estimation. *Electronic Journal of Statistics*, **2**, 494–515.

Rudelson, M. 1999. Random vectors in the isotropic position. *Journal of Functional Analysis*, **164**, 60–72.

Rudelson, M., and Vershynin, R. 2013. Hanson–Wright inequality and sub-Gaussian concentration. *Electronic Communications in Probability*, **18**(82), 1–9.

Rudelson, M., and Zhou, S. 2013. Reconstruction from anisotropic random measurements. *IEEE Transactions on Information Theory*, **59**(6), 3434–3447.

Rudin, W. 1964. *Principles of Mathematical Analysis*. New York, NY: McGraw-Hill.

Rudin, W. 1990. *Fourier Analysis on Groups*. New York, NY: Wiley-Interscience.

Samson, P. M. 2000. Concentration of measure inequalities for Markov chains and Φ-mixing processes. *Annals of Probability*, **28**(1), 416–461.

Santhanam, N. P., and Wainwright, M. J. 2012. Information-theoretic limits of selecting binary graphical models in high dimensions. *IEEE Transactions on Information Theory*, **58**(7), 4117–4134.

Santosa, F., and Symes, W. W. 1986. Linear inversion of band-limited reflection seismograms. *SIAM Journal on Scientific and Statistical Computing*, **7**(4), 1307—1330.

Saulis, L., and Statulevicius, V. 1991. *Limit Theorems for Large Deviations*. London: Kluwer Academic.

Schölkopf, B., and Smola, A. 2002. *Learning with Kernels*. Cambridge, MA: MIT Press.

Schütt, C. 1984. Entropy numbers of diagonal operators between symmetric Banach spaces. *Journal of Approximation Theory*, **40**, 121–128.

Scott, D. W. 1992. *Multivariate Density Estimation: Theory, Practice and Visualization*. New York, NY: Wiley.

Seijo, E., and Sen, B. 2011. Nonparametric least squares estimation of a multivariate convex regression function. *Annals of Statistics*, **39**(3), 1633–1657.

Serdobolskii, V. 2000. *Multivariate Statistical Analysis*. Dordrecht, The Netherlands: Kluwer Academic.

Shannon, C. E. 1948. A mathematical theory of communication. *Bell System Technical Journal*, **27**, 379–423.

Shannon, C. E. 1949. Communication in the presence of noise. *Proceedings of the IRE*, **37**(1), 10–21.

Shannon, C. E., and Weaver, W. 1949. *The Mathematical Theory of Communication*. Urbana, IL: University of Illinois Press.

Shao, J. 2007. *Mathematical Statistics*. New York, NY: Springer.

Shor, N. Z. 1987. Quadratic optimization problems. *Soviet Journal of Computer and System Sciences*, **25**, 1–11.

Silverman, B. W. 1982. On the estimation of a probability density function by the maximum penalized likelihood method. *Annals of Statistics*, **10**(3), 795–810.

Silverman, B. W. 1986. *Density esitmation for statistics and data analysis*. Boca Raton, FL: CRC Press.

Silverstein, J. 1995. Strong convergence of the empirical distribution of eigenvalues of large dimensional random matrices. *Journal of Multivariate Analysis*, **55**, 331–339.

Slepian, D. 1962. The one-sided barrier problem for Gaussian noise. *Bell System Technical Journal*, **42**(2), 463–501.

Smale, S., and Zhou, D. X. 2003. Estimating the approximation error in learning theory. *Analysis and Its Applications*, **1**(1), 1–25.

Spirtes, P., Glymour, C., and Scheines, R. 2000. *Causation, Prediction and Search*. Cambridge, MA: MIT Press.

Srebro, N. 2004. *Learning with Matrix Factorizations*. Ph.D. thesis, MIT. Available online: http://ttic.uchicago.edu/ nati/Publications/thesis.pdf.

Srebro, N., Rennie, J., and Jaakkola, T. S. 2005a (December 2004). Maximum-margin matrix factorization. In: *Advances in Neural Information Processing Systems 17 (NIPS 2004)*.

Srebro, N., Alon, N., and Jaakkola, T. S. 2005b (December). Generalization error bounds for collaborative prediction with low-rank matrices. In: *Advances in Neural Information Processing Systems 17 (NIPS 2004)*.

Srivastava, N., and Vershynin, R. 2013. Covariance estimation for distributions with $2 + \epsilon$ moments. *Annals of Probability*, **41**, 3081–3111.

Steele, J. M. 1978. Empirical discrepancies and sub-additive processes. *Annals of Probability*, **6**, 118–127.

Steinwart, I., and Christmann, A. 2008. *Support vector machines*. New York, NY: Springer.

Stewart, G. W. 1971. Error bounds for approximate invariant subspaces of closed linear operators. *SIAM Journal on Numerical Analysis*, **8**(4), 796–808.

Stewart, G. W., and Sun, J. 1980. *Matrix Perturbation Theory*. New York, NY: Academic Press.

Stone, C. J. 1982. Optimal global rates of convergence for non-parametric regression. *Annals of Statistics*, **10**(4), 1040–1053.

Stone, C. J. 1985. Additive regression and other non-parametric models. *Annals of Statistics*, **13**(2), 689–705.

Szarek, S. J. 1991. Condition numbers of random matrices. *J. Complexity*, **7**(2), 131–149.

Talagrand, M. 1991. A new isoperimetric inequality and the concentration of measure phenomenon. Pages 94–124 of: Lindenstrauss, J., and Milman, V. D. (eds), *Geometric Aspects of Functional Analysis*. Lecture Notes in Mathematics, vol. 1469. Berlin, Germany: Springer.

Talagrand, M. 1995. Concentration of measure and isoperimetric inequalities in product spaces. *Publ. Math. I.H.E.S.*, **81**, 73–205.

Talagrand, M. 1996a. New concentration inequalities in product spaces. *Inventiones Mathematicae*, **126**, 503–563.

Talagrand, M. 1996b. A new look at independence. *Annals of Probability*, **24**(1), 1–34.

Talagrand, M. 2000. *The Generic Chaining*. New York, NY: Springer.

Talagrand, M. 2003. *Spin Glasses: A Challenge for Mathematicians*. New York, NY: Springer.

Tibshirani, R. 1996. Regression shrinkage and selection via the Lasso. *Journal of the Royal Statistical Society, Series B*, **58**(1), 267–288.

Tibshirani, R., Saunders, M. A., Rosset, S., Zhu, J., and Knight, K. 2005. Sparsity and smoothness via the smoothed Lasso. *Journal of the Royal Statistical Society B*, **67**(1), 91–108.

Tropp, J. A. 2006. Just relax: Convex programming methods for identifying sparse signals in noise. *IEEE Transactions on Information Theory*, **52**(3), 1030–1051.

Tropp, J. A. 2010 (April). *User-friendly tail bounds for matrix martingales*. Tech. rept. Caltech.

Tsybakov, A. B. 2009. *Introduction to non-parametric estimation*. New York, NY: Springer.

Turlach, B., Venables, W.N., and Wright, S.J. 2005. Simultaneous variable selection. *Technometrics*, **27**, 349–363.

van de Geer, S. 2000. *Empirical Processes in M-Estimation*. Cambridge University Press.

van de Geer, S. 2014. Weakly decomposable regularization penalties and structured sparsity. *Scandinavian Journal of Statistics*, **41**, 72–86.

van de Geer, S., and Bühlmann, P. 2009. On the conditions used to prove oracle results for the Lasso. *Electronic Journal of Statistics*, **3**, 1360–1392.

van der Vaart, A. W., and Wellner, J. A. 1996. *Weak Convergence and Empirical Processes*. New York, NY: Springer.

Vempala, S. 2004. *The Random Projection Method*. Discrete Mathematics and Theoretical Computer Science. Providence, RI: American Mathematical Society.

Vershynin, R. 2011. *Introduction to the non-asymptotic analysis of random matrices*. Tech. rept. Univ. Michigan.

Villani, C. 2008. *Optimal Transport: Old and New*. Grundlehren der mathematischen Wissenschaften, vol. 338. New York, NY: Springer.

Vu, V. Q., and Lei, J. 2012. Minimax rates of estimation for sparse PCA in high dimensions. In: *15th Annual Conference on Artificial Intelligence and Statistics*.

Wachter, K. 1978. The strong limits of random matrix spectra for samples matrices of independent elements. *Annals of Probability*, **6**, 1–18.

Wahba, G. 1990. *Spline Models for Observational Data*. CBMS-NSF Regional Conference Series in Applied Mathematics. Philadelphia, PN: SIAM.

Wainwright, M. J. 2009a. Information-theoretic bounds on sparsity recovery in the high-dimensional and noisy setting. *IEEE Transactions on Information Theory*, **55**(December), 5728–5741.

Wainwright, M. J. 2009b. Sharp thresholds for high-dimensional and noisy sparsity recovery using ℓ_1-constrained quadratic programming (Lasso). *IEEE Transactions on Information Theory*, **55**(May), 2183–2202.

Wainwright, M. J. 2014. Constrained forms of statistical minimax: Computation, communication and privacy. In: *Proceedings of the International Congress of Mathematicians*.

Wainwright, M. J., and Jordan, M. I. 2008. Graphical models, exponential families and variational inference. *Foundations and Trends in Machine Learning*, **1**(1–2), 1—305.

Waldspurger, I., d'Aspremont, A., and Mallat, S. 2015. Phase recovery, MaxCut and complex semidefinite programming. *Mathematical Programming A*, **149**(1–2), 47–81.

Wang, T., Berthet, Q., and Samworth, R. J. 2014 (August). *Statistical and computational trade-offs in estimation of sparse principal components*. Tech. rept. arxiv:1408.5369. University of Cambridge.

Wang, W., Wainwright, M. J., and Ramchandran, K. 2010. Information-theoretic limits on sparse signal recovery: dense versus sparse measurement matrices. *IEEE Transactions on Information Theory*, **56**(6), 2967–2979.

Wang, W., Ling, Y., and Xing, E. P. 2015. Collective Support Recovery for Multi-Design Multi-Response Linear Regression. *IEEE Transactions on Information Theory*, **61**(1), 513–534.

Wasserman, L. A. 2006. *All of Non-Parametric Statistics*. Springer Series in Statistics. New York, NY: Springer.

Widom, H. 1963. Asymptotic behaviour of Eigenvalues of Certain Integral Operators. *Transactions of the American Mathematical Society*, **109**, 278–295.

Widom, H. 1964. Asymptotic behaviour of Eigenvalues of Certain Integral Operators II. *Archive for Rational Mechanics and Analysis*, **17**(3), 215–229.

Wigner, E. 1955. Characteristic vectors of bordered matrices with infinite dimensions. *Annals of Mathematics*, **62**, 548–564.

Wigner, E. 1958. On the distribution of the roots of certain symmetric matrices. *Annals of Mathematics*, **67**, 325–327.

Williams, D. 1991. *Probability with Martingales*. Cambridge, UK: Cambridge University Press.

Witten, D., Tibshirani, R., and Hastie, T. J. 2009. A penalized matrix decomposition, with applications to sparse principal components and canonical correlation analysis. *Biometrika*, **10**, 515–534.

Woodruff, D. 2014. Sketching as a tool for numerical linear algebra. *Foundations and Trends in Theoretical Computer Science*, **10**(10), 1–157.

Wright, F. T. 1973. A bound on tail probabilities for quadratic forms in independent random variables whose distributions are not necessarily symmetric. *Annals of Probability*, **1**(6), 1068–1070.

Xu, M., Chen, M., and Lafferty, J. D. 2014. *Faithful variable selection for high dimensional convex regression*. Tech. rept. Univ. Chicago. arxiv:1411.1805.

Xu, Q., and You, J. 2007. Covariate selection for linear errors-in-variables regression models. *Communications in Statistics – Theory and Methods*, **36**(2), 375–386.

Xue, L., and Zou, H. 2012. Regularized rank-based estimation of high-dimensional nonparanormal graphical models. *Annals of Statistics*, **40**(5), 2541–2571.

Yang, Y., and Barron, A. 1999. Information-theoretic determination of minimax rates of convergence. *Annals of Statistics*, **27**(5), 1564–1599.

Ye, F., and Zhang, C. H. 2010. Rate minimaxity of the Lasso and Dantzig selector for the ℓ_q-loss in ℓ_r-balls. *Journal of Machine Learning Research*, **11**, 3519–3540.

Yu, B. 1996. Assouad, Fano and Le Cam. *Research Papers in Probability and Statistics: Festschrift in Honor of Lucien Le Cam*, 423–435.

Yuan, M. 2010. High dimensional inverse covariance matrix estimation via linear programming. *Journal of Machine Learning Research*, **11**, 2261–2286.

Yuan, M., and Lin, Y. 2006. Model selection and estimation in regression with grouped variables. *Journal of the Royal Statistical Society B*, **1**(68), 49.

Yuan, M., and Lin, Y. 2007. Model selection and estimation in the Gaussian graphical model. *Biometrika*, **94**(1), 19–35.

Yuan, X. T., and Zhang, T. 2013. Truncated power method for sparse eigenvalue problems. *Journal of Machine Learning Research*, **14**, 899–925.

Yurinsky, V. 1995. *Sums and Gaussian Vectors*. Lecture Notes in Mathematics. New York, NY: Springer.

Zhang, C. H. 2012. Nearly unbiased variable selection under minimax concave penalty. *Annals of Statistics*, **38**(2), 894–942.

Zhang, C. H., and Zhang, T. 2012. A general theory of concave regularization for high-dimensional sparse estimation problems. *Statistical Science*, **27**(4), 576–593.

Zhang, Y., Wainwright, M. J., and Jordan, M. I. 2014 (June). Lower bounds on the performance of polynomial-time algorithms for sparse linear regression. In: *Proceedings of the Conference on Learning Theory (COLT)*. Full length version at http://arxiv.org/abs/1402.1918.

Zhang, Y., Wainwright, M. J., and Jordan, M. I. 2017. Optimal prediction for sparse linear models? Lower bounds for coordinate-separable M-estimators. *Electronic Journal of Statistics*, **11**, 752–799.

Zhao, P., and Yu, B. 2006. On model selection consistency of Lasso. *Journal of Machine Learning Research*, **7**, 2541–2567.

Zhao, P., Rocha, G., and Yu, B. 2009. Grouped and hierarchical model selection through composite absolute penalties. *Annals of Statistics*, **37**(6A), 3468–3497.

Zhou, D. X. 2013. Density problem and approximation error in learning theory. *Abstract and Applied Analysis*, **2013**(715683).

Zou, H. 2006. The Adaptive Lasso and its oracle properties. *Journal of the American Statistical Association*, **101**(476), 1418–1429.

Zou, H., and Hastie, T. J. 2005. Regularization and variable selection via the elastic net. *Journal of the Royal Statistical Society, Series B*, **67**(2), 301–320.

Zou, H., and Li, R. 2008. One-step sparse estimates in nonconcave penalized likelihood models. *Annals of Statistics*, **36**(4), 1509–1533.

Subject index

Author index

Printed in the United States
by Baker & Taylor Publisher Services